AF832015

Date Palm Biotechnology

Shri Mohan Jain • Jameel M. Al-Khayri
Dennis V. Johnson
Editors

Date Palm Biotechnology

Springer

Editors
Shri Mohan Jain
Department of Agricultural Sciences
University of Helsinki
PL-27
Helsinki
Finland
mohan.jain@helsinki.fi

Jameel M. Al-Khayri
Department of Agricultural Biotechnology
King Faisal University
Al-Hassa 31982
Saudi Arabia
jkhayri@kfu.edu.sa

Dennis V. Johnson
3726 Middlebrook Ave
Cincinnati, OH 45208
USA
djohn37@aol.com

ISBN 978-94-007-1317-8 e-ISBN 978-94-007-1318-5
DOI 10.1007/978-94-007-1318-5
Springer Dordrecht Heidelberg London New York

Library of Congress Control Number: 2011934965

© Springer Science+Business Media B.V. 2011
No part of this work may be reproduced, stored in a retrieval system, or transmitted in any form or by any means, electronic, mechanical, photocopying, microfilming, recording or otherwise, without written permission from the Publisher, with the exception of any material supplied specifically for the purpose of being entered and executed on a computer system, for exclusive use by the purchaser of the work.

Printed on acid-free paper

Springer is part of Springer Science+Business Media (www.springer.com)

Preface

Date palm (*Phoenix dactylifera* L.) is a dioecious fruit tree native to the hot arid regions of the world, mainly grown in the Middle East, and North Africa. Through germplasm exchange, date palm agriculture has expanded to Australia, Southern Africa, South America, Mexico and the United States of America. Since ancient time this majestic plant has been recognized as the "tree of life" because of its integration in human settlement, wellbeing, and food security in hot and barren parts of the world, where only a few plant species can flourish. Date palm trees continue to provide the most sustainable agro-ecosystems in harsh dry environments providing raw materials for housing, furnishings, and many handcrafts in addition to supplying nutritious delicious fruits that can be consumed fresh, dried, or processed, providing a nutritious source of sugars, minerals, and vitamins. Economically, date palm provides a major source of income for local farmers and associated industries in communities where it is grown.

Expansion of date palm agriculture is faced with challenges stemming from propagation and genetic improvement limitations. The heterozygous nature of this dioecious species hampers the use of seeds which produce off type seedlings, and normally are not used to propagate known elite cultivars. The limited availability of offshoots and the difficulties of establishing propogules from offshoots render this traditional propagation method inadequate, particularly for large-scale propagation. Based on recent advances in plant tissue culture, micropropagation technique has been developed for the rapid mass propagation of date palm. Some limitations associated with genetic improvement have been circumvented by taking advantage of tissue culture applications and molecular methodologies.

Overall, this book discusses the major developments in date palm biotechnology during the last few decades highlighting genetics and germplasm, tissue culture methodologies and applications, genetic engineering, genomics, and molecular techniques. The book contains an introductory chapter: Date palm biotechnology from theory to practice that gives a plausible background for the 33 review chapters which highlight current research status relevant to various aspects of date palm biotechnology. The book is divided in five parts. Part I discusses the research development, methodology, and commercial application of micropropagation in seven chapters: Potential of date palm micropropagation for improving small farming systems; Date palm tissue culture: a pathway to rural development; Date palm micropropagation via somatic embryogenesis; Date palm micropropagation via organogenesis; Micropropagation of date palm using inflorescence explants; Bioreactors and automation in date palm micropropagation; Commercial date palm tissue culture procedures and facility establishment.

Somaclonal variation is quite common in micropropagated plants occurring spontaneously in many plant species, but can be controlled by in vitro culture practices. However, it can be extremely useful to select somaclones exhibiting desirable traits like enhanced tolerance to biotic or abiotic stress agents. In fact, scientists often resort to mutagens to induce mutations to gain a broader genetic pool for more efficient in vitro selection. The maintenance of genetic fidelity of in vitro plantlets is highly desirable and that can be achieved by phenotypic characterizations, which is quite slow; however, molecular techniques are reliable to identify somaclones rapidly. In this book, Part II deals with the research development and applications of somaclonal variation and mutation in date palm, covered in seven chapters: Somaclonal variation in date palm; Growth abnormalities associated with micropropagation of date palm; Molecular detection of somaclonal variation in date palm; In vitro selection for abiotic stress in date palm; *Fusarium oxysporum* f. sp. *albedinis* toxins characterization and use for selection of resistant date palm to Bayoud disease; Radiation induced mutations for date palm improvement; Magnetic field induced biochemical and growth changes in date palm seedlings.

Genetic diversity of date palm is threatened by human development and reduction of suitable arable lands, in addition to ecosystem changes, monoculture, and biotic invasions. Successful utilization of date palm genetic resources requires characterization and preservation of date palm germplasm biodiversity using various conservation methods based on in situ and ex situ collections. Date palm field gene banks are difficult to maintain and the storage of offshoots under biocontrolled environment is impracticable. Seed conservation is not viable due to genetic heterozygosity. To augment traditional conservation methods, in vitro technologies have been applied in date palm. Molecular techniques to characterize date palm germplasm biodiversity are modern tools that have proved useful. The status of date palm germplasm and current techniques employed in conservation and molecular characterization are described in seven chapters in Part III of this book: Date palm germplasm; In vitro conservation of date palm germplasm; Molecular markers in date palm; Biodiversity in date palm: molecular markers as indicators; Polymorphism and genetic relationship in date palm using molecular markers; Date palm genome project at the kingdom of Saudi Arabia; Potential of arbuscular mycorrhizal technology in date palm production.

Plant tissue culture techniques offer several tools to plant breeders based on an understanding of genetic principles. For instants, embryo rescue, in vitro fertilization, in vitro flowering, somatic hybridization. An obstacle of date palm breeding is sex identification of offsprings since first flowering requires at least 3 years. Molecular technologies can effectively identify sex at early growth stages and produce molecular markers which can drastically reduce breeding cycles. Genetic improvement is necessary to enhance the resistance to numerous date palm diseases and insects, improve yield and fruit quality, and increase tolerance to abiotic stresses. However, traditional breeding methods are inapplicable due to inherently slow growth nature and the long generation time of the date palm. Genetic engineering has proved invaluable to genetically improve many plant species. Although in its infancy, date palm genetic transformation and genomic studies to identify genes coding for useful traits are witnessing great interests. Part IV of this book contains eight chapters addressing research progress made in these areas: Date palm genetics and breeding; Development of new Moroccan selected date palm varieties resistant to Bayoud and of good fruit quality; Molecular markers for genetic diversity and Bayoud disease resistance in date palm; Towards sex determination of date palm; Interspecific hybridization and embryo rescue in date palm; In vitro flowering of date palm; Date palm cell and protoplast culture; Transgenic date palm.

The potential of producing secondary metabolites, enhancing date utilization and the possibility of producing biofuels from date palm-associated microorganisms paves the way to industrial biotechnological applications. Part V of this book describes up-to-date related progress made in three chapters: Secondary metabolites of date palm; Industrial biotechnology: date palm fruit applications; Date palm as a source of bioethanol producing microorganisms.

It is evident that biotechnology has significantly influenced date palm agriculture. Although research in date palm biotechnology is relatively limited, achievements accumulated thus far have inspired us to collect this valuable information under one cover to provide an updated source for beginners in the field of date palm biotechnology as well as a reliable reference for specialists. This book is beneficial to students, researchers, scientists, commercial producers, consultants, and policy makers interested in agriculture or plant science particularly in date palm biotechnology. It is highly recommended for plant biotechnology courses especially in date palm biotechnology graduate courses and training.

The chapters in this book were authored and reviewed by prominent specialists demonstrating distinct research contributions to date palm biotechnology, invited from industry, universities, and research institutes. Their contribution to the quality of this book is gratefully acknowledged.

Shri Mohan Jain
Jameel M. Al-Khayri
Dennis V. Johnson

Contents

Contributors

L. Abahmane Laboratoire de Biotechnologie Végétale, INRA, Centre Régional de la Recherche Agronomique de Marrakech, 533, Marrakech, Morocco, abahmanel@yahoo.fr

A.A. Abul-Soad Horticulture Research Institute, Agricultural Research Center, 9 Cairo Univ. St., Orman, Cairo, Egypt, adelaboelsoaud@gmail.com

S.M. Aleid Date Palm Research Center of Excellence, King Faisal University, Al-Hassa, Saudi Arabia, seid@kfu.edu.sa

N.S. Al-Khalifah King Abdulaziz City for Science and Technology, KACST, P. O. Box 6086, Riyadh 11442, Saudi Arabia, abujawad@kacst.edu.sa

J.M. Al-Khayri Department of Agricultural Biotechnology, College of Agricultural and Food Sciences, King Faisal University, P.O. Box 420, Al-Hassa 31982, Saudi Arabia, jkhayri@kfu.edu.sa

I.S. Al-Mssallem Date Palm Genome Project, King Abdulaziz City for Science and Technology, Riyadh, Kingdom of Saudi Arabia, imssallem@kacst.edu.sa

Y. Al-Shayji Biotechnology Department, Food Resources and Marine Sciences Division, Kuwait Institute for Scientific Research, P.O. Box 24885, Safat 13109, Kuwait, yshayji@kisr.edu.kw

E. Askari King Abdulaziz City for Science and Technology, KACST, P.O. Box 6086, Riyadh 11442, Saudi Arabia, easkari@kacst.edu.sa

A. Assani Department of Plant Agriculture, University of Guelph, Guelph, ON, N1G 2W1, Canada, aassani@bell.net

C. Bayoudh Laboratory of In Vitro Culture, Regional Center of Oases Agriculture Researches, 2260 Degache, Tunisia, civpalmier@yahoo.fr

S.A. Bekheet Plant Biotechnology Department, National Research Center, El-Tahrir Str., Dokki 12622, Cairo, Egypt, shawky005@yahoo.com

N. Bouguedoura Laboratoire de Recherche sur les Zones Arides (LRZA), Université des Sciences et de la Technologie Houari Boumediene (USTHB), BP 31 El Alia Bab Ezzouar, Algiers, Algeria, nadiaboug@voila.fr

G.M. Brown Department of Biology, Sultan Qaboos University, Al-Khod, Sultanate of Oman, gmarbrown@yahoo.co.uk

D. Chabane Laboratoire de Recherche sur les Zones Arides (LRZA), Université des Sciences et de la Technologie Houari Boumediene (USTHB), BP 31 El Alia Bab Ezzouar, Algiers, Algeria, chabanedj@yahoo.fr

E. Cherif Laboratoire de génétique moléculaire, immunologie et biotechnologie, Faculté des Sciences de Tunis, Université de Tunis El Manar, Campus Universitaire, 2092 El Manar Tunis, Tunisia

Y. Cohen Department of Fruit Tree Sciences, Institute of Plant Sciences, Volcani Research Center, Bet Dagan, 50250, Israel, vhyuvalc@volcani.agri.gov.il

C. Cullis Department of Biology, Case Western Reserve University, 10900 Euclid Avenue, Cleveland, OH 44106-7080, USA, cac5@case.edu

F. Daayf Department of Plant Science, University of Manitoba, 222, Agriculture Building, Winnipeg, MB R3T 2 N2, Canada

A.J. Dakheel International Center for Biosaline Agriculture, Dubai, United Arab Emirates, a.dakheel@biosaline.org.ae

F. Dhawi Department of Agricultural Biotechnology, College of Agricultural and Food Sciences, King Faisal University, P.O. Box 420, Al-Hassa 31982, Saudi Arabia, faten.dhawi@live.com

S. Dkkhil-Dakhlaoui Laboratoire de génétique moléculaire, immunologie et biotechnologie, Faculté des Sciences de Tunis, Université de Tunis El Manar, Campus Universitaire, 2092 El Manar Tunis, Tunisia

N. Drira Laboratory of Plant Biotechnology, Faculty of Sciences of Sfax, 3018 Sfax, Tunisia, drira_riadh@yahoo.fr

A. El Hadrami OMEX Agriculture Inc. Canada, P.O. Box 301, 290 Agri Park Road, Oak Bluff, Manitoba, R0G 1N0, Canada, abdele@omexcanada.com

I. El Hadrami Department of Biology, Faculty of Sciences Semlalia, Laboratory of Biotechnologies, Cadi Ayyad University, B.P. 2390, 40 000 Marrakech, Morocco

B. El-Korchi Date Palm Research and Development Unit, Plant Tissue Culture Laboratory, United Arab Emirates University, 81908, Al Ain, United Arab Emirates

S. Elshibli Department of Agricultural Sciences, University of Helsinki, FI-00014 Helsinki, Finland, sakina.elshibli@helsinki.fi

M. Ferry INRA, Phoenix Research Station, Apartado 996, 03201 Elche, Spain, m.ferry@telefonica.net

L. Fki Laboratoire des Biotechnologies Végétales Appliquées à l'Amélioration des Cultures, Université de Sfax-Faculté des Sciences de Sfax, Route de Sokra Sfax-Tunisie, lotfi.fki@fss.rnu.tn

R. Gargouri-Bouzid Department of Biology, Institute for Engineering Studies, University of Sfax, Road Soukra BPW, 3038 Sfax, Tunisia, gargouriradhia@yahoo.fr

N. Gupta Department of Microbiology and Biotechnology, Faculty of Science, M S University, Baroda, Gujarat, India

Department of Molecular Biology and Genetic Engineering, College of Basic Sciences and Humanities, G.B. Pant University of Agriculture and Technology, Pantnagar, Uttarakhand 263 145, India, nidhigupta_2411@yahoo.co.in

M.S. Hanafy Plant Biotechnology Department, National Research Center, El-Tahrir Str., Dokki, 12622, Cairo, Egypt, mshanafy@yahoo.com

S.M. Jain Department of Agricultural Sciences, University of Helsinki, PL-27, Helsinki, Finland, mohan.jain@helsinki.fi

D.V. Johnson 3726 Middlebrook Ave, Cincinnati, OH 45208, USA, djohn37@aol.com

H. Korpelainen Department of Agricultural Sciences, University of Helsinki, FI-00014 Helsinki, Finland, helena.korpelainen@helsinki.fi

W. Kriaâ Laboratory of Plant Biotechnology, Department of Biology, Faculty of Sciences Sfax, Road Soukra BP. 1171, 3000 Sfax, Tunisia, maswkri@yahoo.fr

R.R. Krueger USDA-ARS National Clonal Germplasm Repository for Citrus & Dates, 1060 Martin Luther King Blvd, Riverside, CA 92507, USA, Robert.Krueger@ars.usda.gov

H. Kushwaha Department of Molecular Biology and Genetic Engineering, College of Basic Sciences and Humanities, G.B. Pant University of Agriculture and Technology, Pantnagar, Uttarakhand 263 145, India

B.H. Lazrek Laboratoire de chimie bio-organique, Département de Chimie, Faculté des sciences Sémlalia, Université Cadi Ayyad, Boulevard Moulay Abdellah B.P S 15, Marrakech, Morocco

A. Mahjoub Laboratoire des Biotechnologies Végétales Appliquées à l'Amélioration des Cultures, Université de Sfax-Faculté des Sciences de Sfax, Route de Sokra Sfax-Tunisie

M. Marrakchi Laboratoire de génétique moléculaire, immunologie et biotechnologie, Faculté des Sciences de Tunis, Université de Tunis El Manar, Campus Universitaire, 2092 El Manar Tunis, Tunisia

R. Masmoudi Laboratoire des Biotechnologies Végétales Appliquées à l'Amélioration des Cultures, Université de Sfax-Faculté des Sciences de Sfax, Route de Sokra Sfax-Tunisie

F. Masmoudi-Allouche Laboratory of Plant Biotechnology, Department of Biology, Faculty of Sciences Sfax, Road Soukra BP: 1171, 3000 Sfax, Tunisia, faizamasmoudi@yahoo.fr

B. Meziou Laboratory of Plant Biotechnology, Department of Biology, Faculty of Sciences Sfax, Road Soukra BP: 1171, 3000 Sfax, Tunisia, besmameziou@yahoo.fr

A. Mliki Laboratoire de Biotechnologie Végétale, Centre de biotechnologie de Borj Sedria, Borj Sedria Tunis-Tunisie

R. Mzid Laboratoire de Biotechnologie Végétale, Centre de biotechnologie de Borj Sedria, Borj Sedria Tunis-Tunisie

Laboratory of Plant Molecular Physiology, Biotechnology Center of Borj Cédria, 901 Hammam-Lif, Tunisia, rmzid@yahoo.fr

A. Othmani Laboratory of Molecular Genetics, Immunology & Biotechnology, Faculty of Sciences of Tunis, 2092, El Manar Tunis, Tunisia

Laboratoire de génétique moléculaire, immunologie et biotechnologie, Faculté des Sciences de Tunis, Université de Tunis El Manar, Campus Universitaire, 2092 El Manar Tunis, Tunisia, degletbey@yahoo.fr

K. Rajmohan Atul Rajasthan Date Palms Ltd., Chopasni, Nandanvan P.O., Jodhpur 342008, Rajasthan, India, rajmohan33@yahoo.com

S. Rhouma Laboratoire de génétique moléculaire, immunologie et biotechnologie, Faculté des Sciences de Tunis, Université de Tunis El Manar, Campus Universitaire, 2092 El Manar Tunis, Tunisia

M.C. Rillig Institut für Biologie, Plant Ecology, Freie Universität, Berlin, Germany, rillig@zedat.fu-berlin.de

A. Rival UMR DIADE, CIRAD, F-34398, Montpellier, France

M.M. Saker Department of Plant Biotechnology, National Research Center, Dokki, Cairo, Egypt, msakr@asrt.sci.eg; sakrmahmoud@yahoo.com

A.O.M. Salem Département de Biologie, Faculté des Sciences et Techniques, Université de Nouakchott, B.P 5026, Nouakchott, Mauritania

MyH Sedra Laboratoire de phytopathologie, Génétique et de Lutte Intégrée, Centre Régional de Marrakech, Regional Center of Agriculture Research, Institut National de Recherche Agronomique (INRA), BP. 533, Ménara 40000, Marrakech, Morocco, sedramh@hotmail.com; mhsedra@yahoo.fr

B. Sghaier Laboratoire des Biotechnologies Végétales Appliquées à l'Amélioration des Cultures, Université de Sfax-Faculté des Sciences de Sfax, Route de Sokra Sfax-Tunisie

G. Shabbir Department of Field and Forage Crops, International Center for Biosaline Agriculture (ICBA), P.O. Box 14460 Dubai, UAE

Institut für Biologie, Plant Ecology, Freie Universität, Berlin, Germany, g.shabbir@biosaline.org.ae

H. Shittu Department of Molecular and Cellular Biology, University of Guelph, Guelph, ON, N1G 2W1, Canada, hshittu@uoguelph.ca

C. Sudhersan Biotechnology Department, Food Resources and Marine Sciences Division, Kuwait Institute for Scientific Research, P.O. Box 24885, Safat 13109, Kuwait, schellan@kisr.edu.kw

J. Tan Date Palm Genome Project, King Abdulaziz City for Science and Technology, Riyadh, Kingdom of Saudi Arabia

Institute of Bioinformatics, Chongqing University of Posts and Telecommunications, Chongqing, China, tanjun0528@gmail.com

M. Trifi Laboratory of Molecular Genetics, Immunology & Biotechnology, Faculty of Sciences of Tunis, 2092, El Manar, Tunis, Tunisia

Laboratoire de génétique moléculaire, immunologie et biotechnologie, Faculté des Sciences de Tunis, Université de Tunis El Manar, Campus Universitaire, 2092 El Manar Tunis, Tunisia

H.J. Visser Date Palm Research and Development Unit, Plant Tissue Culture Laboratory, United Arab Emirates University, 81908, Al Ain, United Arab Emirates

M. Yang Date Palm Genome Project, King Abdulaziz City for Science and Technology, Riyadh, Kingdom of Saudi Arabia

Key Laboratory of Genome Science and Information, Beijing Institute of Genomics, Chinese Academy of Science, Beijing, China, yangmeng@big.ac.cn

Y. Yin Date Palm Genome Project, King Abdulaziz City for Science and Technology, Riyadh, Kingdom of Saudi Arabia

Key Laboratory of Genome Science and Information, Beijing Institute of Genomics, Chinese Academy of Science, Beijing, China, yinyx.1@gmail.com

J. Yu Date Palm Genome Project, King Abdulaziz City for Science and Technology, Riyadh, Kingdom of Saudi Arabia

Key Laboratory of Genome Science and Information, Beijing Institute of Genomics, Chinese Academy of Science, Beijing, China

Plant Stress Genomics Research Center, King Abdullah University of Science and Technology, Thuwal, Kingdom of Saudi Arabia, junyu@big.ac.cn; jun.yu@kaust.edu.sa

A. Zaid Date Palm Research and Development Unit, Plant Tissue Culture Laboratory, United Arab Emirates University, 81908, Al Ain, United Arab Emirates, zaid@uaeu.ac.ae

S. Zehdi-Azouzi Laboratoire de génétique moléculaire, immunologie et biotechnologie, Faculté des Sciences de Tunis, Université de Tunis El Manar, Campus Universitaire, 2092 El Manar Tunis, Tunisia

X. Zhang Date Palm Genome Project, King Abdulaziz City for Science and Technology, Riyadh, Kingdom of Saudi Arabia

Key Laboratory of Genome Science and Information, Beijing Institute of Genomics, Chinese Academy of Science, Beijing, China, zhangxw@big.ac.cn

Chapter 1
Introduction: Date Palm Biotechnology from Theory to Practice

D.V. Johnson

Abstract Date palm has been a cultivated tree crop for at least 5,000 years. Early date palm-specific technologies were developed to enhance crop productivity and fruit quality by means of selecting superior female palms and their propagation by offshoots. Other old innovations included crop and water management, segregation of trees by gender, artificial pollination, naming of cultivars and the characterization of fruit development stages, fruit flesh texture and fruiting seasonality. Modern biotechnology techniques are carrying forward date palm development in attempts to understand the genetic basis of the palm, to produce tissue-cultured plantlets on a large scale to more rapidly expand planting and replanting of date groves and to employ molecular breeding of new cultivars for increased fruit yield and resistance to pests and pathogens. This volume surveys the current state of date palm biotechnology through contributions by leading researchers in the field.

Keywords Cultivar • Domestication • Gender • Genome • Germplasm • Molecular biology • Pathogens and pests • Pollination • Propagation • Tissue culture

1.1 Introduction

This comprehensive volume covers a broad range of highly-technical subjects demonstrating how modern scientific theory and practice are being applied to overcome agronomic impediments to achieving the long-term goal of sustained, expanded and enhanced quality of date fruit production worldwide. Disease and pest problems threaten date growing in several countries of North Africa and the Middle East and biotechnology has the potential to make significant contributions

D.V. Johnson (✉)
3726 Middlebrook Ave, Cincinnati, OH 45208, USA
e-mail: djohn37@aol.com

S.M. Jain et al. (eds.), *Date Palm Biotechnology*,
DOI 10.1007/978-94-007-1318-5_1, © Springer Science+Business Media B.V. 2011

to their effective management. DNA studies which are able now to confirm the identity of date palm cultivars, have also provided evidence of genetic resistance to some diseases and pests in certain cultivars.

Rather than attempt to describe the technical contents of this book, my intention is to provide some background and context to the chapters which follow. This I believe can be accomplished by sketching major technical breakthroughs associated with the date palm over its long history. Crop domestication is a lengthy process, and we need to remind ourselves that virtually all of the major world crops were brought into cultivation by innovative food providers long before the advent of modern agricultural sciences. The date palm serves as a general example of the necessary sequence of technologies that had to be devised to domesticate, propagate and manage the palm, and to achieve fruit yields of high quantity and quality, sufficient to justify the effort in terms of human inputs.

The remainder of this introduction is divided into three sections. First is a discussion of selected aspects of the date palm in terms of its biology, domestication and agronomy. The second section summarizes the technological innovations in early date palm growing, especially in Mesopotamia. Finally, the third section attempts to summarize the rapid technological advances in recent years with regard to tissue culture and molecular biology research.

Before proceeding, it is worthwhile to review familiar terminology. The term *technology* refers to the application of science or the scientific method to achieve a targeted objective. These days we may think of it as referring almost exclusively to intricate sophisticated techniques or to machines that can perform remarkable tasks. But we must remember that the ancient world had its own technological obstacles and breakthroughs, many of them of a mechanical nature, but which had profound importance in their time. The term *biotechnology* incorporates the use of living organisms to the meaning of technology; it is a relatively new word coined in the early 1940s. A widely accepted definition of biotechnology is: "Any technological application that uses biological systems, living organisms, or derivatives thereof, to make or modify products or processes for specific use." (UN 1993). Science and technology represent, in human history, a hypothetical pyramid constructed such that each successful block laid down becomes a new foundation and provides an opportunity to erect yet another more sophisticated block above it.

1.2 Biology, Domestication and Agronomy

The date palm (*Phoenix dactylifera* L.) is one of the 14 recognized species of the genus *Phoenix*, which is itself one of the 183 palm genera currently known. *Phoenix* belongs to the tribe Phoeniceae, the subfamily Coryphoideae and family Arecaceae (Palmae) and has a wide natural distribution in the Old World. *Phoenix dactylifera* is the type specimen for the genus. Species of *Phoenix* range in size from stemless to tall, the date palm being the largest, reaching over 30 m in height. *Phoenix* spp. are solitary or clumping and dioeceous; inflorescences are interfoliar with male and female flowers borne on separate trees. The leaves are pinnate, induplicate (V-shaped) and erect, with basal leaf spines. In date palm the leaves are 3–6 m in length. *Phoenix* fruits are ovoid to oblong, smooth and have a fleshy mesocarp. Domesticated date palm fruits have quite variable shapes and range in size from 18 to 110 × 8 to 32 mm with the weight of an individual fruit varying from 2 to 60 g (Zaid 2002); they are significantly larger than in other *Phoenix* species. Date palm has 36 chromosomes (n = 18; 2n = 36).

Worldwide, about 3,000 date-palm cultivar names exist; however, some names are probably synonyms, the result of a local or national name given to a cultivar which also occurs in another country, but under a different name. Apart from being classified by cultivar or variety according to fruit characteristics, two other sets of overriding fruit designations typically are employed: flesh consistency and ripening time. Date cultivars are grouped into three classes: *soft*, *semidry* or *dry*, based upon the texture of the fruit under normal ripening conditions. Cultivars are also classified according to the length of time needed to produce mature fruit, as *early*, *midseason* and *late*.

Date palm is one of the world's first cultivated fruit tree. It is one of the classical Old World fruits, and, along with the olive and fig, the three represent an ancient group of fruit trees closely associated with the beginnings of agriculture. Date palm was domesticated in Mesopotamia, modern-day Iraq, 5,000 or more years ago. Until recently, it was believed that there were no wild ancestors of the date palm. However, in recent decades, archaeological research coupled with contemporary botanical field and laboratory studies have revealed that the cultivated date is closely related to wild and feral populations in North Africa and the Middle and Near East. Wild dates are considered to be the same species, can hybridize with the named cultivars, and are morphologically similar to the domesticated form; the main distinguishing feature of wild dates is their much smaller fruits (Zohary and Hopf 2000).

The value of the date palm within a subsistence or market economy represents much more than a source of nutritious high-energy fruit which can be eaten fresh, and also is easily stored by sun drying and used as supplementary foodstuff throughout the year. Fruits can be pressed into an easily-transportable date cake, made into syrup and fermented into date wine, vinegar and serve as a source of bioethanol; processing of the latter generates a feedstock by-product. When a tree is felled, the heart can be extracted and eaten. The pits (seeds) are eaten by livestock. At present, the fruits are a source of raw materials for agrofood industries and the production of secondary metabolites are important in human diet. In an oasis, the trees create a microenvironment suitable for other forms of agriculture and livestock raising, and provide much-needed shade. Stems of trees can be split for construction wood, the leaf midribs provide fencing material and the leaves can be woven into baskets, mats, hats and so forth. Date palm indeed is a multipurpose species from the past and of the present.

Date palms are commonly distinguished between seed-derived (*seedling* or *khalt*) plants, the result of sexual propagation, and those derived from offshoots removed from desirable female or male cultivars. Seedling dates exhibit characters inherited from both parents and may be female or male, with gender only revealed at the initiation of flowering at 5–7 years of age. In contrast, an offshoot-derived plant is the gender of the parent plant and in other ways true-to-type. Date palms produce only a limited number of offshoots; estimates range from 20 to 30, most of them when the palm is relatively young. The number of offshoots produced varies with the cultivar and in certain ones is very low or even nil. Date palms reach their full fruit-bearing potential at about 10 years of age (Zaid 2002). On average, a date palm plantation has an economic life of 40–50 years. Some estimates are that the palm can continue to bear fruit until the age of 100 or even older, but no documented long-term studies in this regard appear to have been done. In practical terms, there are strong inducements for replacement planting of very old palms because of the difficulty in climbing ever-taller trees, declining productivity and greater susceptibility, with age, to pests, diseases and blow-down.

Date palms cultivated in the typical desert environment flower once a year, in spring in the Northern Hemisphere. This habit appears to be governed by climate. In years of exceptional rainfall in the desert, a second flowering in the same year can occur. In the wild, date palms are pollinated by wind or insects. In extreme northern Chile, date palms grown under ideal stress-free conditions of plentiful irrigation water, high temperatures with very little annual temperature fluctuation, the palms flower and fruit continuously like coconuts. The fact that date palm has the genetic capability to flower and fruit continuously, very likely would not result in a greater annual fruit production per tree than occurs in a single annual flowering season. However, the potential of this genetic trait may be worth exploiting to cultivate dates, where similar ideal environments exist, for off-season marketing.

Diseases and pests are, to varying degrees depending upon location, ever-present threats to date fruit production, and in certain cases to the life of the trees themselves. Among the various maladies affecting date palm, two are of contemporary prominence. Bayoud disease, a fatal vascular wilt, is caused by a soil-borne fungus disease (*Fusarium oxysporum*). This disease first appeared in Morocco in the late nineteenth century and has spread widely in that country and into neighboring Algeria. In particular it attacks the prized Moroccan Medjool cultivar; it is dispersed chiefly by the transfer of offshoots. From the onset of symptoms of leaf withering, death of the palm follows in several months. No chemical or biological control is known; the only present solution appears to be the selection of cultivars which are resistant to the disease (Zaid 2002).

A second major threat to date palms is the red palm weevil (*Rhychophorus ferrugineus*). Native to tropical Asia, in recent decades the insect has spread westward steadily into South Asia, the Middle East, North Africa and southern Europe. It attacks a number of commonly-cultivated palms. The larvae burrow into the palm trunk, but infestation is not obvious until serious damage has been done and leaf wilt occurs. At that point the palm is beyond recovery and dies when the terminal bud is attacked. Following good plantation practices to avoid plant stress and attention to sanitation appear to lessen the incidence of infestation. A variety of chemical and biological controls are being studied to control this insect but a protocol for early detection and effective treatment of infested trees has not been achieved.

Increasing fruit yield is the key objective of date improvement efforts. Current yield figures can provide an approximate benchmark against which to measure future success, but average fruit yields per tree are very difficult to determine. For example, according to FAOSTAT, world date fruit production in 2008 reached 7,048,089 mt, from a harvested area of 1,264,611 ha. Calculated data gives an average yield of 5.57 mt/ha. The basic problem with such a calculation is that the number of trees planted per hectare varies widely depending upon whether the plantings in a country are traditional and without fixed spacing, or modern plantations with fixed spacing, or, a combination of the two. To demonstrate the reported range: in Morocco 50 trees/ha is typical, whereas in Somalia it reportedly can be to 577 palms/ha. Another difficulty is that yields measured by weight vary considerable among cultivars because of fruit size and weight. The most reliable yield figures come from two popular high-value cultivars. Medjool and Barhee, grown under modern plantation conditions in the USA and Israel, achieve yields of 80–120 and 200 kg per tree, respectively (Zaid 2002).

1.3 Early Date Palm Technology

Date fruits, along with grapes and cereals such as barley, played a key role in Mesopotamia and Ancient Egypt to make fermented alcoholic beverages. In fact, the word *alcohol* is derived from Arabic. The development of fermentation technology, therefore, is directly related to date palm. At some later time, the technique of tapping date palm inflorescences and trunks for sap was perfected; sap also ferments naturally into a mild alcoholic beverage. The consumption of *laqmi*, fermented from either sap or fruit, at certain locations in North Africa, represents a contemporary practice of an ancient technology.

At least seven ancient technologies, directly or indirectly related to date palm, were developed in Mesopotamia and are summarized below (Dowson 1921, 1923, 1982; Popenoe 1973; Pruessner 1920):

1. Plant sexuality recognized; male and female date palms distinguished.
2. Date groves organized using fixed tree spacing and with catch cropping of annual crops in initial years of establishment.
3. Date palms under cultivation segregated by gender.
4. Free flow irrigation and other water management practices of date palm groves.
5. Artificial pollination of date palms.
6. Offshoot separation propagation devised.
7. Stages of fruit development were recognized and given specific designations, as follows:

From pollination to final date-fruit ripening takes about 200 days. Date palm flowering and fruiting were recognized to have five distinct stages over the ripening period. *Hababauk*, female flowers and immediate post-pollination period when the very young fruits are creamy white in color; *kimri*, green fruit undergoing rapid growth; *khalal*, fruit grows slowly to full size, sugar content increasing while moisture content decreases, hard, glossy, red or yellow in color; *rutab*, fruits ripening to a soft stage, brown in color; *tamar*, fruits fully ripened, wrinkled, brown or black in color. In Arabic, *tamar* is also a generic term for dates. These designations, especially the last four, are basic technical terms in use today. The terms appear in the literature with variable English spellings.

Apparently, the naming of date palm cultivars also originated in Mesopotamia to distinguish different fruit types. Popenoe (1973), writing about the ancient Zahidi cultivar of Iraq, reports on a comment to him by V.H.W. Dowson in the 1920s to the effect that Zahidi was a Basrah variety that, according to local tradition, was the first female variety known and that all other varieties were derived from it. Pruessner (1920) made calculations of fruit productivity of date palms in Mesopotamia through deciphering stone tablets from about 4,000 years ago, and estimated that annual yields per tree were in a range of 105–180 kg. These could be considered moderate as compared to contemporary yields.

Two issues that early date palm technology could not deal with effectively were rapid cultivar propagation and an efficient method of breeding improved cultivars. Seed propagation, intentional or spontaneous, has a major drawback of producing mixed populations of female and male palms, and those progeny, because of cross-pollination, did not fully replicate the desirable fruit characteristics sought. Offshoot propagation was able to solve the true-to-typeness problem, but with a major limitation because relatively small numbers of offshoots are available, far short of the numbers needed for rapid expansion of date palm cultivation.

Since domestication, sexually propagated date palms, past and present, have produced many new forms and some of those with desirable fruit or other characteristics have become named cultivars, in turn being perpetuated and propagated by offshoots. But this did not contribute much to achieving positive changes in certain cultivars that only plant breeding could accomplish.

The Department of Agriculture was a major promoter of date cultivation in the USA beginning in the early twentieth century with the importation of some 20,000 offshoots, representing about 150 cultivars, from Iraq, Pakistan, Egypt, Algeria, Tunisia and Morocco. Once the industry was well established, a breeding program was initiated at a date palm research station in Indio, California, which ran from 1948 to 1978. The primary goal of the breeding program, using backcrossing, was to obtain female Deglet Noor cultivars with fruit adapted to mechanical harvesting and processing, but without any loss of fruit quality. Deglet Noor has always been the most important cultivar in California and even today accounts for 70% of total production (Fig. 1.1). The program ended before that goal could be achieved. Some promising progeny from those efforts are preserved in the National Date Palm Germplasm Repository, Thermal, California. Genetic improvement through date palm breeding is not feasible because of the length of time for a palm to reach sexual maturity (5–7 years). Moreover, offshoot propagation cannot produce large numbers of progeny.

Traditional date cultivation typically carried out on a small scale, faces special difficulties, often because of inaccessibility to modern technologies developed for and utilized by large commercial orchards. Cost is most often the main obstacle.

Fig. 1.1 Mature stand of Deglet Noor cultivar in Indio, California. Note the metal ladders permanently affixed to the upper portion of the trees to facilitate access to the crown for date culture and management

However, traditional producers can achieve enhanced productivity by adopting inexpensive manual practices relative to watering, drainage, fertilizer, weed control, tree spacing, etc.

1.4 Tissue Culture and Molecular Biology

The last half century has been characterized by scientific efforts to advance agricultural production through new methods of propagation and to gain insights into the molecular structure of plants to select progeny with enhanced productivity and resistance to specific pests and diseases. Both subjects treated in this section are being studied intensively as evidenced by the numerous recently-published references cited in the chapters which follow.

1.4.1 Tissue Culture

Research on date-palm propagation via tissue culture (also called *in vitro* propagation or micropropagation) began in about the 1970s. At about the same time, palm tissue culture research also was undertaken with coconut (*Cocos nucifera*) and the

African oil palm (*Elaeis guineensis*). Research progress made since then in date palm micropropagation has been reviewed by Al-Khayri (2005, 2007). Zaid (2002) provides a comprehensive discussion of all aspects of date palm propagation. In it he summarized the general advantages of tissue culture techniques. In slightly modified form they are as follows:

- Production of healthy disease- and pest-free cultivars of female or male palms with desirable qualities, e.g. bayoud resistance in females; superior pollen in males;
- Large-scale multiplication of plantlets at any season of the year;
- Production of genetically-uniform progeny;
- Ability to propagate elite cultivars which lack offshoots and seed-only derived plants;
- Facilitate exchange of plant materials among laboratories for research purposes without risk of spreading diseases or pests and avoiding often cumbersome plant quarantine regulations;
- Reliable source of a large quantity of plantlets, if required.

Two tissue culture techniques are used. The first is somatic embryogenesis which involves plantlet production by generating embryos from cells not originating from reproductive organs. Tisserat (1981) is credited with developing the technique for date palm. In this technique, derived cells are cultured in a medium of growth regulators to produce a mass of disorganized cells called *callus*. Subsequently, the culture medium is modified to induce embryos from the callus. Key advantages of this method are that it produces bipolar structures ostensibly able to germinate, avoiding distinct phases of shoot and root induction. Continuous production is feasible to produce large numbers of plantlets at relatively low cost per unit. The main disadvantage is that high hormone levels in the medium make the plantlets prone to mutation which yields off-types, which are only manifested a few years later after field planting. Off-types produced by somatic embryogenesis, however, do represent induced genetic variability which may have desirable genetic traits of value in molecular breeding.

The second technique, direct organogenesis, utilizes meristematic cells and low concentrations of plant growth regulators in the medium. Axillary bud, root tip or floral bud cells can also be used. The advantage of this technique is that it avoids callus formation and produces plantlets directly, which are true-to-type and not susceptible to mutation. On the other hand, a comparatively small number of plantlets are produced because many do not survive the rooting stage. Organogenesis is slow and expensive as compared to somatic embryogenesis.

Research related to tissue culture is also investigating cell and protoplasm culture to produce callus and radiation to induce mutations *in vitro* to create promising genetic variants which then can be multiplied.

In several countries, commercial plantings have been and are being carried out using tissue-cultured plants derived by both techniques. Private companies and government-supported facilities are engaged in plantlet production. Tissue culture laboratories are expensive to establish and run, and need to produce large quantities of plantlets to achieve an economy of scale. The future potential of tissue culture appears to be greatest in countries with large areas in production needing rehabilitation or replacement, or to establish extensive new areas. As research reveals more about the date palm genome, tissue culture represents the best technique for selecting and propagating new improved cultivars.

Tissue culture also has a large potential role to play in date palm germplasm conservation. Long-term storage of tissue cultured material can be achieved by *in vitro* cryopreservation. Storage requires maintaining *in vitro* material at temperatures approaching –196°C, the temperature of liquid nitrogen. At that temperature all biological processes cease and the material remains genetically stable. Cryopreservation requires expensive refrigeration equipment and a reliable source of electrical power, but the actual storage space needed is minimal.

The conservation of date palm germplasm would be well served if a website were created and maintained to provide a means for researchers and others to post and retrieve data and information on the conservation status of cultivars. Such a website could function as a clearinghouse to identify cultivars in particular need of cryopreservation. It could also provide valuable information on the existence and status of cultivars maintained as *in vivo* germplasm collections. As research on the molecular biology of date palms progresses, *in vivo* palms will provide a means to compare the DNA, for example, of a Medjool cultivar infected with bayoud with one that is healthy.

1.4.2 Molecular Biology

Complementary to advances in tissue culture techniques are the development of new biotechnologies which can accelerate plant breeding efforts. DNA, or genetic, fingerprinting provides a genetic profile to reveal molecular markers linked to ancestry, cultivar and relationship to related plants, morphology, disease and pest resistance, drought tolerance, soil adaptability and so on. There are four major fingerprinting techniques:

- Restriction Fragment Length Polymorphisms (RFLPs);
- Randomly Amplified Polymorphic DNAs (RAPDs);
- Amplified Fragment Length Polymorphisms (AFLP);
- Simple Sequence Repeats (SSR), often referred to as microsatellites.

These techniques are discussed in detail in the chapters which follow. An important achievement in understanding the taxonomy of genus *Phoenix* palms was realized through a study of microsatellite markers which affirmed the validity of 14 species within the genus, and that *P. dactylifera* was initially domesticated from wild populations of the same species, and not derived from other species of *Phoenix* (Pintaud et al. 2010). Previously, there was uncertainty about the validity of certain taxonomic names, which was complicated by the fact that *Phoenix* species readily hybridize naturally in the wild where species distributions overlap or in cultivation when grown near each other.

DNA fingerprinting research is being carried out to confirm the identity of date palm cultivars in various date-growing countries including Algeria (Benaceur et al. 1991);

Egypt (El-Assar et al. 2005); Iraq (Khierallah et al. 2010); Morocco (Sedra et al. 1998); Oman (Al-Ruqaishi et al. 2008); Qatar (Ahmed and Al-Qaradawi 2010); Saudi Arabia (Al-Khalifah and Askari 2003); Sudan (Elshibli and Korpelainen 2008); Tunisia (Zehdi et al. 2004); UAE (Al Kaabi et al. 2007); and USA-California (Johnson et al. 2009). These selected references show the breadth of research activities under way, which are using the various techniques available. Genetic analysis can also reveal more about individual cultivars, such as the work on the Medjool cultivar which confirmed its status as a landrace variety (Elhoumaizi et al. 2006).

A milestone in molecular research on date palm was achieved with the sequencing of the draft genome of the Khalas cultivar. (http://qatar-weill.cornell.edu/research/datepalmGenome/download.html). This resource is freely-available to researchers on the internet and should bring about an increase in the use of molecular markers. Reflective of the rise of genomic studies of the date palm, there were four papers on the subject presented at the March 2010 Fourth International Date Palm Conference in Abu Dhabi.

A current key research issue is to determine date palm gender at a very early stage of development through genetic analysis. Some progress along these lines reveals the potential of molecular biology to differentiate gender. However, thus far molecular markers linked to gender have not been identified and this represents a key impediment to improvement of date palm. When this particular problem is solved it will represent a significant breakthrough and speed crop improvement. Future DNA studies also may reveal the genetic markers for fruit flesh being soft, semidry or dry, and for early, midseason and late ripening.

References

Ahmed TA, Al-Qaradawi AY (2010) Genetic diversity of date palm genotypes in Qatar as determined by SSR and ISSR markers. Paper presented at fourth international date palm conference, Abu Dhabi. Abstract Book p 80

Al Kaabi HH, Zaid A, Shephard H, Ainsworth C (2007) AFLP variation in tissue culture-derived date palm (*Phoenix dactylifera* L.) plants. Acta Hort 736:135–159

Al-Khalifah NS, Askari E (2003) Molecular phylogeny of date palm (*Phoenix dactylifera* L.) cultivars from Saudi Arabia by DNA fingerprinting. Theor Appl Genet 107:1266–1270

Al-Khayri JM (2005) Date palm *Phoenix dactylifera* L. In: Jain SM, Gupta PK (eds.) Protocols of somatic embryogenesis in woody plants. Springer, Netherlands, pp 309–319

Al-Khayri JM (2007) Date palm *Phoenix dactylifera* L. micropropagation. In: Jain SM, Häggman H (eds.) Protocols for micropropagation of woody trees and fruits. Springer, Netherlands, pp 509–526

Al-Ruqaishi IA, Davey M, Alderson P, Mayes S (2008) Genetic relationships and genotype tracing in date palms (*Phoenix dactylifera* L.) in Oman, based on microsatellite markers. Plant Gen Res Char Util 6:70–72

Benaceur M, Lanaud C, Chevallier M-H, Bounaga N (1991) Genetic diversity of the date palm (*Phoenix dactylifera*) from Algeria revealed by enzyme markers. Plant Breed 107:56–57

Dowson VHW (1921, 1923) Dates and date cultivation of the 'Iraq. Parts I, II, III. Heffer & Sons, Cambridge

Dowson VHW (1982) Date production and protection. FAO, Rome

El-Assar AM, Krueger RR, Devanand PS, Chao CT (2005) Genetic analyses of Egyptian date (*Phoenix dactylifera* L.) accessions using AFLP markers. Genet Res Crop Evol 52:601–607

Elhoumaizi MA, Devanand PS, Fang J, Chao CT (2006) Confirmation of 'Medjool' date as a landrace variety through genetic analysis of 'Medjool' accessions in Morocco. J Am Soc Hort Sci 131:403–407

Elshibli S, Korpelainen H (2008) Microsatellite markers reveals high genetic diversity in date palm (*Phoenix dactylifera* L.) germplasm from Sudan. Genet 134:251–260

Johnson C, Cullis TA, Cullis MA, Cullis CA (2009) DNA markers for variety identification in date palm (*Phoenix dactylifera* L.). J Hort Sci Biotech 84:591–594

Khierallah HSM, Bader SM, Baum M, Hamwich A (2010) Assessment of genetic diversity for some Iraqi date palms (*Phoenix dactylifera* L.) using AFLP markers. Paper presented at fourth international date palm conference, Abu Dhabi. Abstract Book p 84

Pintaud J-C, Zehdi S, Couvreur T (2010) Species delimitation in the genus *Phoenix* (Arecaceae) based on SSR markers, with emphasis on the identity of the date palm (*Phoenix dactylifera* L.). In: Seberg A, Peterson G, Barfod A, Davis J (eds) Diversity, phylogeny, and evolution in the monocotyledons. Aarhus Univ Press, Denmark, pp 267–286

Popenoe PC (1973) The date palm. Field Research Projects, Coconut Grove, Miami

Pruessner AH (1920) Date culture in ancient Babylonia. Am J Sem Lang Lit 36:213–232

Sedra MH, Lashermes P, Trouslot P, Combes M-C (1998) Identification and genetic diversity analysis of date palm (*Phoenix dactylifera* L.) varieties from Morocco using RAPD-markers. Euphy 103:75–82

Tisserat B (1981) Date palm tissue culture. Agricultural Research Service, Advances in Agricultural Technology, Western series, No. 17. Oakland, California

UN (1993) The United Nations convention on biological diversity. Concluded at Rio de Janeiro 5 June 1992, into force 29 Dec 1993

Zaid A (ed.) (2002) Date palm cultivation. Rev. ed. FAO, Rome

Zehdi S, Trifi M, Marrakchi M, Pintaud J-C (2004) Genetic diversity of Tunisian date palms (*Phoenix dactylifera* L.) revealed by nuclear microsatellite polymorphism. Hered 141:278–287

Zohary D, Hopf M (2000) Domestication of plants in the old world, 3rd edn. Oxford University Press, Oxford

Part I
Micropropagation

Chapter 2
Potential of Date Palm Micropropagation for Improving Small Farming Systems

M. Ferry

Abstract The date palm multiplication by tissue culture is very helpful in some definite cases including the propagation of cultivars resistant to diseases such as bayoud, propagation of rare and elite cultivars in high demand and propagation of new cultivars adapted to particular ecological and socioeconomic conditions. However, several essential precautions must be taken to prevent counterproductive and damaging results. It is especially necessary to adopt measures to avoid accelerating erosion of date palm agro-diversity by propagating a small number of cloned cultivars. It is also indispensable to employ a regeneration process assuring the genetic conformity of tissue-culture palms that are eventually distributed to farmers. It is recommended to avoid the use of the somatic embryogenesis approach which may produce seriously abnormal plants; whereas, strict organogenesis assures genetic stability. Finally, an assessment of the date palm sector must be achieved in relation to environmental, economic and social issues before a decision is made to initiate any massive tissue culture production program. Indeed, in many date palm growing countries, the limiting factors in the development of new date palm plantations are shortage of water and unfavorable socioeconomic trends, rather than the shortage of offshoots.

Keywords Agro-diversity • Bayoud • Flowering • Organogenesis • Socioeconomics • Somaclonal variation • Somatic embryogenesis • Sustainable development

2.1 Introduction

The first research studies on date-palm propagation by tissue culture go back to the 1970s (Ammar et al. 1977; Eeuwens 1978; El-Hennawy et al. 1978; Poulain et al. 1979; Reuveni et al. 1972; Schroeder 1970; Smith 1975; Tisserat 1979).

M. Ferry (✉)
INRA, Phoenix Research Station, Apartado 996, 03201 Elche, Spain
e-mail: m.ferry@telefonica.net

S.M. Jain et al. (eds.), *Date Palm Biotechnology*,
DOI 10.1007/978-94-007-1318-5_2, © Springer Science+Business Media B.V. 2011

Both regeneration techniques, organogenesis and somatic embryogenesis, were investigated. At that time, the applied interest of these investigations was focused on obtaining date palm resistant to bayoud disease, in addition to meeting the strong demand for date palm shoots of a few high-quality commercial cultivars. Lack of offshoots in various production regions (California, Middle East) was a problem and prices were very high.

Assessment of the development and behavior of the first obtained tissue culture-derived date palms were realized during the 1980s (Ferry et al. 1986, 1988). The first fruit harvest of tissue culture-derived date palms occurred in 1989 (Ferry et al. 1993). The conformity of the date palms produced by organogenesis was established (Al-Ghamdi 1996; Booij et al. 1993; Ferry et al. 1990). Date palm plants obtained by somatic embryogenesis were already showing abnormalities during the 1980s, especially dwarfism (late J. Carpenter, 1985, personal communication). During the same time period, a serious problem of abnormality was observed in plantations of oil palms produced by somatic embryogenesis (Corley et al. 1986). Starting from 2000 onwards, many research papers have described similar abnormalities with date palms obtained by somatic embryogenesis (Al-Wasel 2001; Cohen et al. 2004; McCubbin et al. 2000).

Some 40 years after the first demonstration of date palm propagation by tissue culture, it is essential to assess the benefits, limitations and perspectives of this technique, especially for small-farming systems.

2.2 Benefits of Micropropagation

The main benefit of date palm micropropagation is to respond more quickly than with offshoots to the needs of the farmers. From a date palm, an average of 10 offshoots can be obtained by the traditional technique. The propagation rate with this technique does not exceed 10 every 10–15 years. By micropropagation based on organogenesis, this rate is of 3 every 2 months; more than half million shoots can be obtained within 2 years from only one proliferating explant.

But, it is important to note that the very low potential of propagation with offshoots can be considered quite sufficient as demonstrated by the two following examples. First, California planters continue to use offshoots for the renewal of their plantation even though it was in California, in 1979, where the first propagation of date palm by somatic embryogenesis from offshoots was achieved (Tisserat 1979). Second, in Tunisia, the considerable development of cv. Deglet Noor plantations from the 1970s has entirely been based on the use of offshoots (Ferry 1996).

In fact, it is mainly in some countries of the Arabian peninsula that plantations of tissue culture-derived date palms have reached relative importance due to the very high purchasing power in this region. Indeed, the usual price for tissue culture-derived shoots is very high and not at all accessible to small farmers. Recently plantlets have

Fig. 2.1 Date palm attacked by bayoud disease

been proposed to farmers in the Sahel at the excessive price of 30 Euros per unit (Rédaction 2010). This is more than 30 times the price of a tissue culture-derived banana plantlet in Africa (Bauer et al. 2009). The prices presently proposed for date palm tissue culture derived plantlets are much too high for small farmers. This situation is in fact similar to the one that predominates with other species of tissue culture-derived shoots that represent elite material (Ahloowalia 2004).

In vitro multiplication of date palm has both certain advantages and disadvantages. Since *in vitro* propagation of date palm makes available a large number of cloned plants in a short time, there is a risk to accelerate agro-diversity erosion. It could lead to reducing the number of cultivated varieties to a smaller group of cultivars of international reputation propagated by a few commercial tissue-culture laboratories.

Contrary to this risk, date palm *in vitro* propagation allows the multiplication of very rare quality genotypes or of genotypes without offshoots. In this last case, the operation is feasible with various types of explants: (1) Use of undifferentiated buds (Ferry et al. 1999), apex or very young leaves. However, collecting this type of explants means the sacrifice of the palm that is the unique specimen of the selected genotype; (2) Young leaves collected and cultivated, according to the technique used for the oil palm tissue culture (Noiret et al. 1988); (3) Young inflorescences. There is no need to sacrifice the palm when the flowering pattern is well established and the right extraction technique is used.

Tissue culture multiplication technique has another important interest. This technique allows a secure exchange of healthy material. This issue is crucial regarding date palm to avoid the spread of at least two organisms deadly for the date palm: the *Fusarium oxysporum* f. sp. *albedinis* responsible for bayoud disease (Fig. 2.1) and the red palm weevil, *Rhynchophorus ferrugineus* (Fig. 2.2). The present dramatic spread of both organisms has been due to the exchange of infested offshoots and the importation of ornamental palms.

Fig. 2.2 Red palm weevil (*Rhynchophorus ferrugineus*)

2.3 Limits of Micropropagation

Tissue-culture palms are to produce date fruits, but of what quality and for which outlets? Since the first research on date palm tissue culture, interest in it has created an infatuation with the technique that has not always been justified by the actual and reasonable needs of the growers. In a larger way, this infatuation reflects the general one associated with the accelerated and unsustainable development of date palm plantations over about the past 30 years.

Since the beginning of 1980s, there has been a spectacular increase in the date production worldwide (Fig. 2.3) (Ferry 2005). However, there has been a decline in the palm yield during the last 20 years as compared with the previous period (Fig. 2.4) (Ferry 2005).

Therefore, the accelerated increase of world production is not the result of a yield increase but of an exponential increase in the number of planted palms. Several countries have increased their date palm plantations, but in one in particular, United Arab Emirates, this increase has been really enormous: from a population of 2–3 million palms trees in the 1970s, the UAE has reached today more than 40 million trees and now occupies first place in the world for the number of date palms.

On the worldwide scale, the main date-growing problem is the very poor yield per palm and not the lack of palms. The worldwide production of dates is only an average of 30–35 kg per palm a year, whereas the normal potential of date production per palm is about 100–120 kg per year. The low productivity of palms is not due to genetic limitations but rests with agronomic and socioeconomic causes. Among these is the general abandonment of oases by the younger generation in most of the date palm growing countries (Ferry 1996).

The primary agronomic reason for poor palm productivity is a high deficit of irrigation water. In date palm-growing regions, water resources are generally limited whereas evapotranspiration is high. For the last 30 years, the development

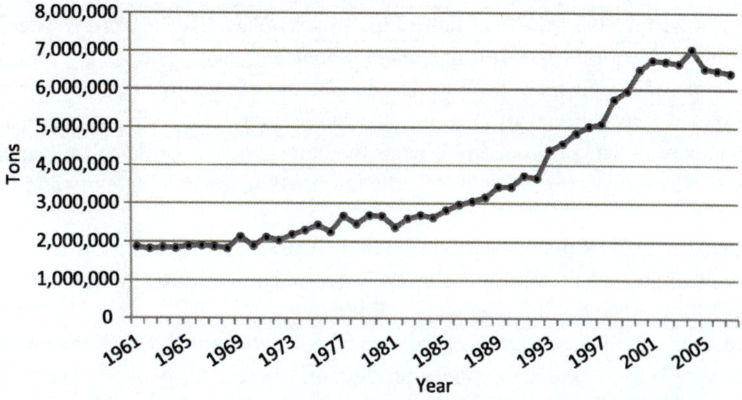

Fig. 2.3 Evolution of the world total yield (tons) of dates during the last five decades

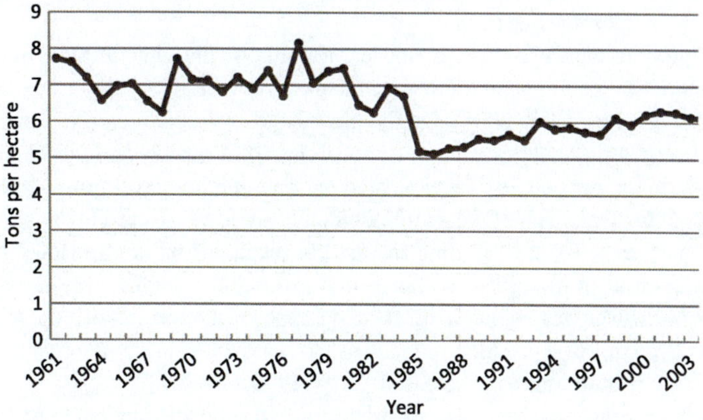

Fig. 2.4 Evolution of average world yield per area (tons/ha) date palm during the last five decades

of agriculture in these regions has contributed to the emergence of an extremely serious water-shortage situation in many countries. In the majority of date palm-growing countries, water-resource exploitation exceeds the natural renewal capacities, or relies on nonrenewable fossil water resources which are more and more problematic and costly to extract (Ferry 2005).

With the development of new technologies for water resource exploitation, the management of these resources, that was previously well balanced, has evolved into a mining exploitation with disastrous consequences: artesian water source depletion, rapid water-table lowering, intrusion of salty water of marine or underground origin, and both water table and soil salinization (Bouhlassa et al. 2008; Lemarchand 2008). The situation has become so precarious in many countries that total drying-up of some palm groves can be observed. This phenomenon is clearly visible in the countries where the extension of the palm groves has been considerable, for example, in the United Arab Emirates (Dakheel 2003).

This unsustainable growth of date palm groves raises other questions, beside the irreversible deterioration of soil and water resources: what has been the purpose to plant so many palms and to produce so much larger quantities of dates? For which national and international market and at what selling price, corresponding to the huge increase in date production? What is the long-term economic and social profitability of this growth for the date farmers, for others in the date industry and for the governments?

More than 90% of the dates produced in the world are consumed domestically in the countries of production and chiefly in the areas of production (Ferry 2005). Reliable future date export expansion is limited.

Date fruits are consumed mainly by two types of consumers: First, the producers and the people living in the regions of production where dates constitute an essential part of the diet; second, people of the Moslem religion or culture. The consumption of dates is traditionally associated with breaking of the fast during Ramadan and major religious and domestic feasts. In the western world, the importation and consumption of dates are most important in those countries where there is an immigrant population from Islamic countries.

Even in predominantly Islamic countries, nowadays, the general tendency is a gradual decrease in date consumption with the development of an urban way of life and increased income (Ferry 2005). On the other hand, with the increase of income, consumers become more demanding in terms of date fruit quality and presentation.

Therefore, to the opposite of the explosive palm plantation development that has been fomented for these last 30 years, the objective should be to produce dates in a sustainable manner without degrading the natural resources while offering farmers sufficient economic profitability. To reach this profitability, market prices for the dates must compensate for farmer's efforts and expenses. Unfortunately, the present situation of date fruit overproduction leads to the opposite result and explains why there is a manpower shortage in the oases (Ferry 1999).

Another issue must be taken into consideration. To produce dates of better quality does not mean only to promote cultivars of superior quality. Other factors are also decisive: proper cultural and fruit storage techniques, and efficient trade and marketing to offer healthy and attractive fruit.

Tissue culture technology application has a role to play in this perspective but its use will indeed only be useful provided a full set of the other components are taken into consideration.

2.4 Prospective

The need for tissue culture-derived date palms exists to contribute to the solution of very specific problems. A well-known problem pertains to the temporary situation in locations where offshoots of commercial date palm cultivars are nonexistent or scarce. This was the reason for the initiation of tissue culture research in California 40 years ago. In those localities where the idea is to introduce for the first time elite commercial cultivars, it is absolutely necessary to ensure beforehand the adaptability of these cultivars to local environmental conditions. Unfortunately, this obvious preliminary experimental path is often forgotten. Such is presently the situation in the Sahel and it was the case with the cv. Deglet Noor. Several laboratories undertook industrial tissue culture multiplication of this internationally very well-known cultivar in the hope to sell shoots in the Middle East. However, this cultivar requires strict ecological conditions and its introduction in many places, especially in the Middle East, failed.

In the fight against bayoud disease, tissue-culture multiplication of resistant and high-quality cultivars continues to represent the only feasible solution. However, results obtained until now are far from satisfying the needs. Some 25 years after the start-up of the tissue culture multiplication program in Morocco, only 300,000 date palms have been produced, 90% of which are sensitive to the bayoud disease. Practically, the only genotype called Nejda, presenting both characteristics of bayoud resistance and high fruit quality, has been distributed so far to the farmers. The need to reconstitute groves destroyed by bayoud disease will require several millions of date palms. It is necessary to underscore that the somatic embryogenesis technique in Morocco has been rejected in favor of organogenesis. This choice was been made in order to avoid all risk of distribution of abnormal tissue culture-derived date palms that would have been catastrophic for small farmers, who constitute the majority of the date palm growers in the country.

With date palm, the control of the process at the industrial stage is much more difficult to achieve when using organogenesis rather than somatic embryogenesis. On the other hand, organogenesis guarantees the conformity of the material offered to the farmers, as has been verified once more in Morocco (Aaouine 2000).

For the extension of the date palm culture beyond its traditional area, the use of the tissue-culture process represents considerable interest. As an example, the situation in the Sahelian region is pertinent.

In this region, for some tens of years, many projects have, one after the other, sought to introduce date palm culture, starting with a small stock of offshoots from the few palm groves originating in the northern Sahelian area. In spite of the serious efforts committed to this objective, none of the projects has succeeded. Even if socio-economic considerations largely explain these failures, the difficulties associated with the acquisition and transportation of the offshoots also hampered success. To make available date palm planting material from tissue culture constitutes a complete change of perspective.

However, before distributing tissue-cultured date palms in the Sahel, it is imperative to conduct preliminary behavior tests of the cultivars to be introduced. The extension of date palm culture in this region is facing a limiting factor of crucial importance: it is imperative that the maturation of the fruits takes place before the start of the rainy season otherwise the dates will rot before being harvested. The more one descends in latitude, from the Sahelo-Saharian strip toward the Sahelo-Sudanese strip, the greater the risk of this occurring. The first behavior trials were conducted in the area of Gao, Mali, 20 years ago (Fig. 2.5).

This type of preliminary testing is absolutely indispensable when introduction is planned to take place at the southern limits of the Sahelo-Saharian strip. For some

Fig. 2.5 Tissue culture date palm behavior trial at Gao, Mali, in 1988

years, several cultivars have been distributed in regions at the latitude of Niamey, Niger, without such trials having been done. Such a decision constitutes a serious mistake. Furthermore, among the cultivars distributed, cv. Barhee is of the very soft type and ripens very quickly. Even though this date cultivar can be eaten before it is fully ripe (in the khalal stage), Barhee fruit storage will be impossible for the small farmers. The selection of cv. Barhee represents *a priori* a bad candidate for cultivation in Sahel and, in any case, it is seriously irresponsible to distribute the cultivar without preliminary behavior trials.

In the palm groves of the southern part of the Sahelian strip, many date palms flower twice a year (Fig. 2.6). Important research work has been accomplished on this phenomenon (Jahiel 1996). On completion of this work, it was observed that the abnormal flowering of some of these palms was predominant. As an example, Table 2.1 shows the observations made of a date palm in Kojimeri, Niger, on 15 May 1995.

All bunches and inflorescences in Table 2.1 were localized after the establishment of the phyllotaxis pattern and according to their precise rank on the chronological spiral starting from the oldest (n°1). Without any aborted axillary production, this palm carries, at the same time, 14 bunches exhibiting abnormal flowering (mature dates or about to ripen) and 7 inflorescences exhibiting normal flowering of the year.

These genotypes are ideal for developing new cultivars after proving that this character is not influenced by the environmental conditions. Indeed, these cultivars would flower totally or principally during the abnormal period and consequently would mature their fruit long before the rainy season.

Fig. 2.6 Date palm flowering twice per year in Niger

Table 2.1 Axillary production of a predominant twice-flowering genotype

Chronological rank	Type of axillary production	Chronological rank	Type of axillary production
1	Not harvested bunch	12	Harvested bunch of 4.5 kg
2	Harvested bunch of 5 kg	13	Bunch with dates at khalal stage
3	Harvested bunch of 5.5 kg	14	Bunch with dates at khalal stage
4	Harvested bunch of 5.5 kg	15	Inflorescence
5	Harvested bunch of 4.5 kg	16	Inflorescence
6	Harvested bunch of 7 kg	17	Inflorescence
7	Harvested bunch of 7.5 kg	18	Inflorescence
8	Harvested bunch of 6 kg	19	Inflorescence
9	Harvested bunch of 7.5 kg	20	Inflorescence
10	Harvested bunch of 6.5 kg	21	Inflorescence
11	Harvested bunch of 7.5 kg	22	Inflorescence

Apart from these ecological constraints, extension of date palm culture in the Sahel is also made difficult due to the very specific socio-cultural context. A large proportion of the population living in the Sahelian strip is essentially made up of nomadic herders, without any tradition of farming, except in some cases of the practice of seasonal rainfed cropping. The recurrent droughts of recent years, however, have stimulated emergence of agro-pastoralism which is far from being well controlled. Even though, for cultural reasons, these agro-pastoralists express an interest in principle in date palm culture, this interest does not always correspond to a sincere will or to a valid capability to dedicate time and means to this crop. The identification of people who are truly ready to grow date palms requires a very good

Fig. 2.7 Date palm micropropagation by organogenesis

understanding of the social environment and a very progressive intervention. Experience acquired in this domain demonstrates that the adoption of the date palm has no chance of success unless it is integrated into diversified farming systems, including various other crops (especially food crops) and associated with livestock-raising activities (Ferry et al. 2001).

Candidate farmers ready to grow tissue culture-derived date palms need complete training as to the different specific practices of date palm culture. In addition, they have to receive regular and prolonged technical support until at least the third or fourth harvests. Initiatives involving the massive distribution of tissue culture date palm plants without any extended program of assistance and training are bound to fail (Ferry 1998).

As to the economic issue, it is essential that the proposed cultivars, in addition to their agro-climatic adaptability which has already been emphasized, yield dates competitive with those that are usually imported. The selection of the cultivars to propagate by tissue culture must be based, therefore, on a multiple-criteria approach before their multiplication and distribution to candidate agro-pastoralists to become date-palm growers.

A last point of very high importance must be respected: it is essential to provide the candidate famers, not only well adapted cultivars, but also they must be true-to-type tissue culture-derived date palms that do not present any abnormalities. This true-to-typeness is guaranteed when the tissue culture palms have been produced by strict organogenesis (Fig. 2.7), without any callus formation as minimal it may be.

Fig. 2.8 Date palm micropropagation by somatic embryogenesis

Fig. 2.9 Abnormal flowering of *in vitro* date palms propagated by somatic embryogenesis

It has been well-documented that the date palms obtained by indirect somatic embryogenesis (based on a maintained callogenesis phase) (Fig. 2.8) may present serious flowering abnormalities (Fig. 2.9). The most important abnormalities are infertility of the female flowers and dwarfism. For the cv. Barhee, the batches produced by some laboratories were 100% abnormal (Al-Wasel 2001). One or two laboratories now have modified their protocol (in particular the important reduction in the number of tissue culture cycles) and have succeeded to reduce this risk of anomaly. A mixed embryogenesis/organogenesis method could also eliminate the risk of abnormalities development (Ferry et al. 2000). But, for distribution to small farmers, this risk still remains too high. It is interesting to note that the commercial laboratories that produce oil palms shoots by somatic embryogenesis do not distribute these shoots to small farmers (Durand-Gasselin 2009). This distribution concerns only large companies that can economically run the risk to plant an unknown percentage of abnormal shoots.

2.5 Conclusion

It is clear that tissue culture multiplication technique can be a very efficient tool for the development of the date palm culture in defined cases: to reconstitute the groves destroyed by bayoud disease by planting resistant and high fruit quality cultivars, to improve the quality of the date fruit production with elite cultivars, to extend the date culture into new areas with adapted cultivars and to facilitate safe germplasm exchanges.

One of the main technological factors limiting the use of this technique is the production of abnormal plants, when plantlets are obtained by somatic embryogenesis. If this problem can be solved, this method of propagation could reduce the cost of production and, consequently, the selling prices of date palm tissue culture-derived shoots. At the moment, the shoots that are supplied with a true-to-typeness guarantee are produced by organogenesis. However, this technique does not offer the same propagation speed. It is a high labor-consuming process and consequently a costly one. Furthermore, very few laboratories control organogenesis at an industrial scale.

Nevertheless, date palm organogenesis-derived shoots represents, even for the small farmers, a high interest to obtain selected mother plants. From these plants, the farmers can increase or renew their plantations from the offshoots produced by the mother plants. This process leads to a relatively slow palm grove development or renewal.

This slow development associated with a participative-approach strategy is usually better adapted to the small farmer capacity and, quite preferable to the uncontrolled, unreasonable and unsustainable date palm plantations growth that has been adopted in some countries over the last 30 years.

References

Aaouine M (2000) Production of date palm vitroplants: the Moroccan experience. Proceedings of the date palm international symposium, Namibia, 22–25 Feb 2000. pp 46–52

Ahloowalia BS (2004) Integration of technology from lab to land. Proceedings of a technical meeting on low cost options for tissue culture technology in developing countries, Vienna, 26–30 Aug 2002. IAEA, pp 87–89

Al-Ghamdi A (1996) True to type date palm produced through tissue culture techniques. Proceedings of the third symposium on date palm. Al Hassa, 17–20 Jan 1993. Mars Publishing House, pp 39–53

Al-Wasel S, Abdelrahman S (2001) Field performance of somaclonal variants of tissue culture-derived date palm (*Phoenix dactylifera* L.). Plant Tissue Cult Biotech 11:97–105

Ammar S, Benbadis A (1977) Multiplication végétative du palmier dattier par la culture de tissus de jeunes plantes issues de semis. CR Acad Sci 284:1789–1793

Bauer V, Burkart S, Abele S, Kahangi E (2009) Financing, handling, hardening and marketing of tissue culture-derived planting material through nurseries: the case of banana in Kenya, Uganda and Burundi. Poster at Tropentag 2009, University of Hamburg, 6–8 Oct

Booij I, Piombo G, Risterucci JM et al (1993) Sugar and free amaino acid composition of five cultivars of date palm from offshoots or vitroplants in open field. J Agric Food Chem 41:1553–1557

Bouhlassa S, Alechcheikh C, Kabiri L (2008) Origine de la minéralisation et détérioration de la qualité des eaux souterraines de la nappe phréatique du quaternaire du bassin-versant de Rheris (Errachidia, Maroc). Sécher 19:67–75

Cohen Y, Korchinsky R, Tripler E (2004) Flower abnormalities cause abnormal fruit setting in tissue culture-propagated date palm (*Phoenix dactylifera* L.). J Hort Sci Biotech 79:1007–1013

Corley R, Lee C, Law L et al (1986) Abnormal flower development in oil palms clones. Planter 62:233–240

Dakheel A (2003) Date palm and biosaline agriculture in the United Arab Emirates. In: The date palm from traditional resource to green wealth. ECSSR, Abu Dhabi, pp 199–211

Durand-Gasselin T (2009) Production de vitroplants de palmier à huile à grande échelle. 3ème séminaire du Réseau, Biotechnologies du Palmier Dattier, Montpellier, 18–20 Nov 2008

Eeuwens CJ (1978) Effects of organic nutrients and hormones on growth and development of tissue explants from coconut and date palms cultured *in vitro*. Physiol Plant 42:173–178

El-Hennawy HM, Wally YA (1978) Date palm bud differentiation in vitro. Egypt J Hort 5:81–82

Ferry M (1996) La crise du secteur phoenicicole dans les pays méditerranéens. Quelles recherches pour y répondre? Opt Méditer. CIHEAM Série A 28:129–156

Ferry M (1998) Le développement du palmier dattier en zone semi-aride au Sahel. Flamboy 46:27–30

Ferry M (2005) Production et commercialisation de la datte dans le monde. Situation et perspectives. Symposium international sur le développement agricole durable des systèmes oasiens. 08–10 mars 2005, Erfoud, Maroc, INRAM, p 18

Ferry M, Ruiperez E (1999) Organogenesis *in vitro* a partir de las producciones axilares de palmeras datileras adultas de Elche. VIII Congreso Nacional de Ciencias Horticolas, Murcia, pp. 147–151

Ferry M, Louvet JM, Desfonds V (1986) Resurgence of juvenile foliage characters among *in vitro* date palm plants. Second symposium on date palm. King Faisal University, Mar 1986, Abstract

Ferry M, Toutain G, Desfonds V et al. (1988) Morphological differences in the development of date palm vitroplants of various varieties. Symposium Irako-Français sur la culture de tissus de palmier dattier, Baghdad, Iraq, Sept 26–29, p 5

Ferry M, Toutain G, Al Faihani A et al. (1990) Behaviour of date palm vitroplants from *in vivo* transfer to the first date production. Seventh international congress on plant tissue and cell culture, Amsterdam, 24–29 June, Poster A3

Ferry M, Toutain G, Monfort S (1993) La multiplication du palmier dattier. In: Physiologie des arbres et arbustes en zones arides et semi-arides. Groupe d'études de l'arbre, Paris, pp 353–363

Ferry M, Navarro J, Ruiperez E, et al. (2000) An embryogenesis-organogenesis mixed system for *in vitro* date palm durable mass propagation. In: Proceedings of date palm international symposium, Windhoek, Namibia, Feb 22–25, p 27

Ferry M, De Saint SA, Maha I (2001) Le palmier dattier dans les systèmes de production pastoro-oasiens du Nord-Niger. L'acacia 23:23–24

Jahiel M (1996) Phénologie d'un arbre méditerranéen acclimate en région tropical :le dattier au sud du Niger et son appropriation par la société Manga. Thèse Université de Montpellier 2

Lemarchand F (2008) Les nappes fossiles du Sahara. La Rech 421:82–87

McCubbin MJ, Van Staden J, Zaid A (2000) A southern African survey conducted for off-types on date palms produced using somatic embryogenesis. Proceedings of the date palm international symposium, Windhoek, Namibia, 22–25 Feb, pp 68–72

Noiret JM, Gascon JP, Pannetier C (1988) La production de palmier à huile par culture in vitro. Oléag 40:365–372

Poulain C, Rhiss A, Beauchesne G (1979) Multiplication végétative en culture in vitro du palmier dattier. CR Séan Acad Agric 11:1151–1154

Rédaction (2010) Des dattiers au Sahel. Spore no 145, pp 6

Reuveni O, Adato Y, Lillien-Kipnis H (1972) A study of new and rapid methods for the vegetative propagation of date palms. Date Grow Inst Rep 49:17–24

Schroeder CA (1970) Tissue culture of date shoots and seedlings. Date Grow Inst Rep 47:25–27

Smith SN (1975) Vegetative propagation of the date palm by root tip culture. Bull Agron Sahar 1:67

Tisserat B (1979) Propagation of date palm (*Phoenix dactylifera* L.) in vitro. J Exp Bot 30:1275–1283

Chapter 3
Date Palm Tissue Culture: A Pathway to Rural Development

K. Rajmohan

Abstract Date palm, being a crop of the arid and marginal lands, has an integral relationship with the life of the rural people. It has excellent potential for improving the rural sector, and contributing to the economic, social and cultural aspects of rural areas. Various primary and secondary products of date palm add to the economic and social security of the people. Date palm cultivation helps generate considerable opportunities for rural employment, ensures livelihood and food security of the rural masses and facilitates eco-restoration. It can act as a catalyst for the development of the rural sector in the arid regions. Large-scale scientific cultivation of date palm in traditional and non-traditional areas necessitates quality planting materials in sufficient quantity. Tissue-culture propagation ensures rapid multiplication and establishment of true-to-type plants of elite cultivars. It is an effective and efficient alternative for conventional vegetative propagation. Research and development work to design viable protocols for the tissue-culture propagation of leading commercial cultivars of date palm has been undertaken at various centers. Several commercial agencies are involved in the large-scale multiplication and distribution of date palm using *in vitro* techniques. Tissue-culture technology, its potential and impact on rural development and the efforts for large-scale cultivation of date palm using tissue culture plants in the arid regions of India are discussed.

Keywords Food security • Rural employment • Rural development • Tissue culture

K. Rajmohan (✉)
Atul Rajasthan Date Palms Ltd., Chopasni, Nandanvan P.O., Jodhpur 342008, Rajasthan, India
e-mail: rajmohan33@yahoo.com

S.M. Jain et al. (eds.), *Date Palm Biotechnology*,
DOI 10.1007/978-94-007-1318-5_3, © Springer Science+Business Media B.V. 2011

3.1 Introduction

Date palm is a traditional crop and in recent years has gained acceptance in 40 countries including the USA, South Africa, and Australia. It is well-known for its ability to withstand extremes of temperature and can tolerate very high summer temperatures as well very cold winters. Date palm is best suited to semi-arid and arid climates and can withstand long drought periods without irrigation and is well adapted to saline conditions.

Date palm, when propagated by seedlings, takes 4–7 years after planting to bear fruit (Zaid 2002). Mature date palms can produce 80–200 kg of fruit per year. Dates ripen in four stages: *kimri* (unripe), *khalal* (full-size, crunchy), *rutab* (ripe, soft) and *tamar* (ripe, sun-dried). Fresh date is a premium source of vitamin C and supplies 230 kcal (960 kJ) of energy per 100 g fruit. It is rich in iron, potassium and calcium and is an abundant source of sugar.

Three main cultivar groups of dates exist; soft (e.g. cvs. Barhee, Halawy, Khadrawy, Medjool), semi-dry (e.g. cvs. Dayri, Deglet Noor, Zahidi) and dry (e.g. cv. Thoory). Date palm is dioecious, having separate male and female plants. There are more than 3,000 cultivars grown worldwide (Zaid 2002).

The date palm is a multipurpose tree and the fruit has high nutritional value and a long storage life. Products from date fruits include syrup, jam and beverages. Building materials and handicrafts are produced from the trunks and leaves. Date seeds are used for animal feed. The palm requires only limited inputs and can be productive for about 65 years. Besides direct consumption, date fruits are traditionally used to prepare a wide range of products such as date juice concentrates (spread, syrup and liquid sugar), fermented date products (wine, alcohol, vinegar, organic acids) and date pastes for different uses in baking and confectionary. Also, by-products arising from date processing can be used for different purposes. Within agricultural systems date press cake (a by-product of date juice production) as well as date pits can be used as animal feed.

Worldwide date production has increased exponentially over the last three decades (FAO 2008). Annual expansion of about 7% was recorded. The total number of date palms has been estimated to be approximately 100 million, distributed in 40 countries. Most of the major date-producing countries have steadily expanded production over the last few decades.

Date palm is an important subsistence crop of the desert regions and is a rich source of nutrition, contributing to food security. Being a crop of arid, semi-arid and marginal lands it has an integral relationship with the life of the rural people. It has excellent potential for improving the rural sector fostering holistic development. It can make significant contributions to the economic, social and cultural aspects of rural areas (Safwat 2007). There are ample opportunities for rural employment and development involved in large scale date palm cultivation. Scientific and systematic cultivation requires quality planting materials in sufficient quantity. Tissue-culture propagation ensures rapid multiplication and establishment of true-to-type plants of choice cultivars.

3.2 Methods of Date Palm Propagation

3.2.1 *Traditional Propagation*

Date palm can be easily grown from seed. However, the seedlings are variable and take 4–7 years to fruit. Furthermore, 50% of the seedlings may turn out to be males. Also, date fruits from seedling plants are often smaller and of poorer quality. The common means of propagation is by transplanting suckers or offshoots when they are 3–5 years old. Plants grown from offshoots will fruit 2–3 years earlier than seedling plants. However, offshoots develop slowly and the numbers produced per tree are limited. Also, they are produced only during the early years of life of the mother palm. Offshoots will not survive in the field unless they are large enough. Hence the propagation efficiency of date palm via offshoots is low. Basal offshoots with a well-developed root system and 10–15 kg in weight are being used for planting. However, after separation from the mother palms they have to be maintained in a nursery for about 1 year for better development of the root system. Aerial offshoots are not used.

3.2.2 *Tissue Culture Propagation*

The production and productivity of date palms can be substantially improved by using quality planting materials of superior genetic stock. *In vitro* clonal propagation is an effective and efficient alternative for conventional vegetative propagation, to ensure rapid multiplication and establishment of true-to-type plants of elite cultivars (Al-Khayri 2005, 2007). The annual demand for date palm planting materials has been estimated to be seven million plants. Protocols have been developed for tissue-culture propagation of date palm. Tissue-culture techniques proved useful for commercial propagation of date palm (Hoop 2000). Several cultivars are being multiplied worldwide (Larbi 2010). There are a number of public and private sector agencies involved in commercial tissue-culture propagation of date palm. The UAE University in Al Ain, Al Rajhi in Saudi Arabia, Green Coast Nurseries in Fujairah, Al Wathba Marionnet in Iran, Date Palm Developments in UK, etc. are some of the leading agencies in this regard, producing millions of tissue-culture plants annually. UAE University is reported to have tissue-culture propagation protocols for about 50 cultivars of date palm. Atul Rajasthan Date Palms Ltd, has initiated actions for the massive greening of the desert areas of India, producing tissue-culture date palm plants of suitable cultivars. Tissue-culture techniques help to achieve precise production targets. Stringent specifications can be met, for increasing the bargaining power of the grower. The process can deliver planting materials to suit pre-determined mass production programs to meet well-planned market demands.

3.2.2.1 An Overview of Date Palm Tissue Culture

The earliest reports on date palm tissue culture were made by Tisserat (1979) and Reuveni (1979). Subsequently several publications dealt with standardizing and refining the technique with respect to various date palm cultivars, adopting different methods. There are mainly two methods of *in vitro* propagation adopted for date palm tissue culture viz. direct organogenesis and somatic embryogenesis (Raj Bhansali 2010). Explants like zygotic embryos, root segments (Sharma et al. 1980), young leaves (Sharma et al. 1984), shoot apices (Al-Khayri 2001; Raj Bhansali et al. 1988; Zaid and Tisserat 1983), inflorescence (Bhaskaran and Smith 1992; Drira and Benbadis 1985; Loutfi and Chlyah 1998) and adventitious buds (Bouguedoura et al. 1990) have been used for this purpose.

During the past three decades there were about 50 publications on the *in vitro* propagation of date palms. Murashige and Skoog (1962) medium is the most widely used for date palm tissue culture (Raj Bhansali 2010). The basal medium is supplemented with amino acids (arginine, asparagines, glycine, adenine and glutamine), vitamins (inositol, biotin, pyridoxine, nicotinic acid and thiamine), sucrose, etc. Activated charcoal and polyvinylpyrrolidone have been observed to reduce the problem of phenolic interference. Various auxins (2,4-dichlorophenoxy acetic acid, indole acetic acid, naphthoxy acetic acid, naphthalene acetic acid) and cytokinins (6-benzyl aminopurine, kinetin, 2-iP) are also being used.

Protocols for *in vitro* somatic embryogenesis in date palm have been developed by several researchers (El Hadrami et al. 1995; Raj Bhansali et al. 1988; Sharma et al. 1986; Sudhersan et al. 1993). The explants are incubated in complete darkness for 3–6 months, in culture rooms for the production of embryogenic callus. The phenolics interference is overcome by the use of activated charcoal, polyvinylpyrrolidone, cysteine-HCl, ascorbic acid and citric acid (Dass et al. 1989; Raj Bhansali 1988; Raj Bhansali and Kaul 1991; Zaid and Tisserat 1983). Frequent subculturing is also adopted to overcome a browning problem. Regeneration of somatic embryos occurs when the callus is subcultured using suitable media, often hormone-free. Suspension culture of date palm friable callus for rapid somatic embryogenesis has been established (Bhaskaran and Smith 1992; Fki et al. 2003; Othmani et al. 2009a; Sharma et al. 1986). Hundreds of somatic embryos can be developed from suspension culture within a short time.

Refinement of protocols for direct organogenesis in date palm has been accomplished by several researchers (Sudhersan et al. 1993; Zaid et al. 2002). However, the rate of multiplication is less as compared to somatic organogenesis (Raj Bhansali 2010). The explants enlarge in the culture establishment medium within 2–3 months and then regenerate into adventitious buds. These further develop into shoots within another 6 months. The shoots develop into plantlets on transfer into rooting medium. There are recent reports on shoot proliferation from shoot apex culture (Eshraghi et al. 2005; Taha et al. 2001).

3.2.2.2 Improvement of Technology

The rate of *in vitro* multiplication needs to be improved further. A major problem in tissue-culture propagation of date palm, at present, is the long gestation period for the production of plantlets as compared to other crops. This needs to be reduced. Also, comparative evaluation of field performance needs to be done for the various methods of *in vitro* propagation. Somaclonal variation under field conditions has to be assessed properly (Al-Wasel 2001; Othmani 2009b). The present cost of production needs to be reduced drastically for popularizing tissue-culture propagation of date palm. The cost of date palm plantlet production is notoriously high as compared to that of other horticultural crops. For example, it is more than 100 times that of banana. The comparatively slow and less efficient regeneration process contributes to a major share to this. The average sale price of tissue culture plants is USD 20–23 (Zaid 2002). The price varies with cultivar. Elite cultivars may cost even USD 50–60 each. Hence, there is a pressing need to improve the protocols, to bring down the plant price to affordable levels of the ordinary farmers. A higher rate of *in vitro* proliferation can bring down the cost. Somatic embryogenesis-based protocols may be desirable in this respect, as the rate of multiplication is comparatively high.

3.2.2.3 Production and Distribution of Planting Materials

The present world population of date palms is about 100 million. The annual increase in area is about 7%, based on statistics for the last 10 years. Hence the annual demand for planting materials could be around seven million plants. Tissue-culture propagation of date palm needs to be popularized for improving the production and productivity of this crop, using superior genotypes. Sufficient tissue-culture production units have to be established in order to meet the demands for planting materials. As the cost of plantlets is comparatively high, they have to be made available to the farmers at subsidized rates. Concerned governments have to take policy decisions in this regard.

3.3 Establishment and Management of Demonstration Plantations

Demonstration plantations have to be established in key locations to convince the farmers of the advantages of tissue-culture date palm plants. The apprehension of the farmers about tissue-culture plants needs to be addressed effectively. They have to be convinced regarding the earliness, uniformity and true-to-type nature of tissue-culture plants. In addition, the administrators, planners and other decision makers also have to be convinced about the advantages. The advantage of tissue culture in a dioecious crop like date palm, for ensuring the sex of the plant, has to be convincingly demonstrated to farmers. In seedling propagation about 50% of the plants may turn out to be males and this will be known only after 4–7 years, when the seedling plants flower. This is a major drawback of traditional cultivation. In tissue-culture date palm plantations, optimum male: female ratio among the plants can be maintained. Tissue-culture plants will not be harboring any disease-causing organisms. They may even be free from viruses. They will come to bearing in the 3rd year onwards. There will be uniformity in flowering and bearing habits. This will facilitate the pollination and harvesting processes. Established plantations must serve as models to understand scientific cultural operations and management.

3.4 Harvesting, Processing and Marketing

Technological upgrading of the harvesting and processing of dates is required. Optimum stages of harvesting in relation to market demands have to be identified for each agro-climatic zone or location. Value addition, product diversification and packaging are very much important from the marketing point of view. The awareness about quality and cultivar superiority is increasing rapidly. There is an increasing demand for sorted and packed fruits of superior cultivars. Hygienic packing is important. Previously dates were sold loose, without any packaging. Later polythene bags and baskets were used. The type of container and packaging material used in the national and international date trade is varied. They can be traditional bulk packs, export bulk packs and retail packs.

Bulk packs for dates have been traditionally jute bags, baskets woven of palm leaflets or tins. These are still in use, especially the basket for export of lower- quality dates. Dates are sold straight from the baskets or bags in the local markets. The process of pressing dates has at least partially been mechanized by the introduction of semi-automatic presses.

However, of late, dates are packed in attractive consumer packets of convenient size and shape. Size grading and cultivar labeling are positive developments in quality upgradation of the date trade. Impregnation of dates with nuts and dry fruits is another recent development.

Glove boxes, originally made of wood strips, have been used for packing Deglet Noor cv., in which 250 g of dates are packed in a fish-bone arrangement along a central piece of spikelet (Barreveld 1993). Date fruits are also marketed while still attached to the spikelets, mostly in cellophane bags or window cartons. Polythene bags and cellophane bags are used for pressed dates. Small moulds with a sheet of cellophane and label underneath are filled in layers with 0.1–1.0 kg dates and compressed. The mould is removed and the cellophane is wrapped around the block and heat sealed. For loose dates a closed window carton and transparent plastic cup are popular.

The moisture content, storage temperature and size and shape of fruits, etc. are critical. New products and value-addition techniques are being tried. In this regard, it is important to establish processing units, preferably near the production centers.

3.5 Transfer of Technology

In most of the date-producing countries farmers are unaware of scientific cultivation practices. Traditional cultivation is more or less unscientific and unsystematic. This leads to low productivity, more disease and pest incidence, poor quality of the produce, etc. Production and productivity of dates can be substantially increased by the adoption of quality planting materials, scientific cultivation and processing. Sufficient information has been generated in this respect by various research and development centers. However, effective technology transfer is essential. Collection and dissemination of information on production and planting, marketing, post-harvest handling and processing are critical. Efforts of agencies like the Food and Agricultural Organization (FAO), the Regional Project for Palm and Dates Research Centre in the Near East and North Africa, the Date Palm Research and Development Network, the Maghreb Date Palm Regional Project, the North African Regional Date Palm Network, the Biocontrol Project of Red Palm Weewil in the Arabian Gulf Region, etc. are remarkable in this respect. Still, most of the farmers are less aware of the advantages of quality planting materials and scientific management practices, the influence of soil and water properties on the yield and quality of dates, pollination methods, efficient harvesting and post-harvest handling processes, etc. Further concerted efforts of the concerned agencies and governments are required in the area of transfer of technology, utilizing mass media, publications, training programs and demonstration plots.

3.6 Support of Multicropping Systems

Date palm is very well-suited for multiple cropping systems, which can accommodate compatible intercrops such as pomegranate, papaya, vegetable crops, medicinal plants, etc. Intercropping is the practice of growing two or more crops in close proximity, in order to produce more yield per unit area, making use of resources like nutrients, water and sunlight, otherwise not utilized by a single crop. Examples of intercropping strategies are planting a deep-rooted crop with a shallow-rooted crop, or planting a tall crop with a shorter crop that requires partial shade. This will significantly enhance unit-area productivity and returns, making the best use of available inputs in a sustainable manner. In addition to the positive influences on soil health, multicropping systems will be helpful for the efficient use of limited resources.

3.7 Food and Nutritional Security

Dates already have the reputation of reducing vulnerability to famine and hunger, as well as contributing to food and nutritional security. There are date palm cultivars yielding 100–200 kg per palm per year, resulting in a productivity of 16–32 mt of fruit per ha. Fresh date is a premium source of vitamin C (30 mg) (Panhawar 2006).

Date fruits of 20% moisture content will provide about 3,000 Kcal per kg of date flesh (Barreveld 1993). They contain 1–3% protein and 2.5–7.5% fat. Palmitic, capric and caprylic acid are the major free fatty acids in date flesh, followed by linoleic, lauric, pelargonic and myristic acids. Moreover, it is rich in iron (6 mg), phosphorus (350 mg), potassium and calcium and is an abundant source of sugar (average 70%). Dates contain 110–170 mg of beta-carotene. It is a good source of fiber also and it has been reported to have medicinal properties. It is believed to be useful in controlling cold, fever, edema, bronchial catarrh and gastric problems.

3.8 Economic Impact

Scientific cultivation of date palms is highly remunerative, as compared to many of the major perennial crops. A comparative example is coconut cultivation. Even on a very conservative estimation, date palm can provide 7 times more unit area income to farmers, the average per hectare income from coconut being Rs. 55,360, (173 plants per hectare; 80 nuts per palm per year; Rs. 4 per nut) while that of date palm is Rs. 390,000 (156 plants per hectare; 100 kg fruit per palm per year; Rs. 25 per kg). No other crop in the arid zone can provide a comparable productivity or profitability. Economic returns can be expected from the 4th year onwards, in cultivation using tissue-culture plants and can offer stable returns for about 70 years. In addition to fruit, other plant parts also have economic significance. Until the commencement of economic returns from the palms, the farmers can make some reasonable income through intercropping with vegetable crops and fruit crops like papaya, pineapple, pomegranate, etc. Date palm-based markets and industries have already been proven economically viable and sustainable. The introduction of scientific date palm development programs in rural and marginal areas can have a significant impact on the national and international economic security. The global initiative on date palm cultivation is certain to bring about substantial impact in the coming years.

The date fruit, which is a product of the hot arid regions of Southwest Asia and North Africa, is marketed all over the world as a high-value confectionery and fruit crop. It is also an extremely important subsistence crop in most desert regions. There is an increasing trend in the export of dates. Processed dates are generally more popular than natural dates. Although Tunisia accounts for only 2% of world date production, its share of global exports in value is 21%. Main importers of dates are India, United Arab Emirates, Pakistan, Malaysia and Europe. France, United Kingdom, Germany, Italy and Spain are the major importers within the European Union.

3.9 Employment Generation and Poverty Alleviation

A date palm development program offers significant opportunities for employment generation in rural as well as urban sectors.

3.9.1 Production and Distribution of Planting Materials

The productivity of date palms can be substantially improved by using quality planting materials of superior genetic stock. *In vitro* clonal propagation is an effective and efficient alternative for conventional vegetative propagation, to ensure rapid multiplication and establishment of true-to-type plants of choice cultivars. Precise production targets can be achieved and stringent specifications met for increasing the bargaining power of the grower. The process can deliver planting materials to suit pre-determined mass production programs and meet well-planned market demands, especially for international markets. It is obvious that there is increasing demand for date palm tissue culture plants, of superior cultivars, in UAE as well as other countries. Establishment of a number of tissue-culture production units is necessary, requiring a sizable number of technicians, both male and female.

3.9.2 Establishment and Management of Plantations

There is broad potential for expanding the area under date palm cultivation, all over the world, both in the traditional and non-traditional areas. More and more plantations are being established every year in areas with congenial climatic conditions, soil and water quality and water availability. Suitable cultivars for each locality have to be multiplied and made available at reasonable cost and in a short period of time. There is substantial difference in economic returns between scientific cultivation and conventional cultivation. Expansion of area necessitates skilled as well as unskilled manpower. Operations like fencing, clearing, leveling, lay-out, pitting, pit filling, manuring, planting, shading, irrigation, intercropping, weeding, intercultural operations, disposal of debris, composting and assisted pollination demand labor. Mechanization can reduce manpower requirements, but cannot replace them. The average cost of cultivation per hectare, supported by mechanization, from the 4th year onwards, will be about Rs. 100,000. There can be an escalation of about 40%, in conventional cultivation without mechanization. It has been reported that the operational costs can be covered in the 1st year of production (Zaid 2002) i.e. 4 years after planting.

3.9.3 Harvesting, Processing and Marketing

Dates ripen in four stages, *kimri* (unripe), *khalal* (full-size, crunchy), *rutab* (ripe, soft), *tamar* (ripe, sun-dried). Several harvests are required. There is specific harvesting and packing considerations for each date cultivar and the form in which they will be consumed (Zaid 2002). Experienced persons are essential for harvesting.

Rain can cause damage to the fruit and affect its quality and may cause rotting, fermentation and lead to insect infestation of fruit. The fruit must be protected against rain. Beside direct consumption of the whole dates the fruits are traditionally used to prepare a wide range of different products such as date juice concentrates (spread, syrup and liquid sugar), fermented date products (wine, alcohol, vinegar, organic acids) and date pastes for different uses (bakery and confectionary). Also, the by-products arising from date processing can be used for different purposes. Within agricultural systems date press cake (by-product of date juice production) as well as date pits can be used as animal feed. Harvesting, transportation, grading, packing and processing necessitate manpower. Domestic and international marketing usually require different standards. Special care is required to meet these standards. Marketing also offers employment opportunities. Domestic retail market as well as international market demands manpower for transportation, sales, promotion of diversified products and business development.

3.10 Support to Marginalized Farmers and Weaker Sections

Date palm cultivation in rural areas can help in the empowerment of marginalized farmers, as well. They have practically no opportunities for self-sufficiency and survive on charity or welfare and are resource poor and depended on the favor of climatic conditions. They do not have access to scientific cultivation practices. Deficiencies related to land, credit, planting materials, marketing, etc. are common. Crop loss due to biotic and abiotic stress conditions is common, worsening their conditions. Non-governmental organizations and self-help groups can help marginalized farmers to make use of the opportunities of large-scale date palm cultivation, processing and marketing programs. A perennial crop like date palm can ward off many of their problems.

3.11 Female Empowerment

Economic insecurity, gender inequalities and discriminations, insufficient employment opportunities, food and nutritional insecurities, etc. are the serious concerns of rural women. Date palm cultivation has enough scope for fostering the skills and providing employment opportunities for the rural womenfolk, who are marginalized, poor and socially excluded. Female empowerment is the process of enabling access to opportunities, by encouraging the development of skills for self-sufficiency. Date palm cultivation provides ample opportunities for the female workers, in aspects of land preparation, nursery maintenance, planting and aftercare, cultural practices like weeding, irrigation, fertilizer application, plant protection operations, harvesting, packing and processing. This offers year- round employment potential for them.

3.12 Social Security and Rural Livelihood Security

Being a perennial crop that grows well under adverse conditions where many other crops fail, qualifies date palm to ensure security to the farmers. This indirectly provides social security, as observed in the case of plantation crops like rubber. Since most of the date palm areas belong in the rural sector, it contributes to rural livelihood security as well.

3.13 Environmental Improvement

Bringing large areas under date palm cultivation can definitely help in environmental improvement and sustainable development. It can improve soil health by root activities, supporting microbial action. Soil structure and texture can be improved and can bring about changes in microclimate and alleviate the negative impacts of stressed ecosystems. It results in efficient use of limited resources and helps thwart desertification. Deforestation leads to environmental degradation by promoting floods, drought, landslides and soil erosion. Moreover, the survival of traditional flora and fauna will come under threat. Extensive tree planting can counteract these effects. Trees give off valuable oxygen and help in the conservation of water, prevention of drought and ensure the survival and protection of fauna.

3.13.1 Carbon Sequestration and Carbon Credit

Carbon sequestration is the process of removing carbon from the atmosphere and depositing it in a reservoir. All crops absorb CO_2 during their growth and release it after their death as they are consumed, destroyed or incorporated into the soil. The goal of agricultural carbon removal is to use the crop itself and its relation to the carbon cycle to sequester carbon within the soil. The average rates of accumulation for forest or grassland establishment are 33.8 gC m^{-2} y^{-1} and 33.2 gC m^{-2} y^{-1}, respectively (Post and Kwon 2000). Date palm plantations in desert climates can greatly help in carbon sequestration. Carbon credit is a key component in emission-trading schemes and is useful to mitigate global warming. It is coming under the purview of the Clean Development Mechanism (CDM), which is an arrangement under the Kyoto Protocol allowing industrial countries to invest in projects that reduce emission in developing countries as an alternative to more expensive emission reductions in their own countries. Credits can be exchanged between businesses or bought and sold on international markets at the prevailing market price. Establishment of large plantations of date palm, especially in deserts and uncultivated areas, can amass carbon credit.

3.13.2 Combating Desertification

Desertification is the degradation of land in arid and dry sub-humid areas due to climatic variations and human activities. It results from overgrazing, overexploitation of groundwater and diversion of water from rivers for human consumption and industrial use. There are intrinsic linkages between desertification, land degradation and climate change. A major impact of desertification is reduced biodiversity and diminished overall productivity. It has been considered essential to provide local populations with a central role in determining a course of action to fight desertification based on their knowledge and experience. Date palm, being a crop of the desert, can have a key role in this respect. It can slow down the desertification process to a great extent. It can act as wind breaks, prevent sand dune shifting, add to the organic content of the soil, change the microclimate and support other flora as well as fauna. The effort of the UAE government in expanding the area under date palms during the last decade, facilitating the fight against desertification, is a good example, in this regard.

3.14 Cultural Impacts

Date palms represent history, culture and values. They are an integral part of the culture and tradition of many countries in the Gulf Region. Evidence of date palm cultivation goes back to at least 4,000 B.C. The earliest form of date palm cultivation coincided with the oldest civilizations. Date palm development, in a sense, represents the extension of culture and tradition and can result, even, in the betterment of international relations. There have been a number of exchanges of offshoots between date producing countries. An example of the commercial introduction of the date palm into new lands was the importation of a large number of offshoots into California and Arizona in the USA, during the recent past. Global partnerships in date palm development will have far-reaching cultural impacts and can foster communal harmony and integration.

3.15 Adoption of Tissue Culture Technology in the Indian Rural Context – An Example

Atul Ltd., a member of Lalbhai Group of companies, is one of the oldest business houses of India. It has made great strides in extending date palm cultivation in the deserts on the western border of India, using date palms in a systematic and scientific approach in a phased manner. This has had far-reaching impact on rural development

Fig. 3.1 Date palm demonstration plantation at Jaisalmer, Rajasthan in an area of 100 ha

in this underdeveloped area of India. So far, about 50,000 tissue-culture plants of promising cultivars have been distributed to the small and marginal farmers of Rajasthan. A well-knit strategy has been adopted for effective implementation of the project using a participatory public/private partnership mode. There are four phases for the implementation of the project. The first part envisages the establishment of model demonstration plantations in key locations to convince the farmers of the superior performance of tissue-culture raised date palms. A plantation consisting of seven superior cultivars has been established in an area of 100 ha at Jaisalmer, Rajasthan (Fig. 3.1). Another plantation of 40 ha has been established using five cultivars at Bikaner, Rajasthan (Fig. 3.2). The second phase aims to mobilize quality planting materials of superior cultivars, through tissue-culture laboratories for the establishment of plantations.

Atul has linked-up with internationally-reputed firms in the Gulf Region to source tissue-cultured plants. Primary hardened tissue-culture date palm plants of promising cultivars have been imported from Arab nations and subjected to secondary hardening at Jodhpur, Rajasthan (Figs. 3.3 and 3.4). The third phase targets capacity-building for the generation of tissue-culture date palm plants in India, adopting the best available protocol. A tissue-culture laboratory at Jodhpur has been established with overseas technology. The objectives of the fourth phase include large-scale scientific cultivation of date palms in the arid regions of Western India. The fifth phase encompasses setting up co-operatives which in turn will do buy-back arrangements with farmers to purchase and market date fruit. Infrastructure for collection of date fruit, grading, processing, packing, storage, logistics, branding, distribution and marketing will be established. Atul will

Fig. 3.2 Date palm demonstration plantation at Bikaner, Rajasthan in an area of 40 ha

Fig. 3.3 Primary hardened tissue culture date palm plants in the greenhouse of Atul at Jodhpur

provide training modules and other support services to the farmers for adopting the best scientific practices of date palm cultivation. Atul has developed a team of extension personnel, farm managers, scientists, academicians and marketing personnel for this purpose. Technical bulletins for promoting scientific date palm cultivation are being distributed to farmers. Classes and field training are being organized at the demonstration farms. Atul has assumed a major leadership role

Fig. 3.4 Secondary hardened tissue culture date palm plants in the net house of Atul at Jodhpur

for date palm development in North West India, providing help for rural development, creating employment opportunities, empowering women and fostering environmental integrity.

3.16 Conclusion and Prospective

Date palm cultivation helps generate considerable opportunities for rural employment, ensures livelihood and food security of the rural masses and facilitates substantial eco-restoration. It can act as a catalyst for the overall development of the rural sector in the arid regions. Large-scale scientific cultivation of date palm in traditional and non-traditional areas necessitates quality planting materials in sufficient quantity. This can be ensured through tissue-culture technology. Research and development programs need to be strengthened to develop reliable protocols for the large-scale commercial tissue culture propagation of the leading date cultivars. Facilities for ensuring clonal fidelity using molecular techniques should be established. Concerted efforts of public and private agencies are critical in deciding the success of date palm cultivation projects for rural development. Sustainable and scientific cultivation of date palms, supporting the marginal farmers and weaker sectors of the society, has to be promoted. Generation of date palm-based employment opportunities in the rural sector has great potential.

References

Al-Khayri JM (2001) Optimization of biotin and thiamine requirements for somatic embryogenesis of date palm (*Phoenix dactylifera* L.). In Vitro Cell Dev Biol Plant 37:453–456

Al-Khayri JM (2005) Date palm *Phoenix dactylifera* L. In: Jain SM, Gupta PK (eds.) Protocol for somatic embryogenesis in woody plants. Springer, Netherlands, pp 309–319

Al-Khayri JM (2007) Micropropagation of date palm *Phoenix dactylifera* L. In: Jain SM, Haggman H (eds.) Protocols for micropropagation of woody trees and fruits. Springer, Berlin, pp 509–526

Al-Wasel AS (2001) Field performance of somaclonal variants of tissue culture-derived date palm (*Phonix dactylifera* L.). Plant Tissue Cult 11:97–105

Barreveld WH (1993) Date palm products. Food and Agriculture Organization of the United Nations, Rome

Bhaskaran S, Smith RH (1992) Somatic embryogenesis from shoot tip and immature inflorescence of *Phoenix dactylifera* cv Barhee. Plant Cell Rep 12:22–25

Bouguedoura N, Michaus-Ferriere N, Bompard JL (1990) Comportment in vitro de bourgeons axillaries de type indéterminé du palmier dattier (*Phoenix dactylifera*). Can J Bot 68:2004–2009

Dass HC, Kaul RK, Joshi SP, Raj Bhansali R (1989) In vitro propagation of *Phoenix dactylifera* L. Curr Sci 58:22–24

Drira N, Benbadis A (1985) Multiplication vegetative du palmier dattier (*Phoenix dactylifera*) par reversion on culture in vitro debauches florales de pieds femelles. J Plant Phys 119:227–235

El Hadrami I, Cheikh R, Baaziz M (1995) Somatic embryogenesis and plant regeneration from shoot tip explants in *Phoenix dactylifera* L. Biol Plant 37:197–203

Eshraghi P, Zarghami R, Ofoghi H (2005) Genetic stability of micropropagated plantlets in date palm. J Sci Iran 16:311–315

Food and Agricultural Organization of the United Nations (FAO) (2008) FAOSTAT Prod STAT, Crops

Fki L, Masmoudi R, Drira N, Rival A (2003) An optimised protocol for plant regeneration from embryogenic suspension cultures of date palm (*Phoenix dactylifera* L.) cv. Deglet Nour. Plant Cell Rep 21:517–524

Hoop BM (2000) Date palm micropropagation in Saudi Arabia: policies and technology transfer. Int J Biotech 2:333–341

Loutfi K, Chlyah H (1998) Vegetative multiplication of date palm from the in vitro cultured inflorescence: effect of some growth regulator combinations and organogenetic potential of various cultivars. Agron 18:573–580

Larbi A (2010) Micropropagation of date palm (*Phoenix dactylifera* L.) selected genotypes from inflorescence tissues by using somatic embryogenesis technique. Fourth international date palm conference, Abu Dhabi. Abstracts Posters. p 49

Murashige T, Skoog F (1962) A revised medium for rapid growth and bioassay with tobacco tissue cultures. Phys Plant 15:473–497

Othmani A, Bayoudh C, Drira N et al (2009a) Somatic embryogenesis and plant regeneration in date palm *Phoenix dactylifera* L., cv. "Boufeggous" is significantly improved by fine chopping and partial desiccation of embryogenic callus. Plant Cell Tissue Org Cult 97:71–79

Othmani A, Bayoudh C, Drira N et al (2009b) Regeneration and molecular analysis of date palm (*Phœnix dactylifera* L.) plantlets using RAPD markers. Afr J Biotech 8:813–820

Panhawar F (2006) Date nutrition and medicinal values. Third international date palm conference, Abu Dhabi. Abstracts. p 156

Post WM, Kwon KC (2000) Soil carbon sequestration and land-use change: processes and potential. Glob Change Biol 6:317–327

Raj Bhansali R (2010) Date palm cultivation in the changing scenario of Indian arid zones: challenges and prospects. In: Ramawat KG (ed.) Desert plants. Springer Verlag, Berlin, pp 423–459

Raj Bhansali R, Kaul RK (1991) Into future – date through tissue culture. Ind Hort 36:6–10

Raj Bhansali R, Kaul RK, Dass HC (1988) Mass cloning of date palm plantlets through repetitive somatic embryogenesis. J Plant Anat Morph 5:73–79

Reuveni O (1979) Embryogenesis and plantlet growth of date palm (*Phoenix dactylifera* L.) derived from callus tissues. Plant Phys 63:138

Safwat MSA (2007) Organic farming of date palm and recycling of their wastes. In: African crop science conference proceedings 8:2109–2111

Sharma DR, Dawra S, Chowdhuri JB (1984) Somatic embryogenesis and plant regeneration in date palm (*Phoenix dactylifera*) cv Khadrawy through tissue culture. Ind J Exp Biol 22:763–766

Sharma DR, Deepak S, Chowdhuri JB (1986) Regeneration of plantlets from somatic tissues of date palm (*Phoenix dactylifera*). Ind J Exp Biol 24:763–766

Sharma DR, Kumari R, Chowdhuri JB (1980) In vitro culture of female date palm (*Phoenix dactylifera*) tissue. Euphy 29:169–174

Sudhersan C, El-Nil A, Al-Baiz A (1993) Occurrence of direct somatic embryogenesis on the sword leaf of in vitro plantlets of *Phoenix dactylifera* cultivar Barhee. Curr Sci 65:887–888

Taha HS, Bekheet SA, Saker MM (2001) Factors affecting in vitro multiplication of date palm. Biol Planta 44:431–433

Tisserat B (1979) Propagation of date palm (*Phoenix dactylifera*) in vitro. J Exp Bot 30:1275–1283

Zaid A (2002) (ed.) Date palm cultivation. Rev. ed. FAO, Rome

Zaid A, Tisserat B (1983) In vitro shoot tip differentiation in *Phoenix dactylifera* L. Date Palm J 2:163–182

Zaid A, de Wet PF, Djerbi M, Oihabi A (2002) Diseases and pests of date palm. In: Zaid A (ed.) Date palm cultivation. Rev. ed. FAO, Rome. pp 227–281

Chapter 4
Date Palm Micropropagation via Somatic Embryogenesis

L. Fki, R. Masmoudi, W. Kriaâ, A. Mahjoub, B. Sghaier, R. Mzid, A. Mliki, A. Rival, and N. Drira

Abstract During recent years, different approaches have been designed for the micropropagation of elite date palm cultivars. The most commonly used technology approach is somatic embryogenesis which presents a great potential for the rapid propagation and genetic resource preservation of this species. Considerable progress has been made in the development and optimization of this regeneration pathway through the establishment of embryogenic suspension cultures. However, several problems still need to be solved and are currently under study, such as the abnormal differentiation of somatic embryo, the proliferation of endophytic bacteria within *in vitro* cultured material and the occurrence of somaclonal variants in regenerated offspring. The present review is aimed at providing updated and innovative information on recent progress, applications and prospects for somatic embryogenesis in the date palm.

Keywords Conservation • Date palm • Micropropagation • Somatic embryogenesis

---

L. Fki (✉) • R. Masmoudi • A. Mahjoub • B. Sghaier
Laboratoire des Biotechnologies Végétales Appliquées à l'Amélioration des Cultures, Université de Sfax-Faculté des Sciences de Sfax, Route de Sokra Sfax-Tunisie
e-mail: lotfi.fki@fss.rnu.tn; maswkri@yahoo.fr

W. Kriaâ • N. Drira
Laboratory of Plant Biotechnology, Department of Biology, Faculty of Sciences Sfax, Road Soukra BP. 1171, Sfax 3000, Tunisia
e-mail: maswkri@yahoo.fr; drira_riadh@yahoo.fr

R. Mzid • A. Mliki
Laboratoire de Biotechnologie Végétale, Centre de biotechnologie de Borj Sedria, Borj Sedria Tunis-Tunisie
e-mail: rmzid@yahoo.fr

A. Rival
UMR DIADE, CIRAD, F-34398, Montpellier, France

S.M. Jain et al. (eds.), *Date Palm Biotechnology*,
DOI 10.1007/978-94-007-1318-5_4, © Springer Science+Business Media B.V. 2011

4.1 Introduction

The date palm, *Phoenix dactylifera* L., is one of the most economically important plants in arid and hot regions, especially in the Middle East and southern Mediterranean countries. Conventionally, this plant is propagated from offshoots which availability is often limited: indeed a given palm produces about 20 offshoots during its lifetime. Seed-derived palms show variable field performance because of their genetic heterogeneity. Using classical breeding methods, the genetic improvement of date palm is hampered by its long life cycle. More than 30 years are necessary to complete three backcrosses and to obtain the first offshoots from a given inter-varietal cross.

Bayoud disease caused by *Fusarium oxysporum* fungus has destroyed two thirds of the Moroccan palm plantations (more than ten million trees) and more than three millions palms in Algeria, causing considerable economic, ecological as well as social damage (Fernandez et al. 1995). Today a new lethal threat, brittle leaf disease, is spreading at a fast pace, as more than 36,000 trees are infected in Tunisia (Triki et al. 2003). In view of this critical situation, the integration of biotechnology in date palm propagation, breeding and conservation strategies has become essential.

Plant regeneration through tissue culture is able to provide technologies for the large-scale propagation of healthy true-to-type plants. Plant tissue culture is also an essential tool for plant breeding programs (Parveez et al. 2000) and the conservation of plant genetic resources (Engelmann and Dussert 2000). Somatic embryogenesis, which leads to embryo differentiation from somatic cells and not from fertilized ovules, is often employed because of its numerous advantages (Carlos and Martinez 1998). Embryogenic cell suspension cultures have shown higher morphogenetic capacity when compared with other methods of *in vitro* propagation. Well developed somatic embryos from efficient embryogenic cultures are highly critical to produce synthetic seeds (Prewein and Wilhelm 2003). In addition, embryogenic cells are considered as choice material for (i) the isolation of protoplasts with a highly regeneration capacity (Ling and Iwamasa 1994), (ii) genetic engineering (Cabrera et al. 1996) and (iii) the cryopreservation of plant genetic resources (Engelmann 2004).

Since 1970, extensive efforts have been undertaken to mass-propagate date palms using *in vitro* techniques (Al-Khalifa 2000; Al-Khateeb 2008a,b; Al-Khayri and Al-Bahrani 2001; Ammar and Benbadis 1977; Bekheet et al. 2001; Bhaskaran and Smith 1992; Bouguedoura et al. 1990; Daguin and Letouze 1988; Drira 1983; Drira and Benbadis 1985; Eeuwens 1978; El Hadrami et al. 1995; Fki et al. 2003; Masmoudi et al. 1999; Othmani et al. 2009; Poulain et al. 1979; Reuveni 1979; Reuveni et al. 1972; Reynolds and Murashige 1979; Schroeder 1970; Sharma et al. 1980, 1984, 1986; Sharon and Shankar 1998; Taha et al. 2001; Tisserat 1979, 1982, 1984; Tisserat and Demason 1980; Veramendi and Navarro 1996, 1997; Zouine et al. 2005). A number of excellent reviews have also been published (Al-Khayri 2005, 2007; Bhaskaran and Smith 1995; Branton and Blake 1989; El Hadrami and El Hadrami 2009).

To establish aseptic cultures, various different explants have been used, including zygotic embryos (Ammar and Benbadis 1977; Reynolds and Murachige 1979), shoot tips (Veramendi and Navarro 1996), lateral buds (Bouguedoura et al. 1990; Drira 1983), leaves (Bhaskaran and Smith 1992; Fki et al. 2003) or inflorescences (Bhaskaran and Smith 1992; Drira and Benbadis 1985; Fki et al. 2003). Different types of disinfecting solutions have been used, among them mercuric chloride ($Hg Cl_2$) was found to be the most efficient (Drira and Benbadis 1985). Some authors have given more importance to somatic embryogenesis, mainly because of its higher potential for mass propagation (Al-Khateeb 2008a; Al-Khayri and Al-Bahrani 2001; Bhaskaran and Smith 1992; Daguin and Letouzé 1988; El Hadrami et al. 1995; Fki et al. 2003; Masmoudi et al. 1999; Othmani et al. 2009; Poulain et al. 1979; Reynolds and Murachige 1979; Tisserat and Demason 1980). Several research groups have opted for date palm regeneration through adventitious organogenesis, this approach being reputed to be slow, but less risky in terms of somaclonal variation (Al-Khateeb 2008b; Bekheet et al. 2001; Drira and Benbadis 1985; Taha et al. 2001; Tisserat 1984). The simple development of lateral buds has also been explored for the successful regeneration of whole plants (Drira 1983).

The culture medium devised by Murashige and Skoog (1962) has been commonly used for date palm tissue culture. The 2,4-D auxin is the most popular for callogenesis (Fki et al. 2003). The use of liquid media for plant regeneration has been widely used (Bhaskaran and Smith 1992; Daguin and Letouze 1988; Fki et al. 2003; Sharma et al. 1986; Veramendi and Navarro 1996; Zouine et al. 2005). The present chapter is aimed at describing research work on date palm somatic embryogenesis. Its potential for large-scale propagation, conservation and breeding is discussed.

4.2 Somatic Embryogenesis for Large-Scale Propagation

Somatic embryogenesis is one of the most important technologies for plant regeneration. There are two morphogenetic pathways ensuring the production of somatic embryos. The first pathway is direct somatic embryogenesis, which is yet to be fully developed for massive plant regeneration in date palm (Sudhersan et al. 1993). The second pathway is an indirect method which is based on the induction of embryogenic calli (Al-Khayri 2005).

4.2.1 Initiation of Embryogenic Calli

4.2.1.1 Factors Controlling Callus Induction

Different types of explants have been used to induce embryogenic calli, such as zygotic embryos, shoot tips, leaves, lateral buds or inflorescences. However, the successful induction of calli requires very specific physicochemical conditions which are essential for the dedifferentiation of cells. Previous research work has shown that the 2,4-D auxin is the most suitable plant growth regulator for the initiation of callogenesis in date palm; indeed, concentrations such as 5 mg/L and more (El Hadrami and Baaziz 1995; Tisserat 1979) and even lower than 1 mg/L were found to be efficient (Fki et al. 2003; Masmoudi et al. 1999). Cytokinins such as 2iP (2-isopentyl adenine) seems not to be necessary, as it is for many monocotyledonous plants (Magnaval et al. 1997). Our studies showed that Picloram (0.2–0.5 mg/L) induced callogenesis although it generated non-embryogenic calli or abnormal somatic embryos (Fki 2005). In date palm, callogenesis is a very slow process which may require 4–8 months. This seems to be a generic characteristic of *in vitro* cultivated *Arecaceae*, as it was also described for the coconut palm (Verdeil and Buffard-Morel 1995) and the African oil palm (*Elaeis guineensis* Jacq.) (Duval et al. 1995). This very slow culturing step has obviously a negative impact on the cost of vitroplants.

Calli generally appear on either upper or lower surfaces of the leaves; the latter side is clearly more productive and generates the major part of the calli population (Fki et al. 2003). When an immature inflorescence is used as a primary explant, only calli which originated from the proliferation of floral tissues showed embryogenic competency. Comparatively, calli deriving from axial tissues bearing flower primordia showed a lower embryogenic potential. The callogenic capacity of inflorescences was generally found to be higher than that of leaves (Drira and Benbadis 1985; Fki et al. 2003).

Mature and immature zygotic embryos produce calli with a low embryogenic capacity (Fki 2005; Reynolds and Murashige 1979). This was an unexpected result considering the immense morphogenetic capacity of this explant in other species like the African oil palm (Teixeira et al. 1994). Calli resulting from zygotic embryos could not produce a high number of embryos, and since date palm is a dioecious and heterozygotic plant, calli cannot be used to propagate desired cultivars. On the other hand, the combination of this propagation method, *in vitro* flowering (Ammar et al. 1987; Masmoudi-Allouche et al. 2010) and *in vitro* fertilization technologies may be an innovative approach for date palm breeding.

Furthermore, Al-Khayri (2001) reported that biotin and thiamine could improve the quality of the embryogenic calli. Silver nitrate was also found to promote somatic embryogenesis (Al-Khayri and Al-Bahrany 2001, 2003). In a study to determine the effects of date palm syrup on somatic embryogenesis induction, Al-Khateeb (2008a) found that such a natural extract could be used at a 6% concentration as a replacement for sucrose. Date palm meristematic tissues extract also enhanced date palm somatic embryogenesis (El-Assar et al. 2004).

The genotype has a strong influence on callus induction. Indeed, using the same concentration of 2,4-D (1 mg/L), the frequencies of callus induction obtained from juvenile leaves from cvs. Klasse and Barhee were found to reach, respectively, 10% and 40% (Fki 2005).

The fragmentation of explants, which cuts morphogenetic correlations and enables a direct contact of totipotent cells with the culture medium, improved rates in the initiation of embryogenic calli (Fki 2005). Nevertheless it was found that explant fragmentation often led to an early proliferation of endophytic bacteria.

Fig. 4.1 Date palm embryogenic callus

The massive release of phenolic compounds, which was found in many date palm cultivars, can be overcome by culturing the explant in media supplemented with activated charcoal (Fki 2005). Other anti-oxidative substances, such as citric acid or ascorbic acid, have been successfully used for the reduction of excessive browning and eventual necrosis of date palm tissues (Zaid and Tisserat 1984).

The physiological state of the donor plant, the period in which primary explants were sampled and their position on the culture medium are also important. For example, in the case of callogenesis from leaves, the period ranging from September to January seems to be the most favorable under the Tunisian climate (Fki 2005).

Recent basic studies by Gueye et al. (2009a) have shown that callus initiation by 2,4-D requires polar auxin transport and is characterized by the reactivation of fascicular parenchyma cells followed by the dedifferentiation of perivascular sheath cells. Both callogenesis and rooting are initiated from perivascular sheath cells, the ultimate developmental fate depending upon auxin concentration (Gueye et al. 2009b).

4.2.1.2 Characterization of Embryogenic Calli

In vitro propagation of date palm is a very slow process. Finding cyto-morphological, biochemical and molecular markers characterizing embryogenic cultures would be a gain. At the cyto-morphological level, embryogenic calli are friable, showing small (<2 mm) white nodules, slightly connected to the rest of the whitish callus matrix (Fig. 4.1). Non-embryogenic cells are prenchymatous with a small nucleus

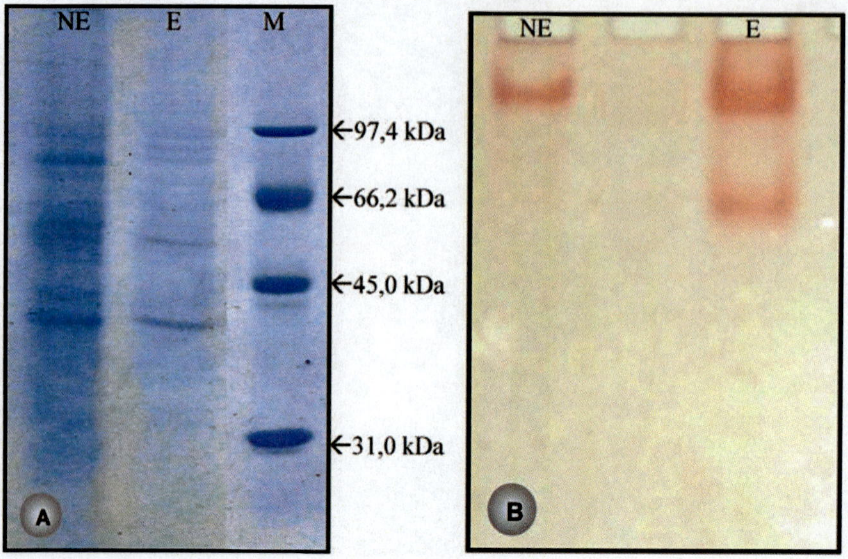

Fig. 4.2 Molecular analyses. **(a)** SDS-PAGE of total soluble proteins extracted from non-embryogenic (NE) and embryogenic (E) date palm calli. Lane M: molecular weight markers. **(b)** *In situ* detection of peroxydase activity after PAGE separation of Total Soluble Proteins extracted from non-embryogenic (NE) and embryogenic (E) date palm calli

and a large centrally-located vacuole. In contrast, embryogenic cells are small with a large centrally-located nucleus and dense cytoplasm (Veramendi and Navarro 1997). Morphological and cytological analyses are often insufficient to characterize embryogenic calli; it is then of paramount interest to find biochemical and molecular markers.

At the biochemical level, total soluble protein contents were found to be higher in embryogenic calli than in their non-embryogenenic counterparts (El Hadrami and Baaziz 1995; Fki 2005; Masmoudi et al. 1999). Some proteins, as revealed by one-dimensional SDS-PAGE, can be used to distinguish between the two types of calli (Fig. 4.2a). According to studies from our group and El Hadrami and Baaziz (1995), peroxydase activity is much higher in embryogenic callus than in the non-embryogenic, and some isoforms of the enzyme can be used to identify embryogenic calli (Fig. 4.2b).

While describing major trends of carbon metabolism during the initiation and expression of somatic embryogenesis in date palm cv. Deglet Noor, Masmoudi et al. (1999) determined PEPC patterns in embryogenic and non-embryogenic calli. Detection of PEPC activity on polyacrylamide native gels after electrophoresis revealed the presence of three active isoforms in crude extracts from the embryogenic callus strain, whereas only a single band was present in the non-embryogenic one.

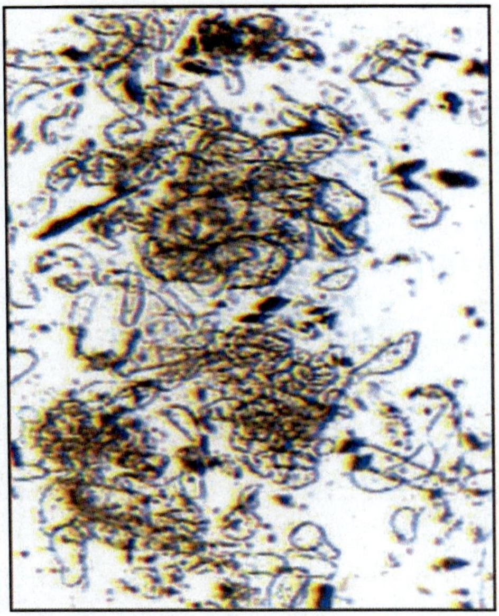

Fig. 4.3 Microscopic view of a date palm embryogenic suspension

4.2.2 Multiplication, Development and Maturation of Somatic Embryos

Solid media have been usually and extensively employed to produce somatic embryos from embryogenic calli. This is obviously an important step in the control of the different stages of somatic embryogenesis in date palm (El Hadrami et al. 1995; Masmoudi et al. 1999; Reynolds and Murashige 1979; Tisserat 1979). However, solid media could not be used to ensure large-scale propagation. For this reason, extensive efforts have been deployed to establish embryogenic suspension cultures with high morphogenetic potentialities (Bhaskaran and Smith 1992; Daguin and Letouzé 1988; Fki et al. 2003; Sharma et al. 1986; Veramendi and Navarro 1996; Zouine et al. 2005).

An optimized protocol for plant regeneration from embryogenic suspension cultures of date palm cv. Deglet Noor has been described by our group (Fki et al. 2003). Embryogenic cell suspensions established after 500 mm mesh filtration were studied for cell shape, size and cell cluster formation (Fig. 4.3). The resulting suspensions were highly heterogeneous, containing cells at various stages of differentiation. Chopping the callus into pieces favored the formation of PEMs (Fig. 4.4). These results confirmed those obtained by Kreuger et al. (1995) with cultures of *Cyclamen persicum*. Sané et al. (2006) and Othmani et al. (2009) recently confirmed the positive effect of callus chopping on the differentiation of date palm somatic embryos. A twofold dilution of liquid MS medium (Murashige and Skoog 1962) showed a positive effect on somatic embryo differentiation. Furthermore, the addition of 1 mg/L 2,4-D and the presence of activated charcoal

Fig. 4.4 Mass production of date palm proembryos through liquid suspension culture

at a low concentration (300 mg l^{-1}) in the liquid medium promoted the differentiation of somatic embryos. This result is in contrast to a previous report on the Barhee cv. by Bhaskaran and Smith (1992), which described the occurrence of embryogenesis only in the absence of 2,4-D. Furthermore, De Touchet et al. (1991) described for African oil palm the use of a washing step before embryo differentiation, with the aim of eliminating any traces of PGRs in the culture medium. The simultaneous presence of embryos at various differentiation stages in the same suspension culture facilitated the study of somatic embryogenesis. Indeed, various stages of embryo development, namely: spherical, elongated (after the cotyledon appears) and cotyledonary leaf formation could be observed in a given flask (Fig. 4.5).

It is worth noting that mature embryos were sampled continuously at each transfer. By contrast, the protocol described for African oil palm by De Touchet et al. (1991) involves the total conversion of a given culture towards the differentiation pathway at a given time, this step being initiated by a washing treatment and completed with the plating of the culture on a solid medium.

The use of liquid medium was found to drastically increase the productivity of somatic embryogenesis. Whereas only 10 ± 2 embryos were recovered from 100 mg callus (FW) on solid medium, the same amount of callus produced up to 200 ± 10 embryos in liquid medium after a 1-month cultivation period (Fig. 4.6). Suspension cultures were maintained up to 6 months, thus enabling the continuous production of mature embryos. The overall production rate was 10,000 ± 45 typical embryos per liter per month. Bhaskaran and Smith (1992) reported a lower productivity and a high proportion of abnormal embryos for the Barhee cv. De Touchet et al. (1991) indicated that when using a PGR-free liquid medium,

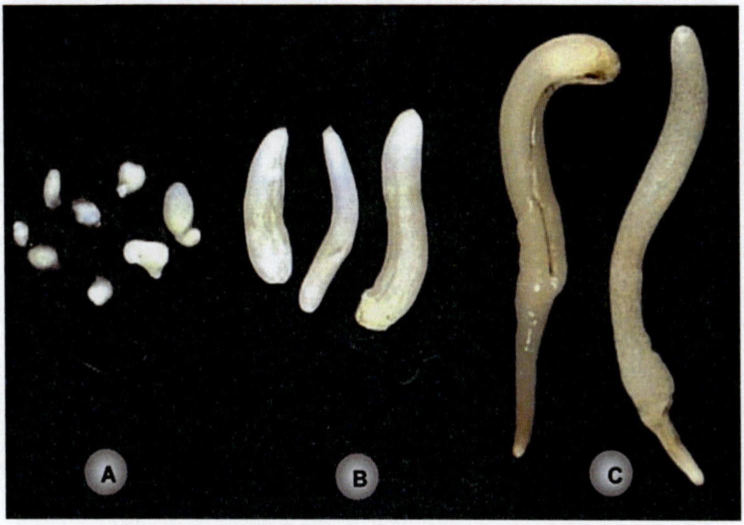

Fig. 4.5 Various different stages of date palm somatic embryo development: (**a**) globular juvenile, (**b**) Mature, and (**c**) Beginning of germination

Fig. 4.6 Mass production of date palm somatic embryos from embryogenic suspension culture

fully developed somatic embryos exhibiting both a gemmule and a radicule were obtained at a very low frequency; they did not germinate and become necrotic after their transfer to a solid medium.

Microscopic examination showed that the embryogenic aggregates initiated adventitious nodules, which separated from each other with agitation. Each culture was made of nodules showing various different sizes. Some nodules continued

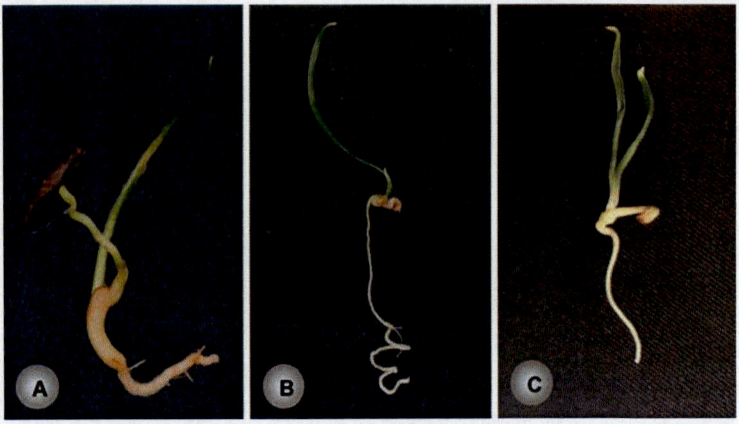

Fig. 4.7 Developing date palm plantlets from various origins: (**a**) Seed-derived seedling; (**b**) zygotic embryo-derived vitroplant; (**c**) somatic embryo-derived vitroplant

to proliferate, the others differentiated into embryos. No marked growth lag was noticed after the transfer of PEMs into a fresh medium. During the first 5 days of subculture, the growth rate was slow. After that period, a steady increase in fresh weight was measured up to day 25. From that point onwards, there was a decrease in growth. Thus fresh biomass increased about fourfold in 1 month, a performance comparable to that previously reported for the African oil palm (De Touchet et al. 1991).

In order to determine the origin of somatic embryos, filters of different mesh sizes were used to sieve cell suspensions (Fki et al. 2003). These experiments suggest that somatic embryos seem to originate from cell aggregates and not from single cells.

Some embryogenic callus lines were not able to generate vigorous somatic embryos. This is certainly a consequence of poor accumulation of storage compounds such as proteins, lipids and sugars. In this respect, a comparative study was carried out between date palm somatic and zygotic embryos (Fki 2005; Sghaier et al. 2008; Sghaier-Hammami et al. 2009) in order to improve the quality of the somatic embryos.

In date palm, the zygotic embryo taken from seed is at a developmental stage corresponding to the post-globular stage. The full development of this embryo takes place after germination, which begins with the elongation of the cotyledon. Morphologically, these embryos are longer and thicker than the somatic ones. Besides, no differences exist between somatic embryos and zygotic embryos which were developed on a PGR-free medium (Fki 2005). Total soluble protein contents were much higher in the zygotic embryos taken from seeds a few days after germination than in somatic embryos. These differences reflect on the regenerated plants. Indeed, seed-derived plants are more vigorous than somatic embryo- and zygotic embryo-derived plants (Fig. 4.7). The vigor of the plants resulting from seeds could be explained by the specific role of the endosperm which contains important storage compounds (Besbes et al. 2004).

Sucrose and ABA were evaluated at different concentrations for their respective effects on somatic embryo maturation (Fki 2005; Sghaier et al. 2009; Sghaier-Hammamia et al. 2009; Zouine et al. 2005). Mature somatic embryos developed on half strength MS medium enriched with high concentration of sucrose (60 g/L) and ABA (2 mg/L) were found to be thicker, longer and richer in proteins than the control (Fki 2005). The promotive effect of abscissic acid (ABA) is mainly exerted during the development of the cotyledon. Indeed, ABA induces the accumulation of storage proteins and prevents precocious germination.

4.2.3 *Somatic Embryos Germination*

The duration of the cultivation period in liquid medium was found to be very important for the balanced germination of somatic embryos. Indeed, the cultivation of mature embryos in liquid medium for more than 1 month led to hyperhydration (Fki et al. 2003). We have demonstrated that partial desiccation of mature somatic embryos, corresponding to a decrease in water content from 90 down to 75%, did significantly improve germination rates on modified MS medium deprived of PGRs (from 25% to 90%).

In the same way, cutting back the cotyledon leaf to about half its length was found to stimulate embryo germination. This kind of physiological response has been observed on zygotic embryos from various plant species. In our experiments, embryos with a shortened cotyledon leaf showed an 80% germination rate, compared to 25% with those with an intact cotyledon leaf. The cotyledon leaf seemed to inhibit the apical meristem development of some embryos, but did not influence the growth of the root meristem. The transfer of germinated embryos onto a medium supplemented with 1 mg/L NAA and 1 mg/L BAP enabled the production of vigorous plantlets showing a balanced shoot and root development. The positive effect of BAP on the germination of African oil palm somatic embryos derived from embryogenic suspensions was also reported by Aberlenc-Bertossi et al. (1999).

Al-Khayri (2003) showed a positive effect of IBA (0.2–0.4 mg/L) on germination rates of somatic embryos which were produced on solid medium. Media containing an additional source of inorganic phosphate (170 mg/L sodium dihydrogen phosphate + 100 mg/L potassium dihydrogen phosphate) resulted in faster germination (Sharon and Shander 1998).

4.2.4 *Physiology of the Somatic Embryo-Derived Plants and Their Acclimatization*

Acclimatization is the last stage of micropropagation; when not properly controlled, it leads to high loss rates. Plants transplanted into the greenhouse should progressively resist to: (i) higher luminosity; (ii) lower relative humidity; (iii) fluctuation of

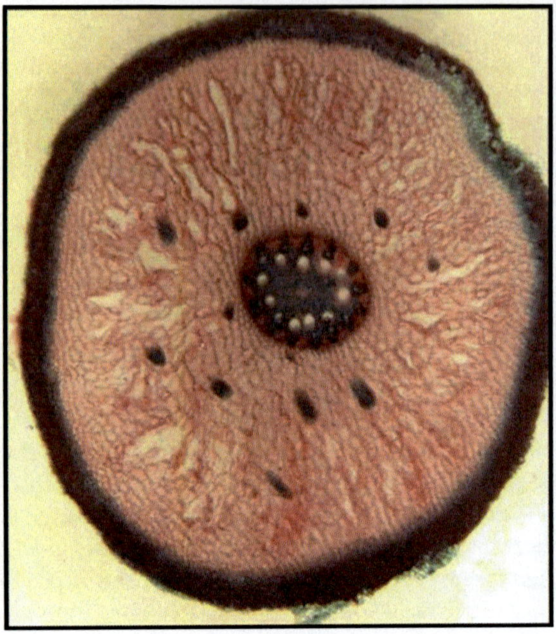

Fig. 4.8 Cross section of *in vitro* grown primary root from a 3 month-old plantlet derived from suspension-cultured somatic embryo

temperature and (iv) biotic stresses. For a better control of acclimatization, we have studied several factors, including the physiology of vitroplants and the physico-chemical conditions of acclimatization (Fki 2005). Our anatomical analyses showed that *in vitro* grown roots have a structure similar to those sampled from acclimatized plants. This proves that roots of vitroplants are virtually functional at all developmental stages (Fig. 4.8). With regard to the photosynthetic capacity of vitroplants, we found that only after an *in vitro* hardening period of 12 months does the photochemical activity of Photosystem II become close to that measured in an already acclimatized plant.

Scanning electron microscopy (SEM) examination of detached leaf surfaces showed the regulation of stomatal aperture, preventing excess transpirational vapor loss in 12-month-old vitroplants. Indeed, all stomata were closed in this material, which was not the case for 3-month-old vitroplants (Fig. 4.9). In this respect, Zaid and Hughes (1995) reported that a polyethylene glycol treatment of vitroplants increased the amount of wax deposition on leaf surfaces and as a consequence it was able to decrease water losses which were observed during acclimatization.

When exploring the role of carboxylases in date palm acclimatization (Masmoudi et al. 1999), it was found that the PEPC/RubisCO ratio decreased (from 17.7 to 0.2) throughout the *in vitro* development of plantlets, due to a substantial depletion of PEPC activity. Concomitantly, RubisCO activity assumed greater importance and became the main route for inorganic carbon fixation. Western blot analysis using polyclonal antibodies raised against PEPC and RubisCO purified from tobacco leaves confirmed this trend in terms of relative enzyme abundance.

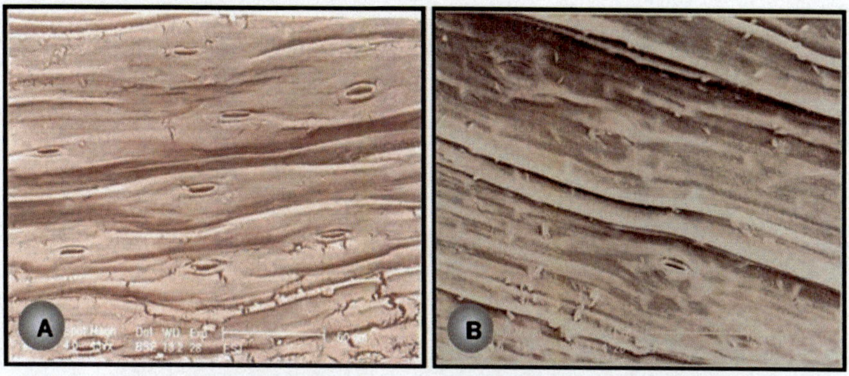

Fig. 4.9 SEM observation of leaf epidermis: **(a)** 3-month-old vitroplant (400x); **(b)** 12 month-old vitroplant

According to our experiments, the optimal conditions for acclimatization of date palm vitroplants, ensuring high survival rates (70%) for 12-month-old plantlets, can be listed, as follows: (i) temperature: 25–30°C (ii) moderate light intensity (iii) a draining substrate and (iv) a *plant by plant* instead of *whole batch* acclimatization procedure. In order to promote plant growth in the greenhouse, Awad (2008) suggested the utilization of a 5-aminolevulinic acid-based fertilizer.

4.2.5 Genetic Stability and In-Field Behavior of Somatic Embryo-Derived Plants

Several strategies have been used to assess the genetic integrity of somatic embryo-derived plants. In the case of date palm, reports about somaclonal variation are still controversial. Using flow cytometry, we have estimated the nuclear genome size of cv. Deglet Noor and analyzed the stability of this parameter in regenerated plants (Fki et al. 2003). Our experiments showed that all the analyzed adult plants and *in vitro* regenerated plantlets were diploid. No variation in genome size that could be linked to the micropropagation protocol could be detected. Our estimation of 2 C DNA content of cv. Deglet Noor was 1.96 ± 0.05 pg and therefore approximately 1.77×10^9 bp. Date palm was found to have a much smaller genome than the African oil palm, which was reported to be $2C = 3.76 \pm 0.09$ pg; 3.4×10^9 bp by Rival et al. (1997).

RAPD and ISSR analyses of leaf genomic DNA reportedly are able to distinguish between date palm cultivars (Corniquel and Mercier 1994; Sedra et al. 1998; Zahdi et al. 2002) although according to studies from our group, they failed to reveal polymorphism in genomic DNA which could be associated with somaclonal variations in tissue culture-derived date palm plants (Fki 2005; Mahjoub 2003; Mzid 1999; Othmani 1998). Similarly, Rival et al. (1998) reported that a RAPD approach could not differentiate between normal and variant oil palms, as the fraction of the genome explored through this technology was clearly not large enough

Fig. 4.10 Normal fruit bunches from a somatic embryo-derived date palm

to draw any valuable conclusion. However, Saker et al. (2000) reported that RAPD and isozyme analyses were reliable techniques which could be used to detect somaclonal variations in date palm.

The use of juvenile explants and low doses of 2,4-D can minimize the risks of somaclonal variation and generally ensure regeneration of morphologically uniform plants, as proven by the fruit quality of palms that have been planted in the field by our group a few years ago (Fig. 4.10). On the other hand, high concentrations of 2,4-D could induce somaclonal variation and the most genetically unstable cultivar seems to be Barhee, since some somatic embryo derived-plants now produce parthenocarpic fruits (unpublished results). The genetic instability of this cultivar also has been mentioned by Cohen et al. (2004) after a multi-seasonal analysis of fruit setting and by Zivdar et al. (2008) according to isoenzyme analysis.

In the same way, morphological abnormalities in somatic embryo-derived date palm cv. Sukkari have been described by Al-Mazroui et al. (2007). However, it has been reported that date palm cv. Barhee derived from somatic embryos and natural offshoots produced fruits bearing the same characteristics (Smith and Aynsley 1995).

4.2.6 Major Constraints and Remedies

According to our investigations, the three major constraints in date palm somatic embryogenesis are endophytic bacterial contamination, abnormal somatic embryo differentiation and somaclonal variation (Fki 2005). Concerning endophytic bacterial contamination, only juvenile explants could be used to establish clean *in vitro* tissue culture, since antibiotics such as cefotaxim have only a bacteriostatic effect. Immaturity of vascular tissue in these explants may explain the absence of endophytic contaminants (Fki 2005). Both abnormal somatic embryo differentiation and somaclonal variation were especially associated with the utilization of high concentrations of 2,4-D. Reducing its concentration significantly minimized the number of abnormal somatic embryos and somaclonal variants (Fki 2005).

4.3 Applications of Somatic Embryogenesis

4.3.1 Synthetic Seed Technology

Bio-encapsulation of somatic embryos is most appropriate, especially for dioecious and vegetatively-propagated plant species such as the date palm. As we are able to produce a high number of somatic embryos, our research group is now focusing on the production of date palm synthetic seed. Somatic embryos were placed in sterilized 3% sodium alginate, dissolved in MS medium without calcium. Each one was picked up using a micropipette and dropped into a 75 mM $CaCl_2.2\,H_2O$ solution in MS medium. After 30 min, the calcium chloride solution was decanted and the alginate beads, each containing a single somatic embryo, were rinsed with sterile distilled water. Each encapsulated somatic embryo can be considered a synthetic seed, provided it survives further conservation. Furthermore, alginate was not toxic to somatic embryos and the *in vitro* germination rate was unaffected (Fig. 4.11). This is in accordance with results obtained by Daikh and Demarly (1987) and Bekheet et al. (2002).

4.3.2 Protoplasts

Chabane et al. (2007) reported for the first time the formation of a callus from protoplasts in date palm. Protoplasts were isolated from young offshoot leaves and embryogenic calli. Different combinations of enzyme solutions were tested for protoplast isolation and a cocktail of 1.5% Cellulase RS (Yakult Pharmaceutical Ind. Co., Ltd, Tokyo Japan), 0.15% Pectolyase (Kyowa Chemical Products Co., Ltd, Osaka Japan) and 0.2% Hemicellulase (Sigma, USA) was found to be the most efficient. Cell division was induced in both liquid culture and nurse culture. Nevertheless these proto-calli failed to regenerate plants.

4.3.3 In Vitro Mutagenesis

Mutagenesis is a very interesting approach for crop improvement since many superior genotypes have been obtained using diverse mutagenic agents. In our laboratory, date palm embryogenic calli were exposed to gamma irradiation (30 Gy) and plantlets material. In order to select bayoud-resistant cell lines, a toxin extracted from *Fusarium*

Fig. 4.11 Synthetic seeds of date palm

oxysporum f. sp. *albedinis* was added to the culture media at several concentrations (5, 10 and 20 μg/mL) (Fki 2005). This experiment suggests that a toxin can be used as an *in vitro* selection agent to screen for resistance. The putatively resistant plantlets are now under investigation at the biochemical and molecular levels and resistance will be field-assessed in bayoud-contaminated zones.

4.3.4 Genetic Transformation

In our laboratory, embryogenic cultures of date palm were bombarded with tungsten particles coated with a plasmid encoding β-glucuronidase (GUS) and neomycin phosphotransferase (NPTII) which results in kanamycin resistance (Fki 2005). Expression of these proteins is under the control of a CaMV 35 S promoter. Expression of the reporter gene was histochemically confirmed in embryogenic cultures of cvs. Barhee and Deglet Noor (Fig. 4.12). Helium pressure of 2 bars and a distance of 6 cm between the particle launch and the target were found to increase the number of GUS-expressing colonies. The number of GUS-expressing colonies decreased with time and further studies are being carried out to obtain stable transformation.

4.3.5 Cryopreservation

Studies on cryopreservation of date palm embryogenic cultures are scarce. In most of cases, a classical freezing protocol, which is relatively time consuming, has been used. It consists in a slow cooling of plant material down to a temperature range of −30°C to −40°C, followed by the immersion of the vials containing tissues and a cryoprotectant in liquid nitrogen (Bagniol and Engelmann 1991, 1992; Fki 2005; Mycock et al. 1995; Ulrich et al. 1982).

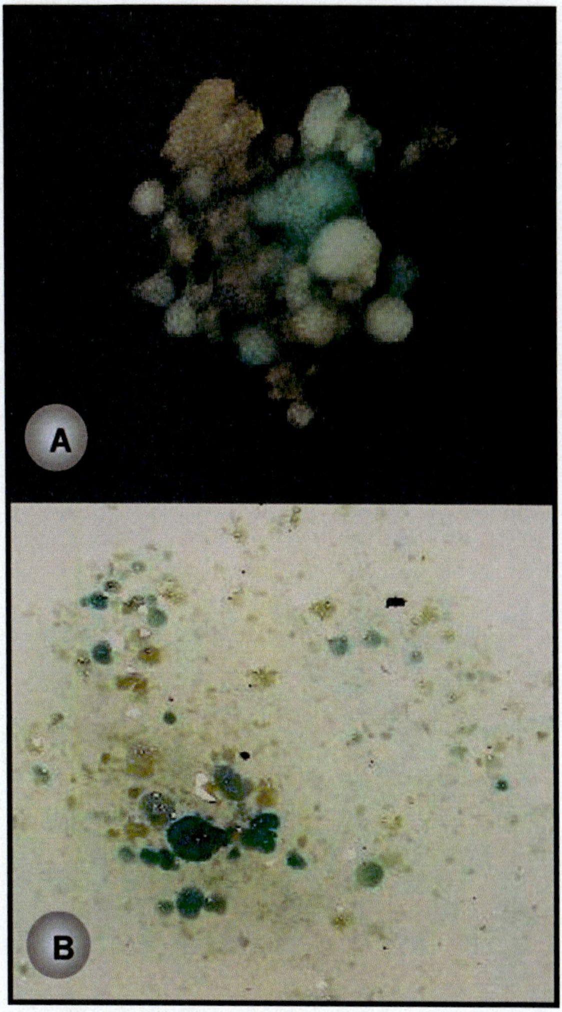

Fig. 4.12 Transient *GUS* gene expression in date palm embryogenic cultures. (**a**) Callus. (**b**) Suspension culture

Cryopreservation of date palm tissue cultures is a major research topic in our group and optimized cryopreservation protocols based on the vitrification process have been established for date palm embryogenic cultures (unpublished results).

4.4 Conclusion and Prospective

Somatic embryogenesis has a tremendous potential for date palm development. This technology can be used for large-scale propagation, thereby opening the way for the production of synthetic seeds. However, the factors controlling callus induction still need to be mastered and the quality of embryogenic calli requires further improvement.

Moreover, high-frequency callus induction needs to be obtained from explants originating from various organs/genotypes. Further research is needed to find the most appropriate biochemical and molecular markers of embryogenesis in date palm. In date palm, most of the molecular methods used to assess somaclonal variation have shown some limitations. Indeed, cytogenetic analysis proved unable to reveal any alteration in genome structure, and isozyme markers are subject to large ontogenic variation. Molecular markers are able to investigate only a small part of the genome and they are useless in the case of epigenetic changes. Somaclonal variation was found to be from epigenetic origin in oil palm and differences in DNA methylation rates could be linked to the occurrence of epigenetic instability (Jaligot et al. 2000). This research opens new ways for the investigation of somaclonal variation in date palm. To date, field evaluation, even if it is a long and costly process, remains the most reliable strategy to assess the genetic integrity of regenerated date palms.

Very limited work has been carried out on the cryopreservation of date palm embryogenic cultures and therefore the development of innovative procedures is needed for the efficient preservation of genetic resources and the management of commercial propagation. Preliminary studies revealed that embryogenic cultures provide a choice plant material for *in vitro* mutagenesis experiments and the further selection of useful mutants, the generation of protoplasts and somatic cell hybridization, and for genetic engineering. Further research is required to overcome problems such as abnormal somatic embryo differentiation, endophytic bacteria proliferation and somaclonal variation.

Acknowledgements This work was supported by the Ministry of Higher Education, Scientific Research and Technology of Tunisia, the International Atomic Energy Agency (TC Project RAF/5/049) and the European Commission (COST Action 871).

References

Aberlenc-Bertossi F, Noiret M, Duval Y (1999) BA enhances the germination of oil palm somatic embryos derived from embryogenic suspension cultures. Plan Cell Tissue Org Cult 56:53–57

Al-Khalifa NS (2000) In vitro culture studies on date palm (*Phoenix dactylifera* L.) cv. "Mosaifah" "Nabtat Sultan". Plant Tissue Cult 10:1–8

Al-Khateeb A (2008a) Comparaison effects of sucrose and date palm syrup on somatic embryogenesis of date palm (*Phoenix dactylifera* L.). Am J Biochem Biotech 4:19–23

Al-Khateeb A (2008b) Regulation of in vitro bud formation of date palm (*Phoenix dactylifera* L.) cv. "Khanizi" by different carbon sources. Bioresour Technol 99:6550–6555

Al-Khayri JM (2001) Optimization of biotin and thiamine requirements for somatic embryogenesis of date palm (*Phoenix dactylifera* L.). In Vitro Cell Dev Biol Plant 37:453–456

Al-Khayri JM (2003) In vitro germination of somatic embryos in date palm: effect of auxin concentration and strength of MS salts. Curr Sci 84:101–104

Al-Khayri JM, Al-Bahrany AM (2001) Silver nitrate and 2-isopentyladenine promote somatic embryogenesis in date palm (*Phoenix dactylifera* L.). Sci Hort 89:291–298

Al-Khayri JM, Al-Bahrany AM (2003) Genotype-dependent in vitro response of date palm (*Phoenix dactylifera* L.) cultivars to silver nitrate. Sci Hort 99:153–162

Al-Khayri JM (2005) Date palm *Phoenix dactylifera* L. In: Jain SM, Gupta PK (eds.) Protocols for somatic embryogenesis in woody plants. Springer, Netherlands, pp 309–319

Al-Khayri JM (2007) Date palm *Phoenix dactylifera* L. micropropagation. In: Jain SM, Haggman H (eds.) Protocols for micropropagation of woody trees and fruits. Springer, Netherlands, pp 509–526

Al-Mazroui HS, Zaid A, Bouhouche N (2007) Morphological abnormalities in tissue culture-derived date palm (*Phoenix dactylifera* L). Acta Hort 736:329–335

Ammar S, Benbadis A (1977) Multiplication végétative du palmier dattier (*Phoenix dactylifera* L.) par la culture de tissus de jeunes plantes issues de semis. Comp Rend Biol 284:1789–1791

Ammar S, Benbadis A, Krishna TB (1987) Floral induction in date palm seedlings (*Phoenix dactylifera* var. Deglet Nour) cultured in vitro. Can J Bot 65:137–142

Awad MA (2008) Promotive effects of a 5-aminolevulinic acid-based fertilizer on growth of tissue culture-derived date palm plants (*Phoenix dactylifera* L.) during acclimatization. Sci Hort 118:48–52

Bagniol S, Engelmann F (1991) Effects of pregrowth and freezing conditions on the resistance of meristems of date palm (*Phoenix dactylifera* L. var. "Bou Sthammi Noir") to freezing in liquid nitrogen. Cryo Lett 12:279–286

Bagniol S, Engelmann F (1992) Effect of thawing and recovery conditions on the regrowth of meristems of date palm *Phoenix dactylifera* L. var. "Bou Sthammi Noir" after cryopreservation in liquid nitrogen. Cryo Lett 13:253–260

Bekheet SA, Taha HS, Saker MM (2001) Factors affecting in vitro multiplication of date palm. Biol Plant 44:431–433

Bekheet SA, Taha HS, Saker MM, Moursy HA (2002) A synthetic seed system of date palm through somatic embryogenesis encapsulation. Ann Agric Sci 47:325–337

Besbes S, Blecher C, Deroanne C et al (2004) Date seeds: chemical composition and characteristic profiles of the lipid fraction. Food Chem 24:577–584

Bhaskaran S, Smith R (1992) Somatic embryogenesis from shoot tip and immature inflorescence of *Phoenix dactylifera* L. cv. Barhee. Plant Cell Rep 12:22–25

Bhaskaran S, Smith R (1995) Somatic embryogenesis in date palm *Phoenix dactylifera* L. In: Jain SM, Gupta PK, Newton RJ (eds.) Somatic embryogenesis in woody plants, vol 2. Springer, Netherlands, pp 461–470

Bouguedoura N, Michaux-Ferrière N, Bompar JL (1990) Comportement in vitro de bourgeons axillaires de type indéterminé du palmier dattier (*Phoenix dactylifera* L.). Can J Bot 68:2004–2009

Branton RL, Blake J (1989) Date palm. In: Bajaj YPS (ed.) Biotechnology in agriculture and forestry, vol 5. Springer Verlag, Berlin, pp 161–175

Cabrera PJL, Vegas GA, Herrera M (1996) Regeneration of transgenic papaya plants via somatic embryogenesis induced by *Agrobacterium rhizogenes*. In Vitro Cell Dev Biol Plant 32:86–90

Carlos M, Martinez FX (1998) The potential uses of somatic embryogenesis in agroforestry are not limited to synthetic seed technology. Rev Bras Fis Veg 10:1–12

Chabane D, Assani A, Bouguedoura N et al (2007) Induction of callus formation from difficult date palm protoplasts by means of nurse culture. Comp Rend Biol 330:392–401

Cohen BY, Korchinsky R, Tripler E (2004) Flower abnormalities cause abnormal fruit setting in tissue culture-propagated date palm (*Phoenix dactylifera* L.). J Hort Sci Biotech 79:1007–1013

Corniquel B, Mercier L (1994) Date palm (*Phoenix dactylifera* L.) cultivar identification by RFLP and RAPD. Plant Sci 101:163–172

Daguin F, Letouze R (1988) Régénération du palmier dattier (*Phoenix dactylifera* L.) par embryogenèse somatique: amélioration de l'efficacité par passage en milieu liquide agité. Fruits 43:191–194

Daikh H, Demarly Y (1987) Résultats préliminaires sur l'obtention d'embryons somatiques et la réalisation de semences artificielles de palmier dattier (*Phoenix dactylifera* L.). Fruits 42:593–596

De Touchet B, Duval Y, Pannetier C (1991) Plant regeneration from embryogenic suspension cultures of oil palm (*Elaeis guineensis* Jacq.). Plant Cell Rep 10:592–532

Drira N (1983) Multiplication végétative du palmier dattier (*Phoenix dactylifera* L.) par la culture in vitro de bourgeons axillaires et de feuilles qui en dérivent. Comp Rend Biol 296:1077–1082

Drira N, Benbadis A (1985) Multiplication végétative du palmier dattier (*Phoenix dactylifera* L.) par réversion, en culture in vitro, d'ébauches florales de pieds femelles adultes. J Plant Phys 119:227–235

Duval Y, Engelmann F, Durand-Gasselin T (1995) Somatic embryogenesis in oil palm (*Elaeis guineesis* Jacq). In: Bajaj YPS (ed.) Somatic embryogenesis and synthetic seed, vol 1. Springer-Verlag, Berlin/Netherlands, pp 235–352

Eeuwens CJ (1978) Effects of organic nutrients and hormones on growth and development of tissues explants from coconut and date palms cultured in vitro. Phys Plant 42:173–178

El-Assar AM, El-Messeih WM, El-Shenawi MR (2004) Applying of some natural extracts and growth regulators to culture media their effects on "Sewi" cv. date palm tissues grown in vitro. Assuit J Agric Sci 35:155–168

El Hadrami I, Baaziz M (1995) Somatic embryogenesis and analysis of peroxydases in *Phoenix dactilifera* L. Biol Plant 37:179–203

El Hadrami I, Cheikh R, Baaziz M (1995) Somatic embryogenesis and plant regeneration from shoot-tip explants in *Phoenix dactylifera* L. Biol Plant 37:205–211

El Hadrami I, El Hadrami A (2009) Breeding date palm. In: Jain SM, Priyadarhan PM (eds.) Breeding plantation tree crops. Springer, Netherlands, pp 191–216

Engelmann F (2004) Plant germplasm cryopreservation: progress and prospects. In Vitro Cell Dev Biol Plant 40:427–433

Engelmann F, Dussert S (2000) Développement de la cryoconservation pour la conservation des ressources génétiques végétales. Agric 9:237–244

Fernandez D, Lourd M, Quinten M et al (1995) Le Bayoud du palmier dattier, une maladie qui menace la phoeniciculture. Phytoma 469:36–39

Fki L (2005) Application des suspensions cellulaires embryogenes au clonage et à l'amélioration *in vitro* du Palmier dattier. Thèse de doctorat, Faculté des Sciences de Sfax-Tunisie

Fki L, Masmoudi R, Drira N, Rival A (2003) An optimised protocol for plant regeneration from embryogenic suspension cultures of date palm (*Phoenix dactylifera* L.) cv. Deglet Nour. Plant Cell Rep 21:517–524

Gueye B, Morcillo F, Collin M et al (2009a) Acquisition of callogenic capacity in date palm leaf tissues in response to 2,4-D treatment. Plant Cell Tissue Org Cult 99:35–45

Gueye B, Said-Ahmed H, Morcillo F et al (2009b) Callogenesis and rhizogenesis in date palm leaf segments: are there similarities between the two auxin-induced pathways? Plant Cell Tissue Org Cult 98:47–58

Jaligot E, Rival A, Beulé T et al (2000) Somaclonal variation in Oil Palm (*Elaeis guineensis* Jacq.): the DNA methylation hypothesis. Plant Cell Rep 19:684–690

Kreuger M, Postma E, Brouwer Y, Holst GJ (1995) Somatic embryogenesis of *Cyclamen persicum* in liquid medium. Phys Plant 94:605–612

Ling JT, Iwamasa M (1994) Somatic hybridization between *Citrus reticulata* and *Citroptis gabunensis* through electrofusion. Plant Cell Rep 13:493–497

Magnaval C, Noirot M, Verdeil JL et al (1997) Specific nutritional requirements of coconut calli (*Cocos nucifera* L.) during somatic embryogenesis induction. J Plant Phys 159:719–728

Mahjoub A (2003) Mutagenèse induite par des rayons gamma sur le palmier dattier en culture in vitro et perspectives de sélection de mutants résistants au bayoud. Mastère. Université de Sfax, Tunisie

Masmoudi-Allouche F, Meziou B, Kriaâ W et al (2010) In Vitro flowering induction in date palm (*Phoenix dactylifera* L.). J Plant Growth Regul 29:35–43

Masmoudi R, Rival A, Nato A et al (1999) Carbon metabolism in in vitro cultures of date palm: the role of carboxylases (PEPC and RubisCO). Plan Cell Tissue Org Cult 57:139–143

Murashige T, Skoog F (1962) A revised medium for rapid growth and bioassays with tobacco tissue culture. Phys Plant 15:473–479

Mycock DJ, Wesley-Smith J, Berjak P (1995) Cryopreservation of somatic embryogenesis of four species with and without cryoprotectant pre-treatment. Ann Bot 75:331–336

Mzid R (1999) Effet d'un rayonnement type gamma sur la variabilité génétique induite sur des tissus de palmier dattier (*Phoenix dactylifera* L.) en culture in vitro. Mastère, Université de Sfax, Tunisie

Othmani A (1998) Etude de l'influence d'un facteur hormonal sur la variabilité génétique chez le palmier dattier (*Phoenix dactylifera* L.). Mastère, Université de Tunis, Tunisie

Othmani A, Bayoudh C, Drira N et al (2009) Somatic embryogenesis and plant regeneration in date palm *Phoenix dactylifera* L., cv. "Boufeggous" is significantly improved by fine chopping and partial desiccation of embryogenic callus. Plant Cell Tissue Org Cult 97:71–79

Parveez GH, Masri MM, Zainal A et al (2000) Transgenic oil palm: production and projection. Biochem Soc Trans 28:969–972

Poulain C, Rhiss A, Beauchesne G (1979) Multiplication végétative en culture in vitro du palmier dattier (*Phoenix dactylifera* L.). Comp Rend Acad Agric Fr 11:1151–1154

Prewein C, Wilhelm E (2003) Plant regeneration from encapsulated somatic embryos of pedunculate oak (*Quercus robur* L.). In Vitro Cell Dev Biol Plant 39:613–617

Reuveni O (1979) Embryogenesis and plantlets growth of date palm (*Phoenix dactylifera* L.) derived from callus tissues. Plant Phys 63(Suppl):138

Reuveni O, Adato Y, Lilien-Kipnis H (1972) A study of new and rapid methods for the vegetative propagation of date palms. Date Grow Inst Rep 49:16–24

Reynolds J, Murashige T (1979) Asexual embryogenesis in callus cultures of palms. In Vitro 15:383–387

Rival A, Beulé T, Barre P et al (1997) Comparative flow cytometric estimation of nuclear DNA content in embryogenic calli and seed-derived oil palm (*Elaeis guineensis* Jacq). Plant Cell Rep 16:884–887

Rival A, Bertrand L, Beulé T et al (1998) Suitability of RAPD analysis for the detection of somaclonal variants in oil palm (*Elaeis guineensis* Jacq). Plant Breed 117:73–76

Saker MM, Bekheet SA, Taha HS et al (2000) Detection of somaclonal variations in tissue culture-derived date palm plants using isoenzyme analysis and RAPD fingerprints. Biol Plant 43:347–351

Sané D, Aberlenc-Bertossi F, Gassama-Dia YK et al (2006) Histocytological analysis of callogenesis and somatic embryogenesis from cell suspensions of date palm (*Phoenix dactylifera*). Ann Bot 98:301–308

Schroeder CA (1970) Tissue culture of date shoots and seedlings. Date Grow Inst Rep 47:25–27

Sedra H, Lashermes P, Trouslot P et al (1998) Identification andgenetic diversity analysis of date palm (*Phoenix dactylifera* L.) varieties from Morocco using RAPD markers. Euphy 103:75–82

Sghaier B, Bahloul M, Gargouri R, Drira N (2008) Development of zygotic and somatic embryos of *Phoenix dactylifera L.* cv. "Deglet Nour": comparative study. Sci Hort 116:169–175

Sghaier B, Kriaa W, Bahloul M et al (2009) Effect of ABA, arginine and sucrose on protein content of date palm somatic embryos. Sci Hort 120:379–385

Sghaier-Hammamia B, Drira N, Jorrín-Novob JV (2009) Comparative 2-DE proteomic analysis of date palm (*Phoenix dactylifera* L.) somatic and zygotic embryos. J Prot 73:161–177

Sharma DR, Dawra S, Chowdhury JB (1984) Somatic embryogenesis and plant regeneration in date palm (*Phoenix dactylifera* L.) cv. 'Khadravi' through tissue culture. Indian J Exp Biol 22:596–598

Sharma DR, Deepak S, Chowdhury JB (1986) Regeneration of plantlets from somatic tissues of date palm (*Phoenix dactylifera* L.). Indian J Exp Biol 24:763–766

Sharma DR, Kumari R, Chowdury JB (1980) In vitro culture of female date palm (*Phoenix dactylifera* L.) tissues. Euphy 29:169–174

Sharon M, Shankar C (1998) Somatic embryogenenesis and plant regeneration from leaf primordial of *Phoenix dactylifera* cv Yakubi. Indian J Exp Biol 36:526–529

Smith RJ, Aynsley JS (1995) Field performance of tissue cultured date palm (*Phoenix dactylifera* L.) clonally produced by somatic embryogenesis. Prin 39:47–52

Sudhersan C, Abo El-Nil MM, Al-Baiz A (1993) Occurrence of direct somatic embryogenesis on the sword leaf of in vitro plantlets of *Phoenix dactylifera* L., cultivar Barhee. Curr Sci 65:887–888

Taha HS, Bekheet SA, Sakr MM (2001) Factors affecting in vitro multiplication of date palm. Biol Plant 44:431–433

Teixeira JB, Sondahl MR, Nakamura T, Kirby EG (1994) Establishment of oil palm cell suspensions and plant regeneration. Plan Cell Tissue Org Cult 40:105–111

Tisserat B (1979) Propagation of date palm (*Phoenix dactylifera* L.) in vitro. J Exp Bot 30:1275–1283

Tisserat B (1982) Factors involved in the production of plantlets from date palm callus cultures. Euphytica 31:201–214

Tisserat B (1984) Propagation of date palms by shoot tip cultures. Hort Sci 19:230–231

Tisserat B, Demason D (1980) A histological study of development of adventive embryos in organ cultures of (*Phoenix dactylifera* L.). Ann Bot 46:465–472

Triki MA, Zouba A, Khouldia O et al (2003) "Maladie des feuilles cassantes" or brittle leaf disease of date palms in Tunisia: biotic or abiotic disease. J Plant Pathol 85:71–79

Ulrich J, Finkle BJ, Tisserat BH (1982) Effects of cryogenic treatment on plantlet production from frozen and unfrozen date palm callus. Plant Phys 69:624–627

Veramendi J, Navarro L (1996) Influence of physical conditions of nutrient medium and sucrose on somatic embryogenesis of date palm (*Phoenix dactylifera*). Plant Cell Tissue Org Cult 45:159–164

Veramendi J, Navarro L (1997) Influence of explant sources of adult date palm (*Phoenix dactylifera* L.) on embryogenic callus formation. J Hort Sci 72:655–671

Verdeil J, Buffard-Morel J (1995) Somatic embryogenesis in coconut (*Cocos nucifera* L.). In: Bajaj YPS (ed.) Somatic embryogenesis and synthetic seed, vol 1. Springer, Berlin, pp 299–317

Zahdi S, Trifi M, Ould Mohamed Salem A et al (2002) Survey of inter simple sequence repeat polymorpisms in Tunisian date palms (*Phoenix dactylifera* L.). J Genet Breed 56:77–83

Zaid A, Hughes H (1995) Water loss and polyethylene glycol-mediated acclimatization of in vitro grown seedlings of 5 cultivars of date palm (*Phoenix dactylifera* L.) plantlets. Plant Cell Rep 14:385–388

Zaid A, Tisserat B (1984) Survey of the morphogenetic potential of excised palm embryos in vitro. Crop Res 24:1–9

Zivdar S, Mousawi M, Ansari NA (2008) Genetic stability in date palm micropropagation. Asian J Plant Sci 7:775–778

Zouine J, El Bellaj M, Meddich A et al (2005) Proliferation and germination of somatic embryos from embryogenic suspension cultures in *Phoenix dactylifera* L. Plant Cell Tissue Org Cult 82:83–92

Chapter 5
Date Palm Micropropagation via Organogenesis

L. Abahmane

Abstract The world demand for date palm vitro-plants has increased in recent decades. The use of all available techniques of rapid multiplication to fulfill this demand is an objective of great interest. However, the technique used should permit maximum fidelity in terms of genetic stability of regenerated plants. Hence, organogenesis presents an advantage of use of low concentrations of plant growth regulators and consequently the callus phase is avoided. Direct regeneration of vegetative buds minimizes the risk of somaclonal variation among regenerants. Moreover, the duration of culture period is limited by frequent renewal of the plant material. Actually, there are few laboratories that use this technique to produce date palm vitro-plants at the commercial level. Hundreds of thousands of date palm vitro-plants have been produced using this technique mainly in Morocco, Saudi Arabia and the United Arab Emirates. Research has permitted development of this micropropagation process and to adapt it to the most important date palm cultivars. In the present study, a review of the literature and our research results on date palm tissue culture, with emphasis on the organogenesis technique, are presented. The entire multiplication process from offshoot removal to plantlet acclimatization is fully described.

Keywords *In vitro* • Plant growth regulators • Regeneration • Tissue culture

---

L. Abahmane (✉)
Laboratoire de Biotechnologie Végétale, INRA, Centre Régional de la Recherche Agronomique de Marrakech, 533, Marrakech, Morocco
e-mail: abahmanel@yahoo.fr

S.M. Jain et al. (eds.), *Date Palm Biotechnology*,
DOI 10.1007/978-94-007-1318-5_5, © Springer Science+Business Media B.V. 2011

5.1 Introduction

Date palm, *Phoenix dactylifera* L. (Family: Palmae) is a perennial long-lived, dioecious, monocotyledon which is highly heterozygous. It is considered the key species in the preservation of oasis ecosystems. Date palm cultivation also is one of the most economically important activities in the arid zones of the Middle East and North Africa where it is grown not only for its valuable fruits but also to produce fuel, fiber and to provide shelter for ground crops. In this region, 62 million of the 105 million date palm trees worldwide grow on an area of over one million ha (1,264,611 ha). World production of dates is approximately 7.04 million metric tons and generates important commercial activities (FAO 2008).

In many countries, date palms suffer from serious biotic and abiotic constraints. Among the first are bayoud disease, a soil-borne fungus (*Fusarium oxysporum* f. sp. *albedinis*), and the most dangerous threat to date palm groves, mainly in North Africa. In this region, bayoud disease has destroyed, since its appearance in 1870, more than ten million palms in Morocco and three million in Algeria. It attacks the most renowned cultivars which are susceptible and produce high-quality fruit (Medjool, Deglet Noor, Boufegouss, etc.).

The second important biotic threat is the red palm weevil (*Rhynchophorus ferrugineus* Oliver) which has become the most serious pest of date palm in the world (Gómez and Ferry 1999). It is very difficult to detect early infestation, since the larva begins its life inside the palm and normally never comes to the surface. The rotting of tissue due to infestation by this pest produces a characteristic odor. In addition, wilting or yellowing of leaves can be easily observed on infected trees. The high rate of spread of this pest is human-caused, by transporting infested young or adult date palm trees and offshoots from contaminated to uninfected areas (Ferry and Gómez 2002).

Among the abiotic constraints, desertification is the most serious factor reducing palm groves in most of the countries where date palm is cultivated. This phenomenon is aggravated by drought conditions which hamper successful production of the crop. Taking into account those constraints, the need to select palms to replace senescent and destroyed date palm groves has been steadily increasing in decades. Hence, the exploitation of all available means of date palm propagation is required to fulfill the huge demand for palm vitro-plants.

5.2 Means of Date Palm Propagation

5.2.1 Conventional Techniques

Date palm is conventionally propagated sexually by seeds or vegetatively by offshoots. However, the two techniques are inefficient when huge numbers of plant material are needed or when some superior genotypes are selected and need to be cloned vegetatively. Furthermore, attempts to micro-graft shoots on seedling rootstocks did not succeed to regenerate complete plantlets because no vascular connection between the two parts (scions and rootstocks) was established (Loutfi and El Hadrami 2005). As a result of these natural constraints, only biotechnologic tools (tissue culture) can be used to fulfill the increasing world demand of date palm vitro-plants. Advantages and disadvantages of each one of the conventional techniques are discussed below.

5.2.1.1 Sexual Propagation

The seed method is the oldest means of date palm propagation. Its main advantage is that it is simple in practice and also enlarges date palm genetic diversity. Hence, the technique is very useful in breeding programs and selection from among the progeny can lead to developing some elite palms with interesting traits. In some countries the number of date palm trees originating from natural hybrids is important; in Egypt, for example, there are 3.5 million and in Morocco more than two million. In other countries (United Arab Emirates, Kuwait, Pakistan, Yemen, etc.) propagation by seed is still practiced (Ferry et al. 1998).

However, this method cannot be used to propagate elite or select genotype since the progeny will be quite variable because of the highly heterozygous character of date palm (Tisserat 1982). Moreover, half of the progeny will be composed of male trees which cannot be distinguished before flowering. Female seed-derived plants will produce variable fruits which are generally of inferior quality (Al Khateeb 2006; Eke et al. 2005; Tisserat and DeMason 1980; Zaid and de Wet 1999).

5.2.1.2 Offshoot Propagation

Date palms produce shoots from axillary shoot meristems and inflorescences from floral meristems (Sudhersan et al. 2001). The use of offshoots for date palm propagation is the most conventional vegetative technique at the farm level (Al Khateeb and Ali-Dinar 2002). Use of this method permits the preservation of true-to-typeness of multiplied genotypes. However, the average number of suckers per palm per lifetime is very low and restricted to the juvenile stage (Tisserat 1983). Some cultivars do not produce suckers or produce limited numbers (10–30) of transplantable offshoots (Heselmans 1997). In addition, offshoots are difficult to root (Asemota et al. 2007; Eke et al. 2005) and the success in the field transfer is usually less than 60% (Saaidi et al. 1979). Some researchers have reported a high mortality of suckers transferred to soil (Al Khateeb 2006). One more disadvantage of this method is the spread of dangerous diseases and pests such as bayoud disease or red palm weevil which can be transported by contaminated offshoots.

The selection of suckers to be planted is of great importance since it determines the success in the field. The most important parameters to be considered are (Zaid and de Wet 1999): (1) weight: 10–25 kg; (2) age: 3–5 years; (3) base diameter: 20–30 cm; (4) formation of own roots; (5) signs of maturity: production of own roots, first fruit set, production of offshoots of second generation (Nixon and Carpenter 1978); (6) well connected to the mother tree.

To enhance the rate of survival of transplanted offshoots, some researchers propose alternative solutions such as use of exogenous plant growth regulators mainly 25 mM indole butyric acid (IBA) (Qaddoury and Amassa 2004). They observed that IBA-treated offshoots rooted earlier and at a much higher frequency while untreated ones rooted poorly, later and tended to produce few roots. Hodel and Pittenger (2003a) reported that offshoots possessing more roots when removed from the mother palm have a greater ability to regenerate a root system and to become established more successfully and rapidly because over two-thirds of all new roots grow from existing cut roots. In addition, offshoots 10–35 cm in diameter have the highest survival rates, probably because they have more roots when initially removed from the mother palm, more stored carbohydrates to provide energy for root growth and increased levels of naturally occurring root-promoting substances (Hodel and Pittenger 2003b).

Ground offshoots of large size are usually used for date palm propagation. The use of small size and aerial (high and unrooted) offshoots is impractical due to their low rate of survival (Al-Ghamdi 1988). Therefore, they are usually discarded during the separation process of large-sized ones (Mohammed 1978). However, these offshoots could regenerate roots and be used successfully if an intermittent mist system is used (El Hamady et al. 1992). In fact, small offshoots, weighing 5 kg and less, can be used but they should initially be kept, for at least 2 years, in a nursery, mist bed, greenhouse or a shade net structure (Reuveni et al. 1972). Fungi are usually a serious problem in a mist bed, and offshoots must be treated twice a month with a wide spectrum fungicide (Zaid and de Wet 1999). In a study of the rooting of ground and aerial date palm offshoots, Al Mana et al. (1996), showed that rooting medium was an important factor in determining the extent of offshoot root formation. The highest rooting percentages were obtained using the following media: perlite: peat moss (3:1) medium followed by the wood shavings : peat moss (1:1) and perlite: peat moss (1:1) media. Sand medium was inferior to the others. It was also found that neither rooting percentage nor root weight of ground offshoots were increased by NAA and/or catechol treatments. On the other hand, NAA and/or catechol treatments appeared to be essential for good root formation and development of aerial offshoots.

5.2.2 *In Vitro Propagation*

Since conventional propagation techniques are limited in terms of providing insufficient numbers of date palm plants, biotechnology has provided a promising alternative to meet the increasing demand for date palm vitro-plants in the last decades. Plant tissue culture techniques have been used to clone a wide range of economically-important palms (coconut, oil palm and date palm) (Al Kaabi et al. 2001). Using these techniques, date palm can be micropropagated either by somatic embryogenesis in which embryos are produced from embryogenic callus and then germinated to form complete plantlets (Letouze et al. 2000; McCubbin et al. 2000) or through organogenesis in which plantlets are produced from multiplied buds without passing through the callus stage (Al Khateeb 2006). Since buds come directly from mother plant tissue, the plantlets produced are identical to the mother tree (Aaouine 2000; Al Khateeb 2008b; Beauchesne 1983; Beauchesne et al. 1986).

The application of tissue culture techniques for date palm micropropagation has many advantages, particularly: (1) large-scale multiplication of commercial cultivars; (2) propagation of elite and select cultivars with desirable characters (bayoud resistance, males with superior metaxenia characteristics, high yielding, etc.); (3) production of homogeneous, vigorous and disease-free plants; (4) no seasonal effect on plant production; (5) enables exchange of plant material without any risk of the spread of diseases or pests (Zaid and de Wet 1999).

5.2.2.1 Somatic Embryogenesis

In this pathway, cells or callus cultures on solid media or in suspension cultures on liquid media form embryo-like structures called somatic embryos, which on germination media, produce complete plants. The primary somatic embryos are also capable of producing more embryos through secondary somatic embryogenesis (Ahloowalia et al. 2004). Somatic embryogenesis is one of the successful methods widely used for mass propagation of date palm throughout the world (Kunert et al. 2003). This technique can be started from any meristematic part of the date palm tree, particularly from axillary buds, shoot tips, immature inflorescences and immature embryos (Al Khayri 2003; Bhaskaran and Smith 1995). Many reported successful results in this field have been published: (Al Khayri 2003; Ammar and Benbadis 1977; Aslam and Khan 2009; Badawy et al. 2005; Bhaskaran and Smith 1992; Daguin and Letouze 1988; El-Bahr et al. 2004; Othmani et al. 2009; Sharma et al. 1984; Tisserat 1979; Zouine and El Hadrami 2007). For more details see chapter on somatic embryogenesis in this book.

5.2.2.2 Direct Organogenesis

The organogenesis technique, based on the exploitation of meristematic tissue potentialities to form new shoots, avoids callus formation and does not use 2,4-D. The growth regulators incorporated in the media are used at the lowest possible concentration. The organogenesis technique consists of four steps: initiation of vegetative buds; bud multiplication; shoot elongation; and rooting. The success of this technique is highly dependent on the success of the first step (initiation) which requires a well-trained staff. Furthermore, most of problems encountered in the succeeding stages (multiplication, elongation, rooting) may have their origin at the initiation phase (Zaid and de Wet 1999). Furthermore, since shoots are directly initiated from mother tissue without passing through a callus phase, the plantlets produced are supposed to be true-to-type (Kunert et al. 2003).

5.3 Organogenesis Protocols

5.3.1 Selection and Removal of Offshoots

Suitable offshoots for *in vitro* culture may have an average weight of 2.5–6 kg (Badawy et al. 2005; Beauchesne et al. 1986; Mohamed et al. 2001), 3–5 years old, 60–80 cm high or bearing 8–12 leaves (Badawy et al. 2005). Suckers must be disease free and selected from a well-known elite adult date palm tree. When removing offshoots, special care should be given to its attachment to the mother tree. In fact, seeds can germinate within adhering leaf bases and produce a small plant which cannot be distinguished from an offshoot until verification of the connection between the offshoot and the mother palm. Skilled laborers are required to cut and remove an offshoot properly without damage to its base. Once removed, the offshoot should be cleaned of soil then all roots and leaves severely cut back before transportation to the lab. The best time period for starting *in vitro* culture from offshoots is between the end of date fruit harvest and the start of the next flowering stage (Beauchesne et al. 1986).

5.3.2 Offshoot Preparation

Offshoot preparation can be done with a sharp knife by removing, gradually (one by one), outer leaves and fibrous tissues at their bases until exposure of the shoot tip zone (Fig. 5.1 a, b). Careful handling is recommended to avoid damage to the brittle meristematic region. The shoot tip can then be excised by cutting a circle around the base of the cylindrical shoot tip at an angle of 45°. The ultimate size of the excised shoot tip should be about 3–4 cm in width and 6–8 cm in length. A sheathing leaf base enclosing the very young leaves of the heart of the offshoot should be left in place to protect it from disinfection solutions. A chain saw can also be used and makes easier the extraction of the shoot tip. Once removed, the shoot tip should be transferred to an antioxidant solution containing 100 mg ascorbic acid and 150 mg citric acid to avoid tissue browning due to the phenolic compounds.

5.3.3 Shoot Tip Disinfection

The excised shoot tips can be disinfected according to the following steps: (1) clean the shoot tips with distilled water to remove any organic debris; (2) soak in a fungicide solution (benomyl, mancozeb) for 10–15 min; (3) rinse three times with sterile distilled water; (4) soak again, for 20 min, in a commercial Clorox solution (sodium hypochlorite) supplemented with 0.3 g/l potassium permanganate; (5) rinse three times with sterile distilled water under aseptic conditions (Abahmane et al. 1999).

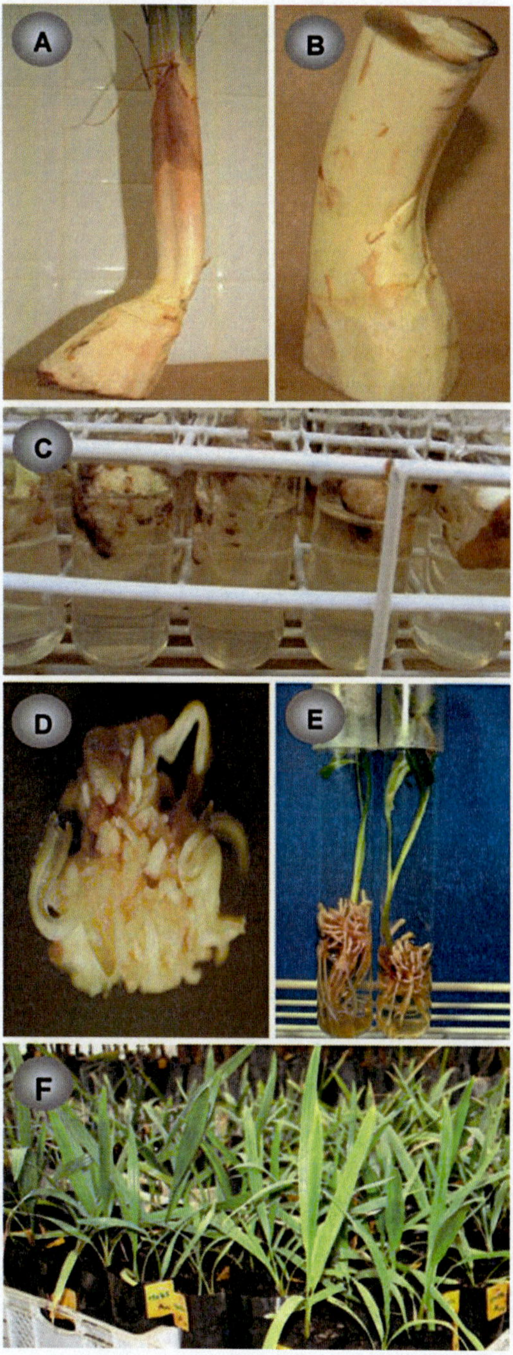

Fig. 5.1 Date palm micropropagation via organogenesis technique from offshoot shoot tip. (**a**) Offshoot after removing external leaves; (**b**) Shoot tip extracted from offshoot and ready to be disinfected; (**c**) Explants from shoot tip on culture medium in the starting stage; (**d**) Vegetative buds in the multiplication stage; (**e**) Well-formed plantlets ready for transfer under greenhouse; (**f**) Acclimatized vitro-plants under greenhouse

Other disinfection protocols are available and can be used according to Al-Khayri (2007), Badawy et al. (2005), Beauchesne et al. (1986), Bekheet and Saker (1998), Othmani et al. (2009), and Rao and Ganapathi (1993).

5.3.4 Explant Removal

After sterilization, the root tip can be dissected to extract cultured explants. Using scalpel and forceps, the young leaves surrounding the apical dome are gradually removed. The explants consist of the bottom of the excised leaves. The very young leaves closely surrounding the apical dome are difficult to separate; the entire shoot tip can be cut into 4–6 pieces and transferred to a culture medium. At the axils of young leaves, axillary buds can be found and are also suitable as explants. To avoid desiccation, explants must be immediately transferred to a prepared culture medium. In general, an average of 15–25 explants can be extracted from each offshoot shoot tip.

5.3.5 Incubation Conditions

After transfer into culture media, explants are incubated for 3–6 months in the dark so as to enhance bud initiation and also to prevent oxidation of phenolic compounds which occurs under light conditions. Explants should be transferred to fresh media each month. After initiation, shoots are transferred to lighted conditions with a photoperiod of 16 h. Air temperature in the growth chamber is maintained at 27 ± 1°C during the illuminated period and 22 ± 1°C during the dark period (Abahmane et al. 1999; Anjarne et al. 2005).

5.3.6 Culture Media

Generally a basal medium of Murashige and Skoog (1962) (MS) inorganic salts is used for micropropagation of date palm (Table 5.1). Based on the multiplication stage, it can be used at full strength or diluted usually to half strength due to its high level of mainly ammonium salts. A formulation for macro elements was proposed by Beauchesne et al. (1986) and can be used primarily at the initiation step. Beyond the previously-mentioned mineral salts, the following chemicals are usually added to the culture medium, especially NaH_2PO_4 (170 mg/l), myo-inositol (100 mg/l), adenine (30 mg/l), glutamine (200 mg/l), nicotinic acid (1 mg/l), pyridoxine-HCl (1 mg/l), Biotin (1 mg/l), calcium pantothenate (1 mg/l), sucrose (30 g/l), and agar (7 g/l) (Al Khateeb 2006; Beauchesne 1983). Plant growth regulators are added according to the multiplication stage (Table 5.1).

Table 5.1 Hormones added to MS medium during initiation and multiplication stages of date palm micropropagation (values in parentheses are in mg/l)

Date palm cultivars	Explant source	Initiation media	Growth responses	Multiplication media	References
–	Shoot tips	NAA(1), IBA (1), NOA(1–5.5), 2-iP(0.1)	Buds	–	Poulain et al. 1979
Boufegous, Bouskri, Black Bou Sthammi	Shoot tips, young leaves, axillary buds	2-iP(0.1), IBA(0.5), IAA(0.5), NOA(2)	Vegetative buds	–	Beauchesne 1983
–	Shoot tips	NAA(1), IAA (1), NOA(1–5.5), 2-iP(0.1–3)	Buds	NOA(2), NAA(1), IAA(1), BA(0.5), 2-iP(1), Kin(1–5)	Beauchesne et al. 1986
–	Shoot tips	2-iP(2), NAA(1)	Shoot buds	2-iP(3), NAA(0.5)	Taha et al. 2001
Maktoom	Shoot tips	2-iP(2), BA(1), NAA(1), NOA(1)	Buds	2-iP(4), BA(1), NAA(1), NOA(1) (Liquid medium)	Khierallah and Bader 2007
Sewy	Shoot tips, leaflets	2-iP(1), Kin (1) NAA(0.5–1)	Buds	MS + Zeatin (1), NOA(0.5)	Wanas et al. 1999
Khenezi	Shoot tips, apical meristem	IAA(1.6) or IAA(0.4), NAA(0.4) and BA(1.6) or 2-iP(3.2)	Bud generative tissue	IAA(0.4), NAA(0.1), Kin(0.1), 2-iP(1.5)	Al Kaabi et al. 2001
Yakubi	Shoot tips	BA(5), 2-iP(1), IAA(1.6), NAA(0.1)	Multiple shoots	–	Sharon and Shankar 1999
Barhi, Maktoom	Shoot tips	BA(1), 2-iP(1), NAA(1), NOA(1)	Vegetative buds	–	Bader and Khierallah 2007
Zagloul	Shoot tips	2-iP(2), NAA(0.1)	Shoot proliferation	2-iP(4), BA(4), NAA(0.5)	Bekheet and Saker 1998
Sukry	–	–	–	Kin(0.2), 2-iP, BA, IAA, NOA, NAA(0.1 each)	Al Khateeb 2006
Zaidi, Hussain, Asil	Shoot tips	IBA(0.1–6), BA(1)	Shoot proliferation	2-iP(1), TDZ(0.5)	Hussain et al. 2001
–	Axillary and apical buds, leaflets bases	NOA(3–5), NAA(1), 2-iP(0.5–1)	Buds from callus	–	Saka et al. 1998
Khalas	Shoot tips	–	Shoots from callus	BA(2)	Aslam and Khan 2009

The culture media are dispensed in test tubes (150 × 25 mm) at 15–20 ml or in culture jars (250 ml flasks, 170 ml baby food jars or magenta containers) at 50 ml. Media are then autoclaved at 121°C and under 1 bar pressure for 15–30 min according to the culture medium volume in the containers.

5.3.7 *Shoot Formation*

Depending on the genotype, shoot formation generally requires 6–12 months (Fig. 5.1 c, d). Bud initiation is controlled by several factors that may act in concert. Those factors in particular are the culture media components, genotype and time period of plant material collection. As shown in Table 5.1, different culture media are used to obtain shoots from explants, depending on the date palm cultivar. Concerning the time period of plant material collection, it was shown that the best period coincides with the dormancy stage in advance of flowering (Amin 2001). According to a study by Al Maari and Al Ghamdi (1998) on the effect of seasonal variation on date palm micropropagation, it was shown that the best growth and bud regeneration, and lowest rate of tissue browning were obtained when cultures were established between November and April (Northern Hemisphere).

5.3.8 *Shoot Multiplication*

Once initiated, vegetative buds should be transferred gradually to lighted conditions with a photoperiod of 16 h. Basal medium used for shoot multiplication is usually MS at full- or half-strength. Plant growth regulators are added in low concentrations as compared with the initiation stage and this is in accord with previous research on date palm micropropagation (Al Khateeb 2006; Zaid and de Wet 1999). Some authors have reported that shoot multiplication occurs on media characterized by auxin/cytokinin ratios > 1 (Loutfi and El Hadrami 2005).

According to a study of the multiplication stage of date palm cv. Sukry, Al Khateeb (2006) reported that low hormone concentrations promoted formation of new buds while high concentrations resulted in abnormal growth and without any observed sign of budding or shoot formation; the best combination that gave a good multiplication rate was (mg/l): Kin (0.2), 2-iP (0.1), BA (0.1), IAA (0.1), NOA (0.1) and NAA (0.1). Working on cv. Khalas, Aslam and Khan (2009) reported that the best shoot multiplication rate was obtained with 7.84 μM of BA. In addition, they reported that when the concentration of BA and KIN increased, respectively, to above 7.84 and 9.28 μM, the shoot regeneration rate decreased. In addition they showed that BA was more effective on shoot multiplication as compared to KIN. In their experiments, the authors found that the highest frequencies of shoot regeneration and number of shoots per an explant were obtained on solid MS medium as compared to liquid medium.

The influence of carbon sources and concentrations on *in vitro* shoot multiplication of date palm Khanezi cv. was investigated by Al Khateeb (2008a). He reported that 30 and 60 g/l of sugar was optimal for either qualitative or quantitative shoot growth, while abnormal growth was observed at 90 and 120 g/l, possibly due to osmotic stress. The author also reported that maltose, fructose or glucose were almost equally effective as a carbon source for date palm tissue culture as compared to sucrose. In addition, root formation was enhanced by increasing the sugar concentration above 60 g/l. This shoot rooting inhibits formation of new buds and hence decreases their ability to multiply during multiplication stage.

In general, the most common plant growth regulators used at this multiplication stage are auxins NAA, IAA and NOA and cytokinins BA, 2-iP and Kinetin (Table 5.1). However, other plant hormones can be used depending on the micropropagated genotype. Hussain et al. (2001) reported that TDZ plays an important role in the multiplication stage because it enhances horizontal rather than vertical growth.

As in the previous stage, the shoot multiplication rate is also genotype dependent (Beauchesne 1983; Hussain et al. 2001). In fact, the culture media of different compositions shown in Table 5.1 clearly elucidate this phenomenon encountered in date palm micropropagation. In addition, it was observed that *in vivo* behavior of date palm is also reflected *in vitro* i.e. cultivars producing more suckers in the field, show higher rate of multiplication *in vitro* (Hussain et al. 2001; Krikorian and Cronauer 1984). Concerning the transfer of cultures onto fresh media, it should be done at 4–6 week intervals depending on growth of genotypes in multiplication.

5.3.9 Shoot Elongation and Rooting

At this stage, the basal medium MS salts are used at either full- or half-strength (Fig. 5.1 e). Generally, shoot elongation requires transfer of shoots from the multiplication medium to another medium with a high auxin/cytokinin ratio (Beauchesne et al. 1986; Loutfi and Chlyah 1998). The most effective plant growth regulators at this stage are NAA or IBA, Kin and 2-iPA or BA. A combination of NAA (1 mg/l), BA (0.5 mg/l) and Kin (0.5 mg/l) enhances shoot growth and elongation. Gibberellins at 2 mg/l can also be incorporated into the culture medium in this stage but for no more than 15 days (Beauchesne et al. 1986). A study conducted on Zaghloul and Sewi cvs. showed that NAA at 0.1 mg/l has a pronounced effect on shoot length as compared with IBA and IAA (El Sharabasy et al. 2001). However, Aslam and Khan (2009) found that a rooting percentage of 87.34% was obtained on solid MS medium supplemented with 24.6 μM of IBA. They also reported that root length was higher when MS liquid medium was used.

Among rooting auxins, NAA added at 0.1 mg/l gave the maximum percentage of root formation, numbers and length (Al Kaabi et al. 2001; Taha et al. 2001). They also reported that use of MS salts at three-fourths strength gave the best results on root formation as compared to one-fourth, one-half and full strength. In the same study, light intensity of 8,000 lx and sucrose at 40 g/l produced the best results on root number and length. On the other hand, some authors have proposed alternatives to promote plant quality. Sidky et al. (2007) reported that plantlets transferred onto MS medium at half strength supplemented with 0.1 mg/l of NAA, 1 g/l of activated charcoal, 40 or 50 g/l of sucrose and 4 mg/l of paclobutrazol, increased thickness of plantlets, accelerated root formation and promoted secondary root formation.

Depending on the protocol used and genotype multiplied, some authors recommended media free of hormones for 1 month during the elongation stage (Al Kaabi et al. 2001). In the subsequent rooting stage, they used NAA at 1 mg/l coupled with sucrose at 30 g/l.

5.3.10 Plant Acclimatization

The specialized conditions which exist during *in vitro* culture can result in the formation of plantlets of abnormal morphology, anatomy and physiology (El Bahr et al. 2003; Saker et al. 2000). After *ex vitro* transfer (Fig. 5.1f), they may be easily impaired by sudden changes in environmental conditions and so need a period of acclimatization to overcome these abnormalities (Pospisilova et al. 1999). Accordingly, the acclimatization phase is the most important stage in the protocol of date palm micropropagation, because if not optimized, the whole process will be inefficient. Hence, it is not acceptable to produce a huge number of plantlets and lose them at this final stage. In fact, major differences exist between the environment of plants growing in tissue culture and those in the greenhouse; particularly, differences in light, both quantity and quality, relative humidity, nutrients and other growth promoters, the gaseous composition and the medium substrate (Seelye et al. 2003).

Previous studies have focused on optimization of factors such as soil mixtures, relative humidity, temperature and other greenhouse conditions. Studies in this field have led to significant enhancement (70%) in the percentage of survival of tissue culture derived plantlets (Quraishi et al. 1997). However, successful plant acclimatization of date palm should be started at the rooting stage. Plantlets to be transferred to the acclimatization stage should have certain important characteristics that enable them to succeed in the greenhouse. In fact, plantlets must be at least 12–15 cm in length with a well-formed and closed crown, two or three fully opened leaves and more than three roots. Such plantlets, if acclimatized under suitable conditions, have the maximum chance to survive when transferred to the greenhouse. However, plant quality is usually genotype dependant; some cvs. (Medjool, Boufeggous, Nejda, Aguellid, etc.) produce vigorous and well-formed plantlets of desired characteristics while others produce plants of inferior quality (open crown, curved leaves, weak plantlets, etc.). In such a situation, plantlets should be kept in the lab until they have acquired the desired characteristics. In fact, Al Salih et al. (1986) reported that death or failure of date palm plantlets during the acclimatization stage may be attributed to the lack of root cell differentiation caused by a deficiency in growth regulators or sugar balance. In this context, El Bahr et al. (2003) pointed out that leaf, root and stomata morphology of date palm *in vitro* plantlets were different in structure and shape as compared with those produced from acclimatized ones.

In order to improve the survival rate at the acclimatization stage, some osmoticums are used prior to the transfer of plantlets to soil. The addition of polyethylene glycol (PEG) at 6 or 8 g/l to the culture medium enhances plantlet survival during the acclimatization stage. Histological studies have shown that PEG increased the epidermal wax layer of leaves as compared to control plants (Sidky et al. 2007). In fact, non-acclimatized *in vitro* plantlets had an average of 15% of the amount of wax of greenhouse plants. Treatment of *in vitro* derived plantlets with PEG during the acclimatization stage resulted in an increase of wax deposition and as a consequence water loss was drastically reduced (Zaid and Hughes 1995a). It was reported that PEG-treated plantlets showed a water loss of only about 27% (similar to that of greenhouse grown plants) as compared to an average of 40% in control plants (Zaid and Hughes 1995b).

In addition, the application of gamma aminobutyric acid (GABA) at 10 mM during the acclimatization stage increased the chlorophyll concentration and then improved the survival percentage of date palm plantlets (Awad 2007). Moreover, other substances were tested at the rooting stage. When the growth retardant paclobutrazol was added to the rooting medium, shoots developed fewer and smaller leaves. This reduction in plant growth makes them more wilt tolerant and increases their survival when transferred to the greenhouse (Seelye et al. 2003). On the other hand, a reduction in relative humidity leads to increased plant respiration with associated development of functional stomata for controlling plant water loss (Seelye et al. 2003).

Multiple soil mixtures have been used to transfer plantlets *ex vitro*. The main mixture characteristic that influences plant growth is moisture which should not be excessive to avoid fungi attacks (rots) and not too low to avoid plantlet desiccation. Tisserat (1981) reported that the best survival rate was recorded for 10–12 cm date palm plantlets transferred to peat moss: vermiculite mixture (1/1:v/v) and covered with transparent plastic. El Sharabasy et al. (2001) reported that the best results were obtained with a planting medium containing equal parts of peat, sand and vermiculite. The survival percentage was 80% after 18 months.

In order to maintain high relative humidity around plantlets newly transferred to the greenhouse, plastic micro-tunnels can be used during the first 3 weeks of acclimatization. In addition, during acclimatization plantlets should be protected against fungi that cause crown and leaf rot. Wide-spectrum fungicides should be used twice weekly to avoid plantlets lost. Insecticides also may be needed to control insects inside the greenhouse.

Water supply must be monitored very carefully during the first month of acclimatization. Too much can lead to plantlet rot and too little moisture in the substrate can decrease the relative humidity around the plants and cause their rapid wilt.

5.4 Date Palm Micropropagation from Inflorescence Tissue

In the case of certain rare or select date palm genotypes, micropropagation is hampered by the absence of suckers needed to start their multiplication. In such situation, the use of inflorescence tissue remains the only way to micropropagate those genotypes. Many publications exist about the use of this technique (Abahmane 1998, 2003, 2005a,b 2007; Abul-Soad 2007; Drira 1985; Drira and Benbadis 1985; Loutfi 1989, 1999; Loutfi and Chlyah 1998). The application of this technique to date palm micropropagation will be fully described in the chapter: "Micropropagation of date palm (*Phoenix dactylifera* L.) using inflorescence explants" in this book.

5.5 Problems Encountered in Organogenesis

Date palm micropropagation is hampered by certain constraints that affect its efficiency. Among these problems, tissues browning, vitrification, early rooting and bacterial contaminants are the most important factors affecting date palm tissue culture. Except for internal bacterial contamination, the other problems listed are encountered when the plant material source is from either offshoot shoot tips or inflorescence tissue.

5.5.1 Tissue Browning

Date palm tissue is known to contain high levels of caffeoylshikimic acids ranging from 190 to 430 μg/g fresh weight depending on the cultivar (Loutfi and El Hadrami 2005). The released polyphenols accumulate in the culture medium which turns brown over time. These substances are oxidized by polyphenoloxydases and form quinons which are highly toxic to cultured tissue. Their secretion is enhanced by injuries due to explant preparation for *in vitro* inoculation or to their manipulation when transferred to fresh media.

Many suggestions have been made to reduce the incidence of this phenomenon. Among these are: pre-soaking of tissue in antioxidant solutions (100 mg/l ascorbic acid and 150 mg/l citric acid) before their transfer to culture media as reported by Murashige (1974) and Zaid and Tisserat (1983). Some researchers report that addition to culture media of some adsorbents like adenine; glutamine and citrate reduce date palm tissue browning (Rhiss et al. 1979).

Activated charcoal is among the compounds widely used to avoid tissue browning. However, a great amount of added growth regulators is also absorbed by activated charcoal and this is why high hormone concentrations are commonly used with this adsorbent (50–100 mg/l coupled with 3 g/l of activated charcoal). In addition, Beauchesne (1983) reported that use of Polyvinylpirrolidone (PVP-40) in the culture medium at a concentration of 2 g/l reduced date palm tissues browning.

Some other measures have also been suggested including use of explants of small size, juvenile tissue (tissue with less lignification) and frequent transfers onto fresh media. Besides, a period of incubation in darkness during the first months in the starting stage is also recommended in date palm micropropagation in order to reduce the incidence of tissue browning, which is enhanced under light conditions. In addition, Al Khateeb (2008b) stated that culturing of explants during winter and spring seasons can reduce this phenomenon.

5.5.2 *Early Rooting*

During the starting stage, roots are formed on cultured explants instead of buds. The appearance of roots in this stage inhibits bud formation and leads to culture elimination. At the multiplication stage, early rooting decreases the rate of bud multiplication by diverting most tissue nutrients to root formation rather than shoot formation (Al Khateeb 2008b). Studies of this phenomenon have shown that root initiation requires high auxin concentrations, especially NAA (3 mg/l). In contrast, root elongation occurs on culture media with low auxin concentrations (0.1 mg/l). Moreover, the time of year when explants are introduced *in vitro* seems to have an important effect on root formation. Rooting percentages of 16% and 40% are observed when explants were cultured respectively in July and December (Northern Hemisphere) (Anjarne and Zaid 1993). Furthermore, low concentrations of mineral nutrients in culture media and incubation of cultures in darkness for a long period also lead to early rooting of buds (Al Khateeb 2008b).

5.5.3 *Tissue Vitrification*

Date palm tissue cultures are susceptible to a vitrification phenomenon. This physiological disorder is characterized by development of tissue with lignification deficiency. It is due to the accumulation of water in the cultured tissues (Al Khateeb 2008b). Many factors that enhance this disorder are cited in the literature. The more important among them are high levels of growth regulators (mainly cytokinins) and some mineral salts (especially ammonium ions) in the culture medium. A study conducted on the Aguellid cv. showed that media containing high concentrations of ammonium nitrate allowed rapid growth and consequently high levels of vitrification, rising from 46% to 53%. In contrast, on culture media containing a moderate concentration of ammonium nitrate, these percentages were reduced to 14–19% (Bougerfaoui and Zaid 1993). Other factors have been reported by Al Khateeb (2008b) including the presence of high humidity levels and gases, particularly ethylene, inside culture tubes and the use of liquid media. Measures to reduce this phenomenon are numerous. Among them the most effective are increasing agar concentration, use of container covers that allow proper release of gases, reduction of hormonal and ammonium concentrations and the use of solid instead of liquid media (Al Khateeb 2006).

5.5.4 *Tissue Contamination*

Date palm tissue cultures are sometimes highly contaminated with internal bacteria. These contaminants are introduced *in vitro* with cultured explants even if they were well disinfected. Many studies have confirmed the existence of internal bacteria in apparently healthy offshoot tissue. Isolation and identification of these contaminants have shown that they belong to the genus *Bacillus* (Leary et al. 1986). Their most important characteristic is formation of endospores which can survive at 80°C for 30 min! The presence of these contaminants in date palm tissues culture can lead to culture elimination at any time. In fact, their appearance is generally observed after 1 month of culture. However, this appearance can also occur even after three or five subcultures. The incidence of this contamination can increase from 20% to 50% of cultures. Control of these contaminants can be done by the use of some antibiotics: tetracycline (30 μg/ml), streptomycin (10 μg/ml), neomycin (20 μg/ml) and chloramphenicol (30 μg/ml) (Leary et al. 1986). To ensure the efficiency of bacterial control when using antibiotics, the following measures are recommended: cleaning contaminated tissue by washing in sterile distilled water, dipping in antibiotic solution before culturing in the physiological medium, sterilizing tools at 180°C and using young date palm tissue in culture (Benjama et al. 2001). However, in practice, the best way to overcome this problem is to screen for contaminated cultures at the beginning of the multiplication stages and discard them.

5.5.5 *Availability of Plant Material*

In the case of some rare or select genotypes, micropropagation is hampered by limited numbers of available offshoots. In fact, some selected genotypes with good fruit quality and presumed resistance to bayoud disease are usually represented by only one or two palms. Hence, few plant materials (shoot tip explants) are available for development of their micropropagation protocol and for plant distribution to farmers. In such a situation, explants excised from emerged inflorescences have been successfully used to multiply genotypes that no longer produce offshoots (Abahmane 2003, 2005a,b, 2007; Loutfi 1999; Loutfi and Chlyah 1998). In fact, inflorescences are produced almost every year and plant material is abundant. In our experience with this plant material, no internal bacteria contaminants have been observed in tissue cultures. In addition, the mother tree is preserved when plant material is removed if some technical cares are taken.

5.6 Conclusion and Prospective

Date palm tissue culture techniques have permitted successful micropropagation of a large number of commercial varieties worldwide. The availability of date palm vitroplants has had a tremendously positive impact on the agricultural sector, mainly in oasis ecosystems. Normally, date palm can be micropropagated through two main methods. One is the embryogenesis technique in which somatic embryos are regenerated from embryogenic callus. By using this method, a large number of plants can be obtained in a short period. The second method is the organogenesis technique in which vegetative buds are regenerated from cultured explants without passing through a callus stage. Since plantlets are obtained directly from mother tissue, typically they will be identical to the mother tree. Hence, this method can be considered more secure and the risk of somaclonal variations can be easily avoided.

Explants are commonly excised from offshoot shoot tips. This plant material is the most suitable for micropropagation of date palm varieties. However, in the case of some select genotypes and rare cultivars, the availability of offshoot is quite limited. In such a situation, explants excised from inflorescences remain the only source of plant material for micropropagation.

The protocols described for date palm micropropagation by organogenesis, can be used both at the research and commercial levels. In fact, many research and commercial labs are actually micropropagating date palm using this technique, chiefly in Morocco, Saudi Arabia and the United Arab Emirates. In Morocco, this protocol has been developed by the National Institute for Agronomic Research (INRA) and transferred to the private sector. The collaborative partnership between INRA-Morocco and the private sector has permitted the production of more than 550,000 date palm vitro-plants. These plants have been distributed by the agricultural ministry extensive services to farmers of the southern oases free of charge.

In the present study, the entire organogenesis process has been described step by step and recommendations related to each stage have been formulated according to the literature and our experience in this field. Some problems facing date palm tissue culture are also discussed in order to help individuals using the technique to overcome them.

More research activities must be undertaken to reduce the time needed to produce date palm vitro-plants via organogenesis. The initial stage remains the main step where more research work has to be undertaken to produce vegetative buds in a shorter time. In addition, protocols have to be refined so as to obtain a routine procedure which can be used at the commercial level to produce more date palm vitro-plants at low cost within a short time.

References

Aaouine M (2000) Production of date palm vitro-plants: the Moroccan experience. Proceedings date palm international symposium, Windhoek, Namibia, pp 46–52

Abahmane L (1998) Utilisation des tissus inflorescentiels comme explants pour la micropropagation du palmier dattier (Phoenix dactylifera L). Proceedings of international conference on date palm, Marrakech, Morocco, Feb 16–18, pp 256–260

Abahmane L (2003) Multiplication du palmier dattier (Phoenix dactylifera L.) à partir des tissus floraux. Proceedings international conférence on date palm, College of Agric. & Vet. Med., King Saud University, pp 911–924

Abahmane L (2005a) Les tissus inflorescentiels: Une nouvelle source de matériel végétal pour la micropropagation des clones sélectionnés de palmier dattier (*Phœnix dactylifera* L.).

Proceedings international symposium on sustainable agricultural development of Oasian systems. Erfoud, Morocco, Mar 8–10, pp 99–105

Abahmane L (2005b) Micropropagation par tissus inflorescentiels du palmier dattier (Phœnix dactylifera L.): un outil efficace pour la sauvegarde des génotypes rares. Al Awamia 113:49–60

Abahmane L (2007) Micropropagation of selected clones from inflorescence tissues and its role in the date palm (*Phoenix dactylifera* L.) improvement programme. Proceedings fourth international symposium on the date palm, King Faisal University – Al Hassa, Saudi Arabia, 5–8 May 2007, p 145

Abahmane L, Bougerfaoui M, Anjarne M (1999) Use of tissue culture techniques for date palm propagation and rehabilitation of palm groves devastated by bayoud disease. Proceedings international symposium on date palm, Assiut University, Assiut (Egypt) Nov 9–11

Abul-Soad AA (2007) Inflorescence tissue culture utilisation for date palm (*Phoenix dactylifera* L.) micropropagation. Abstracts of fourth symposium on date palm in Saudi Arabia, King Faisal University, Al-Hassa, 5–8 May, p 144

Ahloowalia BS, Prakash J, Savangikar VA (2004) Plant tissue culture, In: AIEA-TECDOC-1384: low cost options for tissue culture technology in developing countries, pp 3–10

Al-Ghamdi AS (1988) Rooting of date palm offshoots as affected by offshoot size, cultivar and butyric acid injection. Acta Hort Tech Comm Int Soc Hort Sci (ISHS) 379–388

Al Kaabi HH, Rhiss A, Hassan MA (2001) Effect of auxins and cytokinins on the in vitro production of date palm bud generative tissues and on the number of differentiated buds. Proceedings second international conference on date palm Al Ain, UAE, pp 47–86

Al Khateeb AA (2006) Role of cytokinin and auxin on the multiplication stage of date palm (*Phoenix dactylifera* L.) cv. Sukry. Biotech 5:349–352

Al Khateeb AA (2008a) Regulation of in vitro bud formation of date palm (*Phoenix dactylifera* L.) cv. Khanezi by different carbon sources. Biores Techn 99:6550–6555

Al Khateeb AA (2008b) The problems facing the use of tissue culture technique in date palm (*Phoenix dactylifera* L.). Sci J King Faisal Univ 9:85–104

Al Khateeb AA, Ali-Dinar HM (2002) Date palm in Kingdom of Saudi Arabia: cultivation, production and processing. Translation, Authorship and Publishing Center, King Faisal University, Kingdom of Saudi Arabia, p 188

Al Khayri JM (2003) In vitro germination of somatic embryos in date palm: effect of auxin concentration and strength of MS salts. Curr Sci 84:680–683

Al-Khayri JM (2007) Micropropagation of date palm Phoenix dactylifera L. In: Jain SM, Haggman H (eds.) Protocols for micropropagation of woody trees and fruits. Springer, Berlin, pp 509–526

Al Maari KW, Al Ghamdi AS (1998) Effect of seasonal variation on the multiplication of date palm through tissue culture. Proceedings international conference on date palm. 16–18 Feb 1998, Marrakech, Morocco, pp 244–248

Al Mana, FA, El-Hamady MA, Bacba MA, Abdelrahman AA (1996) Improving root development on ground and aerial date palm offshoots. Res Bul N° 60 Agric Res Cent King Saud Univ., pp 5–19

Al Salih AA, Badr SM, Jarrah AZ, Al Qadi MT (1986) A comparative morphlogical and anatomical study of seed and embryo culture derived seedlings of Phoenix dactilifera L. Date Palm J 4(2):153–162

Amin T (2001) In vitro propagation of date palm (Phoenix dactylifera L.) by adventive buds. Proceedings second international conference on date palm. Al Ain, UAE, pp 568–587

Ammar S, Benbadis A (1977) Multiplication végétative du palmier dattier (*Phoenix dactylifera* L.) par la culture de tissus de jeunes plantes issues de semis. C R Acad Sci Paris 284:1789–1792

Anjarne M, Zaid A (1993) Effets de certains équilibres hormonaux sur l'enracinement précoce des tissus du palmier dattier (*Phoenix dactylifera L.*). Al Awamia 82:197–210

Anjarne M, Abahmane L, Bougerfaoui M (2005) Les techniques de micropropagation du palmier dattier (*Phoenix dactylifera L.*) : Expérience de l'INRA–Maroc. Actes du Symposium International sur le Développement Agricole Durable des Systèmes Oasiens. Erfoud, Morocco, Mar 8–10, pp 86–93

Asemota O, Eke CR, Odewale JO (2007) Date palm (*Phoenix dactylifera* L.) in vitro morphogenesis in response to growth regulators, sucrose and nitrogen. Afr J Biotech 6:2353–2357

Aslam J, Khan SA (2009) In vitro micropropagation of khalas date palm *(Phoenix dactylifera L.)*, an important fruit plant. J Fruit Orn Plant Res 17:15–27

Awad MA (2007) Effect of antistress substances and elemental sulphur application on growth and stress tolerance of tissue culture derived date palm *(Phoenix dactylifera L.)* plantlets cv. Khadrawy during acclimatization. Proceedings fourth international symposium on date palm, King Faisal University, Al Hassa, Saudi Arabia, May 5–8, p 153

Badawy EM, Habib AMA, El Bana A, Yosry GM (2005) Propagation of date palm *(Phoenix dactylifera L.)* plants by using tissue culture technique. Arab J Biotech 8:343–354

Bader SM, Khierallah HSM (2007) Effect of silver thiosulphate and glutamine on direct organogenesis of date palm (*Phoenix dactylifera* L.). Proceedings fourth symposium on date palm. King Faisal University, Al Hassa, Saudi Arabia, May 5–8, p 131

Beauchesne G (1983) Vegetative propagation of date palm *(Phoenix dactylifera L.)* by in vitro culture. Proceedings first symposium on date palm. King Faisal University, Saudi Arabia, pp 698–700

Beauchesne G, Zaid A, Rhiss A (1986) Meristematic potentialities of bottom of young leaves to rapidly propagate date palm. Proceedings second symposium on date palm. King Faisal University, Saudi Arabia, pp 87–94

Bekheet SA, Saker MM (1998) In vitro propagation of Egyptian date palm: II- Direct and indirect shoot proliferation from shoot tip explants of *Phoenix dactylifera* L. cv Zaghlool. Proceedings first international conference on date palm, Al Ain, UAE, pp 150–157

Benjama A, Cherkaoui B, Al Maii S (2001) Origin and detection of *Bacillus* contaminating date palm vitro culture and importance of manipulation conditions. Al Awamia 104:73–74

Bhaskaran S, Smith RH (1992) Somatic embryogenesis from shoot tip and immature inflorescences of *Phoenix dactylifera* L. cv. Barhee. Plant Cell Rep 12:22–25

Bhaskaran S, Smith RH (1995) Somatic embryogenesis in date palm *(Phoenix dactylifera* L.*)*. In: Jain SM, Gupta PK, Newton RJ (eds.) Somatic embryogenesis in woody plants: angiosperms. Kluwer, Dordrecht, pp 446–470

Bougerfaoui M, Zaid A (1993) Effet de la teneur du milieu de culture en amoniaque sur la vitrification des tissus du palmier dattier cultivés in vitro. Al Awamia 82:177–196

Daguin F, Letouze R (1988) Régénération du palmier dattier (*Phoenix dactylifera* L.) par embryogenèse somatique: amélioration de l'efficacité par passage en milieu liquide agité. Fruits 43:191–194

Drira N (1985) Multiplication végétative du palmier dattier (*Phoenix dactylifera* L.) par les néoformations induites en culture in vitro sur des organes végétatifs et floraux prélevés sur la phase adulte. Thèse de doctorat d'Etat Es-Sciences Naturelles Fac Sci Tunis

Drira N, Benbadis A (1985) Multiplication végétative du palmier dattier (*Phoenix dactylifera* L.) par réversion, en culture in vitro, d'ébauches florales de pieds femelles. J Plant Phys 119:223–235

Eke CR, Akomeah P, Asemota O (2005) Somatic embryogenesis in date palm (*Phoenix dactylifera* L.) from apical meristem tissues from 'Zbia' and 'Loko' landraces. Afr J Biotech 4:244–246

El Bahr MK, Ali ZA, Taha HS (2003) In vitro propagation of Egyptian date palm cv. Zaghlool: II. Comparative and anatomical studies between direct acclimatized and in vitro adapted pre-acclimatized plantlets. Arab Univ J Agric Sci 11:701–714

El-Bahr MK, Ali ZA, Saker MM (2004) A comparative anatomical study of date palm vitro-plants. Arab J Biotech 7(2):219–228

El Hamady M, Al Mana FA, Bacha MA (1992) Greenhouse rooting of date palm offshoots using an inverted mist system. Annu Agric Sci Ain Shams Univ Cairo 37:523–529

El Sharabasy SF, Bosila HA, Ibrahim IA et al. (2001) Micropropagation studies on Zaghlool and Sewi cvs. of date palm (*Phoenix dactylifera L.*): III. Plantlet acclimatization. Proceedings second international conference on date palm, Al Ain, UAE, pp 523–530

FAO (2008) http://faostat.fao.org

Ferry M, Gómez S (2002) The red palm weevil in the Mediterranean area. Palms 46:172–178

Ferry M, Bouguedoura N, El Hadrami I (1998) Patrimoine génétique et techniques de propagation in vitro pour le développement du palmier dattier. Spécial Oasis. Sécheresse 9(2):139–146
Gómez S, Ferry M (1999) Attempts at biological control of date palm pests recently found in Spain. In: Canard M, Beysset-Arnouty V (eds.) Proceedings first symposium for applied biological control in Mediterranean countries, Cairo, 25–29 Oct 1998. Sacco Toulouse, France, pp 121–125
Heselmans M (1997) Setting research priorities through an international date palm network. Biotech Dev Monit 30:1820–1822
Hodel DR, Pittenger DR (2003a) Studies on the establishment of date palm (*Phoenix dactylifera*) cv 'Deglet Noor' offshoots. Part I: observations on root development and leaf growth. Palms 47:191–200
Hodel DR, Pittenger DR (2003b) Studies on the establishment of date palm (*Phoenix dactylifera*) 'Deglet Noor' offshoots. Part II. Size of offshoots. Palms 47:201–205
Hussain I, Rashid H, Muhammad A, Quraishi A (2001) In vitro multiplication of date palm. Proceedings second international conference on date palm. Al Ain, UAE, pp 432–438
Khierallah HSM, Bader SM (2007) Micropropagation of date palm (*Phoenix dactylifera* L.) var. Mektoom through direct organogenesis. Acta Hort 736:213–224
Krikorian AD, Cronauer SS (1984) Banana. In: Sharp W, Evans DA, Ammirato PV, Yamada Y (eds.) Handbook of plant cell culture, vol 2. McMillan, New York, pp 327–348
Kunert KJ, Baaziz M, Cullis CA (2003) Techniques for determination of true-to-type date palm (*Phoenix dactylifera* L.) plants: a literature review. Emirates J Agric Sci 15:1–16
Leary JV, Nelso N, Tisserat B, Allingham EA (1986) Isolation of pathogenic Bacillus circulans from callus cultures and healthy offshoots of date palm (*Phoenix dactylifera L.*). Appl Envir Microbiol 52:1173–1176
Letouze R, Daguin F, Hamama L et al. (2000) Mass propagation of date palm through somatic embryogenesis. Histological study of embryo formation and cultivar identification by RAPD markers. Proceedings date palm international symposium. Windhoek, Namibia, pp 55–64
Loutfi K (1989) Multiplication végétative du palmier dattier (*Phoenix dactylifera* L.) à partir de la culture in vitro d'explants inflorescentiels. Thèse Doctorat 3ème cycle. Université Cadi Ayyad, Marrakech
Loutfi K (1999) Organogenèse et embryogenèse somatique à partir des tissus floraux du palmier dattier (*Phoenix dactylifera L.*) cultivés in vitro. Aspects histologiques et caryologie des vitroplants. Thèse doctorat Es-Sciences, Université Cadi Ayyad Marrakech
Loutfi K, Chlyah H (1998) Multiplication végétative du palmier dattier à partir de segments d'inflorescences cultives in vitro: effet de différentes combinaisons hormonales et capacités organogénétiques de divers cultivars. Agron 18:573–580
Loutfi K, El Hadrami I (2005) *Phoenix dactilyfera* date palm. In: Litz RE (ed.) Biotechnology of fruit and nut crops. CAB International, Wallingford, pp 144–157
McCubbin MJ, Van Staden J, Zaid A (2000) A southern African survey conducted for off-types on date palm production using somatic embryogenesis. Proceedings date palm international symposium, Windhoek, Namibia, pp 68–72
Mohamed SM, El Sharabasy SF, Bosila HA (2001) Micropropagation studies on Zaghloul and Sewi cvs of date palm (Phoenix dactylifera L.), 1- Callus initiation and formation. Proceedings second international conference on date palm, Al Ain, UAE, pp 500–512
Mohammed S (1978) Problem in date palm propagation. Indian Hort 23:15–31
Murashige T (1974) Plant propagation through tissue culture. Biologist 21:87–93
Murashige T, Skoog F (1962) A revised medium for rapid growth and bioassays with tissue cultures. Phys Plant 15:473–497
Nixon RW, Carpenter JB (1978) Growing dates in the United States. U.S. Dept. of Agriculture, Agric. Information Bulletin No. 207. USDA, Washington DC
Othmani A, Bayoudh C, Drira N, Trifi M (2009) In vitro cloning of date palm *Phoenix dactylifera* L. cv. Deglet Bey using embryogenesis suspension and temporary immersion bioreactor. Biotech Biotech Equip 23:1181–1188

Pospisilova J, Ticha I, Kadlecek P (1999) Acclimatization of micropagated plants to *ex vitro* conditions. Biotech Plantar 42:481–497

Poulain CA, Rhiss A, Beauchesne G (1979) Multiplication végétative en culture in vitro du palmier dattier (Phoenix dactylifera L.). C R Acad Agric 11:1151–1154

Qaddoury A, Amassa M (2004) Effect of exogenous indole butyric acid on root formation and peroxidase and indol-3-acetic acid oxydase activities and phenolic contents in date palm offshoots. Bot Bull Sin 45:127–131

Quraishi A, Hussain M, Latif M (1997) Sustained long term embryogenic cultures of date palm and their field performance. Pak J Bot 19:135–141

Rao PS, Ganapathi TR (1993) Micropropagation of palms. In: Ahuja MR (ed) Micropropagation of woody plants. Kluwer, Dordrecht, pp 395–424

Reuveni O, Adato Y, Kipnis HL (1972) A study of new and rapid methods for the vegetative propagation of date palms. Date Grow Inst 49:17–24

Rhiss A, Poulain C, Beauchesne G (1979) La culture in vitro appliquée à la multiplication du palmier dattier (*Phoenix dactylifera* L.). Fruits 34:551–554

Saaidi M, Duvauchelle G, Toutain G (1979) Multiplication du palmier dattier: Etude de quelques facteurs conditionnant la reprise végétative des rejets de palmier dattier. Fruits 34:55–61

Saka G, Yahawi N, Abid F (1998) In vitro techniques as applied to the regeneration of different Algerian date palm cvs. Through somatic embryogenesis and organogenesis and perspectives of improvement. Proceedings international conference on date palm, Marrakech, Morocco, Feb 16–18, pp 225–229

Saker M, Bekheet M, Taha HS (2000) Detection of seasonal variations in tissue culture derived date palm plants using isosyme analysis and RAPD fingerprints. Biol Plant 43:347–351

Seelye JF, Burge GK, Morgan Ed R (2003) Acclimatizing tissue culture plants: reducing the shock. Comb Proc Int Plant Prop Soc 53:85–90

Sharma DR, Dawra S, Choudhury JB (1984) Somatic embryogenesis and plant regeneration in date palm (*Phoenix dactylifera L.*) cv. Khadrawi through tissue culture. Indian J Exp Biol 22:596–598

Sharon M, Shankar C (1999) Regeneration of date palm *(Phoenix dactylifera L.)* through direct organogenesis. Indian J Plant Phys 4:323–326

Sidky RA, Zaid ZE, El-Bana A (2007) Optimized protocol for in vitro rooting of date palm (*Phoenix dactylifera* L.). Proceedings fourth international symposium on date palm, King Faisal University, Al Hassa, Saudi Arabia, pp 454–467

Sudhersan C, Abo El-Nil M, Hussain J (2001) Hapaxanthic axillary shoots in date palm plants grown in vitro and in vivo. Palms 45:84–89

Taha HS, Bekheet SA, Saker MM (2001) Factors affecting in vitro multiplication of date palm. Biol Plant 44:431–433

Tisserat B (1979) Propagation of date palm *(Phoenix dactylifera L.)* in vitro. J Exp Bot 30:1275–1283

Tisserat B (1981) Production of free-living date palms through tissue culture. Date Palm J 1:43–54

Tisserat B (1982) Development of new tissue culture technology to aid in the cultivation and crop improvement of date palms. Proceedings first symposium on date palm. King Faisal University, Saudi Arabia, pp 126–140

Tisserat B (1983) Tissue culture of date palm- A new method to propagate an ancient crop and a short discussion of the California date industry. Prin 27:105–117

Tisserat B, DeMason DA (1980) A histological study of development of adventive embryos in organ cultures of *Phoenix dactylifera L.* Ann Bot 46:465–472

Wanas WH, El Hammady AM, Abo Rawash M, Awad AA (1999) In vitro propagation of date palm. 1- Direct organogenesis as affected by cytokinin and auxin levels in the medium. Ann Agric Sci 44:19–31

Zaid A, de Wet PE (1999) Date palm propagation. In: Zaid A, Arias-Jimenez EJ (eds.) Date palm cultivation, FAO Plant Production and Protection Paper 156 Rev. 1

Zaid A, Hughes H (1995a) In vitro acclimatization of date palm plantlets (*Phoenix dactylifera* L.): a quantitative comparison of epicuticular leaf wax as a function of polyethylene glycol treatment. Plant Cell Rep 15:111–114

Zaid A, Hughes H (1995b) Water loss and polyethylene glycol mediated acclimatization of in vitro grown seedlings of 5 cultivars of date palm (*Phoenix dactylifera* L.) plantlets. Plant Cell Rep 14:385–388

Zaid A, Tisserat B (1983) In vitro shoot tip differentiation in *Phoenix dactylifera L.* Date Palm J 2:163–182

Zouine J, El Hadrami I (2007) Effect of 2,4-D, glutamine and BAP on embryogenic suspension culture of date palm (*Phoenix dactylifera* L.). Sci Hort 112:221–226

Chapter 6
Micropropagation of Date Palm Using Inflorescence Explants

A.A. Abul-Soad

Abstract Inflorescence-based micropropagation holds great potential for the multiplication of recalcitrant male and female date palm individual trees and cultivars of commercial interests with limited populations. This can be accomplished in a short time with minimal effort as compared to the traditional practice of using shoot-tip explants. The aim of this technique is to pave the way to use inflorescence explants to micropropagate date palm by direct formation of organs (somatic embryos or shoots) and avoiding most of the constraints that face the shoot tip like high percentage of contamination in the establishment stage, heavy browning, a long time for first cluster initiation, using a considerable number of offshoots and inability to micropropagate the elite palms in case no offshoots are available. The way to excise the immature inflorescence without damage to the mother tree, composition of the nutrient medium for direct organ initiation have remained hindrances to this technique over recent decades, in addition to the technicalities for proper handling of cultures inside the laboratory and successful shifting of plants to the greenhouse. We report for the first time an innovative method used to excise the immature inflorescence at a suitable stage for successful culture initiation. Spikelet explants are induced to produce shining globular structures without a callus phase. Also, explants were exceptionally able to develop direct shoots. Two types of organs developed on the differentiation medium, green shoots and intact somatic embryos. Only green shoots and multiple somatic embryos were subjected to proliferation at the multiplication stage. Well-rooted plantlets were hardened and successfully established in soil.

Keywords Acclimatization • *In vitro* • Inflorescence • *Phoenix dactylifera* L. • Regeneration

---

A.A. Abul-Soad (✉)
Horticulture Research Institute, Agricultural Research Center,
9 Cairo Univ. St., Orman, Cairo, Egypt
e-mail: adelaboelsoaud@gmail.com

S.M. Jain et al. (eds.), *Date Palm Biotechnology*,

DOI 10.1007/978-94-007-1318-5_6, © Springer Science+Business Media B.V. 2011

6.1 Introduction

The date palm (*Phoenix dactylifera* L.) is a dioecious, perennial monocotyledonous plant species belonging to the family Arecaceae. It is one of the oldest fruit tree crops and is primarily cultivated in North African and Middle Eastern countries. Flowers are borne on stalks growing among the leaves. The date palm has compound, branched flower stalks (inflorescences) bearing small whitish or creamy flowers (Fig. 6.1a). Traditionally, shoot tip explants from offshoots are used for various micropropagation protocols for both research and commercial production

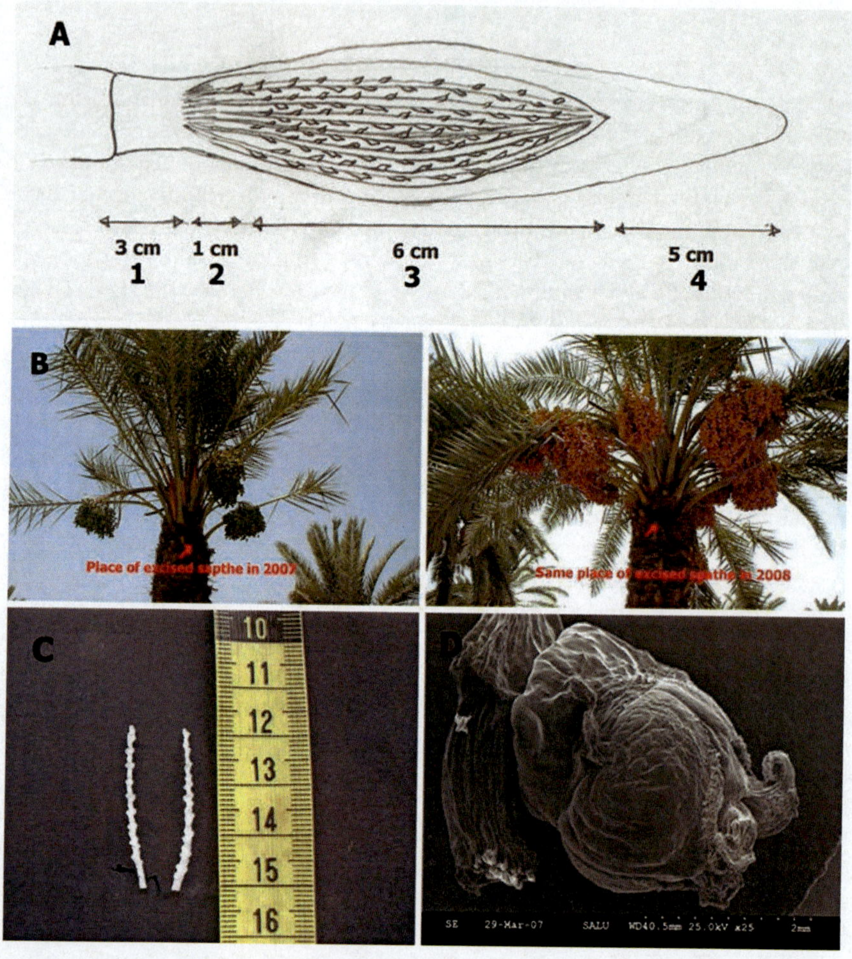

Fig. 6.1 Establishment of the inflorescence cultures. (**a**) Schematic illustration of 15 cm spathe components (*1*: Inflorescence base, *2*: Spikelet base, *3*: Spikelets, *4*: Protective sheath), (**b**) The mother tree of excision (*left*), and subsequent season carrying regular bunches (*right*), note the number of bunches, (**c**) Initial spikelet explant, (**d**) Scanning Electron Microscopic of one floret 25×

purposes (Abul-Soad et al. 2002a,b; El-Hadrami and Baaziz 1995; Tisserat 1984b). The major disadvantage of this practice is scarification of the entire offshoot. Ultimately, this will hinder the micropropagation of superior individuals without offshoots or of commercially viable cultivars with limited numbers of available individual specimens.

Inflorescence explants have proven promising and afford a needed alternative explant source for micropropagation of elite cultivars and rare individuals of date palm. We describe here a stepwise micropropagation procedure using inflorescence explants and compare their benefits with the use of shoot-tip explants; innovative ways to safely excise an immature inflorescence from an adult tree at the appropriate time; establishment of the initial explants on a starting medium after surface sterilization of the spathe; other details on rooting, hardening and field performance in fruiting.

6.2 Morphogenic In Vitro Responses of Date Palm Inflorescence

Inflorescences of several palm species have been cultured *in vitro*. Since 1973, several researchers have attempted to culture palm inflorescences. Explants of female and male oil palm inflorescences were grown on culture media and usually plants regenerated normally without callus formation (Smith and Thomas 1973), whereas in date palm, a high auxin level was required (Eeuwens and Blake 1977). The addition of auxins in the culture media increases the frequency of visible expanded carpels developing from supposedly male date palm (Tisserat and DeMason 1980). Vestigial female date carpels on surviving male flowers become enlarged and quite prominent (Tisserat 1979). White friable callus usually was initiated from the floral bud strand (Tisserat et al. 1979). In some cases, roots and embryoids were initiated from explants of coconut inflorescence rachillae (Eeuwens 1978) and date palm (Tisserat 1979).

Date palm ovules, carpel tissue, parthenogenetic endosperm and fruit stalks turned black within 24 h after culturing on nutrient media and subsequently died. Cultures of date palm floral bud reproductive tissue, especially male anthers, usually turned brown and died after a few weeks in culture (Reuveni et al. 1972).

The tissue of choice in palms for obtaining embryogenic callus is either the actively growing lateral bud (when available) or the shoot tip (Tisserat 1981). Older tissues and organs do not respond well in culture, and callus produced from them usually has limited morphogenetic potential; e.g., mature leaf and flower bud callus produce only roots (Abul-Soad 2003; Tisserat 1984a).

Tissue culture studies of date palm inflorescences have demonstrated the varied potential of female floral initials. Firstly, these initials were able to change from the floral state to vegetative outgrowths by different methods, all of which were correlated to the stage of differentiation at the time of excision. Secondly, they changed according to the composition of the initial nutrient medium, especially to the plant growth regulator formula used. Also, the sequence of nutrient media used is decisive

(Abul-Soad 2003). These floral initials were able to develop directly into complete plantlets. Floral initials also had the potential for induction of adventitious buds or somatic embryos, which in turn developed into complete and separated plantlets (Drira and Al-Sha'ary 1993). Therefore, Abahmane et al. (1999) suggested that a micropropagation technique using floral tissues of floral spikes at an early stage of growth on culture media aided in converting these tissues from floral to vegetative. Moreover, this early stage can vary by cultivar, climatic conditions and nutritional status of the mother tree (Abul-Soad 2007a). On the contrary, Kriaa et al. 2007 envisaged a protocol to be based on the use of mature female flowers, taken at the latest developmental stage, before opening of the spathe. This strategy has an advantage of avoiding damage to the mother plant.

The types of formations of well-responding inflorescence explants are of three types: (a) direct green shoots from stressed or long-term incubated initial explants (Abul-Soad et al. 2011), (b) direct embryogenic callus which will differentiate after maturation into direct somatic embryos or direct shoots (Abul-Soad et al. 2004a; Abul-Soad 2007a, 2009; Kriaa et al. 2007) and (c) unfriable callus (Abul-Soad et al. 2005; Feki and Drira 2007). In another trial carried out on male inflorescences, although the callus formed, it failed to form somatic embryos. This is perhaps related to the age of the explant source (Al-Khayri et al. 2007).

In addition, cultured female flower initials developed into hermaphrodite flowers having stamens and anthers filled with pollen, besides the normal three carpels (Drira and Benbadis 1985) or green flowers without pollen (Abul-Soad 2003). In that way, morphologically-typical hermaphrodite flowers were obtained (Masmoudi et al. 2007).

Thus far no laboratory has succeeded in commercially propagating date palms (male or female) from an inflorescence except in Al Ain City, UAE where male date palm has been commercially propagated (El-Korchi 2007). Also, female inflorescence was successfully used for commercial micropropagation of 12 date palm varieties in Date Palm Research Institute, Shah Abdul Latif Univ., Pakistan (Abul-Soad and Mahdi 2010).

Most of the trials have faced some unknown hindrances at the experimental level. Some of these hindrances are the reproducible protocol by which huge numbers can be produced in a short time. The procedure which can be used for the inflorescence excision without any damage or dangerous for the mother tree at an early stage is sensitive. Producing plantlets through the callus phase can produce somaclonal variation, rather than the lengthy procedure and growth vigor of produced plants from callus. High auxin concentration can be used to induce the callus from inflorescence explants in the case of using mature or late age explants. An appropriate formula is not known which can induce the direct organs, either somatic embryos or green shoots, within a few months for the overwhelming majority of cultured explants. However, the manner of applying this technique is of paramount importance. It is conceivable that more than 100 cultivars can be managed and initially established within 1–2 months in one laboratory, with no wasted offshoots and significantly less efforts. On a commercial level, the Biotechnology Laboratory of Date Palm Research Institute, Khairpur, Pakistan succeeded in producing seven commercial Pakistani cultivars in the first run. These cvs. were Dhakki, Gulistan, Aseel, Kurh, Kashowari, Dedhi and Gajar, most of which were healthy and of commercial interest to the date industry of Pakistan (Abul-Soad 2007b; Markhand and Abul-Soad 2010).

One or two immature inflorescences were excised from targeted cultivars of the major date-growing areas in Pakistan (Khairpur and Dera Ismail Khan). These cultivars are represented by the early cvs. (Gajar and Kashowari), the major export cv. (Aseel), a cultivar with high price and international standards cv. (Dhakki) and low population cv. (Gulistan). For example, cv. Gulistan at Dera Ismail Khan was multiplied by traditional methods late in the last century, amounting to a few thousands and there is growing demand for its offshoots which cannot be met by the traditional offshoot-detachment method. Tens of thousands of plantlets were micropropagated within 2 years from only two small inflorescences (Abul-Soad and Mahdi 2010).

6.3 Benefits of Using Inflorescence Explants

These benefits can be divided into two groups based on considerations related to *in vitro* and *in vivo* growth behavior.

6.3.1 Considerations Related to In Vitro Tissue Culture

- Fungal and bacterial contamination is very low. Despite surface sterilization with chemical substances, 90% of shoot-tip explants can be lost through contamination, particularly from endogenous bacteria. In contrast, inflorescence explants which are contaminant-free can reach up to 100% success (Abul-Soad 2007a; Abul-Soad et al. 2008).
- Minimal browning occurs. Wounding an explant in the shoot-tip technique causes secretion of phenolic compounds into the culture medium and prevents the growth of the initial explant and ultimately leads to death. Nevertheless it can be controlled by incubation in complete darkness and activated charcoal in the media especially during the initial stage. But the whole organ (spikelet) is typically used in this technique.
- Juvenility of inflorescence explants used which have high potential to differentiate quickly into organs. At the initial stages of growth, the explant has masses of meristematic tissue making it easy to express morphogenetic responses (Abul-Soad et al. 2011).
- A time-efficient protocol in which the shorter starting stage is only 2–6 months as compared to at least 6–18 months for organogenesis from shoot tip explants (direct pathway).
- Consequently, inflorescence explant cultures reduce the time scale of the entire production to 1–2 years as compared to 2–3 years with shoot tips. Also, the formation of organs with a minimal intervening callus phase avoids the risk of somaclonal variation.

6.3.2 Considerations Related to In Vivo Open-Field Characters

- Micropropagation of recalcitrant cultivars, especially those existing in limited numbers (Kriaa et al. 2007) or rare individual female lines (Abdallah 2002), facilitate large scale production of plants for commercial uses; including some elite cultivars without any or with few offshoots. No method is known to reproduce such cultivars and lines except through inflorescence explants.
- Micropropagation of highly-selected and extremely-useful male lines (Al-Ghamdi et al. 2007) can be achieved. Fruit quality properties such as size, color and sugar content are dependant mainly on the pollen grain source (the metaxenia phenomenon). This means that not all heterozygous males are equal in their impact on fruit quality. There have been many studies and evaluation programs to select ideal male trees. These studies are without potential if there are no offshoots on the tree, i.e. no source for further propagation. The inflorescence technique is the only way currently available to reproduce such vigorous males (El-Korchi 2007).
- Micropropagation of cultivars resistant to epidemic diseases such as bayoud (*Fusarium oxysporum* f. sp. *albdinis*) and pests like the red palm weevil (*Rhynchophorus ferrugineus* F.). There are a few cultivars and lines exhibiting natural resistance to these maladies, but the trees typically produce no offshoots. It is reported that well-acclimatized plants derived from inflorescences of INRA-J19 and INRA-A6 clones were tested against bayoud disease (Abahmane 2007).

6.4 Excision of Immature Inflorescence

Date palm is an important crop plant in arid and semiarid areas with a hot dry summer season and little rain in winter. Due to these conditions, date palm flowering occurs each year at the beginning of spring (Northern Hemisphere) and facilitates availability of immature female inflorescence (spathe) for tissue culture. The growth and development of small floral buds to reach the usual mature size of the female spathe is rapid. Their appearance on the tree is synchronized with increasing air temperatures in the early spring season. Typically, the flowering season takes 2–6 weeks for complete emergence of all spathes. The appropriate excision time for each cultivar should be investigated to determine a precise time for excision of spathe within that timeframe. The excision time is equally crucial to avoid sexual organ formation which occurs at the late stage of development. In view of the fact that carpels are the last phase of flower differentiation, an early stage of growth is required to avoid any sexual cells interrupting the organogenesis from somatic cells alone.

The age of the explant plays an important role in this protocol. Some reports suggest using mature female flowers at the latest developmental stage of growth, just before the opening of the protective sheath (Kriaa et al. 2007). Others refer to using the inflorescence as soon as it emerges from among the leaves and the brown tip of the spathe becomes visible (Abahmane 2007). In a trial with a male tree from

Al-Ain City, the youngest possible spathes, 51–54 cm long, were excised, (El-Korchi 2007). Also, they referred to the source of inflorescences, offshoots of cv. Barhee approximately 1–1.2 m high; the inflorescence if present in the axil of the offshoot, was removed with its protective sheath (spathe) intact and refrigerated at 4°C until used (Bhaskaran and Smith 1992).

By contrast, the current innovative protocol recommends excision of the spathe while it is still hidden among the leaves (15 cm long). A method developed based on empirical experience was able to estimate the location of young spathes on the tree with a success rate 80–90%. The flower buds that will develop into inflorescences begin to emerge in reciprocal positions around the head of the tree, mainly above 4–5 older frond whorls. Thereafter, spathes emerge among the next 3–4 frond whorls (Abul-Soad 2003). Based on a preliminary study, one frond is selected and the adjacent 4–5 fronds peeled away. In the case of not finding the spathe, the operation should be aborted; otherwise this can cause the crown of the tree to fall off. Only a skilled person can induce the same tree to provide a maximum of 1–2 spathes.

In another proven method, pruning of the outer mature fronds is carried out until reaching the first outer spathe. By cutting 36–40 mature fronds the loss of a paramount source of spathes occurs in the current season. Therefore, bunch thinning should be done to reduce the number of fruiting bunches from 15 to 20 down to only 4 bunches. The mother tree entirely recovered in the following season and the location of excision had almost disappeared (Fig. 6.1b). In both methods, the time of excision, cultivar and climatic conditions, particularly temperature, are variable factors that control the appropriate excision time, i.e. age of the spathe.

The mother tree from which the spathe was excised lost only one bunch in that same year and entirely recovered the next year. Care must be taken to treat the wound with fungicide and pesticide to prevent any further infection or infestation from diseases or pests, in particular the red palm weevil.

The inflorescence was excised from the mother plant in early spring, placed in a clean plastic cover and handled carefully from the open field to the laboratory. In the event of considerable distance between the laboratory and the location of the source material (spathes), the spathe can be conserved for 1–2 days in an ice chest. This step had no adverse impact on the survivability of the spathes.

6.5 Surface Sterilization of Spathes

One of the worthwhile benefits of the inflorescence technique is the simple sterilization protocol to decontaminate explants. As soon as the entire inflorescence (spathe) has been brought into the laboratory, immediate surface sterilization should be carried out according to the following steps:

- Dipping into fungicide solution for 30 s, without any shaking. The most commonly used fungicide is Tobsin or Bennlit (systematic fungicide) solution 2 g l^{-1}.
- Handle carefully spathe while washing under running tap water for 30–60 s. The water stream should be directed at the basal part of the spathe to remove dirt.

- Using 30% sodium hypochlorite (NaOCl) solution (16%) for 1–2 min.
- Washing with sterilized distilled water for 30–60 s one time without shaking.

Cracks or cuts in the outer protective sheath of the spathe due to mishandling during excision, transfer to the laboratory or surface sterilization can increase the contamination rate to as much as 100%. Maximum care must be given to handling the spathe through all stages. Keeping the spathe undamaged results in definitely highly survival and of explants free of contaminants (Abul-Soad 2003, 2007a, 2009). This protocol avoids any direct contact between the surfactants and the spikelet explants. That is why the explants respond well to the nutrient medium. Most of other protocols (El-Korchi 2007) call for surface sterilization of the spikelet explants but not for the entire spathe (Abul-Soad 2007a; Abul-Soad and Mahdi 2010).

6.6 Establishment of Initial Cultures

In a schematic illustration (Fig. 6.1a), the spathe can be divided histologically into four sections: inflorescence base, spikelet base, spikelet and protective sheath.

The spikelet was the only explant component which responded well to different nutrient media (Fig. 6.1c). Other sections failed to produce even callus onto the callus induction medium (Abul-Soad 2003). The procedure of surface sterilization of the intact spathe involves the following steps:

- The outer protective sheath surface is sterilized and the spathe cut longitudinally from the middle like a T from only one side. The cut may be made in the central swollen portion of the spathe due to its softness.
- Spikelets cut, 3–4 cm long, from their base can be cultured. Spikelets are cut 2–3 cm long and placed horizontally on the sloping surface of the nutrient medium. Each of these should have 2-4 immature florets (Fig. 6.1d).
- All cultured tubes are incubated in a controlled growth room at 25±2°C in the dark. Incubated explants are subcultured every 3–4 weeks on the same starting culture medium, as described in Table 6.1.
- Well-responding explants are transferred to a maturation medium for 1-2 subcultures.
- Matured globular structures which are ready to differentiate and early-differentiated explants growing in the darkness are shifted onto the differentiation medium for 16 h photoperiod and under 2,000–3,000 lm m^{-2} conditions for 1-2 subcultures each for 4–6 weeks. Subsequently differentiated cultures are shifted to the multiplication stage or may be rooted directly.

The age of a spikelet explant does not necessarily determine its length. The length of cultivar spikelets is variable within the same inflorescence or even among different inflorescences on a tree. Moreover, the spikelets are located within the spathe in a pyramid-like shape. The longest spikes are found in the center and are surrounded by shorter ones on the periphery.

It seems that the endogenous hormones of the explant control the number of responding florets on the explant. The most suitable culture vessel is a test tube

Table 6.1 Nutrient media composition used for inflorescence protocol and its sequence

	Composition (mg l^{-1})			
Medium	Salts	Additives	Auxins	Cytokinins
1. Starting	Macro of B5[a]+Micro of MS[b]	30000 Suc.[c]+2200 Agar+1400 Gel+Vit.[d] of MS+170 KH_2PO_4+100 Glutamine+40 Ad.[e]	0.1 2.4-D+0.1 IAA+5.0 NAA	–
2. Maturation	Macro of B5+Micro of MS	30000 Suc. + 2200 Agar+1400 Gel+Vit. of MS+170 KH_2PO_4+100 Glutamine+40 Ad. + 1500.0 AC[f]	5.0 2,4-D	1.0 2iP
3. Differentiation	MS	30000 Suc. + 2200 Agar+1400 Gel+Vit.of MS +	0.1 NAA	0.1 Kinetin
4. Proliferation	MS	30000 Suc.+ 2200 Agar+1400 Gel+Vit. of MS +	0.1 NAA	0.05 BA
5. Rooting	¾ MS	50000 Suc. + 2200 Agar+1400 Gel+0.1 Ca-panthothianate+Vit. of MS+with 3000.0 AC.	0.1 NAA	

[a]*B5*: Gamborg et al. (1968) Nutrient Medium
[b]*MS* Murashige and Skoog Medium (1962)
[c]*Suc.* Sucrose
[d]*Vit.* Vitamins
[e]*Ad.* Adenine Sulfate
[f]*AC* Activated Charcoal

Fig. 6.2 Inflorescence morphogenesis. (**a**) Pro-embryogenic masses (PEMs) induction on the spikelet explant of cv. Dedhi after 3 weeks in culture onto modified MS basal medium supplemented with 1.0 mg l^{-1} NAA+0.1 mg l^{-1} IAA+0.1 mg l^{-1} 2,4-D, (**b**) Shining globular creamy structures after 2–3 months in culture, (**c**) Maturation on activated charcoal medium in darkness, (**d**) Caulogenesis during differentiation process

(25×150 mm). However, there was no significant difference according to the type of culture vessel, especially at the starting stage of culture initiation. Mostly there was one explant per tube; however, more than one explant can be cultured in a single tube but that may result in poor growth of cultures. Moreover, only rarely did one explant have all immature florets on it responding. One type of positive response is the formation of a globular shining structure, instead of the immature floret or cleaved from the floret. The color typically is creamy and sometimes bright white (Fig. 6.2a). It is worth mentioning that the sequence and manipulation of cultures is very important. This is related to the proper direction of each morphogenetic type to its subsequent suitable stage of growth and nutrient medium. For example, if the globular structures shift directly to a differentiation medium without maturation, most of them will not differentiate; nevertheless, there is no significant change in appearance before and after transfer to the maturation medium. As well as, transfer of the differentiated cultures to the multiplication or rooting stage directly according to its shape to maximize the production. Individual cultures will go directly to the rooting stage to reduce the production time and give a whole plant, but multiplied cultures will go on to the multiplication stage to proceed with normal growth and produce more cultures. The time schedule of organogenesis from well-established initial explants was as follows:

- Shining globular creamy structure formation was at about after 2 months of culture, through 1–2 subcultures (Fig. 6.2b).
- Maturation of initial structures after 2–3 months, through 2–3 subcultures (Fig. 6.2c).
- Differentiation process at 1–2 months, through 1–2 subcultures (Fig. 6.2d).
- Proliferation phase of individual shoots and multiple somatic embryos.
- Multiplication and rooting stages. This can be extended from a few months to 1–2 years depending on simultaneously required plantlet numbers and their health.

The first and second steps must be done in complete darkness for induction and maturation processes of organs, while other further steps can be done under illuminated conditions. It was observed that the presence of light during the initial phase disrupted the growth and increased the degree of browning. In the case of matured cultures failing to develop into organs, they must be returned to the second step under darkness which is an indication that they were not yet mature. It is necessary to mention that a successful maturation phase could produce differentiated organs on the same maturation medium in darkness. At such time it must be taken out to pursue their growth in the proliferation phase under illumination.

As a fundamental principle, one meristematic cell or group of cells may be induced to develop single or multiple somatic embryos (George 1996). In date palm, typically there are two types of common somatic embryos: individual and multiple (Abul-Soad et al. 2004b). Both multiple somatic embryos and green shoots only were subjected to the proliferation phase and through the multiplication stage, whereas the individual somatic embryos were preferably left to grow by themselves. The latter type of somatic embryos grow rapidly to an intact shoot-root plantlet within 2–3 months in the rooting stage. Relying on such types of embryos can minimize the entire production-cycle time as compared to traditional shoot tip explants. In the case of rare male or female individuals, it is preferable to quickly regenerate plants rather than consume time in the multiplication stage of individual embryos to expedite mass production.

6.7 Direct and Indirect Formation of Organs by the Differentiation Process

Friable callus is formed from initial cultured explants until the beginning of differentiation stage. Typically two structures were identified during the initial stage. Most prevalent were the globular structures which developed into green shoots or somatic embryos on the differentiation medium, as discussed above. These globular structures are like granules which are different in size; each is 1–3 mm in diameter and usually existing in clusters. Each cluster emerged from one floret. In the beginning, most of spikelet florets produced pro-embryonic masses (Fig. 6.2a) which are crystal white or often brownish yellow. These small masses multiplied and increased in number and became more organized. The growth and development was so quick as to make the initial explant of the spikelet appear as a swelled aggregate of these shiny globular structures (Fig. 6.2b). Also, the shiny brown color increased over time and the maturation of these globular structures. Therefore, the size of the globular structure clusters was gradual. The lower cluster on the spikelet explant was the largest, and then the size gradually decreased up to explant tip. It seems the endogenous hormonal content of each floret on the spikelet is different and due to this balance the nearest floret to the spikelet base produced more globular structures. Rarely all florets responded and produced these globular structures which indicate a nutritive competition among different florets on the spikelet explant leading to the death of un-responded florets. The second organized structure was direct shoot formation. Formation of direct shoots was rare in the inflorescence protocol. Moreover, produced shoots were unable to continue their growth and development to intact plantlets with roots. In addition, growth was slow compared to regular shoots which were produced from the globular structures after differentiation. However, it occurred when the initial explants were subjected to a long-incubation term in the state of physiological stress. Likewise this has been done with shoot-tip explants to induce direct organs by osmotic stress (Abul-Soad et al. 2007; Sidky et al. 2007). But the direct shoots of the shoot-tip protocol are so powerful and they can continue to growth vigorously.

One of the advantages of the inflorescence technique is the normal character of the organized somatic embryos. Normal embryos (Fig. 6.3a) can be recognized by the white embryo coat which encloses a shoot and is connected with a thin white primary root. This coat splits longitudinally from the top producing a new green leaf. By continuing the embryo growth, the coat shrinks and turns brown. Sometimes, this coat produces embryogenic callus (Fig. 6.4). Mostly the normal embryos are twisted around themselves. There are two shapes for normal embryos: non-repeated embryos (Fig. 6.3b) and repeated embryos (Fig. 6.3c). Off-types of somatic embryogenesis are reported with shoot tip callus (Abul-Soad et al. 2004b). During the differentiation process of shoot-tip embryogenic callus, many somatic embryo shapes were differentiated. Determination of these shapes may be effective to improve the quality of produced plants and avoid abnormalities. Two shapes of somatic embryos were differentiated from the embryogenic callus (pro-embryos). One of them named *repeated embryo* (RE) and the other given the designation *non-repeated embryo* (NRE), or in other words, individual embryos. The RE is a cluster of embryos having the self-capability to proliferate additional secondary embryos and shoots. But the NRE directly differentiates into a single embryo which continues to grow into a distinct plantlet or can be used to form secondary embryos on its base or the periphery of the embryo coat (Fig. 6.3d). Growth depends on the nutrient medium composition. Using MS medium supplemented with 0.1 mg l^{-1} ABA and 0.5–1.0 mg l^{-1} BA produced secondary embryos on the NRE (Zaid 2003).

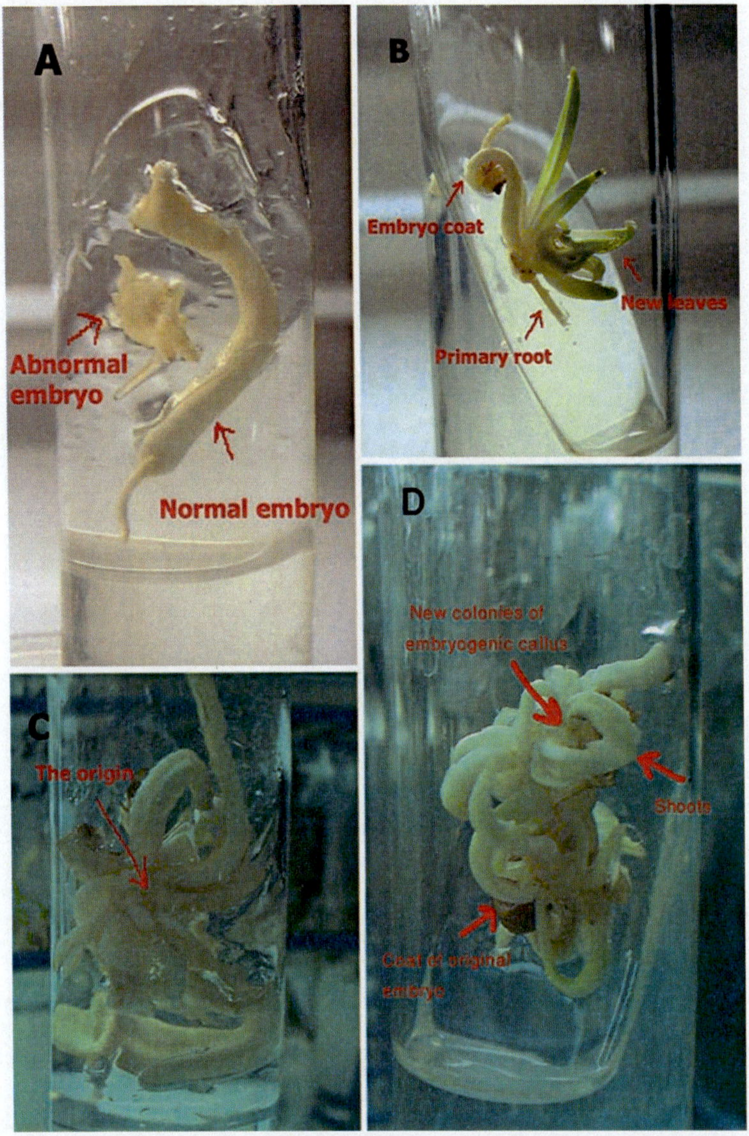

Fig. 6.3 Somatic embryos of inflorescence. (**a**) Normal and abnormal embryos, (**b**) Growth of an individual normal embryo. (**c**) A repeated embryo comprises more than 10 synchronized embryos. (**d**) Formation of secondary embryos on NRE having in their axils new colonies of embryogenic callus

The origin of these secondary embryos could be the coat enclosing the embryo (Fig. 6.4a) or the base of an embryo at the connection area of enclosed shoot and primary root (Fig. 6.4b) or at any part of the embryo coat (Fig. 6.4c). It was observed that the outer white coat of a differentiated embryo splits vertically from the middle

Fig. 6.4 Origin of secondary embryos. (**a**) On the coat split, (**b**) The connected area of enclosed embryo leaf and primary root, (**c**) Originates from the embryo coat. (**d**) Growth and development of secondary embryos on the base of embryo

and white or creamy bright granules, which are in fact pro-embryos, were directly initiated on this split to form the secondary embryos. Secondary embryos develop from granules to somatic embryos with a white coat within 6–8 weeks (Fig. 6.4d).

Mostly, abnormal embryos are smaller than normal (Fig. 6.3a). Some of these off-types were collected during the differentiation process and traced through their different stages of growth and development *in vitro* (Ibrahim 1999; Zaid 2003).

These off-types are designated to characterize their morphology: multi-tail, fused and multi-head (Zaid 2003). Therefore, using abscisic acid (ABA) in the maturation medium of date palm prevented maturing embryos from germinating precociously. The addition of 0.1–0.8 mg l^{-1}ABA to the proliferation medium stimulated the process of secondary embryo formation, while increasing the concentration to 1.6 mg l^{-1} ABA was not effective (Hassan 2002).

These off-types were not able to develop into complete plantlets and died (Abul-Soad et al. 2004b; Zaid 2003). Using such off-types in the micropropagation led to embryogenic callus formation. Expectedly the embryos produced from such embryogenic callus can produce the somaclonal variation and originate off-types in the field. Thus, it is recommended that during the differentiation process, especially when shoot tip explants are used, these abnormalities be discarded to avoid any probability of variants formation (Abul-Soad et al. 2004b).

Therefore, variety and subculture number during differentiation stage showed significant variance in abnormal production of shoot-tip embryogenesis. The percentage of abnormal shapes in somatic embryos was increased by augmenting the re-culture number. The Egyptian cvs. Sewi and Bent-Esha were higher than cv. Hiane in developing more abnormal shapes of somatic embryos. Moreover, the percentage of abnormalities increased from 64.78% in the first subculture of differentiation stage to 90.27% in the fourth subculture (Zaid 2003).

Somatic embryos, which usually appear during the differentiation stage of the shoot-tip technique, were determined in a schematic illustration (Fig. 6.5). This classification is based on their morphology during the differentiation stage (Abul-Soad et al. 2004b). It can be concluded that the micropropagation protocol utilized two subsequent subcultures (12 weeks) for the differentiation stage during which different shapes of somatic embryos were recognized. In the first subculture, the embryogenic callus (pro-embryos) which appearing as free crystal-white granules differentiated into predominant shapes: repeated embryos and non-repeated embryos. The NRE was an individual embryo 5–10 mm in length and fast in growth and development to green leaves (Fig. 6.3b). On the other hand, the RE is a granule developed in a cluster of embryos, all of them gathered at the same point of origin. These embryos often number 10–15 and are synchronized, i.e. having different sizes indicating different ages (Fig. 6.3c). These two shapes are typically similar to the differentiated somatic embryos of the inflorescence technique.

In the second subculture, the RE and NRE were transferred to the same nutrient medium to trace their growth and development. It is worth mentioning that picking up these germinating embryos in the first subculture stimulated further germination of remaining granules into new somatic embryos as well as new granules. This may indicate the synchronization in the differentiation process, i.e. earlier differentiated embryos prohibited the development of other matured pro-embryos to proceed to grow and become intact somatic embryos.

The second subculture allowed fast growth of individual embryos to whole plantlets (5–10 cm in length) after 6 weeks in culture, which were transferred directly to the pre-rooting stage (Abul-Soad et al. 1999).

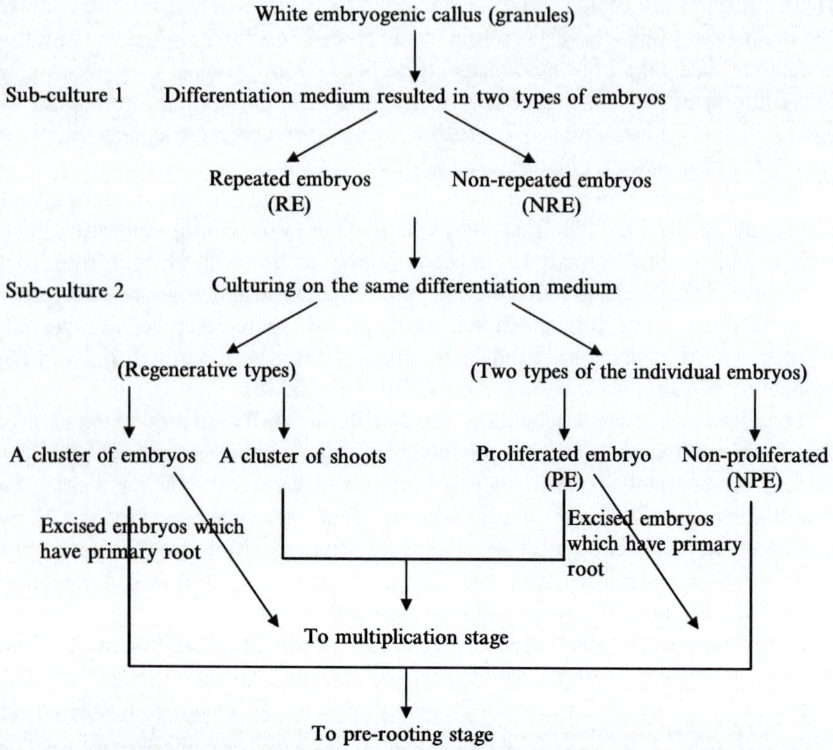

Fig. 6.5 Schematic illustration of shoot-tip embryogenic callus differentiation during two subcultures. The differentiation medium was composed of MS-basal medium supplemented with NAA 0.1 mg l^{-1}, 2iP 0.1 mg l^{-1} and activated charcoal 100 mg l^{-1} (Abul-Soad et al. 2004b)

This type of NPEs were healthy and rapidly growing to intact plantlets compared to the plantlets which will be separated from multiplying clusters at the end of multiplication stage. Sometimes, an embryogenic callus produced on the basal part of this individual embryo (NRE), is called a *proliferated embryo* (PE) (Fig. 6.4b). The PE of the second subculture was used in the multiplication stage with the RE of the first subculture as multiplying types in the multiplication stage (Fig. 6.3c). Classification of somatic embryos into these categories during the differentiation stage and directing them into the multiplication stage or the rooting stage is crucial for commercial production of *in vitro* date palms. No morphological difference was observed between the differentiated embryos of shoot tip or inflorescence explants. Moreover, abnormal embryos were rarely discovered during the differentiation process of the inflorescence culture (Fig. 6.3a).

According to the inflorescence technique used, the overwhelming majority of emerged somatic embryos were normal (Fig. 6.3a). The abnormal somatic embryos must be discarded to avoid somaclonal variations. The intervening callus phase in the shoot-tip technique could be the reason for abnormalities as compared to the current inflorescence technique without an unfriable callus phase. More details on types of somatic embryos and abnormality in date palm are provided by Abul-Soad et al. (2004b).

Differentiation of the new organs, i.e. green shoots (Fig. 6.2d) and somatic embryos are associated with direct emergence of new embryogenic callus. Typically the color of this embryogenic callus is white or sometimes creamy. Its organogenic potential is high by continuous subculture onto the same differentiation medium resulted in new somatic embryos as well as production of embryogenic callus without loss of the organogenic potential. This process occurred through 6–7 subcultures each from 3–4 weeks in semi-log plot, i.e. the amount of friable callus was low in the beginning of the differentiation stage, then increased to the highest level by increasing the subculture up to number 7, then decreased and eventually no additional friable callus formed, and the differentiation stage was ended at subculture 13 (Abul-Soad and Mahdi 2010). In embryogenic callus derived from shoot-tip explants, the events occur just the opposite at the differentiation stage. The newly differentiated callus rapidly loses organogenic potential after only 4–6 subcultures. This could be due to habituation of callus cells on the same composition of the nutrient medium. Eventually after 4–6 re-cultures of the shoot-tip embryogenic callus and side process of new callus formation, a big amount of unfunctional friable callus was achieved. The new clones of callus were morphologically similar to the ordinary embryogenic callus, white granules and friable but lost the ability to development to intact somatic embryos. Sometimes this type of callus produces tiny and abnormal embryos.

6.8 Proliferation During the Multiplication Stage

As mentioned earlier, multiple embryos and green shoots after differentiation would were transferred onto the proliferation medium at the multiplication stage (Fig. 6.6a). During the proliferation process no secondary unfriable callus was observed. This callus is globular and variable in size (1–2 mm), always initiating in leaf axils as a source of new shoots. It is total caulogenesis (shoot forming) and sometimes with a minimally intervening embryogenic callus phase (the same shining globular structures). Simultaneously proliferation via caulogenesis was associated with secondary somatic embryo formation. The origin of this was the regular shining globular structure which formed in the axils of juvenile leaves. The proliferated somatic embryos easily separate from the multiplied shoot cluster (Fig. 6.6a). Then these individual somatic embryos will be transferred directly to the rooting stage. According to the knowledge of the author, no research has been performed to determine the maximum number of subcultures in the multiplication stage. However, it is recommended that a lower number of subcultures will always be better. In fact, the maximum number of subcultures in the multiplication stage is not definite. However, as a specific example, the cultures of inflorescence were multiplying well throughout 13 subcultures done at the Date Palm Research Institute, Khairpur, Pakistan, during the period 2008–2010. Also, there were produced from one inflorescence of cv. Gulistan; it provided 10,000 healthy plantlets in the rooting stage and partially transplanted

Fig. 6.6 Multiplication, rooting and acclimatization stages. (**a**) Proliferation of multiple embryos and individual embryos, (**b**) Plantlets during rooting stage, (**c**) Direct shoot formation on spikelet initial explant after 11 subcultures, (**d**) 3, 6, 12 and 18 month stages of growing plants

into the greenhouse. It is very important to mention that for over one century the population of recalcitrant cv. Gulistan at the Dera Ismail Khan District, stood at 1,000–2,000 trees due to the limitation of new offshoots through traditional propagation. Notwithstanding, within only 2 years it became possible to produce that number and more from a single inflorescence. This is strong justification for using the promising inflorescence technique.

6.9 Formation of Healthy Plantlets

In this protocol, three steps were taken to produce a good root-shoot system (Fig. 6.6b), whereas the fourth step is to pre-acclimatize (*in vitro* hardening protocol) the plantlets *in vitro* to reduce *ex vitro* stress in the greenhouse.

- Well-rooted plantlets are achieved on a medium comprised of ¾ MS salt strength and 50 g l^{-1}- sucrose (Abul-Soad et al. 1999; Ibrahim et al. 1999b). This nutrient medium supplemented with 0.1 mg l^{-1} NAA to improve the root system and 0.05 mg l^{-1} BA to enhance leaf formation (Omar 1988a). Also, 1.0–2.0 mg l^{-1} Ca-panthothianate was added in order to prevent necrosis and vitrification. The base of culture tubes may be covered with aluminum foil to reduce light emission by which growth and development of root system can be improved. The base covering can be done for the entire rack of tubes (Abul-Soad et al. 1999).
- Plantlets are transferred from standard small tubes (25 × 150 mm diameter and height, respectively) to larger tubes (28 × 250 mm) to allow erect shoot growth. Plantlets should be maintained on this medium until they reach a suitable height with a good root system and 2–3 leaves. Typically this occurred at 3 months (2 re-cultures) at least on such rooting medium. About 80–90% of cultured plantlets on this medium produced a fair root system (Abul-Soad et al. 1999). Whereas, it is reported that 30–40% of plantlets cultured on agar media, containing only 0.1 mg l^{-1} NAA only produced prolific adventitious root growth (Tisserat 1982).
- In the third step, better growth responses of date palm plantlets are associated with the addition of 3 g l^{-1} AC to the medium (Abul-Soad et al. 2006). Generally, low rates of photosynthesis in plantlets grown *in vitro* have been attributed to low light conditions. Thus, increasing the light level to 9,000 lm m^{-2} strongly enhanced growth (Ibrahim et al. 1999a).

6.10 In Vitro Hardening Protocol

The process of acclimatization can begin while plantlets are still *in vitro*. *In vitro*-grown plantlets grow under the controlled growing conditions such as low light levels, aseptic conditions, medium containing sugar and nutrients to allow for heterotrophic growth and in high humidity. They are unable to survive when transferred directly to the greenhouse where low humidity levels and opened stomata of newly-transplanted plantlets allow intensive transpiration which leads to rapid drying of leaves. *Ex vitro* date palm plants which were newly shifted to the greenhouse without keeping the humidity above 90% could not resist the humidity deficiency more than half an hour. After that, leaves shrink and became like thin thread. Therefore, leaves could not be able to return back to their natural state after shifting them to high humidity conditions (85–95%). Thus it is advisable to pre-acclimatize the plantlets gradually while they are *in vitro* (Abul-Soad et al. 1999).

- In this step, plantlets are rinsed with sterile distilled water in order to remove excess adherent media under aseptic conditions, and then cultured on a medium free of sugar, half strength of MS and in culture tubes capped with aluminum foil. Cultures are kept under a high light intensity (9,000 lm m^{-2}).
- In this procedure, ventilation between the inside and outside of the tube was increased through gradually punching holes in the aluminum foil, followed by complete removal a few days before transplanting. This gradual approach is beneficial to reduce the relative humidity in the tubes and to increase epicuticular wax development on the leaves.

At the same time, omission of sucrose from the nutrient medium brings about a shift from heterotrophy to autotrophy (Abul-Soad et al. 1999). Hereafter, high rates of photosynthesis along with a high light level and adequate gas exchange become paramount for survival in the greenhouse. The same *in vitro* hardening procedure has been used in other crops (Kozai et al. 1987). All of these factors help to acclimatize date palm plantlets gradually to ensure survival in the greenhouse. Date palm plants are given the chance to rely on themselves to respond like natural seedlings in the field and produce all their requirements from carbohydrates without any addition of chemical elements during the first 1–2 months in the greenhouse.

Another method of *in vitro* hardening in date palm has been described wherein when plantlets reach a desired size (15 cm in length), they are hardened by subculturing into minimal organic medium (MS salts; inositiol, 100 mg l^{-1}; thiamine-HCl, 0.4 mg l^{-1}; sucrose, 30 g l^{-1}) for 2 weeks with incubation at 10,000 lm m^{-2} illumination (Omar 1988b).

It was observed that the typical twisted shoots became erect, and the pale green color became dark green. Moreover, in the case of strong plantlets which have 4–5 adventitious roots and wide, erect and dark green leaves 15–20 cm in length, the closure may be pushed up by plantlet tip. At the end of the pre-acclimatization stage, the wilting of the leaves when they are in uncovered tubes represents a positive sign for early *in vitro* detection of inadequate plants. Such plantlets must be discarded to avoid further effort in transplanting and acclimatization in the greenhouse. Plantlets with mostly erect leaves were shifted to soil beds under a low transparent plastic tunnel to keep the relative humidity at 80–90%. The reason for the change is to reduce fungal infection as compared to the more common range of 90–100%.

6.11 Transplanting

Specialized environmental conditions, high humidity and low photosynthesis of *in vitro* culture can result in the formation of plantlets exhibiting abnormal morphology, anatomy and physiology. After *ex vitro* transfer, plantlets might easily be impaired as well by sudden changes in environmental conditions and therefore a period of acclimatization is needed to overcome some abnormalities (Pospisilova et al. 1999).

Healthy shoot-root system plantlets of date palm are transplanted on different soil beds in the greenhouse for acclimatization. Many soil mixtures have been used as a substrate for date palm acclimatization (Abul-Soad et al. 1999; Al-Jibouri et al. 1988; Hegazy et al. 2006; Madhuri and Shankar 1998). The highest plantlet survival rate after 3 months was 93%, planted under the current procedure of transplanting (Fig. 6.6d). Plantlets selected for this procedure should have 2–3 erect leaves of an average 10–20 cm height with well-developed adventitious root systems. These plantlets then were transplanted into special pots (5 cm diameter × 18 cm height) containing vermiculite as the soil bed. Placement of small stones at the bottom of pots promoted root growth due to effective drainage of irrigation water.

6.12 Acclimatization

6.12.1 Procedures

- Ideal plantlets within long tubes are transferred from the laboratory to a greenhouse and left for at least 1–2 h under the greenhouse environment which has a higher temperature of around 30°C, relative humidity of 40–60% and natural illumination (daylight without direct sun) 10,000–25,000 lm m^{-2}. The nutrient media stuck around the root system was washed off by immersion in sterilized water and carefully shaken.
- Plantlets are dipped in 3–5 g l^{-1} systematic fungicide (Topsin M 70) solution for 1–2 min.
- Direct planting in suitable pots (5 × 18 cm) filled with autoclaved vermiculite and making note on the container of the associated data of the cultivar name.
- Planted pots placed under low transparent plastic tunnels one by one within a few min to avoid humidity decrease inside the tunnel associated with spontaneous opening of the tunnel to put in the plants. It is preferable to construct a tunnel to contain 250 plants as a maximum number which can be handled within 1–2 h. Growing conditions within the tunnel are maintained at 30–35°C and 90–100% relative humidity. These conditions are gradually reduced by placing the plants within the tunnel in parallel with the ventilation process until the plants are fully exposed to the natural growing conditions within the open greenhouse after 1–2 months when the plastic sheet is removed.
- The tunnel is kept closed after placement of the pots and left undisturbed for 3–7 days. During this period irrigation is prohibited.
- The tunnel is opened after 7 days for 10–15 min of ventilation and a Copper OxyChloride spray treatment of 3 g l^{-1} solution fungicide applied. Also, dead plants are removed and leaves infected with fungus are cut off.
- Gradual opening of the tunnel after 2 weeks of planting for the subsequent 2 weeks.

- Fertilization and irrigation process after 2–3 months of planting as required (Fig. 6.6d). Until now no work has been done to determine the actual requirements of these growing date palm plants from different elements. However, 2–3 g l^{-1} N-P-K fertilizer (17-17-17) gave good results. Also, using the mycorrhiza (Bouamri et al. 2006) and humic acid which is a natural chelating soil mobilizing agent was effective.
- Transfer successful plants into larger pots 20×25 cm. Fertilizer dose increased slightly. Growing plants must be fertilized with irrigation water containing gradual doses of a chemical fertilizer to encourage the plant growth. The program of fertilization beginning after 2–3 months of transplanting with light dose 2–3 g l^{-1} N-P-K fertilizer (17-17-17, Zarkhaze) every 1–2 weeks. Every 1–2 months an additional application of Copper OxyChloride 3 g l^{-1} over 6–18 months (Fig. 6.6d). Plants 50 cm in height are then transferred to the nursery or directly planted in the field. Plants at 12–18 months old begin to produce feathered (compound) leaves. When plants push thick white roots through the pores at the pot bottom after 5–6 months, they should be transferred to a larger pot. Growing plants can be placed on a wooden base or plastic sheet to prevent root extension into underlying soil.

6.12.2 *Most Important Factors in Date Palm Acclimatization*

- Pre-acclimatization on a low-nutrient liquid medium while plantlets are kept under laboratory conditions.
- Use of a sterilized soil bed and fungicide Topsin M 70 application before transplanting in the greenhouse.
- Maintain 85–90% relative humidity and 30–35°C temperature in ambient environment during the first 7–10 days of transplanting to avoid leaf wilting and fungus infection.

6.12.3 *Expected Problems and Their Solutions*

- Quick wilting of leaves while transplanting. Rapid planting process within the shortest time 1–2 min and pre-acclimatizes the plants while they are in the lab.
- Fungal infection of basal part of the plant and leaves separation while their root may be still healthy, white and inside the soil. Always wash the plants thoroughly with sterilized water before planting. Also, keep the basal part of the plant completely above the pre-sterilized soil surface. Treatment with a systematic fungicide (Topsin M 70 or any other fungicide) solution is beneficial after 1 week of transplanting.
- Fungal infection on leaves, mostly during the first 2 weeks of acclimatization. It is caused by condensation of water drops falling inside the tunnel. Tunnel should not be opened during the first 3–7 days. In addition, dipping the plants in a systematic

fungicide solution for 1–2 min before planting, without washing after treatment. Also, spraying with Copper OxyChloride solution is helpful after 1 week of transplanting.

- Dead and infected outer leaves during first month. Cut them off then spray with Copper OxyChloride solution 5 g l^{-1}.
- Low humidity in the tunnel. Keep the temperature in the range of 30–35°C.
- Leaves with late fungal infection. Relative humidity should be gradually reduced after a maximum 20 days of transplanting and kept at 50–60%.

6.13 Fruit Type in Open-Field

There are some reports from Tunisia and Morocco referring to fruiting of date palms derived from the inflorescence. INRA-J19 a selected date palm genotype which was field planted in Marrakech (Morocco) began to produce fruits in 2005. This is providing information on true-to typeness and consequently the applicability of such a technique for mass propagation of date palm (Abahmane 2007).

The Plant Tissue Culture Laboratory, Al-Ain City, UAE announced in 2005 the early emergence of male spathes of an extraordinary male date palm micropropagated from an inflorescence. Furthermore, the average number of spathes per plant was about 2–3, while the percentage of spathe production was 96%, i.e. 67 male plants produced spathes out of the original 70 plants (El-Korchi 2007). However, more investigations are needed to ensure the genetic stability of inflorescence *in vitro*-derived plantlets. Also, evaluation of fruit quality in the field should be carried out similar to work performed with shoot tip *in vitro*-derived plantlets (Abd-Allah et al. 2008).

6.14 Conclusion and Prospective

Research work has been ongoing for a few decades to use inflorescence explants in micropropagation of date palm. However, commercial laboratories are using shoot tip explants despite the competitive benefits of inflorescence buds. This chapter explores these benefits and gives a detailed procedure for how to use inflorescences in the micropropagation of date palm. Immature inflorescence culture involves sterilization of small spikes and culturing segments on nutrient media according to the culture stage described in Table 6.1. Direct formation of globular structures occurs, pro-embryos or direct shoots formation (Fig. 6.6c). A histology study is required to further clarify the developmental process involved. Resultant somatic embryos can be classified into two categories, individual and multiple somatic embryos. Directing these embryos to the multiplication or rooting stage is obligatory. The multiplying cultures are used in the multiplication stage to get the maximum number of shoots which are then directed to the rooting stage. Rooted plantlets are subjected to an *in vitro* hardening procedure before transplanting in the greenhouse. The question of when to stop the cycle of *in vitro* multiplication to produce the maximum number of high quality true-to-type plantlets remains unanswered. Studying the genetic stability of *in vitro* plantlets could enhance consumer confidence in relation to commercial production through avoiding somaclonal variation which may result of in prolonged cultures. Also, it would be of importance in studying genetic stability in relation to various culture stages. Once its applicability is demonstrated in different cultivars the inflorescence explant is expected to gain more acceptance as an alternative to producing shoot-tip explants.

References

Abahmane L (2007) Micropropagation of selected clones from inflorescence tissues and its role in the date palm (*Phoenix dactylifera* L.) improvement program. Fourth symposium on date palm in Saudi Arabia, King Faisal University, Al-Hassa, 5–8 May, Abstracts Book, p 145

Abahmane L, Bougerfaoui M, Anjarne M (1999) Use of tissue culture techniques for date palm propagation and rehabilitation of palm groves devastated by bayoud disease. In: Proceedings of the international symposium on date palm, Assiut University, Assiut, Egypt, 9–11 Nov, pp 385–388

Abdallah AS (2002) Evaluation of three date palm seedling types grown at El-Dakhla Oasis New Valley, Egypt. In: Proceedings of Minia first conference for agriculture and environmental science. Minia, Egypt, 25–28 Mar, pp unknown

Abd-Allah BM, Mohamed AY, Abul-Soad AA, El-Kholy LA (2008) Behavior study of some dry-date palm (*Phoenix dactylifera* L.) cultivars derived from tissue culture under Aswan conditions. In: Proceedings of third international conference on date palm. Faculty of Agriculture and Environmental Science, Suez Canal University, North Sinai, Egypt, 25–27 Apr, pp 231–240

Abul-Soad AA (2003) Biotechnological studies of date palm: micropropagation of inflorescence, molecular biology, and secondary metabolites. Ph.D. dissertation, Pomology Department, Faculty of Agriculture, Cairo University

Abul-Soad AA (2007a) Inflorescence tissue culture utilization for date palm (*Phoenix dactylifera* L.) micropropagation. Fourth symposium on date palm in Saudi Arabia, King Faisal University, Al-Hassa, 5–8 May, Abstracts Book, p 144

Abul-Soad AA (2007b) Study of processing development problems of dates in Egypt. Fourth symposium on date palm in Saudi Arabia, King Faisal University, Al-Hassa, 5–8 May, Abstracts Book, p 406

Abul-Soad AA (2009) Inflorescence utilization in tissue culture for date palm (*Phoenix dactylifera* L.) micropropagation. In: Proceedings of the international dates seminar, 28 July. Date Palm Research Institute, Shah Abdul Latif Univ., Khairpur, Pakistan, pp 33–43

Abul-Soad AA, El-Sherbeny NR, Baker SI (2004a) Organogenesis in female inflorescence of date palm (*Phoenix dactylifera* L. *cv*. Zaghloul). In: Proceedings of second international conference on date palm. Suez Canal University Faculty of Environmental Agricultural Sciences, El-Arish, Egypt, 6–8 Oct, pp 139–163

Abul-Soad AA, El-Sherbeny NR, Baker SI (2005) Date palm (*Phoenix dactylifera* L. *cv.* Zaghloul) propagation using somatic embryogenesis of female inflorescence. In: Third conference on recent technologies in agriculture. Cairo University, Egypt, 14–16 Nov, 3:423–441

Abul-Soad AA, El-Sherbeny NR, Baker SI (2007) Effect of basal salts and sucrose concentrations on morphogenesis in test tubes of female inflorescence of date palm (*Phoenix dactylifera* L.) cv. Zaghloul. Egyp J Agric Res 85(1B):385–394

Abul-Soad AA, Ibrahim AI, El-Sherbeny NR, Baker SI (1999) In vitro and *ex vitro* optimization for rooting and acclimatization of date palm. In: Proceedings of first international conference in Egypt on plant tissue culture and its application, 12–14 Sept, Egypt, pp 227–241

Abul-Soad AA, Ibrahim AI, El-Sherbeny NR, Baker SI (2002a) In vitro optimization for plant regeneration of date palm (*Phoenix dactylifera* L.). Minia J Agric Res Develop MCAES 1st Egypt 22(2D):2265–2282

Abul-Soad AA, Ibrahim AI, El-Sherbeny NR, Baker SI (2004b) Improvement and characterization of somatic embryogenesis in date palm (*Phoenix dactylifera* L.). In: Proceedings of the international conference of genetic engineering and its applications. Egyptian Society of Genetics and Suez Canal University, Sharm El Sheikh City, South Sinai, Egypt, 8–11 Apr, pp 359–373

Abul-Soad AA, Mahdi SM (2010) Commercial production of tissue culture date palm (*Phoenix dactylifera* L.) by inflorescence technique. J Genet Eng Biotech 8(2):39–44

Abul-Soad AA, Mahdi SM, Markhand GS (2011) A comparison between the organogenesis potentiality of shoot tip and inflorescence explants in *Phoenix dactylifera* L. In: Second international meeting of date palm and its related subjects. Qassim University, Kingdom of Saudi Arabia (in press)

Abul-Soad AA, Markhand GS, Shah SA (2008) Effect of Naphthaleneacetic acid and Indole-3-acetic acid on somatic embryogenesis of female inflorescence explants of date palm (*Phoenix dactylifera* L.) cv. Aseel. In: Proceedings of the third international conference on date palm. Faculty of Agriculture and Environmental Science, Suez Canal University, North Sinai, Egypt, 25–27 Apr, pp 222–231

Abul-Soad AA, Zaid ZE, Salah A, Sidky R (2002b). Tissue culture of date palm (*Phoenix dactylifera* L.). In: Proceedings of the third scientific conference of agricultural science, Assiut, Egypt, 15–17 Oct, pp 327–341

Abul-Soad AA, Zaid ZE, Sidky R (2006) Improved method for the micropropagation of date palm (*Phoenix dactylifera* L.) through elongation and rooting stages. Bull Fac Agric Cairo Univ Egypt 57:791–801

Al-Ghamdi, AS, Al-Bahrany AM, Al-Khayri JM (2007) Evaluation of male date palm *Phoenix dactylifera* L. as a pollination source for Al-Hassa commercial cultivars. Fourth symposium on date palm in Saudi Arabia, King Faisal University, Al-Hassa, 5–8 May, Abstracts Book, p 233

Al-Jibouri AJM, Salman RM, Omar MS (1988) Transfer of in vitro-regenerated date palms to the soil. Date Palm J 6:390–400

Al-Khayri JM, Al-Ghamdi AS, Al-Bahrany AM (2007) Callus induction from inflorescence of elite males of date palm *Phoenix dactylifera* L. Fourth symposium on date palm in Saudi Arabia, King Faisal University, Al-Hassa, 5–8 May, Abstracts Book, p 147

Bhaskaran SH, Smith RH (1992) Somatic embryogenesis from shoot tip and immature inflorescence of *Phoenix dactylifera* L. cv. Barhee. Plant Cell Rep 12:22–25

Bouamri R, Dalpe Y, Serrhini MN, Bennani A (2006) Arbuscular mycorrhizal fungi species associated with rhizosphere of *Phoenix dactylifera* L. (date palm) in Morocco. Afr J Biotech 5:510–516

Drira, N, Al-Sha'ary A (1993) Analysis of date palm female floral initials potentials by tissue culture. In: Proceedings of the third symposium on date palm, Saudi Arabia. King Faisal University, Al-Hassa, pp 161–170

Drira N, Benbadis A (1985) Multiplication végétative du palmier dattier (*Phoenix dactylifera* L.) par réversion en culture in vitro d'ébauches florales de pieds femelles. J Plant Phys 119:227–235

Eeuwens CJ (1978) Mineral requirements for growth and callus initiation of tissue explants excised from mature coconut palms (*Cocos nucifera*) and cultured in vitro. Phys Plant 36:23–28

Eeuwens CJ, Blake J (1977) Culture of coconut and date palm tissue with a view to vegetative propagation. Acta Hort 78:268–277

El-Hadrami I, Baaziz M (1995) Somatic embryogenesis and plant regeneration from shoot-tip explants in *Phoenix dactylifera* L. Biol Plant 37:205–211

El-Korchi B (2007) Large scale in vitro propagation of a rare and unique male date palm (*Phoenix dactylifera* L.) using inflorescence technique. Acta Hort 736:243–254

Feki L, Drira N (2007) Development of an efficient plant regeneration system for date palm. Fourth symposium on date palm in Saudi Arabia, King Faisal University, Al-Hassa, 5–8 May, Abstracts Book, p 143

Gamborg OL, Miller RA, Ojima K (1968) Nutrient requirements of suspension cultures of soybean root cells. Exp Cell Res 50:151–158

George EF (1996) Plant propagation by tissue culture. Part 2, the technology. Exegetics Limited, Edington, pp 591–601

Hassan, MM (2002) In vitro studies on somatic embryogenesis conservation of date palm. Ph.D thesis, Faculty of Agri., Cairo Univ. Egypt

Hegazy AE, Kansowa OA, Abul-Soad AA, Nasr MI (2006) Growing behaviors of *ex vitro* date palm plants after acclimatization. In: Second international conference of genetic engineering and its applications. Sharm El-Sheik City, South Sinai, Egypt, 14–17 Nov, p 69–75

Ibrahim AI (1999) Somaclonal variation during micropropagation of date palm via embryogenesis. In: The first international conference on plant tissue culture and its applications, University of Zagazig, Egypt, 12–14 Sept, pp 189–199

Ibrahim AI, Abul-Soad AA, Bakr EI, El-Sherbeny NR (1999a) Effect of charcoal and light intensity on rooting and acclimatization stages of regenerated date palm (*Phoenix dactylifera* L.) plantlets. In: Proceedings of the first international conference on plant tissue culture and its applications, University of Zagazig, Egypt, 12–14 Sept, pp 219–226

Ibrahim AI, Abul-Soad AA, El-Sherbeny NR, Bakr EI (1999b) Effect of nutrient media on the growth and development of date palm grown in vitro and *ex vitro* plantlets. In: Proceedings of the first international conference on plant tissue culture and its applications, University of Zagazig, Egypt, 12–14 Sept, pp 209–217

Kozai T, Hayashii M, Hirosawa Y et al (1987) Environmental control for acclimatization of *in vitro* cultured plantlets (1) Development of acclimatization unit for accelerating the plantlet growth and test cultivation. J Agric Met Jap 42:349–358

Kriaa W, Masmoudi F, Drira N (2007) In vitro culture of date palm using mature female flower explants. Fourth symposium on date palm in Saudi Arabia, King Faisal University, Al-Hassa. 5–8 May, Abstracts Book, p 146

Madhuri S, Shankar PC (1998) Somatic embryogensis and plant regeneration from leaf primordia of *Phoenix dactylifera* L. cv. Tokubi. Ind J Exp Bio 36:526–529

Markhand GS, Abul-Soad AA (2010) Fruit characterization of Pakistani dates. Pak J Bot 42(6):3715–3722

Masmoudi F, Chaari CB, Drira N (2007) Date palm floral initials and perspectives in tissue culture cloning and improvement. In: Fourth symposium on date palm in Saudi Arabia, King Faisal University, Al-Hassa, 5–8 May, Abstracts Book, p 142

Murashige T, Skoog FA (1962) A revised medium for rapid growth and bioassays with tobacco tissue cultures. Phys Planta 15:473–479

Omar MS (1988a) Callus initiation, asexual embryogenesis and plant regeneration in *Phoenix dactylifera* L. Date Palm J 6:265–275

Omar MS (1988b) In vitro response of various date palm explants. Date Palm J 2:371–389

Pospisilova J, Ticha I, Kadlecek P et al (1999) Acclimatization of micropropagated plants to *ex vitro* conditions. Bio Planta 42:481–497

Reuveni O, Adato Y, Kipins HL (1972) A study of new and rapid methods for the vegetative propagation of date palms. Date Grow Inst Rep 49:17–24

Sidky RA, Zaid ZE, Abul-Soad AA (2007) Direct somatic embryogenesis of date palm (*Phoenix dactylifera* L.) by osmotic stress. Egypt J Agric Res 85(1B):573–582

Smith WK, Thomas JA (1973) The isolation and in vitro cultivation of cells of *Elaeis guineensis*. Oleag 28:123–127

Tisserat B (1979) Propagation of date palm (*Phoenix dactylifera* L.) in vitro. J Exp Bot 30:1275–1283

Tisserat B (1981) Date palm tissue culture. U.S. agriculture research serves, advanced agriculture technology. ATTW 17:1–15

Tisserat B (1982) Factors involved in the production of plantlets from date palm callus cultures. Euphyt 31:201–214

Tisserat B (1984a) Date palm. In: Sharp WR, Evans DA, Ammirato PV, Yamada Y (eds.) Handbook of plant cell culture, vol 2. Macmillan, New York, pp 505–545

Tisserat B (1984b) Propagation of date palms by shoot tip cultures. Hort Sci 19:230–231

Tisserat B, DeMason DA (1980) A histological study of development of adventive embryos in organ cultures of *Phoenix dactylifera* L. Ann Bot 45:465–472

Tisserat B, Foster G, DeMason DA (1979) Plantlet production in vitro from *Phoenix dactylifera* L. Date Grow Inst Rep 54:19–23

Zaid ZE (2003) comparative studies on the production of date palm cultivars via tissue culture technique. Ph.D. thesis, Faculty of Agri., Cairo Univ., Egypt

Chapter 7
Bioreactors and Automation in Date Palm Micropropagation

A. Othmani, R. Mzid, C. Bayoudh, M. Trifi, and N. Drira

Abstract Researchers have successfully micropropagated several cultivars of date palm (*Phoenix dactylifera* L.) using *in vitro* tissue culture techniques based on somatic embryogenesis and shoot organogenesis. However, major hindrances exist in scale-up including low productivity of cultures, synchronization of embryogenesis, conversion of matured somatic embryos, low multiplication rates of *de novo* shoots and high production cost of plantlets. Recently, tremendous success has been achieved in automation of micropropagation steps of many plant species using liquid culture systems. This achievement constitutes an alternative for resolving all the preceding problems. For example, yield of cotyledonary somatic embryos produced in suspension cultures were 17-fold greater than on agar-solidified medium. The transfer of shoot clusters in temporary immersion bioreactor clearly improved the yield of regenerated shoots 5.5-fold in comparison with that regenerated on agar-solidified medium. The aim of this chapter is to critically outline the potential of liquid culture systems (suspension culture and temporary immersion system, TIB) to large scale and automation of date palm micropropagation. The principles, advantages and disadvantages of these methods are also described.

A. Othmani (✉) • M. Trifi
Laboratory of Molecular Genetics, Immunology & Biotechnology, Faculty of Sciences of Tunis, 2092, El Manar Tunis, Tunisia
e-mail: degletbey@yahoo.fr

R. Mzid
Laboratory of Plant Molecular Physiology, Biotechnology Center of Borj Cédria, 901 Hammam-Lif, Tunisia
e-mail: rmzid@yahoo.fr

C. Bayoudh
Laboratory of In Vitro Culture, Regional Center of Oases Agriculture Researches, 2260 Degache, Tunisia
e-mail: civpalmier@yahoo.fr

N. Drira
Laboratory of Plant Biotechnology, Faculty of Sciences of Sfax, 3018 Sfax, Tunisia
e-mail: drira_riadh@yahoo.fr

S.M. Jain et al. (eds.), *Date Palm Biotechnology*,
DOI 10.1007/978-94-007-1318-5_7, © Springer Science+Business Media B.V. 2011

Keywords Cell suspension • *In vitro* • Shoot organogenesis • Somatic embryogenesis • Suspension culture • Temporary immersion systems

7.1 Introduction

Date palm (*Phoenix dactylifera* L.) (2n=36) is an out-breeding perennial monocotyledon cultivated chiefly in arid and desert areas of North Africa and the Middle East. This fruit crop is the mainstay of agriculture in oases zones where many other food crops can be grown in combination. Conventional propagation by offshoots restricts establishment of new date palm plantations. Moreover, several genotypes do not produce offshoots while others are difficult to root. In addition, seed-propagated palms do not bear true to type due to heterozygosity and require up to 7 years reaching fruiting stage.

Many reports have described the use of *in vitro* tissue culture techniques, with the potential to multiply genotypes of superior value on a massive scale in a shorter period than the conventional method. Several protocols of *in vitro* propagation in date palm have been optimized successfully, including somatic embryogenesis using apical meristem (Rhiss et al. 1979; Zouine and El Hadrami 2007), development of adventitious buds from juvenile leaves and axillary bud culture (Othmani et al. 2009a), culture of zygotic embryos (Ammar and Benbadis 1977), reversion of female flowers (Drira and Benbadis1985; Othmani 1998), cell suspension culture (Fki et al. 2003; Othmani et al. 2009b) and protoplast culture (Chabane et al. 2007). These protocols have used agar-solidified media. However, these protocols are difficult to automate and entail a high production cost unsuitable for mass clonal propagation (Vinocur et al. 2000). Besides, the low multiplication rate of shoots induced in agar solidified-medium is a critical problem facing the success of commercial tissue culture in date palm propagation. Liquid cultures are more amenable to automation necessary in commercial scaling-up production systems.

7.2 Micropropagation of Date Palm on Solid Media

7.2.1 *Regeneration Approaches*

Since the mid 1970s, extensive efforts have been undertaken by several research groups to mass clonal propagation of date palm via *in vitro* tissue culture using agar-solidified media (Drira 1983; Drira and Benbadis 1985; Poulain et al. 1979; Tisserat 1979; Tisserat and DeMason 1980). Conventionally, explants were inoculated on Murashige and Skoog (1962) basal medium (MS) incorporating growth regulator(s), a source of carbon (sucrose) and were solidified with agar (0.5–0.8%).

Callogenesis is obtained in darkness at 28±2°C on an initial culture medium consisting of MS agar-solidified medium supplemented with an auxin. After a few months, the calli and explants showing shoot organogenesis are transferred to light (16/8 h photoperiods) on a second culture medium with a lower auxin concentration. After embryogenic structures develop on the surface of the calli, they are isolated and cultured on a plant growth regulator-free medium (Al-Khayri 2005). In shoot organogenesis the entire expanding explants with resultant shoots are transferred to MS agar-solidified medium enriched with cytokinin. Vigorous somatic embryos and single shoots are transferred on an α-naphthalenacetic acid (NAA)-enriched medium to allow the root development (Al-Khayri 2007). Then they are transplanted to the acclimatization stage, and subsequently to the pre-nursery.

Our experiments showed many kinds of morphogenetic responses from leaf explants of 13 cultivars of date palm when cultured on MS agar-solidified media (M_1, M_2, M_3, M_4 and M_5) supplemented with 0.0, 0.5, 1.0, 10, and 50 mg l^{-1} of 2,4-Dichlorophenoxyacetic acid (2,4-D), respectively (Othmani 2010). All the media contained 30 g l^{-1} sucrose and 0.3 gl^{-1} activated charcoal (AC) and were solidified with 0.7% agar. The pH of all media was adjusted to 5.7 prior to autoclaving at 120°C and 1.4 kg cm^{-2} for 20 min.

Regeneration is involved with indirect somatic embryogenesis that is the induction of somatic embryos by passing through the callus phase. Regeneration is also the case of explants of cvs. Deglet Bey (Fig. 7.1a,b); Cheddakh, Ksebba, Tezerzeit Kahla, Boufeggous, Tezerzeit Safra, Gharess Mettig, Kenta and Deglet Noor when cultured on M_3 medium and T10; Bith Hmem and Khwat Ftimi when cultured on M_5 medium involving direct embryogenesis, induction of somatic embryos without passing through the callus phase. As well as explants of cvs. Ksebba and Boufeggous when cultured on M_3 medium and Deglet Bey when cultured on M_4 medium (Fig.7.1c). Indirect caulogenesis, induction of shoots by passing via the callus phase, such as in explants of cvs. Cheddakh, Ksebba, Gharess Mettig (Fig. 7.1d) and Tezerzeit Kahla when cultured on M_2 medium and cvs. Tezerzeit Safra and Gondi when cultured on M_3 medium. Also, direct caulogenesis, induction of shoots with no callus phase, such as in Tezerzeit Kahla (Fig. 7.1e) and Kenta when cultured on M_2 medium. In addition, rhizogenesis was recorded, for example, on explants of cv. Boufeggous (Fig. 7.1f) when cultured on M_5 medium.

7.2.2 *Effect of Desiccation on Somatic Embryo Production*

To study the effect of partial desiccation, embryogenic calli were transferred to sterile empty Petri dishes containing three sterile Whatman filter paper disks. The dishes were sealed with parafilm and kept at 27±2°C in the dark for 6, 12, 24 and 48 h desiccation. The relative water content of embryogenic calli was calculated according to Malabadi et al. (2004). After the desiccation treatments, the partially desiccated calli were transferred to MS medium supplemented with 0.1 mg l^{-1} abscisic acid (ABA) for light culture. The 12 h partial desiccation treatment resulted in a decrease in water content from 95% to 75% with more somatic embryos forming.

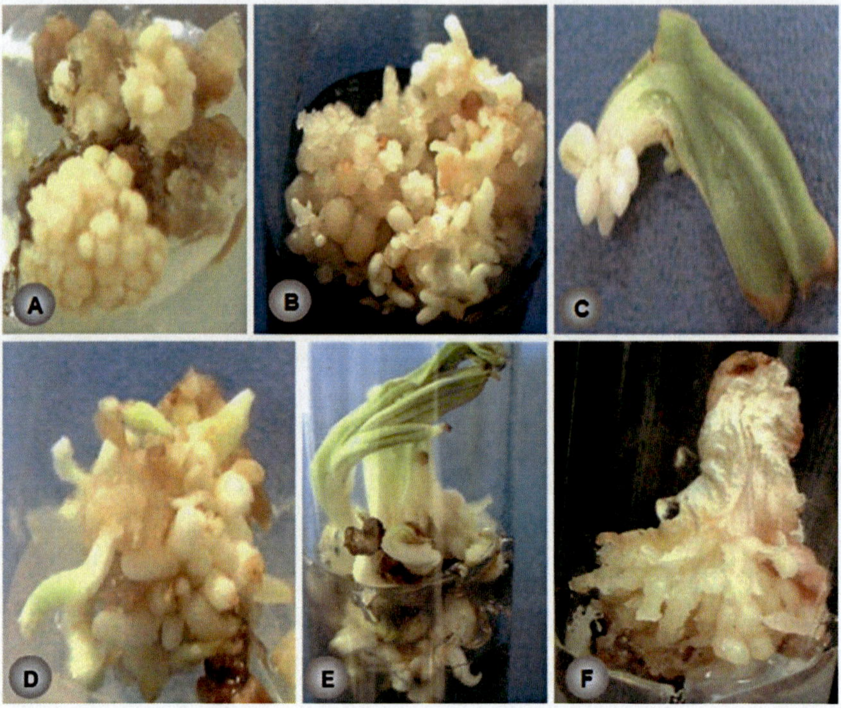

Fig. 7.1 Morphogenetic responses observed from leaves of date palm. (**a**) Embryogenic callus on MS medium supplemented with 10 mg l^{-1} 2,4-D (cv. Deglet Bey). (**b**) Development of somatic embryos from embryogenic callus after culture on MS medium deprived of growth regulators (cv. Deglet Bey). (**c**) Direct formation of somatic embryos on MS medium supplemented with 10 mg l^{-1} 2,4-D (cv. Deglet Bey). (**d**) Indirect formation of shoots on MS medium supplemented with 0.5 mg l^{-1} 2,4-D (cv. Gharess Mettig). (**e**) Direct formation of shoots on MS medium supplemented with 0.5 mg l^{-1} 2,4-D (cv. Tezerzeit Kahla). (**f**) Rhizogenesis on MS medium supplemented with 50 mg l^{-1} 2,4-D (cv. Boufeggous)

Under these conditions somatic embryos started their development 7 days after desiccation treatment (Fig. 7.2a), and after 38 days of culture, elongated embryos matured into cotyledonary embryos (Fig. 7.2b). The maturation frequency increased with increasing desiccation duration up to 12 h (Table 7.1), but decreased at 24 h. In contrast, an increase in the time of desiccation treatment beyond 24 h resulted in increased necrosis and death of nodular somatic embryos.

As in the cases of *Oryza sativa* (Chand and Sahrawat 2001) and *Pinus kesiya* (Malabadi et al. 2004), it was found that desiccation treatment of embryogenic calli was necessary to induce somatic embryo morphogenesis from date palm embryogenic callus. The results suggested that a complex relationship exists between the water content of embryogenic calli and development of somatic embryos. Of the five desiccation treatments evaluated for their effect on date palm somatic embryo maturation, 12 h induced significantly more somatic embryos than 6, 24 or 48 h desiccation. According to Rance et al. (1994), the desiccation treatment might

Fig. 7.2 Improved somatic embryo production and plant regeneration of date palm cv. Boufeggous. (**a**) Development of somatic embryos derived from 12 h desiccated callus cultured for 10 days on MS medium containing 0.1 mg l^{-1} AC. (**b**) Matured somatic embryos derived from 12 h desiccated callus cultured for 38 days on MS medium without AC. (**c**) Proliferation of sterile blade chopped callus after 1 month of culture on MS meduim supplemented with 0.1 g.l^{-1} AC. (**d**), (**e**) and (**f**) Development stages of somatic embryos derived from finely chopped callus within 3, 20 and 30 days of culture on MS media with or without 0.1 mg l^{-1} AC, respectively. (**g**) Plantlet obtained from a converted somatic embryo. (**h**) Potted plants 1 month after transfer to a greenhouse

trigger rapid biochemical changes in the calli and under water stress specific enzymes or polypeptides probably appear in callus culture. Attree and Fowke (1993) reported that desiccation treatment stimulates the accumulation of storage reserves and triglycerides that have a positive effect on somatic embryo development. The desiccation treatment was also found to significantly improve germination rates of somatic embryos of date palm cv. Deglet Noor (Fki 2005). The effect of partial drying may cause a breakdown of endogenous ABA or decreased sensitivity of embryos to ABA which could release embryos from development constraints and allow germination to proceed (Kermode et al. 1989).

7.2.3 Effect of Callus Chopping on Somatic Embryo Production

Two methods of chopping were used. The first consisted of chopping primary calli with a scalpel blade according to the method described by Teixeira et al. (1995). The second method was finer, consisting of straining embryogenic callus through a 40 mesh screen (Sigma S-0070) with a pore size of 380 μm using a tissue grinder glass pestle (Sigma T-8279) (Othmani et al. 2009a). The small pieces obtained were

Table 7.1 Effect of desiccation and fine chopping of embryogenic callus of date palm cv. Boufeggous on time taken by somatic embryos to be matured, production number of somatic embryos per 500 mg FW of embryogenic callus, percentage of germination and frequency of conversion of germinated somatic embryos into plantlets

Callus treatment	Time (day)	No. embryos	% Germination	% Conversion
No treatment	55.33a	50.66d	83.50a	94.50c
Fine chopping	34.667e	422.33a	84.73a	97.26a
Desiccation (6 h)	42.66c	123.66c	83.46a	95.63bc
Desiccation (12 h)	38.33d	306.33b	84.50a	96.10ab
Desiccation (24 h)	51.00b	45.66d	81.20b	94.93bc
Desiccation (48 h)	0.00f	0.00e	0.00c	0.00d

transferred aseptically to MS medium supplemented with 0.1 mg l^{-1} ABA for light culture. Chopping primary calli with a scalpel blade and cultivating on MS medium supplemented with 0.1 g l^{-1} ABA resulted in a 3.5-fold increase in callus fresh weight (FW) per month compared to non-chopped calli, and gave rise to friable, granular calli composed of embryogenic cells. Under these conditions several somatic embryos were observed after 1 month of culture (Fig. 7.2c). Otherwise, fine chopping of embryogenic callus into small pieces (<380 μm), not only improved callus growth, but also induced the development of somatic embryos (Table 7.1) which started within 3 days (Fig. 7.2d). Under these conditions, embryos further differentiated into elongated embryos within 2 weeks (Fig. 7.2e) of culture, and cotyledonary embryos were formed in the next 2 weeks (Fig. 7.2f). Similar to the desiccation treatments, finely chopped embryogenic callus produced relatively synchronized somatic embryo development.

Cotyledonary somatic embryos transferred to MS medium with 1 mg l^{-1} NAA developed into healthy plants with both shoots and roots (Fig. 7.2g) at a frequency of 83% after 1 month of culture. There was no statistically significant difference between embryogenic callus treatments for the germination and conversion into plantlet frequencies (Table 7.1). Nevertheless, the matured somatic embryos induced from finely chopped callus developed roots sooner and converted into plantlets quicker than those obtained from desiccated and nontreated embryogenic calli.

The growth of about 1,200 somatic plants is currently being monitored in growth chambers (Fig. 7.2h). During the acclimatization step, 60% of the plantlets survived in the greenhouse. The results of this study showed that chopping callus into small pieces (<380 μm) is one of the most effective techniques in terms of numbers and synchronized somatic embryo development. Furthermore, this technique has improved the productivity of calli and shortens the period of maturation of embryos. Therefore, this technique could be integrated into other procedures to improve maturation frequency of somatic embryos. These include culturing embryogenic tissue on media supplemented with maltose, polyethylene glycol, ABA, and AC (Maruyama et al. 2007), partial desiccation of embryogenic calli and higher concentration of gellan gum (Malabadi et al. 2004), use of embryogenic suspension cultures (De Touchet et al. 1991), culture of cell suspensions of embryogenic callus on temporary immersion bioreactor (Teisson and Alvard 1995) and addition of thidiazuron to the nutrient medium (Jones et al. 2007).

The reason for the improvement with fine chopping embryogenic calli has not been investigated at the physiological or molecular level and may be attributable to an inhibitory substance(s) released by surrounding embryogenic cells. Hence, finely chopped callus might raise this inhibition; consequently, embryos enter into the development phase. Also, it is possible that the physical stress caused by fine chopping of callus stimulates *de novo* protein synthesis allowing a rapid division and differentiation of embryogenic cells. In addition, chopping of embryogenic calli into small pieces minimizes the intercellular connections between cells and contributes to a partial isolation of cells. According to Lowe et al. (1985), the isolation of embryogenic cells may be essential to the cells for the expression of the embryogenic potential. In agreement with Teixeira et al. (1995), chopping primary calli with a scalpel blade was necessary to enable the appearance and the growth of granular secondary calli after 1 month of culture on the same medium. Fki et al. (2003) and Sané et al. (2006) reported that when globular and compact primary calli of date palm were chopped with a scalpel blade and cultivated on the same medium, they divided actively and after 6–8 weeks of culture gave rise to friable calli composed of embryogenic cells.

7.2.4 Scaling-Up Difficulties

The scaling-up of tissue culture production on solid media has revealed some difficulties. Problems have generally occurred in the following issues: First, micropropagation remains a costly production technology; second, tissue culture of date palm is generally time-consuming, and biological events involved in each step of the procedure progress very slowly, as has previously reported for coconut (Buffard-Morel et al. 1995), oil palm (Rival et al. 1997) and date palm (Drira and Benbadis 1985); third, maintaining synchrony during the induction of embryogenesis remains unsolved (Masmoudi 1999); fourth, abnormal development, low maturation and germination frequencies of somatic embryos are serious limitations (Daguin and Letouzé 1997; Feher et al. 2003; Vieitez 1995); and fifth, time-consuming cleaning, filling and handling of a large number of containers (Maene and Debergh 1985).

7.3 Micropropagation of Date Palm in Liquid Media

7.3.1 Advantages of Liquid Culture

The literature shows that liquid culture media are known to be more suitable for eliciting micropropagation processes (embryogenic and organogenic regeneration pathways) and the adequate solution for decreasing plantlet production costs and for considering automation in a number of plant systems (Aitken-Christie 1991; Debergh 1988; Kumar et al. 2006; Ziv 1995). In reality, liquid culture systems provide much more uniform culturing conditions, the media can be easily renewed without changing the vessel, and disinfection is possible by ultrafiltration and vessel cleaning after a culture period is much easier (Etienne et al. 2006). In addition, liquid culture media provide better plant performance by allowing a direct contact of the medium throughout the plant material (Zobayed and Saxena 2003).

7.3.2 Embryogenic Cell Suspension Culture

Embryogenic suspension culture is defined as single cells or small cell aggregates in agitated liquid media (Preil 2005). Under these conditions, plantlets were developed when subculturing embryogenic callus in rotating vessels and afterwards placing the cultures on stationary medium free of plant growth regulators. The use of suspension culture for the regeneration of date palm has been reported by several authors and the liquid media used are MS or ½MS enriched or not with low amount of auxins alone or associated with cytokinin: MS enriched with 0.1 mg l^{-1} NAA (Sharma et al. 1986), 1/2 MS without auxins (Daguin and Letouzé 1988), MS without plant growth regulators (Bhaskaran and Smith 1992), ½MS enriched with 1 mg l^{-1} 2,4-D (Fki et al. 2003), MS enriched with 2 mg l^{-1} 2,4-D (Sané et al. 2006), ½MS enriched with 0.1 mg l^{-1} 2,4-D and 0.5 mg l^{-1} 6-Benzylaminopurine (BAP) (Zouine and El Hadrami 2007) and 1/2 MS enriched with 2 mg l^{-1} 2,4-D (Othmani et al. 2009b).

According to Fki et al. (2003), by using this method, the productivity of the cultures of cv. Deglet Noor increased 20-fold (from 10 to 200 embryos per month per 100 mg fresh weight of embryogenic callus) when embryogenic suspensions were used instead of cultures on agar-solidified media. The authors reported that the overall production rate was 10,000 ± 45 typical embryos per liter per month. Similarly, Othmani et al. (2009b) demonstrated that the yield of cotyledonary somatic embryos produced were greater than in agar-solidified medium, namely 501 versus 29 per 0.5 g fresh weight of embryogenic callus of cv.Deglet Bey. Otherwise, Sané et al. (2006) reported a lower productivity (80 embryos per 100 mg of cell suspension).

In order to initiate suspension cultures, 0.5 g of friable callus was minced, filtered through a 500 µm mesh filter and transferred to 250 ml Erlenmeyer flasks containing 50 ml half-strength MS liquid media (L_1, L_2, L_3, L_4 and L_5) supplemented with 0.4 mg l^{-1} AC and incorporating 0.0, 0.5, 1, 1.5 and 2 mg l^{-1} 2,4-D, respectively (Othmani et al. 2009b). Subculturing was done every 5 days by decanting off the old medium and replacing it with fresh medium of the same composition. Every 3 days, several samples from suspension cultures were isolated and observed under an inverted microscope for the development of somatic embryogenesis. Morphological characteristics of somatic embryogenesis at different development stages were recorded.

Following transfer into liquid medium cells from callus exhibited an intense division and gave rise to a highly heterogeneous suspension comprising cells showing

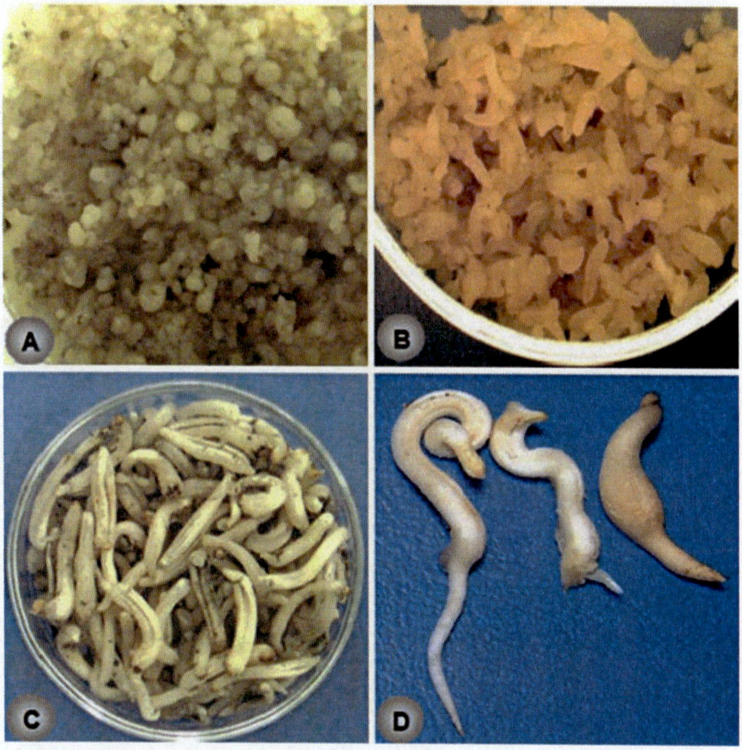

Fig. 7.3 Different stages of somatic embryogenesis in date palm cv. Deglet Bey in suspension culture on half-strength MS liquid medium with 2 mg l^{-1} 2,4-D. (**a**) Globular-stage somatic embryos observed after culturing for 7d. (**b**) Elongated embryos observed after 12 days in culture. (**c**) Cotyledonous-stage embryos observed within 26 days in culture. (**d**) Germination of embryos on MS medium deprived of growth regulators

differences in shape, size and characteristics (Othmani et al. 2009b). Microscopic observations of suspension cultures showed that somatic embryos developed from typical globular embryos (Fig. 7.3a) over a period of 7 days, globular-stage embryos further differentiated into elongated embryos within 12 days (Fig. 7.3b). Cotyledonary embryos (Fig. 7.3c) were formed in the next 26 days. In the liquid culture conditions, all embryo stages were observed in the same culture, this heterogeneity could be explained by the setting of an active and precocious secondary embryogenesis (Zouine and El Hadrami 2007). Additionally, our data showed that increasing the 2,4-D concentration in the liquid media from 0.5 to 2 mg l^{-1} was beneficial to induce somatic embryo proliferation (Table 7.2), a performance better than that previously reported for date palm cv. Deglet Noor (Fki et al. 2003) and for oil palm (De Touchet et al. 1991).

In this research, germination of cotyledonary somatic embryos (Fig. 7.3d) medium and subsequent conversion into plantlets with fully-developed shoots and roots was possible after complete removal of 2,4-D from the culture medium.

Table 7.2 Effect of medium type and 2,4-D concentration on the number of cotyledonary somatic embryos obtained from 0.5 g of embryogenic callus after 45 days culturing on liquid media L_1, L_2, L_3, L_4 and L_5 followed by 2 months of culture on MS solidified medium deprived of growth regulators

Medium type	Code	2,4-D concentration, mg l^{-1}	Number of cotyledonary somatic embryos
Liquid medium	L_1	0.0	0f*
	L_2	0.5	341d
	L_3	1.0	397c
	L_4	1.5	407.66b
	L_5	2.0	499.66a
Agar solidified medium	–	0.0	29e

* Means followed by the same letters are not significantly different as indicated by Newman-Keuls' test at $P=0.05$

Similar results have been reported by Thiruvengadam et al. (2006) with somatic embryos of *Momordica charantia* L. and by Aberlenc-Bertossi et al. (1999), with somatic embryos of oil palm. Al-Khayri and Al-Bahrany (2004) reported that date palm somatic embryos developed into plantlets on MS medium free of plant growth regulators and containing $AgNO_3$. It should be noted that culture inducing differentiation of embryogenic calli in agar-solidified or liquid medium strongly affected the germination and conversion rate of cotyledonary somatic embryos into plantlets when transferred to agar-solidified MS medium deprived of 2,4-D. Indeed, 65% of cotyledonary somatic embryos derived from agar-solidified medium germinated and converted into plantlets rather than 22% from suspension cultures. The low germination rate of cotyledonary somatic embryos derived from suspension cultures is probably due to the inhibitory action of hyperhydricity. It is known that the hyperhydricity phenomenon is a major problem for regeneration of diverse species cultured in liquid media (Piatczak et al. 2005). Kevers et al. (1984) reported that hyperhydration is due to a deficient lignification caused by a serious reduction of peroxydases and phenylalanine ammoniaclyase activities.

For date palm, Othmani et al. (2009b) reported that partial desiccation of mature somatic embryos, corresponding to a decrease in water content from 90.33% to 69% followed by their culture on regeneration medium during 4 weeks, resulted in a significantly higher germination rate (from 21% to 80.66%). These results confirm those from Fki et al. (2003) on the germination of date palm cv. Deglet Noor somatic embryos derived from embryogenic suspensions. Monsalud et al. (1995) reported that mango (*Mangifera indica*) somatic embryos regenerated from suspension cultures germinated precociously when partially dehydrated. According to Jens (1997), partial desiccation of somatic embryos induce a substantial decrease in endogenous levels of ABA, a plant growth regulator with a positive effect in the maintenance of embryos in a maturation stage through preventing precocious germination (Pliego-Alfaro et al. 1996). Additionally, a high frequency of deformed somatic embryos, including deformed cotyledon, absence of cotyledon or plumules, abnormalities in the histodifferentiation of the apical meristem, the fusion of embryos or those with excessive root that can attain in sometimes a yield of 50% (Othmani 2010), constitutes a real constraint to mass propagate date palm by using embryogenic suspension culture.

7.4 Date Palm Micropropagation in Bioreactors

7.4.1 Bioreactor Systems

A bioreactor is a self-contained, sterile environment which capitalizes on liquid nutrient or liquid/air inflow and outflow systems, and is designed to provide most favorable growth conditions by enabling a high degree of control over chemical and/or physical factors such as (pH, oxygen, ethylene, carbon dioxide concentrations, aeration rate and temperature) and are compatible with the automation of micropropagation procedures and a reduction of labor costs (Etienne et al. 2006).

Bioreactor use was first developed for culturing microorganisms and later for plant cell suspensions to accumulate cell biomass for secondary metabolite production (Preil 2005). Currently, bioreactors play a crucial role in scaling up the production for commercialization of somatic embryogenesis and multiplication of clusters of meristem- and bud-based plant micropropagation.

As reported by Etienne et al. (2006), bioreactors can be classified into four categories. First, mechanically agitated bioreactors, including aeration-agitation bioreactors, rotating drums and spin-filter bioreactors. Second, pneumatically agitated bioreactors, including air-lift bioreactors, bubble column bioreactors and simple aeration bioreactors. Third, non-agitated bioreactors, including gaseous phase (mist) bioreactors, oxygen permeable membrane bioreactors, overlay aeration bioreactors and perfusion bioreactors. Fourth, temporary-immersion bioreactors, including systems with temporary complete immersion by pneumatic-driven transfer of liquid medium (RITA® and TIB® systems).

Reports on plant regeneration in bioreactors include a wide range of taxonomically diverse species, e.g. banana (Kosky et al. 2002), potato (Teisson and Alvard 1999), coffee (Etienne and Berthouly 2002; Etienne-Barry et al. 1999), citrus (Cabasson et al. 1997), pineapple (Escalona et al. 1999), tea (Akula et al. 2000), apple (Zhu et al. 2005), strawberry (Hanhineva and Karenlampi 2007) and date palm (Othmani et al. 2009b). In these experiences many kinds of materials were used including micro-cuttings, somatic embryos, apical shoots, flowers, roots, leaves, seeds, rootstocks or shoot clusters.

7.4.2 Temporary Immersion Bioreactors

Temporary immersion bioreactors (RITA™, Vitropic-Cirad, France) contain 15 cm high cylindrical vessels that consist of two compartments (Fig. 7.4); the upper part holds the plant material on a polyurethane filter (1 cm) and the lower part contains

Fig. 7.4 Temporary immersion bioreactor container (RITA™, Cirad, France) consisting of two compartments: the lower part holds the medium and the plant material is located on a polysulfone foam disc in the upper part of the bioreactor

the culture medium. The container is connected to an automated air pump which sets overpressure to the lower part of the container pushing the medium to the upper part through the filter. The overpressure escapes through an air vent in the lid of the container. The air pump is controlled with a timer that sets the duration and frequency of the liquid immersion.

In date palm as in various other species, bioreactor culture is used as a part of the propagation process, especially in scaling-up of shoot organogenesis for further development on agar-solidified media. Othmani et al. (2009b) were the first to report the production of date palm plantlets in TIB-bioreactors. In this experiment, temporary immersion cultures were established with immersion of cultures for 5 min every 8 h for a 6 week culturing period. There were three vessels for each treatment and each vessel contained 10 g of shoot clusters. The experiment was conducted three times.

Our research determined in preliminary experiments (data not shown) that the embryogenic calli of cv. Deglet Bey failed to grow in the TIB and usually turned brown and died. This completely differed from the previous study by Kosky et al. (2002) in which calli of banana, cv. FHIA-18 (AAAB) differentiated into plantlets when cultured in TIB. Therefore, we focused our study on the use of shoots that appeared at the base of young leaf explants (Fig. 7.5a) cultured on MS agar-solidified medium supplemented with 10 mg l^{-1} 2,4-D for 8 months in this system in order to generate plantlets.

After their transfer to the TIB-bioreactor with 200 ml MS liquid medium containing 0.04 mg l^{-1} NAA, 0.2 mg l^{-1} BAP and 0.02 mg l^{-1} Kinetin (KT) (multiplication medium: MM), at an immersion frequency of 3, 5 or 7 min every 8 h for a 6 week

Fig. 7.5 Developmental stages of shoot clusters after transfer in the TIB system. (**a**) Shoot formation at the base of a young leaf explant after 8 months of culture on MS medium supplemented with 10 mg l^{-1} 2,4-D. (**b–d**) Evolution of shoot clusters after transfer in TIB. (**b**) Induction of *de novo* shoots after 2 weeks. (**c**) Shoots development after 4 weeks of culture. (**d**) Vigorous shoots obtained after 6 weeks of culture. (**e**) Rooting of 2 shoots after 1 month of culture on rooting medium. (**f**) Potted plants 18 months after transfer to a greenhouse

culturing period, date palm regenerates as dense clusters of emerging shoots at the base of the shoots after a short callus phase (Fig. 7.5b). Next, shoots elongated faster to become vigorous and gradually produced new shoots (Fig. 7.5c and d) as long as they were kept in the multiplication medium for 6 weeks, in comparison with those cultured on agar-solidified medium particularly when immersion was applied for 5 min every 8 h (Table 7.3).

According to Dufour et al. (1995), immersion times play a crucial role in morphogenesis. Indeed, 1 min immersion every 6 h led to an optimum cell proliferation, whereas for an identical culture medium, immersions of 15 min every 6 h led to complete regeneration of somatic embryos. The high ability of shoots to proliferate in the TIB was previously reported by Zobayed and Saxena (2003). Otherwise, Hanhineva and Karenlampi (2007) mentioned that TIB can be used to clonal propagation by culturing leaf piece explants.

The current result confirmed earlier observations made by Paek and Hahn (2001), who reported that TIB technology provides the potential for producing large numbers of plants economically and efficiently. This improvement of plant performance in the

Table 7.3 A comparison of agar solidified- medium and temporary immersion (TIB) system at three immersion frequencies in relation to the number of shoots, percent hyperhydration and percent shoot necrosis obtained from 10 g (FW) of shoot clusters after 6 weeks of culture on MM

Medium type	Immersion frequency	Number of shoots	Hyperhydration%	Necrosis %
TIB, Liquid medium	3 min/8 h	255b*	3.1d	5.0b
	5 min/8 h	284a	6.0c	5.0b
	7 min/8 h	220c	10.2b	7.0b
Agar solidified medium	–	54d	18.0a	29.7a

*Means followed by the same letter in the same column are not significantly different as indicated by Newman-Keuls' test at $P=0.05$

TIB system is due to a direct contact of the medium with the plant material, a renewing of the culture atmosphere on each immersion and to a reduction of asphyxia, necrosis and tissues hyperhydration phenomena (Etienne and Berthouly 2002).

In the present study, roots were scarcely produced from adventitious shoots (10%) when kept in the TIB container. But, after their transfer to MS agar-solidified medium supplemented with 0.1 mg l^{-1} NAA (rooting medium), elongated shoots exhibited prolific adventitious rooting that reached 96% after 1 month of culture (Fig. 7.5e). This result is in contrast with that reported by Hanhineva and Karenlampi (2007) on strawberry cv. Jonsok for which rooting was optimal when shoots were kept in the TIB container. After acclimatization, plantlets have grown actively in the greenhouse with an apparent normal leaf and shoot morphology (Fig. 7.5f).

7.4.3 Constraints of Micropropagation Bioreactors

Despite many promising results, culturing in bioreactors proves to be trickier than culturing on agar-solidified media in terms of hyperhydricity (vitrification) and contamination risks. These are further difficulties to be considered before using bioreactors on an industrial level. Albarran et al. (2005) observed that hyperhydricity increased with the immersion frequency. They report that 15 min immersions applied two or six times per day led to hyperhydric embryo frequencies of 64% and 90% respectively. Thus, as temporary immersion bioreactors make it possible to adjust the time spent by the plant material immersed in the liquid nutrient medium, it is possible to avoid significantly the hyperhydricity (Etienne et al. 2006). Otherwise, Ziv (2000) reported culturing clusters of buds and meristems as an alternative propagation system for bioreactors and to surmount hyperhydricity, providing a biomass with limited leaf elongation. Concerning the expression of exogenous contamination, it can often be controlled by good sterile technique; however, endogenous contamination cannot be easily controlled in repeated subcultures.

To avoid both hyperyhidricity and contaminations in *Coffea arabusta*, Afreen et al. (2005) developed a scale-up bioreactor called temporary root zone immersion bioreactor (TRI-bioreactor). The system is based on photoautotrophic micropropagation, in a sugar-free medium, under high photosynthetic photon flux (PPF: 100–150 μmol m^{-2} s^{-1}) and an increased CO_2 concentration (1,100 μmol mol^{-1}) and thus reduce the probability of microbial contaminations. Besides, the system can improve both the growth and the quality of regenerated plants.

For date palm, micropropagation in TIB system is associated with endophytic bacteria that proliferated enormously and affected at once the growth and the rooting ability of shoots after their transfer to rooting medium (Othmani et al. 2009b). The disease can be controlled using shoot clusters pro-tested to be free from endophytic bacteria. Nevertheless, subculture of affected shoots in antibiotic-amended liquid medium, did not avoid this problem. The antibiotics tested were cefotaxim (250 mg l^{-1}) and streptomycin (500 mg l^{-1}).

7.5 Conclusion and Prospective

This chapter demonstrates that liquid culture systems, especially temporary immersion bioreactors are well suited for date palm regeneration through shoot organogenesis with many advantages over other systems with solid or semi-solid medium. It follows from this study that the system can be automated and thus workload and cost can be reduced significantly. Results presented in this chapter illustrate the importance of pilot scale step in scaling-up strategies, and its capacity to motivate important research programs involving a fundamental approach with modern research tools. The process of somatic embryogenesis based on suspension cultures could probably adapted to bioreactors for the large-scale multiplication of date palm, thereby opening the way for both the production and solving problems in accumulating secondary metabolites and the production of artificial seeds. Also, without doubt it will supply valuable target tissue for genetic transformation, cell selection and many other practical applications of this species.

References

Aberlenc-Bertossi F, Noiret M, Duval Y (1999) BA enhances the germination of oil palm somatic embryos derived from embryogenic suspension cultures. Plant Cell Tissue Org Cult 56:53–57

Afreen F, Zobayed SMA, Kozai T (2005) Development of photoautotrophy in *Coffea* somatic embryos enables mass production of clonal transplants. In: Hvoslef-Eide AK, Preil W (eds.) Liquid culture systems for in vitro plant propagation. Springer, Netherlands, pp 323–335

Aitken-Christie J (1991) Automation. In: Debergh PC, Zimmerman RJ (eds.) Micropropagation: technology and application. Kluwer, Dordrecht, pp 363–388

Akula A, Becker D, Bateson M (2000) High-yielding repetitive somatic embryogenesis and plant recovery in a selected tea clone, TRI-2025 by temporary immersion. Plant Cell Rep 18:1140–1145

Albarran J, Bertrand B, Lartaud M, Etienne H (2005) Cycle characteristics in a temporary immersion bioreactor affect regeneration, morphology, water and mineral status of coffee (*Coffea arabica* L.) somatic embryos. Plant Cell Tissue Org Cult 81:27–36

Al-Khayri JM (2005) Date palm *Phoenix dactylifera* L. In: Jain SM, Gupta PK (eds.) Protocols of somatic embryogenesis in woody plants. Springer, Berlin, pp 309–318

Al-Khayri JM (2007) Micropropagation of date palm *Phoenix dactylifera* L. In: Jain SM, Haggman H (eds.) Protocols for micropropagation of woody trees and fruits. Springer, Berlin, pp 509–526

Al-Khayri JM, Al-Bahrany AM (2004) Genotype-dependent in vitro response of date palm (*Phoenix dactylifera* L.) cultivars to silver nitrate. Sci Hort 99:153–162

Ammar S, Benbadis A (1977) Multiplication végétative du palmier dattier (*Phœnix dactylifera* L.) par la culture de tissus de jeunes plantes issues de semis. C R Biol 284:1789–1791

Attree SM, Fowke LC (1993) Somatic embryogenesis and synthetic seeds of conifers. Plant Cell Tissue Org Cult 35:1–35

Bhaskaran S, Smith R (1992) Somatic embryogenesis from shoot tip and immature inflorescence of *Phoenix dactylifera* L, cv. Barhee Plant Cell Rep 12:22–25

Buffard-Morel J, Verdeil JL, Pannetier C (1995) Embryogenèse somatique du cocotier (*Cocos nucifera* L.) à partir d'explant foliaire: étude histologique. Can J Bot 70:735–741

Cabasson C, Alvard D, Dambier D et al (1997) Improvement of *Citrus* somatic embryo development by temporary immersion. Plant Cell Tissue Org Cult 50:33–37

Chabane D, Assani A, Bouguedoura N et al (2007) Induction of callus formation from difficile date palm protoplasts by means of nurse culture. C R Biol 330:392–401

Chand S, Sahrawat AK (2001) Stimulatory effect of partial desiccation on plant regeneration in Indica rice (*Oryza sativa* L.). J Plant Biochem Biotech 10:43–47

Daguin F, Letouzé R (1988) Régénération du palmier dattier (*Phoenix dactylifera* L.) par embryogenèse somatique: amélioration de l'efficacité par passage en milieu liquide agité. Fruits 43:191–194

Daguin F, Letouzé R (1997) Le clonage végétal. Biofutur 163:1–12

De Touchet B, Duval Y, Pannetier C (1991) Plant regeneration from embryogenic suspension cultures of oil palm (*Elaeis guineensis* Jacq.). Plant Cell Rep 10:522–532

Debergh P (1988) Improving mass propagation of in vitro plantlets. In: Kozai T (ed.) Horticulture in high technology era. International symposium on high technology in protected cultivation, Tokyo, pp 45–57

Drira N (1983) Multiplication végétative du palmier dattier (*Phoenix dactylifera* L.) par la culture in vitro de bourgeons axillaires et de feuilles qui en dérivent. C R Acad Sci 296:1077–1082

Drira N, Benbadis A (1985) Multiplication végétative du palmier dattier (*Phoenix dactylifera* L.) par reversion, en culture in vitro, d'ébauches florales de pieds femelles adultes. J Plant Physiol 119:227–235

Dufour M, Berthouly M, Alvard D et al. (1995) El sistema de immersion temporal: un nuevo metodo de propagacion in vitro del cafeto en medio liquido. In: Proceedings of the XVII Símposio Sobre Caficultura Latinoamericana, San Salvador, El Salvador, pp 1–8

Escalona M, Lorenzo JC, González B et al (1999) Pineapple (*Ananas comosus* L. Merr) micropropagation in temporary immersion systems. Plant Cell Rep 18:743–748

Etienne H, Berthouly M (2002) Temporary immersion systems in plant micropropagation. Plant Cell Tissue Org Cult 69:215–231

Etienne H, Dechamp E, Etienne-Barry D, Bertrand B (2006) Bioreactors in coffee micropropagation. Braz J Plant Physiol 18:45–54

Etienne-Barry D, Bertrand B, Vasquez N, Etienne H (1999) Direct sowing of *Coffea arabica* somatic embryos mass-produced in a bioreactor and regeneration of plants. Plant Cell Rep 19:111–117

Feher A, Pasternak TP, Duditis D (2003) Transition of somatic plant cell to an embryogenic state. Plant Cell Tissue Org Cult 74:201–228

Fki L (2005) Application des suspensions cellulaires embryogènes au clonage et à l'amélioration in vitro du palmier dattier *Phoenix dactylifera* L. Thèse de Doctorat. Université de Sfax, Faculté des Sciences de Sfax. 183 pp

Fki L, Masmoudi R, Drira N, Rival A (2003) An optimised protocol for plant regeneration from embryogenic suspension cultures of date palm, *Phoenix dactylifera* L., cv. Deglet Nour. Plant Cell Rep 21:517–524

Hanhineva K, Kärenlampi S (2007) Production of transgenic strawberries by temporary immersion bioreactor system and verification by TAIL-PCR. BMC Biotechnol 7:11

Jens F (1997) Changes in endogenous ABA levels in developing somatic embryos of Norway spruce (*Picea abies* (L.) Karst.) in relation to maturation medium, desiccation and germination. Plant Sci 128:75–83

Jones MPA, Yi Z, Murch SJ, Saxena PK (2007) Thidiazuron-induced regeneration of *Echinacea purpurea* L.: micropropagation in solid and liquid cultures systems. Plant Cell Rep 26:13–19

Kermode AR, Dumbroff EB, Bewley JD (1989) The role of maturation drying in the transition from seed development to germination. J Exp Bot 40:303–313

Kevers C, Coumans M, Coumans-Gilles MF, Gaspar TH (1984) Physiological and biochemical events leading to vitrification of plants cultured in vitro. Physiol Plant 61:69–74

Kosky RG, Silva MF, Pérez LP et al (2002) Somatic embryogenesis of the banana hybrid cultivar FHIA-18 (AAAB) in liquid medium and scaled-up in a bioreactor. Plant Cell Tissue Org Cult 68:21–26

Kumar V, Satyanarayana KV, Indu IP (2006) Stable transformation and direct regeneration in *Coffea canephora* by *Agrobacterium rhizogenes* mediated transformation without hairy root phenotype. Plant Cell Rep 25:214–222

Lowe K, Taylor DB, Ryan P, Paterson KE (1985) Plant regeneration via organogenesis and embryogenesis in the maize inbred line B73. Plant Sci 41:125–132

Maene L, Debergh P (1985) Liquid medium additions to established tissue cultures to improve elongation and rooting in vivo. Plant Cell Tissue Org Cult 5:23–33

Malabadi R, Choudhury H, Tandon P (2004) Initiation, maintenance and maturation of somatic embryos from thin apical dome sections in *Pinus kesiya* (Royle ex. Gord) promoted by partial desiccation and Gellan gum. Sci Hort 102:449–459

Maruyama E, Hosoi Y, Ishii K (2007) Somatic embryogenesis and plant regeneration in yakutane-goyou, *Pinus armandii* Franch.var. *amamiana* (Koidz.) Hatusima, an endemic and endangered species in Japan. In Vitro Cell Dev Biol Plant 43:28–34

Masmoudi R (1999) L'embryogenèse somatique chez le palmier dattier (*Phœnix dactylifera* L.) : induction et maintien des structures embryogènes, caractérisation biochimique. Thèse de Doctorat. Université de Sfax. 113 pp

Monsalud MJ, Mathews H, Litz RE, Gray DJ (1995) Control of hyperhydricity of mango somatic embryos. Plant Cell Tissue Org Cult 42:195–206

Murashige T, Skoog F (1962) A revised medium for rapid growth and bioassays with tobacco tissue cultures. Physiol Plant 15:473–497

Othmani A (1998) Etude de l'influence du facteur hormonal sur la variabilité génétique chez le palmier dattier (*Phoenix dactylifera* L.). DEA, Université de Tunis El Manar, Faculté des Sciences de Tunis, 68 pp

Othmani A (2010) Analyse du polymorphisme isoenzymatique et étude des potentialités de régénération chez 13 cultivars rares et/ou de valeur du palmier dattier (*Phœnix dactylifera* L.), Thèse de Doctorat, Université de Tunis El Manar, Faculté des Sciences de Tunis, 199 pp

Othmani A, Bayoudh C, Drira N, Trifi M (2009a) Somatic embryogenesis and plant regeneration in date palm *Phoenix dactylifera* L., cv. Boufeggous is significantly improved by fine chopping and partial desiccation of embryogenic callus. Plant Cell Tissue Org Cult 97:71–79

Othmani A, Bayoudh C, Drira N, Trifi M (2009b) *In vitro* cloning of date palm *Phœnix dactylifera* L., cv. Deglet bey by using embryogenic suspension and temporary immersion bioreactor (TIB). Biotech Biotech Equipa 23:1181–1188

Paek KY, Hahn SH (2001) Application of bioreactors for large-scale micropropagation system of plants. In Vitro Cell Dev Biol Plant 37:149–157

Piatczak E, Wielanek M, Wysokinska A (2005) Liquid culture system for shoot multiplication and secoiridoid production in micropropagated plants of *Centaurium erythraea*. Rafn Plant Sci 168:431–437

Pliego-Alfaro F, Monsalud MJR, Litz RE et al (1996) Effect of abscisic acid osmolarity and partial desiccation on the development of recalcitrant mango somatic embryos. Plant Cell Tissue Org Cult 44:63–70

Poulain C, Rhiss A, Beauchesne G (1979) Multiplication végétative en culture in vitro du palmier dattier. C R Biol 65:1151–1154

Preil W (2005) General introduction: a personal reflection on the use of liquid media for in vitro culture. In: Hvoslef-Eide AK, Preil W (eds.) Liquid culture systems for in vitro plant propagation. Springer, Netherlands, pp 1–18

Rance IM, Tian W, Mathews H et al (1994) Partial desiccation of mature embryo-derived calli, a simple treatment that dramatically enhances the regeneration ability of *Indica rice*. Plant Cell Rep 13:647–651

Rhiss A, Poulain C, Beauchesne G (1979) *In vitro* culture for the vegetative propagation of date palms *Phoenix dactylifera* L. Fruits 34:551–554

Rival A, Beulé T, Lavergne D et al (1997) Development of photosynthetic characteristics in oil palm during in vitro micropropagation. J Plant Physiol 150:520–527

Sané D, Aberlenc-Bertossi F, Gassama-Dia YK et al (2006) Histocytological analysis of callogenesis and somatic embryogenesis from cell suspensions of date palm (*Phoenix dactylifera* L.). Ann Bot 98:301–308

Sharma DR, Deepak S, Chowdhury JB (1986) Regeneration of plantlets from somatic tissues of date palm (*phoenix dactylifera* L.). Indian J Exp Biol 24:763–766

Teisson C, Alvard D (1995) A new concept of plant in vitro cultivation liquid medium: temporary immersion. In: Terzi M, Cella R, Falavigna A (eds.) Current issues in plant molecular and cellular biology. Kluwer, Dordrecht, pp 105–110

Teisson C, Alvard D (1999) In vitro propagation of potato microtubers in liquid medium using temporary immersion. Pot Res 42:499–504

Teixeira JB, Sondahl MR, Nakamura T, Kirby EG (1995) Establishment of oil palm cell suspensions and plant regeneration. Plant Cell Tissue Org Cult 40:105–111

Thiruvengadam M, Varisai Mohamed S, Yang CH, Jayabalan N (2006) Development of an embryogenic suspension culture of bitter melon (*Momordica charantia* L.). Sci Hort 109:123–129

Tisserat B (1979) Propagation of date palm (*Phoenix dactylifera* L.) in vitro. J Exp Bot 30:1275–1283

Tisserat B, DeMason DA (1980) A histological study of development of adventive embryos in organ culture of *Phoenix dactylifera* L. Ann Bot 46:465–472

Vieitez AM (1995) Somatic embryogenesis in *Camellia spp*. In: Jain SM, Gupta PK, Newton R (eds.) Somatic embryogenesis in woody plants angiosperms. Kluwer, Netherlands, pp 235–276

Vinocur B, Carmi T, Altman A, Ziv M (2000) Enhanced bud regeneration in aspen (*Populus tremula* L.) roots cultured in liquid media. Plant Cell Rep 19:1146–1154

Zhu LH, Li XY, Welander M (2005) Optimization of growing conditions for the apple rootstock M26 grown in TIB containers using temporary immersion principle. Plant Cell Tissue Org Cult 8:313–318

Ziv M (1995) The control of bioreactor environment for plant propagation in liquid culture. Acta Hort 393:25–38

Ziv M (2000) Bioreactor technology for plant micropropagation. Hort Rev 24:1–30

Zobayed SMA, Saxena PK (2003) *In vitro*-grown roots: a superior explant for prolific shoot regeneration of St. John's wort (*Hypericum perforatum* L. cv 'New Stem') in a temporary immersion bioreactor. Plant Sci 165:463–470

Zouine J, El Hadrami I (2007) Effect of 2,4-D, glutamine and BAP on embryogenic suspension culture of date palm (*Phoenix dactylifera* L.). Sci Hort 112:221–226

Chapter 8
Commercial Date Palm Tissue Culture Procedures and Facility Establishment

A. Zaid, B. El-Korchi, and H.J. Visser

Abstract Tissue culture techniques are commonly used for the purpose of mass vegetative propagation in many plant species. Important aspects of *in vitro* techniques are mainly the nature of plant material, suitable nutrient media composition, controlled growth conditions and maintaining strict aseptic conditions. Tissue culture provides a means for rapid mass clonal propagation of desired cultivars, and a mechanism for somatic hybridization and *in vitro* selection of novel genotypes. The need for date palm tissue culture came about because offshoots that are traditionally used for propagation cannot satisfy the urgent needs for large quantities of planting material. Several laboratories worldwide initiated *in vitro* propagation of date palm using either organogenesis or somatic embryogenesis techniques. This chapter includes a description of *in vitro* regeneration of date palm through the organogenesis pathway with a detailed guideline to establish a commercial plant tissue culture facility suitable for mass propagation of various plant species including date palm. Procedures including the source of explants material, disinfection, initiation, multiplication media and incubation conditions are well detailed. The focus made by the authors is on the laboratory initial establishment and nature of needed equipment and supplies. The quality control aspect is also well detailed along with the required safety measures. The establishment implementation plan and estimated budget of both the laboratory and the greenhouses are also presented.

Keywords Acclimatization • *In vitro* • Micropropagation • Somatic embryogenesis

A. Zaid (✉) • B. El-Korchi • H.J. Visser
Date Palm Research and Development Unit, Plant Tissue Culture Laboratory, United Arab Emirates University, 81908, Al Ain, United Arab Emirates
e-mail: zaid@uaeu.ac.ae

S.M. Jain et al. (eds.), *Date Palm Biotechnology*,
DOI 10.1007/978-94-007-1318-5_8, © Springer Science+Business Media B.V. 2011

List of Abbreviations Used

°C	Degrees celsius
2,4-D	2,4-Dichlorophenoxyacetic acid
AC	Air conditioning
AFLP	Amplified fragment length polymorphism
BA	6-Benzyl adenine
BAP	6-Benzylaminopurine
CFU	Colony forming unit
cm	Centimeter
dH	Deutsche Härte
EDTA	Ethylenediaminetetraacetic acid
FOB	Free on board
g	Gram
g/l	Gram per liter
GA_3	Gibberellin A3
GI	Galvanized iron
HEPA	High efficiency particulate absorbing
IAA	Indole acetic acid
kg	Kilogram
KVA	Kilovolt ampere
KW	Kilowatt
l	Liter
l/h	Liter per hour
m	Meter
m^2	Square meter
m^3	Cubic meter
mg/l	Milligram per liter
ml	Milliliter
mm	Millimeter
MS	Murashige and Skoog
NAA	Naphtaleneacetic acid
NOA	Naphthoxyacetic acid
pcs	Pieces
pH	Percentage hydrogen
PVC	Polyvinyl chloride
RAPD	Random amplified polymorphic DNA
RFLP	Restriction fragment length polymorphism
SSR	Simple sequence repeat
UK	United Kingdom
UV	Ultraviolet
USD	United States dollar
V	Volt
v/v	Volume to volume ratio
W	Watt

8.1 Introduction

The techniques used to propagate date palm are seed propagation, offshoot propagation and tissue culture (micropropagation). Seed propagation is by far the easiest and quickest method; however, half of the seed progeny are male and half are female, with no easy and reliable way of determining sexuality at an early stage. Furthermore, female plants originating from seedlings usually produce late maturing fruits, generally of inferior quality compared to established clonal trees. Seedlings also differ considerably with regard to production potential, fruit quality and harvesting time, making them very difficult to market as a single harvest.

Offshoots develop from axillary buds on the trunk of the mother plant; consequently the fruits produced are of the same quality as the mother tree. The time between the differentiation of an axillary leaf bud into an offshoot and growing outwards may be up to 3 years (18–36 months), with another 3–4 years before it reaches the necessary size for its separation and subsequent planting. Although 20–30 offshoots are produced by a tree during its life span, only three or four offshoots are suitable for planting out in a single year and these must still progress through a nursery for 1–2 years before field planting.

Although offshoot propagation supports production of true-to-type trees, it is not commercially practical for the following reasons: First, offshoot production is limited to a relatively short vegetative phase of about 10–15 years. Second, only a limited number of offshoots are produced, depending on the cultivar. Third, offshoot survival rate is low. Fourth, the use of offshoots enhances the spread of date palm diseases and pests. Fifth, offshoot propagation is difficult, laborious and therefore expensive.

The application of tissue culture propagation (micropropagation) of date palm enables large-scale mass multiplication, production of genetically-uniform plants, propagation of healthy selected female cultivars or males having superior pollen with useful metaxenia characteristics, removal of any seasonal effects, and simple and fast exchange of plant material within or between countries, without the risk of spreading diseases and pests. Micropropagation generates a large number of clonal plants in a relatively short period of time. Date palm plantlets can be regenerated from cells or tissues through organogenesis or embryogenesis, the route depends on the plant genotype, explant developmental stage, medium composition and incubation conditions.

Various laboratories worldwide have made attempts to propagate date palm through tissue culture techniques. Success has been achieved by only a few international laboratories. This is due to the nature of the date palm which is a woody species without cambium. Some of these laboratories were established within the last 3 years, while others have been functioning for approximately 15 years. There are currently seven functional laboratories worldwide, found in England, France, Israel, Morocco, UAE, Oman, and KSA (Table 8.1).

Most of these laboratories' efforts were focused on cv. Medjool and recently cv. Barhee with an average sale (FOB) of about USD 24–26 per plant. Delivered plants have only juvenile leaves and still need to be hardened-off by the buyer before field planting. The selling price depends on the cultivar, number of plants ordered and growth stage at delivery.

Table 8.1 List of some international date palm commercial laboratories

Country	Laboratory name	Address
France	Marionnet G.F.A.	211 Rue de Courmemin 41230 Soings, France Tel: +33 254 987 103 Fax: +33 254 987 523
Israel	Rahan Meristem	Propagation Nurseries Kibbutz Rosh Hanikra Western Galilee 22825, Israel Tel: 972 4 985 7100 Fax: +972 4 982 4333
KSA	Al Rajhi Tissue Culture Laboratory	P.O. Box 55155, Riyadh 11534, Kingdom of Saudi Arabia Tel: +966 156 21 308 Fax: +966 554 64 482
KSA	Sapad Tissue Culture Date Palm Co.	P.O.Box 1806, Dammam 31441, Kingdom of Saudi Arabia Tel: +966 382 23 850 Fax: +966 382 10 385
Morocco	Domaine Agricole El Bassatine	B.P. 299, Meknes, Morocco Tel: +212 5 50 0493 Fax: +212 5 50 0730
Oman	Date Palm Research Center	Ministry of Agriculture, P.O. Box 467, Muscat 113, Sultanate of Oman. Tel: +968 2 4696 300 Fax: +968 2 4696 271
UAE	UAE University Date Palm Tissue Culture Laboratory	P.O. Box 81908, Al Ain, UAE Tel: +971 3 7832334 Fax: +971 3 7832472
United Kingdom	Date Palm Developments	Baltonsborough, Somerset BA6 8QG, United Kingdom Tel: +44 1458 850576 Fax: +44 158 851104

8.2 Tissue Culture Procedures

Date palm plantlets may be produced *in vitro* through either asexual (somatic) embryogenesis, i.e. initiation and germination of somatic embryos from callus, or organogenesis, i.e. rooting and division of shoot tips and lateral buds. The organogenesis technique for date palm utilizes meristematic tissues, avoids callus formation and excludes 2, 4-D. Furthermore, growth substances included in the media are used in low concentrations.

Limited observations related to the biochemical regulatory mechanisms of organogenesis coupled with the differences of *in vitro* growth requirements among the numerous date palm cultivars necessitate further research to improve this technique. In general, the efficiency of organogenesis in date palm is considered low. This is due to low numbers of explants responsive to *in vitro* culture, a long time required for the initiation phase, low multiplication rates and strong influence of cultivar and medium. Despite avoiding a callus phase, genetic variation can occur as a result of low subculture frequency or excessive use of growth regulators to maximize shoot formation.

Organogenesis procedures consist of four consecutive steps: meristematic bud initiation, multiplication, elongation and rooting. The success of cultures is highly dependent upon the success of the initiation step. Furthermore, various problems encountered during *in vitro* propagation may originate at the initiation step.

8.2.1 Source of Explant Material

Offshoots, 5 kg in weight, separated from adult palm trees of selected cultivars are the preferred sources of explant material (Fig. 8.1a). Using a knife, tapestry maker, or sometimes a hatchet the bottom of young palms (and offshoot roots if existing) are carefully cut off and discarded. This should result in an offshoot 15–50 cm in length (Fig. 8.1b). The best period for starting the *in vitro* culturing of date palm offshoots is from the onset of flowering (it varies from one cultivar to another; however, between April and July is a safe time in the Northern Hemisphere). When separating offshoots from the mother plant, care must be taken to avoid damaging the offshoot base or roughly pulling it off. In such cases, microorganisms such as bacteria can easily penetrate the *heart* of the offshoot (the meristematic area). Explants from damaged offshoots usually become contaminated regardless of disinfection procedures.

When the meristematic area (soft white part) is reached, careful handling is necessary to avoid breaking the offshoot tip (mainly between the lignified part of the base and the soft tissue). Hence, one must leave a small lignified base in order to avoid such damage. The final size of the shoot tip area isolated for disinfection should be about 3–4 cm in width and 6–8 cm in length (Fig. 8.1c).

8.2.2 Disinfection Technique

The soft part, with its lignified base, is then soaked in an anti-oxidant solution of 150 mg/l citric acid and 100 mg/l ascorbic acid until culturing to avoid tissue browning. Disinfection of offshoot *heart* or tip is based on the use of 4 g/l Mancozan (or any other fungicidal solution) for a duration of 20 min, then it should be rinsed three times with autoclaved distilled water and dipped in a 9% NaOCl (sodium hypochlorite) solution (with few drops of Tween 20) under a weak vacuum, for 5 min, with intermittent breaks of the vacuum, and then under normal pressure for 20 min.

8.2.3 Explant Culture

Under an air laminar flow hood, without rinsing it off, the offshoot tip is dissected and put on an initiation medium. The bottom of young leaves, mostly the axillary

Fig. 8.1 Date palm explant preparation for *in vitro* culture. (**a**) Offshoot mother plant, (**b**) Preparing offshoots, (**c**) Offshoot sterilization, (**d**) Piece of tissue for introduction, (**e**) Introduction stage

part between the leaf and underlying tissue, is cultured. The top ends of these leaves are cut off and discarded.

At the bottom of the young leaves there are some very small axillary buds that are often visible. It is common to find some axillary buds when the older leaves are cut out. All these buds are cultured, and after cutting into two parts.

When the cluster of small leaves near the apex is reached, it may be difficult, without the use of magnification, to remove individual leaves. So, at this point of the process, they are removed several at a time. At the end, the cluster of leaves becomes so small that it is better to cut it at right angles in two or four parts and placed into culture. One offshoot as described here may give young leaves or axillary buds, sufficient for starting a batch of 15–25 culture tubes (Fig. 8.1d, e).

8.2.4 *Initiation Medium*

The success of organogenesis is entirely dependent on the initiation success. Furthermore, various problems encountered at other steps may have their origin at the initiation phase. With the initiation, Murashige and Skoog nutrient medium is at full strength; the high level of ammonium salts causes several problems with date palm mainly concerning the formation of vitrified buds.

The objective is to stimulate a flush of young buds from meristematic areas. All growth regulators (NAA, IBA and 2IP) used should have their concentration adjusted between 0.1 and 3 mg/l. Each cultivar of date palm requires slight modifications concerning the balancing of growth regulators. The pH adjusted to 5.7 after adding the agar prior to autoclaving.

Normally, after 4–6 months depending on the cultivar, the base of cultured young leaves gives some growth signs that budding has started (not complete buds yet) (Fig. 8.2a,b). At this stage, cultures are transferred to multiplication medium (Fig. 8.2c).

The response to the medium composition depends on several factors including genotype, offshoot age and weight, time of introduction to culture, and browning phenomenon. Delay in the subculturing process (even 1 week) may result in severe browning and consequently death of the culture. Explants with an elongated and enlarged leave, originating from a primordial leaf isolated from shoot tip meristematic area, are cut and discarded to save nutrients to tissue potentially more conducive of bud and shoot development.

8.2.5 *Browning Phenomenon*

During the course of growth and development, date palm tissues release phenolic compounds (quinones) that can accumulate in the medium. Such phenols have profound negative physiological effects on the initiation and multiplication stages. To alleviate browning problems researchers have engaged in presoaking of explants in an anti-oxidant solution of 150 mg/l of citric acid and 100 mg/l of ascorbic acid, employed small explants and frequent re-culturing and added activated charcoal. Charcoal, however, reduces the availability of hormones and thus must be supplied at high concentrations to compensate for this limitation.

8.2.6 *Contamination and Mite Problems*

It is worth noting that the entire production of a tissue culture laboratory could be lost as a result of bacterial contamination. The use of large-spectrum antibiotics at

Fig. 8.2 Some *in vitro* culture stages of date palm. (**a**) and (**b**) Buds at initiation media, (**c**) Buds at multiplication stage

very low concentrations to control bacteria and fungi is advised, but in a careful manner to avoid selection of resistant strains of pathogenic bacteria.

Another important source of general contamination are microscopic mites which are airborne, or carried by clothing, utensils, or explants. Mites pass easily from the outside environment (field) to inside the laboratory through air conditioner filters which have relatively large pores. Doubling such filters and frequent filter cleaning is highly recommended. Air-borne mites can be controlled by installing air sterilizers. Laboratory workers, or any person entering the facility after visiting field plantations, should do so only after showering and changing clothes.

Plant materials should always be checked for mites under a stereomicroscope; when under suspicion, treating with a mitecide is highly advisable. When a great percentage of explants contaminated by fungi are observed, the source of such a spread may be due to mites because they can survive disinfectants. Contamination appears on the surface of the agar at the base of explants or on the explants, where the mites can be seen under a microscope. If mite spread is detected in the laboratory, all cultures must be removed and the room fumigated. When mites are detected early, contaminated area and shelf surfaces within a 50 cm diameter are rubbed with mitecide and all the flasks in that area autoclaved.

8.2.7 *Incubation Conditions*

At the start of initiation, all cultures are incubated in total darkness, but after 3–4 months a low light is provided by 16 h photoperiods. The culture room temperature is adjusted to 27°C ± 1 during 16 h of light, and 22°C ± 1°C during the 8 remaining dark hours. Cultures favor thermoperiods over a constant temperature. Transferring cultures to a fresh medium each month, especially during the first few months, is recommended even if growth changes were not noticed.

8.2.8 *Multiplication Medium*

Explants showing bud proliferation are transferred to multiplication media that have the same composition as the initiation medium except for the growth regulators, where the type and concentrations are modified. For this stage, NAA, NOA, IAA, BA and Kinetin are used at 0.5–5 mg/l (Fig. 8.3a).

Leaves start sprouting in this medium, but many fail to give elongated shape plantlets, rather they give rise to plantlets in a rosette shape. Such leaves are cut and removed as a measure of quality control of multiple bud formation. Cutting gently between buds is recommended while dividing the explant covered with buds into 2–4 pieces. Such pieces are again placed on multiplication medium for further multiplication and subsequent subculturing at a monthly interval. When a well-developed plantlet seems to dominate growth, it is necessary to separate it from the remaining budding tissues, because enlarged buds (plantlets) seem to inhibit other parts of budding tissues.

8.2.9 *Rooting and Elongation of Date Palm In Vitro Plantlets*

When enough buds are obtained and most of them show a resemblance to a young plantlet, it is necessary to transfer the budding tissues onto rooting–elongation medium supplemented with NAA, BAP, Kinetin and GA_3 at 1, 0.5, and 0.5 mg/l,

Fig. 8.3 Some *in vitro* culture stages of date palm. (**a**) Plants at multiplication stage, (**b**) Plants at Elongation stage, (**c**) Plants at rooting stage

respectively (Fig. 8.3b,c). It is highly recommended to keep young plantlets on this medium for only 10–15 days. A longer period on this medium causes leaf vitrification (grass-like appearance). Then the budding tissues are again transferred on a *swelling* medium which is the same as the multiplication medium with the same hormones, except that the sucrose level is adjusted at 100–150 g/l. Elongated shape plantlets may be left on this swelling medium for several months without transfer where they grow vigorously and often produce roots. No special rooting medium is used since date-palm plantlets root easily and evenly. Without roots, they may be transferred to air conditioned greenhouses as un-rooted cuttings and dipped in a solution of rooting hormones made of 0.01 mg/L NAA for 2 min.

8.2.10 Acclimatization and Soil Transfer

Several techniques have been used to acclimatize date plantlets and improve their survival during establishment under greenhouse conditions. The effectiveness of these methods depend upon ambient conditions, and most methods have involved environmental modification.

Date palm plantlets are ready for transplanting only when they gain the following characteristics:

- Two to three healthy and enlarged leaves with no curling phenomenon;
- A shoot length of at least 10–1 cm from stem base to the highest point of the leaves;
- A shoot base with an onion bulb-like form (also called pear-shaped crown);
- A well developed root system with an average of 5 cm in length. Adventitious rooting is obtained by trimming these primary roots to 1–1.5 cm in length and reculturing the plant to an agar nutrient medium containing 0.01/0.1 mg/l NAA without charcoal.

Plants are then rinsed in distilled water to remove adhering agar and residual sucrose. A spray with Benlate solution at 0.5% (or any wide spectrum fungicide) is important since it protects the plant from fungal attack. In order to achieve the above, and consequently produce a well pre-acclimatized date plant that will survive the transplanting stress, it is recommended that the following be ensured:

- Do not transplant any plant until it gains the abovementioned characteristics;
- Enhance a root-elongation process by using auxins at the last *in vitro* stage;
- Increase the light intensity during the last 4–6 weeks; and
- Create an artificial osmotic stress (at the nutrient medium level).

The transplanting operation should be done as quickly as possible to avoid plant dehydration and avoid root damage as far as possible. The soil medium must always be sterilized and usually consisting of equal parts peat and vermiculite (v/v) mixture. Sterile sand with a large grain size could also be added to improve drainage. Bark is to be avoided because it dries out rapidly and causes a water-stress situation. To summarize, the substrate should be a well-drained one, yet with good water-retention capacity. The adequate pH to work with should be about 6.5. Plastic pots (7.5–12.5 cm), jiffy peat pots or trays (25 plants; in case of commercial production) are often used for date-palm transplanting.

Plants are immediately irrigated with 50% Hoagland's solution or 10% MS solution before their incubation into a micro-tunnel located in an environmentally controlled glasshouse (or a large plastic tunnel). These environmental conditions will ensure a high relative humidity (90–95%) and a constant temperature ±25–26°C during the night. Bottom heating of the micro tunnel (±23°C) was found to be very helpful.

To ensure a high survival rate, date palm tissue culture-derived plants should be adapted to gradually decreasing humidity and gradually increasing light. The light intensity is important during the first 3–4 weeks in the glasshouse around (10,000 lx) with a 16 hr photo period. Benlate is to be applied to the foliage once a week, and irrigation using 10% MS solution (or 50% Hoagland) every 3rd or 4th day depending on the hygrometry level of the micro-tunnel.

Four to six weeks later, the plastic of the micro-tunnel is gradually opened to decrease humidity and prepare the plants to adapt to the large glass house (or tunnel) conditions which preferably should have a fog system. Plantlets are now ready to be transplanted into larger plastic bags.

It is worth mentioning that at all stages; water should never be sprayed from the top of the plant. Plants could stay in the glass house (or a tunnel) for a period of 3–4 months before their transfer to a less environmentally controlled nursery, which is usually at the farm level, for their further hardening-off process.

8.2.11 *Nursery Hardening of Acclimatized Date Plants*

Plants are transferred to larger bags (7–10 l capacity) with an adequate substrate, usually sand (soil), vermiculite and gravel at a ratio of 1:1:1, respectively. Transplanting should be done properly with no disturbance to the root system. Original substrate around the roots should stay intact. Plants are then left in the nursery for approximately 8–12 months depending on surrounding conditions and care given, until most of them reach the 4-pinnae leaf stage.

The nursery size and type are related to the number of plants to be hardened-off. An average size of 150 m^2 will be adequate for 1,000 plants. An ultra-violet resistant shade net of 80% is recommended during the first 6 months. During summer, the top of the nursery should have a double layer of the shade net for heat insulation purposes. The nursery should be well located (close to several trees to benefit from their shade) but also in a protected area to avoid sandstorms and severe wind. A water tap should be installed inside the unit for easy irrigation and the unit must be enclosed to avoid animals getting in and eating the plants.

Irrigation is an important factor and must be implemented once a week in winter and at least twice a week during summer. Water should never be sprayed on top of the plant; soil is to be mounded around the base of the plant so water cannot get into its heart.

Fertilization is to be applied once per month: apply 5 g of ammonium sulphate/plant bag (5% nutrient solution; thus 15 kg diluted per 63 l water for 650 plant). Apply 120 ml of solution per plant bag.

Control of diseases and pests is also recommended and the use of Benlate (or any other large spectrum fungicide) has proven to be highly efficient. Foliar spray of Benlate is to be applied every 3–4 weeks.

Close monitoring is advised as mistakes could be disastrous. If all above recommendations and advices are respected, the date grower could expect a survival rate of 90–95%.

8.3 Laboratory Establishment

8.3.1 *Surface Area and Layout*

The total laboratory surface area is estimated at approximately 1,350 m^2 and consists of two technical departments, one administrative unit and annexes, as presented in Tables 8.2, 8.3 and 8.4. Additional areas for hardening greenhouses and nurseries would also be required.

Table 8.2 Surface area for plant tissue culture facilities

Sections	Area (m^2)	Remarks
Media preparation section	216	6 Rooms
Transfer and selection section	172	5 Rooms
Growth room section	288	6 Rooms
Offices and administration section	216	5 Offices and 1 Meeting room
Other facilities	188	Toilets (2), Storage room (1), Rest and cloths change rooms (2), Kitchen (1) and Electrical room (1)
Corridors and passages	270	Several corridors inside and outside the aseptic area
Total	1,350	

Table 8.3 Divisions of the media preparation section of the tissue culture laboratory

Type of room	Number	Dimension (m × m)	Surface area (m^2)
Transfer rooms (2)	4	6 × 6	71
Selection room	1	6 × 4	24
Growth room	6	6 × 8	288
Autoclaving room	1	6 × 6	36
Media preparation room	1	6 × 6	36
Chemicals storage room	1	6 × 6	36
Media and glassware storage	1	6 × 6	36
Weight and fridges room	1	6 × 3	18
Distillation room	1	6 × 3	18
Glassware cleaning room	1	6 × 6	36
Rest and cloths change room	1	6 × 6	36
Total			635

Table 8.4 Details of the administrative unit and related services

Type of room	Number	Dimension (m × m)	Surface area (m^2)
Director's office	1	6 × 8	48
Meeting room	1	6 × 6	36
Secretarial office	1	6 × 4	24
Accounting	1	8 × 6	48
Standard offices	2	6 × 3	36
Storage room	1	6 × 3	18
Toilets	2	6 × 3	36
Electrical cabinet room	1	8 × 2	16
Corridors	5	Various	270
Total			532

8.3.2 General Laboratory Specifications

- External walls and laboratory roofs are to be heat-insulated.
- Surfaces of floors, walls and partitions should be constructed from impermeable materials that have no toxic or adverse effects on staff or handled plants.

- Walls and partitions should have smooth surfaces up to a height that appropriately suits laboratory activities.
- Windows should be of a double-glass type and with the ability to insulate heat.
- All doors of the laboratory are to be well sealed and insulated from drastic temperature changes. Doors should also have smooth, non-absorbing surfaces, and be easy to clean.
- All main entrances of the laboratory are to be equipped with positive air pressure in order to avoid air from outside the aseptic area entering the facility.
- Ceilings, overhead fittings and air conditioning systems should be constructed and refined to prevent the shedding of particles and minimize the build-up of dirt, dust and moisture.
- The main laboratory entrance doors as well as the growth chamber doors are to be equipped with security locks (preferably with numeric combinations).
- All doors and passages/corridors should be large enough to allow passage and circulation of laboratory trolleys (±2 m width).
- Floors of the unit are to be constructed of sheet vinyl (similar to hospital floors) allowing adequate drainage and continuous cleaning/washing with detergents and bleaches.
- All benches/working surfaces are to be water and chemical resistant (marble or granite is recommended). They must also be equipped with sinks with a hot and cold water tap feed and drainage system.
- All sinks must be designed to resist the chemical actions of acids and alkalis.
- All painted surfaces inside the facility must be of excellent quality, covered with a soft PVC coating that tolerates water cleaning (waterproof paints) and also resists moisture and fungal growth. This will assure optimal sterile conditions.
- The laboratory is to be equipped with a safety generator (500 KVA) to ensure power to the laboratory in case of a general power failure.
- The nutrient media preparation room is to be equipped with benches in three directions and should contain cupboards and drawers according to laboratory standard specifications and needs, along with the electrical outlets (220 V) for the installation of needed equipment. The media preparation unit is to have 330 V outlets at the side without benches.
- Transfer rooms are to be equipped with Air Laminar Flow Hoods, and need to be equipped with electrical and gas outlets.
- The air-conditioning system of the entire facility is to be equipped with good quality filters (such as 0.2 μm HEPA filters) in order to ensure the purity of air within the facility.
- Power voltage is to be 220 V, unless specified otherwise.
- Water delivery pumping system should be of the booster-pressure type.
- Extension of the required surface area should always be possible by simply adding other rooms to the existing ones.
- A high standard of cleanliness requires not only a high standard of work by technicians and scientists alike, but also adequate daily cleaning of the laboratory.

The facility can be divided into two distinctive parts, namely septic- and aseptic areas. The septic areas would include structures such as the greenhouses, nurseries, maintenance rooms and general storage areas. These are characterized by the lack of a highly-controlled environment in terms of sterilization needs with a high bacterial and fungal presence. On the other hand, the aseptic (laboratory) area is characterized by having a high degree of sterility, constant maintenance of cleanliness, strict aseptic protocols for staff and possesses a highly-controlled environment regarding temperature and lighting. The aseptic area commonly consists of structures such as the growth chambers, transfer rooms, nutritive media preparation section, sterilization chambers, washing rooms and special storage areas. For a commercial date palm tissue culture laboratory, the following basic installations are required:

• Media preparation room	• Growth rooms
• Distillation room	• Chemicals storage room
• Glassware cleaning room	• Weighing room
• Cold room	• Dark room
• Media and clean glassware storage room	• Reception room
• Autoclaving room	

8.3.3 Media Preparation Room

The media preparation room serves as laboratory space where the four major nutritive agar-based media: initiation, multiplication, elongation and rooting media, upon which the plant cultures are grown, is prepared. The operational activities include the basic media preparation, heating and filling procedures. This room is to be furnished with the following:

• All necessary chemicals	• Filling (pumping) devices
• Water availability	• Work benches
• Necessary glassware	• Electrical outlets (3) of 220 V

The whole section measures 16.35 m long, 7.4 m wide and 3.27 m high. The section is divided into different rooms. The extra autoclave room measures 5.75 × 5.65 m, and houses an additional autoclave (Brand: Webeco; 800 l capacity) that is used facultatively as a backup sterilization system.

The media preparation room measures 6.82 × 5.65 m and is furnished with ample bench top workspace. A built-in bench top stretches along the wall between the two entrances, and extends 70 cm from the wall. Various drawers and storage cabinets are located beneath the bench top. The room also possesses a centrally located bench top *island* with additional workspace, with cabinet and drawer storage below. The *island* measures 158 × 55 cm, and has four sinks with hot and cold water tap feeds at one end. All of the built-in benches are made from high quality aluminum frames and plastic paneling, while the top working surface is made from marble slabs. The various solutions that make up the different nutritive media are prepared here. Additionally the prepared off-shoot cuttings are also sterilized here before their introduction into *in vitro* culture. For this reason the media preparation room should possess electrical supply sockets in order to house equipment such as pH-meters, magnetic stirrers for solution and media preparation, vacuum pumps for sterilization purposes and drying ovens to dry glassware and utensils.

The cooking area measures 16.35 m in length by 1.75 m in width. The cooking section has two entrances, one leading to the rest of the aseptic facility, and the other one serving as an emergency exit leading towards the reception and entrance area. In the cooking area the nutritive media is heated to help dissolve the added agar powder and sucrose. The liquid media is then dispersed into the various glassware containers using a pumping device, before sterilization. It possesses two separated working bench tops along the one wall, constructed in the same manner as the rest of the media section. The first bench top measures 310 cm by 70 cm and houses the gas-powered cookers as well as the media dispensers (pumps). The second bench top work space measures 10.63 m by 70 cm, and houses an additional media dispenser and provides work space for media labeling and packing, utensil wrapping and consumables preparation.

8.3.4 Distillation Room

The distillation room should be provided with ample sinks, hot and cold water supplies, which could be distilled water and/or de-mineralized water. Well designed workbenches are essential. The room may form part of the glassware cleaning room.

- Benches in this room are to be only one-sided with a sink and inlet of water (hot and cold) and electricity supply sockets (220 V).
- All walls are to be covered with white ceramics.
- Purification units are recommended to be of the Millipore (MilliQ) Brand.
- Have a water drainage system for the purifying and distillation units.
- The room is to be equipped with at least three electrical outlets of 220 V.

8.3.5 Glassware Cleaning Room

The glassware cleaning room should be provided with several large sinks designed to resist the corrosive action of acids and alkalis; be supplied with hot and cold water and large workbenches. Additionally, a washing machine and oven to dry instruments and glassware could be housed. Wire-mesh baskets and trolleys are useful for transferring washed and dried glassware and instruments to the media and glassware storage or transfer rooms.

- The room is to have two large benches equipped with several sinks of large size, deep enough to submerge the test tubes and other glassware and vessels. These sinks must have cold and hot water supplies. Drainage systems are to be proportional to the size of these sinks.

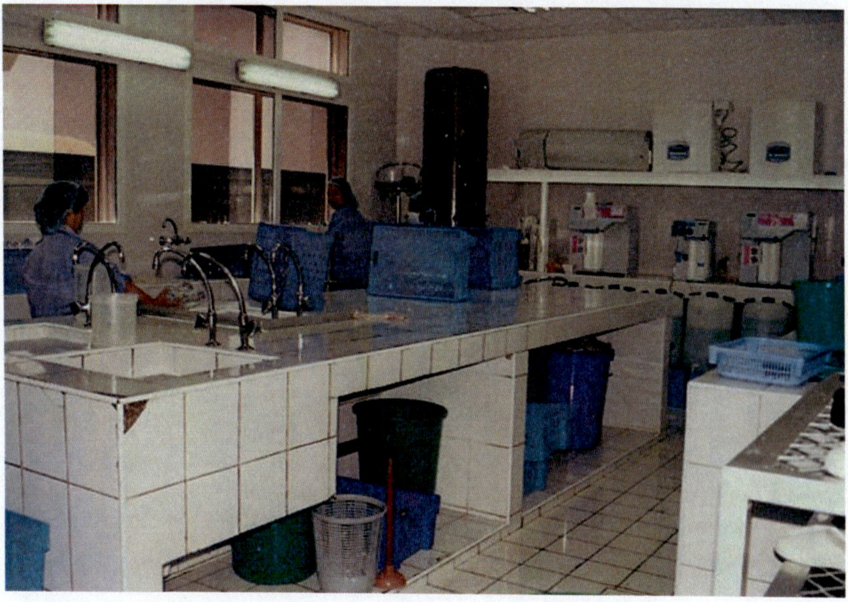

Fig. 8.4 Glassware cleaning room

- All walls are to be of white ceramics.
- Several points of water drainage/sewage.
- Several electrical supply sockets of 220 V.
- Have space to house one or two glassware cleaning machines (Fig. 8.4).

The washing room serve as the location where all cleaning and washing of glassware and disposal of old and contaminated cultures (after autoclaving) takes place. The room measures 8.7 × 5.5 m and is 3.27 m high, with two entrances, one door leading into the rest of the aseptic facility, and the second entrance with two separate doors leading to the outside of the building to serve as an emergency exit. The floor and walls of the washing room are finished with high quality, heavy duty ceramic tiles that are smooth, non-porous, washable and chemically resistant. The room should have a solid ceiling with a smooth finish. The room should support two washing areas/surfaces. One washing area is a continuous surface running along the length of the walls, 80 cm from the walls and at a height of 70 cm above the floor. The second washing area is located on an isolated *island*, measuring 3 m in length, 1.5 m in width and 70 cm in height, in the center of the room. All of the washing surfaces are tiled using the same tiles as the floor and walls. All of the washing sinks are built-in and also tiled, and they measure 60 cm in width (sinks on the *island* measure 55 cm in width), 85 cm in length and 30 cm in depth.

The washing surface running along the length of the room nearest the emergency exit supports four washing sinks, of which the middle two lie adjacent to each other and the other two are placed 80 cm on each side. The washing surface running along the wall nearest to the main entrance supports three washing sinks, which are 80 cm apart.

The working surface running along the farthest wall (width) of the room, houses the water purification system, as previously mentioned. The system consists of two water purification units (Brand: Millipore/MilliQ) that are connected to a water feed and two reservoir tanks (one for each unit). The two reservoirs tanks are located in the corners of the washing room, one atop the working surface, and the other reaching from the floor to the ceiling. The two water purification units should be housed inside a glass and aluminum cabinet for protection.

8.3.6 Cold Room

The cold room serves as a refrigerated walk-in storage area attached to the media preparation room, for the chilled storage of pre-prepared stock solutions and some of the nutrient media. The room could also be used for hormone and vitamin storage. The temperature of the room should be fixed at approximately 5°C ± 1°C. The cold room should measure 2.9 × 2.64 m, and must have a fitted alarm button inside for emergency use.

8.3.7 Media and Clean Glassware Storage Room

- Must be well equipped with storage shelves all around the room (Fig. 8.5 a,b);
- These shelves are to be 30 cm above the floor surface, and should consist of four shelves that are 50–60 cm apart;
- Walls are to have electrical outlets (220 V);
- The room is to be equipped with ultra-violet (UV) lights, connected to a timer in order to sterilize the room;
- This room as well as the six other growth chambers are to be equipped with temperature control (±26°C ± 1°C) and an alarm system in the case of air-conditioner failure.

8.3.8 Autoclaving Room

Before the introduction or cultivation of date palm explant material on a nutrient medium, the medium should first be sterilized to eliminate both fungi and bacteria. Sterilization is mostly done in autoclaves, sometimes by filtering and occasionally chemically or by irradiation. The autoclaving room (Fig. 8.5c) should have the following technical specifications:

- All walls should be constructed from heavy duty white ceramics;
- One side of the room should have a bench with a sink with cold and hot water tap feeds;
- Three 220 V and four 330 V electrical sockets are necessary for the autoclaves to be installed in this room;

Fig. 8.5 Tissue culture facilities. (a) Media storage room, (b) Media filling area, (c) Autoclave rooms

- A water-drainage system must be provided for the autoclaves;
- HEPA filters are also to be installed in this room;
- One steam supply (low and high pressure steam);
- Windows in this room are designed to be opened only in case of emergency.

Successful sterilization may be achieved by:

- Physical destruction of micro-organisms by dry hot air, steam or irradiation (such as gamma-rays). Pressurized steam (wet heat) will destroy all micro-organisms if correctly dosed.
- Chemical destruction of micro-organisms by sterilizing agents such as ethylene-oxide, alcohol or hypochlorite solutions.

- Physical removal of micro-organisms by filtration and/or washing. After sterilization, the nutrient media, glassware, etc. should not be kept in contact with open air in the laboratory, but in sterile areas, a closet for example which has been surface-sterilized with 96% alcohol.

The most common method of sterilizing nutrient media is autoclaving. An autoclave is a container with an automatically closing cover. In practice it is an apparatus in which sterilization is done under high pressure and heat. Nutrient media are put under high pressure for 20 min at 120°C in saturated steam. There are horizontal type autoclaves which open from the front and vertical autoclaves which open from the top. The horizontal autoclave is easier to operate, but more expensive.

Autoclaving is done within a temperature range of 115–135°C. The following factors determine successful sterilization: duration, pressure, temperature and quality of the steam as well as the size (contents) of the object to be sterilized. Advantages of autoclaving are: a quick and simple method and destruction of viruses without adsorption. Disadvantages of autoclaving are: possible pH-changes may occur and components may denature and lose their effectiveness. Basic programs for autoclaving are (timing started when effective temperature is reached):

- Test tubes with nutrient medium: 15 min at 121°C or 20 min at 115°C.
- Flask containing 100–500 ml nutrient medium: 30 min at 121°C.
- Flasks containing 500–5,000 ml nutrient medium: 30–40 min at 112°C.
- Empty test tubes, flasks, etc. and filter paper: 30 min at 130°C.
- Nutrient media and empty objects such as glassware, paper, etc. should be sterilized separately.

The autoclaving room houses large-capacity autoclaves that are used to sterilize utensils, consumables, glassware, and prepared growth media (already poured into glassware) as well as to kill any contaminated cultures (bacterial and fungal) before their final disposal. The room itself measures 6.75 m in length, 5.22 m in width and is 3.00 m high. The wall and floor finishes should be of good quality, highly durable ceramic tiles that are non-porous and washable. It is also recommended that sufficient floor-drainage should be provided in an area of the room that will not be occupied by autoclaving equipment (to dispose of excess water spilled after the autoclaving procedure).

The autoclaving room should house two large capacity autoclaves, and a third smaller autoclave. The first two recommended autoclaves are LTE Touchclave System 1,185 l capacity models (192 cm high), while the recommended third smaller autoclave is a Falcon 400 l capacity model (159 cm), also from LTE Scientific.

8.3.9 Transfer Room

Transfer rooms should be kept as sterile and dust-free as possible. The rooms should be entered only through an air-lock. It should be fitted with sliding doors and airtight windows to ensure that no drafts occur. The rooms should be air conditioned and equipped with laminar flow cabinets.

The cutting (preparation) of a piece of tissue (an organ etc.) in a sterile petri-dish or between layers of sterile filter papers is usually performed in a laminar airflow cabinet, since the risk of bacterial or fungal contamination will be considerably higher in non-sterile circumstances, which will result in a high percentage of lost cultures. The laminar airflow cabinet provides a sterile working surface.

- The four transfer rooms can each accommodate four Air Laminar Flow Hoods each (a total of 16 hoods); and each hood can accommodate two technicians; yielding a work force of 32 people at the same time. Essentially, a laminar flow cabinet has a small motor to circulate air, which first passes through a coarse filter, and subsequently through a fine filter, and ultra clean air, free of fungal or bacterial contaminants, flows through the work area. All contaminants, such as hairs, salts, and flakes, are removed by the ultra clean air flow, and while the cabinet is in operation a completely aseptic environment is maintained on the working surface. The cabinet should never be placed facing a door or a window which is used frequently.
- Electrical outlets (220 V) in six points in each transfer rooms;
- Air conditioning system is to be provided with HEPA filters;
- Each room is to be equipped with UV lights with a timer in order to sterilize the room during the night;
- Each door of the four rooms is to be provided with positive air pressure;
- Windows must be fully sealed with no openings;
- Each room is to be fitted with cupboards to store instruments for inoculation or transfer;
- The rooms should be kept clean; no infected material or contaminated media should be brought into the rooms;
- The doors should be kept closed (no air movement);
- Tables and walls should be cleaned with 96% alcohol;
- Floors should be regularly disinfected. All traces of spilled nutrient medium should be removed. Objects not belonging in the sterile room are to be kept outside;
- Hands and arms must be habitually and thoroughly washed;
- Clean laboratory-coats must be used;
- Used instruments are placed in a disinfectant solution immediately after use.

Inside the culture transfer rooms the *in vitro* plant cultures are continuously transferred onto fresh media and advanced through different propagation stages, under sterile conditions. Introductions of new cultivars are also done in one specifically designated transfer room. All of the culture transfer rooms are virtually identical. The rooms must have durable, high quality wall and floor finishes that are easily washed and chemically resistant. It is suggested to make use of either high quality heavy duty ceramic tiles, or heavy duty sheet vinyl flooring or a combination of both. The transfer rooms have a solid ceiling with a smooth surface. The physical room itself measures 6.76 m long, 5.17 m wide and 3.27 m high.

General lighting in the Transfer Rooms are provided by 12 sets of four fluorescent tube lights (PHILIPS TL-D 18 W/54-765), set up in a 60×60 cm configuration. Specific lighting for working inside the laminar flow cabinets/ hoods is provided by built-in lights of the cabinets themselves. Each transfer room has 8 gas access valves/ points for Bunsen burners which are operated inside the laminar flow cabinets. All gas pipes and valves are monitored by a gas leakage detection system. The use of a zonal electronic Eds Shield System is suggested, which controls gas leakage detection in all rooms. This system is also connected to an automatic shut-down system for the main gas supply valve. Additionally, all rooms are equipped with two fire detectors, one for smoke detection and the other for temperature. The smoke detectors are connected to an alarm system according to specifications and requirements of the local Civil Defense Department.

All of the laminar flow cabinets suggested for use are manufactured by ADS Laminaire (France) (Fig. 8.6). The specific model of laminar cabinets is identifiable through its serial number: 504579. The specific filter used by the laminar cabinets is an Absolute filter M 6/18, also manufactured by ADS Laminaire. There is no specific brand of Bunsen burners used inside the laminar cabinets, but it is recommended that micro-burners or torch-burners not be used, since they consume large amounts of flammable gas. The gas piping used inside the transfer rooms are not brand specific, but must be approved by the local Civil Defense Department.

Temperature control in the transfer rooms is maintained through a central air conditioning unit (a split A/C system is not recommended), and makes use of a HEPA filter system. The air-conditioning system is coupled to a control panel/ thermostat where the desired temperature is set. The temperature is maintained at a comfortable level for the technicians working in the rooms (about 18°C).

In addition to the transfer rooms, a selection (or screening) room is needed. The selection room serves as a contamination screening area, before plant cultures are moved to transfer rooms for handling. The purpose is to identify any cultures with bacterial or fungal contamination and to eliminate these cultures, avoiding the further spread of such contamination. The dimensions of the selection room are 6.76 m long by 5.17 m wide and 3.27 m high. In terms of floor-, wall- and ceiling finishes the selection room is identical to the transfer rooms. The only differences between the two rooms are that the selection room does not need a gas supply, and that the selection room should possess ample built-in bench top workspace.

The key to a good selection room is adequate lighting. For general lighting purposes the selection room has the same setup as the transfer rooms. Supplemental illumination (if needed) can be provided by ordinary bench top lamps (no specific recommended brand). It is therefore advisable to construct the selection room with enough 220 V electrical sockets at bench top height. The temperature-control/ air-conditioning system is identical to that described for the transfer rooms. The surface of the bench top workspace should be constructed of a durable and washable material such as marble or granite.

8.3.10 Growth Rooms

The growth rooms serve as the structures housing the date palm cultures through the entire *in vitro* propagation cycle beginning with initiation and continuing through

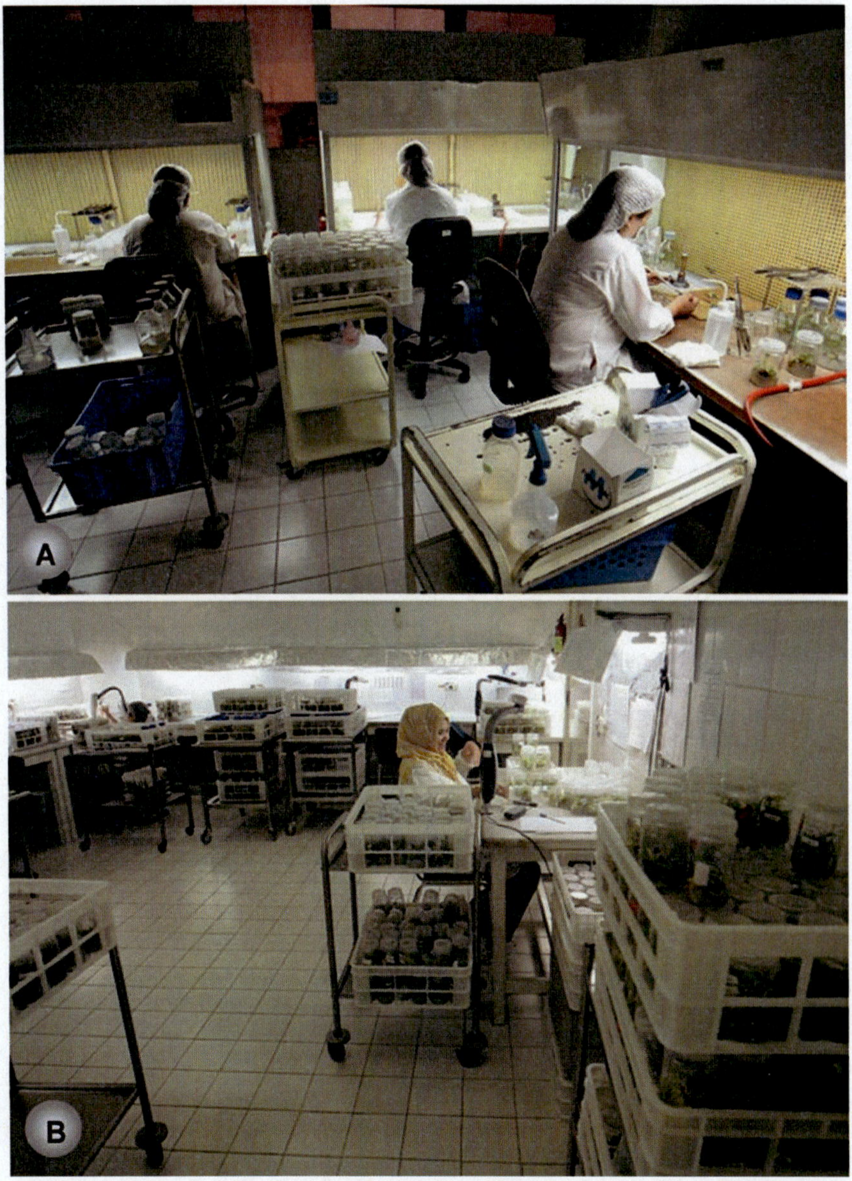

Fig. 8.6 Tissue culture facilities. (**a**) Sub-culturing room (**b**) Selection and Transfer room

multiplication, elongation and finally rooting (Fig. 8.7). It provides optimal growing conditions for the cultured plants, and is therefore equipped with light, temperature and humidity controls. A growth room may house several thousand cultures in various stages of the propagation process, or it may house only cultures of a specific stage,

Fig. 8.7 Growth room shelves full of date palm cultures

such as the initiation growth room (or dark room), which houses all of the newly introduced date palm cultures. The growth rooms' temperature and light/dark photoperiod are strictly controlled according to the various propagation stages. The specific dimensions and setups of the growth chambers play an important role in their efficiency.

Growth rooms should be isolated completely so that constant ambient temperature can be maintained. Air conditioners and heaters (a reversible system) are essential. Illuminated shelves should be installed in a way to prevent a build-up of hot air in the shelves due to the lamps; the rooms should be well ventilated. With regard to lighting, control must be exercised over intensity, quality and photoperiod. Hence, all regimes should be controlled by thermostats and time clocks to maintain uniform conditions. A temperature recorder to detect fluctuations in temperature is necessary in each growth room. It is desirable to have an emergency power point connected to a generator in order to maintain both lighting and temperature in the culture rooms in case of a major power breakdown which may destroy all cultures.

- The rooms must not have windows;
- Air conditioning of these rooms is of extreme importance and the temperature should be 27°C ± 1°C during the day time (with light) and 24°C ± 1°C during night time (16/8 h light / dark photoperiod);
- An extra split AC system is to be provided for each growth room. This is in case of a failure in the main/centralized air-conditioning system;
- Each of the six growth rooms is to be equipped with HEPA filters;
- Entrance to these rooms is to be designed with positive air pressure;

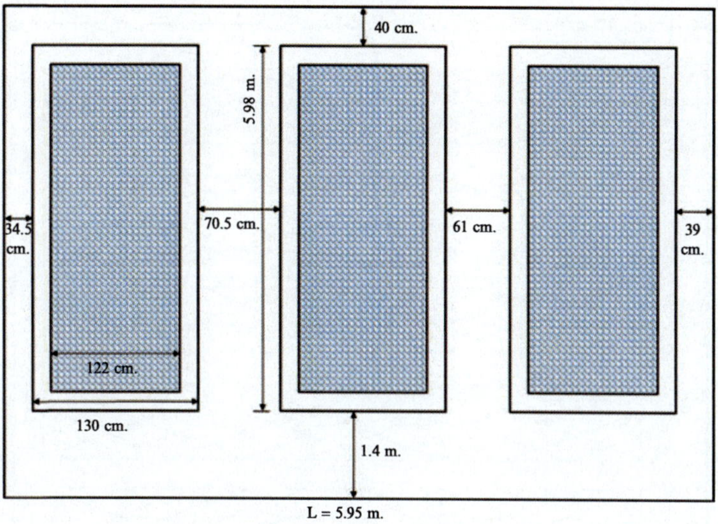

Fig. 8.8 Representation of the floor plan of a suggested growth room, depicting dimensions and shelf spacing

- Growth rooms should be secured (preferably with a combination lock);
- Lighting in these rooms is to be directly connected to the air-conditioning system in order to turn off the light whenever the AC system is not working;
- Shelves are to be equipped with fluorescent bulbs (cool white) specific to the growth of plants and able to provide 2,500–3,000 lx (except in the case of a elongation/rooting room which needs approximately 10,000 lx); lighting should be uniform for all shelves and well distributed inside the room;
- Each room is to be equipped with aluminum shelves;
- Electrical cabinets and ballasts are to be installed outside the aseptic area. This is to reduce heat production inside the laboratory and also to minimize the entering of maintenance staff into the aseptic facility;
- All regimes should be controlled by thermostats and time clocks to maintain uniform conditions;
- A temperature recorder to detect fluctuations in temperature is necessary.

Figure 8.8 represents the floor plan of a suggested growth room with all the respective measurements. Each of the growth rooms are, for all practical reasons, identical. Each growth room can house up to 90,000 cultures. The length of the room is 7.67 m, width 5.98 m and height 3.27 m. Each growth chamber possesses a double door in the middle of the short sidewall (5.98 m).

Each growth room contains three wrought aluminum alloy shelves. Each of the shelves has the following dimensions: total height 2.26 m, width 1.31 m and length 6.1 m. Each shelf has five racks (storage levels), which are made from aluminum grids. Each rack is divided into three equal sections (aluminum grids) each measuring 1.92 m in length. Between each section (grid) there is a wrought aluminum beam

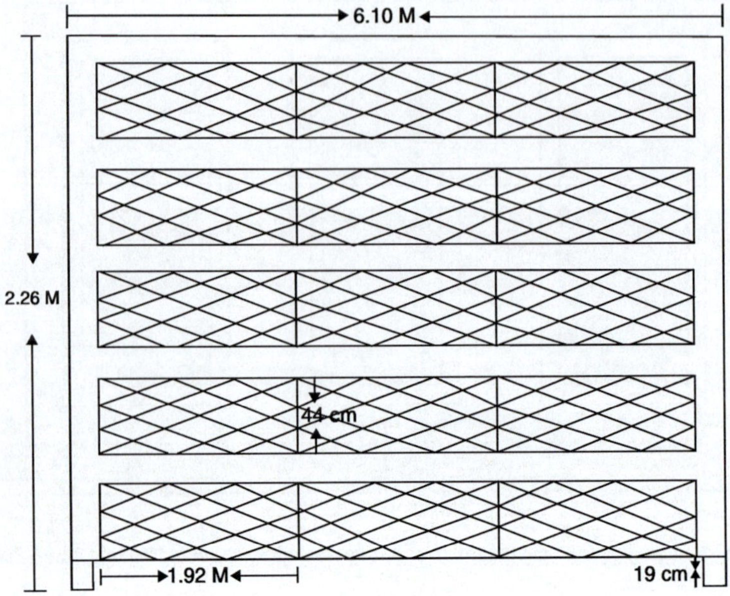

Fig. 8.9 Side-view representation of a growth room shelf

forming part of the structural frame of the entire shelf. The first (lowest) rack is located 19 cm from floor level. Each of the racks is separated by 44 cm intervals. The total shelve surface of each growth room is about 120 m^2.

The spacing of the individual shelves within the growth chamber is shown in Fig. 8.8, and is as follows: The first shelf is placed 33 cm from the long side wall, 25 cm from the back wall and 1.32 m from the front wall (with the double door entry). The second shelf stands directly facing the double door entry. It stands 71 cm spaced from the first shelf, 25 cm from the back wall and 1.32 m from the front wall (or entry door). The third and last shelf stands 61.5 cm spaced from the middle shelf and 38.5 cm from the other side wall. It is also 25 cm removed from the back wall and 1.32 m from the front wall. A detailed side view depicting shell specifications is provided in Fig. 8.9.

Doors should be constructed of a heavy duty material, able to withstand physical damage from culture trolleys. The doors should also have well-sealing bottom- and side-hinges, with a hard plastic flap covering the joining slit to prevent the entry of dust and small foreign objects. The floor of the room should be finished using heavy duty sheet vinyl flooring, or may be covered with a heavy duty epoxy coating, both of which are non-absorbent, durable, with a smooth surface and easy to clean. The growth chambers possess a solid ceiling with a smooth finish. Drop ceilings with suspension systems are not advisable, since this promotes a buildup of dirt and dust which eventually sheds onto the cultures, posing a contamination risk.

Lighting inside the growth chamber can be divided into two categories, namely critical growth lighting and non-specific, general purpose illumination. For general illumination purposes two sets of four fluorescent tube lights (Philips TL-D 18 W/54-765) are used. Each set of four lights are set up in a 60×60 cm configuration.

The critical lighting for the growth of the *in vitro* cultures is provided by numerous fluorescent tube lights. The specific lights used in the growth rooms are either Philips TL-D 90 De Luxe Pro 36 W/950 or Osram Lumilux De Luxe Daylight L 36 W/954 fluorescent lamps (depending on availability and price), and they are 1,200 mm in length. The tube lights are mounted parallel to each other and to the width of the shelves. The tube mountings are 40 cm from the rack below, and 4 cm from the above overhead rack. The individual tubes are 20 cm apart and there are 30 tube lights per rack level. Thus, 150 tube lights per shelf [5 racks × 30 tube lights], and 450 tube lights total per growth room (3 shelves × 150 lights). The ballasts regulating the flow of electrical current through each of the fluorescent bulbs are located in a separate, designated room. All of the growth lights for each growth room are connected to an adjustable automatic timer (Brand: Hager) which maintains a 16 h/8 h light/dark photoperiod.

Supplemental growth lighting for the *in vitro* cultures is provided by four mobile stands of growth lights. These lights are only used facultatively. One mobile, supplemental lighting stand is placed in each corner of the growth room. A full-length, normal cosmetic mirror is mounted against the wall or in the corner closest to the supplemental lights, to improve light distribution. Each mobile supplemental light stand consists of an aluminum frame with five mounted Colombo Illuminazione /50 W 240 V RX7S-24MD lamps.

Temperature control inside the growth rooms are maintained using a central air-conditioning system that is coupled to a thermostat (Model: SIEMENS Landis and Staefa REV II). The growth room temperature is maintained at 27–28°C. The use of a specific brand of air conditioner is not important, and preferences of certain brands such as York, Mitsubishi and Hitachi are at the discretion of the user. The air-conditioning system makes use of a primary filter. It is a built-in 0.2 μm HEPA filter that comes standard with the air conditioner. A heating system coupled to the thermostat is also installed for use during very cold periods. Each growth room also hosts a thermohygrograph (no specific brand), which constantly and independently monitors growth room temperature and relative humidity.

8.3.11 Chemical Storage Room

This storage area is to be equipped with storage cupboards and shelves along three of the walls; positive air pressure needs to be maintained; electrical outlets (4) of 220 V; no windows are to be installed in this room. The room should measure 6.86 m long, 6.75 m wide and 3.27 m high.

8.3.12 Receiving Room

Because of the need to work in semi-sterile conditions, this room should be separated from the aseptic laboratory area since mainly external plant material grown under field conditions (e.g. date palm offshoots) will be dealt with there. The receiving room should be furnished with a sink, hot and cold running water, and a cupboard to store sterilizing agents, glassware and instruments. This room should be kept very clean because of the high presence of possible contaminants and non-sterilized plant material.

8.3.13 Weighing and Refrigeration Room

The weighing room should be provided with a special weighing table, constructed of granite or another natural stone. It should be shock proof and the top precisely horizontal. The room houses all electronic scales and balances for weighing of dry reagents used for making up solutions. The room measures 3.78 × 2.9 m, and it also serves as storage space for opened chemical containers. Along one wall it possesses a bench top work space extending 60 cm from the wall, similar to the media preparation room (indicated in Fig. 8.2). It also houses refrigerators and freezers (4°C and −20°C respectively), where temperature sensitive chemicals and solutions are stored. The microscope and optical equipment could also be housed in this room.

8.3.14 Dark Chamber

The dark chamber, which serves as the initiation culture growth room, should be equipped identical to the growth rooms with two exceptions (no lighting and a smaller room area). A separate, isolated space may be used in one of the six growth rooms.

8.3.15 Air and Traffic

The areas located between the bathroom/changing room and the autoclaving room, the area between the chemicals storage and the transfer rooms, and the area around the laboratory main entrance, are all to be equipped with positive air pressure systems. Every room in the laboratory has its own unique function and unnecessary traffic between rooms should be avoided in order to reduce the risk of widespread and also occasional contamination.

8.3.16 Utilities and Water Quality

The estimated monthly electricity need is 205,000 KW. The monthly water requirement of 4,000 m^3 for the entire facility can be proportioned as 200 m^3 per day for an average of 20 working days per month. The facility's daily water consumption of 200 m^3 can be split into two equal parts: One half for the laboratory (aseptic) area

Table 8.5 Water quality specifications adequate for use in plant tissue culture facility

Property	Specification limit
Water temperature	Max. 25°C
NH_4	0–0.2 mg/l
NO_2	0–0.5 mg/l
NO_3	10–30 mg/l
CI	10–30 mg/l
$KMnO_4$ requirement	6–12 mg/l
Urochrom	0–0.1 mg/l
PO_4	0.01–0.02 mg/l
SO_4	25–50 mg/l
Fe	0.05–0.1 mg/l
Mn	0.01–0.03 mg/l
CaO	50–100 mg/l
MgO	50–100 mg/l
1° dH (German hardness)	100 mg/l CaO
Total hardness	5–10 ° dH
Carbon hardness	5–8 ° dH
Non–carbon hardness	5–10 ° dH
pH value	7.1–8.00
H_2S	0 mg/l
Extract after evaporation	500–1,000 mg/l
Total CFU (colony forming units) / ml water	100
Coliformae	Nil
Escherichia coli presence	Not demonstrable

and the other half for the external (non-sterile) area and general irrigation of the surrounding site. The water quantity used by the laboratory can be described as follows: 30 m^3 as hot water for glassware cleaning, rinsing and for the autoclaving process. Autoclaves should be equipped with their own steamer to produce the needed steam; 50 m^3 to be used as tap water at room temperature for the cleaning of glassware and the laboratory; 20 m^3 is used by the double-distillation/water purification systems to obtain pure water for media preparation (about 1,000 l). The yield of pure water is approximately 1.0 m^3 out of 20 m^3; a rate of 5%.

Water analysis provided for facilities performing micropropagation via *in vitro* tissue culture should be free of solid suspensions such as sand and sludge and should correlate to the specifications as provided in Table 8.5. With regards to the facility's run-off sewage, the waste water that is drained is mostly mixed with Chlorox (NaOCl) and traces of micro-elements (N, P, K, Zn and Cu).

8.3.17 Personnel

By using *in vitro* techniques for mass propagation, a tissue culture facility can easily produce up to 100,000 date palms annually. An aspiring mass propagation tissue

Table 8.6 List of culture vessels for the annual production of 100,000 plants

Item	Quantity
Test tubes, dimension 24 × 200 × 1.2 mm, pyrex, rimless	150,000 pcs
Test tubes plastic closures, transparent natural autoclavable	150,000 pcs
Racks, electro polished stainless steel, each holds 48 tubes	4,000 pcs
370 ml Pickle Jar, model JR0060-10	57,600 pcs
Twist-off cup, diameter 63 mm, polypropylene transparent	45,000 pcs
Media bottles, 500 ml capacity, with polypropylene screw cap	30,000 pcs

culture laboratory should initiate a comprehensive feasibility study in order to plan the multiplication of high-quality cultivars and the training of local staff. Both the multiplication and the training should take place in the completed laboratory. During the training period the staff should be completely familiarized with all techniques and procedures. The training should focus on the acquisition of practical knowledge, skills and basic techniques and consist of basic working procedures, media preparation and filling, sterilization procedures, and administrating a tissue-culture laboratory.

8.4 Equipment and Supplies

A list of culture vessels needed which should be of autoclavable grade (Table 8.6); List of laboratory equipment needed (Table 8.7); Basic nutrients and reagent chemicals (Table 8.8); and List of other accessories and supplies (Table 8.9);

8.5 Quality Control

One of the major drawbacks of mass propagation using plant tissue culture techniques is the appearance of undesired plant phenotypes (somaclonal variants). Somaclonal variants are clearly different from the mother plant. The production of genetically uniform and stable plants in plant tissue culture is, therefore, of prime importance in the efficiency and success of *in vitro* mass-propagation. Understanding the nature and mechanism underlying somaclonal variation is consequently a vital prerequisite for developing new methods to minimize its occurrence.

There are on-going disputes amongst date growers, technicians and scientists about the true-to-typeness of plants produced *in vitro*. It is worth mentioning that tissue culture-derived plants of many species are subject to somaclonal variation in particular, and to genetic variation in general. The origin of meristematic buds is important in relation to the uniformity of organogenesis-derived plants. If buds are pre-existent, then plants will be true-to-type, but if they are newly formed, the possibility exists for plant variation.

Factors causing variations in plant tissue culture are: the technique used for propagation; nature of mother plant (possible chimeras); type of growth regulators

Table 8.7 List of laboratory equipment for the annual production of 100,000 plants

Equipment	Quantity
Autoclave, horizontal, 1,200 l capacity	2
Automizer, to spray sterilizing agent in transfer room	25
Balances, normal and sensitive with accessories	6
Burners with electronic ignition	25
Cart, stainless steel, 20 gauge, 3 shelves, 16½ × 26½ × 32¼ in.	15
Cart wire with baskets and removable drip pan	15
Ceramic lab plates	25
Deep freezer for long term storage of stock solutions	2
Dish washing machine (for test tubes and bottles).	2
Electric power generator, 500 KVA, for emergency	1
Gas taps	25
Hot plate, gas	4
Illuminated shelves with lamp fixture for Gro lux lamp.	18
Ladder, 3-step with casters	6
Laminar air flow, ADS horizontal H 18 type	20
Lamps, fluorescent, TLD 36/W 950, 600 pcs. per growth room	4,000
Media distribution unit, 10–500 ml	4
Media mixer, vortex – genie	2
Microscope, light and a binocular	2
Oven to dry instruments and glassware	2
Oven, sterilization of instruments	2
pH meter, corning 12, with buffer solutions	3
Pipet washing assembly	1
Quadrupeds stands / Duran stainless steel squared, 175 × 175 mm	25
Refrigerator for storage of stock solutions and chemicals	4
Safety thermostats and timers for photoperiod regulations	6
Stirrer, magnetic, without heating	4
Stirrer, magnetic, heating	4
Temperature and humidity recorder	6
Vacuum pump, for disinfection of explant material	2
Vacuum source for filter sterilization	2
Venting systems to remove chemical fumes	4
Water bath, 5–100°C	2
Water distillation, demineralization unit, 4 l/h, Corning 440207	2

used; type of explant used (ploidy gradients: apex to root); age of culture (older than 1 year); medium composition; and incubation conditions.

Most commercial laboratories strive to ensure the true-to-typeness of the produced date plant material, and have various quality control measurements in place. Various molecular biology-based techniques are used to attest the conformity of the plants (histocytology, isoenzyme analysis, RFLP, screening with molecular genetic markers such as RAPDs, AFLPs or SSRs). In most cases, genetic fingerprinting (such as the AFLP technique) is the most reliable scientific method of choice, but we feel that the field response is the only reliable way to confirm if the palms derived from tissue culture are true-to-type to the mother plant.

Table 8.8 A list of basic nutrients and chemical reagents used in date palm tissue culture

Common name	Formula	Quantity
2,4-Dichloro phenoxy acetic acid	2,4 D ($C_8H_6O_3Cl_2$)	1 × 100 g
6-Furfuryl aminopurine	Kinetin ($C_{19}H_{22}O_6$)	10 × 10 g
Adenine hemisulfate salt	$C_5H_5N_5.1/2.H_2SO_4$	4 × 1 kg
Aga agar powder; Gum agar	Plant TC Tested	10 × 25 kg
Ammonium nitrate	NH_4NO_3	3 × 25 kg
Benzylamino purin	BAP ($C_{12}H_{11}N_5$)	10 × 5 g
Biotin (vitamin H) 99%	$C_{10}H_{16}N_2O_3S$	2 × 500 mg
Boric acid	HB_3O_3	1 × 5 kg
Calcium chloride	$CaCl_2.2H_2O$	10 × 1 kg
Calcium nitrate 4-hydrate	$Ca(NO_3)_2.4H_2O$	25 kg
Cobalt chloride	$CoCl_2.6H_2O$	1 × 100 g
Cupric sulfate penta hydrate	$CuSO_4.5H_2O$	1 × 250 g
D-Pantothenic acid calcium salt	$C_9H_{16}NO_5 1/2.Ca$	1 × 100 g
Gibberellin acid	GA3 ($C_{12}H_{11}N_5$)	7 × 10 g
Indole-3-Acetic acid	IAA ($C_{10}H_9NO_2$)	3 × 25 g
Inositol (Meso)	–	10 kg
Isopentenyl adenine	2-I P ($C_{10}H_{13}N_5$)	2 × 25 g
L-Glutamine	$C_5H_{10}N_2O_3$	7 × 1 kg
Magnesium sulphate.7hydrate	$MgSO_4.7H_2O$	25 kg
Manganese sulfate one hydrate	$MnSO_4.H_2O$	5 kg
Naphthalene acetic acid	NAA ($C_{12}H_{10}O_2$)	3 × 25 g
Naphtoxy acetic acid	NOA	3 × 25 g
Nicotinic acid	$C_6H_5NO_3$	2 × 100 g
Potassium nitrate	KNO_3	6 × 25 kg
Pyridoxine HCl (B6)	$C_8H_{11}NO_3.HCL$	1 × 25 g
Sodium dihydrogen ortho-phosphate	$NaH_2.PO_4.\ H_2O$	10 × 1 kg
Thiamine Hcl (B1)	$C_{12}H_{17}ClN_4OS.HCl$	2 × 100 g
Vitamin C	$C_6H_8O_6$	1 × 100 g
Zinc sulfate	$ZnSO_4.7H_2O$	5 kg
Ammonium nitrate	$NH_4.NO_3$ AR	1 × 1 kg
Potassium nitrate	KNO_3	1 × 1 kg
Calcium chloride dihydrate	$CaCl_{2.}2H_2O$	1 × 1 kg
Magnesium sulphate.7hydrate	$MgSO_{4.}7H_2O$	1 × 1 kg
Potassium dihydrogen ortho-phosphate	$KH_{2.}PO_4$	1 × 1 kg
Sodium dihydrogen ortho-phosphate	$NaH_2PO_4.H_2O$	1 × 1 kg
Boric acid	H_3BO_3	1 × 1 kg
Ethylene diaminetetra acetic acid disodium salt di hydrate	Na_2 EDTA	1 × 1 kg
Ferrous sulfate	$FeSO_4.7H_2O$	1 × 1 kg
Manganese sulfate one hydrate	$MnSO_4.H_2O$	1 × 1 kg
Cupric sulfate penta hydrate	$CuSO_4.5H_2O$	1 × 1 kg
Sodium molybdate	$Na_2MoO_4.2H_2O$	1 × 1 kg
Zinc sulfate	$ZnSO_4.7H_2O$	1 × 1 kg
Cobalt chloride .6hydrate	$CoCl_2.6H_2O$	1 × 1 kg
Potassium iodide	KI	1 × 1 kg
Sodium hydroxide pellets	NaOH	1 × 1 kg

(continued)

Table 8.8 (continued)

Common name	Formula	Quantity
Potassium hydroxide pellets	KOH	1 × 1 kg
Hydrochloric acid 37%	HCl, 37%	1 × 1 kg
pH 7 buffer conc.	–	10 packets
Sucrose	–	50 kg / month
Ethanol (95% purity)	–	50 gal/month

Table 8.9 A list of accessories, consumables and supplies

Items	Quantity
Alcohol lamp	2 pcs
Aluminum foil (25 m per roll)	50 rolls
Asbestos gloves	2 prs
Aseptic filling bell	1 pc
Aseptic filling bell, Bellco 5611-0038	10 pcs
Beaker, 10 ml, 100 ml, 250 ml, 400 ml, 1,000 ml	10 each
Bottle, reagent, glass stopper, 250 ml, 500 ml, 1,000 ml	10 each
Brush, buret	3
Brush, test tube	3
Buret, Kimble 17024, 25 ml	10
Buret, Kimble 17080 F, 250 ml	6
Cheesecloth	5
Clamp holder	10
Clamp, pinch	10
Clamp, screw	10
Clamp, versatile, vinyl jaws, large and micro	10 each
Clock with timer	1
Corkborer, stainless steel, size # 2 or 3	2
Cotton, non-absorbent	2 rolls
Counter, hand tally	1
Coverglass	1,000
Filter paper, Whatman, , 9 cm #42, #50	5 boxes each
Flask, delong 50 ml, 125 mm, 250 mm	100 each
Flask, Erlenmeyer, narrow mouth, 1,000 ml, 2,000 ml, 4,000 ml	5 each
Flask, Erlenmeyer, narrow mouth, 50 ml, 125 ml, 500 ml	50 each
Flask, media storage, 500 ml, Bellco 2536-00500	10
Flask, media-storage, 2,000 ml	1
Flask, volumetric, 10 ml , 50 ml, 100 ml, 500 ml, 1000 ml	5 each
Forceps, 9½ Pott-Smith, dressing	4
Forceps, micro-dissecting (Clay-Adams # 6441)	4
Forceps, standard dissecting	4
Funnel, powder, Nalgene, 65 – mm diameter	10
Funnel, produce	1
Graduated cylinder, 10 ml , 100 ml, 250 ml, 1,000 ml, 2,000 ml	5 each
Haemocytometer (Maemacytometer)	2
Lamp, Gro Lux regular bulbs and VHO bulbs	1,800 each
Markers, permanent, Edding 3000	10

(continued)

Table 8.9 (continued)

Items	Quantity
Membrane filter syringe and accessories	5
Microslide	500
Morton stainless-steel closure, with fingers, Bellco 2005-00015	100
Morton stainless-steel closure, with fingers, Bellco 2005-00025	200
Needle, dissecting	4
Needle, inoculating, platinum	4
Paper towels	
Parafilm	5 boxes
Petri dish, 15 – × 100 – mm, Pyrex	100
Pipet box	2
Pipet, Mohr, 1 ml , 10 ml, 5 ml	10 boxes each
Pipet, serological, large orifice, 1 ml, 5 ml, 10 ml	5 each
Prunning shear	1
Ring stand and clamp set for rinse water	1
Ring stand, rectangular base, 36	10
Scalpel blade, surgeon's # 10, # 11, # 15	100 boxes each
Scalpel handle, surgeon's # 7, #11	100 boxes each
Scalpel, solid steel	4
Scoop	10
Spatula, micro, double bladed, stainless steel	10
Test tube, 18 × 150 –mm	24
Thermometer, 0–150°C	1
Tray, cafeteria, autoclavable	10
Tubing, silicon, Dow-Corning, 1/4 × 3/8	50 ft
Wash bottle, polyethylene, 250 ml	5

Should a tissue culture facility wish to implement a technique such as AFLP (Amplified Fragment Length Polymorphism) analysis, as a type of quality control to ensure produced plants are true-to-type, they can follow one of two options. The first option would be to establish an internal fingerprinting facility that would routinely test randomly selected produced plants. This would mean the facility would have to set up an available room inside the aseptic area as a molecular biology laboratory. The laboratory would have to be equipped with expensive, specialized analytical equipment (Table 8.10), depending on what analysis technique is chosen, and expert technicians hired to manage it. Although the initial establishment costs are high, this is the most cost-effective option in the long-term.

The second approach would be to routinely send random plant samples for analysis by a reliable analytical facility offering genetic fingerprinting. Such facilities are difficult to find, but eliminates the need for expensive equipment, highly-trained technicians and additional work. However, over the long term this option would certainly become the most expensive option, since analytical facilities charge premium prices for such analyses.

Table 8.10 List of equipments, chemicals for molecular biology laboratory

No.	Equipment / Chemicals	Quantity
1	Refrigerated centrifuge / Microcentrifuge	1 unit each
2	Water bath	1 unit
3	Tissue homogenizer/Disruptor (Tissuelyzer)	1 unit
4	Nanodrop photospectrometer (DNA quantification)	1 unit
5	Plant DNA extraction kit	(Pack of 96) – as needed
6	PCR thermal cycler	1 unit
7	Flaked – ICE making machine	1 unit
8	AFLP Plant mapping kit for small plant genomes (applied biosystems)	(100 reactions) as needed
9	PCR components (TAQ, DNTPs PCR buffer, $MgCl_2$)	As needed
10	DNA capillary sequencer (8 capillaries)	1 unit
11	Computer with data analysis software (Genemapper)	1 unit
12	Sequencing reagents (Polymer, Buffer, Formamide, Standards)	As needed

8.6 Greenhouses and Nurseries

8.6.1 General Acclimatization

Acclimatization presents challenges at least equal to those posed by the initiation of cultures because it marks the end of artificially-controlled cultivation and the beginning of autonomous plant growth.

The culturing of plant tissue *in vitro* with almost 100% relative humidity inside the culture vessel can lead to various abnormalities in the plant structure. Plants of many species produced *in vitro* often show morphological, physiological, structural and biochemical differences from ones produced conventionally. These include reduced cuticular wax deposits, altered leaf anatomy, excessive water loss and stomatal abnormalities in comparison with greenhouse-grown plants. It is worth mentioning that the loss of viability is almost always attributed to poor control of transpirational water loss from the date plants and their heterotrophic nature.

Even when gradual hardening off has been concluded, poor survival and slow growth of date plantlets is commonly reported. Such a low survival rate (sometimes reaching below 50%) is caused by several factors. These are mainly the physiological immaturity of transferred plantlets, an inadequate root system, incorrect irrigation schedule and the lack of technical care at the *in vitro* laboratory stage.

Several techniques have been used to acclimatize date plantlets and improve their survival during establishment under greenhouse conditions. The effectiveness of these methods depends upon ambient conditions, and most methods involve environmental modification.

Date-palm plantlets are ready for transplanting only when they gain the following characteristics: 2–3 healthy and enlarged leaves with no curling phenomenon, a shoot length of 10–15 cm from the stem base to the highest point of the leaves, a shoot base with an onion bulb-like form (also called pear-shaped crown), and a well developed root system with an average length of 5 cm. Adventitious rooting is obtained by trimming these primary roots to 1–1.5 cm in length and reculturing the plant to an agar nutrient medium containing 0.01–0.1 mg/l NAA without charcoal.

Plants are then rinsed in distilled water to remove adhering agar and residual sucrose. A spray with Benlate solution at 0.5% (or any wide spectrum fungicide) is important since it protects the plant against fungal attack. In order to produce a well pre-acclimatized date plant that will survive transplanting stress, it is recommended that the following be ensured: avoiding transplanting any plant until it gains the above mentioned characteristics, enhancing the root-elongation process using auxins at the last in vitro stage, increasing the light intensity during the last 4–6 weeks and creating an artificial osmotic stress (at the nutrient medium level).

The transplanting operation should be done as quickly as possible to avoid plant dehydration and root damage. The soil medium must always be sterilized and usually consisting of 1 part peat: 1 part vermiculite (v/v) mixture. Sterile sand with a large grain size could also be added to improve drainage. Bark is to be avoided because it dries out rapidly and causes a water-stress situation. To summarize, the substrate should be well drained, yet with good water retention capacity. The average working pH should be about 6.5. Plastic pots (7.5–12.5 cm), jiffy peat pots or trays (25 plants; in case of commercial production) are often used for date palm transplanting (Fig. 8.10a,b).

Plants are immediately irrigated with 50% Hoagland's solution or 10% MS solution before their incubation into a micro-tunnel located in an environmentally controlled glasshouse (or a large plastic tunnel). These environmental conditions will ensure a high relative humidity (90–95%) and a constant temperature ±25–26°C during the night. Bottom heating of the micro-tunnel (±23°C) has been found highly beneficial.

To ensure a high survival rate, date palm tissue culture-derived plants should be adapted to gradually decreasing humidity and gradually increasing light. The light intensity is important during the first 3–4 weeks in the glasshouse, around 10,000 lx with a 16 h photo period. Benlate is to be applied to the foliage once a week, and irrigation using 10% MS solution (or 50% Hoagland) every 3rd or 4th day depending on the hygrometry level of the micro-tunnel.

Four to six weeks later, the plastic of the micro tunnel is gradually opened in order to decrease humidity and prepare the plants to adapt to the large glasshouse (or tunnel) conditions which preferably should have a fogging system. Plantlets are now ready to be transplanted into larger plastic bags.

It is worth mentioning that at all stages water should never be sprayed from the top of the plant. Plants could stay in the glass house (or a tunnel) for a period of 3–4 months before transfer to a lower environmentally-controlled nursery, which is usually at the primary production level, for further hardening-off (Fig. 8.11).

Fig. 8.10 (a) Transplanted plants in a micro-tunnel, (b) Transplanted plants in plastic bags

Fig. 8.11 Tissue-culture produced plants at different stages. (**a**) VP1 stage (2 months after acclimatization); (**b**) VP2 stage (6 months from VP1 stage), (**c**) VP3 stage (5 months from VP2 stage)

8.6.2 Nursery Hardening-Off of Acclimatized Date Plants

Plantlets received from a laboratory are usually about 35–45 cm long with 4–5 leaves among which are up to 2 pinnae leaves (called also permanent leaves). The plant must have a thick shoot system and the base must be similar in form to a large onion bulb (pear-shaped). As stated above, the plant must have a well-developed root system.

Transportation of these plants must be done in a proper manner and plants must preferably not be stacked on top of each other to avoid stem breakage and/or leaf damage. Transport must preferably also be in a single stage and if plants/truck are to stay over somewhere, it must be in a shaded area; watering should not be neglected if transport takes several days.

It is recommended that, upon reception of this material by the date grower, plants are transferred to larger bags (7–10 l capacity) with a suitable substrate, usually sand (soil), vermiculite and gravel at a ratio of 1:1:1, respectively. Transplanting should be done properly with no disturbance to the root system. The original substrate around the roots should stay intact. Plants are then left in the nursery for approximately 8–12 months depending on surrounding conditions and care given, until most of them reach the 4-pinnae leaf stage. The date grower is advised to co-ordinate the purchase and the hardening-off period, to ensure that planting can be timely performed (during February/March in the Southern Hemisphere and September/October in the Northern).

The nursery size and type is related to the number of plants to be hardened-off. An average size of 150 m^2 will be adequate to accommodate 1,000 plants. A ultra-violet resistant shade net of 80% is recommended during the first 6 months. During summer, the top of the nursery should have a double layer of the shade netting for heat insulation purposes. The nursery should be well located (close to several trees to benefit from their shade) but also in a protected area to avoid sandstorms and severe wind. A water tap should be installed inside the unit for easy irrigation and the unit must be fenced to avoid animals getting in and feeding on the plants.

Irrigation is an important factor and must be implemented once a week in winter and at least twice a week during summer. Water should never be sprayed on top of the plant; soil is to be mounded around the base of the plant so water cannot get reach its heart. Fertilizer is to be applied once per month: apply 5 g of ammonium sulphate per plant bag through 120 ml of made-up solution. Control of diseases and pests is also recommended and the use of Benlate (or any other large spectrum fungicide) has proven to be highly effective. The foliar spraying of Benlate is to be performed every 3–4 weeks. Close monitoring is advised as mistakes could be disastrous. If all above recommendations and specifications are respected, the date grower could expect a survival rate of 90–95%.

For the acclimatization of 100,000 date palm plantlets, it is necessary to start with five greenhouses and five nurseries during the first year and then to double it in year 3 or 4. The greenhouses are to be constructed with the following dimensions: 40 m long × 8 m wide with a surface area of 320 m^2. Requirements for the greenhouses are as follows: light regulation, temperature regulation, humidity control, pathogen-free concrete floors, filtered air, insect exclusion, intermittent fogging-system, automated irrigation and fertilizer system, sterilized soil mixes, storage area for greenhouse supplies, and potting and repotting benches.

Figure 8.12 illustrates the recommended greenhouse model with galvanized iron (GI pipe) –to be used with 2.5 in. diameter for ground pole and 2 in. diameter for roof. The distance between poles is to be 4 m with a height of poles from the ground of about 2.4 m. Poles to be fixed 0.5 m deep from the ground level and on cement base of 40 × 40 cm.

8.7 Implementation Plan and Budget

8.7.1 Project Implementation Timetable

- Building of the laboratory facilities according to agreed layouts (including air conditioning).
- Purchase of all needed equipment and its installation in respective rooms of the laboratory.

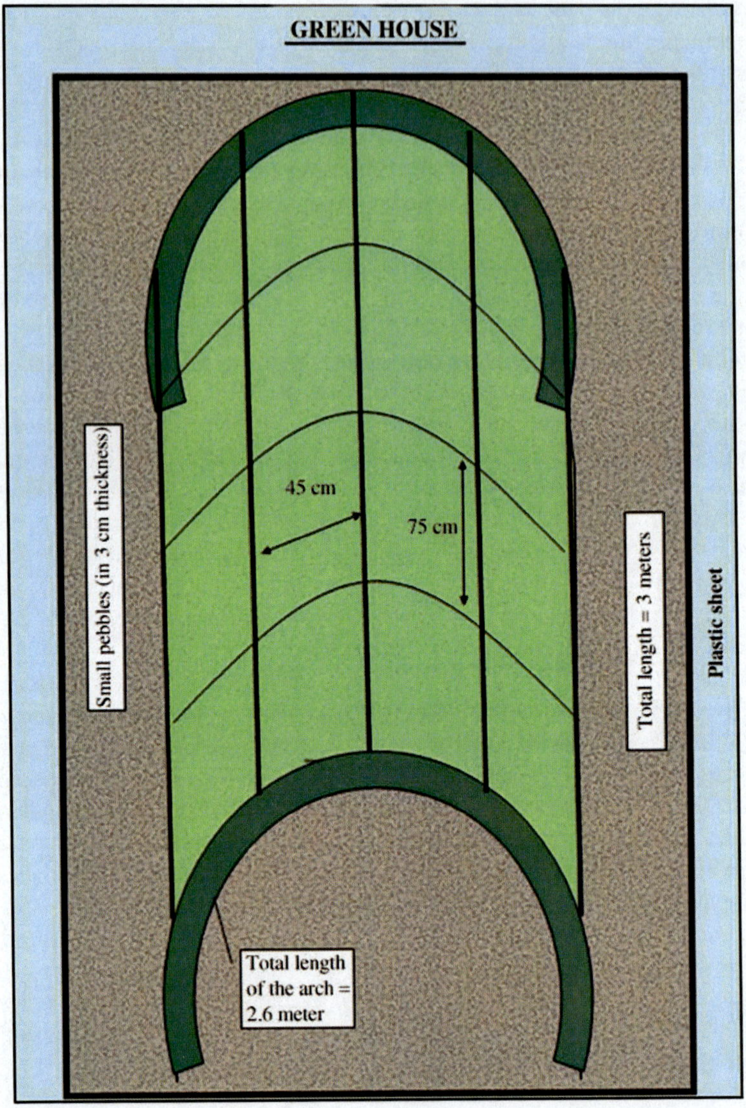

Fig. 8.12 Greenhouse specifications (length: 65 m, and width: 28 m: total area: 1,820 m^2)

- Appointment of the project director as well as hiring the scientific and technical staff.
- Training of recruited personnel in all operations.
- Start-up of the production system by the establishment of mother-/explants multiplying tissues of various selected date palm cultivars.
- Initiate the large-scale propagation of disease- and pest-free plant material.

Table 8.11 Estimated project budget for first 5 years

Item	Year 1	Year 2	Year 3	Year 4	Year 5	Subtotal
Buildings	250,000	–	–	–	–	250,000
Glassware (Table 8.6)	250,000	100,000	100,000	50,000	50,000	550,000
Laboratory equipment (Table 8.7)	50,000	50,000	50,000	50,000	50,000	250,000
Nutrients and reagents (Table 8.8)	50,000	50,000	50,000	50,000	50,000	250,000
Accessories and supplies (Table 8.9)	60,000	20,000	20,000	15,000	15,000	130,000
Greenhouses (16) and nurseries (11)	35,000	–	–	35,000	–	70,000
Personnel						
Technical director (1)	130,000	130,000	135,000	135,000	140,000	670,000
Supervisors (2)	12,000	12,000	14,000	15,000	15,000	68,000
Technicians and laborers (10)	20,000	20,000	25,000	25,000	30,000	120,000
Subtotal	857,000	382,000	394,000	375,000	350,000	2,358,000

- Approximately 5 years are required from the start of construction until the first field planting. It will take 18–24 months for the *in vitro* introduction, while the acclimatization will take approximately 16–18 months.

8.7.2 Estimated Budget

The estimated establishment budget for the first 5 years of the project is about USD 2.4 million and can be found in Table 8.11.

8.8 Laboratory Safety

8.8.1 General Laboratory Safety

The laboratory should be equipped with fire alarms and gas leakage detection systems. It is essential that all staff be alert and aware of their surroundings at all times and be safety-conscious and to anticipate possible hazardous situations. Staff should be made familiar with all safety equipment through regular professional training. It is advisable to have safety rules and guidelines printed as a separate booklet and distributed to staff. The laboratory should provide adequate safety facilities, ensuring both the provision of items to counteract specific emergencies (emergency escapes, fire extinguishers, eye-baths, and emergency showers) and by enforcing regular maintenance of various instruments.

Special rules should be stipulated for hazardous chemical use and equipment. The use of particularly dangerous or complicated apparatus should be entrusted to specialized technicians. General cleaning, utensil-washing, electrical ballast rooms, cold rooms and growth chambers are examples of sectors to be entrusted to specific responsible officers.

Equipment running more or less permanently (such as refrigerators, incubators) should be labeled *Continuously On*. The consumption of food and drink in the laboratory can be highly dangerous and should be forbidden. Individuals allergic to certain substances should inform lab management.

8.8.2 Fire Safety

If the fire alarms sound, all persons not involved in fighting the fire must leave the building as soon as possible through the nearest emergency exit. When possible, all equipment should be turned off and gas burners extinguished. Doors must be closed only when it is determined that everyone has left the room. Personnel should not attempt to re-enter the building until permission is issued. Once clear of the building, everyone should assemble at a previously designated point to determine if anyone is missing.

Laboratories should be kept clean, tidy and free from waste paper and packing material. Solvents or other inflammable materials should not be hoarded. Firefighting equipment must not be moved from its appointed place. All personnel should be familiarized with the position of alarm buttons, fire extinguishers, fire blankets and emergency exits. It is vitally important to ensure that fire extinguishers and all other safety equipment (first aid boxes, showers, eye-wash bottles etc.) are always readily accessible. Emergency exits to be always free of obstructions.

Many laboratory fires start with ignition of flammable solvents or their vapors by Bunsen burner flames or electric hotplates. Wherever possible, use non-flammable solvents and chemicals. If it is unavoidable, beware of nearby possible ignition sources on adjacent benches. Keep only minimum quantities of solvents needed in the laboratory at the appointed place and avoid solvent bottles spills. A non-smoking environment must be enforced.

8.8.3 Work-Related Accidents

All accidents, however insignificant, must be reported as soon as possible, including dangerous occurrences not involving injury, to lab management and recorded in a log book.

There should be a dedicated First Aid Room and all personnel should be familiar with its location, layout and inventory. First aid kits must be available in all laboratory sections and restocked upon use.

Many accidents occur through inexperience; anyone uncertain of any technique should ask for help or advice. When there is a risk of injury, e.g. handling a particularly toxic or corrosive material, operating an instrument, or using a new technique, make certain that a skilled staff member is present to provide suitable assistance in case of an emergency.

Most chemicals used in the laboratory are potentially dangerous. Always read the warning information printed on the label and proceed accordingly. The following are examples of dangerous chemicals: benzene and benzidine, carbon tetrachloride, sodium and potassium cyanides, mercury and mercury compounds, pyridine, osmium tetroxide (osmic acid), phenol, bromide, dioxine, ethers, chromic acid and chromates.

No liquid chemical or solution should be pipetted by mouth. Never transfer cyanide solutions of any strength, or any solutions which may be toxic or pathogenic by pipette. There is specialized equipment available for such purposes. Carry Winchester bottles or other large bottles using the special carriers provided and never carry a bottle by its neck.

A flame- and chemical-retardant laboratory coat should always be worn and kept buttoned up. Non-slips soles, low heels, closed footwear should be worn. The location and operation of emergency showers and eyebaths should be known to all personnel and present in several strategic locations.

Most cuts due to broken glass, which can be very serious, occur when the users are trying to force a glass tube through a hole in a rack, rubber tubing or rubber grommets. All injuries should be tended to immediately, and any broken glass shards removed and the area cleaned. Broken glass or sharp instruments should be placed in a separate Sharps Box and disposed of suitably.

In order to prevent physical injuries no running should be allowed in the laboratories or the corridors. Keep floors dry, clean and tidy. If there is a spillage, attend to it immediately in an approved way or place a warning sign. If heavy equipment must be moved, seek assistance and use a trolley.

Equipment operated at increased or reduced pressures must be guarded by a suitable protective shield. An eye- or facial-shield must be worn while handling any apparatus which carries the risk of exploding or imploding.

Gas cylinders must be kept in a protected stand, strapped or chained, outside the laboratory, and moved on a specialized trolley. Whenever gas is tapped from a cylinder, the cylinder must be fitted with the correct reducing valve. When not in use, the main valve on the cylinder must be turned off. Never use a hammer or blunt object to loosen a valve and never use grease anywhere on a cylinder or on the reducing valve.

8.8.4 Electrical Safety

Always use a properly fused electrical socket when connecting any apparatus to the power supply. If the instrument has a ground wire, make sure that it is properly connected. Do not overload an electrical outlet in excess of the designed current.

Electric repairs should be conducted only by certified personnel. If any electrical appliance appears to be malfunctioning in any way, switch off the main. Never try to service an electrical appliance before unplugging the power supply. Contact the instrument supplier for service.

If someone has sustained an electrical shock and is still in contact with the source of electrical current, switch off the appliance first then use a non-conducting item such as a blanket or wooden broomstick to separate the person from the device before rendering first aid.

8.8.5 *Facility Security*

Access to the laboratory outside working hours should be restricted to authorized technicians with special permits. Permitted staff should report any unauthorized persons to the security department. A list of emergency telephone numbers should be prominently placed at all telephone points in the laboratory.

8.9 Conclusion and Prospective

Lately *in vitro* tissue culture techniques have received important attention because of its inherent advantages and applicability to the rapidly developing fields of molecular biology, biotechnology and agricultural industry. Tissue culture-based endeavors are based on the establishment of aseptic culture conditions, multiplication of initiated cultures through the organogenesis process or the development of embryos, and hardening-off of tissue culture-derived plants before field planting. This chapter presents a comprehensive guide, with sufficient information and specifications, for the establishment of a tissue culture facility focused on the mass-propagation of any suitable plant species. It may interest and encourage individuals or entities to plan, initiate and contribute to the continuous development of similar units. There is currently a large demand for date palm plant material. It appears that both Iraq and Iran will need approximately ten million date trees each. Egypt and Sudan are also continuously requesting plant material, which will certainly strengthen this growing demand.

Part II
Somaclonal Variation, Mutation and Selection

Chapter 9
Somaclonal Variation in Date Palm

A. El Hadrami, F. Daayf, S. Elshibli, S.M. Jain, and I. El Hadrami

Abstract The exploration of somaclonal variation is an approach that could provide date palm breeding programs with new genotypes. Naturally occurring or induced variants may have superior agronomic quality and/or enhanced performance but could also harbor new traits such as tolerance to drought and salinity or resistance to major diseases i.e. bayoud. This chapter summarizes recent progress in terms of studying and exploring date palm somaclonal variation, and provides an outlook about future applications of this biotechnology in this socioeconomically important crop.

Keywords Biotic and abiotic stress • Bayoud • Conventional breeding • Date palm • Drought • *In vitro*-selection • *Phoenix dactylifera* L. • Salt • Somaclonal variation

---

A. El Hadrami (✉)
OMEX Agriculture Inc. Canada, P.O. Box 301, 290 Agri Park Road, Oak Bluff, Manitoba, R0G 1N0, Canada
e-mail: abdele@omexcanada.com

F. Daayf
Department of Plant Science, University of Manitoba, 222, Agriculture Building, Winnipeg, MB R3T 2N2, Canada

S. Elshibli • S.M. Jain
Department of Agricultural Sciences, University of Helsinki, FI-00014 Helsinki, Finland
e-mail: sakina.elshibli@helsinki.fi; mohan.jain@helsinki.fi

I. El Hadrami
Department of Biology, Faculty of Sciences Semlalia, Laboratory of Biotechnologies, Cadi Ayyad University, B.P. 2390, 40 000 Marrakech, Morocco

S.M. Jain et al. (eds.), *Date Palm Biotechnology*,
DOI 10.1007/978-94-007-1318-5_9, © Springer Science+Business Media B.V. 2011

9.1 Introduction

Genetic variation is essential to fulfill the needs of conventional and non-conventional date palm-breeding programs. Sufficient genetic diversity and variation can be found among cultivated germplasm, and in wild relatives such as species of *Phoenix* and *Sabal* palms. Once a trait is bred-in or introgressed into a new cultivar, several years of selection follow in the field to assess its stability and affect on agronomic performance and yield. Date palm can be propagated sexually and recombination becomes the main genetic event that allows for the incorporation of genetic variability into segregating populations. However, this is a lengthy process that may take decades before introgressing the new desirable traits. To accelerate the process, the species can be vegetatively propagated, which allows the introgression of the new traits through mechanisms other than the recombination. Induced mutagenesis and *in vitro* techniques represent some of the alternatives to alter desired genetic traits, at the same time as preserving the integrity of the genome and the clone characteristics.

Somaclonal variation and *in vitro*-selection represent useful biotechnology tools in date-palm breeding for tolerance to biotic and abiotic stresses i.e. drought, salinity, diseases and pests. These techniques also offer an improvement of the value-added of the new genotypes with traits such as an increase in the number and/or size of fruits or improved texture or taste, or modification in flower structure (Ahloowalia and Maluszynski 2001; Pedrieri 2001; Witjaksono 2003). Somaclonal variation represents a real advantage in widening the genetic basis of this species, relying more or less solely on vegetative propagation. By applying specific selective agents or providing particular conditions to *in vitro*-propagated tissues, somaclones with desired traits can be produced at a high frequency (Karp 1995). The causes of somaclonal variation during multiplication are diverse and tightly dependent upon the genotype, its level of ploidy, the growth conditions and duration of selection (Maluszynski and Kasha 2002). Studies of the determinants of such variation have revealed that it can be due to changes at the gene level through genetic events such as duplication, translocation, mutation by insertion or deletion of transposable elements, or methylation. It can also occur at the chromosome level through instability, inversion and transient or permanent ploidy changes (Dennis 2004; Kumar and Marthur 2004; George and Sherington 1984; Phillips *et al.* 1990). These phenomena often lead to irreversible pleiotropic and epigenetic events and the production of variants called chimera. Commonly used mutagens include microbial synthetic toxins i.e. crude fungal-culture filtrates; fusaric acid (El Hadrami et al. 2005); chemicals such as ethylene scimine (ES), diethyl sulphonate (DES), ethyl methane-sulphonate (EMS), and the azida group (i.e. NaN3) or physical mutagens such as χ- and γ-rays (Co60), fast and thermal neutrons (nf and Nth).

Depending on the selective agent, *in vitro*-selection could be conducted using regenerative and embryogenic calli, cell suspensions, zygotic and somatic rescued embryos, fused protoplasts and cybrids, but also at later stages during the regeneration of shoot and root meristems. The method of choice often depends on the advanced control of the micropropagation technique as well as the ease of application and the efficiency of the selective agent in inducing high levels of variation.

The regeneration method of tolerant cells is also important in order to preserve the inheritance of the desired trait or traits.

9.2 Somaclonal Variation in Date Palm

Somaclonal variation is an essential component of date-palm breeding in which variation regenerated from somatic cells can be used for the introduction of new agronomic, tolerance or quality traits (El Hadrami and El Hadrami 2009; Jain 2001). Variation in the somaclones has often been associated with changes in chromosome numbers and/or structure, punctual mutations or DNA methylation or other epigenetic events (Brown et al. 1993; Larkin and Scowcroft 1981). Somaclonal variation is undesirable from an industrial production stand point of view but may provide an enrichment of the gene pool. It also provides additional advantages such as the mass production of plants, opportunities for synthetic seeds, cryopreservation and direct delivery system for genetic variation. Its frequency depends, among other factors, on the genotype and the length of the proliferation process.

Jain (2007) reported that rapid shoot proliferation can be achieved from various parts of the plant including shoot tips, stem cuttings, auxiliary buds and roots. He also pointed out that the selection of the genotype and the number of sub-culture cycles help limit the appearance of somaclones after the step of plant regeneration. Many off-type plants and abnormal dwarf phenotypes with low fruit sets may still be observed among the *in vitro*-propagated populations with high frequencies. These phenotypes are not always detectable at seedling stages and often become apparent a few years after planting. However, the technological advances and the development of molecular markers have made it possible, in recent years, to early and accurately detect these variants and eliminate them from the mass production (Baaziz et al. 1994; Corniquel and Mercier 1994; Cullis et al. 1999; Powell et al. 1996; Saker et al. 2000; Salman et al. 1988). These off-types and somaclones can be further investigated to enrich the genetic pool.

9.2.1 Somaclonal Variation

The concept of somaclonal variation was introduced in the early 1980s to describe any variation observed in tissue culture (Larkin and Scowcroft 1981). Somaclones spontaneously develop during tissue culture due to the plasticity of the genome and its ability to restructure in response to exo- or endogenous conditions encountered *in vitro*. Although this can lead to novel sources of genetic variation for breeding programs, it is a phenomenon that needs to be regulated to guarantee uniformity during multiplication.

In comparison with the findings on African oil palm, it can be assumed that the use of higher concentrations of auxins during the redirection of calli to regenerate plantlets has a dramatic effect on the rearrangement of the genome and often leads to epigenetic events and the formation of somaclones. Some auxins such as 2,4-D, more than others, have been shown to be highly effective in inducing this phenomenon. Date palm genotypes also differ in terms of their responses to these growth regulators. Some cultivars may exhibit a genotype-fidelity and produce a high rate of true-to-type plants while others are more prone to variations. It is noteworthy that most date palm cultivars remain to be tested for their *in vitro* propagation abilities through tissue culture. Among those tested, a certain degree of recalcitrance has been observed, especially for those exhibiting agronomic, production, and quality or resistance traits (El Bellaj 2000; El Hadrami 1995; Gueye et al. 2009; Zouine et al. 2005).

The body of knowledge regarding the potential use of somaclonal variation in date palm breeding remains to be fulfilled and most available data are either sporadic or lack conclusive thoroughness. On the other hand, it is usually described and acknowledged in many other systems that micropropagation via somatic embryogenesis may lead, as compared to organogenesis, to higher percentages of somaclonal variation. Comparison between the two systems is lacking in date palm due to the lengthy process of achieving tissue cultures and producing vitroplants.

9.2.2 Sources of Variation

Besides the microchanges involved in the variation in the expression of specific genes, macrochanges such as deletion/addition of chromosomes or chromosome breakage may occur during tissue culture. These modifications remain dependent upon the explant origin, the concentrations of plant growth regulators used and the type of multiplication technique used to propagate the tissues. Undifferentiated cells induced under *in vitro* culture conditions are often genetically variable and unstable, with variations in chromosome numbers and ploidy levels. The causes of such chromosomal variation include irregular cell events that can occur during the induction of calli from explants i.e. endoreduplication, amitosis and DNA amplification (D'Amato 1985). Other factors also include higher ploidy levels and polysomaty. In date palm, the high frequency of retroelements and the sensitivity to certain auxins often triggers variations during tissues culture (Jain 2007).

The relationship between polysomaty and the level of diploidy remains unclear in date palm. Studies using tomato vitroplants, for instance, have shown that the frequency of polyploidy production is likely to be higher using hypocotyl segments as compared to leaf and cotyledon explants, which predominantly produce diploid seedlings (Van den Bulk et al. 1990). On the other hand, the use of older cotyledons as explants had led to either a poor regeneration of plants or to the regeneration of tetraploid or mixoploid seedlings with a high polysomaty status (Colijn-Hooymans et al. 1994).

Chromosomal variation has been previously reported in date palm. Various studies showed a chromosome number varying from 26 to 36 (Al-Salih and Al-Jarrah 1987; Al-Salih and Al Rawi 1987; Al-Salih et al. 1987; Beal 1937; Ibrahim et al. 1998; Loutfi 1999; Nemec 1910). Such variation may be linked to phenomena such as polysomaty, chromosome breakage or nucleolar heterochromatin aggregation/disaggregation that occur in *in vitro*-propagated material. In addition, somaclonal variation can occur in micropropagated tissues as a result of a gameto-, proto- or soma-clonal origin.

9.2.2.1 Gametoclonal Variation

Gametoclonal variation refers to variants derived from gametic and gametophytic cells (Evans et al. 1984). The value of gametoclonal variation in plant breeding comes from the development of double-haploids after anther culture. Gametes, as a product of meiosis, receive according to Mendel segregation laws half of the genetically segregating alleles as opposed to somatic cells, which divide their genetic material equally during mitosis. Gametoclones are able to express both recessive and dominant alleles while crossovers may occur during meiosis, creating new sources of variability. Once doubled for stability, gametoclones can be examined for residual heterozygosis (Evans and Sharp 1986) and used in breeding programs. This strategy has been explored in date palm using double haploids derived from micro- and macrospores (Chaibi et al. 2002; Zouine and El Hadrami 2004).

Anther and ovule culture recovery is difficult in date palm. Investigations in this regard have led to the achievement of cell divisions and to the formation of globular embryoids from uninucleate microspores. Successful attempts report the importance of cold treatment combined with the use of two auxins and one cytokinin to generate embryoids (Chaibi et al. 2002), that unfortunately were unable to develop further. Other studies, using various treatments and exogenous factors, did not provide any major improvements apart from production of weak and short-living calli from these propagules. It is also important to mention that one of the main difficulties in developing these studies is related to the short duration of the flowering period in date palm, preventing the harvest of fresh anthers with uninucleate microspores. These anthers are also likely to turn brown and die a few weeks after their culture. Chaibi et al. (2002) reported that treating anthers with a thermal shock at 37–38°C is suitable only within a narrow window of time prior to their *in vitro* culture. Combined with that the use of MS medium amended with 2,4-D and 2-isopentenylaminopurin (2-iP), as well as activated charcoal, often help prevent tissue browning, and increase the percentage of microspore division.

Some haploid recovery attempts have also been conducted using unfertilized date-palm ovules. Due to the small size of these ovules, browning and necrosis were the main limits encountered by these cultures. Although, the carpel enlarged and became quite prominent when cultured, the use of activated charcoal was required to ensure a much longer survival, and root or callus formation. Until now, the best results obtained were from flowers taken from closed spaths in which the embryo sacs were formed and contained undifferentiated cells. Recently, Masmoudi-Allouche et al. (2009) also reported on the potential induction of hermaphrodism in date palm inflorescences.

In spite of its current unsuccessful use in date palm, gametoclones offer a great opportunity for breeding of this plant. Clonal propagation generates identical copies of the selected genotype but applying specific stresses at given times may result in loosing-up the control mechanisms, guaranteeing the stability of the genome of the gametoclones. Long-term propagation of multiple shoots *in vitro* and excessive sub-cultures of the same stock may expose the tissue to an environment where their genetic stability gets altered, creating internal repeats of genetic sequences or transposition of retroelements. All these events may affect essential genes required for the growth and development or implicated in the differentiation/de-differentiation process, consequently leading to polymorphism. Methylation/demethylation is also likely to occur under these particular conditions, leading to variation among the propagules. Being haploids, the gnome of these gametoclone can be doubled to contribute to the extension of the genetic variation among the germplasm.

9.2.2.2 Protoclonal Variation

Protoplasts are cells derived from within the cell membrane. Their genetic integrity remains intact. Their manipulation offers certain flexibility in terms of creating variation and recovering protoclones with desired traits. In date palm, serious difficulties were encountered generating and maintaining protoplasts, although several attempts have been carried out in many laboratories around the world. An early browning followed by a rapid death of the protoplasts occurs in the suspensions. Alternatively, most reports describe the use of cell suspensions for any genetic manipulation of date palm (Fki 2005).

9.2.2.3 Somaclonal Variation

Somatic cells represent a source of variability that could be utilized within a breeding program to generate new clones with desired traits. Date palm micro-cultures were previously irradiated *in vitro* (Ahloowalia and Maluszynski 2001) to induce mutations. This strategy is thought to be an effective way of introducing variability into either wild or bred stock (Jain 2007; Maluszynski and Kasha 2002; Szarejko et al. 1995). Besides, it can be combined with somatic embryogenesis or organogenesis to regenerate new material. Among the traits affected by somaclonal variation, dwarfism and abnormal floral development are the most observed (Al-Kaabi et al. 2007; Zaid and Al-Kaabi 2003).

Somatic embryos starting initially as single cells, represents the most appropriate material for induced mutagenesis and are less prone to chimerism (Jain 2002). However, their poor rate of germination makes large-scale multiplication very complicated (Jain 2002). Progress is tentatively being made to control their germination and increase recovery rates.

9.2.3 *Factors Affecting Somaclonal Variation*

Many factors contribute to the somaclonal variation observed in plant tissues, propagated *in vitro*. These include the effect of the explant and its source; the combined effect of culture age and number of subculture cycles used; the effect of endo- and exogeneous growth regulators; the genotype fidelity and flexibility; the abundance of retroelements and post-transcriptional events; as well as other factors such as atypical duplication of DNA.

9.2.3.1 The Effect of the Explant and Its Source

In date palm, many studies have revealed that the explants from leaves and apical/lateral meristems were less prone to produce somaclones when organogenesis is considered. However, it has also been described that date palm propagated by organogenesis exhibits a certain level of genetic variation when tested in the field several years after culture. At present, it is too difficult to estimate the percentage of this genetic variation and determine its positive or negative impacts on this long-living tree. For instance, this variation cannot be observed unless fruits are set, which often occurs 5–7 years after field planting.

It is widely acknowledged that explants with highly differentiated tissues (i.e., roots, leaves, stems) are more likely to produce variations than explants with meristematic tissues (i.e. auxiliary buds, shoot/root tips) (Duncan 1997). This seems to be related to the fact that DNA at the initial stages of development in somatic cells is less prone to methylation events as compared to later stages when the differentiated cells are metabolically active (i.e. Munksgaard et al. 1995). This change in the methylation status of DNA could be due to the alteration of the balance between *de novo* synthesis and/or activation of enzymes catalyzing methylation/demethylation reactions, or changes in the concentrations of substances or co-factors involved in these reactions (Munksgaard et al. 1995). Methylation is the process that controls gene expression during somatic embryogenesis through inactivation of transcription (Duncan 1997). All these general observations hold true in date palm tissue culture, suggesting that the methylation constitutes the main factor controlling variation during *in vitro* propagation.

The hypomethylation status of somatic cells suggests a link with the state of differentiation and recalls what is observed in the early stages of zygotic embryos (Herman 1991; Lo Schiavo et al. 1989). Recently, Sgheir-Hammama et al. (2009) compared the proteomic map of somatic and zygotic embryos of date palm cv. Deglet Noor. The authors showed significant differences among the two types of embryos with regard to their proteins content and function. The abundant protein pool in zygotic embryos was carbohydrate biosynthesis and storage/stress-related proteins while in somatic embryos, glycolysis-related proteins were predominant. Having no link with other complementary genomic studies, one would expect to see that some of these differences are related to methylation mechanisms. The recent unveiling of the entire genome (Al-Dous et al. 2009) will in the future shed light on some of these questions.

9.2.3.2 The Combined Effect of Culture Age and Number of Subculture Cycles

Aging cultures are more likely to lead to variants than freshly cultured tissue. Changes in the *in vitro* growth conditions and the quickness of the multiplication process can also increase the mutation rate per cell generation, as well as the number of mutations accumulating in the tissue over time (Duncan 1997).

This may affect the genetic stability of the plant tissue and lead to somaclonal variations (Martinez et al. 1998). For example, studies in banana, using 3, 5, 7, 9, and 11 subcultures, revealed no variation up to third subcultures. However, somaclones were observable starting from the fifth subculture, and their frequency kept on increasing with the increase in subcultures number (Rodrigues et al. 1998). The frequency of numerical and structural errors in chromosomes was also high when cultures were kept for a longer time or propagated over five sub-culture cycles (Shepherd et al. 1996).

Similar data were reported in date palm, where somaclonal variants have been shown in cvs. Barhee and Khalas (Al-Kaabi et al. 2007; Zaid and Al-Kaabi 2003). These two cultivars exhibited abnormal floral development and dwarfism as the main problems inherent in somaclonal variation. Higher percentages of variation were recorded in Barhee as compared to Khalas. Palms with albino stripes at their mid-ribs and albino and variegated leaflets were described, at low frequency, in cv. Khalas. In cv. Barhee, the impact of the variation was much more dramatic with palms unable to set fruit (Al-Wasel 2005). Saker et al. (2000) showed that the frequency of somaclonal variations was age-dependent and that the abnormal shoots showed genetic variations at the molecular level when compared to normal genotypes.

In another study involving cv. Khanizi and the use of RAPD markers to examine the genetic stability of the somatic embryogenesis-derived plants, Eshraghi et al. (2005a) reported the occurrence of some genetic variation, up to the sixth generation. The genetic similarity between the mother and calli-derived plants ranged between 94% for R1-2 and 83% for R5, suggesting an increase in the accumulation of genetic changes among generations of micropropagules. Assuming that R5 had accumulated the maximum genetic change, the percentage of variation is about 17% with an average rate of 3.4% per multiplication cycle.

9.2.3.3 The Effect of Growth Regulators

Endo- and exogenous hormonal homeostasis is a key element in the stability of *in vitro* cultures. Disturbance of the cells cycle often occurs after exogenous application of hormones and may lead to variability. Concentration of 6-benzylaminopurine (BAP) as high as 22, 44 or 66 mM were reported to be responsible of variability in many plant systems (Trujillo and Garcia 1996), including date palm (El Hadrami et al., unpublished). The combination of BAP with adenine at higher concentrations could also alter the number of chromosomes. Besides, the choice of growth regulators should be adjusted depending on the genotype and the explant used to initiate the culture. In many systems, kinetin seems to be required for bud propagation while root tips preserve their mitosis under a very low concentration of BA, or in media completely deprived of BAP. Although, tissue culture instability was never clearly correlated to the increase in BAP concentration, variation in ploidy level has been observed in many *in vitro* cultures, including those of date palm (Loutfi 1999). Likewise, other growth regulators such as the auxin DICAMBA, often added to regenerative growing calli to induce somatic embryogenesis, can increase variability. Omar and Novak (1990) reported on the effect of DICAMBA and PICLORAM on date palm calli growth and embryogenesis. Both auxins induce similar effects to the ones described using 2,4-D (El Bellaj and El Hadrami 2004; Zouine et al. 2005). Recently, Gueye et al. (2009) emphasized that while callogenesis in date palm can be initiated by culturing immature leaf segments on medium containing 2,4-D, it is still quite difficult to obtain calli from certain genotypes.

Calli induction requires independent spatial and temporal events. Upon culturing explants on media amended with 2,4-D, cells from the fascicular parenchyma that is perpendicular to the vascular axis are subjected to dramatic structural and ultrastructural changes that recall meristematic cells. Later during the early stages of culture, modifications also occur in the adjacent perivascular sheath cells, leading to callogenic abilities. These cells often reinitiated their cycle and undergo a series of division forming calli. In date palm, the most callogenic segments are obtained from within the leaf elongation zone that requires a polar auxin transport to initiate callogenesis. This zone is also reported to contain the highest content of free endogenous indole-3-acetic acid. El Bellaj et al. (2000) also showed the importance of the IAA oxidase as a regulator for the somatic embryogenesis processes in date palm.

Endogenous gibberellins levels or exogenously applied to induce dwarfing could also affect somaclonal variations. The combination of abscisic acid with indole-3-acetic acid, and cytokinins was also reported to be responsible for endogenous plant growth regulator homeostasis and to lead to variants during the late phases of culture. In date palm, only a few studies such as the one by Khierallah and Bader (2007) examined the effect of these factors and not even under the context of inducing variations but rather avoiding it.

9.2.3.4 Genotype Fidelity and Flexibility

Technically all genotypes can be used in tissue culture. However, some genotypes are more prone to variability than others. This could be attributed to the microenvironmental influence on cell behavior or the stability of the genotypes; some being more stable than others. Variation linked to the genotype fidelity can lead to variation in height, and bunch characteristics i.e. atrophied fruits. Somaclones derived from cvs. Barhee and Khalas, for example, were reported to develop abnormally, especially at the flowering stage, besides being dwarf (Al-Kaabi et al. 2007; Zaid and Al-Kaabi 2003).

Within a commercial operation of micropropagation, the primary achievable objective is to massively produce true-to-type clonal planting material. Representational difference analysis (RDA) is usually used to detect culture-induced variations (Lisitsyn et al. 1993). This approach helps in delineating DNA changes between off-types induced in tissue culture and true-types producing normal plants. It is a PCR-based system that monitors the genomic integrity and changes occurring in a batch of tissue culture-derived plants. This approach is able to reveal the extent of stress exerted toward plant cells, hence triggering mutagenic processes during explant establishment, calli induction or maintenance, embryo development, and plant regeneration.

Plant genomes, as in many other Eukaryotes, have two components (Capy 1998). The plastic genome is the labile and highly instable portion of the genetic material, and consequently prone to variability and rearrangement. This segment often guarantees certain flexibility to the genotypes in terms of altering their cell cycle *in vitro*, and could explain the variability observed in tissue culture using RDA. It also supports the fact that somaclonal variation is not a random phenomenon but occurs only in loci with higher mutation rates. The other fraction of the genome is composed of a much more stable genetic material called the *core* genome, responsible for ensuring the fidelity of the genotype.

9.2.3.5 The Abundance of Retroelements and Post-Transcriptional Events

Variation in tissue culture is highly dependent on the level of ploidy, karyotypic changes as well as on specific post-transcriptional events. Date palm being a diploid species still achieves high levels of variation in tissue culture that may be controlled karyotypically, epigenetically or post-transcriptionally. Somaclonal variants, if genetically stable, could contribute to breeding programs in terms of introgressing new traits. For instance, chromosome breakage during rearrangement is a phenomenon that is often observed in conventional sucker-grown plants such as date palm. This can lead to an alteration in the number of chromosomes, resulting in the regeneration of plants different from the original clone. The same phenomenon occurs at a higher rate under tissue-culture conditions, and can be explored to generate variability. In addition, tissue-culture techniques may increase the rate of specific mutations involving certain DNA sequences, making variants segregate in a homozygous fashion. It may also affect the methylation of DNA depending on the genotype, age of the culture and its stage of growth.

Transposable elements are also drivers for variation in tissue culture because they can create either stable or unstable gene mutation, changes in the DNA methylation, or even lead to chromosomes breakage (Pechke and Philips 1992). In many plant species, it has been shown that culture conditions create an environment that is conducive to autonomous transposition of DNA sequences (Larkin and Scowcroft 1981). In date palm, a high frequency of retroelements has been already detected, especially in certain cultivars prone to variation (Jain 2007).

9.2.3.6 Other Factors

Many other factors contribute to variation during tissue culture. For instance, to help maintain the diploid status of the cells in meristematic tissues, the cell cycle is carried out in a normal way, where duplication of DNA is immediately followed by caryokinesis and cytokenesis. However, in differentiated tissues, cells could undergo DNA duplication and autoreduplication with division and mitosis. Polyploidization may occur in these tissues, leading to a phenomenon called polysomaty, where chromosomes undergo duplication within the same nuclear membrane (endomitosis).

Chromatids of each chromosome may also be duplicated during the interphase with no subsequent changes in the final chromosomes number (endoreduplication) leading to chromosomes with four, eight or 2^x chromatids (diplo-, quadriplo-, poly-chromosomes). These phenomena often increase in frequency in aging tissues subjected to mutagenic treatments. Date palm, as a diploid species, does not exhibit all these genetic events during tissue culture as would polyploid species. However, the current lack of in-depth examination of some of the somaclones observed in date palm does not exclude the occurrence of some of these DNA and chromosomes changes, which could explain the observed growth and development abnormalities reported (Al-Kaabi et al. 2007; Smith and Ansley 1995; Zaid and Al-Kaabi 2003).

In date palm micropropagation, genotype fidelity and stability represents a prime goal. Somatic embryogenesis offers these features and was considered for the last two decades as the most economical method to commercially produce vitroplants. However, occurrence of up to 100% off-types at maturity has raised concerns and uncertainties about genotype fidelity. Optimization of the protocols used during date palm somatic embryogenesis needs, hence, to be revised. For instance, during somatic embryo maintenance and organ proliferation in some procedures in the literature, which is applicable for commercial large-scale production, proliferation of plantlets takes place before separating somatic embryos, then individual plants moved to maintenance cultures (e.g. Eke et al. 2005; Eshraghi et al. 2005b; Othmani et al. 2009a; Taha et al. 2007). The formation of discrete mature somatic embryos at maintenance the stage of somatic embryogenesis is well documented in some forest tree species, such as Norway spruce, *Picea abies*, (Vagner et al. 2005), white spruce, *P. glauca* (Yeung and Thrope 2005) and Douglas Fir, *Pseudotsuga menziesii*, (Gupta and Holmstrom 2005). Although many other reasons have proved to cause somatic variation in date palm and other plant species, such as the number of sub-cultures and other potential stress conditions; however, we suggest that some of these acute variations also may take place due to organ proliferation, mostly shoots, from other calli cells adjacent to somatic embryos which are not maintained separately. Calli morphogenesis and proliferation to shoots or roots is known to be directed by the plant hormones cytokinin and auxin extensively used for plant propagation. Although, both shoot/root proliferation as well as somatic embryogenesis from callus cultures are indirect forms of organogenesis, each is a different method and has different culture requirements such as subcultures and manipulation of growth regulators in the culture media (George and Debergh 2008). Genetic stability within somatic cells depends on the competence of these cells (George and Debergh 2008); while, a strong selection in favor of genetically normal cells during somatic embryo development is more expected (Vasil 1994). It would be more valuable to study these two methods separately for somaclonal variation at DNA and sequence level, as well as for DNA methylation and developmental processes of *in vitro* date palm plants which may provide a tool for early detection of some phenomenon of these off-types. On the other hand, separating individual somatic embryos to the maturation and germination phase has good potential for further production of secondary somatic embryos with minimum hormonal usage and would be a preliminary check point for applying molecular markers to inspect random samples for high genetic deviation from the mother tree or detection of specific genotypes before proceeding with further efforts. The procedure using embryogenic suspension with bioreactor aid described by Othmani et al. (2009b) constitute a promising update to somatic embryogenesis protocols for date palm, especially in terms of the direct development of somatic embryos which can be applied for repetitive embryogenesis with less expected genetic variation.

9.2.4 *Induction of Somaclonal Variation*

The induction of somaclonal variation can be carried out by adjusting some of the factors contributing to the regulation and redirection of cell cycle during micropropagation, such as the explant and its source, the age of the culture, the number of subcultures and the growth hormones. Other sources of creating this variability rely on the use of chemical or physical mutagens such as microbial or synthetic toxins, sulphonate, χ- and γ-rays (Co60). Combined with various *in vitro* cultures, induced mutagenesis represents a fast and efficient way for introducing desirable traits into date palm, such as resistance/tolerance to biotic or abiotic stresses, or the improvement of yield and fruit quality. Some morphological features were also reported to be altered through this process (Jain 2002). The improvement of *in vitro* techniques applied to date palm in recent years has made it possible to irradiate large populations of vitroplants, and to maintain somaclones within the same collection (Jain 2007). Maluszynski and Kasha (2002) and Ahloowalia et al. (2004) reported that more than 1,800 cultivars from the collection maintained by the FAO/IAEA-Nuclear Techniques in Agricultures division are/were direct mutants or derived from crosses involving genotypes subjected to induced mutagenesis. Some of these cultivars have been released in over 50 countries for various tests of agronomic, tolerance/resistance, and yield performances or simply to monitor their growth and development habits.

9.2.4.1 Mechanisms and Potential Application in Breeding

Throughout the domestication process of date palm, whenever productivity has been increased it depended upon a narrowing of the genetic base. One of the modern breeding challenges is to return to the wild ancestors of contemporary cultivars and explore some of the diversity that has been lost during their domestication and/or breeding. However, it is usually challenging to retrieve the wild variation to be used in a breeding program because most of it has a negative impact on growth and adaptation. In recent years, the development of DNA markers in many plant species, allowed for the saturation of genetic maps established based of segregating populations and the establishment of quantitative trait loci (QTL) that could improve the yield or other continuous traits. To the contrary, date palm is still lagging behind due to many technicalities linked the regeneration and establishment of segregating populations and the conduction of backcrossing. Besides, the introgression of wild alleles could also mask the magnitude of some favorable effects, resulting in minimal promotion of yield or improvement of the phenotype. This along with other factors have led to a worldwide intensive monoculture of certain bred varieties, which reduces the diversity among the grown germplasm and puts at risk the sustainability of this sub-Saharan crop.

Changes in the methods used to analyze plant genomes in the last few decades allowed for a dramatic shift from extensive breeding experiments involving crosses and backcrosses to the use of molecular markers that identify differences in the DNA sequences among genomes. These can now be depicted more intensely using methods such as the so-called representational difference analysis or RDA (Lisitsyn et al. 1993). This method relies on subtractive hybridization to isolate differential genomic regions between compared genotypes. Applied in date palm, RDA revealed sequences that were dispersed and repetitive (Lisitsyn et al. 1993). Some of these sequences were more abundant in one genome as compared to the others included in the study, suggesting their involvement as stress-related sequences (Lisitsyn et al. 1993).

Kunert et al. (2002) investigated the use of DNA microarrays to address the problem of screening off-types. The authors designed a diagnostic *DNA microchip* for the detection of off-types derived from micropropagation by tissue culture. This approach offers a high throughput screening and can be used as a quality control measure in the plant tissue-culture industry. Hybridization with fluorescence-tagged DNA from tested plant DNA to microarrays carrying chemically homogenous plant off-type-derived hybridization targets is also possible (Lemieux et al. 1998). These targets will be isolated by RDA (Lisitsyn et al. 1993), and monitored in the plant genome of micropropagated lines. Pilot studies reported the usefulness of RDA to identify targets from tissue culture-derived date palm plants produced via the process of somatic embryogenesis. Evaluation of the RDA technology on date palm somatic embryogenesis showed that it can be used along with field evaluations to increase the accuracy and reproducibility of detection of off-types.

9.2.4.2 Selection of Somaclones

Somaclonal variation, even if it is a burden for the propagation of true-to-type clones for commercial plantations, has a specific interest in non-conventional date palm breeding. This variation, if selected for, could lead to the creation of new varieties with traits affecting nutrient uptake and overall agronomic performance, dwarfism, the characteristics and arrangement of leaves on the stem, fertilization proprieties, tolerance/resistance to abiotic or biotic stresses, as well as yield, fruit size, shape, and texture. Somaclones can result *in vitro* due to genetic changes triggered by the growth conditions or simply due to hardening errors. During growth, somaclones are often associated with the effect of growth regulators added to the culture media during initiation of calli or redirecting dedifferentiated cells toward regeneration of somatic embryos (El Bellaj et al. 2000; El Hadrami 1995; El Hadrami and Baaziz 1995; El Hadrami and Coumans 1994; El Hadrami et al. 1995; Tisserat 1979). Auxins in particular are known to induce genetic instability and DNA rearrangement, hence leading to somaclonal variations (Cullis 1999; Karp 1989; Phillips et al. 1994).

Other epigenic events independent of auxins also may be involved in triggering the expression of specific genes or the activation of retroelements (Cassells et al. 1999; Cullis et al. 1999; Karp 1989, 1993; Phillips et al. 1994; Skirvin et al. 1994). These often lead to changes in certain traits of the plant that can be carried over to the offspring only if selected for through DNA amplification, methylation or activation of retroelements (Brar and Jain 1998).

Once a variant is selected, a series of sub-cultures are required to test its stability. Depending on the selected somaclones, tests can be conducted either at random or targeting certain traits such as tolerance to abiotic stresses or resistance to diseases and pests, either *in vitro*, or under controlled condition or in the field. The use of molecular techniques could be also used at this stage to accelerate the process of testing.

A recent study by Sghaier-Hammami et al. (2009) examined, at the proteome level, the differential changes that occur in somatic embryos derived from cv. Deglet Noor versus their zygotic counterparts. Several qualitative and quantitative differences were observed between the two tested embryos. Among them, somatic embryos were most likely to activate their glycolysis battery of enzymes while the zygotic counterparts use storage and stress-related proteins in their early stages of development. These findings may suggest new perspectives for research on developing synthetic seeds with somatic embryos.

9.2.4.3 Molecular and Biochemical Characterization of Somaclonal Variation

In vitro production of date palm using either organogenesis or somatic embryogenesis has been established in recent years as a routine procedure to satisfy the large demands of commercial plantations. Rigorous controls are executed to ensure the quality, homogeneity and genetic identity of the production. Unfortunately, variants occur under these intensive propagation conditions. Keeping in mind that, once a plant is transferred into the field, it requires at least 10–15 years to bear fruits, selecting those variants can lead to serious economical losses. Therefore, it is of a primary importance to develop biochemical and molecular tools to early detect and discriminate off-types, genetically non-identical to the propagated mother plant.

A heated controversy exists in the literature regarding the occurrence of off-types in micropropagated date palm. An abnormal fruiting of cv. Barhee (80–100% pathenocarpy with 1–3 carpels) occurred upon propagation through somatic embryogenesis (Djerbi 2000). These abnormalities were observed in trees that were planted in 1990s after the pollination of over 100,000 trees. It is still not clear whether the observed off-types are associated with DNA changes that occurred during their *in vitro*-propagation or due to non-optimal environmental conditions associated with unusual cultural practices. To the contrary, Smith and Ansley (1995) and Al-Ghamdi (1996) reported that plants derived from the same cv. Barhee through somatic embryogenesis did not show any abnormalities. Disregarding that this is a case of cultivar mis-naming/designation or a sampling size matter, it is still unclear whether the cultivar shows growth and development abnormalities or not. Investigating two other cvs., Thoory and Zahdi, Al-Ghamdi (1996) reports that there were no significant differences in flowering or fruit set while McCubbin et al. (2000) reported certain abnormalities such as leaf variegation, seedless fruit, broader leaves, different spine structure, bending of stems and a compact growth form.

Over the years, several markers have been developed to scrutinize off-types. The first markers were morphological characters that are still used by certain small tissue culture-producing companies (El Houmaizi et al. 2002; Hussain and El-Zeid 1978; Sedra et al. 1998). Trained nursery technicians can visually inspect and discard any variants. This technique does not include the variation due to age, and the interaction between the genotype and the environment. In addition, certain important criteria of selection such as fruit set and quality, pollination potential, and disease resistance cannot be assessed at a juvenile stage, which make this technique, although economic, unsuitable for a successful breeding program.

The recent development of molecular markers is currently replacing the use of morphological screening for off-types. These markers rely on the variation in the genomic DNA or the expression of specific genes. Isozymes were among the earliest adopted markers (Kephart 1990) because they were thought to exhibit differential patterns between true-to-type and off-type genotypes, hence leading to a fingerprint for the latter (Baaziz et al. 1994; Saker et al. 2000; Salman et al. 1988). Commonly used isoenzymatic systems include peroxidase, polyphenol oxidase, phosphoglucose isomerase, esterase, glutamate oxaloacetate transaminase, endopeptidase, alcohol dehydrogenase, phosphoglucose mutase and cytosolic leucine aminopeptidase. Studying four date palm cultivars derived from calli cultures, Salman et al. (1988) reported a variation in the isoenzymatic patterns of the esterase, glutamate oxaloacetate transaminase and cytosolic leucine aminopeptidase. Similar results were reported by Saker et al. (2000) analyzing peroxidase, polyphenol oxidase, and glutamate oxaloacetate transaminase, at the isoenzymatic pattern level and enzymatic activity. Although successful in depicting differences between true-to-types and off-types, isozymes lack coverage of the genetic variation across the genome that may be due to re-arrangement of other epigenetic events.

As alternatives to isozymes, a number of DNA-based markers have been developed in the last two decades including RAPD, RFLP, AFLP, microsatellites and RDA (Corniquel and Mercier 1994; Cullis et al. 1999; Powell et al. 1996). These techniques revealed various levels of polymorphism among date palm cultivars (e.g. Corniquel and Mercier 1994; Sedra et al. 1998), and were also used to tentatively detect somaclonal variations (Saker et al. 2000). RAPD analysis, although easy to perform, is not reproducible and results can vary among laboratories and even among users in the same laboratory (Jones et al. 1997; Skroch and Nienhaus 1995).

Another set of molecular markers often used to study the diversity among date palm germplasm, or to assess somaclonal variability relies on the use of RFLP. This technique requires a large amount of high quality DNA to be digested with restriction enzymes to screen for polymorphic probes among tested genotypes (Ait-Chitt et al. 1993; Corniquel and Mercier 1994, 1997; Ouenzar et al. 1998). The transferability limitations of the technique between laboratories led to its replacement by PCR-RFLP.

AFLP analysis represents a combination of the RFLP and RAPD markers, and often generates a large number of fragments that could be used in genotype fingerprinting. This technique has been used for cultivar discrimination in date palm (e.g. Lacaze and Brackpool 2000) as well as for the study of the genetic fidelity of the tissue-culture propagated genotypes in comparison with the original offshoots. Its downside comes from the fact that it samples only a small portion of the genome, consequently leading to misses in terms of genomic variation. To overcome these limitations, microsatellites were developed in recent years. These markers are small in size (~100 bp) with di- or tri-nucleotides repeats (Scriber and Pearce 2000), and vary in length, number of repeats, and can be highly polymorphic. As RFLP markers, these microsatellites have to be isolated and characterized. Pre-screening of a genomic library with labeled repeats often leads to the identification of polymorphic microsatellites. Sequencing of the regions containing the microsatellite allows for the development of specific PCR primers for the flanking regions of the microsatellite, which becomes a sequence-tagged marker. Polymorphic microsatellites with multiple alleles could then be applied to discriminate commonly-grown varieties of date palm. Given the absence of evidence that these markers could be present in the highly variable parts of the genome, and to the recent information about the low frequency of repeats in the whole genome of date palm (Al-Dous et al. 2009), these markers may be limited in terms of identification of somaclone variants.

Another technology that has made some progress in date palm breeding programs is the use of the RDA analysis. This technique allows for the development of probes to study genomic losses, re-arrangements, amplification or punctual mutations in the studied organism (Lisitsyn et al. 1993). This technique combines the use of a representational sample of the variation with subtractive hybridization and enrichment to focus only on unique and differential sequences between populations or genotypes. This technique has been used in date palm to differentially characterize cvs. Barhee and Medjool (Kunert et al. 2000; Vorster et al. 2002) and believed to cover hypervariable regions, allowing the study of off-type somaclone variants.

The availability of the draft genome of date palm (Al-Dous et al. 2009) has expanded perspectives for developing new molecular markers relying on chip technologies, microarrays or other florescent probes. Already, a set of 850,000 SNPs have been detected to be useful in delineating differences among parental genotypes and screening progenies (Al-Dous et al. 2009). Recent advances have also been made in terms of 2D-proteomics, where zygotic and somatic embryos proteome maps were constructed and unveiled significant differences (Sghaier-Hammama et al. 2009). This technique could also be applied to differentiate true-to-type and off-types somaclone variants.

9.3 Conclusion and Prospective

Micropropagation of date palm has gone a long way and offers tremendous support to breeding programs. It has resolved many technical issues encountered to produce progenies and reduced the time required to produce them. It also allowed the creation of new applications intended to introduce variations (i.e. somaclonal variation) and widen the polymorphism among date palm cultivated germplasm. Progress being made is modest in comparison with other clonally-propagated species due to difficulties that are still encountered generating protoplasts or cultures from anthers and ovules. The recent sequencing of the entire genome will certainly help unravel some of the mechanisms controlling somaclonal variation in this species to hopefully use them in the development of new cultivars with desirable traits.

Acknowledgments This review has been supported by several research grants to Professor Ismail El Hadrami (Ifs/AUF/PRAD/CNRST/AI Morocco-Tunisia/AI Morocco-France).

References

Ahloowalia BS, Maluszynski M (2001) Induced mutations: a new paradigm in plant breeding. Euphytica 119:167–173

Ahloowalia BS, Maluszynski M, Nichterlein K (2004) Global impact of mutation-derived varieties. Euphytica 135:187–204

Ait-Chitt M, Ainsworth CC, Thangavelu M (1993) A rapid and efficient method for extraction of total DNA from mature leaves of date-palm (*Phoenix dactylifera* L.). Plant Mol Biol Rep 11:317–319

Al-Dous EK, George B, Salameh YM et al (2009) Qatar Researchers Sequence Draft Version of Date Palm Genome. URL: http://qatar-weill.cornell.edu/research/datepalmGenome/download.html

Al-Ghamdi AS (1996) Date palm (*Phoenix dactylifera* L.) germplasm bank in King Faisal University, Saudi Arabia. Survival and adaptability of tissue cultured plantlets. Acta Hortic 560:241–244

Al-Kaabi HH, Zaid A, Ainsworth C (2007) Plant-off-types in tissue culture-derived date palm (*Phoenix dactylifera* L.) plants. Acta Hortic 736:267–281

Al-Salih AA, Al Rawi A (1987) A study of the cytology of two female cultivars of date palm. Date Palm J 5:123–124

Al-Salih AA, Al-Jarrah A (1987) Chromosomes number of a date palm male: cultivar Ghannami Akhdar. Date Palm J 5:128–137

Al-Salih AA, Al-Najjar NR, Al-Mashhadani AN (1987) A study on the chromosome number of two specific female date palm cultivars. Date Palm J 5:134–143

Al-Wasel ASAA (2005) Survey study on somaclonal variations in vitro-derived date palm trees. In: Proceedings international workshop on true-to-typeness of date palm tissue culture-derived plants. Institut National de Recherche Agronomique, Morocco, 23–25 May 2005. http://www.inra.org.ma/ist/publications/ouvrages/truetype.pdf

Baaziz M, Aissam F, Brakez Z et al (1994) Electrophoretic patterns of acid soluble proteins and active isoforms of peroxidase and polyphenoloxidase typifying calli and somatic embryos of two reputed date palm cultivars in Morocco. Euphytica 76:159–168

Beal JM (1937) Cytological studies in the genus *Phoenix*. Bot Gaz 99:400–407

Brar DS, Jain SM (1998) Somaclonal variation: mechanism and applications in crop improvement. In: Jain SM, Brar DS, Ahloowalia BS (eds.) Somaclonal variation and induced mutations in crop improvement. Kluwer Academic, Dordrecht, pp 15–38

Brown PTH, Lange FD, Kranz E, Lörz H (1993) Analysis of single protoplast and regenerated plants by PCR and RAPD technology. Mol Gen Genet 237:311–317

Capy P (1998) Evolutionary biology: a plastic genome. Nature 396:522–523

Cassells AC, Joyce SM, Curry RF, McCarthy TF (1999) Detection of economic variability in micropropagation. In: Altman A, Ziv M, Izhar S (eds.) Plant biotechnology and in vitro biology in the 21st century. Kluwer Academic, Dordrecht, pp 241–244

Chaibi N, Ben Abdallah A, Harzallah H, Lepoivre P (2002) Potentialités androgénétiques du palmier dattier *Phoenix dactylifera* L. et culture in vitro d'anthères. Biotechnol Agron Soc 6:201–207

Colijn-Hooymans CM, Hakkert JC, Jansen J, Custers JBM (1994) Competence for regeneration of cucumber cotyledons is restricted to specific developmental stages. Plant Cell Tissue Organ Cult 39:211–217

Corniquel B, Mercier L (1994) Date palm (*Phoenix dactylifera* L.) cultivar identification by RFLP and RAPD. Plant Sci 101:163–172

Corniquel B, Mercier L (1997) Identification of date palm (*Phoenix dactylifera* L.) cultivars by RFLP: partial characterization of a cDNA probe that contains a sequence encoding a zinc finger motif. Int J Plant Sci 158:152–156

Cullis C, Rademan S, Kunert KJ (1999) Method for finding genetic markers of somaclonal variation. International publication number WO 99/53100 1999

D'Amato F (1985) Cytogenetics of plant cell and tissue culture and their regenerates. CRC Crit Rev Plant Sci 3:73–112

Dennis ES (2004) Molecular analysis of the alcohol dehydrogenase (adhI) genes of maize. Nucleic Acids Res 12:3983–4000

Djerbi M (2000) Abnormal fruiting of the date palm trees derived from tissue culture. In: Proceedings date palm international symposium, Windhoek, Namibia, 22–25 Feb 2000, p 73

Duncan RR (1997) Tissue culture-induced variation and crop improvement. Adv Agron 58:201–240

Eke CR, Akomeah P, Asemota O (2005) Somatic embryogenesis of Date palm (*Phoenix dactylifera* L.) from apical meristem tissues from "Zebia" and "Loko" landraces. Afr J Biotechnol 4:244–246

El Bellaj M (2000) Etude de quelques mécanismes biochimiques impliqués dans l'acquisition des potentialités embryogènes et la maturation des embryons somatiques chez le palmier dattier (*Phoenix dactylifera* L.). Thèse d'Université, Faculté des Sciences Semlalia, Marrakech, Morocco

El Bellaj M, El Hadrami I (2004) Characterization of two non constitutive hydroxycinnamic acid derivatives in date palm (*Phoenix dactylifera* L.) callus in relation with tissue browning. Biotechnology 3:155–159

El Bellaj M, El Jaafari S, El Hadrami I (2000) L'AIA-oxydase marqueur et régulateur potentiel de l'embryogenèse somatique chez *Phoenix dactylifera* L. Cah Agric 9:193–195

El Hadrami I (1995) L'embryogenèse somatique chez Phoenix dactylifera L. quelques facteurs limitants et marqueurs biochimiques. Thèse de Doctorat d'Etat, Faculté des Sciences Semlalia, Université Cadi Ayyad, Marrakech, Morocco

El Hadrami I, Baaziz M (1995) Somatic embryogenesis and analysis of peroxidases in *Phoenix dactylifera* L. Biol Plant 37:197–203

El Hadrami I, Coumans M (1994) Somatic embryogenesis in *Phoenix dactylifera* L.: relation with phenolic compounds. In: de Kouchkovsky Y, Larher F (eds.) Plant sciences. SFPV-Université de Rennes I, Rennes

El Hadrami I, El Hadrami A (2009) Breeding date palm. In: Jain SM, Priyadarshan PM (eds.) Breeding plantation tree crops. Springer, New York, pp 191–216

El Hadrami I, Cheikh R, Baaziz M (1995) Somatic embryogenesis and plant regeneration from shoot-tip explants in *Phoenix dactylifera* L. Biol Plant 37:205–211

El Hadrami A, El Idrissi-Tourane A, El Hassni M, Daayf F, El Hadrami I (2005) Toxin-based in vitro selection and its potential application to date palm for resistance to the bayoud Fusarium wilt. A review. C R Biol 328:732–744

El Houmaizi MA, Saaidi M, Oihabi A, Cilas C (2002) Phenotypic diversity of date-palm cultivars (*Phoenix dactylifera* L.) from Morocco. Genet Resour Crop Evol 49:483–490

Eshraghi P, Zarghami R, Ofoghi H (2005a) Genetic stability of micropropagated plantlets in date palm. J Sci Islam Repub Iran 16:311–315

Eshraghi P, Zaghami R, Mirabdulbaghi M (2005b) Somatic embryogenesis in two Iranian date palm cultivars. Afr J Biotechnol 4:1309–1312

Evans DA, Sharp WR (1986) Somaclonal and gametoclonal variation. In: Evans DA, Sharp WR, Ammirato PV (eds.) Handbook of plant cell culture, vol 4. Macmillan, New York, pp 43–97

Evans DA, Sharp WR, Medina-Filho HP (1984) Somaclonal and gametoclonal variation. Am J Bot 71(6):759–774

Fki L (2005) Application des suspensions cellulaires embryogènes au clonage et à l'amélioration in vitro du palmier dattier. Thèse d'Université, Faculté des Sciences de Sfax, Sfax, Tunisia

George EF, Debergh PC (2008) Micropropagation: uses and methods. In: George EF, Hall MA, De Klerk G (eds.) Plant propagation by tissue culture, vol 1, 3rd edn. Springer, Dordrecht, pp 29–64

George EF, Sherington PF (1984) Plant propagation by tissue culture: handbook and directory of commercial laboratories. Exegetics Ltd, London

Gueye B, Morcillo F, Collin M et al (2009) Acquisition of callogenic capacity in date palm leaf tissues in response to 2,4-D treatment. Plant Cell Tissue Organ Cult 99:35–45

Gupta PK, Holmstrom D (2005) Douglas - fir (*Pseudotsuga menziesii*). In: Jain SM, Gupta PK (eds.) Protocol for somatic embryogenesis in woody plants. Springer, Dordrecht, pp 25–34

Herman EB (1991) Recent advances in plant tissue culture. In: Regeneration, micropropagation and media 1988–1991. Agritech Consultants, Shrub Oak, pp 6–10

Hussain F, El-Zeid A (1978) Studies on physical and chemical characteristics of date varieties of Saudi Arabia. Ministry of Agriculture and Water, Riyadh

Ibrahim AMF, El Kobbia AM, Kitat FM, Abd El Kawy MM (1998) Cytological studies on date palm. I. Chromosomal behavior during meiosis of two date palm (*Phoenix dactylifera* L.) male types. Alex J Agric Res 43:237–246

Jain SM (2001) Tissue culture-derived variation in crop improvement. Euphytica 118:153–166

Jain SM (2002) Tissue culture and induced mutations useful tools for floriculture industry. In Vitro Cell Dev Biol Plant 38:643

Jain SM (2007) Recent advances in date palm tissue culture and mutagenesis. Acta Hortic 736:205–211

Jones CJ, Edwards KJ, Castaglione S et al (1997) Reproducibility testing of RAPD, AFLP and SSR markers in plants by a network of European laboratories. Mol Breed 3:381–390

Karp A (1989) Can genetic instability be controlled in plant tissue cultures? Newsl Int Assoc Plant Tissue Cult 58:2–11

Karp A (1993) Are your plants normal? – Genetic instability in regenerated and transgenic plants. Agro-Food-Industry Hi-Tech May/June 7–12

Karp A (1995) Somaclonal variation as a tool for crop improvement. Euphytica 85:1–3

Kephart SR (1990) Starch gel electrophoresis of plant isozymes: a comparative analysis of techniques. Am J Bot 77:693–712

Khierallah HSM, Bader SM (2007) Micropropagation of date palm (*Phoenix dactylerefa* L.) var. Maktoom through organogenesis. Acta Hortic 736:213–223

Kumar PS, Mathur VL (2004) Chromosomal instability is callus culture of *Pisum sativum*. Plant Cell Tissue Organ Cult 78:267–271

Kunert KJ, Vorster J, Bey E, Cullis CA (2000) Representational difference analysis as a DNA-based quality assurance procedure for date palm micropropagation. In: Date palm international symposium, Windhoek, Namibia, 22–25 Feb 2000, pp74–80

Kunert KJ, Vorster J, Bester C, Cullis CA (2002) DNA microchip technology in the plant tissue culture industry. In: Rajasekaran K, Jacks TJ, Finley JW (eds.) Crop biotechnology, vol 829, ACS symposium series. American Chemical Society, Washington DC, pp 86–96

Lacaze P, Giraud-Henry I, Brackpool A (2000) Molecular fingerprinting of date palm cultivars using AFLP. In: Date palm international symposium, Windhoek, Namibia, 22–25 Feb 2000, pp 89–90

Larkin PJ, Scowcroft WR (1981) Somaclonal variation: a novel source of variability from cell cultures for plant improvement. Theor Appl Genet 60:197–214

Lemieux B, Aharoni A, Schena M (1998) Overview of DNA chip technology. Mol Breed 4:277–289

Lisitsyn N, Lisitsyn N, Wigler M (1993) Cloning the differences between two complex genomes. Science 259:946–951

Lo Schiavo F, Pitto L, Giuliano G et al (1989) DNA methylation of embryogenic carrot cell cultures and its variations as caused by mutation, differentiation, hormones and hypomethylating drugs. Theor Appl Genet 77:325–331

Loutfi K (1999) Organogenèse et embryogenèse somatique à partir des tissus floraux du palmier dattier (*Phoenix dactylifera* L.) cultivés *in vitro*. Aspects histologiques et caryologie des vitroplants. Thèse de Doctorat d'Etat, Faculté des Sciences Semlalia, Marrakech, Morocco

Maluszynski M, Kasha KJ (2002) Mutations, in vitro and molecular techniques for environmentally sustainable crop improvement. Kluwer Academic, Dordrecht, 246 p

Martinez O, Reyes LM, Beltran M (1998) Chemovariability in the genus *Musa*: similarities and differences. Infomusa 7:16–20

Masmoudi-Allouche F, Mezioul B, Kriaâ W, Gargouri-Bouzid R, Drira N (2009) In vitro flowering induction in date palm (*Phoenix dactylifera* L.). J Plant Growth Regul 29(1):35–43

McCubbin MJ, van Staden J, Zaid A (2000) A southern African survey conducted for off-types on date palms produced using somatic embryogenesis. In: Proceedings of date palm international symposium, Windhoek, Namibia, 22 Feb 2000, pp 68–72

Munksgaard D, Mattsson O, Okkele FT (1995) Somatic embryo development in carrot is associated with an increase in levels of *S*-adenosylmethionine, *S*-adenosylhomocysteine and DNA methylation. Physiol Plant 93:5–10

Nemec B (1910) Das problem der befruchtungsvorgänge und andere zytologische Fragen. Gebrüder Borntraeger, Berlin 532 p

Omar MS, Novak FJ (1990) In vitro regeneration and ethylmethane sulfonate (EMS) uptake in somatic embryogenesis of date palm. Plant Cell Tissue Organ Cult 20:185–190

Othmani A, Bayoudh C, Drira N et al (2009a) Somatic embryogenesis and plant regeneration in date palm *Phoenix dactylifera* L., cv. Boufeggous is significantly improved by fine chopping and partial desiccation of embryogenic callus. Plant Cell Tissue Organ Cult 97:71–79

Othmani A, Bayoudh C, Drira N, Trifi M (2009b) In vitro cloning of date palm *Phoenix dactylifera* L., CV. Deglet Bey by using embryogenic suspension and temporary immersion bioreactor (TIB). Biotechnol Biotechnol Equip 23(2):1181–1188

Ouenzar B, Hartmann C, Rode A, Benslimane A (1998) Date palm DNA mini-preparation without liquid nitrogen. Plant Mol Biol Rep 16:263–269

Pedrieri S (2001) Mutation induction and tissue culture in improving fruits. Plant Cell Tissue Organ Cult 64:185–210

Peschke VM, Phillips RL (1992) Genetic implication of somaclonal variation in plants. Adv Genet 30:41–75

Phillips RL, Kaeppler M, Peschkke VM (1990) Do we understand somaclonal variation? In: Nijkamp HJJ, van der Plast LHW, van Aartrijk J (eds.) Progress in plant molecular biology. Kluwer Academic, Dordrecht, pp 131–141

Phillips RL, Kaeppler SM, Olhoft P (1994) Genetic instability of plant tissue cultures: breakdown of normal controls. Proc Nat Acad Sci USA 91:5222–5226

Powell WW, Koput K, Smith-Doerr L (1996) Interorganizational collaboration and the locus of innovation: networks of learning in biotechnology. Adm Sci Q 41:116–145

Rodrigues PHV, Tulmann Neto A, Cassieri Neto P, Mendes BMJ (1998) Influence of the number of subcultures on somaclonal variation in micropropagated Nanico (*Musa* spp., AAA group). Acta Hortic 490:469–473

Saker M, Bekheet S, Taha H et al (2000) Detection of somaclonal variation in tissue culture-derived date palm plants using isozyme analysis and RAPD fingerprints. Biol Plant 43:347–351

Salman RM, Al Jibouri AAM, Al Quadhy WK, Omar MS (1988) Isozyme and chromosomal analyses of tissue culture derived date palms. Date Palm J 6:401–411

Scribner KT, Pearce JM (2000) Microsatellites: evolutionary and methodological background and empirical applications at individual, population, and phylogenetic levels. In: Baker A (ed.) Molecular methods in ecology. Blackwell Science, London, pp 235–271

Sedra MH, Lashermes P, Trouslot P et al (1998) Identification and genetic diversity analysis of date-palm (*Phoenix dactylifera* L.) varieties from Morocco using RAPD markers. Euphytica 103:75–82

Sghaier-Hammami B, Drira N, Jorrín-Novo JV (2009) Comparative 2-DE proteomic analysis of date palm (*Phoenix dactylifera* L.) somatic and zygotic embryos. J Prot 73:161–177

Shepherd K, Souza FVD, Da Silva KM (1996) Mitotic instability in banana varieties. IV. BAP concentration and effects of number of subcultures. Fruits 51:211–216

Skirvin RM, McPheeters KD, Norton M (1994) Sources and frequency of somaclonal variation. HortScience 29:1232–1237

Skroch P, Nienhaus J (1995) Impact of scoring error and reproducibility of RAPD data on RAPD based estimates of genetic distance. Theor Appl Genet 91:1086–1091

Smith RJ, Ansley JS (1995) Field performance of tissue cultured date palms (*Phoenix dactylifera*) clonally produced by somatic embryogenesis. Principes 39:47–52

Szarejko Y, Guzy J, Jiménez-Dávalos J et al (1995) Production of mutants using barley DH systems. In: IAEA (ed.) Induced mutations and molecular techniques for crop improvement. (IAEA/FAO Proceedings of an international symposium). IAEA, Vienna, pp 517–530

Taha SH, Hassan MM, El-Bahr MK (2007) Micropropagation of some Egyptian date palm dry cultivars Maturation of somatic embryos. Arab J Biotechnol 10:333–340

Tisserat B (1979) Propagation of date palm (*Phoenix dactylifera* L.) in vitro. J Exp Bot 30:1275–1283

Trujillo I, de García E (1996) Strategies for obtaining somaclonal variants resistant to yellow Sigatoka (*Mycosphaerella musicola*). Infomusa 5:12–13

Vagner M, Fischerova L, Spackova J, Vondrakova Z (2005) Somatic embryogenesis in Norway Spruce. In: Jain SM, Gupta PK (eds.) Protocol for somatic embryogenesis in woody plants. Springer, Dordrecht, pp 141–156

Van den Bulk RW, Löffler HJM, Lindhout WH, Koornneef M (1990) Somaclonal variation in tomato: effect of explant source and a comparison with chemical mutagenesis. Theor Appl Genet 80:817–825

Vasil IK (1994) Automation of plant propagation. Plant Cell Tissue Organ Cult 39:105–108

Vorster JB, Kunert KJ, Cullis CA (2002) Use of representational difference analysis for the characterization of sequence differences between date palm varieties. Plant Cell Rep 21:271–275

Witjaksono (2003) Peran bioteknologi dalam pemuliaan tanaman buah tropika. *Seminar Nasional Peran Bioteknologi dalam Pengembangan Buah Tropika*. Kementerian Riset dan Teknologi RI & Pusat Kajian Buah Buahan Tropika, IPB. Bogor

Yeung EC, Thrope TA (2005) Somatic embryogenesis in Picea glauca. In: Jain SM, Gupta PK (eds.) Protocol for somatic embryogenesis in woody plants. Springer, Dordrecht, pp 47–58

Zaid A, Al-Kaabi H (2003) Plant-off types in tissue culture-derived date palm. (*Phoenix dactylifera* L.). Emir J Agric Sci 15:17–35

Zouine J, El Hadrami I (2004) Somatic embryogenesis in *Phoenix dactylifera* L: effect of exogenous supply of sucrose on proteins, sugars, phenolics and peroxidases activities during the embryogenic cell suspension culture. Biotechnology 3:114–118

Zouine J, El Bellaj M, Meddich A et al (2005) Proliferation and germination of somatic embryos from embryogenic suspension cultures in *Phoenix dactylifera* L. Plant Cell Tissue Organ Cult 82:83–92

Chapter 10
Growth Abnormalities Associated with Micropropagation of Date Palm

N.S. Al-Khalifah and E. Askari

Abstract Traditionally date palm (*Phoenix dactylifera* L.) is propagated through offshoots but limited numbers bar their wide distribution. With the advent of tissue culture massive expansion of elite cultivars was made possible, but commercially *in vitro* plants were introduced only in the last two decades. Propagation of date palm through tissue culture generally results in true-to-type plants but some off-types with abnormal phenotypes also develop due to somaclonal variation caused by genetic and/or epigenetic changes during the tissue culture process. In tissue culture-derived date palm many off-type phenotypes are detected that include stunted or severely retarded growth, leaves with wide leaflets, variegated leaves, malformation of inflorescence, abnormal flowers, abnormal multiple carpels and low levels of fruit set. Off-types with genetic changes may be produced due to excessive plant multiplication cycles and growth regulators used to get a maximum number of plants from a single mother plant. Interestingly many of these off-types appear in the plants at maturity thus escaping the selection and screening at an early stage. It is therefore important that appropriate quality assurance tests are performed both at the molecular and morphological levels to ensure true-to-typeness. Since phenotypic identification of off-types is not reliable at early growth stages of tissue culture-derived plants, identification at the molecular level, isozyme and DNA, is inevitably the solution. The use of hormones as a remedy to abnormalities is also discussed.

Keywords Molecular analysis • Mutation • Off type • Somaclonal variation • Tissue culture

---

N.S. Al-Khalifah (✉) • E. Askari
King Abdulaziz City for Science and Technology, KACST, P. O. Box 6086, Riyadh 11442, Saudi Arabia
e-mail: abujawad@kacst.edu.sa; easkari@kacst.edu.sa

S.M. Jain et al. (eds.), *Date Palm Biotechnology*,
DOI 10.1007/978-94-007-1318-5_10, © Springer Science+Business Media B.V. 2011

10.1 Introduction

Date palm (*Phoenix dactylifera* L.) (2n = 2x = 36) production is one of the important economic supports for the agricultural sector and also one of the major sources of income for farmers and related industries. A vast genetic diversity is found in date palm and is maintained as cultivars by local farmers. It is presumed that most of the present-day cultivars of date palms evolved through hybridization with wild species in both Africa and southwest Asia. Human selection also helped in maintaining superior cultivars through vegetative propagation, which was otherwise not possible due to a high degree of cross-pollination and heterozygosity. The natural hybridization process followed by human selection brought about many cultivars, which were given folk names and grown in local orchards. These cultivars were later transplanted to other areas, through off-shoot multiplication.

Date palm belongs to the family Arecaceae and genus *Phoenix* (Moore 1973). The genus *Phoenix* contains about 13 species (Barrow 1998) which are mostly wild and are also grown as ornamental palms both indoors and outdoors. Only *P. dactylifera* is cultivated for its fruit. Pollen of one species can fertilize other species thus they are all interfertile (Muirhead 1961). Cultivated date palms are, for example, successfully pollinated with *P. reclinata* in East Africa and in Northeast Africa with *P. atlantica*. In Pakistan and India cultivated date palm hybridizes with *P. sylvestris* and in Spain and Northwest Africa with *P. canariensis* (Benbadis 1992; Oudejans 1979). If the hypothesis of a unique progenitor in the origin of *Phoenix dactylifera* is true, it must have been hybridized with other wild species of the genus (Munier 1981).

Date palm is a dioecious, woody, perennial, monocot, and its out-breeding habit makes its seedling progeny heterogeneous (Munier 1981). Dioecism of the date palm has given rise to artificial pollination by hand, and the necessity for vegetative propagation from offshoots which has facilitated cloning of elite cultivars with economically-important traits. Date palms are mainly cultivated in arid regions in the Middle East, where they have been domesticated for at least 500 years and are believed to have originated in Mesopotamia (Wrigley 1995; Zohary and Hopf 2000). The dry climate in the arid regions facilitates pollination and fruit ripening. In early human history date palm offshoots of different cultivars were transplanted, established and adapted to distant areas like North Africa, Egypt, Spain and Pakistan, from their initial area of origin. At the beginning of the twentieth century, offshoots of elite cultivars from Iraq, Saudi Arabia, Algeria, Egypt and Tunisia were also transplanted to California, USA (Nixon and Carpenter 1978).

10.2 Economic Importance and Production

Date palm is of great socioeconomic importance in many countries of the Middle East and North Africa where its cultivation generates a main income source for farmers. The number of known date palm varieties distributed all over the world is approximately 3000 (Zaid and de Wet 1999) out of which about 450 alone are found

Table 10.1 Estimated area (in hectares) and production (in metric tons) of dates crop in different regions of the Kingdom of Saudi Arabia from 2000 to 2004

Region/year	2000	2001	2002	2003	2004
Riyadh	185,680	207,706	209,938	212,398	213,065
	35,352	33,552	33,432	33,128	37,361
Makkah	55,342	63,305	69,734	68,109	69,002
	8,510	9,191	9,072	8,981	9,773
Madinah	82,345	90,697	90,298	98,664	100,906
	16,350	17,181	17,536	18,199	18,819
Qaseem	112,355	126,451	130,041	138,470	160,844
	35,340	33,728	34,402	34,277	35,808
Eastern	95,823	103,470	104,269	1,186,667	129,711
	13,750	12,085	11,766	11,043	11,639
Aseer	75,325	80,914	82,802	80,444	72,666
	8,923	8,061	8,016	8,392	7,154
Tabuk	24,345	27,300	27,816	29,691	30,078
	4,523	4,464	4,715	4,678	4,437
Hail	60,545	69,049	68,232	85,159	97,284
	12,525	12,653	12,924	13,979	14,889
Nothern	56	61	66	84	69
	16	17	8	12	14
Jazan	125	147	130	298	221
	125	318	318	205	166
Najran	12,355	13,426	13,462	14,125	17,922
	3,235	3,078	3,045	3,438	3,390
Baha	2,315	3,512	3,006	2,729	3,547
	345	698	680	712	963
Jouf	28,233	31,849	29,746	35,251	45,978
	3,456	4,073	4,065	4,378	4,389
Kingdom of Saudi Arabia	734,844	817,887	829,540	884,088	941,293
	142,450	139,099	139,979	141,421	148,801

Source: Ministry of Agriculture, Saudi Arabia (2005)

in the Kingdom of Saudi Arabia (Bashah 1996). Hussain and El-Zeid (1978) have reported almost 400 cultivars in Saudi Arabia, based mainly on their fruit characteristics. In Tunisia more than 250 cultivars have been documented but Deglet Noor dominates all over the country (Rhouma 1994). Iraq, Iran, Saudi Arabia, Pakistan and Egypt are the leading producers and represent 69% of the world date production. Morocco, Tunisia, Algeria, United Arab Emirates, Sudan, USA, Israel and Oman are also some of the most important date-producing countries (Botes and Zaid 2002). Iraq formerly led the world in date production with more than 881,000 mt annually. In recent years output has decreased in Iraq due to a trade embargo and subsequent war. Cultivars of outstanding performance are found mostly in the Basra region of Iraq some of which are Zahidi, Halawy, Barhee, Dayri, Khadrawy, Khastawi, Breem and Maktoom. Saudi Arabia produces more than 712,000 mt per year. The Al-Gassim, Al-Madinah and Al-Ahsa regions are famous for good quality and high yielding cvs. like Barhee, Sukkary, Khlass Ajwah and Sirri (Table 10.1).

Iran also produces fine quality dates. More than 900,000 mt of dates are produced per year mainly in the regions of Kerman, Hormozgan and Khuzestan. Some of the selected cvs. in Iran are Mazafati, Rabbi, Zahidi, Kabkab, Piyarom, Shahani, Sayer, Zarday and Farkan. In Pakistan the annual production of dates is more than 550,000 mt (Botes and Zaid 2002). Baluchistan, Sind and part of the Punjab region produce good quality dates; Aseel, Berni, Begun, Jangi, Hillawi, Khadrawy and Zahidi are some of the cvs. grown in these areas.

10.3 Propagation Through Offshoots and Tissue Culture

True-to-type multiplication of elite cultivars with economically important traits is mainly through vegetative propagation from offshoots of the mother trees to fulfill the demand for the date palm industry (Qaddoury and Amssa 2003; Zaid and de Wet 2002). Since conventional propagation through offshoots is slow and is affected by less production and low survival rate, tissue culture (TC) has been used over the last two decades as an alternative for the mass propagation of true-to-type cultivars of date palm (Al-Wasel 2000a, b; Branton and Blake 1989; Zaid and de Wet 2002). With the advent of tissue culture techniques in date palms, massive expansion of elite cultivars was made possible and transportation of the planting materials of these cultivars to distant places became easier. In the past two decades, several small- and large- scale growers have adopted this innovative technique of propagation to expand their plantings. They imported tissue culture-derived plantlets of their choice from well-known tissue culture laboratories of Europe. This eventually helped in improving the yield and spreading of the cultivars to faraway places, which would otherwise have been confined only to certain localities.

Regeneration of the date palm by somatic embryogenesis has been reviewed by Tisserat (1979, 1984) and by Branton and Blake (1989). In 1979 Tisserat established somatic embryogenesis in date palm. Shoot tips and lateral buds of offshoots are used to induce embryogenic callus (Al-Khalifah 2000; Al-Khalifah et al. 2001, 2006; Al-Khayri 2001, 2003; Al-Khayri and Al-Bahrany 2001; Beauchesne et al. 1989; Gabr and Tisserat 1985; Sharma et al. 1984; Tisserat and DeMason 1980; Zaid and Tisserat 1983). For large-scale micropropagation of cv. Deglet Noor, clonal plant regeneration has been reported from somatic embryos derived from highly proliferating suspension cultures (Fki et al. 2003). Embryogenic calli were initiated from leaf and inflorescence explants. Large numbers of somatic embryos were produced by the subculturing of suspensions in liquid media with a low concentration of growth regulators. A 20-fold increase in the production of embryos was achieved by using embryogenic suspensions instead of solid media cultures. An increase in the germination rate of 25–80% was also observed, which may be due to a partial decrease in the water content of the somatic embryos.

Embryo germination was also found to be stimulated by simply cutting the cotyledonary leaf. With somatic embryogenesis it is easy to produce plants on a large scale but the use of a high level of hormones in the media causes somaclonal mutations which are detected only after 6–8 years of plantation growth i.e. at the flowering stage. Thus, Beauchesne and Rhiss (1979) established organogenesis as an alternative for date palm tissue culture. Organogenesis by single cell from organ-like structures, mostly shoots, may not cause mutations in the plants due to the use of a low level of hormones in the media. Many other explants like leaves (Sharma et al. 1980; Zaid 1981), inflorescences (Drira 1981; Tisserat and DeMason 1980), stems (Poulain et al. 1979; Tisserat 1979), roots (Eeuwens 1978; Smith 1975) and zygotic embryos (Reuveni 1979; Zaid and Tisserat 1983) have also been used for this purpose. However, due to somaclonal variations in particular and genetic variations in general, the genome of the tissue-cultured plants cannot be certified to be identical to its mother plant (Al-Khalifah et al. 2006; Evans 1989; Karp 1995; Rhiss 1988; Saker et al. 2000; Zaid 1990). The analysis of genetic stability of tissue-cultured plants is then necessary for micropropagation of true-to-type plants and germplasm conservation.

10.4 Somaclonal Variations in TC-Derived Plants

Propagation either through offshoots or TC techniques like somatic embryogenesis (Al-Khalifah 2000; Al-Khalifah et al. 2006; Al-Khayri 2003; Tisserat 1982; Zaid and de Wet 1999) must result in true-to-type plants. However, off-types not identical to their progenitor are also produced due to somaclonal variation caused by genetic and/or epigenetic changes during the TC process (Al-Khalifah and Askari 2007; Al-Khalifah et al. 2007a; Cullis et al. 1999; Kaeppler et al. 2000; Kunert et al. 2003; Sala et al. 1999). Tissue culture-derived plants of a large number of species are known to exhibit somaclonal variations (Duncan 1997). These changes in the phenotype may be genetic in nature, which are heritable through sexual cycles and may even be used for crop improvement (Jain 2001; Karp 1995). Chromosomal aberrations, like translocation, insertion/deletion, caused by delayed replication of heterochromatin under *in vitro* conditions, and sequence change, singularly or in combination, are believed to be responsible for these stable variations (Dennis et al. 1987; Hao and Deng 2002).

Tissue culture generated variations may also be epigenetic or transient, which are manifested as a carry-over effect of tissue culture and are eliminated with repeated sexual cycles, or after the passage of time in the case of perennial plants (Kaeppler et al. 2000). During *in vitro* culture, phenomena like oxidative stresses (Cassells and Curry 2001; Joyce et al. 2003), altered level of DNA methylation (Bender 2004; Jaligot et al. 2000, 2002; Xu et al. 2004) and activation of transposable elements (Courtial et al. 2001; Peschke and Phillips 1991; Peschke et al. 1991) are associated with epigenetic variations appearing in tissue culture-derived populations.

A common observation in most studies related to date palm is the severity of this problem of variations in tissue culture-derived plants during the initial years of fruiting, but it declines with age of the plant (Cohen et al. 2004; Djerbi 2000; Gurevich et al. 2005; Reuveni 1986). This observation points to the epigenetic nature of this abnormality. Gradual elimination of the epigenetic changes is believed to be due to correction of the DNA methylation pattern and actions of transposons/retrotransposons by cellular repair mechanism (Brettell and Dennis 1991; Hirochika et al. 1996). For example, Oono (1985) obtained a dwarf mutant of rice which bred true on selfing, but showed signs of reversion after out-crossing. When treated with the DNA methylation inhibitor, five-deoxyazacytidine the dwarf plants grew to normal height. These findings suggested that epigenetic changes may be reversed by applying an appropriate agent.

Plant growth hormones, like 2, 4-D, are known to cause mutations. Concentration of plant growth regulators in culture medium has been found to affect the DNA methylation level of callus cells (Arnholt-Schmitt 1995; LoSchiavo et al. 1989) and perhaps result in epigenetic variations as well (Corley et al. 1986). Therefore in tissue-cultured date palm plants physiological disorders like failure to set fruit, leaves with wide leaflets, variegated leaves, deformed and seedless fruits, and formation of abnormal multicarpel fruitlets may be attributed to somaclonal variations (Al-Wasel 2000a,b; Djerbi 2000; McCubbin et al. 2000). Some of these disorders are reversible and are believed to be due to epigenetic changes (Cohen et al. 2004; Gurevich, et al. 2005). In other cases it is hard to differentiate between permanent genetic changes and epigenetic changes (Cullis et al. 1999; Sala et al. 1999). Epigenetic changes are expressed under stress conditions possibly due to DNA methylation, DNA amplification and/or activation of transposable elements (Brar and Jain 1998; Brettell and Dennis 1991; Hirochika et al. 1996; Kaeppler et al. 2000). DNA methylation is reported to increase with the increase in concentration of auxin-type 2, 4-D hormone and to decrease with the increase in concentration of cytokinins-type Kinetin (LoSchiavo et al. 1989). Most of the somaclonal variations are found in plants tissue cultured from callus tissues (Al-Wasel 2000a, 2001; Ramage et al. 2004). In TC-derived cvs. like Barhee, Khalas, Deglet Noor, Nabtat Saif and Medjool, physiological disorders are high (Al-Wasel 2001). Interestingly many of these off-type phenotypes appear in the plants at maturity thus escaping the selection and screening at an early stage which is not economical either for the commercial producer or the grower. It is therefore important that appropriate quality assurance tests be performed to ensure the production of true-to-type plants.

10.5 The Role of Molecular Markers

Since phenotypic identification of off-types is not reliable at the early stages of the TC-derived plants, identification at the molecular or isozyme and DNA level is inevitably the solution. The advantage of using DNA analysis lies in the fact that the DNA is not affected by environment and is also not tissue- or stage-specific. Many molecular techniques like RAPD, RFLP, AFLP and RDA have been applied in the identification of date palm cultivars (Al-Khalifah and Askari 2003, 2006; Al-Khalifah et al. 2006, 2007c; Askari et al. 2003; Cullis et al. 1999; Kunert et al. 2003). Molecular analyses of TC plants showing abnormal phenotypes were also performed using RAPD (Al-Khalifah and Askari 2007; Al-Khalifah et al. 2007a; Saker et al. 2006) and isozymes (Azeqour et al. 2002; Saker et al. 2000). Both techniques detected plantlets that were showing difference in traits at an early stage but were unable to resolve differences in the abnormal traits that appeared later at a mature stage of the plant growth. Saker et al. (2000) reported the occurrence of genetic difference in almost 4% of a total 70 TC-derived plantlets which were 6–12 months old. Gurevich et al. (2005) recently performed AFLP analysis of both normal and abnormal off-types of TC-derived plants of cv. Barhee but were unable to detect any AFLP marker linked to traits of abnormal flowering and low level of fruit set. Al-Kaabi et al. (2005) also performed a comparative study of the two tissue culture methods (embryogenesis and organogenesis) to estimate the frequency of somaclonal variation in the TC plants by using AFLP analysis. They concluded that the frequency of somaclonal variation was relatively high in plants regenerated by embryogenesis. Al-Khalifah et al. (2007a) also found a very low level of DNA polymorphism among 30 TC-derived plants.

Some genetic variation was detected in abnormal and normal fruit bearing TC-derived plants of Barhee. This indicates that at least to some extent the abnormal fruit bearing off-types can be detected from the normal plants through RAPD analysis. They have suggested that the abnormal fruit-bearing traits in most of the TC plants are not only due to genomic changes but also due to some epigenetic changes that might have occurred during tissue culture stages. Due to the low level of genetic variation among the abnormal and normal fruit bearing TC plants, no consistent marker could be identified between the two phenotypes except the low level of similarity observed between three abnormal and two normal plants (Fig. 10.1a–c). Corniquel and Mercier (1994) also detected some genetic variation between different plants of Barhee by applying RAPD and RFLP techniques. New advanced molecular techniques like DNA microarray technology may help to resolve the genetic diversity and detection of markers linked to this trait (Kunert et al. 2002).

10.6 Abnormal Flowers and Low Fruit Set

Abnormal flowers and low levels of normal fruit set, especially in TC plants of cultivar Barhee and some other elite cultivars, are common in Saudi Arabia. Large numbers of tissue-cultured plants introduced in the country in and around 1995 are producing an average of more than 60% abnormal flowers and fruitlets per plant causing low levels of fruit set, which represents a great economic loss to the farmers (Kunert et al. 2003). A plausible reason for low levels of fruit set may be inadequate pollination due to abnormalities in the flowers (carpels, style and stigma). As a consequence all three of the unfertilized carpels develop into abnormal and small fruitlets. In normal date flowers, after pollination, only a single carpel develops into a fruit and the other two carpels degenerate. More than three carpelled florets and fruitlets (4–7 carpels) have been observed (Al-Wasel 2000b, 2001; Cohen et al. 2004; Djerbi 2000; Gurevich et al. 2005; Reuveni 1986). It is possible that the

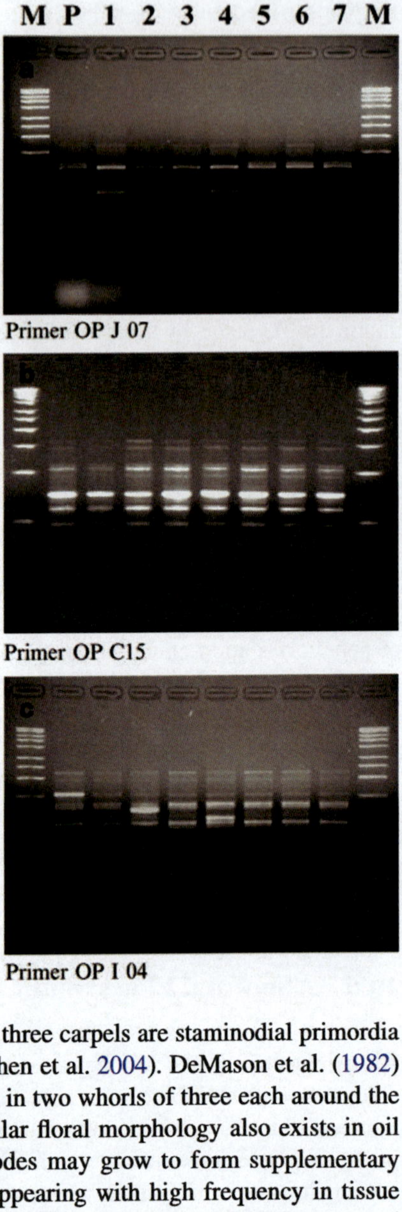

Fig. 10.1 RAPD profiles of Mother Plant and 7 TC derived plants of cultivar Barhee using Operon primers (**a**) OPJ07 (**b**) OPC15 and (**c**) OPI 04. Lanes: *M* molecular weight marker, *P* parent mother plant and 1–7: TC-derived plants

supernumerary carpels other than the main three carpels are staminodial primordia transformed into carpel-like structures (Cohen et al. 2004). DeMason et al. (1982) have shown the presence of six staminodes in two whorls of three each around the pistil in female flowers of date palm. Similar floral morphology also exists in oil palm (Van Heel et al. 1987); and staminodes may grow to form supplementary carpels of mantled fruit, an abnormality appearing with high frequency in tissue culture-derived oil palm plants (Corley et al. 1986). Cohen et al. (2004) suggested that, like oil palm, supernumerary carpels in abnormal fruits of tissue culture-derived date palm may also originate from the staminodes. In flowers from tissue culture derived date palm, structurally deformed stigmas also occur with a high frequency (Cohen et al. 2004). Normal pollination takes place on these stigmas but for unknown reasons, the pollen tube degenerates inside the style-tissue. Continued growth of all three carpels even after the pollination suggests that the degeneration of two

Fig. 10.2 Comparison of the normal and abnormal multicarpel fruits

carpels and growth of single ovary may not be triggered by pollination, but after commencement of fertilization. When fertilization fails to occur, all the existing ovaries may continue to grow to form small fruitlets with rudimentary ovules, instead of two degenerating, as occurs in normal fruit formation (Fig. 10.2a, b).

Development of extra-pistillary fruitlets probably is not associated with failure of fertilization. This suggestion is supported by the absence of such fruitlets in unpollinated flowers of offshoot-derived plants and their abundance in unpollinated rachises of tissue culture-derived plants. Growth of these structures may be an independent phenomenon associated with *in vitro* background of the plants. DeMason and Tisserat (1980) have shown that a high concentration of 2, 4-D in the medium induced bisexuality in male flowers in culture, resulting in the growth of pistillodes into fruit-like structures devoid of an ovule. Many flowers have impaired pollen tube elongation. Limited tube elongation to the stigma or the region around its attachment to the carpel has been observed. Either the pollen tube elongation seizes-up or is unidirectional (Cohen et al. 2004).

The incidence of abnormalities in tissue culture-derived plants of many date palm cvs., especially Barhee, has been reported with varying degree of occurrence (Al-Wasel 2000b; Cohen et al. 2004; Djerbi 2000; McCubbin et al. 2000).

Tissue culture methodology, genotype and age of the plant may have an influence on prevalence of abnormalities in tissue culture-derived date palm populations (Cohen et al. 2004).

10.7 Hormones as a Remedy

Abnormal multicarpel fruits in date palm cause a great reduction in normal fruits and is of concern to date farmers. The abnormal fruits are being produced in abundance mainly by tissue culture-derived plants but have also been observed in some natural trees. This abnormality may also be attributed to ineffective pollination and/or incompatible pollen with female flowers. Multicarpeled fruitlets are of small size, without seeds and are normally shed before ripening. The role of hormones on this abnormality cannot be denied. Cytokinins may have a great effect on reducing the rate of abnormal fruitlets especially in tissue-cultured cv. Barhee. Multicarpel fruits could be suppressed by exogenous application of kinetin on stigmas before pollination.

The process of fruit formation in higher plants is mediated by many hormones produced within the ovule as a result of fertilization (Gillaspy et al. 1993; Gorguet et al. 2005; Vivian-Smith and Koltunow 1999). Generally, in the absence of fertilization, fruit development either fails to take place at all or is severely retarded (Granell et al. 1992; O'Neill and Nadeau 1997; Vercher and Carbonell 1991; Vivian-Smith et al. 2001), probably due to an inadequate level of endogenous hormones (Gillaspy et al. 1993; Pharis and King 1985). It is believed that developing seeds also produce hormones which contribute to ovary wall changes leading to normal fruit formation (Ben-Cheikh et al. 1997; Rodrigo et al. 1997; Talon et al. 1990a). Extraction, purification and identification of such growth regulators may prove effective for exogenous application at different stages of pollination to reduce the rate of abnormal fruits in different economically-important cultivars of date palm.

In some species, however, a failure of fertilization leads to formation of parthenocarpic fruits (Gorguet et al. 2005; Talon et al. 1990a, 1992) suggesting that under natural conditions there are also pathways for ovary wall growth and senescence that may operate without stimuli from fertilization or growing seed.

It has been observed that naturally-occurring parthenocarpic fruits have higher levels of endogenous gibberellins and auxins (Talon et al. 1990b,1992) and parthenocarpic fruits can also be produced by exogenous application of plant growth regulators, including gibberellins, auxins and cytokinins, which apparently make up for inadequacy of the endogenous hormones (Schwabe and Mills 1981). In one study, abnormalities were suppressed by the exogenous application of kinetin on stigmas before pollination (Al-Khalifah et al. 2007b). These reports provide sufficient clues to the involvement of hormones in abnormality formation in date palm. However, hormone profiles of developing fruits of date palm have not yet been worked out and the cause of abnormalities remains unclear, although preliminary information is available on the genesis of this abnormality (Al-Khalifah et al. 2007b; Cohen et al. 2004).

10.8 Conclusion and Future Prospects

This study reviews current efforts to elaborate the causes of, and to find a plausible solution for, the detrimental somaclonal variations in tissue-cultured date palm cultivars. Plant regeneration from embryogenic tissue culture needs more research attention as at present it is not yielding promising results. Gene cloning and transformation technologies (protoplast and biolistics) also need to be explored to develop healthy high-yielding cultivars devoid of abnormalities. Genetic engineering in date palm will prove successful only when the problems faced at present in the tissue culture propagation of cultivars are solved, because tissue culture is a basic requirement of genetic engineering. DNA fingerprints of elite cultivars of date palm can also serve as genetic markers to generate linkage maps of date palm which may in turn facilitate anchoring of genetic markers linked to abnormal traits and also other economically important traits. Marker-assisted selection also shows promise of an efficient transfer of traits from related species and elite cultivars. New advanced molecular techniques such as DNA microarray may help to better resolve the genetic diversity and detection of markers linked to these abnormal traits (Kunert et al. 2002).

References

Al-Kaabi HH, Zaid A, Ainsworth C (2005) AFLP variation in tissue culture-derived date palm plants. In: Proceedings international workshop on true-to-type of date palm tissue culture-derived plants, Morocco, 23–25 May, p 93

Al-Khalifah NS (2000) In *vitro* culture studies on date palm (*Phoenix dactylifera* L.) cv. Mosaifah and Nabtet Sultan. Plant Tissue Cult 10:1–8

Al-Khalifah NS, Askari E (2003) Molecular phylogeny of date palm (*Phoenix dactylifera* L.) cultivars from Saudi Arabia by DNA fingerprinting. Theor Appl Gent 107:1266–1270

Al-Khalifah NS, Askari E (2006) Detection of genetic diversity in date palm (*Phoenix dactylifera*). In: Read MM (ed.) Focus on DNA fingerprinting research. NOVA Science, New York, pp 137–149

Al-Khalifah NS, Askari E (2007) Early detection of genetic variation in date palms propagated from tissue culture and offshoots by DNA fingerprinting. Acta Hort 736:105–112

Al-Khalifah NS, Okawara R, Al-Hafidh Y et al (2001) *In vitro* culture of some cultivars of date palm (*Phoenix dactylifera* L.) in Saudi Arabia. Hort Sci 36:457

Al-Khalifah NS, Khan FA, Askari E et al (2006) *In vitro* culture and genetic analysis of male and female date palm (*Phoenix dactylifera* L.). Acta Hort 725:653–662

Al-Khalifah NS, Askari ES, Hadi S et al (2007a) Genetic analysis of abnormal fruiting in tissue culture-derived trees of date palm 'Barhy' grown in Saudi Arabia. Acta Hort 763:155–165

Al-Khalifah NS, Hadi S, Khan FA et al (2007b) Effect of IAA and kinetin on reversing fruit abnormality of tissue culture-derived date palm (*Phoenix dactylifera* L.). Abstracts. Fourth symposium on date palm in Saudi Arabia, King Faisal University, Al-Hassa, 5–8 May, pp 163

Al-Khalifah NS, Shanavaskhan AE, Askari E (2007c) A morpho-genetic approach to characterize genetic diversity in date palms (*Phoenix dactylifera* L.). 3rd Global Botanic Garden Congress, Wuhan, 16–20 Apr

Al-Khayri JM (2001) Optimization of biotin and thiamine requirements for somatic embryogenesis of date palm (*Phoenix dactylifera* L.). In Vitro Cell Dev Biol Plant 37:453–456

Al-Khayri JM (2003) In *vitro* germination of somatic embryos in date palm: effect of auxin concentration and strength of MS salts. Curr Sci 84:680–683

Al-Khayri JM, Al-Bahrany AM (2001) Silver nitrate and 2- isopentyladenine promote somatic embryogenesis in date palm (*Phoenix dactylifera* L.). Sci Hort 89:291–298

Al-Wasel AS (2000a) Tissue culture technique: is it a safe method to micropropagate elite date palm (*Phoenix dactylifera* L.) cultivars. Arab J Biotechnol 3:245–246

Al-Wasel AS (2000b) Vegetative and fruiting composition of tissue culture-derived and conventionally propagated date palm (*Phoenix dactylifera* L.) cv. Barhy trees. In Vitro Cell Dev Biol Plant 36:1010

Al-Wasel AS (2001) Field performance of somaclonal variants of tissue culture-derived date palm (*Phoenix dactylifera* L.). Plant Tiss Cult 11:97–105

Arnholt-Schmitt B (1995) Physiological aspects of genome variability in tissue culture II. Growth phase-dependent quantitative variability of repetitive BstNI fragments of primary cultures of Daucus carota L. Theor Appl Genet 91:816–823

Askari E, Al-Khalifah NS, Ohmura T et al (2003) Molecular phylogeny of seven date palm (*Phoenix dactylifera* L.) cultivars by DNA fingerprinting. Pak J Bot 35:323–330

Azeqour M, Majourhat K, Baaziz M (2002) Morphological variations and isoenzyme polymorphism of date palm clones from *in vitro* culture acclimatized and established on soil in south Morocco. Euphytica 123:57–66

Barrow S (1998) A monograph of *Phoenix* L. (Palmae: Coryphoideae). Kew Bull 53:513–575

Bashah MA (1996) Date variety in the Kingdom of Saudi Arabia. Guidance booklet, palms and dates. King Abdulaziz University Press, Riyadh, pp 1225–1319

Beauchesne G, Rhiss A (1979) Micropropagation of selected palms. Proc Sym Plant Tissue Culture, FAO, Vienna

Beauchesne G, Zaid A, Rhiss A (1989) Meristimatic potentialities of bottom of young leaves to rapidly propagate date palm. In: Proceedings second symposium on date palm, Saudi Arabia, 3–6 Mar 1986, pp 87–95

Benbadis AK (1992) Coconut and date palm. In: Hammerschlag FA, Litz RE (eds.) Biotechnology of perennial fruit crops. CAB Int, Wallingford, pp 383–400

Ben-Cheikh W, Perez-Botella J, Tadeo FR et al (1997) Pollination increases gibberellin levels in developing ovaries of seeded varieties of citrus. Plant Physiol 114:557–564

Bender J (2004) DNA methylation and epigenetic. Annu Rev Plant Biol 55:41–68

Botes A, Zaid A (2002) The economic importance of date production and international trade. Chapter III. In: Zaid A (ed.) Date palm cultivation. FAO Plant Production Protection Paper 156, Rev. 1. FAO, Rome

Branton RL, Blake J (1989) Date palm (*Phoenix dactylifera* L.). In: Bajaj YPS (ed.) Biotechnology in agriculture and forestry, vol 5, Trees I. Springer, Heidelberg, pp 161–175

Brar DS, Jain SM (1998) Somaclonal variation: mechanisms and applications in crop improvement. In: Jain SM, Brar DS, Ahloowalia BS (eds.) Somaclonal variation and induced mutations in crop improvement. Kluwer, Boston, pp 17–37

Brettell RLS, Dennis ES (1991) Reactivation of a silent Ac following tissue culture is associated with heritable alterations in its methylation pattern. Mol Gen Genet 229:365–372

Cassells AC, Curry RF (2001) Oxidative stress and physiological, epigenetic and genetic variability in plant tissue culture: implications for micropropagators and genetic engineers. Plant Cell Tiss Organ Cult 64:145–157

Cohen Y, Korchinsky R, Tripler E (2004) Flower abnormalities cause abnormal fruit setting in tissue culture-propagated date palm (*Phoenix dactylifera* L.). J Hort Sci Biotechnol 79:1007–1013

Corley RHV, Lee CH, Wong CY (1986) Abnormal flower development in oil palm clones. Planter 62:233–240

Corniquel B, Mercier L (1994) Date palms (*Phoenix dactylifera* L.) cultivar identification by RFLP and RAPD. Plant Sci 101:163–172

Courtial B, Feuerbach F, Eberhard S et al (2001) Tnt1 transposition events are induced by *in vitro* transformation of *Arabidopsis thaliana* and transposed copies integrated into genes. Mol Genet Genom 265:32–42

Cullis C, Rademan S, Kunert KJ (1999) Methods for finding genetic markers of somaclonal variation. International publication number WO 99/53100

DeMason DA, Tisserat B (1980) The occurrence and structure of bisexual flowers in date palm, *Phoenix dactylifera* L. (Arecaceae). Bot J Linn Soc 81:283–292

DeMason DA, Stolte KW, Tisserat B (1982) Floral development in *Phoenix dactylifera* L. Can J Bot 60:1439–1446

Dennis ES, Brettell RIS, Peacock WJ (1987) A tissue culture induced Adhl null mutant of maize results from a single base change. Mol Gen Genet 210:181–183

Djerbi M (2000) Abnormal fruiting of the date palm derived from tissue culture. In: Proceedings date palm international symposium, Windhoek, p 73

Drira N (1981) Multiplication vegetative et micropropagation du palmier dattier (*Phoenix dactylifera* L.) a partir d'organes preleves sur la phase adulte cultives "in vitro". These pour le titre de Docteur de Specialite. Faculte des Sciences de Tunis

Duncan RR (1997) Tissue culture-induced variation and crop improvement. Adv Agron 58:201–240

Eeuwens CJ (1978) Effects of organic nutrients and hormone on growth and development of tissue explants from coconut (*Cocos nucifera*) and date (*Phoenix dactylifera* L.) palms cultured *in vitro*. Physiol Plant 42:173–178

Evans D (1989) Somaclonal variation-genetic basis and breeding applications. Trends Genet 5:46–50

Fki L, Masmoudi R, Drira N, Rival A (2003) An optimized protocol for plant regeneration from embryogenic suspension cultures of date palm, *Phoenix dactylifera* L., cv. Deglet Nour. Plant Cell Rep 21:517–524

Gabr MF, Tisserat B (1985) Propagating palms *in vitro* with special emphasis on the date palm (*Phoenix dactylifera* L.). Sci Hort 25:255–262

Gillaspy G, Ben-David H, Gruissem W (1993) Fruits: a developmental perspective. Plant Cell 5:1439–1451

Gorguet B, van-Heusden AW, Lindehout P (2005) Parthenocarpic fruit development in tomato. Plant Biol 7:131–139

Granell A, Harris N, Pisabarro AG, Carbonell J (1992) Temporal and spatial expression of a thiolprotease gene during pea ovary senescence and its regulation by gibberellins. Plant J 2:907–915

Gurevich V, Lavi U, Cohen Y (2005) Genetic variation in date palms propagated from offshoots and tissue culture. J Am Hort Sci 130:46–53

Hao YJ, Deng XX (2002) Occurrence of chromosomal variations and plant regeneration from long-term-cultured citrus callus. In Vitro Cell Dev Biol 38:472–476

Hirochika H, Sugimoto K, Otsuki Y et al (1996) Retroposons of rice involved in mutations induced by tissue culture. Proc Nat Acad Sci USA 93:7783–7788

Hussain F, El-Zeid A (1978) Studies on physical and chemical characteristics of date palm varieties of Saudi Arabia. Ministry Agriculture Water, Jeddah

Jain SM (2001) Tissue culture-derived variation in crop improvement. Euphytica 118:153–166

Jaligot E, Rival A, Beule T et al (2000) Somaclonal variation in oil palm (*Elaeis guineensis* Jacq.): the DNA methylation hypothesis. Plant Cell Rep 19:684–690

Jaligot E, Beule T, Rival A (2002) Methylation-sensitive RFLPs characterization of two oil palm markers showing somaclonal variation-associated polymorphism. Theor Appl Genet 104:1263–1269

Joyce SM, Cassells AC, Jain SM (2003) Stress and aberrant phenotypes in *in vitro* culture. Plant Cell Tiss Organ Cult 74:103–121

Kaeppler SM, Kaeppler HF, Rhee Y (2000) Epigenetic aspects of somaclonal variation in plants. Plant Mol Biol 43:179–188

Karp A (1995) Somaclonal variation as a tool for crop improvement. Euphytica 85:295–302

Kunert KJ, Vorster J, Bester C et al (2002) DNA microchip technology in the plant tissue culture industry. In: Rajasekaran K, Jacks TJ, Finley JW (eds.) Crop biotechnology, ACS Sym Ser 829. American Chemical Society, Washington, DC, pp 86–96

Kunert KJ, Baaziz M, Cullis CA (2003) Techniques for determination of true-to-type date palm (*Phoenix dactylifera* L.) plants: a literature review. Emirates J Agric Sci 15(1):1–16

LoSchiavo F, Pitto L, Giuliano G et al (1989) DNA methylation of embryogenic carrot callus culture and its variations as caused by mutation, differentiation, hormone, and hypomethylating drugs. Theor Appl Genet 77:325–331

McCubbin MJ, Van-Staden J, Zaid A (2000) A southern African survey conducted for off-types in date palm produced using somatic embryogenesis. In: Proceedings of the date palm international symposium, Windhoek, pp 68–72

Ministry of Agriculture, Saudi Arabia (2005) Agriculture statistical year book. Ministry of Agriculture, Riyadh, Saudi Arabia, 18th issue, p 68

Moore HE (1973) The major groups of palms and their distributions. Gent Herb 11:27–141

Muirhead D (1961) Palms. Dale Stuart King Pub, Phoenix

Munier P (1981) Origine de la culture sur palmier dattier et sa propagation en Afrique. Fruits 36:437–450

Nixon RW, Carpenter JB (1978). Growing dates in the United States. USDA, Agric Inf Bull No. 207, Washington, DC

O'Neill SD, Nadeau JA (1997) Post-pollination flower development. Hort Rev 19:1–58

Oono K (1985) Putative homozygous mutants in regenerated plants of rice. Mol Gen Genet 198:377–384

Oudejans JHM (1979) Date palm, *Phoenix dactylifera* (Palmae). In: Simmonds NW (ed.) Evolution of crop plants. Longman, London, pp 229–231

Peschke VM, Phillips RL (1991) Activation of the maize transposable element suppressor mutant (Spm) in tissue culture. Theor Appl Genet 81:90–97

Peschke VM, Phillips RL, Gengenbach BG (1991) Genetic and molecular analysis of tissue culture-derived Ac elements. Theor Appl Genet 82:121–129

Pharis RP, King R (1985) Gibberellins and reproductive development in seed plants. Annu Rev Plant Physiol 36:517–568

Poulain C, Rhiss A, Beauchesne G (1979) Multiplication vegetative en culture *in vitro* du palmier dattier (*Phoenix dactylifera* L.). CR Acad Agric 11:1151–1154

Qaddoury A, Amssa M (2003) Impact of indole butyric acid on the rooting capacity of young date palm offshoots. Acta Bot Gall 150:213–222

Ramage CM, Borda AM, Hamill SD, Smith MK (2004) A simplified PCR test for early detection of dwarf off-types in micropropagated Cavendish bananas (*Musa* spp. AAA). Sci Hort 103:145–151

Reuveni O (1979) Embryogenesis and plantlet growth of date palm (*Phoenix dactylifera* L.) derived from callus tissue. Plant Physiol Supp 63:138

Reuveni O (1986) Date. In: Monselise SP (ed.) Handbook of fruit set and development. CRC Press, Boca Raton, pp 119–144

Rhiss A (1988) La muiltiplication vegetative *in vitro* appliquee au palmier dattier (*Phoenix dactylifera* L.). Compte Rendu du Premier Groupe de Travail sur le Multiplication Rapide du Palmier dattier par les Techniques de Culture *in vitro*. Project de Lutte Contre le Bayoud, FAO, 84/018, Rome, pp 23–28

Rhouma A (1994) Le palmier dattier en Tunisie. I: le patrimoine genetique. ARABASQUE, Tunis

Rodrigo MJ, Garcia-Martinez JL, Santes C et al (1997) The role of gibberellins A_1 and A_3 in fruit growth of *Pisum sativum* L. and the identification of gibberellins A_4 and A_7 in young seeds. Planta 201:446–455

Saker MM, Bekheet SA, Taha HS et al (2000) Detection of somaclonal variations in tissue culture-derived date palm plants using isoenzyme analysis and RAPD fingerprinting. Biol Plant Prague 43(3):347–351

Saker MM, Adawy SS, Mohamed AA, El-Itriby HA (2006) Monitoring of cultivar identity in tissue culture-derived date palms using RAPD and AFLP analysis. Biol Plant 50(2):198–204

Sala F, Arencibia A, Castiglione S et al (1999) Molecular and field analysis of somaclonal variation in transgenic plants. In: Altman A, Ziv M, Izhar S (eds.) Plant biotechnology and *in vitro* biology in the 21st century. Kluwer, Dordrecht, pp 259–262

Schwabe WW, Mills JJ (1981) Hormones and parthenocarpic fruit set: a literature survey. Hort Abst 51:661–699

Sharma DR, Kumari R, Chowdhury JB (1980) In *vitro* culture of female date palm *(Phoenix dactylifera* L.) tissues. Euphytica 29:169–174

Sharma DR, Dawra S, Chowdhury JB (1984) Somatic embryogenesis and plant regeneration in date palm (*Phoenix dactylifera* L.) cv. Khadravi through tissue culture. Indian J Exp Biol 22:596–598

Smith SN (1975) Vegetative propagation of date palm by root-tip culture. Bull Agron Sahar 1:67

Talon M, Haden P, Primo-Millo E (1990a) Gibberellins in *Citrus sinuses*: a comparison between seeded and seedless varieties. J Plant Growth Regul 9:201–206

Talon M, Zacarias L, Primo-Millo E (1990b) Hormonal changes associated with fruit set and development in mandarins differing in their parthenocarpic ability. Physiol Plant 79:400–406

Talon M, Zacarias L, Primo-Millo E (1992) Gibberellins and parthenocarpic ability in developing ovaries of seedless mandarins. Plant Physiol 99:1575–1581

Tisserat B (1979) Propagation of date palm (*Phoenix dactylifera* L.) *in vitro*. J Exp Bot 90:1275–1283

Tisserat B (1982) Factors involved in the production of plantlets from date palm callus cultures. Euphytica 31:201–214

Tisserat B (1984) Date palm. In: Sharp WR, Evans DA, Ammirato PV, Yamada Y (eds.) Handbook of plant cell culture, vol 2, Crop species. Macmillan, New York, pp 505–545

Tisserat B, DeMason DA (1980) A histological study of development of adventives embryos in organ cultures of *Phoenix dactylifera* L. Ann Bot 46:465–472

Van Heel WA, Breure CJ, Menendez T (1987) The early development of inflorescences and flowers of the oil palm (*Elaeis guineensis* Jacq.) seen through the scanning microscope. Blumea 32:67–78

Vercher Y, Carbonell J (1991) Changes in the structure of ovary tissues and in the ultra structure of mesocarp cells during ovary senescence or fruit development induced by plant growth substances in *Pisum sativum*. Physiol Plant 81:518–526

Vivian-Smith A, Koltunow A (1999) Genetic analysis of growth regulator-induced parthenocarpy in *Arabidopsis*. Plant Physiol 121:437–451

Vivian-Smith A, Luo M, Chaudhaury M, Koltunow A (2001) Fruit development is actively restricted in the absence of fertilization in *Arabidopsis*. Development 128:2321–2331

Wrigley G (1995) Date palm. In: Smartt J, Simmonds NW (eds.) Evolution of crop plants, 2nd edn. Longman Group, Essex, pp 399–403

Xu M, Li X, Korban SS (2004) DNA methylation alterations and exchanges during *in vitro* cellular differentiation in rose (*Rosa hybrida* L.). Theor Appl Genet 109:899–910

Zaid A (1981) Rapid propagation of the date palm through tissue culture. Master of Science thesis prepared at the U.S. Date and Citrus Station, Indio

Zaid A (1990) In *vitro* hardening-off of date palm (*Phoenix dactylifera* L.) plantlets. Ph.D. thesis, Colorado State University, Colorado

Zaid A, de Wet PF (1999) Botanical and systematic description of the date palm. In: Zaid A (ed.) Date palm cultivation. FAO, Rome, pp 1–28

Zaid A, de Wet PF (2002) Date palm propagation. Chapter V. In: Zaid A (ed.), Date palm cultivation. FAO Plant Production Protection Paper 156. Rev. 1. FAO, Rome

Zaid A, Tisserat B (1983) *In vitro* shoot tip differentiation in *Phoenix dactylifera* L. Date Palm J 2:163–182

Zohary D, Hopf M (2000) Domestication of plants in the Old World, 3rd edn. Oxford University Press, New York

Chapter 11
Molecular Detection of Somaclonal Variation in Date Palm

Y. Cohen

Abstract Somaclonal variation is a general phenomenon in tissue culture-regenerated plants of many species. Several typical abnormal phenotypes have been detected in tissue culture-propagated date palms. Variation from the source phenotypes is undesirable in the commercial propagation of specific cultivars, and efforts have been made to eliminate the formation of these off-types. Some of these phenotypes resemble off-type phenotypes occurring in other crop plants, such as banana and oil palm. Genetic variation, and altered DNA methylation patterns and gene expression are associated with these abnormal phenotypes. However, the specific key factors involved in their generation are still unknown and efficient markers for their early identification do not exist. The newly available draft of the date palm genome sequence and high-throughput sequencing techniques will enable genomic and transcriptomic analyses of off-types, which in turn will provide a better understanding of the phenomenon of somaclonal variation.

Keywords DNA methylation • Dwarf • Molecular analysis • Off-type • Tissue culture

11.1 Introduction

Date palm (*Phoenix dactylifera* L.) is the major fruit tree grown in arid areas in the Middle East and North Africa, where the crop has a significant impact on the economy and serves as a major food source (Chao and Krueger 2007; Zaid and De Wet 1999). In the last century, dates have been introduced to new regions in Australia, Southern Africa, South America, Mexico and the USA. Date palms are traditionally propagated through offshoots. This propagation method, however, is limited by the availability

Y. Cohen (✉)
Department of Fruit Tree Sciences, Institute of Plant Sciences,
Volcani Research Center, Bet Dagan, 50250, Israel
e-mail: vhyuvalc@volcani.agri.gov.il

S.M. Jain et al. (eds.), *Date Palm Biotechnology*,
DOI 10.1007/978-94-007-1318-5_11, © Springer Science+Business Media B.V. 2011

of offshoots, usually ranging between 1 and 20 per tree, depending on the cultivar. Expansion of date palm plantations and introduction to new regions are therefore restricted by the availability of offshoots.

Tissue culture (TC) is an efficient method for the rapid mass propagation of plants. This procedure is especially important in monocot crops such as banana, oil palm and date palm, in which clonal propagation is limited. The development of TC propagation methods has enabled the expansion of date palm groves (Al-Khayri 2007; Tisserat 1979a, b). These techniques are based on either embryogenesis or organogenesis. TC propagation has the potential to generate a large number of homogenous plants, has no seasonal effect on plant source and enables easy and safe exchange of plant material. In addition, by generating large numbers of pest- and disease-free trees it serves to fight the spread of date pests (such as the red palm weevil) and diseases (most importantly bayoud disease, caused by the fungus *Fusarium oxysporum*). Attempts are being made to generate transgenic date palm trees with disease resistance or various other functional advantages (Habashi et al. 2008). TC propagation is therefore crucial for the future development of date palm plantations and culture.

11.2 Somaclonal Variation and Formation of Date Palm Off-Types

In vitro propagation requires that plants remain true-to-type, i.e., be similar in every way to their progenitors. However, the process of TC may also result in the production of off-types, which differ from the original propagated plants. Somaclonal variation is a general phenomenon in TC-regenerated plants of many species (Kaeppler et al. 2000). It is associated with several types of genomic and epigenetic changes, including changes in ploidy, chromosomal rearrangements, point mutations, rearrangement of transposable elements and DNA methylation. Somaclonal variation is unpredictable in nature and can be both heritable (genetic) and non-heritable (epigenetic) (El Hadrami et al. 2011; Jain 2001; Joyce et al. 2003; Kaeppler et al. 2000). Somaclonal variation can be beneficial and may be used for varietal improvement: the various different phenotypes generated can be screened for traits of interest (El Hadrami et al. 2011; Jain 2001; Larkin and Scowcroft 1981). A discussion of the specific potential of somaclonal variation for date palm improvement is presented in another chapter of this book (El Hadrami et al. 2011). However, since many off-types exhibit reduced vigor or yield, from a practical point of view, variation from the source phenotypes in the commercial propagation of specific clones is undesirable. Effort has therefore been invested in eliminating the formation of these phenotypes, by identifying and removing them at early stages of development. In *in vitro*-propagated monocot crops such as banana, oil palm and date palm, somaclonal variation is a significant horticultural problem and studies have been performed to characterize the phenomenon, define conditions and protocols that will minimize it, and develop tools for its early detection (Al Kaabi et al. 2007; Cohen et al. 2007; Corley et al. 1986; Kunert et al. 2003; Sahijram et al. 2003).

Several typical off-types have been detected in TC-propagated date palms. These include trees producing seedless parthenocarpic fruits, trees with variegated leaves or with variations in leaf structure, dwarf trees and trees with variations in overall growth patterns (Al Kaabi et al. 2007; McCubbin et al. 2000). The occurrence of off-types is dependent on TC protocols and varies among batches and producers (Al Kaabi et al. 2007; Cohen et al. 2004; Gurevich et al. 2005). While in most cases the fraction of off-types generated is limited, there have been several reports in which most of the generated trees were found to be abnormal (Al-Wasel 2001; Al Kaabi et al. 2007; Cohen et al. 2004; Gurevich et al. 2005). Both producers and growers try to identify and remove the phenotypically abnormal trees as soon as possible. Some of these trees can be detected and removed at early stages of plant growth, many of them during the *in vitro* process, and others during the hardening stage. However, other off-types can only be detected in the field, several years after planting. In some cases, the off-types can only be identified after tree maturation, flowering and fruit set. Here, we summarize characterization studies and molecular analyses of three common off-types: dwarf trees, trees with fruit-setting abnormalities and trees with leaf variegation.

11.3 Somaclonal Off-Types in Other Tree Crops

The occurrence of off-types is not specific to date palm: it is common to a variety of plant species. It is especially important in several monocot fruit crops which are commonly propagated via TC. While eudicot trees can be propagated by grafting, traditional vegetative propagation of monocot trees is dependent on the availability of offshoots (or suckers). Such offshoots are limited in some crops (date palm and banana), and completely absent in others (oil palm and coconut palm): in the latter, TC is the only available method for clonal propagation. Various off-types are detected in these species and interestingly, many of them are similar in various aspects, suggesting common mechanisms underlying their generation (Cohen et al. 2003; Corley et al. 1986; Gurevich et al. 2005; Sahijram et al. 2003). The occurrence of homologous phenotypes in other species and the additional information available on them are discussed in relation to the major date palm off-types.

11.4 Typical Off-Types in Date Palm: Phenotypic Description and Comparative Analysis with Off-types in Other Crops

11.4.1 Variegation

One of the most common off-type phenotypes in date palm is leaf variegation, characterized by light-colored sectors along the rachis and leaflets (Al Kaabi et al. 2007; McCubbin et al. 2000). This phenotype is detected in many plantlets at the hardening stage and although these are removed and not planted in the field, additional variegated trees, which may reach up to 3% of the planted trees, are detected during the first years after planting. The variegated sectors can be relatively small, comprising several leaflets, or cover large regions in a number of leaves. Although in several instances, normal-looking fruits were produced by such variegated trees, farmers usually remove them upon detection of the abnormality. Leaf variegation probably occurs due to mutations in the photosynthetic apparatus or in chlorophyll synthesis. Therefore, many independent mutations can cause the variegated phenotype (Al Kaabi et al. 2007; Gurevich et al. 2005).

11.4.2 *Fruit-Setting Abnormalities*

Many TC-originated date palms have been identified as having a unique abnormal phenotype characterized by very low levels of fruit set. This phenotype is very common in trees of the Barhee and Hallas cvs., but also occurs in others. The same phenomenon has been noted in several countries in the Middle East and North Africa (Al-Wasel 2001; Al Kaabi et al. 2007; Cohen et al. 2004; Djerbi 2000; McCubbin et al. 2000). In Saudi Arabia for example, approximately 100,000 similarly abnormal trees were discovered (Al-Wasel 2000, 2001). In normal trees, following fertilization only a single carpel, of the three carpels present in each flower, develops into a normal fruit (Fig. 11.1a). However, in these off-type trees, most of the flowers turn into three-carpel parthenocarpic fruits (Fig. 11.1b, c). In addition, multi-carpel flowers and fruitlets (with up to six or nine carpels instead of the normal three-carpel flowers) have also been detected (Al-Wasel 2001; Cohen et al. 2004; Djerbi 2000) (Fig. 11.1c–d). These parthenocarpic fruits have no commercial value. Other structural abnormalities in the flowers of these trees include twisting and distortion of the stigmas (Cohen et al. 2004). The severity of this phenotype varies among trees of various sources and ages, as well as among similar trees in the same grove. Moreover, it has been found to be alleviated during maturation of the trees, as detected by reductions in the levels of both parthenocarpic fruits and fruits with supernumerary carpels, in parallel to an increase in the level of normal fruit produced. Many of these trees revert to normal, with yields similar to those of normal trees, within 10 years of planting (Cohen et al. 2004). Nevertheless, the phenomenon reappears in trees originated from offshoots of these off-type trees. The severity of the symptoms correlates with that of the mother tree from which they originated (Cohen et al. 2004).

11.4.2.1 "Mantled" Oil Palm

A phenomenon similar to the abnormal fruit setting of date palms was discovered almost 30 year ago in TC-originated African oil palms (*Elaeis guineensis* Jacq.) (Corley et al. 1986). These abnormal trees share a specific phenotype termed "Mantled", which is characterized by very low fruit-setting efficiency and formation

Fig. 11.1 Description of the two major date palm somaclonal off-types: (**a**–**d**) Fruit setting abnormality: (**a**) Barhee fruit bunch with normal, single carpel fruitlets, (**b**) An off-type bearing mainly three-carpelate fruitlets, (**c**) Hallas spikelet bearing parthenocarpic fruits with supernumerary carpels (*arrows*), and (**d**) fruitlets with various numbers of supernumerary carpels. Bar represent 10 mm in **c** and **d**. (**e**–**f**) Dwarf Medjool date trees (*left*) compared to normal trees of similar age (*right*)

of supernumerary carpels. As with date palms, the severity of the symptoms in most of the "mantled" oil palm trees declines during their maturation, with many trees reverting to normal following several years of growth in the field. The genetic similarities between the date and oil palms and their common abnormal phenotype suggest that the mechanism governing the formation of the fruit-setting phenotype in date palms may be similar to that in the "mantled" oil palm off-types (Cohen et al. 2004).

11.4.3 Dwarfism

Another major off-type phenotype generated through TC propagation, termed *dwarfism*, is characterized by a slow growth rate in the field. This off-type has been detected in thousands of young trees in different countries (Al Kaabi et al. 2007; Gurevich et al. 2005; McCubbin et al. 2000), and is very common among specific batches of the Medjool cv. (Cohen et al. 2007). These trees have a very short trunk and shorter than normal leaves (Fig. 11.1e, f). Interestingly, fruitlets carrying supernumerary carpels have been detected in the inflorescences of these trees (Cohen et al. unpublished). While thousands of these dwarf trees originated from a specific TC laboratory, they have been found among trees produced by other laboratories as well (Al Kaabi et al. 2007; McCubbin et al. 2000).

11.4.3.1 Dwarf Banana

Another monocot that is largely produced through TC is banana. In this crop, as in dates, many off-types are generated in TC. The levels of somaclonal abnormal phenotypes differ among cultivars, TC conditions and batches. Although usually less than 10% of the plantlets planted in the field are later identified as off-types, their occurrence causes major economic damage (Damasco et al. 1998; Israeli et al. 1991; Sahijram et al. 2003; Withers 1993). The most common banana off-types are the dwarf trees (Israeli et al. 1991; Sahijram et al. 2003). Dwarf off-types may account for up to 80% of all somaclonal variants in Cavendish bananas (Bairu et al. 2006; Reuveni and Israeli 1990). TC-propagated plants of other species also tend to form abnormal dwarf plants.

11.5 Genetic Variation and Markers for Varietal Description Among Cultivars

Many molecular studies have compared different cultivars and identified characteristic band patterns. In principle, any available marker can be used to characterize different cultivars. Application of such methods to date palms has resulted in the identification of differences between cultivars by isoenzymes (Baaziz and Saaidi 1988; Bendiab et al. 1998), random fragment length polymorphism (RFLP) (Corniquel and Mercier 1994, 1997), random amplified polymorphic DNA (RAPD) (Corniquel and Mercier 1994; Saker and Moursi 1999; Sedra et al. 1998), amplified fragment length polymorphism (AFLP) (Cao and Chao 2002; El-Assar et al. 2005; Jubrael et al. 2005; Rhouma et al. 2007), simple sequence repeats (SSR) (Hamama et al. 2003; Zehdi et al. 2005) and representational difference analysis (RDA) (Johnson et al. 2009; Vorster et al. 2002). These studies have enabled the genetic characterization of cultivars and the assessment of genetic distances between them.

11.5.1 Variation Among Trees Propagated by Traditional Methods

Since commercial date palms are clonally propagated from offshoots, different individuals of the same cultivar should be genetically identical. Most studies compare only a single accession from each cultivar. However, in some reports, several individual plants from the same cultivar were analyzed, and a considerable amount of sequence variation was found to exist, even in trees propagated from offshoots. For example, polymorphism was detected within the cvs. Barhee, Deglet Noor and Medjool propagated from offshoots using RAPD (Corniquel and Mercier 1994) and AFLP (Devanand and Chao 2003; Gurevich et al. 2005) analyses. In Medjool in particular, a very high level of variation was detected, suggesting that this cultivar is actually a land race rather than a single highly conserved cultivar (Elhoumaizi et al. 2006). Since genetic variation exists within different exemplars of a same cultivar clonally propagated by offshoots, identification of off-types by the existence of genetic variation is problematic, and does not necessarily suggest that an abnormal phenotype can be expected.

11.6 Methods for True-to-Typeness Detection

11.6.1 Morphological Differences

Many off-types are morphologically different at the TC stage or during hardening, and molecular analysis of abnormal-looking plantlets can usually detect genetic variation. Such variation was found by both isoenzyme and RAPD analysis (Saker et al. 2000): in one experiment, all morphologically abnormal shoots showed genetic variations at the molecular level (Saker et al. 2000). However, many off-types are indistinguishable at the early stages. Important characteristics such as vegetative growth, pollination and fruit set can only be observed years after the trees have been planted and grown in the field (Kunert et al. 2003).

In banana, methods to screen and identify dwarf plantlets, based on morphological characteristics at the end of their nursery stage, were developed (Reuveni and Israeli 1990; Smith and Hamill 1993) and are now routinely used by commercial TC companies. However, *in vitro*-grown date palms are planted as plantlets, years before they develop their tree morphology, characterized by a fast-growing trunk. Therefore, screens for the early detection of dwarf off-types are not available for date palms.

11.6.2 Use of Isoenzymes and DNA Markers to Look for Genetic Variation Associated with Specific Phenotypes

Most of the available DNA marker types have been used on date palms, mostly to cluster cultivars into related groups. A detailed description of the use of such markers in date palm is presented in another chapter of this book (Cullis 2011). Various methods have been developed to assess the genetic fidelity and uniformity (true-to-typeness) of micropropagated plants (Kunert et al. 2003; Read and Preece 2003). These methods focus on identifying changes in the DNA sequence, such as point mutations, deletions or insertions and DNA rearrangements. Most conservative methods for the identification of true-to-type date palms have been reviewed by Kunert et al. (2003). Isoenzyme (Azeqour et al. 2002; Zivdar et al. 2008), RAPD (Saker et al. 2000, 2006) and AFLP analyses (Gurevich et al. 2005; Saker et al. 2006) were used to screen for abnormal trees originated from TC, and variation between the different accessions was detected.

In several reports, higher variation was found in TC-originated trees, but variation was also detected among normal trees propagated either from offshoots or through TC (Al Kaabi et al. 2007; Gurevich et al. 2005). However, because most of the variation was not specific to unique markers, such studies have not yet been able to provide effective methods for the early detection of off-types. Variation in DNA sequences is not necessarily restricted to different individual plants. If a high number of mutations accumulate during the TC process, several independent mutations can occur in single developing plants, and different band patterns can be detected between various organs of the same plant. Such variation was detected in TC-originated variegated trees (Gurevich et al. 2005): AFLP analysis detected genetic variation between tissues from variegated and non-variegated regions of the leaves, and from roots of variegated plantlets of the same plants, as well as between root and leaf samples.

11.7 DNA Methylation and Somaclonal Variation

Not all somaclonal variation is due to sequence changes or chromosomal abnormalities. Altered epigenetic control of gene expression, which can cause gene silencing or gene activation, may also contribute to the variation. Epigenetic changes can be unstable or reversible through meiosis or somatically. One of the main mechanisms controlling gene expression is DNA methylation, which plays an important role in plant development and gene regulation (Finnegan et al. 2000). Changes in DNA methylation modify patterns of gene expression, which are associated with developmental processes. Demethylation affects plant morphology, including floral organs, and timing of flowering (Finnegan et al. 1996, 2000). The generation of somaclonal variation is suggested to be associated with epigenetic mechanisms, including altered patterns of DNA methylation (Jain 2001; Kaeppler et al. 2000), and such alterations have been identified in various plant species propagated by TC (Kaeppler et al. 2000). In micropropagated banana plants, for example, changes in DNA methylation patterns were detected, while no polymorphism could be identified in conventionally propagated banana plants (Peraza-Echeverria et al. 2001).

One method of studying DNA methylation patterns is methylation-sensitive amplified polymorphism (MSAP) (Mingliang et al. 2000). This method is a modification of AFLP (Vos et al. 1995), in which DNA is digested by two isoschizomers (*Hpa*II and *Msp*I) with different sensitivities to DNA methylation in their recognition sites. Very little information exists on DNA methylation in date palms, but methylation patterns are known to be cultivar-specific (Cohen et al. 2007; Fang and Chao 2007). Methylation patterns studied by MSAP in mother date palm trees and their offshoots (Fang and Chao 2007) were observed to be generally conserved. However, some of the methylated sites detected in the mother trees were demethylated in some of their offshoots.

As already discussed, TC-propagated "mantled" oil palm off-types share many similarities with the date palm abnormal fruit set off-type, including low levels of fruit set and supernumerary carpels (Al-Wasel 2001; Cohen et al. 2004). The "mantled" phenotype has been found to be associated with reduced global methylation levels and altered DNA-methylation patterns (Jaligot et al. 2004; 2002; 2000; Matthes et al. 2001; Rival et al. 2009). MSAP analysis of date palm off-types has also detected variation in DNA-methylation patterns—in Barhee date palm off-types with reduced fruit set and in dwarf Medjool off-types (Cohen et al. 2007). Attempts are currently being made to identify methylation-specific differences associated with each of these phenotypes (Cohen et al. unpublished). DNA-methylation patterns are established and maintained by DNA methyltransferases. In oil palms, three different methyltransferases were cloned and their expression was followed in calluses that tend to form off-types (fast growing) and in those in which off-types are only rarely generated (nodular compact). A consistent reduction in methylation was found in the fast-growing calluses as compared to the nodular compact ones (Rival et al. 2008, 2009). Some differences were also detected in the expression levels of methyltransferases in immature oil palm inflorescences, although these could not explain the formation of the off-type phenotypes.

11.8 Characterization of Somaclonal Variants Using Candidate Gene Approach

11.8.1 Dwarfing Genes

One approach to characterizing somaclonal off-types involves the exploration of candidate genes. The phenotypes of many off-types resemble those of known mutations, and similar phenotypes can be generated by altering the function or expression of those specific genes. This approach has been applied in several species to understand the phenotype of dwarf trees, as dwarf mutations have been identified in many different plant species. In cereals, use of such mutants led to improved crop yields, in the Green Revolution (Hedden 2003; Hedden and Phillips 2000). Molecular characterization of these mutants identified many of them as defective in gibberellic acid or brassinosteroid metabolism. Characterization of banana dwarf off-types revealed reduced levels of gibberellic acids (GA-3 and GA-20) (Sandoval et al. 1995). However, preliminary analysis of dwarf date palm off-types has not detected differences in the partial sequences of 15 genes related to vegetative growth (Sekar, Lavi and Cohen unpublished).

11.8.2 Flower-Development Genes

Another phenotype that has been extensively studied is the fruit-setting phenotype in dates and the similar "mantled" oil palm. The formation of supernumerary carpels suggests a possible alteration in flower structure. The genes controlling flower development have been extensively studied in recent years (Meyerowitz et al. 1991), and have been found to be conserved in monocot plants (Bommert et al. 2005). Mutations in MADS-box genes are known to cause homeotic transformations in which different whorls of the flower organ change their identities. Formation of the multicarpel phenotype can be generated by a genetic and/or epigenetic change in one of the B-type genes, which may change the identity of stamens to carpel-like organs. Such mutations have been found in monocots, e.g. the Superwoman gene (an AP3 homolog) in rice (Nagasawa et al. 2003) and the Silky1 gene in maize (Ambrose et al. 2000). Flower-development MADS-box genes were identified for oil palms (Adam et al. 2006; Alwee et al. 2006), and the expression levels of some of them were reduced in the female inflorescence of "mantled" oil palms (Adam et al. 2007). The level of *Eg*DEF1, an AP3 ortholog, was specifically reduced during the early stages of flower development (Adam et al. 2007). Preliminary analysis has detected reduced levels of the AP3/DEF date palm ortholog in female inflorescences of trees with reduced levels of fruit set and formation of supernumerary carpels (Shochat, Lavi and Cohen unpublished results).

Several additional genes show differential expression patterns in oil palm calluses that tend to form the "mantled" phenotype. These include a defensin gene (Tregear et al. 2002) and an early auxin-responsive gene (Morcillo et al. 2006). However, these insights into specific gene-expression patterns have not provided a tool for early detection of somaclonal variation in oil palm or date palm, as neither provides any explanation for the mechanism of off-type formation.

11.9 Genomic Analysis of Off-Types

Recent developments in molecular techniques are enabling researchers to look for differences between entire genomes. In several studies, somaclonal variation has been analyzed by RDA. This method can specifically sample and characterize the differences between closely related genomes. Under the stress of TC, specific labile portions of the genome (hot spots) can be modulated. Therefore, RDA can be an efficient approach to identifying these specific labile changes in the genomic DNA during TC propagation. Using this method, DNA sequences of both normal and off-type banana (Oh et al. 2007) and oil palm off-types (Cullis et al. 2007) were identified. The identified genomic regions were highly polymorphic, containing a large number of mutations that arise A frequently during *in-vitro* propagation. Interestingly, use of some of the primers designed to identify somaclonal variation in banana also detected variation between normal and off-type oil palms (Cullis et al. 2007). It was suggested that the high rate of variation in these specific labile regions could make these regions candidate markers for early detection of variation. The same method has proven efficient for the identification of different date palm varieties (Johnson et al. 2009; Vorster et al. 2002). Moreover, primers developed to detect "mantled" off-types were also efficient at discriminating between different date varieties

(Johnson et al. 2009). These results suggest that the same regions that are variable in oil palm off-types are also highly variable in date cultivars and represent labile regions in the palms' genomes. Attempts are currently being made to use the same approach for identification of labile regions that differ between normal and date palms off-types (Cullis unpublished).

In the last few years, high-throughput sequencing has dramatically improved sequencing capabilities, enabling the sequencing of non-model organisms' genomes. A draft of the date palm genome, representing 53X coverage, has been published (Al-Dous et al. 2011; Genbank accession no. ACYX02000000). These data can facilitate molecular understanding of date palm biology, and help in developing specific markers for the identification of somaclonal off-types. Preliminary transcriptome analysis is currently being performed to characterize dwarf date palm off-types (Cohen et al. unpublished). This analysis has detected dozens of differentially expressed genes in normal vs. dwarf trees. Thus, the potential of transcriptome and genome analysis is only beginning to be explored.

11.10 Conclusion

Somaclonal variation has been known for decades. Various tools and techniques have been used to characterize the phenomenon, and several molecular mechanisms have been suggested. During the TC process, changes occur in the DNA that can be either genetic (in the sequence) or epigenetic. Clearly, most of these changes are silent, and only some are specifically related to off-type phenotypes. A candidate gene approach, based on genes that are expected to be involved in specific phenotypes, did show differences in the expression patterns of wild-type and off-type trees. However, most of these altered expression patterns were probably the result of somaclonal variation, and not its primary cause. As-yet unknown master genes are probably altered to cause the specific phenotypes.

Since most changes in the DNA are random, molecular analyses usually detect higher levels of variation in off-types, but do not detect specific conserved changes. Other changes occur in specific hot spots—labile sequences in the genome that are highly variable. While these specific sequences make very good candidate markers, they tend to vary not only among off-types but also among different cultivars and among different individual plants. These labile regions do not transcribe specific genes and as such they cannot provide any new candidate that might initiate and control these processes.

Clearly, many changes occur during TC: some of them result in specific phenotypes while most of them are not related to the expressed phenotypes. The repeated recurrence of similar phenotypes in different cultivars and even in different plant species suggests the involvement of some key factors. However, to date, research has been unsuccessful at identifying these factors. In many ways, finding specific markers can be likened to finding a needle in a haystack. Despite great progress, we are still far from understanding the processes, and from being able to develop efficient markers for the early identification of off-types. The recent availability of the date palm genome draft and the development of new high-throughput sequencing techniques are expected to enable global genomic and transcriptomic analyses of off-types and may provide some new insights for better understanding the phenomenon of somaclonal variation.

Acknowledgments Research on somaclonal variation of date palms was supported by Research Grant Award no. IS-3886-06 from BARD, United States-Israel Binational Agricultural Research and Development Fund. This manuscript is contribution no. 114/2010 from the Volcani Research Center, Israel.

References

Adam H, Jouannic S, Morcillo F et al (2006) MADS box genes in oil palm (*Elaeis guineensis*): patterns in the evolution of the *SQUAMOSA*, *DEFICIENS*, *GLOBOSA*, *AGAMOUS*, and *SEPALLATA* subfamilies. J Mol Evol 62:15–31

Adam H, Jouannic S, Orieux Y et al (2007) Functional characterization of MADS box genes involved in the determination of oil palm flower structure. J Exp Bot 58:1245–1259

Al-Dous EK, George B, Al-Mahmoud ME et al (2011) De novo genome sequencing and comparative genomics of date palm (*Phoenix dactylifera*). Nat Biotech doi:10.1038/nbt.1860

Al Kaabi HH, Zaid A, Ainsworth C (2007) Plant-off-types in tissue culture-derived date palm (*Phoenix dactylifera* L) plants. Acta Hort 736:267–281

Al-Khayri J (2007) Date palm *Phoenix dactylifera* L. micropropagation. In: Jain SM, Häggman H (eds.) Protocols for micropropagation of woody trees and fruits. Springer, Dordrecht, pp 509–526

Al-Wasel AS (2000) Vegetative and fruiting comparison of tissue culture derived and conventionally propagated date palm (*Phoenix dactylifera* L.) cv. Barhi trees. In Vitro Cell Dev Biol 36:P-1010

Al-Wasel AS (2001) Field performance of somaclonal variants of tissue culture-derived date palm (*Phonix dactylifera* L.). Plant Tissue Cult 11:97–105

Alwee S, Van der Linden CG, Van der Schoot J et al (2006) Characterization of oil palm MADS box genes in relation to the mantled flower abnormality. Plant Cell Tissue Organ Cult 85:331–344

Ambrose BA, Lerner DR, Ciceri P et al (2000) Molecular and genetic analyses of the silky1 gene reveal conservation in floral organ specification between eudicots and monocots. Mol Cell 5:569–579

Azeqour M, Majourhat K, Baaziz M (2002) Morphological variations and isoenzyme polymorphism of date palm clones from in vitro culture acclimatized and established on soil in South Morocco. Euphy 123:57–66

Baaziz M, Saaidi M (1988) Preliminary identification of date palm cultivars by esterase isoenzymes and peroxidase activities. Can J Bot 66:89–93

Bairu MW, Fennell CW, van Staden J (2006) The effect of plant growth regulators on somaclonal variation in Cavendish banana (*Musa* AAA cv. 'Zelig'). Sci Hort 108:347–351

Bendiab K, Baaziz M, Majourhat K (1998) Preliminary date palm cultivar composition of Moroccan palm groves as revealed by leaf isoenzyme phenotypes. Biochem Syst Ecol 26:71–82

Bommert P, Satoh-Nagasawa N, Jackson D, Hirano HY (2005) Genetics and evolution of inflorescence and flower development in grasses. Plant Cell Phys 46:69–78

Cao BR, Chao CT (2002) Identification of date cultivars in California using AFLP markers. Hort Sci 37:966–968

Chao CT, Krueger RR (2007) The date palm (*Phoenix dactylifera* L.): overview of biology, uses, and cultivation. Hort Sci 42:1077–1082

Cohen Y, Korchinsky R, Tripler E et al (2003) Fruit setting abnormalities in date palms of the cultivar Barhee originated from tissue culture. Alon Hanotea Hebrew 57:275–279

Cohen Y, Korchinsky R, Tripler E (2004) Flower abnormalities cause abnormal fruit setting in tissue culture propagated date palm (*Phoenix dactylifera* L.). J Hort Sci Biotech 79:1007–1013

Cohen Y, Gurevich V, Korchinsky R et al (2007) Molecular and phenotypic characterization of somaclonal variation in date palm off-types originated from tissue culture. Acta Hort 738:417–423

Corley RHV, Lee CH, Law IM, Wong CY (1986) Abnormal flower development in oil palm clones. Planter KL 62:233–240

Corniquel B, Mercier L (1994) Date palm (*Phoenix dactylifera L.*) cultivar identification by RFLP and RAPD. Plant Sci 101:163–172

Corniquel B, Mercier L (1997) Identification of date palm (*Phoenix dactylifera L.*) cultivars by RFLP: partial characterization of a cDNA probe that contains a sequence encoding a zinc finger motif. Int J Plant Sci 158:152–156

Cullis CA (2011) Molecular markers in date palm. In: Jain SM, Al-Khayari JM, Johnson DV (eds.) Date palm biotechnology. Springer, Dordrecht

Cullis CA, Cullis MA, Ong Abdullah M (2007) Development of markers for the mantled phenotype in oil palm. In: PIPOC 2007. Proceedings agriculture, biotechnology and sustainability conference. Kuala Lumpur. pp 299–312

Damasco OP, Smith MK, Adkins SW et al (1998) Identification and characterisation of dwarf off-types from micropropagated Cavendish bananas. Acta Hort 490:79–84

Devanand PS, Chao CT (2003) Genetic variation within 'Medjool' and 'Deglet Nur' date (*Phoenix dactylifera L.*) cultivars in California detected by fluorescent-AFLP markers. J Hort Sci Biotech 78:405–409

Djerbi M (2000) Abnormal fruiting of the date palm derived from tissue culture. In: The date palm international symposium. Windhoek. p 73

El Hadrami A, Daayf F, Jain SM, El Hadrami I (2011) Somaclonal variation in date palm. In: Jain SM, Al-Khayari JM, Johnson DV (eds.) Date palm biotechnology. Springer, Dordrecht

El-Assar AM, Krueger RR, Devanand PS, Chao CT (2005) Genetic analysis of Egyptian date (*Phoenix dactylifera* L.) accessions using AFLP markers. Genet Res Crop Evol 52:601–607

Elhoumaizi MA, Devanand PS, Fang J, Chao CT (2006) Confirmation of 'Medjool' date as a landrace variety through genetic analysis of 'Medjool' accessions in Morocco. J Am Soc Hort Sci 131:403–407

Fang JG, Chao CT (2007) Methylation-sensitive amplification polymorphism in date palms (*Phoenix dactylifera* L.) and their off-shoots. Plant Biol 9:526–533

Finnegan EJ, Peacock WJ, Dennis ES (1996) Reduced DNA methylation in *Arabidopsis thaliana* results in abnormal plant development. Proc Nat Acad Sci USA 93:8449–8454

Finnegan EJ, Peacock WJ, Dennis ES (2000) DNA methylation, a key regulator of plant development and other processes. Curr Opin Genet Dev 10:217–223

Gurevich V, Lavi U, Cohen Y (2005) Genetic variation in date palms propagated from offshoots and tissue culture. J Am Soc Hort Sci 130:46–53

Habashi AA, Kaviani M, Mousavi A, Khoshkam S (2008) Transient expression of beta-glucuronidase reporter gene in date palm (*Phoenix dactylifera* L.) embryogenic calli and somatic embryos via microprojectile bombardment. J Food Agric Environ 6:160–163

Hamama L, Cornee N, Leclerc V et al (2003) Date palm (*Phoenix dactylifera)* offshoot identification by PCR-ISSR markers. Acta Hort 616:453–457

Hedden P (2003) The genes of the green revolution. Trends Genet 19:5–9

Hedden P, Phillips AL (2000) Gibberellin metabolism: new insights revealed by the genes. Trends Plant Sci 5:523–560

Israeli Y, Reuveni O, Lahav E (1991) Qualitative aspects of somaclonal variations in banana propagated by in vitro techniques. Sci Hort 48:71–88

Jain SM (2001) Tissue culture-derived variation in crop improvement. Euphy 118:153–166
Jaligot E, Rival A, Beule T et al (2000) Somaclonal variation in oil palm (*Elaeis guineensis Jacq.*): the DNA methylation hypothesis. Plant Cell Rep 19:684–690
Jaligot E, Beule T, Rival A (2002) Methylation-sensitive RFLPs: characterisation of two oil palm markers showing somaclonal variation-associated polymorphism. Theor Appl Genet 104:1263–1269
Jaligot E, Beule T, Baurens FC et al (2004) Search for methylation-sensitive amplification polymorphisms associated with the "Mantled" variant phenotype in oil palm (*Elaeis guineensis* Jacq.). Genome 47:224–228
Johnson C, Cullis TA, Cullis MA, Cullis CA (2009) DNA markers for variety identification in date palm (*Phoenix dactylifera* L.). J Hort Sci Biotech 84:591–594
Joyce SM, Cassells AC, Jain SM (2003) Stress and aberrant phenotypes in vitro culture. Plant Cell Tissue Organ Cult 74:103–121
Jubrael JMS, Udupa SM, Baum M (2005) Assessment of AFLP-based genetic relationships among date palm (*Phoenix dactylifera* L.) varieties of Iraq. J Am Soc Hort Sci 130:442–447
Kaeppler SM, Kaeppler HF, Rhee Y (2000) Epigenetic aspects of somaclonal variation in plants. Plant Mol Biol 43:179–188
Kunert KJ, Baaziz M, Cullis CA (2003) Techniques for determination of true-to-type date palm (*Phoenix dactylifera* L.) plants: a literature review. Emir J Agric Sci 15:1–16
Larkin PJ, Scowcroft WR (1981) Somaclonal variation - a novel source of variability from cell cultures for plant improvement. Theor Appl Genet 60:197–214
Matthes M, Singh R, Cheah SC, Karp A (2001) Variation in oil palm (*Elaeis guineensis* Jacq.) tissue culture-derived regenerants revealed by AFLPs with methylation-sensitive enzymes. Theor Appl Genet 102:971–979
McCubbin MJ, Van Staden J, Zaid A (2000) A Southern African survey conducted for off-types on date palms produced using somatic embryogenesis. In: The date palm international symposium, Windhoek, pp 68–72
Meyerowitz EM, Bowman JL, Brockman LL et al (1991) A genetic and molecular model for flower development in *Arabidopsis thaliana*. Dev 1:157–167
Mingliang X, Xiangqian L, Schuyler SK (2000) AFLP-based detection of DNA methylation. Plant Mol Biol Rep 18:361–368
Morcillo F, Gagneur C, Adam H et al (2006) Somaclonal variation in micropropagated oil palm. Characterization of two novel genes with enhanced expression in epigenetically abnormal cell lines and in response to auxin. Tree Phys 26:585–594
Nagasawa N, Miyoshi M, Sano Y et al (2003) SUPERWOMAN1 and DROOPING LEAF genes control floral organ identity in rice. Dev 130:705–718
Oh TJ, Cullis MA, Kunert K et al (2007) Genomic changes associated with somaclonal variation in banana (*Musa* spp.). Phys Plant 129:766–774
Peraza-Echeverria S, Herrera-Valencia VA, James-Kay A (2001) Detection of DNA methylation changes in micropropagated banana plants using methylation-sensitive amplification polymorphism (MSAP). Plant Sci 161:359–367
Read PE, Preece JE (2003) Molecular identification of micropropagated plants. Acta Hort 616:25–47
Reuveni O, Israeli Y (1990) Measures to reduce somaclonal variation in *in vitro* propagated bananas. Acta Hort 275:307–313
Rhouma S, Zehdi-Azouzi S, Ould Mohamed Salem A et al (2007) Genetic diversity in ecotypes of Tunisian date-palm (*Phoenix dactylifera* L.) assessed by AFLP markers. J Hort Sci Biotech 82:929–933
Rival A, Jaligot E, Beule T, Finnegan EJ (2008) Isolation and expression analysis of genes encoding MET, CMT, and DRM methyltransferases in oil palm (*Elaeis guineensis* Jacq.) in relation to the 'mantled' somaclonal variation. J Exp Bot 59:3271–3281
Rival A, Jaligot E, Beulé T et al (2009) The oil palm "mantled" somaclonal variation: a model for epigenetic studies in higher plants. Acta Hort 829:177–182

Sahijram L, Soneji JR, Bollemma KT (2003) Analyzing somaclonal variation in micropropagated banana (*Musa* spp.). In Vitro Cell Dev Biol 39:551–556

Saker MM, Moursi HA (1999) Molecular characterization of Egyptian date palm cultivars: II – RAPD fingerprints. Arab J Biotech 2:71–78

Saker MM, Bekheet SA, Taha HS et al (2000) Detection of somaclonal variations in tissue culture-derived date palm plants using isoenzyme analysis and RAPD fingerprints. Biol Plant 43:347–351

Saker M, Adawy S, Mohamed A, El-Itriby H (2006) Monitoring of cultivar identity in tissue culture-derived date palms using RAPD and AFLP analysis. Biol Plant 50:198–204

Sandoval J, Kerbellec F, Cote F, Doumas P (1995) Distribution of endogenous gibberellins in dwarf and giant off-types banana (*Musa* AAA, Cv. Grand Nain) plants from *in-vitro* propagation. Plant Growth Reg 17:219–224

Sedra MH, Lashermes P, Trouslot P et al (1998) Identification and genetic diversity analysis of date palm (*Phoenix dactylifera* L.) varieties from Morocco using RAPD markers. Euphy 103:75–82

Smith MK, Hamill SD (1993) Early detection of dwarf off-types from micropropagated Cavendish bananas. Aust J Exp Agric 33:639–644

Tisserat B (1979a) Propagation of date palm (*Phoenix dactylifera* L.) *in vitro*. J Exp Bot 30:1275–1283

Tisserat B (1979b) Tissue-culture of the date palm. J Hered 70:221–222

Tregear JW, Morcillo F, Richaud F et al (2002) Characterization of a defensin gene expressed in oil palm inflorescences: induction during tissue culture and possible association with epigenetic somaclonal variation events. J Exp Bot 53:1387–1396

Vorster BJ, Kunert KJ, Cullis CA (2002) Use of representational difference analysis for characterization of sequence between date palm varieties. Plant Cell Rep 21:271–275

Vos P, Hogers R, Bleeker M et al (1995) AFLP: a new technique for DNA fingerprinting. Nucl Acids Res 23:4407–4414

Withers LA (1993) Early detection of somaclonal variation. In: Proceedings of the workshop biotechnology applications for banana and plantain improvement, San Jose, pp 200–248

Zaid A, De Wet PF (1999) Origin, geographical distribution and nutritional values of date palm. In: Zaid A (ed.) Date palm cultivation. FAO, Rome, pp 29–44

Zehdi S, Sakka H, Ould AMS et al (2005) Molecular polymorphism and genetic relationships in a Tunisian date-palm (*Phoenix dactylifera* L.) collection using ISSR amplification fingerprinting. Plant Gen Res Newsl 144:39–44

Zivdar S, Mousawi M, Ansari NA (2008) Genetic stability in date palm micropropagation. Asian J Plant Sci 7:775–778

Chapter 12
In Vitro Selection for Abiotic Stress in Date Palm

A. El Hadrami, F. Daayf, and I. El Hadrami

Abstract *In vitro*-selection of date palm cultivars with an enhanced tolerance to abiotic stress represents a great support to the breeding program. This could lead to variants with superior agronomic quality and/or enhanced performance but also to genotypes harboring new traits such as tolerance to drought and salinity. In an era where a race has started towards functional food and nutraceuticals, this source of variation could also be explored to improve the nutritional value of dates. This chapter summarizes the recent progresses in date palm *in vitro*-selection and provides an outlook about future applications of the technique in this socio-economically important crop. It also highlights some of the challenges faced by this crop with regard to climate change and global warming.

Keywords Biotechnologies • Abiotic stress • Conventional breeding • Date palm • Drought • Excessive heat • *In vitro*-selection • *Phoenix dactylifera* L. • Salinity

---

A. El Hadrami (✉)
OMEX Agriculture Inc. Canada, P.O. Box 301, 290 Agri Park Road, Oak Bluff, Manitoba, R0G 1N0, Canada
e-mail: abdele@omexcanada.com

F. Daayf
Department of Plant Science, University of Manitoba, 222, Agriculture Building, Winnipeg, MB R3T 2N2, Canada

I. El Hadrami
Department of Biology, Faculty of Sciences Semlalia, Laboratory of Biotechnologies, Cadi Ayyad University, B.P. 2390, 40 000 Marrakech, Morocco

S.M. Jain et al. (eds.), *Date Palm Biotechnology*,
DOI 10.1007/978-94-007-1318-5_12, © Springer Science+Business Media B.V. 2011

12.1 Introduction

Abiotic stresses such as drought, salinity, high temperatures, and atmospheric or telluric chemical pollution, cause substantial damage to date palm resulting in annual losses estimated in billions of dollars worldwide. The crop, being a desert plant, has evolved strategies to protect itself against most of these stresses. However, if the combined effect of the intensity of stress and exposure time is too high, major alterations are often observed in terms of plant metabolism, leading to cell death, a substantial change in growth and development habits, and ultimately a reduction in yield. In addition, most deployed cultivars on a commercial scale are very susceptible to abiotic stress and yield poorly if not given the required optimal growth conditions.

In vitro selection is among the biotechnology techniques commonly used in plant breeding for tolerance to abiotic stresses i.e. drought, salinity, Al stress and acidic soils (Yusnita et al. 2005). These techniques often offer a value-added to the regenerated crop such as an increase in the number and/or size of fruits or their texture or taste, or a modification in flower structure (Ahloowalia and Maluszynski 2001; Pedrieri 2001; Witjaksono 2003). Depending on the selective agent, *in vitro* selection is conducted at the protoplast, cell-suspension or callus stages or later during the regeneration of shoot and root meristems. The choice of method often depends on the ease of application and the efficiency of the selective agent in inducing high levels of variation. The regeneration method of tolerant cells/individuals is also important (Widoretno 2003) in order to preserve the inheritance of the sought trait (Yusnita 2005).

Plant tissue and cell cultures such as regenerative and embryogenic calli, cell suspensions, zygotic and somatic rescued embryos, fused protoplasts and cybrids are a great support to many conventional breeding programs. These can be applied at various stages of the breeding process to increase the creation of genetic variation and the efficiency in selecting/multiplying cultivars with desirable traits. For instance, limitation for vegetatively propagated plant species (i.e. date palm) was mainly due to the inability to produce larger populations that could allow for the selection of individuals with desired characters. The *in vitro* techniques can now produce larger populations that can be subjected to chemical or physical mutagens, resulting in soma- or protoclonal variation. Altered genes hence enrich the genetic pool via gene insertion or regular backcrossing.

Traits that are possible to select for *in vitro* include temperature tolerance, photoperiod and day-length, drought and salt tolerance, resistance to microorganisms and pests or their toxic metabolites, resistance to herbicides and antibiotics, heavy-metal tolerance and low NPK inputs (El Hadrami and El Hadrami 2009; Jain and Gupta 2005).

12.2 In Vitro Selection for Abiotic Disorders

12.2.1 The Date Palm Major Abiotic Constraints

A stress by definition is any exogenous factor that negatively impacts on the plant growth and development or has an adverse effect on its biomass or yield. A strain, on the other hand, corresponds to any reduction in plant growth and development or function resulting from a stress. Date palm growth and development, productivity and yield are adversely affected by water stress and drought, salinity, low NPK inputs and many other abiotic constraints. These stresses usually alter various physiological process and metabolic pathways such as the reduction of photosynthetic activity and pigment contents. A decline in photosynthesis increases the importance of the PSII dark reaction, which consequently lead to a decrease in biomass and yield.

12.2.1.1 Drought and Water Stress

The meteorological term *drought* refers to a period of time without significant rainfall (Simpson 1981). Water stress occurs in growing plants as a result of a continuous loss of water through transpiration and/or evaporation while there is a shortage of available water in the soil. Drought stress tolerance is seen in almost all plants but its extent varies from species to species and even within cultivars of the same species. A better understanding of how water stress affects the morphology, the anatomy and physiology of the plant often helps in the creation and selection of new tolerant cultivars (Nam et al. 2001; Shao et al. 2009). Different degrees of water stress can be observed in a plant. A moderate loss of water often leads to stomatal readjustment or closure, which limit gas exchange. Desiccation, on the other hand, occurs when water loss is extensive and often leads to a major disruption of the metabolism, enzymatic activities and cell structure with more of an effect on cells enlargement than on cell division.

Based on the strategies that plants use to tolerate water stress, date palm, being an arid-land plant originally, it does not escape drought and it rather tolerates and/or postpone desiccation. This allows the tissue to maintain a functional level of hydration and to function while dehydrated. Water stress in date palm induces an extensive development of the root system. This allows the increase in water uptake and helps maintain the osmotic pressure through higher proline levels (Djibril et al. 2005). These morphological changes may be due to the activity of the stress hormone abscissic acid (ABA). Most grown date palm cultivars still achieve an acceptable yield under water stress mainly due to the development of the root system to sustain the repartition of the dry matter.

One of the earliest plant responses to water stress is a decrease in leaf area. Date palm leaves are well adapted to the harsh environment where the plant usually grows, which reduces water loss. When water content decreases, cells shrink and the turgor pressure against the cell walls relaxes, which increases cytoplasmic solute concentration. Cell expansion, being a turgor-driven process, is extremely sensitive to water deficit. A decrease in turgor may cause a decrease in the overall growth rate of the plant. This is due to the fact that water stress leads to an alkalinization, and cell wall extensibility is normally greatest when the pH is slightly acidic.

Pigment contents can also be affected under drought conditions. The ratio of chlorophyll (a) and (b) to carotenoid changes in the plants undergoing a water stress. Xanthophyll content is also altered given the importance of these pigments in the dissipation of the excess energy through the xanthophyll cycle. Other enzymatic and non-enzymatic systems are also activated upon stress to reduce damage due to the oxidative burst generated by drought stress. These include β-carotenes, ascorbates (AA), α-tocopherol (α-toc), reduced glutathione (GSH), superoxide dismutases (SOD), peroxidase (POD), ascorbate peroxidase (APX), catalase (CAT), polyphenol oxidase (PPO) and glutathione reductase (GR) (Prochazkova et al. 2001). Carotenes are also a key component of the plant antioxidant system under water stress. They are present in the chloroplasts bound to the core complexes of PSI and PSII (Havaux 1998). Although they are prone to oxidation, their role is very important in protecting these complexes against oxidative damage and shielding the phytochemical process.

Being a desert plant, date palm has to cope with scarcity of surface and ground water and roots to extend deeper for water resources. Date palm is mostly found in region with 75–100 mm of rainfall. Tolerance to water stress is expected in successful cultivars. Many studies are dedicated to the examination of the morphological and physiological responses as well as the photosynthetic activity alteration in date palm plants subjected to water stress under greenhouse conditions. Elshibli and Korpelainen (2009) studied these responses in soft- and dry-type date palm cultivars that were supplied with 10%, 25%, 50%, and 100% of field capacity water and exposed to stepwise changes in CO_2 levels. The authors reported that water stress impacted plant growth and morphology and that the response was cultivar-dependent. Unfortunately, such studies are not often included in the conventional breeding program because they are considered time consuming. Moreover, tolerance to drought seems to be too complex to be assessed solely on the basis of physiological measurements (Jaleel et al. 2009; Razmjoo et al. 2008). Breeders tend to assess achievable yield under drought conditions instead (Blum 1996).

The widespread development of *in vitro* techniques to massively produce date palm clones suitable for planting offers new opportunities to apply *in vitro* selection methods for drought-tolerant cultivars. Sucrose, mannitol or polyethylene glycol (PEG) can be used to achieve selection *in vitro* for drought tolerant cultivars. These molecules could simulate the mechanical stresses caused by the withdrawal of cellular water due to low water availability.

12.2.1.2 Salinity

Sodicity refers to the concentration of Na^+ and salinity to the total salt content; these represent the major abiotic factors that restrict crop production and agricultural capabilities worldwide. Salinity hampers crop production because of the combined effect of water deficit, resulting in an osmotic stress, and the impact of high levels of Na^+ ions on critical biochemical processes. In addition, the climatic changes observed during the last few decades jointly with the steady raise in food demand from an increasingly growing population forces the recourse to an extensive agriculture often with an over-utilization of the arable land, along with over-fertilization under water deficiency conditions.

From the early stages of plant growth and development, date palm is exposed to salt. Evaporation draws water from the soil, causing an accumulation of salt in the upper soil layer where seeds are placed. This often leads to germination under salt concentration higher than the average salt level throughout the soil profile. Date palm is a highly tolerant species for salt. The threshold of average root zone salinity (electrical conductivity of soil saturation extract=ECse in dS/m) of date palm is 6–10 dS/m (Rhoades et al. 1992), which places the plant in the category of tolerant species. Most trees are able to harbor a good production even less than 3,000 ppm of salty water. Some cultivars were reported to be tolerant to up to 22,000 ppm of total dissolved salt but their productivity had been affected (unpublished data).

In date palm as for many other plant species, screening the germplasm and breeding programs for salt tolerance face several limitations. The major one is the growth stage dependency of the phenotypic response to salt, which depends also on the physiological mechanisms involved in tolerance. Besides, genotype x environment and soil physico-chemical proprieties could render the screening rather difficult. Identifying molecular and physiological traits provides the bases for an efficient germplasm screening through conventional or marker-assisted breeding and transgenic approaches. The quantitative nature of salt stress tolerance and the issues linked to having reliable testing set-ups make it difficult to discriminate between salt-sensitive and salt-tolerant lines. Since field selection for salt tolerance is rather a laborious process, implementation of rapid screening procedures under controlled conditions has always been used. Some of these methods rely on the use of *in vitro* tissue culture. Many mechanisms are involved in plant salt tolerance depending on whether the species is a glycophyte, mesophyte or halophyte. These include ion exclusion/sequestration, osmotic adjustment, accumulation of protective macromolecules, and membrane transport system adaptation to saline environment.

Using tissue culture techniques to screen date palm genotypes for salt tolerance, Al Mansoori et al. (2007) studied the effect of sodium chloride (NaCl) on calli derived from immature embryos of four local date palm cultivars. The authors examined the salt effect during the dedifferentiation and the fast-growing stages (exponential multiplication phase). Using fast growing calli, the authors were able to detect significant differences in the response of the compared genotypes to salt. Induced calli, on the other hand, were ineffective in detecting variation among genotypes in response to salinity.

According to Al Mansoori et al. (2007), salinity significantly affects date palm immature embryos. A complete inhibition seemed to occur when NaCl level exceeds 3% in the induction medium. A delay in calli growth was also evident when salt concentration increased. In both types of calli, tissue dehydration symptoms were recorded accompanied by a significant increase in Na^+ and proline content and a dramatic decrease in K^+ content. The increase in endogenous proline content during the induction stage was more pronounced in the progeny of the cultivar exhibiting higher percentage of calli induction. This suggests a role of the free proline content in the regulation of dedifferentiation process by adjusting the intracellular osmotic pressure between the cytoplasm and the vacuole.

12.2.1.3 Excessive Heat

Germination as well as many other growth and development processes are sensitive to or regulated by temperature. The ideal temperature for date palm ranges from

25–32°C. Beyond the high temperature, seeds will still germinate but at a lower rate. When seedlings emerge and are exposed to excessive heat, they may experience a heat shock, which may lead to dramatic consequences especially when the balance between the evapotranspiration and the excess heat is broken. Excessive heat is known to cause many damages at the cellular and sub-cellular levels including the alteration of cellular structures, organelles and cytoskeleton (Buchanan et al. 2000). It is also reported to impair membrane function and to cause a decrease in the overall protein synthesis in favor of the synthesis and accumulation of heat shock proteins. These proteins are represented by chaperones and foldases that refold proteins denatured as a consequence of heat.

Excessive heat, as an interacting stress, is often accompanied by a low availability of water and may increase the impacts of other stresses due to drought or salinity. The steep increases in daytime temperatures, especially in areas where the date palm thrives, are responsible for an increase in the evapotranspiration of the plant and soil evaporation, which consequently lead to high transpiration rates and low plant water potentials (Hall et al. 2000). High temperatures can also damage pollen, especially when combined with drought.

12.2.1.4 Nutrient Uptake Efficiency

The release of macro- and micro-nutrients from fertilizers applied to crops, especially nitrogen and phosphorus, has a dramatic impact on the ecosystems. Cultivars with optimized nutrient uptake efficiencies are in high demand in many crops including date palm. These cultivars often outperform their counterparts currently grown in terms of nutrient uptake and utilization due to a specific absorption, assimilation, translocation and remobilization. These traits often differ among plant species and among cultivars of the same species due to specific morphological and physiological characteristics, and offer an opportunity of improving the plant through breeding.

In the last few decades, an extensive body of knowledge has been gathered with regard to the optimization of date palm cultivation to guarantee better yield and high quality of dates. Unfortunately, most of these investigations did not find a way to be implemented in the plantations. Yet, several cultural practices have made significant contributions to the irrigation and nutrient management in date palm. A common practice is the use of organic manures (5–15 t/ha/year). However, there is no real assessment or guideline regarding the quantities applied. Across date palm growing areas, it seems that sheep manure is the most used. This manure contains 2% nitrogen but spreading of chicken manure would be a better alternative (3–5% nitrogen).

In recent years, inorganic nitrogen and other chemically-synthesized fertilizers have been introduced to date palm cultivation to improve and maintain soil fertility and balance the growth and development. The amounts being applied take into account whether there is a combination with organic manures or not but also the presence or absence of cover crops. Nitrogen uptake in date palm does not exceed 60% according to a few studies, using fast-released nitrogen. According to the most-cited literature, conventional date palm plantations require 500 g of nitrogen,

300 g of phosphorous and 250 g of potassium per tree. Microelements such as Zn, Cu, Fe, B and Mn were also reported to significantly impact on the yield and quality of harvested dates.

In recent years, the development of irrigation in date palm plantations has allowed for the implementation of fertigation strategies. However, date palm is often associated with sandy soils and requires the combination of this strategy with the use of microorganisms to enhance the biological activity involved in nutrient mineralization and recycling, often lacking in these soils. In places where date palm irrigation is a common practice such as in California, flood-basin and drip irrigation represent the main water delivery systems.

In vitro selection of cultivars with better nutrient management efficiency would be an addition to these progresses made in the field in order to direct a breeding program. This is of primary importance to the date palm given the survivability of the plant in some of the harshest conditions, including poor soils and in sites prone to desertification. The latter alone is sufficient reason to promote research in this area to allow a rehabilitation of desertified lands in the arid and semi-arid regions.

A successful selection for nutrient uptake efficiency in date palm should be based on the characterization of the germplasm with respect to components involved in nutrient uptake and utilization, and an investigation of the effects of nitrogen deficiency on the development and physiology of different genotypes. Once interesting genotypes have been identified, a genetic analysis should be carried out to examine their nutrient uptake efficiency, and to develop an appropriate screening and selection procedure. For characterization and identification of date palm genotypes differing in nutrient uptake efficiencies, various cultivars are to be grown *in vitro*, in the greenhouse, and field trials under nutrient –deficient –sufficient conditions (i.e., low vs. high nitrogen levels). A number of criteria can be then be followed among these differential treatments such as the evaluation of the vegetative development, chlorophyll content, protein content and yield under greenhouse or field conditions. If the selection process is operated *in vitro*, vegetative development, chlorophyll content, protein content of shoots, fresh and dry weight of shoots and roots, root architecture as well as the removal of nitrogen from the nutrient solution can be assessed.

12.2.2 Molecular and Biochemical Mechanisms of Abiotic Stress

Salt and drought stresses lead to an *in planta* signal transduction consisting of both osmotic and ionic homeostasis signaling pathways. Osmotic homeostasis involves complex plant reactions that overlap with drought and other stress responses. The ionic aspect, on the other hand, involves the so-called salt overly sensitive (SOS) pathway where a calcium-responsive SOS3-SOS2 protein kinase complex controls the expression and activity of ion transporters such as SOS1. A study of the pathway regulating ion homeostasis in response to salt stress revealed common responses (Xiong and Zhu 2002).

These include (i) temporary changes in the across-membrane influx of calcium (Knight 2000; Sanders et al. 1999); (ii) the activation of mitogen-activated and/or calcium dependent protein kinases (CDPKs) and phosphatases (Merlot et al. 2001; Tahtiharju and Palva 2001; Xiong et al. 2002) as a signaling cascade; (iii) increases in ABA biosynthesis and content in response to stress, which consequently triggers an array of other responses (Xiong et al. 2002). Both ABA-dependent and -independent osmotic stress signaling constitutively change the expression of transcriptional factors, which enhance the expression of early response transcriptional activators, hence activating downstream stress tolerance effector genes; (iv) the accumulation of inositol phosphates that act as a signal molecule for some of the transcriptional changes occurring during stress responses (Xiong et al. 2001); (v) the activation of phospholipases, which in turn generates a diverse array of second messenger molecules, some of which might regulate the activity of stress responsive kinases, i.e., phospholipase D functions in an ABA independent pathway; (Frank et al. 2000); (vi) the induction of late embryogenesis abundant (LEA) type genes including the CRT/DRE responsive COR7RD genes (Xiong and Zhu 2002); (vii) increased levels of antioxidants and compatible osmolytes such as proline and soluble sugars (Hasegawa et al. 2000); and (viii) the accumulation of reactive oxygen species such as superoxide, hydrogen peroxide and hydroxyl radicals (Hasegawa et al. 2000).

Based on the commonality of many aspects of drought and salt stress responses, it seems that genes that would increase tolerance to salt could also lower plant susceptibility to drought. For instance, transcription factors such as AtCBF/DREBl (C-repeat-binding factor/dehydration-responsive element binding protein) and other genes such as OsCDPK7 (Saijo et al. 2000) or AVPl (a vacuolar pyrophosphatase-proton pump) (Gaxiola et al. 2001) were reported to improve both a tolerance to salt or cold stresses as well as to drought in model species such as the thale cress and a limited few crops.

Plant responses to abiotic stress are branched and somewhat complex given the fact that the stress could be perceived in one part of the plant and the response occurring in another part (Tester and Davenport 2003). Facing stress, plants often evolve an avoidance mechanism along with tolerance. For instance, the development of the deep-rooted system during drought represents an avoidance mechanism while the change in genes expression and the regulation of metabolic pathways are part of the tolerance component to enhance or maintain plant functionality under stress condition. Although there is no general mechanism applicable to all plants and that differences can be seen between genotypes from the same species (Welfare et al. 2002), several universal and unique changes occur at transcription level for certain genes (Ho et al. 2001; Yamaguchi-Shinozaki and Shinozaki 2006).

Recently, Rosa et al. (2009) showed that stress perception and signaling pathways share some common mechanisms such as the ones involving sugar-sensing. The authors illustrated several cases where sucrose and hexoses play a dual role in up-regulating growth-related genes and down-regulating stress-related genes through an HXK-dependent pathway.

12.2.3 In Vitro Selection Methods

12.2.3.1 Induction of Stress-Tolerance

To induce stress tolerance in date palm *in vitro* cultures various chemical and physical selective agents could be used. Some of the commonly used osmolytes to simulate drought and water deficit are PEG and mannitol. Sodium chloride, however, has been used to induce calli and other tissue-propagated cultures and assess their tolerance to salinity. The effects of both NaCl (Al-Khayri 2002) and PEG (Al-Khayri and Al-Bahrany 2004) on date palm *in vitro* cultures were investigated to determine the lethal dose and evaluate growth response and proline accumulation as an indicator of salinity and drought stresses. *In vitro* material used often depends on the stress studied, the genotype, and the feasibility of the application and the potential rapid regeneration of tolerant genotypes. These vary from proliferative and regenerative calli, somatic embryos, to cell suspensions.

Another way of inducing stress-tolerance in date palm is through the exploration of somaclonal variation and/or genetic transformation. A number of genes expressed at the extranuclear level in many plants were revealed to enhance resistance to drought, salt, heat, UV-damage and nutrient uptake efficiency, as well as biomass and yield. These approaches have never been explored in date palm and will certainly be in the years to come given the recent completion of sequencing of date palm genome.

12.2.3.2 Isolation and Regeneration of Stable Stress-Tolerant Genotypes

A number of procedures have been developed to regenerate *in vitro* genotypes with potential tolerance to abiotic stress. These methods rely on either an indirect or a direct regeneration of tolerant genotypes to specific abiotic stress situations. For instance, indirect selection to drought uses induced calli at a given stage of their development. This calli are transferred onto a regeneration medium containing increasing concentrations of PEG or mannitol. Direct regeneration, on the other hand, consists of culturing the excised tissues from the explants directly on the regeneration medium supplemented with the drought-selecting agents. In either case, the number of regenerated shoots is assessed according to each applied concentration, after a few weeks of incubation.

To generate genotypes with tolerance to abiotic stress, successive generations of culture are necessary to guarantee the stability of the selected variation. Eshraghi et al. (2005) examined, using RAPD markers, the genetic stability of the somatic embryogenesis-derived plants for up to six generations of cv. Khanizi, and detected an increase in genetic variation. The genetic dissimilarity between the mother and calli-derived plants ranged was of 6% in the R1-2 and 17% in the R5. This suggests an increase in the accumulation of genetic changes as the multiplication cycle number increases.

Since their genetic determinants of avoidance and tolerance to abiotic stress are complex and there is no specific stress-tolerance gene to be targeted, many levels have to be considered in order to select for genotypes harboring stress-related genes and proteins. These include, at the protein level, essential transcriptional factors i.e. CBF/DREBl, signaling cascades and pathways i.e. CDPKs and SOS, Hsps/chaperones, foldases, and LEA proteins, water and ion transporters, ROS scavenging and detoxifying enzymatic and non-enzymatic systems. At the metabolic level, soluble sugars, osmolytes and polyamines could be good markers for stress-tolerance acquisition.

12.2.3.3 Testing Stress-Tolerant Genotypes

Depending on the tolerance to be tested for, protocols and procedures have been developed or on their way to being developed. These include molecular, morphological and physiological techniques either under laboratory conditions using *in vitro*-cultured tissues or under controlled environmental conditions using seedlings or even in the field. For example, material being derived from somatic embryogenesis is currently undergoing extensive testing to detect the frequency and extent of variation that occurs in the somaclones in comparison with the original type (Al-Kaabi et al. 2007; Al-Wasel 2005; Zaid and Al-Kaabi 2003). New molecular tools are providing a tremendous support to these time- and resource-consuming selection and tolerance testing programs.

12.3 In Vitro Selection Within the Breeding Program

12.3.1 Current Status and Future Outlooks

Three main strategies to conduct *in vitro*-selection of date palm towards abiotic stress are: (i) mapping and introgressing quantitative trait loci (QTLs) (Vinocur and Altman 2005); (ii) genetic transformation using genes and sequences intended for extrachromosomal or chromosomal insertions (Apse and Blumwald 2002; Vinocur and Altman 2005); and (iii) a genome-wide targeted approach. Although QTLs and extrachromosomal insertions have been explored with more or less success in many plant systems, they seem far from being applied in date palm due to the absence of genetic maps and the public acceptance of genotypes harboring extra-nuclear DNA from a foreign origin. In addition, the unavailability of defined abiotic stress tolerance-genes has been a major limitation for the latter. Only cases of successful regeneration of partially stress-tolerant plants have been reported and often using short-term assays, which does not allow judging the efficacy and durability of the acquired tolerance (El Hadrami et al. 2005).

Now that a number of tissue culture methods are being optimized in date palm, one would hope to see mature transformed plants, via *Agrobacterium*-mediated gene transfer or direct DNA uptake (electroporation or bombardment), been selected for abiotic stress tolerance testing. In this selection, transformed and wild type plants are exposed to a variety of abiotic stress conditions such as water deficit, salt, high temperature or nutrient deficiency. For example, salt stress can be applied through irrigation using solution with high osmotic value, or by hydroponically cultivating the plants in a hyper-osmotic growth solutions i.e. Hoagland solution, or by transferring seedling onto a high-salt MS medium. The chosen concentration range depends on the cultivar used and should cover mild, moderate to growth-inhibiting or lethal concentrations. The plant undergoing selection should be frequently monitored until substantial physiological and/or morphological effects appear in wild-type plants. Transformed plants exhibiting a better behavior than the wild type and noticeable physiological and/or morphological effects, biomass, should be identified as abiotic stress tolerant plants.

Alternative methods for transformation of date palm tissue with stress-related genes is to use viral-delivered polynucleotides or via introduction into the chloroplast genome. For example, recombinant constructs in which the native coat protein coding sequence from an avirulent strain of a systemically-infecting virus is deleted and replaced with a stress-related protein expressed under an altered viral promoter to be expressed in the plant host. The stress-related polynucleotide sequence or sequences can be transcribed or expressed in the host plant under control of the subgenomic promoter to produce the desired proteins. Constructs can be designed in a way to allow inserted non-native subgenomic promoters to transcribe or express adjacent genes in a plant host without any potential recombination with each other or with native subgenomic promoters. Non-native stress-related polynucleotide sequences may be inserted adjacent to the non-native subgenomic plant viral promoters so that the sequences are transcribed or expressed in the host plant under control of the subgenomic promoters to produce the desired proteins. The viral vectors are then encapsulated by coat proteins encoded by the recombinant plant viral polynucleotide to produce a recombinant plant virus. The recombinant plant viral polynucleotide or recombinant plant virus is used to infect appropriate host plants. The recombinant plant viral polynucleotide is capable of replication and systemic spread in the host, as well as the transcription or expression of stress-related genes of interest.

Exogenous stress-related polynucleotide sequences into cultivars undergoing breeding for tolerance to abiotic stress can be introduced to the genome of the chloroplasts. To achieve that, plant cells are chemically treated to reduce chloroplast numbers to about one per cell. These cells are then bombarded with the desirable exogenous stress-related polynucleotide in a way that a least one copy would be incorporated into the chloroplast. The preparation of the exogenous polynucleotides with at least one polynucleotide stretch derived from the chloroplast's genome guarantees its integration into this sub-cellular genome via homologous recombination. The contrast should also include a selectable marker to trace genotypes harboring at least a copy of the gene of interest.

The exogenous polynucleotides strategy offers the possibility of expressing many stress-related genes in a single host plant at once. This can be achieved by keeping its sequence with its promoter region or by designing constructs with a single promoter sequence under which a polycistronic message from different exogenous polynucleotide sequences can be transcribed. Co-translation of all proteins without interruption can be guaranteed by the insertion of an internal ribosome entry site sequence between the multi-exogenous expressed polynucleotides. Alternatively, constructs can be introduced individually into various plants that can be crossed within a conventional breeding program to generate progenies that could be screened for an enhanced stress-tolerance.

The recent completion of sequencing of the whole genome of date palm (Al-Dous et al. 2009) offers new ways for identifying stress-related genes. This information along with the available expressed sequence tags (ESTs) and cDNA sequences from a number of monocotyledonous species i.e. maize, barley, sorghum and rice could be used to functionally characterize and classify on the scale of whole-genome stress-related homologs whether they are orthologs or paralogs. Orthologous sequences evolved from a common ancestor by specialization and often retain identical function over time while paralogous are related by duplication events and often show a divergence in function. To further investigate and identify putative ortholog genes related to abiotic stress, computational methods such as sequences alignment and clustering or digital expression can be used.

Within newly developed searchable databases containing ESTs and cDNA sequences from a number of monocotyledonous species i.e. maize, barley, sorghum and rice, it is currently possible to generate an e-northern blot that displays the digital expression profile according to all keywords included in the sequence record comprising the sequences cluster. The e-northern blot can provide the expression profile of a cluster in terms of what tissues/organs the gene is expressed, the developmental stage at which the gene is expressed and the physiological conditions under which this gene is expressed i.e. drought, salinity, etc. Given the randomness of ESTs in the different clusters, the digital expression assigns also a probability value of occurrence, hence indicating a specialized or general expression.

12.3.2 In Vitro Selection for Functional Foods and Nutraceuticals

Dates were found to be a high source of energy (278–314 kcal/100 g), due to the high carbohydrate content, mainly fructose and glucose. They were also found to be low in fat (9.0 g/100 g) and proteins (5.1 g/100 g), but rich in dietary fibers and antioxidants. Total dietary fiber content of dates varies from 6.26 to 8.44 g per100 g of dates, and 84–94% of these fibers were insoluble. Dates represent a good source of antioxidant constituents including selenium (0.356–0.528 mg/100 g), total antioxidants (8,212–12,543 μmol of Trolox equiv/g), carotenoids (0.92–2.91 mg/100 g) and phenolics (217–343 mg of ferulic acid equiv/100 g). Date palm cultivars exhibit distinct levels and profiles of phenolic acids (gallic, protocatechuic, *p*-hydroxybenzoic, vanillic, caffeic, syringic, *p*-coumaric, ferulic and *o*-coumaric acids). In their review, Al-Farsi and Lee (2008) summarized information from over 80 references that depicted the content and nutritional value of dates. Ten minerals were reported in dates, the major being selenium, copper, potassium and magnesium. The consumption of 100 g of dates can provide over 15% of the daily-recommended intake. Vitamins B-complex and C were also found in large quantities in dates.

Biglari et al. (2008) analyzed the antioxidant activity of dates using three different methods namely using Trolox equivalent antioxidant capacity (TEAC) method, 2,2′-azinobis (3-ethylbenzothiazoline-6-sulphonic acid) radical cation (ABTS·+) assays, and the ferric reducing/antioxidant power method (FRAP assay), as well as the total phenolic and flavonoid contents and concluded that they have potential use as antioxidant functional food. Al-Farsi et al. (2007) reported on the compositional and functional characteristics of dates, syrups and their by-products such as press cakes and seeds. In this analysis, the authors examined the proximate composition, dietary fibers, total phenolics and antioxidant activities of these food products. The results suggested that date by-products, particularly the seeds, may be considered as a good source for natural antioxidants, and could potentially be used as a functional food or incorporated as an ingredient in a functional food.

All the information gathered on the nutritional value and health benefit of dates and their seeds or other by-products show the value-added of this crop. These will certainly offer new uses and boost the marketability of dates as a source for functional food ingredients. In return new demands upon the breeding programs have to be fulfilled and *in vitro* selection will be one of the key components to screen cultivars with the desirable traits i.e. high content of given phenolics.

12.3.3 In Vitro Selection and Climate Change

Climate change affects plant growth and development through the increase of temperatures as a result of a global warming and through an excess of greenhouse gases in the atmosphere. The impact of increased temperatures is much more important than the increase in the atmospheric CO_2 because they decrease water use efficiency defined as the ratio between the assimilated carbons over the lost water to transpiration. Water leaving the leaf encounters two major resistances opposed by the stomata depending on their degree of openness and the boundary layers (a layer of undisturbed air that envelopes the leaf). Carbon entering the leaf, on the other hand, encounters three resistances exerted by the boundary layer, stomata and mesophyll. Water diffuses faster than CO_2 due to its smaller molecular size. This means that any change in stomatal conductance will have a proportionally greater influence on water loss than carbon gain. Date palm being a plant of drier areas is likely to have a low stomatal conductance to maintain higher water use efficiency.

Global warming will likely have two main impacts on date palm. It will first displace its geographical distribution, and impact on its growth and development habits. Climate change is expected to lead to a substantial decrease in terms of date palm assimilation. This can potentially be restored by modifying canopy architecture or by increasing the net photosynthesis on a leaf area basis provided that the intercepted solar radiation does not change dramatically. Being a C_3 plant one would expect a manipulation of the Rubisco to decrease photorespiration.

12.4 Conclusion and Prospective

For a number of years, date palm tissue culture and *in vitro* selection has been given a considerable attention both in the scientific community and the industrial sector. This places enormous pressure in terms of advancing the technologies and adopting new protocols and refining them to achieve better, faster and reliable clonal multiplication of cultivars with desirable traits such as tolerance to drought, salinity and better nutrient uptake efficiency. Micropropagation and *in vitro* selection offer solutions to many technical issues encountered during the production of progenies and narrowed the time frame required to produce them. They also allowed for new applications intended to introduce variations and widen the polymorphism among date palm cultivated germplasm. Progress being made is still modest in comparison with other clonally propagated species due to difficulties such as generating protoplasts or cultures from anthers and ovules and to the optimization of genetic transformation.

Cultures of date palm from explanted stem tips or lateral buds, organs, tissues, cells, protoplasts and many others all offer varying degrees of potential for clonal multiplication and could more or less easily be subjected to stress conditions to select for stress-tolerance genotypes. When stabilized and understood, this newly introgressed tolerances can increase the polymorphism of the germplasm and help introduce new traits to be selected for within the breeding programs. Such novel traits could be opportunities and solutions for many constraints and challenges that lay on the shoulders of the date palm scientific community, breeders, pathologists, physiologists and agronomists combined to guaranty the sustainability of the earliest domesticated tree.

Acknowledgments This review has been supported by several research grants to Professor Ismail El Hadrami (Ifs/AUF/PRAD/CNRST/AI Morocco-Tunisia/AI Morocco-France). The authors would like to thank Dr. S.M. Jain for the critical review of the first draft of this contribution.

References

Ahloowalia BS, Maluszynski M (2001) Induced mutations- a new paradigm in plant breeding. Euphy 119:167–173

Al Mansoori TA, Alaa El-Deen MN, Caligari PDS (2007) Evaluation of in vitro screening techniques for salt tolerance in date palm. Acta Hort 736:301–307

Al-Dous EK, George B, Salameh YM et al. (2009) Qatar researchers sequence draft version of date palm genome. URL: http://qatar-weill.cornell.edu/research/datepalmGenome/download.html

Al-Farsi MA, Lee CY (2008) Nutritional and functional properties of dates: a review. Crit Rev Food Sci Nutr 48(10):877–887

Al-Farsi M, Alasalvar C, Al-Abid M et al (2007) Compositional and functional characteristics of dates, syrups, and their by-products. Food Chem 104:943–947

Al-Kaabi HH, Zaid A, Ainsworth C (2007) The presence of this insect can be identified through plant-off-types in tissue culture-derived date palm the following symptoms: 1. 'V' cut on the leaf rachis or (Phoenix dactylifera L.) plants. Acta Hort 736:267–281

Al-Khayri JM (2002) Growth proline accumulation and ion content in sodium chloride-stressed callus of date palm. In Vitro Cell Dev Biol Plant 38:79–82

Al-Khayri JM, Al-Bahrany AM (2004) Growth, water content, and proline accumulation in drought-stressed callus of date palm. Biol Plant 48:105–108

Al-Wasel ASAA (2005) Survey study on somaclonal variations in in vitro-derived date palm trees. In: Proceeding of the international workshop on true-to-typeness of date palm tissue culture-derived plants, Institut National de Recherche Agronomique, Morocco, 23–25 May, 2005. http://www.inra.org.ma/ist/publications/ouvrages/truetype.pdf.

Apse MP, Blumwald E (2002) Engineering salt tolerance in plants. Curr Opin Biotech 13:146–150

Biglari F, AlKarkhiand AFM, Mat Easa A (2008) Antioxidant activity and phenolic content of various date palm (*Phoenix dactylifera*) fruits from Iran. Food Chem 107:1636–1641

Blum A (1996) Crop responses to drought and the interpretation of adaptation. Plant Growth Reg 20:135–148

Buchanan B, Gruissem W, Jones R (eds.) (2000) In: Biochemistry and molecular biology of plants. American Society of Plant Physiologists, Rockville

Djibril S, Mohamed OK, Diaga D et al (2005) Growth and development of date palm (*Phœnix dactylifera* L.) seedlings under drought and salinity stresses. Afr J Biotech 4:968–972

El Hadrami I, El Hadrami A (2009) Breeding date palm. In: Jain SM, Priyadarshan PM (eds.) Breeding plantation tree crops. Springer, New York, pp 191–216

El Hadrami A, El Idrissi-Tourane A, El Hassni M et al (2005) Toxin-based in vitro selection and its potential application to date palm for resistance to the bayoud *Fusarium* wilt, a review. CR Biol 328:732–744

Elshibli S, Korpelainen H (2009) Biodiversity of date palms (*Phoenix dactylifera* L.) in Sudan: chemical, morphological and DNA polymorphisms of selected cultivars. Plant Gen Res 7:194–203

Eshraghi P, Zarghami R, Ofoghi H (2005) Genetic stability of micropropagated plantlets in date palm. J Sci Islam Rep Iran 16(4):311–315

Frank W, Munnik T, Kerkmann K et al (2000) Water deficit triggers phospholipase D activity in the resurrection plant *Craterostigma plantagineum*. Plant Cell 12:111–124

Gaxiola RA, Li J, Undurraga S et al (2001) Drought- and salt-tolerant plants result from overexpression of the AVP1 H^+-pump. Proc Natl Acad Sci USA 98:11444–11449

Hall AE, Findell JL, Schaller GE et al (2000) Ethylene perception by the ERS1 protein in *Arabidopsis*. Plant Phys 123:1449–1458

Hasegawa PM, Bressan RA, Zhu J-K et al (2000) Plant cellular and molecular responses to high salinity. Ann Rev Plant Phys Plant Mol Biol 51:463–499

Havaux M (1998) Carotenoids as membrane stabilizers in chloroplasts. Trends Plant Sci 3:147–151

Ho SL, Chao YC, Tong WF et al (2001) Sugar coordinately and differentially regulates growth- and stress-related gene expression via a complex signal transduction network and multiple control mechanisms. Plant Phys 125:877–890

Jain SM, Gupta P (eds) (2005) Protocols of somatic embryogenesis in woody plants. Springer, Dordrecht

Jaleel CA, Paramasivam M, Wahid M et al (2009) Drought stress in plants: a review on morphological characteristics and pigment compositions. Intl J Agric Biol 11:100–105

Knight H (2000) Calcium signaling during abiotic stress in plants. Int Rev Cytol 195:269–324

Merlot S, Gosti F, Guerrier D et al (2001) The ABI1 and ABI2 protein phosphatases 2 C act in a negative feedback regulatory loop of the abscisic acid signaling pathway. Plant J 25:295–303

Nam NH, Chauhan YS, Johansen C (2001) Effect of timing of drought stress on growth and grain yield of extra-short-duration pigeon pea lines. J Agric Sci 136:179–189

Pedrieri S (2001) Mutation induction and tissue culture in improving fruits. Plant Cell Tissue Org Cult 64:185–210

Prochazkova D, Sairam RK, Srivastava GC et al (2001) Oxidative stress and antioxidant activity as the basis of senescence in maize leaves. Plant Sci 161:765–771

Razmjoo K, Heydarizadeh P, Sabzalian MR (2008) Effect of salinity and drought stresses on growth parameters and essential oil content of *Matricaria chamomile*. Intl J Agric Biol 10:451–454

Rhoades JD, Kandiah A, Mashali AM (1992) The use of saline waters for crop production. FAO Irrigation and Drainage Paper 48, Rome

Rosa M, Prado C, Podazza G et al (2009) Soluble sugars-metabolism, sensing and abiotic stress. A complex network in the life of plants. Plant Sign Behav 4:388–393

Saijo Y, Hata S, Kyozuka J et al (2000) Over-expression of a single Ca^{2+}-dependent protein kinase confers both cold and salt/drought tolerance on rice plants. Plant J 23:319–327

Sanders D, Brownles C, Harper JF (1999) Communicating with Calcium. Plant Cell 11:691–706

Shao HB, Chu LY, Abdul Jaleel C et al (2009) Understanding water deficit stress-induced changes in the basic metabolism of higher plants-biotechnologically and sustainably improving agriculture and the ecoenvironment in arid regions of the globe. Crit Rev Biotech 29:131–151

Simpson GM (1981) The value of physiological knowledge of water stress in plants. In: Simpson GM (ed.) Water stress on plants. Praeger, New York, pp 235–265

Tahtiharju S, Palva T (2001) Antisense inhibition of protein phosphatase 2C accelerates cold acclimation in *Arabidopsis thaliana*. Plant J 26:461–470

Tester M, Davenport R (2003) Na^+ tolerance and Na^+ transport in higher plants. Ann Bot 91:503–527

Vinocur B, Altman A (2005) Recent advances in engineering plant tolerance to abiotic stress: achievements and limitations. Curr Op Biotech 16:123–132

Welfare K, Yeo AR, Flowers TJ (2002) Effects of salinity and ozone, individually and in combination, on the growth and ion contents of two chickpea (*Cicer arietinum* L.) varieties. Environ Pollut 120:397–403

Widoretno W (2003) Seleksi in Vitro untuk Toleran terhadap Cekaman Kekeringan pada Kedelai (*Glycine max* (L.) Merr.) dan Karakterisasi var Somaklonal yang Toleran. [Disertasi]. Bogor: Program Pascasarjana IPB

Witjaksono W (2003) Peran bioteknologi dalam pemuliaan tanaman buah tropika. Seminar Nasional Peran Bioteknologi dalam Pengembangan Buah Tropika. Kementerian Riset dan Teknologi RI & Pusat Kajian Buah Buahan Tropika, IPB. Bogor, 9 Mei 2003

Xiong L, Zhu JK (2002) Molecular and genetic aspects of plant responses to osmotic stress. Plant Cell Environ 25:131–139

Xiong L, Lee BH, Ishitani M et al (2001) FIERY1 encoding an inositol polyphosphate 1-phosphatase is a negative regulator of abscisic acid and stress signaling in *Arabidopsis*. Genes Dev 15:1971–1984

Xiong L, Schumaker KS, Zhu JK (2002) Cell signaling during cold, drought, and salt stress. Plant Cell 14:165–183

Yamaguchi-Shinozaki K, Shinozaki K (2006) Transcriptional regulatory networks in cellular responses and tolerance to dehydration and cold stresses. Ann Rev Plant Biol 57:781–803

Yusnita (2005) Induksi Variasi Somaklonal dan Teknik Seleksi in Vitro untuk Mendapatkan Galur Kacang Tanah (*Arachis hypogaea* L.) Resisten Penyakit Busuk Batang Sclerotium. [Disertasi]. Bogor: Program Pascasarjana IPB

Yusnita, W, Sudarsono (2005) In vivo selection of peanut somatic embryos on medium containing culture filtrates of Sclerotium rolfsii and plantlet regeneration. Hayati 12:50–56

Zaid A, Al-Kaabi H (2003) Plant-off types in tissue culture-derived date palm. (*Phoenix dactylifera* L.). Emir J Agric Sci 15:17–35

Chapter 13
Fusarium oxysporum f. sp. *albedinis* Toxin Characterization and Use for Selection of Resistant Date Palm to Bayoud Disease

MyH. Sedra and B.H. Lazrek

Abstract Date palm (*Phoenix dactylifera* L.) is the most economically important food crop in Moroccan oasean agricultural areas, contributing to preserving an arid ecosystem threatened by desertification. The bayoud disease, caused by the fungus *Fusarium oxysporum* f. sp. *albedinis* (Foa), is incontestably the most serious disease affecting date palm in North Africa. The selection for resistance among date palm cultivars was the preferred way to control the disease. New performing cultivars were selected, mass propagated and distributed to farmers. The use of pathogen toxins in *in vitro* selection is an innovative approach for rapid screening for resistance to bayoud disease. This chapter gives an overview of recent knowledge about toxins and other substances produced by plant pathogenic fungi and their applications in *in vitro* and *in vivo* selection for resistance. Foa toxins contain fusaric acid and other toxic fractions. These fractions of toxins have some chemical and biological characteristics that differ from among other fractions isolated from other pathogenic and nonpathogenic strains of *F. oxysporum*. These toxins could be exploited for pre-selection of plants for resistance to bayoud among populations of plants originating from either irradiated tissue culture or conventional breeding programs.

Keywords Date palm • Breeding for resistance • *In vitro* selection • HPLC analysis • Toxins

MyH. Sedra (✉)
Laboratoire de Phytopathologie, Génétique et de Lutte Intégrée,
Centre Régional de Marrakech, Institut National de Recherche
Agronomique (INRA), BP. 533, Ménara, 40000, Marrakech, Morocco
e-mail: sedramh@hotmail.com; mhsedra@yahoo.fr

B.H. Lazrek
Laboratoire de chimie bio-organique, Département de Chimie,
Faculté des sciences Sémlalia, Université Cadi Ayyad,
Boulevard Moulay Abdellah B.P S 15, Marrakech, Morocco

S.M. Jain et al. (eds.), *Date Palm Biotechnology*,
DOI 10.1007/978-94-007-1318-5_13, © Springer Science+Business Media B.V. 2011

13.1 Introduction

Date palm (*Phoenix dactylifera* L.) is the essential element of Saharan and sub-Saharan ecosystems. The problems of protection against pests and diseases constitute major constraints for date palm sector development. Bayoud disease, caused by *Fusarium oxysporum* f. sp. *albedinis*, the abbreviation Foa is used throughout this chapter, is incontestably the most serious and destructive threat menacing date producers in the Maghreb countries. It is one of the wilt diseases which it is so difficult to control. However, to control this disease, the foreseeable control methods are the same as those recommended against *Fusarium* wilts. The adoption of such a method, among others, depends on the countries themselves and the gravity of the disease. An efficient strategy requires the implementation of an integrated management program. In Morocco, the genetic improvement of date palm through a breeding program has been adopted to control the disease for several decades. The diffusion of existing resistant cultivars has been very limited thus far; due to the mediocre quality of date fruits they produced, that lacked the commercial value desired by producers. The research undertaken regarding genetic improvement aimed at the selection of cultivars combining good fruit quality and resistance to bayoud. New performing cultivars were selected from some already mass propagated and distributed to farmers (Sedra 1995, 2003a, b, 2007a, 2010) in order to reconstitute date farms damaged by the disease. The use of toxin pathogens in *in vitro* selection could be one promising approach for a rapid method of screening for resistance to bayoud disease. Before highlighting this approach to the use of toxins as selective agents and to relate the findings about the date palm-Foa pathosystem, an overview is presented to summarize recent knowledge about toxins and other substances produced by plant pathogenic fungi and possible screening and selection for resistance *in vitro* and *in vivo*.

13.2 Overview

Pathosystems are very diverse and there is neither a single model of plant- pathogen interactions nor a simple and common resistance mechanism. Host-pathogen interactions are governed by a complex of molecular and biochemical reactions, which ultimately result in the expression of disease resistance or susceptibility. Passive defense lines such as cell walls, wax layers and chemical barriers confer broad resistance to a wide variety of pathogens (Lebeda et al. 2001). Pathogenic fungi may use several modes of action to penetrate into plant tissue to overcome mechanical, chemical and/or physiological barriers of the host plant. In diseases caused by microorganisms producing toxins, these metabolites could be involved in the pathogenicity (Wolpert et al. 1994; Yoder 1980). Toxins produced by many microorganisms and certain fungi can be grouped by their specific ability to produce toxic substances, but within this group there are large differences in the taxonomic distances between the pathogens, and in their mode of plant-pathogen interactions (Allen et al. 1999). Many papers have been published over the past 20 years, including several valuable review studies on toxins in plant pathogens and plant diseases (Durbin 1981; Hamer and Holden 1997; Hensel and Holden 1996; Huang 2001; Sedra et al. 2002; Svabova and Lebeda 2005; Walton 1996). Toxins are characterized by a high diversity in their chemical structure and physicochemical and biological properties (Wolpert et al. 2002). Diverse chemical classes including proteins, polypeptides, polyketols, terpenes and/or glycoproteins were reported as examples in the literature (Markham and Hille 2001; Svabova and Lebeda 2005; Wolpert et al. 2002). The preceding authors grouped toxic substances by several properties: (1) chemical characteristics (peptide, terpenoid, glycoside, phenol, polysaccharide or other biosynthetic pathways etc.); (2) type of the producing organism (fungus or bacterium); (3) biological activity (enzyme inhibitor, antimetabolite, cell-wall degrading substance etc.); (4) host specificity or nonspecificity (Nedelnik and Repkova 1998). Definitions of toxins are based on the identity of the microorganisms producing them, their chemical nature and mode of action. Therefore, they are considered primary or secondary determinants of pathogenesis according to their involvement level in pathogenesis (Lepoivre 2003; Yoder 1980). Toxins are primary determinants when they act as the key element in infection initiation and symptom development; they are secondary determinants when they only modify the symptom intensity (Lepoivre 2003). Several experimental approaches have been developed in order to evaluate the role of toxins in host-pathogen interactions (Brown et al. 2001; Wolpert et al. 2002). Several authors have assessed their involvement in pathogenesis using five common criteria: (1) host-specificity (Bamburg and Strong 1971; Pringle and Scheffer 1964; Svabova and Lebeda 2005; Wolpert et al. 2002; Yoder 1980); (2) presence in infected plants (Karr et al. 1974; Rudolph 1976; Scheffer 1976; Svabova and Lebeda 2005; Yoder 1980); (3) toxin production at a key step of disease development (Scheffer 1976; Yoder 1980); (4) induction of typical disease symptoms (Daly 1976; Scheffer and Pringle 1967; Svabova and Lebeda 2005; Yoder 1980); and (5) degree of correlation between the quantity of toxin produced *in vitro* and the pathogenicity level (Brian et al. 1952; Kohmoto et al. 1979; Pringle and Scheffer 1964; Rudolph 1976; Sedra et al. 1997; Yoder 1980).

The toxins can be inactivated or specifically eliminated, during the initiation, establishment or expression of the host-pathogen interaction, by antibody and metabolic inhibitors but these may not always have sufficient specificity to neutralize toxin actions (Lepoivre 2003). This neutralization may be possible when the pathogen is affected by natural or induced mutations or the use of resistant plant genotypes (Hodgkin 1990; Knogge 1996; Takken and Joosten 2000). The study of toxins has been dedicated to the aspects of the interactions between plants and toxin-producing pathogens, the mode of phytotoxic action and the host and nonhost-selective toxins. However, they have been focused on host-specific toxins and mycotoxins (Brown and Hunger 1999; Walton and Panaccione 1993) rather than on nonhost specific toxins, which affect a large spectrum of plant species (Chung et al. 2002; Daub and Ehrenshaft 2000; Svabova and Lebeda 2005). There are several types of toxins: mycotoxins, phytotoxins, vivotoxins, pathotoxins and phytoagressins. For fungi, the most studied are the mycotoxins and phytotoxins.

13.2.1 Mycotoxins

The mycotoxins are the product of secondary metabolites, of low molecular weight, produced by fungi and are present in several products of human and animal food and provoking numerous diseases and death (Bennett and Klich 2003); for example, aflatoxins, fumonisines, moniliformines and ochratoxines. The fungi that produce these toxins are, for example, *Aspergillus, Penicillium, Claviceps, Alternaria* and *Fusarium* but certain other fungi cannot produce them; for example, those causing downy mildews (Spencer-Phillips et al. 2002), powdery mildews (Belanger et al. 2002), smuts and rusts (Nedelnik and Repkova 1998). A very important group of *Fusarium* mycotoxins are the trichothecenes.

13.2.2 Phytotoxins

The phytotoxins playing a role in pathogenesis are also called *pathotoxins* or *phytoaggressins* (Graniti 1991); for example, fusaric acid, enniatins and lycomarasmine. Phytotoxins have been recognized as useful tools for the induction and selection of disease-resistant plants using *in vitro* selection (Bhatt et al. 1988; Chawla and Wenzel 1987; Darakov 1995). Phytotoxins are generally classified into two major groups: nonspecific and specific. The possible involvement of fusaric acid, a nonspecific toxin, in disease development has been reviewed by Pegg (1981). Several *Fusarium oxysporum* strains have been reported to produce low molecular-weight phytotoxins with different chemical natures (Hermann et al. 1996; Mepsted et al. 1995; Sutherland and Pegg 1995). In bayoud disease of date palm, Foa produces substantial quantities of fusaric acid in culture filtrate (Mokhlisse 1987; Sedra et al. 2008; Surico and Graniti 1977).

13.2.3 Host Specific and Nonhost Specific Toxins

Fungi can produce one or several toxins capable of acting alone or in synergy (Soler-Rivas et al. 1999). The toxins can be classified in two groups according to host specificity. Host-specific toxins (Park et al. 1994; Wang et al. 1996) are indispensable for disease development and correspond to pathogen pathogenicity factors which constitute, in some cases, the primary determinants of pathogen virulence (Bender et al. 1987; Mo and Gross 1991). Nonhost-specific toxins, for example fusicoccine affect several plant species in the same way; as they can act on a common metabolic process for most plant species.

13.2.4 Selection of Resistant Plants Using Toxins

Svabova and Lebeda (2005) reported that in the last two decades more than 100 publications have focused on selections for the improvement of resistance to plant pathogens.

Over 30 plant species were examined to utilize various selection agents extracted from about 40 plant pathogens. The use of the toxins as selective agents has remained the most adopted method to control some plant diseases caused by pathogens (Bhatt et al. 1988; Branchard 1984; Brettell et al. 1980; Carlson 1973; Daub 1984; Darakov 1995; Megneneau 1994; Sacristam 1982; Sedra et al. 2008). Other studies showed that the toxins can be used for the improvement of the resistance of some plants to pathogenic agents (Svabova and Lebeda 2005). These authors reported this application that successfully resulted in resistant lines in banana, carnation, grapevine, strawberry and wheat and currently, these techniques are an important complement to classical breeding methods.

The *in vitro* selection using a culture filtrate of the pathogen is much more convenient than the use of the pathogenic agents. But this filtrate contains a large amount of undetermined toxins and the chosen resistance could be for the minor phytotoxin constituting the filtrate or for the cytolytic enzymes, or a combination of the two compounds. Consequently, it is necessary to verify the published experiences to the present and the results should be compared with the toxins used to eliminate the selection for the components that are not essential (El Hadrami et al. 2005). Otherwise, Sedra et al. (2008) showed that the culture filtrate of Foa contains fusaric acid and other essential toxic fractions qualitatively capable of differentiating material resistant and susceptible to bayoud disease, knowing that the toxic activity of the fusaric acid and the filtrate is not specific. In spite of this, fusaric acid has been used several times with success to improve resistance in banana (Matsumoto et al. 1995), alfalfa (Remotti and Löffler 1996), barley (Chawla and Wenzel 1987), wheat and potato (Wenzel and Foroughi-Wehr 1990), pineapple (Borras et al. 2001) and tomato (Shahin and Spivey 1986).

13.3 Bayoud Disease and Toxins

Like other pathogenic *Fusarium oxysporum* species, Foa, causal agent of bayoud disease, secretes into a liquid culture media numerous substances like toxins (El Fakhouri et al. 1996a, b; Mokhlisse 1987; Sedra et al. 1993, 1997, 1998; Surico and Graniti 1977) and enzymes (Amraoui et al. 2004; El Modafar et al. 2000). Several special forms of *F. oxysporum* produce various peptidic (Mussel 1972), terpenic (Casinovi 1972), polysaccharidic or glycopeptidic (Lousberg and Salemink 1972) phytotoxic compounds (phytotoxins). Some of them have been chemically and biologically well characterized, such as fusaric acid and its derivatives which were isolated from culture filtrates of *F. oxysporum* f. sp. *elaeidis*, *F. moniliforme* (Mepsted et al. 1995; Porter et al. 1995; Sutherland and Pegg 1995) and from Foa (El Fakhouri et al. 1996a, b; El Hadrami et al. 2005; Mokhlisse 1987; Sedra et al. 1993, 1997; Surico and Graniti 1977). Fusaric acid may play an important role in the early stage of date palm infection by Foa (Bouizgarne et al. 2004; Sedra et al. 2008).

13.4 Isolation, Production and Purification of Toxins of Foa

13.4.1 Fungal Culture, Fungal Culture Filtrate and Toxin Extraction Procedure

In order to produce culture filtrate and toxins, Foa was grown in liquid Czapek's medium according to the protocol determined by Sedra et al. (1993) and El Fakhouri et al. (1996a). This medium was shown to be the best for fungus growth and sporulation in comparison with other media (Sedra 1993a). An improvement of toxin production was achieved when four chemically different media derived from Czapek's medium were used. The best carbon and nitrogen sources were determined for excellent sporulation and fungal growth (Sedra et al. 1998). In fact, Table 13.1 shows that the medium M1, containing mannose and KNO_3 is the more suitable for toxins (195 mg/l) than Czapeck's medium (130 mg/l) production). The nature of carbon and nitrogen plays, therefore, a necessary role in culture medium. Furthermore qualitative analysis by HPLC demonstrated that the composition of the toxin was not been affected by the composition of the culture medium used.

13.4.2 Extraction of Secondary Metabolites

The extraction of secondary metabolites was performed on Foa culture filtrates according to the method of Pringle and Scheffer (1963), modified by El Fakhouri et al. (1996a). During extraction (Fig. 13.1), the adsorbed fractions were diluted by three successive steps of mixtures of 10%, 30%, and 50% pyridine in water (v/v) that, respectively, gave three collected fractions labeled F_I, F_{II}, and F_{III}. These fractions were evaporated until dry and a yellowish powder was obtained for F_I, while whitish ones were obtained for the F_{II} and F_{III} fractions. Recent findings show that the extraction by the butanol/water is the best method to have a good field of fusaric acid, followed by extraction using acetic ether (result not yet published). For further experiments and bioassays, only the fraction F_{II} was used since it showed the highest phytotoxicity (Sedra et al. 1998). The same purification procedure was applied to

Table 13.1 Composition of culture media using different sources of carbon and nitrogen

Culture medium	Carbon source	Nitrogen source
M1	Mannose	KNO_3
M2	Dextrose	$NaNO_3$
M3	Saccharose	NH_4NO_3
M4 (control)	Saccharose	$NaNO_3$

The medium M4 is the control (normal Czapek's medium); the media M1, M2 and M3 are different media derived from Czapek's medium

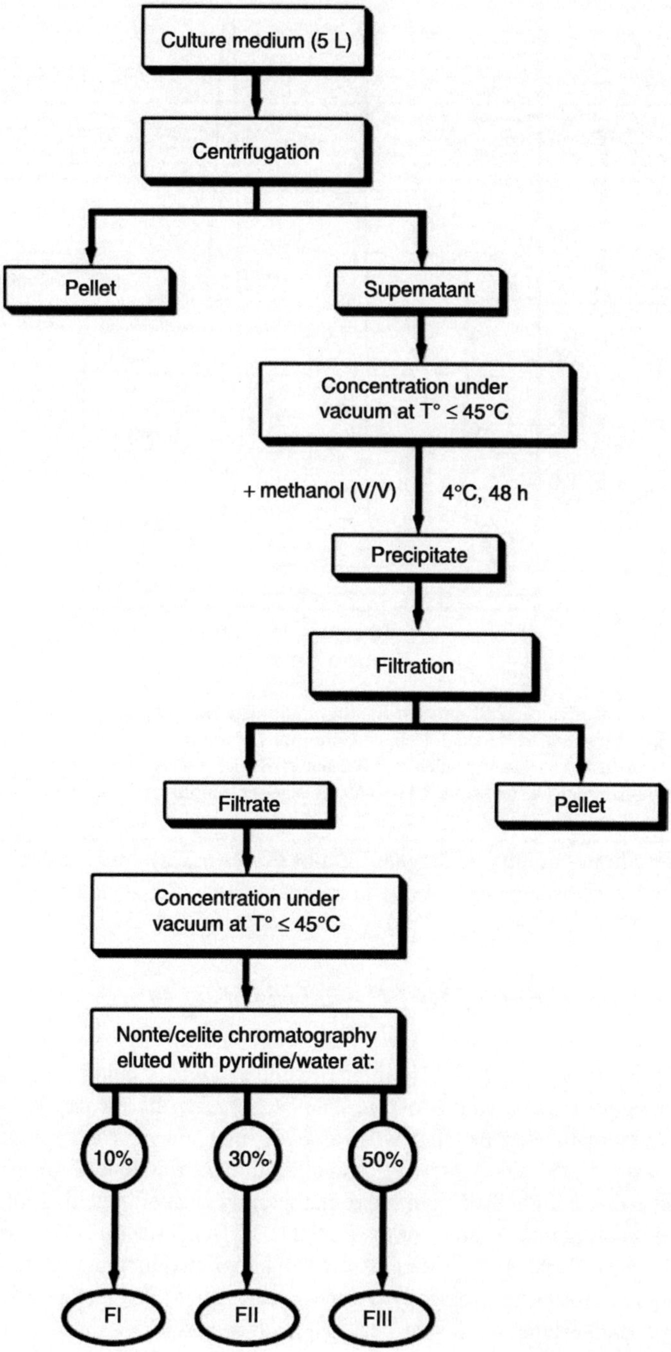

Fig. 13.1 Extraction protocol of secondary metabolites from culture filtrates of Foa and of other pathogenic strains of *Fusarium oxysporum*-like f. sp. *lycopercisi, lini, elaedis* and nonpathogenic *F. oxysporum* strain (AZ4) isolated from soil

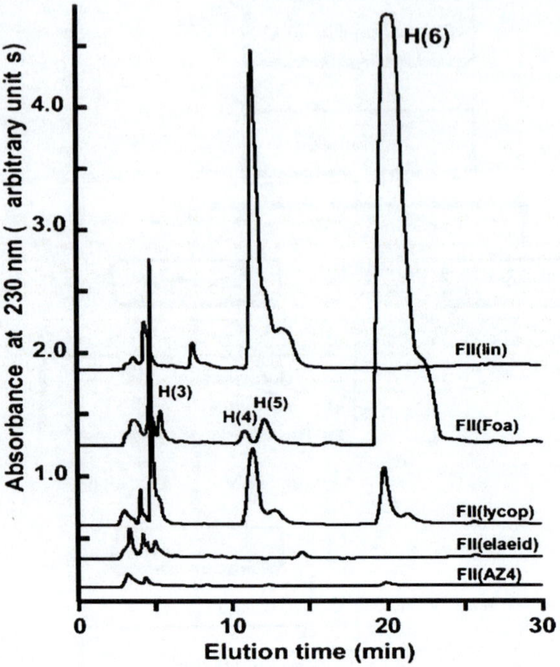

Fig. 13.2 UV absorbance (230 nm) chromatographic pattern of the F_{II} (Foa/133), F_{II} (AZ4), F_{II} (lycop), F_{II} (lin), and F_{II} (elae) fractions. Conditions were: reverse phase C18 Vydac 5 µm (10 × 250 mm) column, linear gradient 5–20% buffer B (0.1% TFA in acetonitrile) in buffer A (0.1% TFA in deionized water), after 10 min 5% B isocratic conditions, 1 ml/min

the culture filtrates of other pathogenic strains of *Fusarium oxysporum*, as indicated below, for comparison (Fig. 13.1) (Amraoui et al. 2005; Sedra et al. 2008).

13.4.3 Comparison of HPLC Analysis of F_{II} Fractions

The F_{II} fractions were obtained from the organic extracts of culture filtrates of Foa and of other pathogenic strains of *Fusarium oxysporum* like *F. oxysporum*. f. sp. *lycopercisi* (tomato wilt), *lini* (flax wilt), *elaeidis* (oil palm wilt) and non pathogenic *F. oxysporum* strain (AZ_4) isolated from soil. All these fractions were analyzed by HPLC on a Hewlett Packard liquid chromatography (Amraoui et al. 2005) and the HPLC chromatographic profiles of the F_{II} (Foa), F_{II} (AZ_4) fractions were very different (Fig. 13.2). No peaks appeared in the chromatogram regarding the AZ4 extracts at 230 nm monitoring spectroscopy (Amraoui et al. 2005; Sedra et al. 2008). In the F_{II} (Foa) HPLC profile, a relevant peak, H(3), appeared at 5.4 min retention time, followed by the peak H(4) at 10.9 min, the peak H(5) at 12.2 min, and then a huge peak H(6) at 20.2 min. Molecular masses of toxins were determined and performed by mass spectroscopy (Amraoui et al. 2005).

13.5 Chemical and Biological Characteristics of *Fusarium oxysporum* Toxins

All of the toxin fractions (F_I, F_{II} and F_{III}) extracted from culture filtrate of Foa is toxic to date palm, but the F_{II} is the most important compound and the most toxic (El Fakhouri et al. 1996b; Sedra et al. 1993, 1997). A laboratory test showed that the F_{II} is a heat-resistant fraction (Sedra et al. 1997, 1998). It has been demonstrated that these toxin fractions can be also be toxic to detached leaves of other plants such as tomato, olive and alfalfa, after 1 week of incubation (Fig. 13.3) (Sedra et al. 2008).

Fig. 13.3 Specificity of different toxin fractions extracted from culture filtrate of Foa, causal agent of bayoud disease. Drying-up and browning caused by toxin fraction as symptoms on detached leaves of other plants: tomato, olive and alfalfa after 1 week of incubation

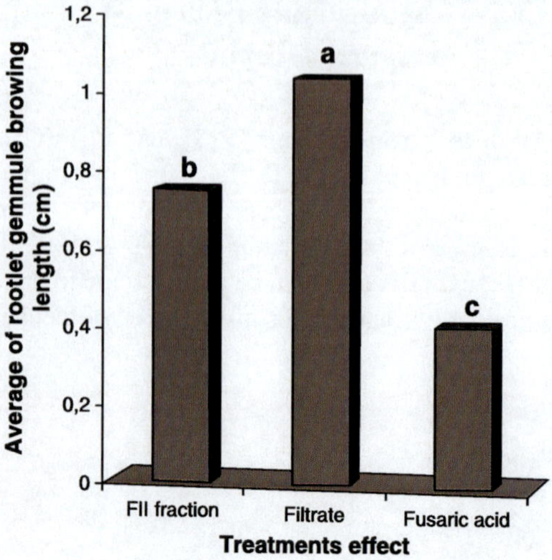

Fig. 13.4 Level of toxic effect of different substances on seedling growth issued from susceptible date palm cv. (Jihel) (different letters mean significant differences between treatments, test of Newman and Keuils, p = 0.05)

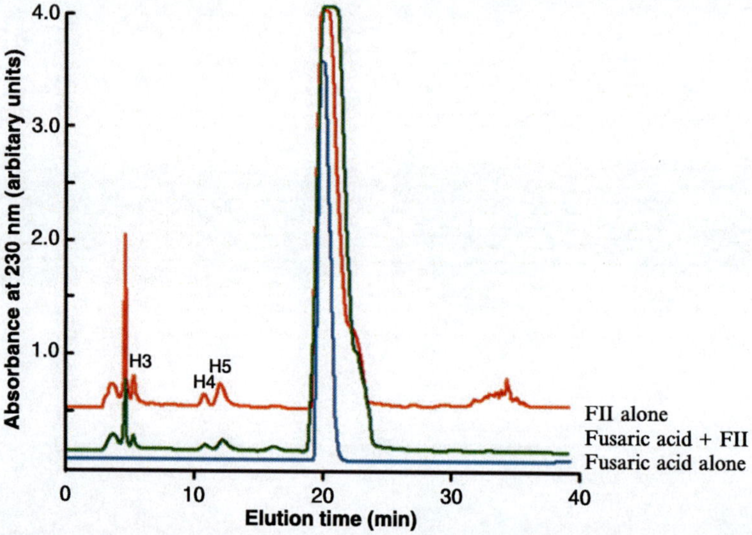

Fig. 13.5 Presence of fusaric acid and other toxic sub-fractions (H3, H4 and H5) in the F_{II} fraction of Foa, causal agent of bayoud (HPLC analysis)

This result may explain the nonspecificity of these different toxin fractions and the dominance of a fusaric acid substance in them. Figures 13.4 and 13.5 reinforce this hypothesis. However, the F_{II} (Foa) fraction contains the fusaric acid and other toxic sub-fractions (H(3), H(4), H(5) and H(6)) (HPLC analysis). These sub-fractions are not only absent in toxin fractions of other pathogenic and nonpathogenic strains of *Fusarium oxysporum* (Fig. 13.2) but showed also their phytotoxicity on date palm (Amraoui et al. 2005; Sedra et al. 2008). In fact, it was demonstrated that fusaric acid and some fractions purified from F_{II} (Foa), i.e. H(3), H(4), and H(5), were toxic to detached leaves. Moreover, H(4) was shown to be more toxic than either the H(6) fraction or authentic fusaric acid, since H(4) induced more severe symptoms more rapidly. Thus, it appears that the F_{II} (Foa) fraction contains new toxins, different from enniatins or fusaric acid by their specificity and their elution time in HPLC, respectively. These molecules could act independently or synergistically with fusaric acid to induce the characteristic symptoms of bayoud disease, but this deserves further studies to be demonstrated. Since these sub-fractions, H(3), H(4), H(5), and H(6) are absent in the F_{II} (AZ_4) HPLC profile, this may explain why the *F. oxysporum* AZ4 strain is not pathogenic on date palm. It would be possible to postulate that the H(3), H(4), H(5) fractions contain new molecules and may play an important role, like fusaric acid, in bayoud disease and the appearance of its symptoms. These toxins can be used bulked or separately as selective agents. Their chemical structure determination and their specificity are currently under investigation.

Previous studies showed that Foa could be distinguished from other special forms or saprophytic isolates according to morphological characters and cultural and pathogenic characteristics (Sedra 1993a, b, 2003a), but also esterase polymorphism (Bounaga 1985; Cherrab 1989), pectinolytic, cellulolytic and proteolytic enzymes (Amraoui et al. 2004), vegetative compatibility and genetic polymorphism (Tantaoui and Boisson 1991; Tantaoui and Fernandez 1993; Tantaoui et al. 1996; Sedra 2003a, 2007b, 2008; Sedra and Zhar 2010). In addition, the use of toxins may allow the noticeable differences observed between Foa and the saprophytic strain (Amraoui et al. 2005; Sedra et al. 1997), as regards their toxicity and chromatographic pattern, suggest that the Foa pathogenicity relies on its ability to produce the F_{II} toxins and others sub-fractions, a property that could alone determine either a pathogenic or a saprophytic strain.

The present results add a new physiological criterion in the Foa characterization by the presence of characteristic HPLC peaks in the F_{II} fraction. Since the F_{II} fraction, as well as the toxic fractions dissolved in HPLC, can readily induce bayoud symptoms, either of these fractions may be used in early selection of new Foa-resistant palm-tree cultivars. Such deeper studies for more of enlightenment are presently under investigation.

13.6 Use of Foa Toxins in Selection of Bayoud Resistant Date Palm

In order to evaluate toxin effects and to develop an efficient and rapid and miniaturized method of selection of cultivars, Sedra et al. (1993, 1998) tested different material of date palm tree: *in vitro* complete plantlet derived from tissue culture, seedlings from germinated seeds, detached leaves of young plants and detached leaflets from adult palm leaves. Symptoms begin to appear during the first week. On seedlings and *in vitro* plants, the toxin produced similar symptoms (leaf root browning, leaflet rolling then withering and the death of plantlets) which occur when plants are inoculated

Fig. 13.6 Drying-up and browning as symptoms on detached date palm leaves that led to their death. Screening of resistant (*right*) and susceptible plants (*left*) after 1 week

Fig. 13.7 Symptoms of the toxic effect characterized by rootlets browning, a delay of the gemmule apparition or inhibition of seedling growth for susceptible genotypes plants to Foa toxin (*right*), resistant genotypes (*left*). Plants from crossed progenies in date palm

with the pathogen. On detached leaves, the symptoms are characterized by chlorotic and necrotic spotting on inoculated leaves, petioles, and stems and a drying-up leading to death. The rapid and most common procedure is using detached leaves (Fig. 13.6) and seedlings (Fig. 13.7); it required minimal toxin concentration which could differentiate resistant and susceptible date palm cultivars is 10 μg/ml for subfraction and 50 μg/ml for fraction F_{II} (Sedra et al. 1993, 1998, 2008). Figure 13.6 showed that after 1 week, this *in vitro* technique can screen resistant and susceptible date palm plants. The susceptible cultivar presents the symptoms of drying-up and browning on detached date palm leaves that led to death. On seedlings derived from crossed progenies, the symptoms of toxic effect are characterized by rootlet browning, a delay of the gemmule apparition or inhibition of seedling growth for susceptible genotypes (Fig. 13.7). Research on *in vitro* selection using toxins in the culture

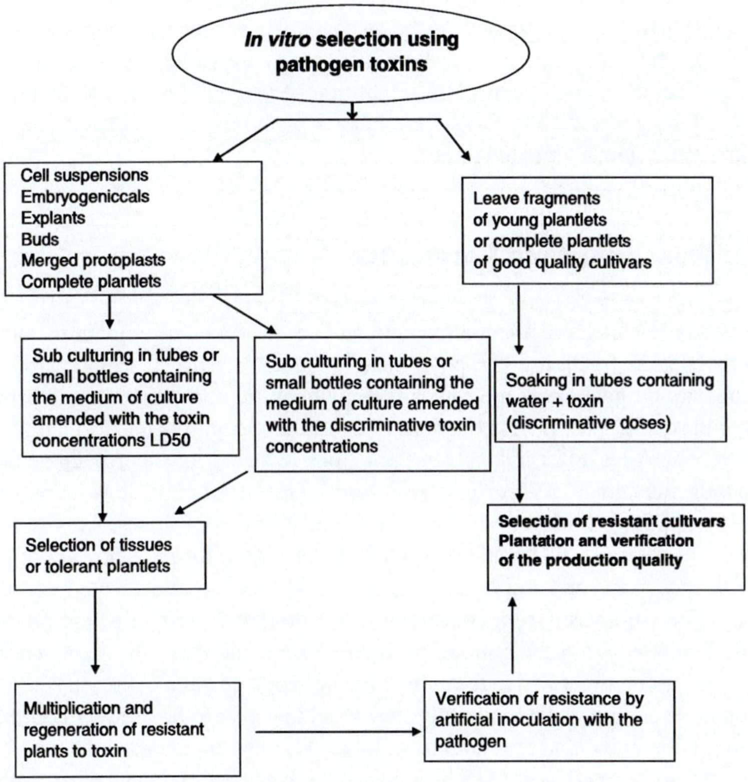

Fig. 13.8 In vitro selection of the resistant cultivars and clones using pathogen toxins

medium of cells was encouraging (result unpublished); these tests need to be more developed for an efficient and economical application method. Figure 13.8 illustrates the different steps of *in vitro* selection procedure in date palm using pathogen toxin (Sedra, 2003a). One the best examples of application of the use of the toxin in a selection program can be found in the research led by INRA (Morocco) in partnership with the International Atomic Energy Agency (IAEA) which permitted *in vitro* creation of some resistant mutants of commercial and susceptible cv. Boufeggous using gamma irradiation and *in vitro* selection using pathogen toxins (Bougerfaoui et al. 2006; Sedra et al. 2008). The selected mutants are under observation in the field at the Zagora Experiment Station (INRA, Morocco). Moreover, within the framework of the regional project IAEA/RAF/5/035 project and/049, the FII fraction of toxins has been produced in our laboratories and distributed to member countries for use in tests of *in vitro* selection in Morocco, Algeria and Tunisia. Important results have been obtained and reported in project countries and IAEA reports.

However, in terms of future prospects, it is important to verify the relationship between resistance to toxins and to the pathogen on the large scale of cultivars. Generally, Graniti (1991) has reported that several plant cultivars showing resistance to selective toxins are also resistant to the pathogen. These interesting and encouraging results open the way to some perspectives of study essentially aiming to determine the sequences of the pathogen's DNA, through the survey of the ARNm that corresponds to the genes coding for the production of the toxin considered one of the primary determinants of pathogenicity.

13.7 Conclusion and Prospective

The toxins of Foa, causal agent of bayoud disease of date palm, contain fusaric acid and others toxic fractions. The phytotoxic compounds purified from the F_{II} (Foa) fractions are probably new molecules that may help in understanding the pathogenesis of bayoud disease. These fractions of toxins have some chemical and biological characteristics that differ among other fractions revealed for other pathogenic and nonpathogenic strains of *Fusarium oxysporum*. This result may allow identification of the Foa pathogen. Moreover, the production strategies of these toxins were performed and determined. These toxins could be exploited for pre-selection of plants for resistance to bayoud among plants originating either from irradiated or nonirradiated tissue culture, or from conventional breeding and mass selection programs. The *in vitro* selection technique and procedure (Fig. 13.8) using the pathogen toxins or metabolites incorporated in the culture medium of date palm cells or tissues could be standardized and made universal. They would be able to be used without risk of contamination in bayoud-free countries. However, other metabolites extracted from culture filtrate of the fungus must be proven as primary determinants of the pathogenicity of the parasite. For the universal use of all these toxic substances, it is important to conduct a survey of their bio-ecological effect beforehand on the environment, to identify their chemical structure and to synthesize them in an industrial way.

Acknowledgements This research was partly supported by Maghreban Project AIEA/RAF/5/049, AOAD Regional Project on Bayoud (2004–8) and PRAD 02–11. Our thanks are also due to individuals in the laboratories at Regional Centre of Agronomic Research of Marrakech, Morocco (INRA, National Institute of Agronomic Research) and at Faculties of Sciences (University of Cadi Ayad, Marrakech), as well as to students and trained staff.

References

Allen DJ, Lenne JM, Waller JM (1999) Pathogen biodiversity: its nature, characterization and consequences. In: Wood D, Lenne JM (eds.) Agrobiodiversity: characterization, utilization and management. CABI, Wallingford, pp 123–153

Amraoui H, Sedra MH, Hamdaoui A (2004) Etude des sécrétions protéiques et enzymatiques du *Fusarium oxysporum* f. sp *albedinis*: agent causal de la fusariose vasculaire du palmier dattier. Al Awamia 109(110):47–62

Amraoui H, Lazrek HB, Sedra MH et al (2005) Chromatographic characterization and phytotoxic activity of *Fusarium oxysporum* f. sp. *albedinis* and saprophytic strain toxins. J Phytopathol 153:203–208

Bamburg JR, Strong FM (1971) 12,13-Epoxytrichothecenes. In: Kadis S, Ciegler A, Ajl S (eds.) Microbial toxins, vol 7. Academic, New York, pp 207–292

Belanger RR, Bushnell WR, Dik AJ, Carver TLW (eds.) (2002) The powdery mildews. A comprehensive treatise. APS Press, St. Paul

Bender CL, Stone HE, Sims JJ, Cooksey DA (1987) Reduced pathogen fitness of *Pseudomonas syringae* pv tomato Tn5 mutants defective in coronatine production. Phys Mol Plant Pathol 30:273–283

Bennett JW, Klich M (2003) Mycotoxins. Clin Microbiol Rev 16:497–516

Bhatt PN, Selvapandiyan A, Tedford E, Mehta AR (1988) Host specific toxin from *Fusarium oxysporum* f. sp. *nicotianae*. In: Abstract of paper 5th international congress of plant pathology, Kyoto, Japan, pp 1–22

Borras O, Santos R, Matos AP et al (2001) A first attempt to use a *Fusarium subglutinans* culture filtrate for the selection of pineapple cultivars resistant to fusariose disease. Plant Breed 120:435–438

Bougerfaoui M, Abahmane L, Anjarne M, Sedra MH (2006) Utilisation de mutagenèse induite pour l'amélioration de la résistance aux toxines du bayoud chez le palmier dattier (*Phœnix dactylifera* L.). Conférence Régionale sur «Mutagenèse Induite et Biotechnologies d'Appui pour la Protection du Palmier Dattier Contre le Bayoud » projet AIEA-RAF/5/049 Alger, 17–18 Juin 2006

Bouizgarne B, Brault M, Pennarun AM (2004) Electrophysiological responses to fusaric acid of root hairs from seedlings of date palm susceptible and resistant to *Fusarium oxysporum* f. sp. *albedinis*. J Phytopathol 152:321–324

Bounaga N (1985) Contribution à l'étude de *Fusarium oxysporum* f. sp. *albedinis* (Killian et Maire) Gordon, agent causal de la fusariose du palmier dattier. Thèse d'état, Alger, Algeria

Branchard M (1984) Application des vitro méthodes à la mise en œuvre de programme de sélections de plantes résistantes à la maladie. Agron 4:905–911

Brettell RIS, Ingram DS, Thomas E (1980) Selection of maize tissue cultures resistant to *Helminthosporum maydis* T-toxin. In: Ingram DS, Helgeson JP (eds.) Tissue culture methods for plant pathologists. Blackwell, Oxford, pp 233–237

Brian PW, Elson GW, Hemming HG, Wright JM (1952) The phytotoxic proprieties of alternaric acid in relation to the etiology of plant diseases caused by *Alternaria solani* (Ell. & Mart.) Jones & Grout. Ann Appl Biol 39:308–321

Brown DA, Hunger RM (1999) Regulation of *in-vitro* Ptr-toxin production by *Pyrenophora tritici-repentis* isolates by environmental parameters and accumulation of Ptr-toxin in culture over time. J Phytopathol 147:25–29

Brown DW, Mc Comrick SP, Alexander NJ et al (2001) A genetic and biochemical approach to study trichothecene diversity in *Fusarium sporotrichioides* and *Fusarium graminearum*. Fung Gen Biol 32:121–133

Carlson PS (1973) Methionine sulfoximine-resistant mutant of tobacco. Science 180:1366–1368

Casinovi CG (1972) Phytotoxins in plant diseases. Academic, London, 105

Chawla HS, Wenzel G (1987) *In vitro* selection of barley and wheat for resistance against *Helminthosporium sativum*. Theor Appl Genet 74:841–845

Cherrab M (1989) Contribution à l'étude morphologique et biochimique de quelques formes spéciales de *Fusarium oxysporum* (Schlecht.). Doct. 3ème cycle, Faculty of science, University Cadi ayyad, Marrakech, Morocco

Chung KR, Ehrenshaft M, Daub ME (2002) Functional expression and cellular localization of cercosporin-resistance proteins fused with the GFP in *Cercospora nicotianae*. Curr Genet 41:159–167

Daly JM (1976) Role of phytotoxins in specificity: a contribution. In: Wood RKS, Graniti A (eds.) Specificity in plant diseases. Plenum Press, New York, pp 231–234

Darakov OB (1995) Gametophyte selection of tomatoes for resistance to early blight disease. Sex Plant Reprod 8:95–98

Daub ME (1984) A cell culture approach for the development of disease resistance: studies on the phytotoxin cercosporin. Hort Sci 19:382–387

Daub ME, Ehrenshaft M (2000) The photoactivated *Cercospora* toxin cercosporin: contribution to plant disease and fundamental biology. Ann Rev Phytopathol 38:491–513

Durbin RD (1981) Toxins in plant diseases. Academic, New York

El Fakhouri R, Lazrek HB, Bahraoui E et al (1996a) Preliminary investigation on a phytotoxic peptide produced *in vitro* by *Fusarium oxysporum* f. sp. *albedinis*. Phytopathol Mediterr 35:121–123

El Fakhouri R, Lotfi F, Sedra MH, Lazrek HB (1996b) Production et caractérisation chimique des toxines sécrétées par *Fusarium oxysporum* f. sp. *albedinis,* agent causal du bayoud. Al Awamia 93:81–92

El Hadrami A, El Idrissi-Tourane A, El Hassni M et al (2005) Toxin-based *in-vitro* selection and its potential application to date palm for resistance to the bayoud *Fusarium* wilt. C R Biol 328:732–744

El Modafar C, Tantaoui A, El Boustani E (2000) Effet de l'acide Caféoylshikimique des racines du Palmier dattier sur L'activité et la Production des enzymes Hydrolytiques de *Fusarium oxysporum* f. sp. *albedinis*. J Phytopathol 1458:715–722

Graniti A (1991) Phytotoxins and their involvement in plant diseases - introduction. Experientia 47:751–755

Hamer JE, Holden DW (1997) Linking approaches in the study of fungal pathogenesis: a commentary. Fung Genet Biol 21:11–16

Hensel M, Holden DW (1996) Molecular genetic approaches for the study of virulence in both pathogenic bacteria and fungi. Microbiology 142:1049–1058

Hermann M, Zocher R, Haese A (1996) Enniatin production by *Fusarium* strains and its effect on Potato tissue. Appl Environ Microbiol 62:393–370

Hodgkin T (1990) *In vitro* pollen selection in *Brassica napus* L. for resistance to phytotoxic compounds from *Alternaria brassicicola* (Schw.) wilts. Sex Plant Reprod 3:1–5

Huang JS (2001) Plant pathogenesis and resistance biochemistry and physiology of plant-microbe interactions. Kluwer, Dordrecht

Karr AL Jr, Karr DB, Strobel GA (1974) Isolation and partial characterization of four host-specific toxins of *Helminthosporium maydis* (race T). Plant Phys 53:250–257

Knogge W (1996) Molecular basis of specificity in host/fungus interactions. Eur J Plant Pathol 102:807–816

Kohmoto K, Scheffer RP, Whiteside JO (1979) Host-selective toxins from *Alternaria citri*. Phytopathology 69:667–671

Lebeda A, Luhova L, Sedlarova M, Jancova D (2001) The role of enzymes in plant-fungal pathogens interactions. J Plant Dis Protect 108:89–111

Lepoivre P (2003) Phytopathologie: bases moléculaires et biologiques des pathosystèmes et fondement des stratégies de lutte. De Boeck & Presses Agronomiques de Gembloux, Brussels

Lousberg JJC, Salemink CA (1972) The chemistry of polysaccharide and glycopeptide phytotoxins. In: Wood RKS, Ballio A, Graniti A (eds.) Phytotoxins in plant diseases. Academic, London, pp 127–137

Markham JE, Hille J (2001) Host-selective toxins as agents of cell death in plant-fungus interactions. Mol Plant Pathol 2:229–239

Matsumoto K, Barbosa ML, Souza LAC, Teixeira JB (1995) Race 1 F*usarium* wilt tolerance on banana plants selected by fusaric acid. Euphy 84:67–71

Megneneau B (1994) Etude de la sensibilité variétale du melon (*Cucumis melo L*) aux toxines du *Fusarium oxysporum* f. sp. *melonis* en culture *in vitro*. Thése de Doctorat d'Université, Paris sud, centre d'Orsay. Paris, France

Mepsted R, Flood J, Paul C et al (1995) A method for rapid selection and investigation of resistance mechanism in *Fusarium* wilt of oil palm. Plant Pathol 44:749–755

Mo YY, Gross DC (1991) Expression *in vitro* and during plant pathogenesis of the syrB gene required for syringomycin production by *Pseudomonas syringae*. Mol Plant Microbe Interact 4:28–36

Mokhlisse N (1987) Contribution à l'identification et à l'étude de la toxicité des différents constituants de la toxine sécrétée par le *Fusarium oxysporum* f. sp. *albedinis*. Thése de 3ême cycle, Faculty des Sciences-Semlalia, Marrakech, Morocco

Mussel HW (1972) Phytotoxins in plant diseases. Academic, London, pp 443–445

Nedelnik J, Repkova J (1998) Plant selection *in vitro* for resistance to some pathogens using secondary toxic metabolites. Czech J Genet Plant Breed 34:69–76

Park SH, Stierle A, Strobel GA (1994) Metabolisme of Maculosin, a host-specific phytotoxine produced by *Alternaria alternata* on spotted knapweed (*Centaurea maculosa*). Phytochemistry 35:101–106

Pegg GE (1981) Biochemistry and physiology of pathogenesis. In: Mace ME, Bell AL (eds.) Fungal wilt diseases of plants. Academic, New York, pp 193–253

Porter JK, Bacon WC, Wray EM, Winston MH Jr (1995) Fusaric acid in *Fusarium moniliforme* cultures, corn and fields toxic to livestock and the neurochemical effects in the brain and pineal gland of rats. Nat Toxins 3:91–100

Pringle RB, Scheffer RP (1963) Purification of the selective toxin of *Periconia circinata*. Phytopathology 53:785–787

Pringle RB, Scheffer RP (1964) Host-specific plant toxins. Ann Rev Phytopathol 2:133–156

Remotti PC, Löffler HJM (1996) The involvement of fusaric acid in the bulb-rot of Gladiolus. J Phytopathol 144:405–411

Rudolph K (1976) Non-specific toxins. In: Heitefuss R, Williams PH (eds.) Physiological plant pathology, vol 4. Springer, Berlin, pp 279–315

Sacristam MD (1982) Resistance response to *Phoma lingam* of plants regenerated from selected cell and embryogenic cultures of haploid *Brassica napus*. App Genet 61:193–200

Scheffer RP (1976) Host-specific toxins in relation to pathogenesis and disease resistance. In: Heitefuss R, Williams PH (eds.) Physiological plant pathology, vol 4. Springer, Berlin, pp 247–269

Scheffer RP, Pringle RB (1967) Pathogen-produced determinants of disease and their effects on host plants. In: Mirocha CJ, Uritani I (eds.) Dynamic role of molecular constituents in plant-parasite interaction. Bruce, St Paul, pp 217–236

Sedra MyH (1993a) Caractérisation morphologique et culturale de *Fusarium oxysporum* f. sp. *albedinis*, agent de la fusariose vasculaire (bayoud) du palmier dattier. Al Awamia 83:209–222

Sedra MyH (1993b) Remarques sur le pouvoir pathogène des isolats de *Fusarium oxysporum* f. sp. *albedinis*, agent de la fusariose vasculaire (bayoud) de palmier dattier. Al Awamia 83:223–240

Sedra MyH (1995) Triage d'une collection de génotypes de palmier dattier pour la résistance au Bayoud causé par *Fusarium oxysporum* f. sp. *albedinis*. Al Awamia 90:9–18

Sedra MyH (2003a) Le bayoud du palmier dattier en Afrique du Nord. FAO, RNE/SNEA-Tunis. Editions FAO sur la protection des plantes. Imprimerie Signes, Tunis, Tunisia

Sedra MyH (2003b) Le palmier dattier base de la mise en valeur des oasis au Maroc. Techniques phoénicicoles et Création d'oasis. Edit. INRA Maroc, Imprimerie Al Watania-Marrakech Morocco

Sedra MyH (2007a) New Morrocan varieties with good agricultural characters and promising to control the bayoud disease. In: Proceedings fourth symposium date palm. King Faisal University, Hofuf, Saudi Arabia

Sedra MyH (2007b) Bayoud disease of date palm in North Africa: recent distribution and remarks about its characterization, diagnosis and origin. In: Proceedings fourth symposium date palm. King Faisal University, Hofuf, Saudi Arabia

Sedra MyH (2008) Pathogenic and molecular characterization of *Fusarium oxysporum* f. sp. *albedenis* strains, causal agent of the bayoud disease of date palm. In: 10th international *Fusarium* workshop and *Fusarium* genomics workshop 2008. Alghero, Sardinia, Italy, Aug 30 to Sept 2, 2008

Sedra MyH (2010) Les nouvelles variétés du palmier dattier sélectionnées pour combattre le Bayoud. Packinfo 89:42–45. Juillet/Aôut 2010, Morocco

Sedra MyH, Zhar N (2010) Genetic variability analysis of populations of *Fusarium oxysporum* f. sp. *albedinis*, causal agent of bayoud disease of date palm and other *Fusarium oxysporum* using molecular techniques. (Proc. 4th Int Date Palm conference United Arab Emirates, Abu Dhabi). ACTA Hort 882:491–504

Sedra MyH, El Fakhouri R, Lazrek HB (1993) Recherche d'une méthode fiable pour l'évaluation de l'effet des toxines secrétées par *Fusarium oxysporum* f. sp. *albedinis* sur le palmier dattier. Al Awamia 82:89–104

Sedra MyH, El Fakhouri R, Lotfi F, Lazrek HB (1997) Activités des toxines secrétées par *Fusarium oxysporum* f. sp. *albedinis*, agent causal du bayoud du palmier dattier et d'autres formes spéciales du *Fusarium oxysporum*. Al Awamia 98:57–65

Sedra MyH, Lazrek H, Lotfi F, Rochat H (1998) *Fusarium oxysporum* f. sp. *albedinis* toxin isolation and use for screening of date palm plants for resistance to the bayoud disesase. In: Proceedings of XXV International Horticultural Congress (IHC), 2–7 Aug 1998, Brussel, Belgium

Sedra MyH, Laouane H, Lazrek HB (2002) Mise en évidence de la présence des toxines dans le filtrat de culture du *Verticillium dahlia*, agent causal de la verticilliose de l'olivier. Al Awamia 105:85–93

Sedra MyH, Lazrek HB, Amraoui H, Nour S (2008) Pathogen toxins of bayoud disease on date palm: *in vitro* selection and biological and specificity activities. In: 10th international *Fusarium* workshop and *Fusarium* genomics workshop 2008. Alghero, Sardinia, Italy, Aug 30 to Sept 2, 2008

Shahin EA, Spivey R (1986) A single dominant gene for *Fusarium* wilt resistance in protoplasts-derived tomato plants. Theor Appl Genet 73:164–169

Soler-Rivas C, Arpin N, Olivier JM, Wichers HJ (1999) The effect of talaasin, the toxin produced by *Pseudomonas talaasi*, on tyrosinase activities and the induction of browning in *Agaricus bisporus* fruiting bodies. Phys Mol Plant Pathol 55:21–28

Spencer-Phillips PTN, Gisi U, Lebeda A (eds.) (2002) Advances in downy mildew research. Kluwer, Dordrecht

Surico G, Graniti A (1977) Produzione di tossine da *Fusarium oxysporum* Schl. f. sp. *albedinis*. Phytopathol Mediterr 16:30–33

Sutherland ML, Pegg GF (1995) Purification of a toxin from *Fusarium oxysporum* f. sp. *lycopersici race 1*. Phys Mol Plant Pathol 46:243–254

Svabova L, Lebeda A (2005) *In vitro* selection for improved plant resistance to toxin-producing pathogens. J Phytopathol 153:52–64

Takken FLW, Joosten MHAJ (2000) Plant resistance genes: their structure, function and evolution. Eur J Plant Pathol 106:699–713

Tantaoui A, Boisson C (1991) Compatibilité végétative d'isolats du *Fusarium oxysporum* f. sp. *albedinis* et des *Fusarium oxysporum* de la rhizosphère du palmier dattier et des sols de palmeraies. Phytopathol Mediterr 30:155–163

Tantaoui A, Fernandez D (1993) Comparaison entre *Fusarium oxysporum* f. sp. *albedinis* et *Fusarium oxysporum* des sols de palmeraies par l'étude du polymorphisme de longueur des fragments de restriction (RFLP). Phytopathol Mediterr 32:235–244

Tantaoui A, Ouinten M, Geiger JP, Fernandez D (1996) Characterization of a single clonal lineage of *Fusarium oxysporum* f. sp. *albedinis* causing bayoud disease of date palm in Morocco. Phytopathology 86:787–792

Walton JD (1996) Host-selective toxins: agents of compatibility. Plant Cell 8:1723–1733

Walton JD, Panaccione DG (1993) Host-selective toxins and disease specificity: perspectives and progress. Annu Rev Phytopathol 31:275–303

Wang H, Li J, Bostock RM, Gilchrist DG (1996) Apoptosis-a functional paradigm for programmed plait cell death induced by host-selective phytoxin and invoked during developement. Plant Cell 8:375–391

Wenzel G, Foroughi-Wehr B (1990) Progeny tests of barley, wheat, and potato regenerated from cell cultures after *in vitro* selection for disease resistance. Theor Appl Genet 80:359–365

Wolpert TJ, Navarre DA, Moore DL, Macko V (1994) Identification of the 100-KD victorin binding protein from oats. Plant Cell 6:1145–1155

Wolpert JT, Dunkle LD, Ciuffetti LM (2002) Host-selective toxins and avirulence determinants: what's in a name? Annu Rev Phytopathol 40:251–285

Yoder OC (1980) Toxins in pathogenesis. Annu Rev Phytopathol 18:103–129

Chapter 14
Radiation-Induced Mutations for Date Palm Improvement

S.M. Jain

Abstract Micropropagation technique is used for rapid shoot proliferation of date palm. Somatic embryogenesis is meant for clonal propagation of date palm and genetic gains can be captured through it, which is rather difficult by zygotic embryo due to its heterozygous nature. Genetic variability is highly desirable for the genetic improvement of crops, which can be either spontaneous or induced by mutagen treatments. Mutation-assisted breeding has been quite successful for the production of new mutant cultivars with desirable traits in both seed and vegetative propagated crops (see: http://www-mvd. iaea.org). In the IAEA date palm project, somatic embryogenic cell cultures were irradiated with gamma radiation, and regenerated plants were transferred to the greenhouse and treated with bayoud toxin, isolated from the causal fungus *Fusarium oxysporum* f. sp. *albedinis*. Several putative mutants tolerant to bayoud disease were initially maintained in the greenhouse and later transferred to the field for further evaluation. Over the last 4 years, these plants have not shown any sign of susceptibility to bayoud disease under field conditions, but they have yet to flower. Thus far, our results suggest that the combination of *in vitro* culture and mutagenesis would be an ideal system for date palm improvement. However, molecular tools are needed to characterize mutants for trait specific gene (s) identification and to develop molecular marker assisted selection and breeding programs. Date palm also has great potential to provide renewable energy or *bioenergy* or *green energy* for producing *bioethanol* and blend it with petrol in the transport industry.

Keywords Genetic variability • *In vitro* selection • Mutation • Tissue culture

S.M. Jain (✉)
Department of Agricultural Sciences, University of Helsinki, PL-27, Helsinki, Finland
e-mail: mohan.jain@helsinki.fi

S.M. Jain et al. (eds.), *Date Palm Biotechnology*,
DOI 10.1007/978-94-007-1318-5_14, © Springer Science+Business Media B.V. 2011

14.1 Introduction

The unique characteristics of date palm (*Phoenix dactylifera* L.) can be truly called a *tree of life* and is considered one of the most ancient plant cultivated in Mesopotamia some 4,000 years ago (Dakheel 2003; Omar and Hameed 2006). Few plant species have been so closely connected with the survival and wellbeing of humans living in hot and arid environments. This plant is basically responsible for the human settlements and expansion in hot and barren parts of the world, and forms the most sustainable agro-ecosystems in harsh dry environments. The date palm is distributed throughout the Middle East, North Africa, South Sahel, areas of East and South Africa, and even in certain parts of Europe and USA (Jain 2007). Dakheel (2003) studied the unique date palm production system under harsh climatic conditions, and described as due to:

- High resilience and tolerance to environmental stresses- high temperature and radiation, low soil and atmospheric moisture, extended periods of drought, high salinity levels, and large diurnal and seasonal fluctuations.
- High resource utilization efficiency and limited input requirement.
- High productivity.
- High nutritional value of date fruit.
- Long productive life, as long as 100 years and multiple uses.
- Creates equable microclimate within oasis ecosystems and thus enables agriculture development (Jain 2007).
- Helpful in the conservation of the fragile environment structure and reduce desertification risks.

Date palm belongs to the monocot family Arecaceae and is an arborescent, dioecious tall evergreen and highly heterozygous plant (Jain 2007). Date fruits are a most important source of human nutrition as well as an export item for many date palm-growing countries. The annual world production of dates was just over seven million mt in 2008 (FAOSTAT). The major bulk of date palm production, 75% of the total world production, comes from Egypt, Iran, Saudi Arabia, UAE, Algeria, Pakistan and Iraq. In the Kingdom of Saudi Arabia, over 200 date palm varieties are grown and date fruit production currently is 980,000 mt, which represents about 15% of world date production.

14.2 Date Palm Fruit as a Staple Fruit

The rich date fruit plays an important role in providing nutrition to humans living under the harsh climatic conditions. Date fruits are the rich source of sweeteners, glucose and fructose (Al-Eid 2006). Date syrup analytical studies showed that it is mainly composed of reduced sugars; glucose and fructose as major source of sugar fraction. Al-Ghamdi and Al-Kahtani (1996a, b, c) made detailed analyses of date palm fruits and described their chemical properties, sugar content and minerals. The sugar content increases as the fruit ripens. The fruit chemical properties and mineral content vary depending on the genotype. However, there was no difference in fruit quality between *in vitro* plants and conventionally grown plants. Furthermore, nutritional analysis of date fruits (Kingdom of Saudi Arabia, Ministry of Agriculture) indicated that they contain:

- High percentage of carbohydrates (total sugars 44–88%).
- Fat (0.2–0.5%).
- Salts and minerals (15 different types).
- Proteins (2.3–5.6%).
- Vitamins and a high percentage of dietary fiber (6.4–11.5%).
- Flesh of dates contains 0.2–0.5% oil.
- Seeds of dates contain 7.7–9.7% oil, which has 14 types of fatty acids.
- Contains elemental fluorine useful for preventing dental decay.
- Contains selenium that prevents cancer and proper function of immune system.

Dates could be considered as an ideal food which provides a wide range of essential nutrients and potential health benefits.

14.3 Major Date Palm Diseases and Pests

Date palm suffers from several diseases and insect pests leading to severe economic lose to growers. There are about 25 diseases and disorders affecting date palm worldwide. Among them, 14 are caused by fungi (Karempour and Pejman 2007). Date bunch fading disorder (DBF) is the most harmful phenomenon damaging both the quality and quantity of date yield. In Iran, this disorder has caused wilting and drying of bunches and finally severe defoliation of date palms over the last 5 years (Karempour and Pejman 2007). In Egypt, 21 fungal species belonging to 15 genera were isolated from diseased date palm samples collected from different Egyptian localities (El-Deeb et al. 2007).

Bayoud disease is a serious threat to date palm plantations in North African Saharan and sub-Saharan regions, which is caused by a soil-borne fungus *Fusarium oxysporum* f. sp. *albedinis* (FOA) (Fig. 14.1). It was first observed in the Draa Valley, Morocco in 1870, and from there it reached the Algerian Central oasis around 1889, and it rapidly advanced into new areas. The disease has destroyed more than ten million palm trees in Morocco and nearly three million trees in Algeria. The rate of destruction by bayoud disease is thus estimated at 5% per year (Oihabi 2003). Presently, the disease is known to occur in Morocco, Algeria and Mauritania. Tunisia takes very drastic quarantine measures and strict surveillance to prevent the introduction of the disease from Algeria. The most popular commercial date palm cvs. such as Deglet Noor, Medjool and Boo Fagoos are under threat from bayoud disease.

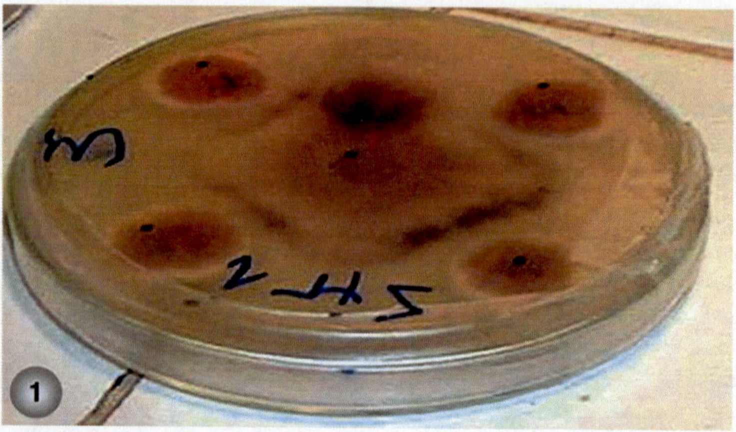

Fig. 14.1 Soil-borne fungus *Fusarium oxysporum* f. sp. *albedinis* (FOA) is the causal agent of bayoud disease

Quenzar et al. (2001) identified two circular plasmid-like DNAs (S and R plasmids) in the mitochondria of date palm. By employing a PCR-based approach, they showed that the presence of R plasmid and absence of S plasmid can be considered as a reliable molecular marker of bayoud disease resistance. In this situation, the presence of S plasmid and the absence of R plasmid are correlated to the susceptibility to bayoud disease. This diagnostic molecular tool could ultimately become a simple, reliable, rapid and efficient approach to identify bayoud resistant and susceptible genotypes from the large pool of date palm lines.

Since 1994, the International Atomic Energy Agency (IAEA) has had a technical cooperation project on the selection of bayoud disease resistant date palm mutants by gamma radiation treatment, working in collaboration with Algeria, Tunisia and Morocco (Jain 2005, 2006), and it continued until 2008. There was a consultant meeting on bayoud disease of date palm, organized by IAEA in March 1996 (RAF/5/035 project). This meeting summarized the status of knowledge on the fungus FOA, its biotypes, its spread in North Africa, disease symptoms and movement of the pathogen in the host, characterization of the fungal toxin, short- and long-term strategies to breed for resistance, induction of variation for resistance in well-established date palm varieties using induced mutations and *in vitro* culture techniques, improvement of screening procedures to select for resistance in breeding programs as well as characterization of variation in the pathogen population and use toxin for *in vitro* selection.

Red palm weevil (RPW) is a major pest in date palm-growing countries in the Near East including the United Arab Emirates, Iran, Egypt and others (Oihabi 2003). It first appeared in the Middle East in 1985 and is of great concern to the date-palm growers in these countries. The control of RPW is mainly done by applying chemical insecticides through direct injection into the trunk of the date palm tree or by fumigation. Pheromone traps are also commonly used to control RPW, but still require more refinement and effectiveness to control this pest. Baculoviruses could be another way to control RPW, especially genetically engineered ones inserted with a set of genes dealing with neurotoxin, light-emission (firefly gene) and heat tolerance.

14.4 *In Vitro* Culture of Date Palm

Date palm is well known to propagate both sexually through seeds and vegetatively by offshoots that are produced from axillary buds situated at the base of the trunk during the juvenile life of the palm tree. Seed propagation of date palm is not appropriate for commercial production due to the heterozygous characteristics of seedlings, which is related to the dioecious nature of the date palm, half of the progeny are generally male and do not produce fruits; also, large phenotypic variation can occur in the progeny (Jain 2006). Currently, there is no known method for sex determination of date palm at the early stage of tree development making it rather difficult to discriminate between productive female and non-productive male trees in the nursery before transplanting them to the field. Furthermore, the seed propagation method has another limitation in that the growth and maturation of seedlings is extremely slow, and only begin to fruit after 8–10 years of planting. The ideal way would be to look for molecular markers for sex determination along the lines of work done in papaya (Deputy et al. 2002).

Offshoot production is slow; their numbers are limited, laborious to separate and cannot meet the rapidly growing demand of varieties. Normally offshoot numbers vary from 10 to 30, depending on the genotype, and are produced only within a certain period time in the mother palm's life (Jain 2007). No field-based methods are yet available for increasing the number of offshoots per plant. There are only a few commercial tissue culture laboratories worldwide micropropagating date palm for large-scale plant production (for more information see: Al Kaabi and Zaid 2003).

In vitro culture techniques such as somatic embryogenesis and organogenesis have been effectively used for large-scale plant multiplication of horticultural crops and forest trees (Jain and Gupta 2005; Jain and Haggman 2007; Jain and Ishii 2003). Plant multiplication via organogenesis is routinely followed in commercial laboratories worldwide especially in ornamental plant industries and also to some extent in fruits and cash crops like coffee, sugarcane etc. The cost of plant production is generally high due to labor and electricity, which reduces the profit margin. Many Western companies have started outsourcing plant multiplication facilities to low labor cost countries such as India, China, Brazil, Kenya, Tanzania and others; in fact, most of the date palm micropropagation commercial laboratories are operating in countries with low labor cost. The performance of *in vitro* propagated plantlets seems to be improving in terms of yield and early flowering. Al-Ghamadi and Al-Kahatani (1996a,b,c) made detailed comparative analyses of fruit quality of micropropagated and conventionally-propagated plants and found no major variation in fruit quality and properties. Results clearly indicate that *in vitro*-grown date palm are quite uniform in terms of fruit quality and physical properties. Smith and Aynsley (1995) reported on field performance of tissue culture-derived date palm clonally produced by somatic embryogenesis. These plants started bearing fruits within 4 years of field planting of small plants with a leaf length of 100 cm and 1.5 cm diameter at the base. Fruit from the tissue culture-derived plants, cv. Barhee, was indistinguishable from the fruits of plants originated from offshoots. These results certainly justify the commercial scale of micropropagation procedures of somatic embryogenesis to provide rapid, cost-effective means of obtaining elite date palm planting material. However, this approach has a major bottleneck in that the plant multiplication rate is highly genotypic dependent, and may require modification of culture medium, depending on the genotype. For more information see Jain (2006) who described the advantages and limitations of date palm micropropagation. Some of the major advantages of micropropagation are year-round availability of plants, quality control, rapid production of plants of elite cultivars and cold storage of elite genetic material.

14.4.1 Techniques for Plant Regeneration

Date palm tissue culture work has revolved around somatic embryogenesis (Al-Khayri 2005; Fki et al. 2003) and organogenesis for plant regeneration (Al-Khayri 2007; Khierallah and Bader 2007). Aaouine (2003) reported plant regeneration from 30 genotypes of date palm via direct shoot organogenesis. Many commercial laboratories in Europe, the Middle East, the United States, North Africa and South Africa are using a combination of somatic embryogenesis and organogenesis. Furthermore, the initiation period for somatic embryogenesis induction is 4–6 months as compared to 8–10 months for organogenesis; the total time from induction phase to plant marketing is 40–44 months via somatic embryogenesis vs. 60 months via organogenesis (Aaouine 2003).

Murashige and Skoog (1962) formulated the most commonly-used culture medium for both somatic embryogenesis and organogenesis of date palm, which also is modified depending on the genotype or cultivar (Jain 2006). Young Maktoom cv. offshoots from 2 to 3 year old date palms were used for direct shoot induction after they were sterilized with commercial bleach and rinsed with sterile distilled water (Khierallah and Bader 2007).

The somatic embryogenesis approach for date palm plant regeneration seems to be more effective for clonal propagation. Fki et al. (2003) improved somatic embryogenesis protocol of date palm cv. Deglet Noor for large-scale clonal propagation. Initially, embryogenic callus cultures were initiated from both leaf and inflorescence explants on MS (Murashige and Skoog) medium containing 0.5 and 10 mg/L 2, 4-D (Figs. 14.2 and 14.3). These cultures were used to develop highly proliferating cell suspension cultures in the liquid medium supplemented with 1 mg/L 2, 4-D (Figs. 14.4–14.6). Somatic embryos were initiated from actively growing cell suspension (Fig.14.7), and finally somatic embryos were germinated (Figs. 14.8 and 14.9) and the whole plantlets regenerated (Figs. 14.10 and 14.11).

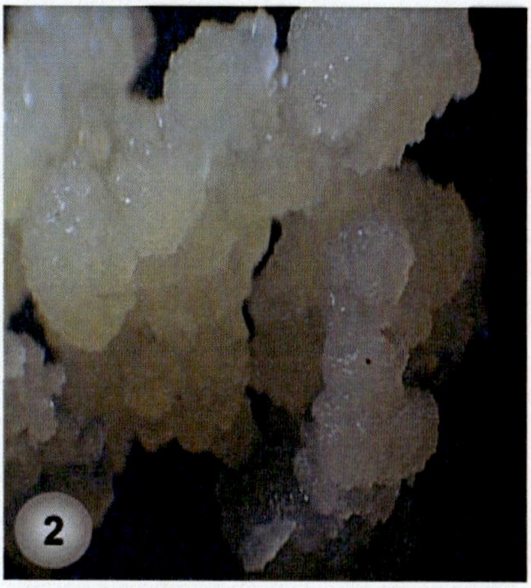

Fig. 14.2 Initiation of somatic embryogenic callus from off shoot of date palm cv. Deglet Noor

Fig. 14.3 Further development of somatic embryos

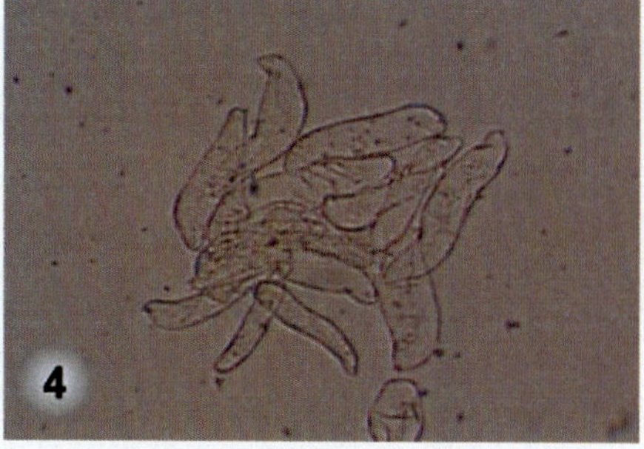

Fig. 14.4 Rapid growing somatic embryogenic cell suspension cultures

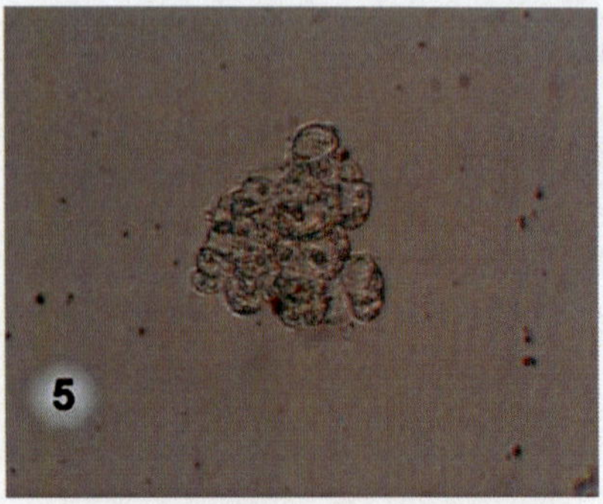

Fig. 14.5 Rapid growing somatic embryogenic cell suspension cultures

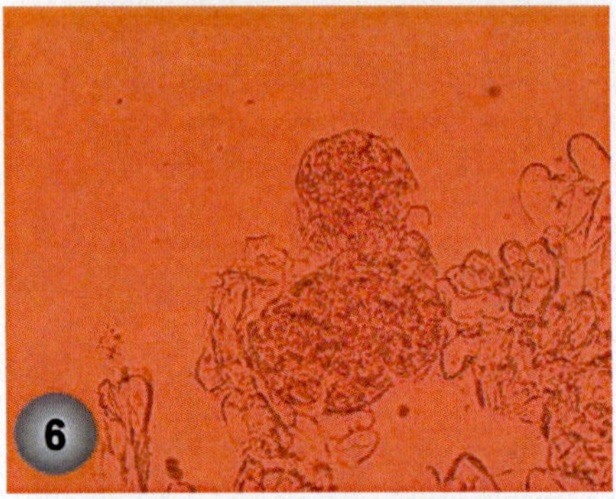

Fig. 14.6 Rapid growing somatic embryogenic cell suspension cultures

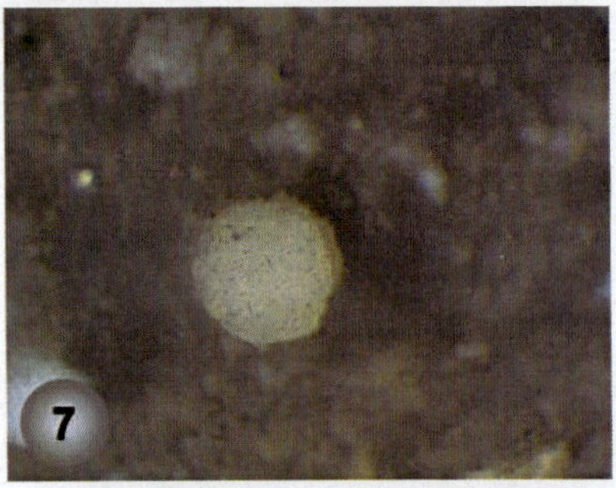

Fig. 14.7 Developing somatic embryo of date palm

Fig. 14.8 Germinating date palm somatic embryos

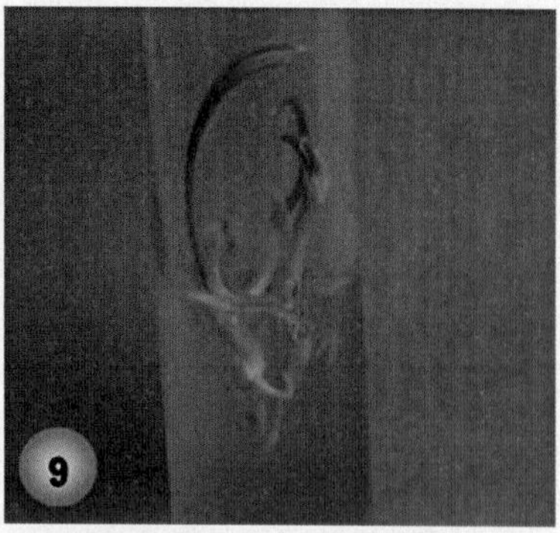

Fig. 14.9 Somatic seedlings of date palm with well-developed shoots and roots

Fig. 14.10 Well-developed somatic seedlings with shoots and roots

Fig. 14.11 Well-developed somatic seedlings with shoots and roots

The overall production of somatic embryos can reach 10,000 units per liter per month. The partial desiccation of the mature somatic embryos significantly improves the somatic embryo germination rate from 25% to 80%. Cutting back of cotyledon leaf was stimulatory to the germination rate. Furthermore, flow cytometry analysis showed no variation in ploidy level of somatic seedlings. Several research groups have modified the culture medium composition by adding vitamins, adenine sulfate, thiamine, glycine, glutamine, myo-inositol and activated charcoal (Al-Khayri 2005). The role of vitamins in date palm tissue culture is not known. For more details see Al-Khayri (2005) and Jain (2006).

14.5 Mutation Induction

The exploitation of genetic variability is essential for the development of new cultivars. Genetic variability can be induced by chemical and physical mutagens, T-DNA insertional mutagenesis, and tissue culture-derived variation or somaclonal variation. The most common physical mutagen used is gamma radiation. In this chapter, we will deal only with physical mutagens. Induced mutations are random changes in the nuclear DNA or cytoplasmic organ, resulting in chromosomal or genomic mutations that enable plant breeders to select useful mutants such as disease resistant, high yield etc. First of all, gamma irradiation breaks DNA into small fragments and secondly DNA starts a repair mechanism. During this second step, new variations develop or mutations occur. In date palm, there is hardly any work done on mutation induction, except that of FAO/IAEA Coordinated Research Project on development of bayoud disease resistant date palm mutant cultivars in North Africa (Jain 2005, 2006). Mutation induction in date palm is feasible now due to a reliable plant regeneration system via somatic embryogenesis and organogenesis. The somatic embryogenesis system is the more preferable approach due to single cell origin of somatic embryos which prevents or reduces the occurrence of chimeras. Moreover, mutant somatic embryos are germinated into direct plantlets in a single step, avoiding the laborious rooting step. The irradiation of multicellular structures, e.g. seed, meristem tissue or offshoots, may result in chimeras in regenerated plants, and that would require a lot of extra work to dissociate chimeras by plant multiplication up to M1V4 generation (Jain 2007).

14.5.1 Determination of Radiosensitive Dose

Plants differ in radio sensitivity and that is why it is important to make a radiosensitive curve to determine LD_{50} dose for mutation induction. High radiation doses are detrimental to the plant genome and cause heavy damage to DNA. This leads to large number of mutations, which are mostly undesirable; it is cumbersome to identify useful mutants, and the handling of the mutant population becomes more difficult.

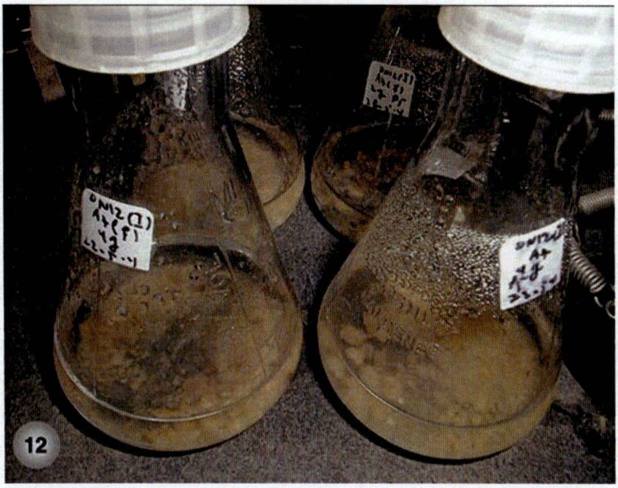

Fig. 14.12 Rapid growing somatic embryo cell suspension ideal for radiation treatment

In some crop plants, a low radiation dose promotes shoot growth, e.g. in citrus 30 Gy dose stimulates shoot growth (Jain personal communication) and 10 Gy maintains somatic embryogenesis nature of date palm up to 3 years (Drira, personal communication). In date palm 20–30 Gy was used for mutation induction depending on the genotype used. For Deglet Noor date palm cv., LD_{50} of somatic embryogenic cell suspension cultures was 20 Gy and used for mutation induction. Actively growing cell suspension in the growth phase (Fig. 14.12) was transferred onto filter paper in the Petri dish. Cell clumps were uniformly spread on the filter paper and the dishes sealed with Para film. Cells were irradiated with different gamma radiation doses and they were transferred for overnight to fresh solid culture medium for recovery from the radiation treatment. The irradiated cells were transferred into liquid medium, and distributed in 50 mL flasks containing 30 mL liquid medium. The cell viability test was made with FDA (fluorescein diacetate) staining to determine the cell survival rate after the different radiation dose treatment. The number of surviving cells was calculated on the basis of cells per mL per radiation dose. After this step, in each flask, 100,000 cells per mL were added as starting material and the cell growth was determined after 1 week. The number of cells per mL per radiation dose treatment was counted with a haemocytometer. These results established the radiosensitive curve and determined LD_{50} radiation dose for mutation induction.

14.5.2 Mutant Isolation

Mutant isolation can be done in two ways, either in a single step or stepwise selection. In the first approach, irradiated cells are put under very high selection pressure for the isolation of mutant cell clumps/lines. The initial selection pressure should be as high as high LD_{75}. Isolated mutant cells are removed and transferred onto fresh culture medium with reduced selection pressure allowing them to recover from the initial selection pressure for about 1 week. The selected lines are put for shoot and root differentiation. Before selected mutant lines are put for shoot differentiation, they should be grown for two generations devoid of selection pressure and then returned to the selection pressure. This step is done to make sure that the selected mutant lines are stable and due to genetic changes rather than epigenetic changes. In the second approach, the selection pressure is reduced stepwise, from high to low concentration. All other steps are more or less similar to the first approach.

With *in vitro* selection of mutants, normally the type of selection pressure varies, e.g. salt concentration, fungal toxin, polyethyl glycol (PEG), herbicide etc. For appropriate selection pressure, it is desirable to determine LD_{50} dose.

A third option is to select mutants at the whole plantlet level, e.g. by spraying herbicide or withholding water for drought-tolerance selection, fungal toxin spraying or injection. In date palm, bayoud disease resistant mutant plants were selected in the greenhouse by treating them with isolated toxin from *Fusarium oxysporum* f. sp. *albedinis* fungus, the causal agent (Jain 2006). These plants have been in the field for the last 4 years and so far are doing just fine.

14.6 Date Palm as a Bioenergy Source

Today the whole of humanity is very much dependent on fossil energy for routine daily life. The surge in the industrialization in the developing world has enhanced fossil energy consumption. Emerging economies like China and India with huge populations are consuming more fossil fuel energy than ever before due to improvement in the socio-economic status of consumers. China is buying 40% more fossil fuel from the international market followed by India 12%. Consequently, the price of fossil fuel is skyrocketing, almost exceeding USD 70.00 per barrel on the international market and this is gradually creating serious problems in the economies of the world. As fossil fuel consumption rises, another point of concern is the gradual depletion of fossil fuels that could last for only 40–50 years more. The rise in fossil fuel energy consumption in the transportation sector, together with other gaseous pollutants, is adversely influencing world climatic conditions. For example, global warming is finally accepted as a cause of concern by the international scientific community and the United Nations. The rise in global temperature even by 1°C will have a negative impact on the earth including reduction in agriculture production, appearance of new pests and diseases, disappearance of some old pests and diseases, water shortage, increase in carbon dioxide and ozone layer depletion. An increase in the ozone layer hole would increase ultra violet-B radiation on the earth that would increase skin cancer incidents and decrease the photosynthesis rate resulting in poor performance of plant productivity.

Renewable energy is the alternate source to the fossil fuel energy, which can be termed as *green energy* or *bioenergy* or *plant-based energy*. Biofuel is gradually replacing petroleum in the transport sector, which is a blend of petrol and bioethanol

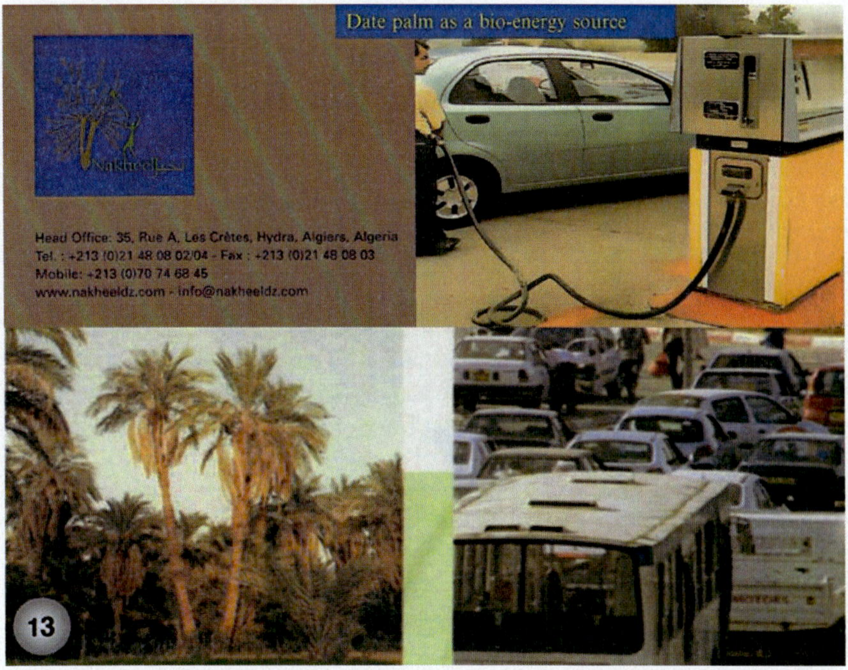

Fig. 14.13 Nakheel, an Algerian biotechnology company, is the first company in the Arab world to promote date palm as a renewable energy crop to produce bioethanol in transport

or biodiesel and fossil diesel. Brazil is the world leader in exploiting biofuel in the transport industry, especially bioethanol produced from sugarcane. The Europeans are leaders in commercial production of biodiesel, mainly from vegetable oil, e.g. *Brassica.* However, the production of biodiesel is comparatively small as compared to bioethanol production in Brazil and USA. Date palm could become a major source of producing bioethanol, since its fruits have a high percentage of carbohydrates (total sugars 44–88%). Millions of date palm trees are grown in the Middle East, North Africa, and South Asia, and they provide food and nutrition to millions of people, and could also become a major source of bioenergy. In Algeria alone, the estimated number date palm trees is over ten million, and production increased from 302,993 mt in 1997 to 526,921 mt in 2007. This increased production provides extra income to date palm growers if price levels are maintained.

The over production of date palm fruits, however, could lead to price reduction and loss of farm income. These problems can be overcome by using date palm for bioethanol production. In Algeria, Nakheel, a biotechnology company, is the first enterprise in the Arab world to promote renewable energy to use in transport. Date palm is the main source for producing bioethanol, which can easily be blended with petrol (Fig. 14.13). In the future, there is a potential to increase date palm-growing areas when the demand is high without compromising dates as a food and nutrition source. Moreover, date palm micropropagation is well established for clonal propagation and the supply of planting material year around.

14.7 Conclusion and Prospective

Date palm is a life-line of people living in Saharan and sub-Saharan regions and also an important source of income in Near Eastern countries. Most of the date palm trees are very old, as much as 70–100 years and probably are becoming more vulnerable to various diseases and pests. One of the reasons could be due to global warming or global climatic change. An increase in global temperature would bring new pests and disease and eliminate some existing types. Since date palm has a long life cycle, it could become more vulnerable to global warming, and that is why it is highly desirable to pay more attention to the genetic improvement of date palm varieties that could withstand natural calamities without compromising yield and quality. The use of chemical insecticide and pesticides is very common to control diseases and pests of date palm. These practices could become a serious hazard to human health and that may also curtail the export market. Innovative techniques need to be applied to the control of disease and pests, and that is where genetic modifications of organisms would be highly effective. Genetic engineering of baculoviruses may be of great help in controlling the RPW by inserting a set of genes including neurotoxin (a gene from the scorpion or snake), light-emitting (fire-fly) and heat tolerance (bacterial gene). The engineered baculoviruses would multiply inside the insects and kill them instantly. One could monitor the rate of viral multiplication inside the insect by light meter. Insertion of *Bt* gene in date palm won't be the proper approach due to long life cycle of date palm and it would be rather difficult to predict the behavior of transgenes in the long run. Moreover, food safety regulations do not permit insertion of the *Bt* gene in food crops.

The progress of *in vitro* culture techniques has enabled date palm micropropagation to become a routine technique for large-scale plant production in many countries. The influence of genotype has handicapped micropropagation of different commercially valuable date palm varieties. This area needs serious attention by modifying, through more empirical work, the culture medium well suited for several date palm cultivars. Now the question arises of how well the molecular approach would assist plant tissue culturists to modify the culture medium and growing conditions or the selection of appropriate explants or pre-conditioning of explants. To answer these questions, plenty of work is foreseen; in other words this area of research is *virgin.*

The date palm shoot multiplication rate could be improved by using a liquid culture system or *bioreactor.* Few groups have started working on liquid culture for *in vitro* propagation of date palm. The RITA bioreactor, based on temporary immersion system, should be tried in date palm shoot multiplication and somatic embryo production. Micropropagation via organogenesis or direct shoot formation is extensive and labor-intensive. Somatic embryogenesis may reduce labor cost and also assist in developing automated somatic embryo production. However, genetic fidelity of micropropagated plants should be maintained with minimal somaclonal variation, otherwise there will be severe economic loses to growers. Molecular marker analysis would be an ideal approach to identify genetic variability at the early stage of plant development. It would be difficult to identify point mutations or any genetic change at the early stage of plant development because it may not express change phenotypically or may express it at the latter stage of plant development. This scenario occurred in African oil palm tissue culture-derived plants in Malaysia and the oil palm industry lost millions of US dollars.

Haploid production in date palm has not yet been accomplished. Inflorescence culture will be one way to induce haploid somatic embryo production. Fki et al. (2003) induced callus from immature inflorescences of date palm cv. Deglet Noor, and the calli originated from the proliferation of floral primordia showed embryogenic potential. The capacity of the inflorescence to form callus was much higher than cultured leaves. They did not determine the ploidy level of callus and regenerated plants from inflorescence-derived callus. In the future, the success of this type of work could revolutionize date palm genetic improvement programs as well as molecular genetics for useful gene identification.

Somatic embryogenic cell suspension is an excellent system for mutation induction and isolates useful mutants of date palm. Direct mutant somatic embryos can be produced and germinated into mutant somatic seedlings. These mutant seedlings can further be micropropagated for large-scale production. The utmost care should be taken when handling somatic embryogenic cultures; failure to do so significantly increases the chances of getting somaclonal variation. This approach is an excellent example of combining mutagenesis and biotechnology for date palm improvement. Transgenic date palms have a long way to go before consumers accept them and consequently export markets initially may also be lost. Therefore, the transgenic approach to modify date palm should be followed with a great caution, even though it has a great potential to overcome several of its problems.

There is a complete lack of date palm molecular biological research to address several issues facing date palm genetic improvement. Molecular marker-assisted selection and breeding need serious attention to identify trait-specific genes from natural or induced genetic variability. Functional genomic date palm breeding could probably become a reality in the future that will speed up genetic improvement of date palm.

Date palm has a great potential as a bioenergy crop, since date fruits have high sugar content and would be ideal for bioethanol production. Molecular genetics will be of great help in gene identification for sugar production and related chemicals. This really needs utmost attention.

Acknowledgement The author acknowledges the contributions of colleagues from Algeria, Morocco, Tunisia, who were involved in FAO/IAEA Technical Cooperation Project on date palm and the International Atomic Energy Agency, Vienna, Austria.

References

Aaouine M (2003) Date palm large-scale propagation through tissue culture techniques. In: The date palm from traditional resource to green wealth. Emirates Centre for Strategic Studies and Research, Abu Dhabi, pp 79–86

Al Kaabi H, Zaid A (2003) Date palm tissue culture in the UAE: research activities and mass propagation. In: The date palm from traditional resource to green wealth. Emirates Centre for Strategic Studies and Research, Abu Dhabi, pp 65–77

Al-Eid SM (2006) Chromatographic separation of fructose from date syrup. Int J Food Sci Nutr 57:83–96

Al-Ghamdi AS, Al-Kahtani MS (1996a) True-to-type date palm (*Phoenix dactylifera* L.) produced through tissue culture. 5. Fruit chemical properties. In: Proceedings third symposium on the date palm, vol 1. Mars Publishing, Riyadh, Saudi Arabia, pp 55–67

Al-Ghamdi AS, Al-Kahtani MS (1996b) True-to-type date palm (*Phoenix dactylifera* L.) produced through tissue culture. 6. Fruit sugar content. In: Proceedings third symposium on the date palm, vol 1. Mars Publishing, Riyadh, pp 67–78

Al-Ghamdi AS, Al-Kahtani MS (1996c) True-to-type date palm (*Phoenix dactylifera* L.) produced through tissue culture techniques. 7. Mineral content. In: Proceedings third symposium on the date palm, vol 1. Mars Publishing, Riyadh, pp 81–90

Al-Khayri JM (2005) Date palm, *Phoenix dactylifera* L. In: Jain SM, Gupta PK (eds.) Protocols for somatic embryogenesis in woody trees. Springer, Dordrecht, pp 309–320

Al-Khayri JM (2007) Protocol for micropropagation of date palm, *Phoenix dactylifera*. In: Jain SM, Haggman H (eds.) Protocols for micropropagation of woody trees and fruits. Springer, Berlin, pp 509–526

Dakheel A (2003) Date palm and biosaline agriculture in the United Arab Emirates. In: The date palm from traditional resource to green wealth. Emirates Centre for Strategic Studies and Research, Abu Dhabi, pp 199–211

Deputy JC, Ming R, Ma H et al (2002) Molecular markers for sex determination in papaya (*Carica papaya* L.). Theor Appl Genet 106:107–111

El-Deeb HM, Lashin SM, Arab YA (2007) Distribution and pathogenesis of date palm fungi in Egypt. Acta Hort 736:421–429

Fki L, Masmoudi R, Drira N, Rival A (2003) An optimised protocol for plant regeneration from embryogenic suspension cultures of date palm, *Phoenix dactylifera* L.cv. Deglet Nour. Plant Cell Rep 21:517–524

Jain SM (2005) Major mutation-assisted plant breeding programmes supported by FAO/IAEA. Plant Cell Tissue Organ Cult 82:113–121

Jain SM (2006) Radiation-induced mutations for developing Bayoud disease resistant date palm in North Africa. In: Zaid A (ed.) Proceedings of the international workshop on true-to-typeness of date palm tissue cultured-derived plants. Plant Tissue Culture Laboratory, UAE University, Al Ain, pp 31–41

Jain SM (2007) Recent advances in date palm tissue culture and mutagenesis. Acta Hort 736:205–211

Jain SM, Gupta PK (eds.) (2005) Protocols for somatic embryogenesis in woody trees. Springer, Dordrecht

Jain SM, Haggman H (eds.) (2007) Protocols for micropropagation of woody trees and fruits. Springer, Berlin

Jain SM, Ishii K (eds.) (2003) Micropropagation of woody trees and fruits. Kluwer, Dordrecht

Karempour F, Pejman H (2007) Study on possible influence of pathogenic fungi on date bunch fading disorder in Iran. Acta Hort 736:431–439

Khierallah HSM, Bader SM (2007) Micropropagation of date palm (*Phoenix dactylifera* L.) var. Maktoom through direct organogenesis. Acta Hort 736:213–223

Murashige T, Skoog F (1962) A revised medium for rapid growth and bioassays with tobacco tissue culture. Phys Plant 15:473–497

Oihabi A (2003) Major pests of the date palm. In: The date palm from traditional resource to green wealth. Emirates Centre for Strategic Studies and Research, Abu Dhabi, pp 143–147

Omar MS, Hameed MK (2006) *In vitro* propagation of date palm. In: Zaid A (ed.) Proceedings of the international workshop on true-to-typeness of date palm tissue cultured-derived plants. Plant Tissue Culture Laboratory, UAE university, Al Ain, pp 86–92

Quenzar B, Trifi M, Bouachrine B et al (2001) A mitochondrial molecular marker of resistance to Bayoud disease in date palm. Theor Appl Genet 103:366–370

Smith RJ, Aynsley JS (1995) Field performance of tissue cultured date palm (*Phoenix dactylifera*) clonally produced by somatic embryogenesis. Principes 39:47–52

Chapter 15
Magnetic Field Induced Biochemical and Growth Changes in Date Palm Seedlings

F. Dhawi and J.M. Al-Khayri

Abstract The long life cycle of date palm (*Phoenix dactylifera* L.) makes genetic improvement through traditional breeding methods a tedious endeavor. Biotechnology offers advanced tools to augment genetic improvement efforts. *In vitro* selection technique, a major application of plant biotechnology, allows the isolation of mutant cells and the regeneration of plants exhibiting desired new traits. Mutations can be induced chemically and physically through the exposure to various forms of radiation at appropriate levels. Studies indicate that magnetic fields have the ability to induce biochemical and physiological changes including enhanced plant growth and nutritional value. Although the use of magnetic fields in genetic improvement of date palm is in its infancy, necessary relevant information is beginning to accumulate. This chapter reviews research achievements in this area and addresses biological effects resulting from magnetic field exposure in date palm, in anticipation of future application in mutation studies. The impact of both static and alternating magnetic fields on the content of proline, DNA, photosynthetic pigments and elements are discussed. In addition, magnetic field-induced changes in growth parameters like weight and water content of both shoot and root systems are described. Exploration of magnetic field in *in vitro* studies of date palm is yet to be realized but encouraged by the regeneration improvements achieved in other plant species. Considering the effectiveness of magnetic field to modify plant systems, the application of magnetic field to genetically improve date palm merits further investigation.

Keywords Biotechnology • Chlorophyll • DNA • Ions • Mutation • Radiation • Proline

F. Dhawi • J.M. Al-Khayri (✉)
Department of Agricultural Biotechnology, College of Agricultural and Food Sciences, King Faisal University, P.O. Box 420, Al-Hassa 31982, Saudi Arabia
e-mail: faten.dhawi@live.com; jkhayri@kfu.edu.sa

S.M. Jain et al. (eds.), *Date Palm Biotechnology*,
DOI 10.1007/978-94-007-1318-5_15, © Springer Science+Business Media B.V. 2011

15.1 Introduction

The date palm (*Phoenix dactylifera* L.) is one of the most drought and salt tolerant fruit tree species acclimatized to the hot arid regions of the world. In addition to combating desertification, date palm production is a major source of income for many farmers. A date palm seedling requires over 3 years to flower and up to 10 years before full fruiting occurs. Moreover, date palm usually requires more than 30 years to complete only three backcrosses and to obtain the first offshoots from an intervarietal cross (Al-Yahya 1995). This long generation cycle along with its heterozygous nature and inability to determine sex at an early stage of development hamper breeding programs of date palm (Moursy and Saker 1996). Biotechnology provides powerful tools that can assist date palm breeders to genetically improve qualitative and quantitative horticultural traits. Biotechnology has already contributed to expanding agricultural areas of date palm by the development of micropropagation techniques (Al-Khayri 2007).

Currently, various laboratories are focusing on exploration of biotechnology in genetic improvement of date palm through mutation induction and *in vitro* selection strategies (Jain 2007). *In vitro* selection accompanied by induced mutation represents an immediate and inexpensive way of generating and selecting plant variants with tolerance to pathogens or environmental factors. These techniques allow introgression of desired traits to economically important cultivars, without dramatically changing other desirable characters (Evans and Sharp 1986; Jain 2001).

The process of *in vitro* selection involves culturing plant cells on a medium containing a selection agent followed by the regeneration of plants from the tolerant cell lines. Cell cultures are exposed to chemical or physical mutagens prior to *in vitro* selection. Common chemical agents used to induce mutations are ethylemethanesulphonate, sodium azide and diethylsulphate. Biophysical methods involve exposure of cell cultures to controlled doses of radiation to induce nonlethal mutations. The use of radiation to induce mutations is less hazardous to the environment and laboratory personnel in comparison to chemical mutagens. Therefore, many scientists recommend using biophysical methods to achieve improvement of agricultural crop to ensure environmental safety (Vasilevski 2003). Various types of radiations can be applied in plant research including electromagnetic waves, magnetic fields, ultrasound and various other forms of radiation.

Considering the effects of radiation, previous reports have shown that seeds of various cultivated plants react strongly to the impact of magnetic radiation when the optimum exposure dose is applied (Phirke et al. 1996). Lower doses of mutagens can stimulate growth, whereas high doses can cause mutations (Rybinski et al. 1993). Several authors have reported that magnetic fields (MF) influence the normal functions of plant systems (Atak et al. 2003; Belyavskaya 2004; Kiranmai 1994), acceleration of plant growth (Atak et al. 2003; Carbonell et al. 2000) and activation of protein synthesis (Phirke et al. 1996). Križaj and Valenči (1989) observed changes in root and shoot growth following magnetic exposures. They proposed possible mechanisms for the observed effects. The MF treatment significantly reduced the number of microorganism (bacteria, fungi) colonies on the naked oat (cv. Akt), sprouts after being treated with the oscillating magnetic field pulses. At the same time, the level of polyphenols slightly increased, and therefore the antioxidant activity (Lipiec et al. 2005).

In date palm radiation studies on growth and development are very limited. One source of radiation studied in date palm is gamma radiation where embryogenic cell suspension cultures were exposed to induce *in vitro* mutations (Jain 2007). Another source of radiation recently studied in date palm is magnetic field which was applied to seedlings (Dhawi and Al-Khayri 2008a, 2008b). This chapter is intended to review research achievements involving the application of magnetic field in date palm. Since the effects of MF is species specific (Reed et al. 1993), utilizing MF in date palm research needs fundamental studies to assess the optimum dose. The study of the MF effect on date palm covers the content of DNA to determine the mutagenic possibility and measure resulted stress associated with proline accumulation. Moreover, water content, fresh and dry weight for both shoot and root were also studied to determine the potential of MF as a growth enhancer. The effects of magnetic field observed in date palm are discussed in the context of results reported in other plant species. The use of magnetic field in *in vitro* plant systems is also described.

15.2 Definition and Source of Magnetic Field

A magnetic field (MF) is an invisible force created by a magnet or as a consequence of the movement of electric charges (flow of electricity). The magnitude (intensity) of a magnetic field is usually measured in Tesla (T) or millitesla (mT), but it can also be measured in Gauss (G), 1 G = 0.1 mT. The intensity of the field decreases with increasing distance away from the field source.

Magnetic fields are generated when electric charge carriers such as electrons move through space or within an electrical conductor. The geometric shapes of the magnetic flux lines produced by moving charge carriers of electric current are similar to the shapes of the flux lines in an electrostatic field. But there are differences in the ways electrostatic and magnetic fields interact with the environment.

Magnetic fields are difficult to shield and can easily penetrate living systems including plant tissue. Penetration causes electron excitation, moving ions in the field direction and heating via induced current (Moulder and Foster 1995). The field around a permanent magnet or wire carrying a steady direct current is a stationary magnetic field or static magnetic field (SMF), while that around an alternating current or changing direct current is an alternating magnetic field (AMF). The utilization of both types of these magnetic fields in date palm research is discussed in this chapter.

15.3 Effects of Magnetic Field

The physiological changes of plants treated with magnetic field can vary according to plant species, field intensity and period of exposure (Aarholt et al. 1981; Belyavskaya 2004; Blank and Goodman 1997; Goodman et al. 1995; Miyakoshi

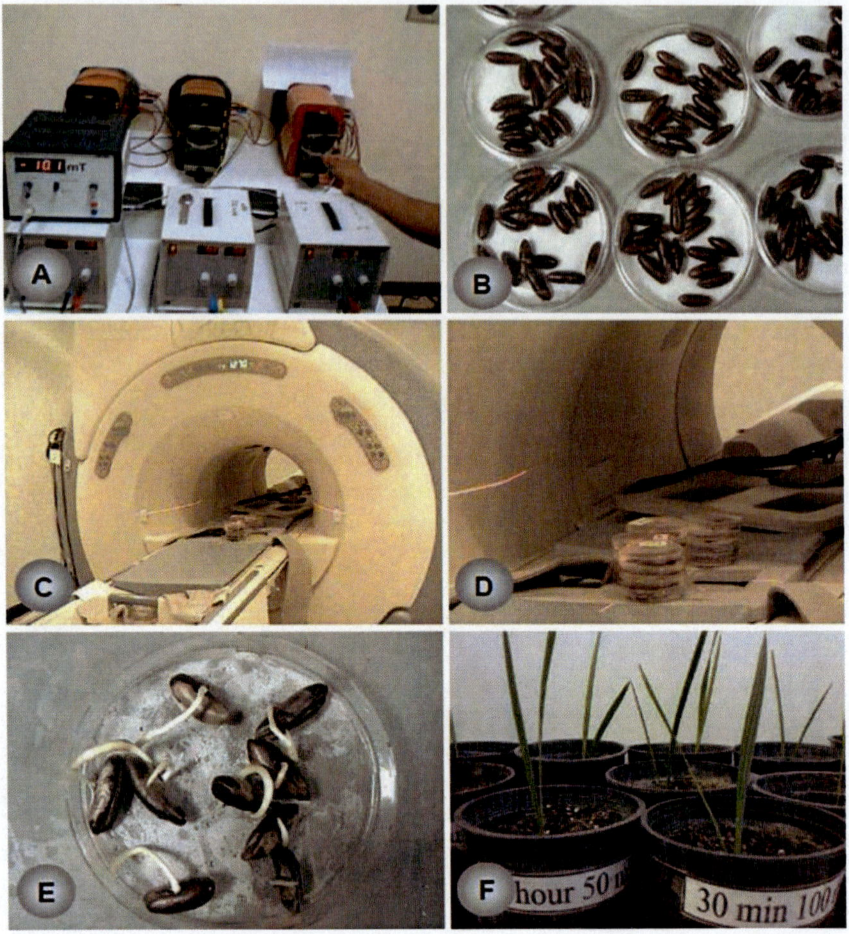

Fig. 15.1 Exposure of date palm seeds to magnetic field. (**a**) The electromagnetic system used to induce static magnetic field, (**b**) seeds prepared for magnetic field treatment, (**c**) the magnetic resonance imaging device used to induce alternating magnetic field, (**d**) seeds exposed to alternating magnetic field, (**e**) seed germination 2 weeks after magnetic field treatment, (**f**) seedlings growing in potted soil 45 days following magnetic field treatment

et al. 1997). Magnetic field changes the characteristics of the cell membrane, cellular functions like mRNA quantity, gene expression and protein biosynthesis and enzyme activities; and cause changes concerning the various functions at the organ and tissue levels (Goodman et al. 1995; Stein and Lian 1992).

Although exploitation of the energy force of MF has been achieved in numerous plant species, in date palm this kind of research is scarce. The first study encountered addressing the use of MF in date palm was reported by Ejraei (2007). Since then, a series of studies were conducted to investigate the behavior of date palm seedlings exposed to MF (both SMF and AMF) (Fig. 15.1). These studies have furthered our understanding of the effect of MF on growth parameters and some biochemical processes essential for plant cell growth and differentiation. In response to MF stress, biochemical modifications studied thus far were related to proline (Dhawi and Al-Khayri 2008a), photosynthetic pigments (Dhawi and Al-Khayri 2008b), DNA (Dhawi and Al-Khayri 2009a) and element composition (Dhawi and Al-Khayri 2009b). Investigations also included the effect of MF on date palm seed germination and seedling growth (Dhawi and Al-Khayri 2009c; Dhawi et al. 2009).

15.3.1 Proline Accumulation

Proline accumulation is a common biochemical indicator for assessing environmental stress in plants. Proline concentration in date palm callus cultures increases significantly under salinity (Al-Khayri 2002) and drought stress (Al-Khayri and Al-Bahrany 2004). Magnetic fields induce electric potential that exerts cellular stress causing biochemical physiological changes in cellular structures and functions in living systems. In higher plants, MF has been shown to induce stress effects (Belyavskaya 2004; Ghanati et al. 2007; Monselise et al. 2003). Another study indicated that exposing pea roots to low MF for 3 days causes a noticeable accumulation of lipid bodies, development of lytic compartment (vacuoles, cytosegresomes and paramural bodies) and reduction of phytoferrritin in plastids. Mitochondria were the most sensitive organelle to low magnetic field; their size and relative volume in cells increased, matrix was electron transparent and cristae reduced (Belyavskaya 2001). Although the role of MF is insufficiently understood, recent studies have shown that MF increases nitrogen and ion uptake by plants (Esitken and Turan 2003; Rochalska 2005; Sharaf El-Deen 2003) which could be indirectly responsible for causing stress.

Proline accumulation induced by MF impact was studied recently by Dhawi and Al-Khayri (2008a). They subjected date palm seedlings (seeds from cv. Khalas female trees) to either static magnetic field (SMF) or alternating magnetic field (AMF). After treatment, seedlings were planted in plastic pots containing potting mix and maintained in a greenhouse for further growth. The SMF was applied for 0, 30, 60, 180, 240 and 360 min using an electromagnetic circuit design to produced 10, 50 and 100 mT. The AMF was applied for 0, 1, 5, 10 and 15 min using a magnetic resonance imaging (MRI) device which has variation from 0.01 to 63,000 Hz, and carried alternating current at 220 V with magnetic flux at 1.5 T (1,500 mT).

Proline concentration was significantly affected by the intensity of the SMF and duration of exposure to SMF. Static magnetic field increased proline significantly, from 1.22 μmol/g in the control to 1.57 μmol/g with seedling exposure to 10 mT for 240 min. With a greater exposure duration proline concentration significantly decreased from its peak concentration, after 360 min of exposure. There was a similar reduction in proline concentration when SMF intensity was increased to 50 or 100 mT, regardless of exposure duration.

Conversely, significant reductions in proline concentration occurred in response to AMF; a greater reduction as the exposure duration increased. AMF reduced proline

Table 15.1 DNA content and proline accumulation of date palm leaves under alternating magnetic field (AMF) impact

AMF duration (min)	DNA content (μg/g)	Proline (μmol/g)
0	48.83 ± 0.97	1.20 ± 0.06
1	52.40 ± 1.51	0.79 ± 0.05
5	46.52 ± 1.80	0.77 ± 0.02
10	32.61 ± 1.02	0.21 ± 0.03
15	32.00 ± 1.01	0.19 ± 0.04

Table 15.2 DNA content and proline accumulation of date palm leaves under static magnetic field (SMF) impact

SMF intensity (mT)	Duration (min)	DNA content (μg/g)	Proline (μmol/g)
10	0	48.80 ± 0.97	1.20 ± 0.06
	30	44.52 ± 0.90	1.20 ± 0.03
	60	36.09 ± 0.75	1.39 ± 0.03
	180	33.95 ± 0.79	1.40 ± 0.04
	240	32.05 ± 0.80	1.57 ± 0.02
	360	32.00 ± 0.56	1.19 ± 0.02
50	0	48.80 ± 0.97	1.20 ± 0.06
	30	31.85 ± 0.70	1.13 ± 0.03
	60	31.80 ± 0.60	1.13 ± 0.04
	180	31.30 ± 0.60	0.96 ± 0.05
	240	31.14 ± 0.97	0.90 ± 0.03
	360	30.93 ± 0.52	0.86 ± 0.04
100	0	48.80 ± 0.97	1.20 ± 0.06
	30	29.40 ± 0.87	0.73 ± 0.02
	60	29.00 ± 1.00	0.70 ± 0.03
	180	29.70 ± 0.98	0.29 ± 0.02
	240	28.77 ± 0.80	0.24 ± 0.02
	360	28.59 ± 0.60	0.20 ± 0.02

concentration with as short as 1 min of exposure of seedlings. MF increased proline accumulation at the beginning of exposure to stress followed by a decrease due to accumulation of its product (Dhawi and Al-Khayri 2008a) (Tables 15.1 and 15.2), when exposure time was prolonged an inverse response was noted analogous to Belyavskaya (2004).

The same pattern was observed in soybean (*Glycine max* L. Merrill) (Atak et al. 2003); low intensity of MF, or short period of exposure to MF, increased chlorophyll concentration, whereas high intensity MF and long exposure to MF reduced the concentration. Magnetic field caused an inconsistency in the function of antioxidant enzymes in tobacco (*Nicotiana tabacum* L.) (Sahebjamei et al. 2007) and decreased phenylalanine ammonialyase activity and phenolic compounds concentration in basil (Ghanati et al. 2007).

The decrease in the cellular concentration of proline at high doses of MF could be explained by proline consumption by the *caged reaction*, whereby proline reacts with the free radicals that are enhanced during exposure to MF (Parola et al. 2006; Scaiano et al. 1994). Proline consumption in the *caged reaction* oxidizes proline to various compounds (Stadtman 1993) and this process can protect plant tissue from potential damage (Matysik et al. 2002). In addition, the proline pathway could be shifted by MF exposure, through oxidation of proline to glutamate or forming glutamic acid *g*-semialdehyde (Matysik et al. 2002). A shift in cellular metabolic pathways following exposure to MF was noted by Ghanati et al. (2007) who reported that under stress of SMF, plants shifted metabolism from biosynthesis of phenolics to the production of essential oils. More studies are necessary to understand and exploit this phenomenon. Although proline accumulation was disturbed, growth of date palm seedlings proceeded normally regardless of the type of magnetic field tested.

This may be explained by the greater ability of young plant seedlings to recover from stress than older plants (Karamanos 1995). The younger more actively growing plants are endowed with a greater ability to recover after stress andthus a greater potential for proline accumulation than plants at later growth stages. It is concluded that proline accumulation in date palm seedlings is affected by exposure to MF. Exposure to MF may be a convenient method to select for tolerance to abiotic stresses as MF induces a stress situation as indicted by the accumulation of proline and thereby can facilitate investigations related to stress physiology.

15.3.2 Modification of DNA Content

Plant stress caused by exposure to magnetic fields induces modifications at the molecular level, particularly in DNA synthesis, structure and function (Goodman and Blank 1999). It is well known that MF can affect singlet-triplet transition of unpaired electrons and prolonged lifetime of free radicals, thereby leading to oxidative stress (Sahebjamei et al. 2007) which causes mutation (Frankel and Liburdy 1996; Kiranmai 1994; Pingping et al. 2007). In a study by Ikehata et al. (1999), a weak effect of SMF on mutation induction was detected. Their results suggest that co-mutagenic effects of reactive chemicals on bacterial DNA increased linearly between 0.07 and 2 T. Stronger static magnetic field up to 13 T have small but detectable mutagenic potential (Ikehata et al. 2009).

Researchers observed the ability of nucleic acids to align with the magnetic field, as well as certain phospholipids. Magnetic effects on protein orientation were originally considered in theoretical calculations of the magnetic anisotropy of peptide bonds (Pauling 1979). Later, alignment of protein crystals in magnetic fields was observed by Astier et al. (1998). Under MF effect total RNA concentration of plantlets regenerated from treated explants significantly increased (Atak et al. 2007). In a study using *in vitro* transcription, Worczak et al. (2006) observed that high doses of MF caused a reduction in the efficiency of RNA polymerase with increasing magnetic field strength, although full-length RNA transcripts were generated. Watanabe et al. (1997) noted an increase of oxidative damage caused by exposure to a strong static magnetic field that would relate to the cause of increase in the mutation frequency. Everything in this universe contains magnetic properties including DNA (Dekker and Ratner 2001). Studying DNA content is important to understand the relation between DNA and plant adaptation (Grime and Mowforth 1982). MF strengths in excess of about 20 T induce expression of the Adh/GUS transgene in roots and leaves of thale cress (*Arabidopsis thaliana* L.) (Paul et al. 2001, 2004). According to Pingping et al. (2007), wheat (*Triticum aestivum*) seeds exposed to 7T MF resulted in 0.29% albino mutant plants which were unable to produce chlorophyll normally. These albino mutants grew weakly but some turned partly green later and produced seeds. After observing three generations and analysis of their genetic stability, seven lines restored their normal state; however, two lines remained in their original albino conditions.

The ability of a magnetic field to play a central role at the molecular level requires validation through extensive studies. Having in mind the future application of the magnetic field as a mutational agent for date palm genetic manipulation, Dhawi and Al-Khayri (2009a) investigated its influence on the DNA content of date palm seedlings. This study showed that DNA content is significantly affected by the SMF through two factors (intensity and exposure duration). Similarly the effect of AMF on DNA content was significantly influenced by exposure time.

The content of DNA in date palm seedlings after SMF exposure significantly decreased in general. A reduction in DNA content is noticed, when increasing SMF intensity from 10 to 100 mT or duration from 30 to 360 min. Static magnetic field treatment at 10 mT decreased DNA significantly to 44.52 μg/g, after 30 min of exposure. Raising SMF intensity to 50 mT decreased DNA significantly to 30.93 μg/g at 360 min, whereas at 100 mT, the lowest level of DNA average was 28.59 μg/g at 360 min. In general, it seems that the reduction in DNA content could stabilize in the same dose of SMF at 50 and 100 mT. The effect of AMF was significantly influenced by exposure time.

Alternating magnetic field treatments increased DNA content after 1 min of exposure to reach its highest level, 52 μg/g. The DNA reduction followed 5 min of AMF exposure to continue with longer duration at 10 min where the DNA content was 32 μg/g, which was the lowest level of DNA content. Exposure to magnetic field caused reduction in the content of DNA. The lowest exposure time tested, 30 min, was sufficient to induce reduction in DNA content. This was true even at the lowest intensity, 10 mT. This dosage caused the DNA content to decrease from 49 to 45 μg/g. Further, increase of the exposure duration to 60 min caused significant reduction in the DNA content. At intensities higher than 10 mT, DNA content decreased significantly even at the shortest exposure of 30 min. At 50 and 100 mT, significant decrease in DNA content was also noted in response to 30 min exposure; whereas the level of DNA increased after 1 min of MRI exposure, then decreased after 5 min. However, longer durations caused no further decrease in the DNA content in response to AMF (Table 15.1) and SMF (Table 15.2).

These observations indicate that magnetic fields interact with DNA processes, probably by inhibiting synthesis or stimulating degradation of DNA. This response merits further exploration as a mutational agent for date palm genetic manipulation. The study has shown that both AMF and SMF treatments caused a reduction in DNA content.

This is comparable to studies by Racuciu et al. (2007, 2008a, b) who reported that low MF reduced the DNA level and has a certain inhibitory effect upon the nucleic acid (DNA and RNA) of seedlings in black locust (*Robinia pseudoacacia*).

Other studies have indicated that MF could lead to decreased genome functional activity (Belyavskaya 2004) and increased mutagenic possibility (Pingping et al. 2007). Koana et al. (1997) showed that magnetic fields enhanced the rate of spontaneous mutation via changes in radical homeostasis, which can indirectly affect DNA integrity. Genotoxic effects of MF may occur from damage to DNA repair mechanisms or by increasing membrane permeability uptake to mutagens (Pingping et al. 2007).

Since magnetic fields interact with moving charges, it is possible that they stimulate stress response by interacting directly with moving electrons in the DNA (Vizcaino 2003). It was shown that even weak magnetic fields have an influence on charged species within DNA (Goodman and Blank 1999, 2002). Moreover, low frequency magnetic fields penetrate the cell and they could theoretically interact directly with the DNA in the nucleus (Goodman and Blank 1999).

Regardless of DNA reduction, growth of date palm seedlings proceeded normally which means that plants could recover after treatment. Pingping et al. (2007) showed that albino plants of magnetically treated wheat turned partly green in later periods of their life. Results indicate that magnetic fields interact with DNA processes. The magnetic field could influence both the charge and electronic properties of the DNA. Another possible explanation is that free radicals induced by MF may block some enzymatic activities, including DNA syntheses enzymes.

15.3.3 Changes in Photosynthetic Pigments Content

Magnetic field has been shown to affect normal plant metabolism (Aladjadjiyan 2007) and impact meristem cell division (Belyavskaya et al. 1992). Additionally, Campbell (1977) found that chloroplasts have paramagnetic properties. That means that in a magnetic field the magnetic movements of the atoms in them are oriented downwards of the field direction. The influence of the magnetic field on plants, sensible to it, increases its energy. Later this energy is distributed among the atoms and causes the accelerated metabolism and, consequently, to better germination (Campbell 1977). Studies carried out by (Pietruszewski 1999a, b) in wheat (*Triticum aestivum* L.) showed increased gluten content of kernels obtained from seeds subjected to the pre-sowing stimulation with magnetic field. Several studies have been done on different plants. MF treatment increased the chlorophyll content in sugar beet leaves (Rochalska 2005). Working with sugar beet (*Beta vulgaris* L.) Pietruszewski and Wójcik (2000) observed increased sugar content in roots in response to magnetic field. In a study by Rakosy-Tican et al. (2005) on the MF effect on potato and wild *Solanum* species, a significant stimulation of leaf growth was noted and the quantity of chlorophyll a and b and carotenoids increased significantly. In *Robinia pseudoacacia* L. seedlings MF affected leaf chlorophyll quantity significantly.

Chlorophyll a as well as chlorophyll b levels were found to decrease with prolonged exposure; while short MF exposure enhanced the two main types of chlorophyll significantly (Sandu et al. 2005). Taia et al. 2007 studied SMF effect on photosynthetic pigments in sweet basil (*Ocimum basilicum* L.). It has been found that an increase occurs in chemical reactions of plants under MF, which has a positive effect on photochemical activity, respiration ratio and enzyme activity in rice (*Oryza sativa* L.) (Carbonell et al. 2000). Plant length and chlorophyll ratio were enhanced under magnetic exposure doses 1, 2 mT whereas, nucleic acid level is diminished quantitatively in maize (Racuciu et al. 2007). Results revealed that photosynthetic pigments were decreased significantly by the exposure to magnetic fields, irrespective of its direction or force and this may be due to the effect of magnetic fields on the reduction in plastids inside the cells.

Additionally, studies by Atak et al. (2000, 2003, 2007) involving MF impact on soybean (*Glycine max* L. Merrill) confirmed that MF significantly increased chlorophyll a, chlorophyll b and total chlorophyll content. Certain stimulation of chlorophyll biosynthesis was noted while inhibitory effect upon the nucleic acid biosynthesis was also revealed. The assimilatory pigments quantity appeared increased in the experimental samples exposed for 0.5 and 1 h, while for longer exposure times a decreased level could be noticed in *Robinia pseudoacacia* (Racuciu et al. 2008a, b). In another study Racuciu et al. 2007 reported that a stimulatory effect on photosynthesis was noted as the magnetic energy dose increasing, suggesting putative stimulation of photosynthesis process.

Dhawi and Al-Khayri (2008b) studied the MF influence on photosynthetic pigments of date palm seedlings. Results indicated that pigment content (chlorophyll a, chlorophyll b, carotenoids and total pigments) was significantly increased under static magnetic field. The highest measurements were recorded at 100 mT, after 360 min of exposure. On the other hand, alternating magnetic field decreased photosynthetic pigment content after 10 min of treatment with 1.5 T. Low magnetic field doses had a stimulatory effect on photosynthetic pigments whereas high doses had a negative effect. Chlorophyll a and carotenoids were more affected than chlorophyll b. This study showed that photosynthetic pigments are significantly affected by the SMF factors (intensity and exposure duration). Chlorophyll a, chlorophyll b, carotenoids and total pigments concentrations increased significantly as SMF intensity increased; however, the significant increase for photosynthetic pigments at low dose treatment at 10 mT began after 180 min of SMF exposure; whereas at 50 mT; short exposure for 30 min was sufficient to increase photosynthetic pigments significantly; the highest values for photosynthetic pigments observed at 100 mT; prolonged exposure time increased the pigments level significantly. The effect of AMF was significantly influenced by exposure time. The highest level of photosynthetic pigment noted at 1 min of AMF treatment, followed by a significant decrease at 5–15 min of AMF exposure. In contradiction to SMF results, increasing exposure time has a negative impact on pigments level under AMF treatments (Tables 15.3 and 15.4).

The electrons with magnetic moments can be oriented in the external MF. As a result of the interaction between the external MF and the magnetic moment of unpaired electrons, the energy is absorbed. Chloroplasts have magnetic moments

Table 15.3 Chlorophyll a, b and total pigments content of date palm leaves under alternating magnetic field (AMF) impact

AMF duration (min)	Chlorophyll a (μg/g)	Chlorophyll b (μg/g)	Carotenoids (μg/g)	Total pigments (μg/g)
0	5.39 ± 0.35	1.94 ± 0.16	2.05 ± 0.21	09.40 ± 0.40
1	7.67 ± 0.56	2.84 ± 0.42	3.09 ± 0.22	13.61 ± 0.76
5	7.62 ± 0.49	2.56 ± 0.20	3.25 ± 0.39	13.44 ± 0.73
10	5.95 ± 0.31	1.96 ± 0.18	2.35 ± 0.16	10.27 ± 0.54
15	5.03 ± 0.21	1.70 ± 0.12	1.72 ± 0.18	8.46 ± 0.25

Table 15.4 Chlorophyll a, b and total pigments content of date palm leaves under static magnetic field (SMF) impact

SMF intensity (mT)	Duration (min)	Chlorophyll a (μg/g)	Chlorophyll b (μg/g)	Carotenoids (μg/g)	Total pigments (μg/g)
10	0	5.39 ± 0.35	1.88 ± 0.04	2.05 ± 0.21	09.34 ± 0.34
	30	5.68 ± 0.19	2.12 ± 0.08	2.13 ± 0.36	09.93 ± 0.39
	60	6.27 ± 0.32	2.18 ± 0.18	2.42 ± 0.33	10.88 ± 0.48
	180	7.17 ± 0.40	2.53 ± 0.23	2.59 ± 0.17	12.31 ± 0.65
	240	7.08 ± 0.46	2.70 ± 0.30	2.59 ± 0.18	12.37 ± 0.84
	360	7.48 ± 0.27	2.86 ± 0.38	2.63 ± 0.16	12.98 ± 0.42
50	0	5.39 ± 0.35	1.88 ± 0.04	2.05 ± 0.21	9.34 ± 0.34
	30	7.8 ± 0.49	3.06 ± 0.14	2.75 ± 0.15	13.62 ± 0.51
	60	8.7 ± 0.35	3.15 ± 0.12	2.93 ± 0.20	14.80 ± 0.47
	180	9.08 ± 0.15	3.04 ± 0.12	3.07 ± 0.28	15.2 ± 0.20
	240	9.3 ± 0.59	3.19 ± 0.34	3.59 ± 0.35	16 ± 1.09
	360	10.6 ± 0.34	4.14 ± 0.33	3.78 ± 0.23	18.52 ± 0.51
100	0	5.39 ± 0.35	1.88 ± 0.04	2.05 ± 0.21	9.34 ± 0.34
	30	11.34 ± 0.37	3.9 ± 0.39	3.57 ± 0.14	18.83 ± 0.56
	60	11.47 ± 0.46	3.8 ± 0.18	3.75 ± 0.24	19.03 ± 0.43
	180	11.8 ± 0.30	3.85 ± 0.28	3.67 ± 0.21	19.33 ± 0.38
	240	13.7 ± 0.27	5.14 ± 0.33	4.5 ± 0.17	23.36 ± 0.53
	360	14.19 ± 0.21	4.87 ± 0.38	4.44 ± 0.18	23.51 ± 0.57

and could be affected by the absorbed energy at a high dose of MF which can disturb the pigments synthesis. Other possible explanations for the decline in pigment content are that carotenoids may be consumed in radical scavenging reactions (Strzalka et al. 2003) or free radicals inhibited the synthesis through affecting photosynthesis enzymes.

15.3.4 Influence on Elemental Composition

The exact mechanism of the effect of static magnetic field on living organisms is still unclear. Theoretically, living cells possess electric charges exerted by ions or free radicals, which act as endogenous magnets. These endogenous magnets can be affected by an exogenous magnetic field, which can orient unpaired electrons. The forces induced by magnetic fields may be large enough to affect any process that can change the rate of movement of electrons significantly (Goodman and Blank 2002). Levels of calcium inside plant cells increase following exposure to MF, which is one of the proposed mechanism by which MF may affect plants. Calcium ions participate in many plant growth processes and responses to stress.

Graziana et al. (1990) studied calcium influx in protoplast isolated from carrot cell suspension in field intensity and frequency-dependent ways. The field-induced calcium uptake during the induction process appeared mainly cumulative as long as the morphology of the protoplasts did not change (up to 10 min). The stimulation elicited fields were effective even after switching the field off, the influx increased for 5 min and then slowed down to its initial value 15 min later. During electrostimulation, an additional amount of ATP was consumed, whereas the plasma membrane was hyperpolarized. Sodium ions were expelled from the protoplast. The electric field stimulates an ATP synthesis-like activity and the consumption of the ATP. Thus formed it elicits an electric potential that drives the influx of calcium and efflux of sodium. A study on tomato plants showed that the application of MF to irrigation water increased nutrient element contents of plants (Duarte Diaz et al. 1997). Owing to plant cells having a negative electrical charge, they take up ions with a positive electrical charge (Marschner 1995). Pea roots exposed to low magnetic field for 3 days, in meristem cells calcium pyroantimonate deposits were observed in all organelles and in the hyaloplasm of cells. Data obtained suggest that the observed LMF effects on the ultra structure of root cells were due to disruptions in different metabolic systems including effects on calcium homeostasis (Belyavskaya 2001). It is suggested that electrophoretic segregation of charged components in the outer leaflet of the cell membrane is responsible for both enhanced adsorption and stimulated uptake macromolecules via changes of the membrane elastic properties that enhance budding and fission processes (Antov et al. 2005). Treatments with magnetic field are assumed to enhance seed vigor by influencing the biochemical processes that involve free radicals, and by stimulating the activity of proteins and enzymes (Kurinobu and Okazaki 1995; Stange et al. 2002). Under SMF (21 G = 0.21 mT) Penuelas et al. (2004) noted inhibition of root growth inthree plant species (*Lens culinaris, Glycine soja*, and *Triticum aestivum*). However, the accumulation of paramagnetic nutrients such as Fe and Mn was different among the three species. The highest Fe concentrations were found in *L. culinaris* and the lowest ones in *T. aestivum*. Although Mn concentrations were lower in the lentils and beans, the total amount of the metal was slightly higher in the legumes than in the cereal. Also higher concentrations of other elements such as S and Cu were noted in the legumes as compared to the cereal. Similarly, a study on tomato plants showed that the application of MF to irrigation water increased nutrient element contents of plants (Duarte Diaz et al. 1997). Numerous studies suggest that magnetic field increases ion uptake and consequently improves nutrition value which could be a good alternative for chemical treatments.

Date palm seedlings were treated with two types of magnetic field (SMF and AMF) in order to evaluate the effect on element uptake (Dhawi and Al-Khayri 2009c;

Table 15.5 Ions accumulation of date palm leaves under alternating magnetic field (AMF) impact

	AMF duration (min)				
	0	1	5	10	15
Ions	Ion concentrations (μg/g)				
Ca	4,160 ± 62	4,897 ± 39	5,376 ± 77	5,477 ± 98	6,312 ± 68
Mg	2,156 ± 44	2,610 ± 54	2,727 ± 39	2,789 ± 39	6,412 ± 69
P	4,528 ± 48	3,792 ± 80	3,755 ± 99	3,709 ± 60	3,743 ± 62
K	3,087 ± 38	3,532 ± 39	3,648 ± 49	3,802 ± 63	3,884 ± 51
Na	1,459 ± 47	1,767 ± 39	1,787 ± 38	1,853 ± 48	1,984 ± 48
Mn	32.70 ± 1.7	50.85 ± 1.7	56.12 ± 1.0	63.30 ± 1.2	72.41 ± 0.9
Fe	32.57 ± 1.4	53.85 ± 1.2	54.15 ± 1.3	58.22 ± 0.1	65.73 ± 1.0
Zn	22.14 ± 1.8	32.57 ± 1.4	35.57 ± 1.3	36.42 ± 0.1	46.53 ± 1.2

Dhawi et al. 2009). The SMF source is produced at three levels of magnetic field intensities (10, 50 and 100 mT). Seedlings were exposed to these magnetic fields for different periods: 0, 30, 60, 180, 240 and 360 min. The AMF Source was magnetic resonance imaging (MRI) with an intensity of 1,500 mT where seedlings treated for 0, 1, 5, 10 and 15 min. Leaf samples were subjected to chemical analysis for elements (Mg, Ca, Na, P, K, Fe, Mn and Zn (Table 15.5). The results revealed that concentrations of Ca, Mg, Mn, Fe, Na, K, and Zn increased, while P concentration decreased with the raising of both types of MF intensities and durations of exposure. Magnetic field has a potential to enhance growth due to the positive effect on the plant major elements such as Ca and Mg, but negative electrical charges on the plants inhibited the uptake of anions such as P. Ion content increased significantly with extended time analogous to the study by Wojcik (1995) who found that MF increased ions if time of exposure was longer. Levels of calcium increased following exposure to MF.

It has been reported that changes in electrical conductivity of $CaCl_2$ solution are caused by exposure to static magnetic fields (Ayrapetyan et al. 1994). Moreover, Mg, K, Fe, Mn, Zn and Na were also affected under MF and increased significantly while P decreased with rising intensity and time of exposure. This is analogous to the Esitken and Turan (2003) findings that increasing MF strength caused increase in contents of N, K, Ca, Mg, Fe, Mn, Na and Zn but reduced P and S content of strawberry leaves. In addition, results may vary according to plant organs; studies indicated that MF increased contents of (Mg, Fe and Cu) in the buckwheat grain var. *Hruszowska* and (P, Ca, K and Zn) in straw (Wojcik 1995). In summary, ions accumulation was affected by magnetic field in date palm plants. Magnetic field may play an important role in cation uptake capacity and has a positive effect on immobile plant nutrient uptake which raises the products nutritional value of date palm.

15.3.5 Effect on Weight and Water Content

Magnetic field could have some beneficial effects on germination as well as fresh and dry weight. The positive effect of the magnetic field treatment was expressed through increasing the germination, height and weight of the shoots compared to the control.

The influence of magnetic field treatment depended on the induction of the field, exposition of the samples and the pre-history of the samples. Alexander and Doijode (1995) found that onion (*Allium cepa* L.) and rice seeds exposed to a weak EMF for 12 h showed significantly increased germination, shoot and root length, and fresh and dry seedling weight. However, MF caused acceleration of plant maturation (Podlesny et al. 2005) and of fruit ripening (Boe and Salunkhe 1963). The fruit ripening was explained earlier by Boe and Salunkhe (1963) that MF could possess an auxin-like character or could activate the auxin or accelerate the enzyme system. Pietruszewski (1999a, b) observed better sprouting of spring wheat seeds, which had been soaked in water before they were subjected to MF, in comparison to dry seeds. He explained this phenomenon by the higher mobility of ions in more humid seeds.

Hypothetically, MF could be a substitute for growth hormones which have been used to increase yield and plant productivity. Several studies have indicated that productivity of some plants increased in response to the MF effect (Aladjadjiyan 2002; Alexander and Doijode 1995). Samy (1998) found earlier flowering and yield increase of cabbage as a result of the treatment with magnetic field at an 8-h exposition.

The variable and contradictory results seem to suggest that the effect of MF on plants may be species-specific. In a study where 16⅔ Hz MF continued exposure of sunflower (*Helianthus annuus*) and wheat (*Triticum aestivum*) seedlings, Fischer et al. (2004) found significant but different effects. Sunflower showed significant increase in total fresh weights, shoot fresh weights, and root fresh weights, whereas dry weights and germination rates remained unaffected under MF exposure. In the same study, treated wheat seedlings showed significant higher fresh and dry weights of roots, total fresh weight and germination rate.

A study of MF effects in tomato (*Lycopersicon esculentum* L. Mill.) seeds during the generative stage, De Souza et al. (2006) showed that leaf area per plant and relative growth rates of fruits of plants from magnetically exposed seeds were greater than those of the control. At fruit maturity stage, all magnetic treatments significantly increased fruit weight, fruit yield per plant, fruit yield per area and equatorial diameter of fruits. At the end of the experiment, total dry matter was significantly higher for plants from magnetically-treated seeds than that of the control. Similar results for rice, sunflower and maize were reported (Carbonell et al. 2000). The positive effect of the magnetic field treatment was expressed in increased germination, height and weight of shoots (Aladjadjiyan 2002). Similarly, Racuciu et al. (2006) reported that young maize plants respond to the magnetic field ranging between 50 and 250 mT. Plant lengths are higher for all exposed samples as it was recorded for the first 11 days of growth. The accumulation of dry substance mass was enhanced logarithmically with the increase of the magnetic induction.

In date palm, Ejraei (2007) subjected seeds to MF and found that 50–200 mT caused a significant increase in germination percentage. After 2 years of growth, plants treated with 50 mT had a significant increase in weight and decrease in length compared to the untreated control. All plants germinated after exposure to 100, 150 and 200 mT had a decrease of weight; the least weight was associated with 200 mT.

In a recent study on date palm, Dhawi and Al-Khayri (2009b) evaluated the effect of magnetic field on seedling growth parameters including fresh and dry

Table 15.6 Growth parameters, leaf fresh and dry weight and water content and root fresh and dry weight and water content, of date palm under alternating magnetic field (AMF) impact

AMF duration (min)	Leaf tissue Fresh weight (g)	Dry weight (g)	Water content (%)	Root tissue Fresh weight (g)	Dry weight (g)	Water content (%)
0	1.82 ± 0.06	1.38 ± 0.05	24.11 ± 0.96	2.17 ± 0.09	1.68 ± 0.03	22.3 ± 0.16
1	2.53 ± 0.58	1.84 ± 0.04	27.04 ± 0.91	2.45 ± 0.05	1.79 ± 0.05	26.8 ± 0.42
5	2.75 ± 0.11	1.85 ± 0.07	32.71 ± 0.55	2.49 ± 0.03	1.80 ± 0.03	26.0 ± 0.20
10	2.74 ± 0.03	1.84 ± 0.03	32.60 ± 0.77	2.61 ± 0.05	1.83 ± 0.02	30.2 ± 0.16
15	2.82 ± 0.03	1.88 ± 0.03	33.32 ± 0.9	2.62 ± 0.06	1.81 ± 0.02	30.2 ± 0.12

weights of both leaves and roots, and calculated water content. This study included both SMF and AMF which were compared. The results showed that magnetic field had a significant impact on date palm seedlings leaf and root fresh weight under SMF; whereas, dry weight for both leaf and root were less affected when AMF was used. Leaf fresh weight increased significantly as influenced by SMF treatment from 10 mT for 30 min to 100 mT for 360 min. The highest leaf fresh weight was obtained at 100 mT for 240 min. Moreover, leaf dry weight significantly increased under the SMF effect. At 10 mT leaf dry weight reached 1.6 g after 360 min of exposure and 1.7 g and 1.8 g at 50 and 100 mT for the same duration, respectively.

Leaf water content was affected significantly by the SMF intensity and duration. At a low dose of 10 mT, leaf water content reached the highest increase of 32% after 180 min of exposure. At 50 mT, the increase in leaf water content reached the same value after 240 min of exposure; whereas this level occurred after only 60 min when 100 mT was applied.

Similarly, SMF affected the increase in root fresh weight. The highest increase at 10 mT and 50 mT was associated with 360 min duration, while at 100 mT the highest root fresh weight was associated with 180 min duration. On the other hand, root dry weight was not affected by SMF treatment. Root water content has increased significantly for all SMF doses. Root water content increased from 26% at 10 mT for 30 min to 32% at 100 mT for period of 60 min.

Increased growth parameters were also noted in response to AMF. The effect of AMF was influenced by exposure time. Root fresh and dry weights as well as water content increased significantly in response to AMF treatments. The highest values for both fresh and dry weights and water content were associated with a dose of 15 min. Leaf fresh weight was influenced by AMF durations (1–15 min). The highest value for fresh weight was after 15 min of exposure. Opposite to the SMF results, AMF caused a significant increase in leaf dry weight. Leaf dry weight highest value was noted at 15 min. Leaf water content significant increase was accompanied with leaf fresh weight and the highest value was after 15 min. In contrast to SMF, AMF increased leaf and root fresh and dry weight as well as water content. Although the exact mechanism of such changes in plant weight is yet to be elucidated, it is likely that MFs may function via the production of reactive oxygen species and free radicals leading to stress reactions. These stress reactions may be responsible for dry weight limited increase or the increase in water content in response to AMF (Tables 15.6) and SMF (Table 15.7).

Table 15.7 Growth parameters, leaf fresh and dry weight and water content and root fresh and dry weight and water content, of date palm under static magnetic field (SMF) impact

		Leaf tissue			Root tissue		
SMF intensity (mT)	Duration (min)	Fresh weight (g)	Dry weight (g)	Water content (%)	Fresh weight (g)	Dry weight (g)	Water content (%)
10	0	1.78 ± 0.12	1.34 ± 0.10	24.83 ± 0.87	2.23 ± 0.15	1.66 ± 0.11	25.41 ± 0.75
	30	2.16 ± 0.05	1.54 ± 0.03	28.70 ± 0.65	2.30 ± 0.07	1.68 ± 0.05	26.90 ± 0.50
	60	2.35 ± 0.09	1.65 ± 0.06	29.71 ± 0.56	2.31 ± 0.04	1.68 ± 0.04	27.22 ± 0.61
	180	2.36 ± 0.08	1.61 ± 0.05	32.21 ± 0.46	2.40 ± 0.03	1.71 ± 0.02	28.50 ± 0.67
	240	2.36 ± 0.05	1.60 ± 0.03	31.91 ± 0.74	2.51 ± 0.05	1.74 ± 0.03	30.86 ± 0.58
	360	2.36 ± 0.10	1.60 ± 0.07	32.18 ± 0.63	2.52 ± 0.08	1.74 ± 0.04	30.87 ± 0.71
50	0	1.78 ± 0.12	1.34 ± 0.10	24.83 ± 0.87	2.23 ± 0.15	1.66 ± 0.11	25.40 ± 0.75
	30	2.44 ± 0.07	1.77 ± 0.05	27.47 ± 0.73	2.58 ± 0.06	1.75 ± 0.04	32.01 ± 0.57
	60	2.44 ± 0.06	1.76 ± 0.06	27.70 ± 1.44	2.60 ± 0.05	1.77 ± 0.04	32.00 ± 0.49
	180	2.46 ± 0.07	1.77 ± 0.07	27.92 ± 1.67	2.61 ± 0.05	1.77 ± 0.03	31.80 ± 0.50
	240	2.56 ± 0.06	1.74 ± 0.05	32.00 ± 0.85	2.62 ± 0.05	1.77 ± 0.03	32.18 ± 0.53
	360	2.56 ± 0.11	1.76 ± 0.06	31.29 ± 0.78	2.64 ± 0.07	1.81 ± 0.04	31.21 ± 0.37
100	0	1.78 ± 0.12	1.34 ± 0.10	24.83 ± 0.87	2.23 ± 0.15	1.66 ± 0.11	25.40 ± 0.75
	30	2.50 ± 0.12	1.74 ± 0.08	30.53 ± 0.81	2.69 ± 0.06	1.82 ± 0.04	32.16 ± 0.63
	60	2.66 ± 0.04	1.79 ± 0.02	32.61 ± 0.43	2.71 ± 0.03	1.81 ± 0.02	33.00 ± 0.31
	180	2.77 ± 0.03	1.88 ± 0.01	32.15 ± 0.31	2.77 ± 0.06	1.86 ± 0.04	32.87 ± 0.36
	240	2.79 ± 0.05	1.89 ± 0.05	32.00 ± 0.63	2.78 ± 0.04	1.89 ± 0.03	32.00 ± 0.49
	360	2.78 ± 0.03	1.89 ± 0.02	32.00 ± 0.41	2.78 ± 0.05	1.88 ± 0.03	32.22 ± 0.32

Reina et al. (2001) studied water absorption by lettuce seeds previously treated in a stationary magnetic field of 0–10 mT. They found significant increase in the rate of water absorption accompanied with an increase in the total mass. The potential explanation for the increase in water content could be due to the variations induced by magnetic fields in the ionic currents across the cellular membrane (Reina and Pascual 2001), which increasing water absorption and water content, thus increasing fresh weight for both leaf and root. Whereas, root dry weight was not affected by SMF treatment analogous to a study by Yano et al. (2004), who found a decrease on CO_2 uptake rate and early growth parameters of radish (*Raphanus sativus* L.) seedlings exposed to magnetic force. Magnetic field is known to decrease the force of water hydrogen bonds (Baran et al. 2006), and lead ions to higher mobility (Pietruszewski 1999a, 1999b), which decreases water potential and leads to increase water absorption and consequently water content.

The effect of SMF on water content in sweet basil was studied by Taia et al. (2007). Their results showed an increased rate of water absorption accompanied with an increase in the total mass in response to increased magnetic forces. This observation coincides with the finding of Monselise et al. (2003) who observed that magnetic fields of low frequency induce a stress on higher plants. Meanwhile Scaiano et al. (1994) assumed that low intensity and low frequency magnetic fields can potentially alter chemical processes in which free radicals are involved. Jones (2000) explained that the effect of magnetic forces on water is due to the electromagnetic fields originated from the atoms and the existence of defects in molecular structure of water. Hong (1995) retained the effect of magnetic fields on water by the bipolar character of water molecules.

In a study by Dhawi and Al-Khayri (2009b), water content increased significantly in both leaves and roots. This is comparable to results obtained by Reina and Pascual (2001). They found a significant increase in the rate of water absorption accompanied by an increase in the total mass in magnetically-treated seeds. Atak et al. (2003), found that fresh weights of soybean (*Glycine max* L. Merrill) seedlings regenerated from treatment explants significantly increased in root formation percent and fresh weight following MF treatment.

Moreover, MF increased the germination of the English oak (*Quercus robur* L.) seeds and their subsequent growth (Celestino et al. 2000), tomato (De Souza et al. 2006) and maize fresh weight and shoot length (Aladjadjiyan 2002). Maize (*Zea mays* L.) seeds were found to grow faster giving taller stature and higher fresh weight than the control plants when exposed to static magnetic field (SMF) (Flórez et al. 2005). This was confirmed by Racuciu et al. (2008b) who also observed stimulated growth of maize seedlings subjected to SMF at 50 mT. This treatment significantly enhanced fresh tissue weight, assimilatory pigment levels, chlorophyll ratio, average nucleic acid levels and plant height. However, the dry substance mass accumulation decreased. At higher level of magnetic fields ranging between 100 and 250 mT all measured growth and biochemical parameters were inhibited. Moreover, MF has been shown to increase seed yield in pea (*Pisum sativum* L.) through a higher number of pods per plant and lower losses of plants per area unit during vegetative growth (Podlesny et al. 2005).

This positive influence of MF on date palm growth parameters may be caused by higher ion mobility due to water ionization and absorption which enhanced the fresh weight. Another possible explanation is that free radicals enhanced by MF treatments (Scaiano et al. 1994) play an important role in electron transfer and in the kinetics of the chemical reactions and absorbed energy (Commoner et al. 1956); this energy is transformed into chemical energy and accelerates the vital processes. In summation, magnetic treatments exerted a positive influence on date palm fresh weight and water content.

15.4 *In Vitro* Responses to Magnetic Field

Magnetic field application in *in vitro* culture of date palm has not been demonstrated yet. In other plant species, a number of studies suggest that the exposure to MFs may results in enhanced *in vitro* responses and improve plant regeneration systems. A study by Celik et al. (2008) on soybean revealed that MF has increased regeneration percentage, chlorophyll a, chlorophyll b and total chlorophyll contents. Plant regeneration and growth from shoot tips exposed to short periods of exposure to MF were positively affected.

In other studies by Atak et al. (2003, 2007), soybean shoot-tip explants exposed to a MF exhibited higher regeneration percentages as compared to the non-treated control. While the regeneration percentage was 28.57% in the control, explants subjected to MF for 2.2 and 19.8 s resulted in 94.33% and 78.18% regeneration, respectively. Root formation occurred after only 9 days in explants exposed to MF as compared to 11–28 day in the control group. While root formation was 4.76% in the control group on the 28th day, this rate increased to 47.17% and 54.54% in explants that passed through the MF with a period of 2.2 and 19.8 s, respectively.

Moreover Atak et al. (2003, 2007) have shown that the total RNA concentration of regenerated plantlets from explants exposed to MF at 2.9–4.6 mT at 2.2 s significantly increased in RNA concentration. This could be due to the increase in expression of enzymes that play a role in shoot formation, chlorophyll biosynthesis and peroxidase biosynthesis. Additionally, the shoot numbers of regenerated explants exposed to MF and total RNA concentration of seedlings that regenerated from explants were also higher than the control.

In vitro cultures of node explants of *Paulownia tomentosa* and *P. fortunei* were exposed to a magnetic flow density of 2.9–4.8 mT and 1 m s^{-1} flow rate for a period of 0, 2.2, 6.6 and 19.8 s. The MF increased the regeneration capability and shortened the regeneration time. Positive effects of MF on plant fresh weight, length, number of leaves and chlorophyll content were observed. It was found that this effect varied with exposure time. When the cultures were exposed to a MF with strength of 2.9–4.8 mT for 19.8 s, the growth of regenerated plants surpassed the control plants (Yaycili and Alikamanoglu 2005).

In a relevant report, Alikamanoglu et al. (2007) studied the effect in tissue culture of MF in *Paulownia tomentosa*. Their results showed that MF increased percentage and accelerated plant regeneration. A positive effect was noted on plant fresh weight, leaf number and chlorophyll amount. In the cultures irradiated with 10 Gy radiations, the regeneration percentage was 83.3% and this percentage was increased from 83.3% to 87.5% when applied together with MF. When the radiation dose increased to 25 Gy, the regeneration percentage has decreased according to control. But regeneration percentage has increased from 79.2% to 83.3% by magnetic field application. The study also tested the effect of 2.9–4.8 mT for 19.8 s MF combined with 10 and 25 Gy gamma radiation to show that MF can repair the damage induced by exposure to gamma rays 10 Gy. However, the repair effect of MF was less pronounced when a higher dose of gamma radiation, 25 Gy, was employed suggesting the occurrence of irreversible damage at high intensities.

15.5 Conclusion and Prospective

Magnetic fields elicit different degrees of alterations in biochemical processes depending on exposure intensity, which is a function of radiation level and duration of exposure. Ultimately, these changes influence plant growth. Chemical analysis has shown that DNA and proline content decreased, whereas an increase was noted in photosynthetic pigments and the content of some ions. These biochemical changes influence growth parameters including fresh weight, dry weight and water content of both shoot and root systems of date palm seedlings.

Radical ions induced by MF are thought to interfere with DNA synthesis and consume proline in stress reactions. While the increased in growth parameter and photosynthetic pigments could be explained by the ability of seedlings to recover through other mechanisms. Photosynthetic pigments, especially carotene, could react as a radical scavenger and protect a plant from potential damages. In the same way calcium participates in many reactions from growth organization to radical ions scavenging and stress reactions. This may explain the increase in photosynthetic pigments and calcium concentrations in response to MF. The increased water content could be explained by the accumulation of ions induced by differences in electric potential across cell membrane thus increasing osmotic potential leading to increased water content.

Low intensities of MFs were found to elicit positive effects on date palm seeds and seedlings. The promoting effect of MF on plant growth may be associated with changes in endogenous production of growth hormones e.g. cytokinins. This possibility requires research initiatives for clarification. The use of MF is a promising technique for various agricultural applications including plant genetic improvement. Research needed in this area in relation to date palm is examining the potential of MF to induce useful mutations in the presence of other mutagens. The potential application of MF to eliminate *in vitro* culture contamination as well as field pathogens is another area that merits investigation.

Studies involving MF in date palm *in vitro* cultures have not been reported yet. Based on the information accumulated thus far about other plant species, research is highly encouraged to characterize the behavior of date palm *in vitro* cultures in response to MF. An area of needed investigation is examining the potential of low doses of MF to accelerate and enhance plant regeneration.

References

Aarholt E, Flinn EA, Smith CW (1981) Effect of low frequency magnetic fields on bacterial growth rate. Phys Med Biol 26:613–621

Aladjadjiyan A (2002) Study of the influence of magnetic field on some biological characters of *Zea mays*. J Cent Eur Agric 4:90–94

Aladjadjiyan A (2007) The use of physical methods for plant growing stimulation in Bulgaria. J Cent Eur Agric 8:369–380

Alexander MP, Doijode SD (1995) Electromagnetic field, a novel tool to increase germination and seedling vigour of conserved onion (*Allium cepa* L.) and rice (*Oryza. sativa* L.) seeds with low viability. Plant Genet Res News 104:1–5

Alikamanoglu S, Yaycili O, Atak C et al (2007) Effect of magnetic field and gamma radiation on *Paulowinia tomentosa* tissue culture. Biotechnol Biotechnol Equip 1:49–53

Al-Khayri JM (2002) Growth, proline accumulation, and ion content in NaCl-stressed callus cultures of date palm (*Phoenix dactylifera* L.). In Vitro Cell Dev Biol Plant 38:79–82

Al-Khayri JM (2007) Micropropagation of date palm *Phoenix dactylifera* L. In: Jain SM, Haggman H (eds.) Protocols for micropropagation of woody trees and fruits. Springer, Berlin, pp 509–526

Al-Khayri JM, Al-Bahrany AM (2004) Growth, water content, and proline accumulation in drought-stressed callus of date palm. Biol Plant 48:105–108

Al-Yahya A (1995) Biotechnology and date palm development. Nakhlatec date palm consultative group, Wye College, University of London. http://aggie-horticulture.tamu.edu/tisscult/biotech/datepalm.html

Antov Y, Barbul A, Mantsur H et al (2005) Electroendocytosis exposure of cells to pulsed low electric fields enhances adsorption and uptake of macromolecules. Biophys J 88:2206–2223

Astier PJ, Veesler S, Boistelle R (1998) Protein crystals orientation in a magnetic field. Acta Crystallogr D 54:703–706

Atak C, Danilov V, Yurttafl B et al (2000) Effect of magnetic field on *Paulownia* seeds. Com JINR Dubna 1–14

Atak C, Emiroglu O, Alikamanoglu S et al (2003) Stimulation of regeneration by magnetic field in soybean (*Glycine max* L. Merrill) tissue cultures. J Cell Mol Biol 2:113–119

Atak C, Celik O, Olgun A et al (2007) Effect of magnetic field on peroxidase activities of soybean tissue culture. Biotechnol Biotechnol Equip 21:166–171

Ayrapetyan SN, Grigorian KV, Avanesian AS et al (1994) Magnetic fields alter electrical properties of solutions and their physiological effects. Bioelectromagnetics 15:133–142

Baran B, Berezyuk O, Golonzhka V (2006) Water systems after magnetic field action. Environ Res Eng Manag 38:19–23

Belyavskaya NA (2001) Ultra structure and calcium balance in meristem cells of pea roots exposed to extremely low magnetic fields. Adv Space Res 28:645–650

Belyavskaya NA (2004) Biological effects due to weak magnetic field on plants. Adv Space Res 34:1566–1574

Belyavskaya NA, Fomicheva VM, Govorun RD et al (1992) Structural-functional organisation of the meristem cells of pea, lentil and flax roots in conditions of screening the geomagnetic field. Biophys Acta 37:657–666

Blank M, Goodman R (1997) Do electromagnetic fields interact directly with DNA? Bioelectromagnetics 18:111–115

Boe AA, Salunkhe DK (1963) Effect of magnetic fields on tomatoes ripening. Nature 199:91–92

Campbell GS (1977) An introduction to environmental biophysics. Springer, New York

Carbonell MV, Martinez E, Amaya JM (2000) Stimulation of germination in rice (*Oryza sativa* L.) by a static magnetic field. Electro Magn Biol 19:121–128

Celestino C, Picazo ML, Toribio M (2000) Influence of chronic exposure to an electromagnetic field on germination and early growth of *Quercus suber* seeds preliminary study. Electro Magn Biol 19:115–120

Celik O, Atak C, Rzakulieva A (2008) Stimulation of rapid regeneration by a magnetic field in *Paulownia* node cultures. J Cent Eur Agric 9:297–304

Commoner B, Heise JJ, Townsend J (1956) Light-induced paramagnetism in chloroplasts. Proc Natl Acad Sci USA 42:710–714

De Souza A, Garcí D, Sueiro L et al (2006) Pre-sowing magnetic treatments of tomato seeds increase the growth and yield of plants. Bioelectromagnetics 27:247–257

Dekker C, Ratner M (2001) Electronic properties of DNA. Phys World 14:29–33

Dhawi F, Al-Khayri JM (2008a) Proline accumulation in response to magnetic fields in date palm (*Phoenix dactylifera* L.). Open Agric J 2:80–83

Dhawi F, Al-Khayri JM (2008b) Magnetic fields induce changes in photosynthetic pigments content in date palm (*Phoenix dactylifera* L.) seedlings. Open Agric J 2:121–125

Dhawi F, Al-Khayri JM (2009a) Magnetic fields-induced modification of DNA content in date palm (*Phoenix dactylifera* L.). J Agric Sci Technol 2:6–9

Dhawi F, Al-Khayri JM (2009b) The effect of magnetic resonance imaging on date palm (*Phoenix dactylifera* L.) elemental composition. Commun Biometeorol Crop Sci 4:11–18

Dhawi F, Al-Khayri JM (2009c) Magnetic field increase weight and water content in date palm (*Phoenix dactylifera* L.). J Agric Sci Technol 2:23–29

Dhawi F, Al-Khayri JM, Essam H (2009) Static magnetic field influence on elements composition in date palm (*Phoenix dactylifera* L.). Res J Agric Biol Sci 5:161–166

Duarte Diaz CE, Riquenes JA, Sotolongo B et al (1997) Effects of magnetic treatment of irrigation water on the tomato crop. Hortic Abst 69:494

Ejraei A (2007) Effect of magnetic field on germination and growth of date palm. Fourth Symposium on Date Palm in Saudi Arabia, King Faisal University, Al-Hassa, 5–8 May. Abstracts. p 231

Esitken A, Turan M (2003) Alternating magnetic field effects on yield and plant nutrient element composition of strawberry (*Fragaria x ananassa* cv. Camarosa). Acta Agric Scand B Soil Plant Sci 54:135–139

Evans DA, Sharp WR (1986) Applications of somaclonal variation. Biotechnology 4:528–532

Fischer G, Tausz M, Köck M, Grill D (2004) Effects of weak 16 ⅔Hz magnetic fields on growth parameters of young sunflower and wheat seedlings. Bioelectromagnetics 25:638–641

Flórez M, Carbonell MV, Martinez E (2005) Exposure of maize seeds to stationary magnetic fields: effects on germination and early growth. Ingen Agrón 59:68–75

Frankel RB, Liburdy RP (1996) Biological effects of static magnetic fields. In: Polk C, Postow E (eds.) Handbook of biological effects of electromagnetic fields, 2nd edn. CRC Press, Boca Raton, pp 149–183

Ghanati F, Abdolmaleki P, Vaezzadeh M et al (2007) Application of magnetic field and iron in order to change medicinal products of *Ocimum basilicum*. Environ 27:429–434

Goodman R, Blank M (1999) Electromagnetic fields may act directly on DNA. J Cell Biochem 75:369–374

Goodman R, Blank M (2002) Insights in to electromagnetic interaction mechanisms. J Cell Physiol 192:16–22

Goodman EM, Greenebaum B, Marron MT (1995) Effects of electromagnetic fields on molecules and cells. Int Rev Cytol 158:279–338

Graziana A, Ranjeva R, Teissie J (1990) External electric fields stimulate the electrogenic calcium/sodium exchange in plant protoplast. Biochemistry 29:8313–8318

Grime JP, Mowforth MA (1982) Variation in genome size – an ecological interpretation. Nature 299:151–153

Hong FT (1995) Magnetic field effects on biomolecules, cells, and living organisms. Biosystems 36:187–229

Ikehata M, Koana T, Suzuki Y et al (1999) Mutagenicity and co-mutagenicity of static magnetic fields detected by bacterial mutation assay. Mutat Res 427:147–156

Ikehata M, Yoshie S, Hirota N et al (2009) Effects of static magnetic field on mutagenesis in *in vitro*. J Phys 156:12–15
Jain SM (2001) Tissue culture-derived variation in crop improvement. Euphy 118:153–166
Jain SM (2007) Recent advances in date palm tissue culture and mutagenesis. Acta Hortic 736:205–211
Jones RR (2000) Modifying atomic architecture. Sci Spectra 22:52–59
Karamanos AJ (1995) The involvement of proline and some metabolites in water stress and their importance as drought resistance indicators. Bulg J Plant Physiol 21:98–110
Kiranmai V (1994) Induction of mutations by magnetic field for the improvement of sunflower. J Appl Phys 75:71–81
Koana T, Okada MO, Ikehata M et al (1997) Increase in the mitotic recombination frequency in *Drosophila melanogaster* by magnetic field exposure and its suppression by vitamin E supplement. Mutat Res 37:55–60
Križaj D, Valenči V (1989) The effect of ELF magnetic fields and temperature on differential plant growth. Electromagn Biol Med 8:159–165
Kurinobu S, Okazaki Y (1995) Dielectric constant and conductivity of one seed in the germination process. Ann Conf Rec IEEE/IAS 1329–1334
Lipiec J, Janas P, Barabasz W et al (2005) Effects of oscillating magnetic field pulses on selected oat sprouts used for food purposes. Acta Agrophys 5:357–365
Marschner H (1995) Mineral nutrition of higher plants, 2nd edn. Academic, London
Matysik J, Alia PSP, Bhalu B et al (2002) Molecular mechanisms of quenching of reactive oxygen species by proline under stress in plants. Curr Sci 82:525–532
Miyakoshi J, Kitagawa K, Takebe H (1997) Mutation induction by high-density, 50 Hz magnetic fields in human MeWo cells exposed in the DNA synthesis phase. Int J Radiat Biol 71:75–79
Monselise E, Parola AH, Kost D (2003) Low-frequency electromagnetic fields induce a stress effect upon higher plants, as evident by the universal stress signal, alanine. Biochem Biophys Res Commun 302:427–434
Moulder JE, Foster KR (1995) Biological effects of power-frequency fields as they relate to carcinogenesis. Proc Soc Exp Biol Med 209:309–324
Moursy HA, Saker MM (1996) Date palm problems and need for biotechnology. Abstracts Fifth International Conference on Desert Development. Texas Tech University
Parola A, Kost K, Katsir G et al (2006) Radical scavengers suppress low frequency EMF enhanced proliferation in cultured cells and stress effects in higher plants. Environ 25:103–111
Paul AL, Daugherty CJ, Bihn EA et al (2001) Transgene expression patterns indicate that spaceflight affects stress signal perception and transduction in *Arabidopsis*. Plant Phys 126:613–621
Paul AL, Schuerger AC, Popp MP et al (2004) Hypobaric biology: *Arabidopsis* gene expression at low atmospheric pressure. Plant Phys 134:215–223
Pauling L (1979) Diamagnetic anisotropy of the peptide group. Biophysics 76:2293–2294
Penuelas J, Llusia J, Martinez B (2004) Diamagnetic susceptibility and root growth responses to magnetic fields in *Lens culinaris*, *Glycine soja*, and *Triticum aestivum*. Electromagn Biol Med 23:97–112
Phirke PS, Patil NN, Umbarkar SP et al (1996) The application of magnetic treatment to seeds: methods and responses. Seed Sci Technol 24:365–373
Pietruszewski S (1999a) Influence of pre-sowing magnetic biostimulation on germination and yield of wheat. Int Agrophys 13:241–244
Pietruszewski S (1999b) Magnetic biostimulation of wheat seeds. Int Agrophys 13:497–501
Pietruszewski S, Wójcik S (2000) Effect of magnetic field on yield and chemical composition of sugar beet roots. Int Agrophys 14:89–92
Pingping Z, Ruochun Y, Zhiyou C et al (2007) Genotoxic effects of superconducting static magnetic fields (SMFs) on wheat (*Triticum aestivum*) pollen mother cells (PMCs). Plasma Sci Technol 9:241–247
Podlesny J, Pietruszewski S, Podlsena A (2005) Influence of magnetic stimulation of seeds on the formation of morphological features and yielding of the pea. Int Agrophys 19:61–68
Racuciu M, Galugaru GH, Creanga D (2006) Static magnetic field influence on some plant growth. Rom J Phys 51:245–251

Racuciu M, Creanga D, Amoraritei C (2007) Biochemical changes induced by low frequency magnetic field exposure of vegetal organisms. Rom J Phys 52:601–606

Racuciu M, Creanga D, Galugaru GH (2008a) The influence of extremely low frequency magnetic field on tree seedlings. Rom J Phys 53:337–342

Racuciu M, Creanga D, Horga I (2008b) Plant growth under static magnetic field influence. Rom J Phys 53:353–359

Rakosy-Tican L, Aurori CM, Morariu VV (2005) Influence of near null magnetic field on *in vitro* growth of potato and wild *Solanum* species. Bioelectromagnetics 7:548–557

Reed DD, Jones EA, Mroz GD et al (1993) Effects of 76 Hz electromagnetic fields on forest ecosystems in northern Michigan: tree growth. Int J Biometeorol 37:229–234

Reina FG, Pascual LA (2001) Influence of a stationary magnetic field on water relations in lettuce seeds, Part I: Theoretical considerations. Bioelectromagnetics 22:589–595

Reina FG, Pascual LA, Fundora IA (2001) Influence of a stationary magnetic field on water relations in lettuce seeds, Part II: Experimental results. Bioelectromagnetics 22:596–602

Rochalska M (2005) Influence of frequent magnetic field on chlorophyll content in leaves of sugar beet plants. Nukleonika 50:25–28

Rybinski W, Patyna H, Przewozny T (1993) Mutagenic effect of laser and chemical mutagens in barley (*Hordeum vulgare* L.). Rom Genet Pol 34:337–343

Sahebjamei H, Abdolmaleki P, Ghanati F (2007) Effects of magnetic field on the antioxidant enzyme activities of suspension-cultured tobacco cells. Bioelectromagnetics 28:42–47

Samy CG (1998) Magnetic seed treatment influence on flowering, siliqua and seed characters of cauliflower. Orissa J Hortic 26:68–69

Sandu DD, Goiceanu C, Ispas A et al (2005) A preliminary study on ultra high frequency electromagnetic fields effect on black locust chlorophylls. Acta Biol Hung 56:109–117

Scaiano JC, Cozens FL, McLean J (1994) Model of the rationalization of magnetic field effects *in vivo*. Application radical pair mechanism biological systems. Photochem Photobiol 59:585–589

Sharaf El-Deen S (2003) Improvement of some characters of edible mushroom with magnetic field. Bull NRC Egypt 28:709–717

Stadtman ER (1993) Oxidation of free amino acids and amino acid residues in proteins by radiolysis catalyzed reactions. Annu Rev Biochem 62:797–821

Stange BC, Rowland RE, Rapley BI et al (2002) ELF Magnetic fields increase amino acid uptake into *Vicia faba* L. roots and alter ion movement across the plasma membrane. Bioelectromagnetics 23:347–354

Stein GS, Lian JB (1992) Regulation of cell cycle and growth control. Bioelectromagn Suppl 1:247–265

Strzalka K, Kostecka-Guga A, Latowski D (2003) Carotenoids and environmental stress in plants: significance of carotenoid-mediated modulation of membrane physical properties. Russ J Plant Phys 50:168–173

Taia KW, Al-Zahrani SH, Kotbi MA (2007) The effect of static magnetic forces on water contents and photosynthetic pigments in sweet basil *Ocimum basilicum* L. (Lamiaceae). Saudi J Biol Sci 14:103–107

Vasilevski G (2003) Perspectives of the application of biophysical methods in sustainable agriculture. Bulg J Plant Physiol Spec Iss 2003:179–186

Vizcaino V (2003) Biological effects of low frequency electromagnetic fields. Radiobiology 3:44–46

Watanabe Y, Nakagawa M, Miyakoshi Y (1997) Enhancement of lipid peroxidation in the liver of mice exposed to magnetic fields. Ind Health 35:285–290

Wojcik S (1995) Effect of the pre-sowing magnetic biostimulation of the buckwheat seeds on the yield and chemical composition of buckwheat grain. Curr Adv Buckw Res 93:667–674

Worczak M, Wadelton K, Davis JC et al (2006) Effects of high magnetic fields on *in vitro* transcription. In: Torbet J, Rivoirard S, Beaugnon E (eds.) Proceedings of the 2nd international workshop on materials analysis and processing in magnetic fields. Grenoble, France, pp 67–70

Yano A, Ohashi Y, Hirasaki T et al (2004) Effects of a 60 Hz magnetic field on photosynthetic CO_2 uptake and early growth of radish seedlings. Bioelectromagnetics 25:572–581

Yaycili O, Alikamanoglu S (2005) The effect of magnetic field on *Paulownia* tissue cultures. Plant Cell Tissue Org Cult 83:109–114

Part III
Germplasm Biodiversity and Conservation

Chapter 16
Date Palm Germplasm

R.R. Krueger

Abstract Date palm (*Phoenix dactylifera* L) belongs to the genus *Phoenix* which is distributed throughout the tropical and subtropical areas of South Asia and Africa. The species are dioecious and the fruits are drupes of varying sizes. *P. dactylifera* apparently originated in the area near the present Iraq and Iran, was domesticated in ancient times and spread into adjacent regions having appropriate climates. Primitive date palms were probably similar to modern date palms but with smaller, less palatable fruits. Genetic diversity of date palms in the center of origin is threatened by such factors as development, desertification, ecosystem change and salinization. Examination of traditional oasis culture of date palms suggests that many of the trees in those oases were *khalts* or mixtures of seedlings. Propagation by offshoots was probably less common but was important in introducing superior varieties from other areas. In general, populations of date palms appear to be mixtures of *khalts* with some elite named varieties present. The introduction of elite varieties, whether from other oases or via tissue culture, may contribute to the genetic erosion of *P. dactylifera* in traditional oases. Farmers should be encouraged to continue cultivating traditional varieties in the oasis cultures, but *ex-situ* collections of date palms should also be established in order to facilitate access and utilization of date palm germplasm. Work is needed to better characterize and evaluate date palm germplasm and to understand the genetic basis for the traits expressed. This is vital to a rational and successful utilization of date palm genetic resources.

Keywords Conservation • Distribution • Genetic diversity • Genetic erosion • Genetic resources • *Phoenix*

R.R. Krueger (✉)
USDA-ARS National Clonal Germplasm Repository for Citrus & Dates
1060 Martin Luther King Blvd, Riverside, CA 92507, USA
e-mail: Robert.Krueger@ars.usda.gov

S.M. Jain et al. (eds.), *Date Palm Biotechnology*,
DOI 10.1007/978-94-007-1318-5_16, © Springer Science+Business Media B.V. 2011

16.1 Introduction

Plant *germplasm* is living tissue from which new plants can be produced. It contains the genetic information that gives plants their individual characteristics and links generations of living plants to one another. The genetic diversity of plants, developed by evolution, hybridization, natural selection, and manipulation by humans, provides the basis for the food production that supports the world's population. Agricultural utilization of many crops usually involves a narrow range of genetic material, making these crops vulnerable to genetic erosion. Genetic diversity in the centers of origin is threatened or severely threatened by habitat loss caused by deforestation, population pressure, fire, hydroelectric development, clearance for agriculture or other development, tourism, etc. (WWF and IUCN 1994–1995). Preservation of the genetic diversity represented in plant ecosystems throughout the world has therefore become a major issue of international concern. The loss of increasingly large numbers of plant species through habitat destruction threatens the availability of a diverse plant germplasm base which will be needed to feed future generations (Brown et al. 1989; Chrispeels and Sadava 1994; Holden and Williams 1984; Holden et al. 1993; National Research Council 1993; Wilkes 1988). Similar losses have in some cases occurred in existing plant collections through inadequate maintenance.

Ideally, genetic resources should be conserved *in situ*. However, the factors mentioned above, especially habitat loss, make maintenance of genetic resources *in situ* somewhat precarious. Consequently, *ex situ* conservation is often necessary to salvage genetic resources. Genetic materials may be lost through disease, weather, etc., and so *ex situ* collections should be maintained in many cases even when there is not an immediate threat of habitat loss. *Ex situ* collections also make germplasm more accessible for researchers. *Ex situ* collections are necessary for characterization and evaluation of germplasm, and in many cases for its utilization in breeding programs and other research activities. Maintenance of germplasm in a disease-free state is also desirable, and this is often possible only in *ex situ* collections.

Assessment of the genetic vulnerability of any crop requires knowledge of the extent and distribution of genetic diversity. This is acquired by systematic sampling and mapping of the flora of the geographical areas in which the species in question are found, as well as an assessment of *ex situ* collections. Unfortunately, information on natural and semi-natural germplasm is often limited on the international level. This is due to the remoteness of some of the material, a lack of resources devoted to assessing these areas and political considerations. In some cases, information may be available at the local or national level, but not to the international genetic resource conservation community. More detailed characterization and evaluation data are needed to adequately assess the actual amount of genetic diversity present. These data should include both descriptive data and molecular level genetic analysis of germplasm existing both *in situ* and *ex situ*.

16.2 The Genus *Phoenix*

The genus *Phoenix*, which includes the date palm (*Phoenix dactylifera* L.), is the sole member of the tribe *Phoeniceae* of the Monocotyledonous family *Palmae* (Bailey and Bailey 1976; Moore 1963). *Phoenix* species are either single trunked or clumping. Trunks range in size from nearly trunkless to over 30 m. *Phoenix* species may be distinguished from other palms having feather-type leaves by the modification of the basal leaflets into spines, the presence of a terminal leaflet and a central fold or ridge on the leaflets, which cause the leaflets to remain erect at all times. *Phoenix* species are dioecious, with the inflorescences arising among the leaves. The small, pale yellowish flowers are borne singly, with the sepals being united into a cupule. There are three petals. Female flowers have three carpels, only one of which matures; male flowers generally have six stamens. The fruits of *Phoenix* species are drupes of variable size, depending on the species, with a single grooved seed.

The status of genetic vulnerability is not well established for most *Phoenix* species. Although many of the species are cultivated as ornamentals, there are probably few pure *Phoenix* in ornamental plantings due to the readiness of *Phoenix* to hybridize. In addition, selection of plants for ornamental uses would exert selection pressure towards characteristics considered aesthetically pleasing. There is apparently little information available about the status of these species in the wild. Their areas of origin are in many cases threatened with habitat loss or ecosystem change due to some of the factors mentioned above. It is quite probable that there is at least some threat of genetic erosion to some of the species. The status of these species needs more attention devoted towards it. The information available has been summarized by Johnson (1996) and Barrow (1998). There is also a lack of characterization and evaluation data for these species that affects their utility as germplasm accessions.

The taxonomy of the genus *Phoenix* has not been well established in the literature until relatively recently. There has been disagreement between various taxonomic treatments and some confusion about species names and validity. *Phoenix* species hybridize readily, which has lead to the suggestion that the genus *Phoenix* is monotypic (Wrigley 1995). The ready inter-specific hybridization of *Phoenix* species can lead to confusion, especially when several species are present, as may occur in *ex situ* collections. In cultivation, *Phoenix* species are often mislabeled in gardens and in some cases are obvious hybrids or off-types (Hodel 1995).

Although 19 species of *Phoenix* have been named, most taxonomic treatments of *Phoenix* list about 12 species as valid, although not necessarily the same 12 (Table 16.1). It is apparent that *P. acaulis*, *P. canariensis*, *P. dactylifera*, *P. paludosa*, *P. reclinata*, *P. rupicola*, and *P. sylvestris* have been widely accepted as valid species. However, there is some confusion over the other species. Miller et al. (1930), Chevalier (1952), Moore (1963), Munier (1973), and Bailey and Bailey (1976) have presented similar taxonomic treatments, but with a few differences. Chevalier (1952) and Munier (1973) accept *P. atlantica* as valid, although it is probably a hybrid of *P. dactylifera* that does not deserve species status. Chevalier (1952), Moore (1963),

Table 16.1 Species of *Phoenix* recognized by several investigators

	Investigator[a]								
Phoenix species	1	2	3	4	5	6	7	8	9
abyssinica Drude						+		+	
acaulis Roxb	+	+	+	+	+	+	+	+	+
andamanensis S Barrow									+
atlantica A Chev				+			+		
caespitosa Chiov									+
canariensis Hort ex Chab		+	+	+	+	+	+	+	+
dactylifera L	+	+	+	+	+	+	+	+	+
farinifera Roxb	+	+		+		+	+		
hanceana Naudin				+			+		
humilis Royle		+	+	+	+		+		
loueirii Kunth	+				+	+		+	+
ouseleyana Griff	+								
paludosa Roxb	+	+	+	+	+	+	+	+	+
pumila Hort			+						
pusilla Gaertn	+	+	+		+	+		+	+
reclinata Jacq	+	+	+	+	+	+	+	+	+
roebelinii O'Brien			+	+		+	+	+	+
rupicola T Anderson		+	+	+	+	+	+	+	+
spinosa FC Schum	+								
sylvestris (L) Roxb	+	+	+	+	+	+	+	+	+
Theophrastus Greuter									+
zeylanica Trimen			+		+			+	

[a]Reference: Martius (1836–1850); Beccari (1890); Miller et al. (1930); Chevalier (1952); Mowry et al. (1952); Moore (1963); Munier (1973, 1974); Bailey and Bailey (1976); Barrow (1998)

and Munier (1973) use the species *P. farinifera*, while Miller et al. (1930) and Bailey and Bailey (1976) prefer *P. zeylanica*. There is some confusion over *P. farinifera*, *P. pusilla*, and *P. zeylanica*. Moore (1963) and Bailey and Bailey (1976) use *P. loueiri* for the *P. hanceana* and *P. humilis* of other workers. Most of the other specific epithets shown in Table 16.1 are more obscure and lack botanical standing. In addition to these, Moore (1963) and Bailey and Bailey (1976) list a number of specific epithets without standing: (*P. andersonii*, *P. cycadifolia*, *P. formosana*, *P. glauca*, *P. hybrida*, *P. leonensis*, *P. natalensis*, *P. porphyrococca*, and *P. tomentosa*. These early treatments of *Phoenix* taxonomy do not deal with the recently described *P. theophrasti* (Greuter 1967).

As stated by Uhl and Dransfield as recently as 1987, "Despite the economic importance of *Phoenix* and the great ease of recognition of the genus, the species remain poorly known and in much need of a careful revision. The most useful account remains that of Beccari (1890)." The relatively recent monograph of Barrow (1998) has greatly elucidated the relationship between the various reported species. This treatment, as summarized in Table 16.2, has become widely accepted and is the system utilized in the USA by the USDA-ARS National Plant Germplasm System (NPGS) (http://www.ars-grin.gov/cgi-bin/npgs/html/tax_search.pl). Barrow (1998) noted that prior to her

Table 16.2 The genus *Phoenix*: a summary (After Barrow 1998)

Phoenix species	Common names	Distribution	Notes	Synonyms
acaulis	–	N India, Burma	Stemless; fruit edible; sometimes confused with *P loureiri*	–
andamanensis	–	Bay of Bengal	Single trunk; semi-dwarf; species status somewhat questionable	–
caespitosa	–	Somalia, Arabian peninsula	Habitat: wadis; stemless; fruit edible; species status somewhat questionable	*arabica*
canariensis	Canary Islands date palm	Canary islands	Wide range of habitats within distribution; single trunk; fruit edible; widely cultivated as ornamental	*cycadiflora, jubae, tenuis*
dactylifera	Date palm	Middle East to W India, N Africa	Habitat: wadis, oases; widely cultivated in suitable climates for fruit; many other plant parts utilized	*atlantica*
loureiri	–	India, China, Indochina, Philippines	Dwarf; fruit edible; other plant parts utilized; taxonomy somewhat confused: two varieties (*Loureiri, humilis*)	*formosana, hanceana, humilis, ousleyana*
paludosa	–	Bay of Bengal, Indochina, Malaysia	Habitat mangrove swamps and estuaries; semi-dwarf	*siamensis*
pusilla	–	S India, Sri Lanka	Fruits edible; other plant parts utilized	*farinifera, zeylanica*
reclinata	Senegal date palm	Tropical and subtropical Africa, Madagascar, Comoro islands	Habitat and morphology variable; fruit edible; other plant parts utilized; widely cultivated as ornamental	*abyssinica, baoulensis, comorensis, madagascariensis, senegalensis, spinosa, zanzibarensis,* etc.
roebelenii	Pygmy date palm	Laos, Vietnam, S China	Rheophytic; dwarf; widely cultivated as ornamental	–
rupicola	Cliff date palm	N India	Single trunk; semi-dwarf; fruits eaten by animals but not humans	–
sylvestris	Indian date palm	India, Pakistan	Wide range of habitats; utilized for sugar, fruit	–
theophrasti	Cretan date palm	Crete, Turkey	Habitat: coastal areas; species status questionable	–

monograph, Beccari (1890) was the most recent monograph dealing with *Phoenix*, recognizing ten species, and that since that publication three additional species had been published: *P. caespitosa* Chiov. from Somalia, *P. atlantica* A. Chev. from the Atlantic Islands, and *P. theophrasti* Greuter from Crete. The treatment of Moore (1963) recognized 12 species. Barrow (1998) pulled Moore's (1963) *P. abyssinica* into *P. reclinata* and his *P. farinifera* into *P. pusilla*, recognized the recently published *P. caespitosa* and *P. theophrasti*, and recognized two varieties of *P. loureiri*, included a new species *P. andamanensis* from the Andaman Islands, and considered *P. atlantica* to be incompletely known. This resulted in 13 species of *Phoenix* being recognized, with 33 synonyms and 57 nomina nuda (invalid or unpublished names).

Phoenix is widespread in the tropical and subtropical areas of southern Asia and Africa (Fig. 16.1). It does not occur naturally in the New World, but most species have been introduced and are cultivated. Uses of *Phoenix* species include food for man and animals, fiber, wood, fuel and as handsome ornamental plants. *Phoenix dactylifera*, the date palm, is the type species for this genus and will be discussed in the following section. Immediately below are short descriptions and comments on the conservation status of the various species recognized by Barrow (1998).

Phoenix canariensis, the Canary Islands date palm, is native to the Canary Islands, and is adapted to more moderate climatic conditions and cooler temperatures than some of the other *Phoenix* species. It has a stout single trunk and can reach heights of 20 m. It is widely planted as an ornamental in the US and the Mediterranean area. Barrow (1998) notes that although *P. canariensis* is widely planted, there is a threat to its genetic integrity due to its ready hybridization with other *Phoenix* species. This is true in its native area of the Canary Islands, to which various *Phoenix* species have been introduced. Due to this threat of genetic erosion, importation of *Phoenix* species to the Canary Islands has been banned. See Morici (1998) for additional information.

Phoenix sylvestris, the Indian wild date palm, is widespread in India. It may also reach 20 m in height, but its single trunk is not as stout as that of *P canariensis*. Its fruits are eaten and its sap used to make sugar and other products and its leaves are used in basketry. *P. sylvestris* is not considered threatened by Johnson (1996) or Barrow (1998).

Phoenix rupicola, the cliff date palm, has a thin trunk to 7 m in height. It is native to northern India. It is considered one of the most attractive for cultivation. *Phoenix pusilla* (syn. *P. farinifera*, *P. zeylanica*), is also from southern India and Sri Lanka, and is about the same size and general appearance as *P. rupicola* but generally shorter in stature. The fruit is edible and the trunk has a floury pith which is also said to be edible. *Phoenix rupicola* is stated by Barrow (1998) to have an uncertain conservation status. The habitat range is restricted but the remoteness of the populations may be beneficial to its survival. Barrow (1998) does not consider *P. pusilla* to be threatened; however Johnson (1996) considers its conservation status to be unclear.

Phoenix reclinata, the Senegal date palm, in which Barrow (1998) included *P. abyssinica*, is a somewhat variable species from the tropical parts of Africa. It has thin, clustering trunks, and may reach a height of 10 m. It is widely planted as an ornamental. *Phoenix paludosa*, the mangrove date palm, is similar in appearance to *P. reclinata*, and is native to swampy areas in southeast Asia. It is also grown as an

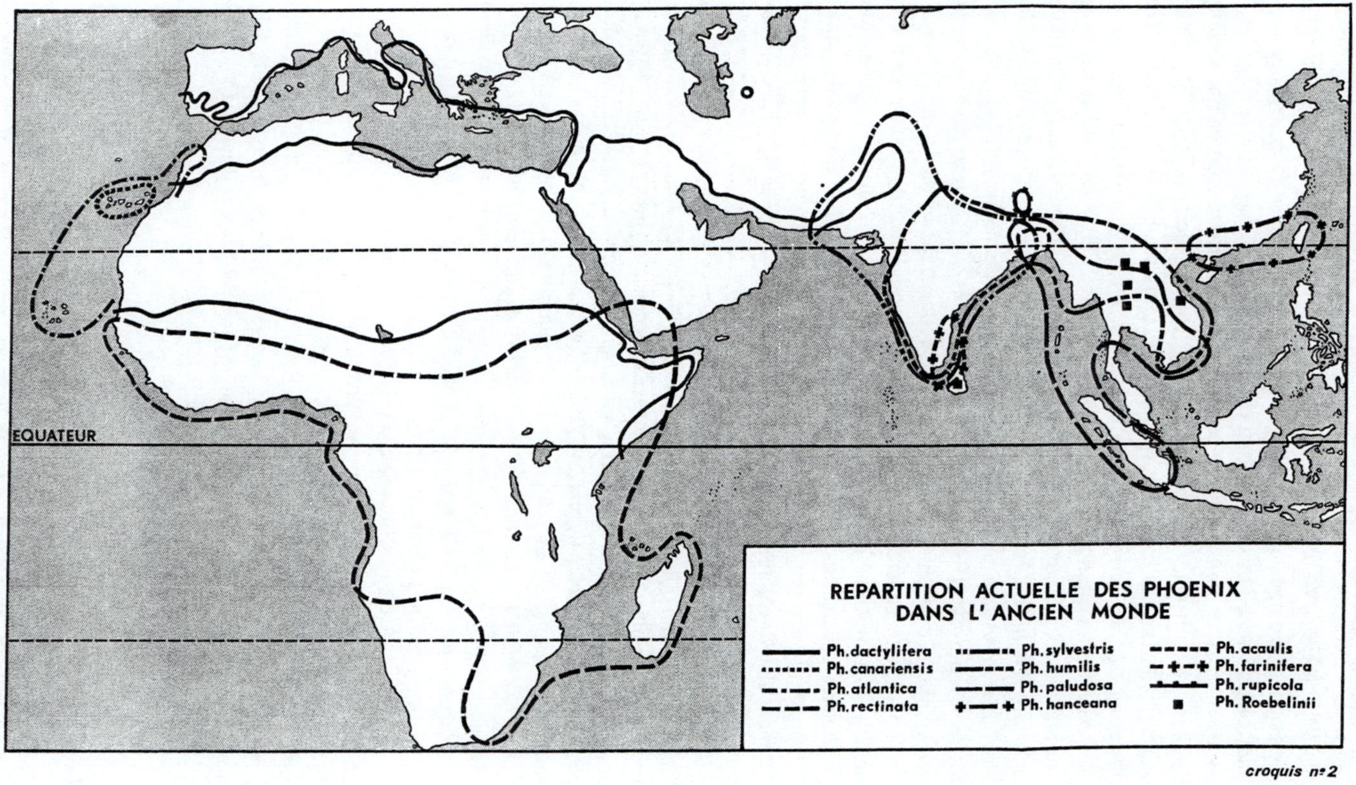

Fig. 16.1 Distribution of *Phoenix* spp. throughout the world (Munier 1973)

ornamental, and the two species are often confused in cultivation. Barrow (1998) does not regard *P. reclinata* to be threatened, although Johnson (1996) regards its synonym *P. abyssinica* to be of unknown conservation status. *Phoenix paludosa* is not considered to be a threatened species, although populations in specific geographic areas may be threatened by urbanization (Barrow 1998).

Phoenix acaulis, *P. loureiri* and *P. roebelenii* are short statured palms, usually less than 2 m in height. *Phoenix acaulis*, which is from northern India and Myanmar, is clumping, whereas the latter two species apparently have both solitary trunked and clustering trunked forms. *Phoenix roebelenii*, the pygmy date palm, is from southeast Asia and is often grown as an elegant ornamental. *Phoenix loueiri*, from northern India and southern China, is a poorly understood species that is sometimes confused with *P. acaulis*, which certain stemless individuals greatly resemble. Older references to *P. hanceana* and *P. humilis* should be understood to refer to *P. loureiri* (Barrow 1998). Barrow (1998) recognizes two varieties of *P. loureiri* based upon the presence or absence of tannin-filled cells in specific areas of the leaves. *Phoenix roebelenii* may be considered threatened upon additional investigation, as its use as an ornamental has resulted in the removal of plants from its native habitat (Barrow 1998). *Phoenix acaulis* has an unclear conservation status that also needs further study. Apparently its range is becoming more restricted due to development. Development also threatens specific populations of *P. loureiri*, although the species as a whole is not considered threatened (Barrow 1998).

A more recently-described species, *Phoenix theophrasti* (Greuter 1967), is not well characterized and not well established as a valid species. It appears to be native to the eastern Mediterranean area (Cyprus, Greece and Turkey). It is similar in appearance to *P. dactylifera*, although apparently somewhat smaller, and before being validated as a species by Barrow (1998) was sometimes thought to be simply a population of distinct *P. dactylifera* native or naturalized in the region. A similar situation exists with *P. atlantica* (Munier 1974), which Barrow (1998) does not accord species status and which may represent only feral populations of *P. dactylifera*. *Phoenix theophrasti* is considered a threatened species due to its restricted range and population pressures (Johnson 1996).

Barrow (1998) also published a new species, *Phoenix andamanensis*. This species resembles and is apparently closely related to *P. rupicola*, being distinguished mostly in seed morphology. Apparently *P. andamanensis* is rare even in the Andaman Islands and might be considered threatened. Barrow (1998) also recognized a previously published but generally unknown species, *P. caespitosa*, from Somalia. This is a stemless palm with sweet fruit that also apparently has a restricted geographic range and thus may be considered potentially threatened.

16.3 The Date Palm *(Phoenix dactylifera)*

The date palm proper (*Phoenix dactylifera* L.) is the tallest *Phoenix* species, reaching heights of more than 30 m. It has clustering trunks smaller in diameter than *P canariensis*, but larger than other *Phoenix* species. In cultivation, it usually appears as a single trunked tree, as the offshoots are usually removed for propagative purposes. The fruit is the largest of any *Phoenix* species, reaching up to 100 mm × 40 mm in size. The fruits are very tasty and nutritious, and are the reason that the date palm is widely cultivated in areas with suitable climates.

Date palms evolved in a unique manner (Wrigley 1995). They have characteristics that adapt them to varied conditions, but differ from many other plants that are found in these conditions. The date palm grows well in sand, but it is not arenaceous. It has air spaces in its roots and may grow well where soil water is close to the surface, but it is not aquatic. It grows well in saline conditions, but it is not a true halophyte and does better in higher quality soil and water. Its leaves are adapted to hot, dry conditions, but it is not a xerophyte and requires abundant water.

The date palm is adapted to areas with long, very hot summers with little rain and low humidity, but with abundant underground water. This is expressed by the saying that the date palm 'must have its feet in running water and its head in the fire of the sky.' These conditions are found in oases and river valleys in the arid sub-tropical deserts of the Middle East. Although there is some question as to where the date palm originated, it most probably arose in the area of northeastern Africa (the Nile delta), northern Arabia, Iraq, and western Iran (Fig. 16.2). This includes the area known as the Fertile Crescent (ancient Mesopotamia), where agriculture in the Old World is thought to have arisen. Indeed, the date palm has been cultivated in this area from ancient times, possibly being one of the first crops domesticated.

The specific geographic area in which *Phoenix dactylifera* originated within this broad area is somewhat unclear. Barrow (1998) has summarized current thinking in this area and further suggested that "a multi-disciplinary approach is advisable, combing evidence from botanical and ecological data with historical, cultural, and archaeological information." Barrow (1998) notes the existence of feral date palms throughout the range of current date cultivation "notably in the southern, warm and dry Near East as well as the northeastern Sahara and north Arabian deserts." Zohary and Hopf (2000) claim that true wild dates have been identified in these geographic areas. The 'wild dates' are said to be similar to cultivated *P. dactylifera* but producing smaller, unpalatable fruits and reproducing sexually (as opposed to the vegetative reproduction typically found in cultivated date palms). The wild dates are contained within *P. dactylifera* and occur over most of its range of cultivation, but "in some areas in the Near East and very probably also in the north-eastern Sahara, in Arabia, and in Baluchistan, dates are genuinely wild and occupy primary niches" (Zohary and Hopf 2000). This viewpoint is supported by the ecological observations of Beccari (1890), which indicated that the conditions of a hot, dry land with wet, saline soils to which many date palm genotypes are adapted are found precisely in the lands adjoining the Arabian Gulf. Further support for this area of origin is found in paleobotanical data and archeological, historical, and cultural information (Barrow 1998; Nixon 1951; Zohary and Hopf 2000). However, there is some uncertainty in this area since the hot, dry climate in this region reduces the amount of vegetative plant material that is intact or identifiable. Barrow (1998) notes that this identification of the Persian Gulf as the natural distribution of *P. dactylifera* places it at the western edge of the natural distribution of *P. sylvestris*. These two species

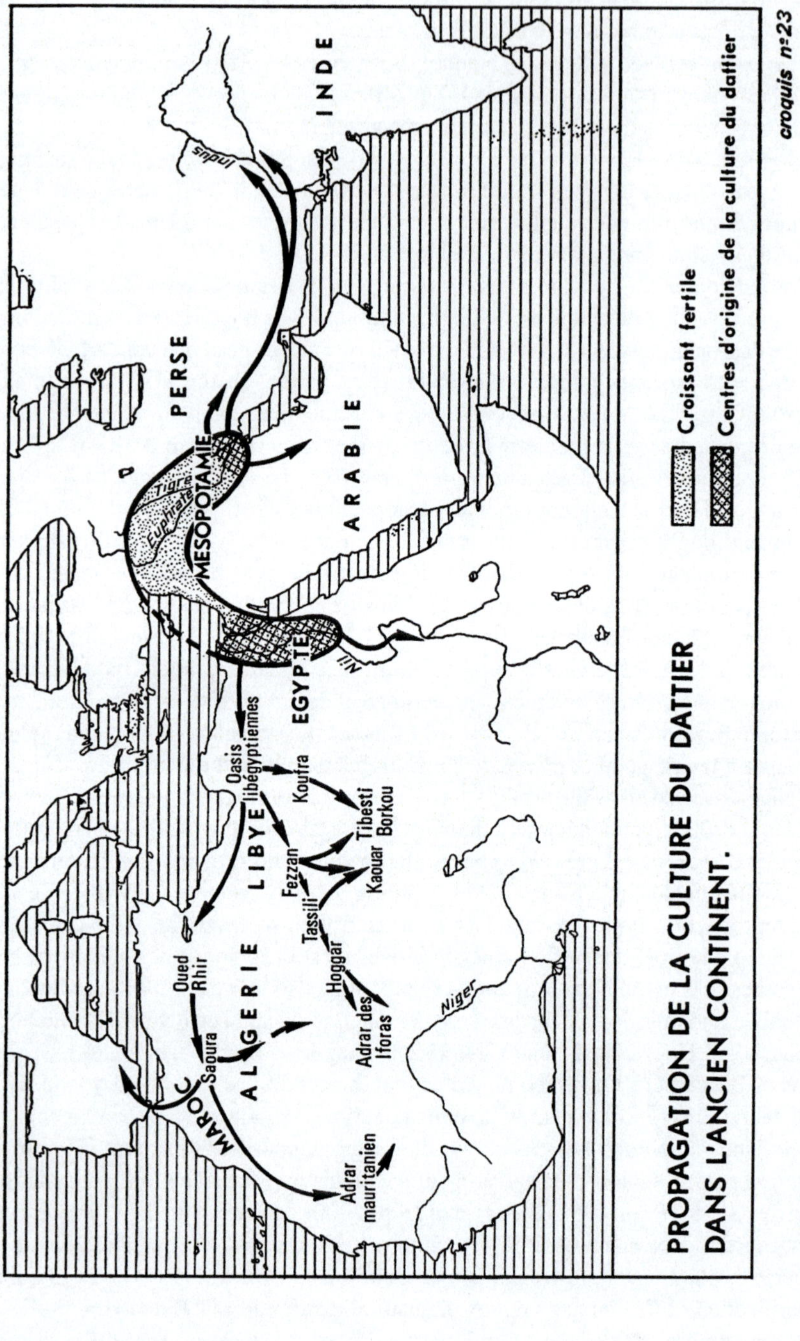

Fig. 16.2 Spread of the date palm throughout the Old World (Munier 1973)

are closely related morphologically and molecularly but their ultimate historical relationship is not clear at this time (Barrow 1998).

Regarding the issue of whether wild date palms might exist, Barrow (1998) stated "A paucity of morphological characters makes differentiation of wild from feral plants difficult. Molecular data from a wide range of domesticated, wild and feral date palms may offer new and useful information. Nuclear and chloroplast DNA regions sequenced in the current study show insufficient variation to be informative at the varietal level within *P. dactylifera*." The cited study was based upon sequencing the 5S intergeneric spacer of the 5S DNA unit. The goal of that research was elucidation of relationships within *Phoenix* rather than within *P. dactylifera*. Use of molecular markers to study feral or wild populations of date palms and use this information to conjecture the origin would be a daunting task that will probably not soon be undertaken. Some indirect evidence for the Middle East as the center of origin comes from AFLP studies made on over 40 date palm accessions collected at different locations in Egypt by El-Assar et al. (2005). They found that almost all the Egyptian collections grouped with varieties from Iraq. This supports but certainly does not prove the concept that the date palm originated near the Arab Gulf and then diffused across North Africa.

The date palm has been cultivated since antiquity, but its wild progenitors (which may or may not still exist) were undoubtedly used by humans long before actual cultivation began. A date palm oasis must have been a welcome sight to those crossing the desert. Here were water, shade, and fresh and dried fruits high in carbohydrates. The dried fruits were easily stored and transported after leaving the oasis. The date palm also supplied building material, fiber, fuel, animal feed, honey (syrup), and wine.

Exploitation of the date palm by man probably began as simple gathering of the fruits and other usable parts of the tree (Nixon 1951; Zohary and Hopf 2000). At this point, there must have been considerable genetic diversity in date palms. Later, trees were probably planted along the network of irrigation canals in ancient Mesopotamia. Selection of trees with superior characteristics probably also originated in ancient times, along with clonal propagation by offshoot planting. Later innovations would include hand pollination, manuring, shading, etc. Spread of date palm germplasm was probably originally as seed, which is much easier to transport than offshoots. After seedling populations were established in other areas, selections could be made and perpetuated by vegetative propagation. This gave rise to the many local varieties that are found in the Middle East. Propagation by offshoots was probably dependent upon the domestication of the camel, due to the weight of the offshoots and the consequent difficulties in transporting them for long distances.

The ancient time of the domestication of the date palm is well documented (Barrow 1998; Munier 1973, 1981a, b, c, d; Nixon 1951; Oudejans 1969; Popenoe 1913; Sauer 1993; Wrigley 1995; Zohary and Hopf 2000). The earliest records of date palm cultivation date from about 7,000 BP in Mesopotamia, but it is generally believed that date culture began thousands of years earlier. Date seeds at least 5,000 years old have been found in the storage godowns at Mohenjo Daro, the ancient city along the Indus river in the Sind, and the date palm was used in the construction of the Temple of the Moon God in Ur (Iraq) some 4,000–5,000 years ago. The date palm is shown in the bas-reliefs at Nineveh (Assyrian Empire). By several millennia BP, date palm culture had spread to Palestine, Arabia, Egypt, North Africa and western India (Fig. 16.2). The date had great spiritual and cultural significance to peoples of the region. The date palm and date culture are depicted on ancient Assyrian and Babylonian tablets, including the famous Code of Hammurabi, which contained laws pertaining to date culture and sales. The date palm is also found in old Egyptian, Syrian, Libyan and Palestinian writings.

It is in Arab culture that the date palm achieves its greatest esteem (Nixon 1951; Popenoe 1913). To Arabic peoples, the date palm is a sacred institution that they have been identified since the dawn of history and which was consecrated by Muhammad in both his public and his private life. There are many references to the date palm in pre-Islamic chronicles, but it becomes more prominent from the time of the Prophet. Muhammad lived in a town at the center of date culture. Dates were the prophet's favorite food, and during his times of poverty he had little to eat but dates and water for months at a time. There are 26 references to dates in the Koran, 16 mentioning them as evidence of God's bounty. According to Muslim tradition, a date palm was said to have sheltered Mary when she was pregnant. There are also references to the date palm in other Islamic writings, including references to its medicinal uses.

The date palm is also mentioned a number of times in Jewish and Christian writings. In the book of Psalms it is said that "the righteous shall flourish like a palm tree" and in the Song of Songs a beautiful woman is likened to a palm tree. The prophetess Deborah sat beneath a palm, and date palms adorned Solomon's temple. In fact, the columnar architecture so common in the Mediterranean area is thought to have been inspired by the use of date palms as building material. Palm leaves were used in temple services during the Feast of Booths and carried in victory parades. Jesus was hailed with palm branches when he entered Jerusalem just before the crucifixion.

This long history of exploitation and selection means that possibly there are no examples of wild date palms. There may be a few apparently wild groves still growing around oases, springs, or seepage areas (as per Zohary and Hopf 2000), but most of the trees that currently exist are the end results of an unknown number of acts of selection over hundreds or even thousands of years. This includes trees which are not currently cultivated, and may appear to be growing wild in oases, abandoned gardens, etc. Because of the length of time of domestication, there has undoubtedly been selection pressure put upon the ancestors of these plants at some time in the past. However, probably evolutionary change due to human selection has been relatively low, so there is a certain amount of genetic diversity present in date palms. This is reflected in the many local varieties, which have been selected for their adaptations to local conditions. Characteristics such as offshoot production, tolerance to humidity and fruit characteristics have been documented. Nixon (1950) has summarized some of the pertinent characteristics of some named date varieties.

This state of affairs makes genetic vulnerability of date palms a less-than-clear-cut matter. Genetic diversity in the region in general is threatened by such factors as population pressure, overgrazing, erosion, dam construction, logging, tourism and other development pressures (WWF and IUCN 1994–1995). In the Middle East and North Africa, desertification and soil salinization are important factors threatening the natural habitat of date palms. Salt water intrusion in certain parts of the Arabian Gulf (e.g., UAE) has caused the loss of entire date palm groves. Due to its extensive cultivation, *Phoenix dactylifera* as a species cannot be considered threatened (Barrow 1998). Moreover, truly wild date palm germplasm may not currently exist (Barrow 1998; Johnson 1996, but see Zohary and Hopf 2000), so cannot be lost due to habitat destruction, etc. However, the genetic diversity that does exist in the cultivated date palm can be lost due to these factors if they result in the loss of local varieties having specific genetic constitutions. Continued selection pressures by man; shifts to fewer, more modern varieties; etc. can also reduce genetic diversity in date palms. Probably the greatest threat to genetic diversity in *P. dactylifera* is the shift from traditional varieties to elite varieties. Modern transportation has facilitated the introduction of new varieties to previously isolated oases. It is also possible that the use of tissue culture to generate large amounts of trees for planting may exacerbate genetic erosion of the date palm since tissue culture production focuses on a relatively small set of cvs., particularly 'Medjool and Barhee'.

To better understand the genetic diversity that is present in the remaining traditional oases, various studies have been done. These have included surveys, morphological observations, and molecular analysis. It is apparent that elite varieties have made their way into at least some traditional oases, but a pool of genetic diversity (reflected by the presence of traditional varieties associated with the oasis as well as seedling populations) remains. A representative sampling of this work has been summarized by Krueger (2008), from which the following illustrative passages are taken:

The spread of date palms from their center of origin near present day Iraq and Iran into other areas of the Middle East, Northern Africa, and Eastern South Asia resulted in the establishment of new date palm oases in these areas. These human-created oases probably resulted from the introduction of a relatively small range of genotypes. Subsequent clonal propagation by the planting of offshoots of desirable types resulted in distinctive fruit types being associated with the various oases. Even in the center of origin, selection pressure resulting from clonal propagation of desired types over a long period of time resulted in a certain amount of genetic erosion and the same association of characteristic varieties with specific oases. The relative isolation of the oases in the days before long-distance transportation was easily accomplished further intensified the association of specific varieties with specific geographical areas. Several oases in close proximity to each other would more likely exchange desirable varieties between themselves than they would with more distant oases. This does not mean that seedling date palms were not present in different oases. Even when seedlings were present, they may have originated from a relatively narrow genetic base since the initial establishment of the palms may have been from only a few seeds or offshoots.

As an example of this, consider the case of Egyptian date culture in the early years of the twentieth century as reported by Mason (1927): "The date varieties are similarly localized, the Samany and Zagloul being largely centered around Edku and Rasheid, the Amri along the eastern Delta border, the Amhat and Saidy (Sewi) in upper Giza Province. The Hayany alone of the commercial varieties enjoying a rather wide range..."

Mason (1927) further described other areas of date production in Egypt farther up the Nile Valley as being centered around seedling dates. In some instances, the "...seedlings of a variety...so nearly resemble the parent variety that they are grown and marketed with it..."

Similarly, Brown and Bahgat (1938) note that: "...most of the well-known varieties confine themselves to certain localities...at Rosetta we find the Zaghloul, the Samani, the Hayâni and BintEisha. Damietta is exclusively a Hayâni district. At Sharqiya, the Amri, the Aglani and Hayâni are most prominent. At Giza and Fayiûm the Siwi and Amhat predominate; while Marg is mainly a Hayâni district. Siwa and Baharia Oases are occupied mostly with Siwi. Aswân is devoted to dry varieties, namely, Bartamuda, Barakawi, Gondeila, Gargouda and Dagana."

In addition to modern industrial production traditional oasis date culture continues in Egypt, apparently with few introductions of new varieties. Nabhan (2007) studied the agricultural crop inventories in Siwa during the period 2004–2006 and compared them with information from 1890 and the early twentieth century. Although it was unclear as to whether or not a few varieties had been lost in this period, it is notable that there have apparently been no introductions of new or elite varieties. Similarly, El-Assar et al. (2005) listed the traditional varieties Amhat, Feryhy, Shakngobil, Siwi, and Taktakt as still existing in Siwa oasis. The Siwi accession from Siwa Oasis, along with other Siwi studied, showed a much different AFLP profile than the other 44 samples studied. Most of the 47 accessions studied fell into one cluster. This suggests that these were similar to the ancestral types from the Middle East (some of which were also included in the molecular study) and that Siwi represents a type that is distinctive genetically as well as morphologically.

Similarly, in Morocco Popenoe (1926) and Service de l'Horticulture (1947) describe various varieties of dates being associated with the various locales of their production. Most of the types described were associated with only one oasis. However, a few types (BouFebouss [BouFeggous], BouIttob. Bouskri, and El Mehjoul [Medjool] were encountered in several locations (Service de l'Horticulture 1947). Bendiab et al. (1998) surveyed 11 palm groves in the Tensift, Draa, Todra, and Ziz river valleys. They observed Bou-Feggous and Jihel to be the most widely distributed. Since cultivars could have different names in different locations, Bendiab et al. (1998) used isozymes to largely support the cultivar names used at the various locations. Mejhoul, a 'fine cultivar', was found only at low levels due to its susceptibility to bayoud. Bou-Feggous, which has some resistance to bayoud, represented the high quality date in most locales. Most date groves showed about 67% seedling mixtures (khalts). This supports the concept that some superior varieties (in this case, Bou-Feggous) tend to be moved between different locations more than inferior varieties and are propagated clonally, whereas seedling mixtures are often left to grow and generally provide lower quality dates that remain local. Some of the locations showed some similarity in cultivar and genetic composition, while the date palms populations of Zagora, M'hamid and Rissani represented unique types as characterized by cvs. Bou-SthammiMoire, Bou-Skri and Iklane, and Bou-Feggous, respectively.

In Tunisia, archival materials from the 1930s and 1940s on file at the USDA-ARS National Clonal Germplasm Repository for Citrus and Dates in Riverside,

California, USA, suggest that many oases have named local varieties present but superior varieties such as Deglet Noor (of Algerian origin) and Bou-Feggous (of Algerian or Tunisian origin) are present in a number of the oases listed. More recently, Mohamed and Rachid (2006) examined several Tunisian littoral oases and 15 cultivars. Specific cultivars were found frequently at the different locations, whereas other varieties were rare. In this case, the elite cultivars were lacking (also in accordance with the archival material). Various studies using isozymes (Ould Mohamed Salem et al. 2001), ISSR (Zehdi et al. 2002), and SSR (Zehdi et al. 2004) markers have examined the genetic diversity within Tunisian oases. The overall picture is that Tunisian date palm groves maintain high levels of genetic diversity both at a specific location and over the entire survey of the three main growing areas. This was despite the long history of domestication and selection pressure. Genetic diversity between locations was low and the cultivars grouped in relation to their geographic origin were not well-defined. This argued for the existence of one Tunisian date palm population rather than separate regional populations.

Krueger (2008) concludes this discussion: These modern studies reinforce the concept that oases away from the center of origin for date palms developed by transport of propagative material from existing oases. In the case of elite or superior selections, movement by offshoot resulted in dissemination of genetically identical or nearly-identical varieties to various areas. In the case of less desirable types, seed dissemination resulted in the establishment of more localized varieties. The selection of superior types by humans would be based mainly upon fruit characteristics and this would be the main selection pressure. Some natural selection pressure may have occurred due to resistance or susceptibility to locally-occurring diseases, pests or environmental conditions. Naturally occurring selection would also be applied to the non-elite varieties that originated from seed.

16.4 Date Palm Germplasm Conservation and Utilization

The preceding discussion suggests that the genetic diversity of date palms should be monitored and *ex situ* collections maintained. There are only a few formal date palm genebanks known. Bettencourt et al. (1992) list only 15, the largest of which were found in Algeria, India, Iraq, Nigeria and the United States. Some of the date palm germplasm collections are far removed from the center of origin and maintain only a few accessions (Brazil, Dominican Republic, France, Taiwan), while others have been destroyed by natural events (Miami, Florida, USA). Except possibly for the Nigerian collections, most date palm accessions maintained in genebanks appear to be elite cultivars or breeding lines, so the genetic diversity is probably rather low. There are other germplasm collections that are either new since the compilation of Bettencourt et al. (1992) or perhaps were simply unknown at the time of that compilation. For instance, a small field genebank was established in UAE at the International Center for Biosaline Studies for salt-tolerant varieties (A. Jaradat, 2010, personal communication). Al-Ghamdi (2001) reports a date palm germplasm bank at King Faisal University, Saudi Arabia, having 18 elite varieties produced by tissue culture in various laboratories around the world. The fact that these were produced by tissue culture raises issues as to genetic integrity. A larger gene bank has reportedly been established at the campus of Al Qasim University in central Saudi Arabia (A. Jaradat, 2010, personal communication). Another example is given by a poster presentation (not appearing in the proceedings) at the Third International Date Palm Conference in the United Arab Emirates, February 2006, concerned the Wadi Quriyat Date Palm Research Station in Bahla, Sultante of Oman. Among the activities of this station is germplasm conservation. There is a collection of 167 pistillate varieties (including 11 important cultivars) and 19 staminate varieties. Evaluation data is taken on some of the varieties. There is also the possibility of the development of a germplasm repository of sorts at the Date Palm Development Center in Al-Ain, UAE (A. Zaid, 2006, personal communication). There may be other genebanks for date palms that are not widely known.

The low number of date palm genebanks and the emphasis on maintenance of elite cultivars in those that do exist lead to two complementary conclusions: First, it is vital that such genebanks be established, and second, that efforts be made to support and maintain the existing traditional oases and the genetic diversity that they contain. Although the oases have endured in some cases for many hundreds or possibly thousands of years, pressures on them have increased greatly in modern times and it cannot be assumed that they will continue to exist or to contain genetic diversity. An example is found in Iraq. Iraq, formerly one of the major date producers in the world, has experienced a great reduction in date production in recent decades. An unfortunate series of events (the Iran-Iraq War, the Gulf War, the draining of the southern marshes, a lack of infrastructure development, UN sanctions, an authoritarian regime and current unrest) has resulted in an unknown but possibly significant portion of the date palms being lost (Walsborn 2008; see also <http://www.usaid.gov/iraq/pdf/AYearInIraq_agriculture.pdf> for one of many web pages dealing with these issues). Undoubtedly, some of the lost trees have contained interesting or useful genes as well as producing fruit for local consumption or export. In addition, the current status of the Iraqi date palm genebank, one of the largest cited by Bettencourt et al. (1992) is currently uncertain. A personal communication (R. Walsborn, 2010, personal communication) indicated that there are still some date palms maintained at the location of the Iraqi genebank but details were not available.

Regarding the vital role of traditional oases in the maintenance of date palm genetic diversity: Support for the traditional oases must balance the need for preservation of their positive aspects with improvement of some of their less desirable characteristics. Traditional oases are often in remote locations, and their inhabitants are often poor and without infrastructure support. A wholesale modernization of the oases, with a shift perhaps to industrial agricultural production and a more modern lifestyle would lose many of the positive aspects of the traditional oases, including the genetic resources of the date palms. At the same time, letting the oases continue on with no changes would also threaten these positive aspects and the genetic resources. A continued impoverished population would not be able to properly care for date palm genetic resources and indeed if the world-wide trend of depopulation of rural areas continues, the oases might be abandoned. This said, coming up with an appropriate plan of action in this area is beyond the capabilities of the writer. In any case, traditional farmers should be encouraged to replant date palms with locally produced offshoots or seedlings. Replacement of old or dead date palms with elite foreign varieties will hasten the genetic erosion of this important world crop. In some cases, a limited number of elite line trees might be planted to supplement the traditional types, as indeed has occurred throughout history.

What is needed is a better understanding of the population structure of date palms found in traditional oases and more extensive evaluation of traits not associated with fruit quality. With increased global interactions and trade, date production whether traditional or industrial faces challenges that can suddenly appear far from their area of origin. Such challenges include increased pressure on water supplies, traditional pest and disease problems such as the dubas bug and bayoud disease, invasive or emerging pests or diseases such as the red palm weevil and others. Genes useful in meeting these challenges may well be present in non-elite date varieties found in traditional oases but their presence is not known. This leads to the need for a concentrated and coordinated effort to survey the oases for these traits and subsequently to understand their genetic basis. The writer knows of only a few studies in this area, such as those cited above. Some of it may be available in journals that are difficult to obtain in the western hemisphere or are written in Arabic or Farsi. However, the writer's sense is that there have been only a few studies examining the population structure of date palm oases and fewer that have looked at important traits other than those associated with fruit quality. Many studies have looked at the genetic relationship between the main named varieties present in an oasis but these have generally looked at only one or a few trees of the purported type. It is possible that many trees bearing the name are mutants or seedling selections (Krueger 2008).

Even with an increased number of *ex situ* collections of date palm germplasm, these collections will be fewer and smaller than for most other crops, due to the relatively limited geographic area in which cultivation is possible and the relatively narrow base of genetic diversity present. In addition to preserving germplasm, *ex situ* collections also increase the efficiency of its utilization. These *ex situ* collections allow a careful preservation of a specific genotype; reduce the chances of disease problems; allow documentation of characterization and evaluation data; and permit easier experimentation to be carried out.

When there is a better understanding of pertinent traits and their genetic bases, important date palm genotypes should be incorporated into regional germplasm repositories. Although the USDA-ARS NPGS is recognized as a world leader in genetic resource conservation, it would be difficult to incorporate important date palm genotypes into the system due to phytosanitary issues on the US side. Also important is that the bulk of date palm research will be done in countries in the Middle East and North Africa, where dates are an important commodity and the date palm is a culturally significant crop. These countries are remote from the US and shipment of germplasm in the form of offshoots from the US is challenging. Therefore, establishment of regional or national date palm gene banks in the Middle East and North Africa would facilitate research into problems important to the local production, including both traditional oasis culture and industrial production. Possibly some important genotypes should also be incorporated into the NPGS as a backup to the regional gene banks.

Germplasm exchange has changed since the era of classic plant exploration in the early years of the twentieth century. During that era, plant genotypes were freely given and taken. Subsequent to that era, germplasm exchange between scientists continued to be done freely with the understanding that plant germplasm represents the common heritage of humanity and that free exchange of germplasm will facilitate its use for the betterment of the human condition. More recently, political considerations have complicated germplasm exchange. Since 1993, the Convention on Biological Diversity (CBD) <http://www.cbd.int>, to which most countries are signees, has recognized rights to plant germplasm at the national level and set various restrictions and conditions on its exchange. The CBD recognized the unique position of agricultural commodities as being international in scope and their exchange being necessary for food security worldwide. Consequently, the International Treaty for Plant Genetic Resources for Food and Agriculture (IT) <http://www.planttreaty.org> was negotiated and approved in 2001. The IT recognizes the importance of germplasm exchange and is designed to facilitate it while promoting a Multi-Lateral System (MLS) promoting benefit sharing, farmer's rights, and other important considerations. The IT places restrictions upon the claiming of Intellectual Property Rights (IPR) on germplasm accessed internationally. While some of the provisions of the IT remain somewhat controversial and the treaty as a whole has not been completely implemented internationally, the IT does provide an explicit framework that defines the manner in which plant genetic resources can be accessed and utilized.

While the IT provides a legal framework for the exchange of plant germplasm, it does not address the logistical or practical aspects thereof. The main considerations in exchanging germplasm are phytosanitary and transportation. In the case of date palms, many countries restrict the introduction of propagative material due to the potential to introduce devastating exotic pests or diseases. For instance, the United States has a small but profitable date industry due largely to having few pests or diseases present in the growing area. Consequently, date palm propagative material is considered 'prohibited' by USDA-APHIS (7 CFR § 319.37-2) and may not enter the United States except under a special Departmental Permit. This is due to the absence in the United States of diseases such as bayoud disease, lethal yellowing and cadang-cadang, and insects such as the red palm weevil. Exchange of germplasm must be accomplished in a manner that minimizes these risks and as such would involve some sort of quarantine and pathogen-testing mechanism. Date palm germplasm can be exchanged as seed or offshoots. Exchange of seed is easier due to the small size of the seed and relative lack of potential pests and diseases accompanying the introduction. However, seedling date palms are variable and as hybrids do not represent the genotype of the mother palm. In addition to the phytosanitary issues, offshoots also present difficulties in transport due to their size, weight and perishable nature. Offshoots are typically 10–20 kg apiece. Consequently, it may be advantageous to exchange date palm germplasm as tissue culture plantlets. This would present distinct advantages in the area of transportation and phytosanitary status.

However, tissue-cultured plantlets of date palms may have some issues with genetic variability (see chapter in this book). Even though tissue cultured plantlets present a lower level of phytosanitary risk, they may still be restricted by a country's plant health organization. Overall, exchange of germplasm as tissue cultured plantlets should be considered a promising method deserving further investigation.

Some preliminary considerations on the implementation of a genetic resource conservation program for date palms will now be presented. The first consideration of the program would be what genotypes to incorporate. This to some extent depends upon the purpose and scope of the project. A genebank having its main objective as supplying propagative material for a local industry would have a more limited range of genotypes than a genebank that has the objective of general genetic resource conservation and utilization. This latter type might incorporate elite varieties, genetic material (local varieties, landraces, genotypes with specific important characteristics), staminate varieties and potentially related taxa. After identification of the varieties to be maintained, an appropriate source will need to be identified. For elite varieties, the best source would be existing collections. Germplasm maintained in formal collections will have the best passport data available, should be properly identified and in most instances will have a higher health standard than commercial date palms. If the variety is not available in a collection, it should be taken from a well-known commercial enterprise. In the case of interesting non-elite varieties, the source may be more obscure. The varieties should be obtained as offshoots, due to the known issues with off-type tissue culture-derived date palms (Aaouine 2003; Al-Wasel 2001; Gurevitch et al. 2005). In some cases, it may be appropriate to obtain material as seeds (this would mostly be genetic material rather than elite varieties).

The accessions should be established with multiple inventory items (i.e., several plants of each type). It is suggested that initially two or three plants of each type be established. The need for offshoots for research and other purposes may mean that some or all accessions need to have additional trees established. This is particularly true for varieties that produce few offshoots, such as Barhee. Regeneration of accessions will also require additional trees to be established (see comments below). The field planting should be at a spacing that allows adequate room for tree growth. A suggested spacing that works well is 9 × 9 m. This spacing allows 125 trees per hectare. If resources permit, backup of the collection via tissue culture or cryogenic storage should be implemented despite any issues with trueness-to-type. Backup at a remote site is desirable as a catastrophic event at the genebank could destroy both the working collection and any backups maintained on site.

Cultural practices appropriate for the area should be established. The cultural care of the valuable collection should be carried out at a high level of quality. This will ensure that the accessions, particularly the fruit, are representative of what would be expected under production conditions. Recommended cultural practices include pollination; thinning (where appropriate or necessary, for instance with Medjool); bunch tie-down; bunch covering; timely harvest; removal of old, senescing leaves and previous season's fruit stalks; fertilization; and irrigation. Irrigation should be by micro-sprinklers. In addition to being more efficient in the use of water, micro-sprinklers allow more flexibility in application rate. This is important since some varieties are smaller statured (e.g. Khadrawy) and might need less water.

An area that needs consideration is the regeneration of the collection. Offshoots will be removed for research and other purposes, but should also be removed if they become too large to be useful. Since date palms produce a limited amount of offshoots during their lifetime, it will be necessary to establish additional trees in order to continue to generate offshoots. The large mature trees may be maintained as examples or they may be removed at the discretion of the curator. Generation of additional offshoots may be by either establishment of new trees in the collection or the establishment of nursery rows.

Rational utilization of genetic resources requires extensive characterization, evaluation and documentation of the germplasm. Characterization and evaluation refer to activities that shed light on the inherent traits of germplasm accessions. *Characterization* is defined as "assessment of the presence, absence, or degree of specific traits that are little influenced in their expression by varying environmental conditions", whereas *evaluation* is defined as "... assessment of plants ... for potentially useful genetic traits, many of which may be environmentally variable (e.g., pest or disease resistance, fruit quality, flavor) (National Research Council 1993)." Thus, characterization generally involves assessment of traits measured on the accessions maintained in a collection (including genetic or molecular constitution), and are generally regarded as a curatorial function. Evaluation activities generally involve assessments made on plants propagated from the accession being maintained and often involve replicated trails and cooperating scientists. In some cases, traits of germplasm accessions are inferred from observations but it is preferred that measurements be made in a scientific manner. For instance, in some cases tolerance of saline soils may be inferred from observations made in the field but this inference does not replace nor have the value of experiments made under controlled conditions. It should be noted that characterization and evaluation activities are usually intense consumers of resources and as such are often neglected in favor of basic maintenance activities.

Most characterization data for date palms has involved fruit characteristics and production. However, there are many other useful traits that should be assessed in germplasm accessions. As examples, Carpenter and Ream (1976) and Barrett (1973) list the following traits that were considered useful in date palm breeding in the USA (for more information on this program, see Krueger 2001):

- Tolerance or resistance to cold, extreme heat, high humidity, rain damage, insect attacks, diseases, saline soil or water, poor drainage and other soil-related problems.
- Adaptation to mechanical harvesting, processing and pest control.
- Rate of vertical growth, number and size of spines, length and flexibility of fruit stalks, distribution and numbers of fruits per bunch to increase size and reduce thinning operations.
- Fruit quality, seed size, uniformity and time of ripening and reduction of skin separation in soft varieties.
- In male palms, metaxenic characters that could be used to manipulate fruit production.

- Identification or discovery of hermaphroditic flowers or monoecious lines.
- Identification, discovery or production of precocious lines.

Among the most important characteristics to be evaluated is resistance or tolerance to diseases and pests. Among the most important diseases to consider are bayoud and lethal yellowing, whereas important insect pests include the dubasbug, the red palm weevil, and the red palm mite. These are currently localized problems in different production areas but may become more widespread in the future. In some cases, specific countries have efforts at the national level to develop date palm lines resistant or tolerant of specific pests or diseases.

Recently, IPGRI expanded its well-known and widely used germplasm descriptors to include extensive descriptors for date palms, including many of those mentioned above (IPGRI et al. 2005). More concise but still extensive and useful descriptors have also been recently published by Rizk and El Sharabasy (2006, 2007). These organized listings of descriptors should be useful to many workers in date palm germplasm conservation and breeding.

In conjunction with characterization and evaluation for horticulturally and phytosanitarily important traits, their genetic basis and control require elucidation. Various marker systems have been used for genetic analysis of date palms, including AFLP, RAPD, RFLP, SSR, etc. SSR markers may be considered the best choice of these systems, but development of SSR markers is resource-intensive compared to some other methods. In any case, use of molecular markers for genetic characterization of date palm populations and differentiation of cultivars or cultivated plants has proven useful, as the discussion above regarding oasis population structure demonstrates. However, development of information on genetic regulation of various date palm traits is still in its infancy. There have also been some important steps towards sequencing of the date palm genome by workers in various countries, including USA/UAE (Donnelly et al. 2007), Spain (L.V. LópezLLorca, 2009, personal communication), and Qatar <http://qatar-weill.cornell.edu/research/datepalmGenome/>. Efforts in this important area will undoubtedly continue in the future and yield interesting and useful information.

16.5 Conclusion and Prospects

Conservation of plant germplasm remains vital for addressing agricultural production problems. The conservation status and genetic vulnerability of date palms are somewhat unclear but deserve attention. The center of origin near present-day Iraq and Iran is threatened by various factors, including development, desertification and salinization. Modernization of traditional oasis culture can also contribute to genetic erosion of date palms due to the importation of more profitable elite varieties. Surveying of existing traditional oasis cultures should be done with the objective of estimating the levels of genetic integrity and erosion present. This will require both morphological and molecular approaches. Farmers in traditional oasis cultures should be encouraged to maintain and plant traditional varieties. However, the vulnerability of the traditional oasis cultures in general makes it important to establish *ex situ* collections of date palm genetic resources to complement the traditional oasis cultures. The *ex situ* collections will allow easier access and more efficient evaluation of date palm germplasm. Although maintenance of date palm genetic diversity is a challenge, the fact that date palms are restricted geographically and climatically may make assessment and conservation of their genetic integrity somewhat easier than for more diverse and widely distributed crops. The date palm will remain an important food source and cultural force in many areas and conservation of date palm germplasm deserves the same attention and action as devoted to other crop species.

References

Aaouine M (2003) Date palm large-scale propagation through tissue culture techniques. In: The date palm: from traditional resource to green wealth. ECSSR, Abu Dhabi, pp 79–86

Al-Ghamdi AS (2001) Date palm (*Phoenix dactylifera* L.) germplasm bank in King Faisal University, Saudi Arabia. Survival and adaptability of tissue cultured plantlets. Acta Horticult 450:241–244

Al-Wasel SAA (2001) Field performance of somaclonal variants of tissue-culture derived date palm (*Phoenix dactyifera* L.). Plant Tissue Cult 11(2)):97–105

Bailey LH, Bailey EZ (1976) Hortus Third. Macmillan, New York

Barrett HC (1973) Date breeding and improvement in North America. Fruit Var J 27:50–55

Barrow SC (1998) A revision of *Phoenix L.* (Palmae: Coryphoideae). Kew Bull 53:513–575

Beccari O (1890) Revista monografica delle specie del genere *Phoenix* Linn. Malesia 3(5):345–416

Bendiab K, Baaziz M, Majourhat K (1998) Preliminary date palm cultivar composition of Moroccan palm groves as revealed by leaf isoenzymephenotypes. Biochem Syst Ecol 26:71–82

Bettencourt E, Hazekamp T, Perry MC (1992) Directory of germplasm collections. 6.1. Tropical and subtropical fruits and tree nuts. IBPGR, Rome

Brown TW, Bahgat M (1938) Date-palm in Egypt. Booklet No 24. Ministry of Agriculture, Egypt, Horticultural Section. Government Press, Cairo

Brown AHD, Frankel OH, Marshall DR et al (eds.) (1989) The use of plant genetic resources. Cambridge University Press, Cambridge

Carpenter JB, Ream CL (1976) Date palm breeding, a review. Date Grow Inst Rep 53:25–33

Chevalier A (1952) Recherches sur les *Phoenix* africains. Rev Intl Bot Appl 32:205–236

Chrispeels MJ, Sadava DE (1994) Plants, genes, and agriculture. Jones & Bartlett, Boston

Donnelly RJ, Yadav J, Zaid A (2007) Sequencing of the date palm genome. Acta Horticult 736:113–115

El-Assar AM, Krueger RR, Devanand PS, Chao CT (2005) Genetic analysis of Egyptian date (*Phoenix dactylifera* L.) accessions using AFLP markers. Genet Res Crop Evol 52:601–607

Greuter W (1967) Beiträge zur Flora der Südägäis 8–9. Bauh 3:243–254

Gurevitch V, Lavi U, Cohen Y (2005) Genetic variation in date palms propagated from offshoots and tissue culture. J Am Soc Hort Sci 130:46–53

Hodel DR (1995) *Phoenix*, the date palms. Palm J 122:14–36

Holden JHW, Williams JT (1984) Crop genetic resources: conservation and evaluation. Allen & Unwin, London

Holden K, Peacock J, Williams JT (1993) Genes, crops, and the environment. Cambridge University Press, Cambridge

IPGRI, INRAA, INRAM, INRAT, FEM, PNUD (2005) Descripteurs du palmier dattier (*Phoenix dactylifera* L.). IPGRI, Rome; FEM, Washington, DC; PNUD, Washington, DC; INRAA, Algiers; INRAM, Rabat; INRAT, Tunis

Johnson DV (1996) (ed.) Palms: their conservation and sustained utilization. Status survey and conservation action plan. IUCN, Gland, Switzerland and Cambridge

Krueger RR (2001) Date palm germplasm: overview and utilization in USA. In: Proceedings: The first international conference on date palms. Al-Ain, March 1998, pp 2–37

Krueger RR (2008) Relationships between oases and germplasm collections. Proceedings II congress on oasis and sustainable tourism. http://www.culturaoasis.com/data/ponencias/Microsoft_Word_-_Robert_R_Krueger-ABS+PON-Oasis_Sustainable_Tourism.pdf

Martius CFP von (1836–1850) Phoeniceae. In: Historia naturalis palmarum: exposition systematica.Vol 3. F. Fleischer, Leipzig

Mason SC (1927) Date culture in Egypt and the Sudan. US Bur Plant Indus Bull 1457, Washington, DC

Miller W, Smith JG, Taylor N (1930) Phoenix. In: Bailey LH (ed.) The standard cyclopedia of horticulture, vol 3. Macmillan, New York

Mohamed BS, Rachid H (2006) Description phenopomologique de 15 cultivars tunisiens de palmier dattier (*Phoenix dactylifera* L.). Plant Genet Res News Lett 148:10–18

Moore HE (1963) An annotated checklist of cultivated palms. Principles 7:119–183

Morici C (1998) *Phoenix canariensis* in the wild. Principles 42(85–89):92–93

Mowry H, Dickey RD, West E (1952) Native and exotic palms of Florida. University of Florida Bulletin 152

Munier P (1973) Le palmier-dattier. Laisonneuve & Larose, Paris

Munier P (1974) Le probleme de l'origine du palmier-dattier et l'Atlantide. Fruits 29:235–240

Munier P (1981a) Origine de la culture du palmier-dattier et sa propagation en Afrique: Notes historiques sur les principales palmeraies africaines. Fruits 36:437–450

Munier P (1981b) Origine de la culture du palmier-dattier et sa propagation en Afrique: Notes historiques sur les principales palmeraies africaines. Fruits 36:531–556

Munier P (1981c) Origine de la culture du palmier-dattier et sa propagation en Afrique: Notes historiques sur les principales palmeraies africaines. Fruits 36:615–631

Munier P (1981d) Origine de la culture du palmier-dattier et sa propagation en Afrique: Notes historiques sur les principales palmeraies africaines. Fruits 36:689–706

Nabhan GP (2007) Agrobiodiversity change in a Saharan desert oasis, 1919–2005: historic shifts in Tasiwit (Berber) and Bedouin crop inventories of Siwa, Egypt. Econ Bot 61:31–43

National Research Council (1993) Managing global genetic resources: agricultural crop issues and policies. National Academy Press, Washington, DC

Nixon RW (1950) Imported varieties of dates in the United States. US Department if Agriculture Circulation 834, Washington, DC

Nixon RW (1951) The date palm: "tree of life" in subtropical deserts. Econ Bot 5:274–301

Oudejans JHM (1969) Date palm. In: Ferwerda FP, Wit F (eds.) Outlines of fruit breeding in the tropic. H Veenman & Zonen NV, Wageningen

Ould Mohamed Salem A, Trifi M, Salhi-Hannachi A et al (2001) Genetic variability analysis of Tunisian date-palm (*Phoenix dactylifera* L.) cultivars. J Genet Breed 55:269–278

Popenoe PB (1913) Date growing in the old world and the new. West India Gardens, Altadena

Popenoe P (1926) Le dattier au Maroc Rev Bot Appl d'Agric Col 6:129–136

Rizk RM, El Sharabasy SF (2006) A descriptor for date palm (*Phoenix dactylifera* L.) characterization and evaluation in genebanks. Am Euras J Agric Environ Sci 1:133–145

Rizk RM, El Sharabasy SF (2007) Descriptors for date palm (*Phoenix dactylifera* L.) characterization and evaluation in genebanks. Plant Genet Res News Lett 150:42–44

Sauer JD (1993) Historical geography of crop plants: a select roster. CRC Press, Boca Raton

Service de l'Horticulture (1947) Contribution a l'étude pomologique des dattes marocaines. Terre Maroc 217:441–454

Uhl NW, Dransfield J (1987) Genera palmarum: a classification of palms based upon the work of Harold E Moore, Jr. Allen Press, Lawrence

Walsborn R (2008) Date sector report and value chain development program. USAID/Iraq/Inma Agribusiness Program Contract No 267-C-00-07-00500-00. USAID. Washington, DC
Wilkes G (1988) Germplasm preservation: objectives and needs. In: Knutson L, Stoner AK (eds.) Biotic diversity and germplasm preservation. XIII Beltsville Symposium. USDA-ARS, Beltsville
Wrigley G (1995) Date palm (*Phoenix dactylifera*). In: Smartt J, Simmonds NW (eds.) The evolution of crop plants, 2nd edn. Longman, Essex
WWF and IUCN (1994–1995) Centres of plant diversity. A guide and strategy for their conservation, 3 vols. IUCN Publications Unit, Cambridge
Zehdi S, Trifi M, Ould Mahamed Salem A et al (2002) Survey of inter simple sequence repeat polymorphisms in Tunisian date palms (*Phoenix dactylifera* L.). J Genet Breed 56:77–83
Zehdi S, Trifi M, Billotte N et al (2004) Genetic diversity of Tunisian date palms (*Phoenix dactylifera* L.) revealed by nuclear microsatellite polymorphisms. Hereditas 141:278–287
Zohary D, Hopf M (2000) Domestication of plants in the Old World, 3rd edn. Oxford University Press, Oxford

Chapter 17
In Vitro Conservation of Date Palm Germplasm

S.A. Bekheet

Abstract Date palm (*Phoenix dactylifera* L.) germplasm is difficult to conserve and store in the form of offshoots or in field collections. Tissue culture technologies have had a major impact on the *ex situ* conservation of plant genetic resources. *In vitro* culture techniques supplement date palm conservation efforts and have been applied to germplasm collection, preservation and rapid clonal multiplication. *In vitro* storage methods have been developed for preservation of date palm germplasm and can be used efficiently for international exchange of germplasm because of their obvious advantages over *in vivo* material. Preservation of plant cells, meristems and somatic embryos has become an important tool for long-term storage of germplasm utilizing minimum space and low maintenance. Short- and mid-term storage is achieved by controlling environmental growth conditions and nutrient media composition. Long-term storage has been reported for *in vitro* cryopreservation of date palm cultures. Encapsulation of plant material in alginate beads has been suggested recently as a possible means of date palm germplasm exchange. Knowledge about germplasm diversity and genetic relationships are highly valuable tools in plant conservation strategies. In this regard, a number of molecular biology methods are currently available for analysis of genetic diversity in date palm genotypes. This chapter discusses the general issues and different aspects of plant biotechnology used for management and conservation of date palm cultivars.

Keywords Conservation • Cryopreservation • Encapsulation • *In vitro*

S.A. Bekheet (✉)
Plant Biotechnology Department, National Research Center,
El-Tahrir Str., Dokki 12622, Cairo, Egypt
e-mail: shawky005@yahoo.com

S.M. Jain et al. (eds.), *Date Palm Biotechnology*,
DOI 10.1007/978-94-007-1318-5_17, © Springer Science+Business Media B.V. 2011

17.1 Introduction

Soil, water and air have long been considered as the essential natural resources. Recently, germplasm has also been added as a fourth essential resource. In recent years there has been an increasing awareness of the holistic view of biodiversity, including agricultural biodiversity and conservation for sustainable utilization and development (Arora 1997). Today, plant germplasm is commanding attention at high government levels in many nations. Several nations have developed extensive systems to carry out their germplasm activities. Scientists also recognize the need to share genetic resources and to take the responsibility to conserve and preserve plant germplasm for the future (Abd El-Wahab et al. 2004). Plant germplasm is defined as a living tissue from which new plants can be grown. It contains the unique genetic information that gives plants their individual characteristics and links generations of living plants to one another. The genetic diversity of plants generated by evolution, hybridization, natural selection and human manipulation, provides the basis for sustainable food production which supports the world's population. This diversity is threatened by habitat loss, development, the change to cultivation of a small number of advanced lines and other factors such as deforestation, fire and hydroelectric power development. Preservation of the genetic diversity represented in all plant ecosystems of the world has become a major issue of international concern. The importance of plant germplasm is universally accepted as a resource of common heritage to be used freely by all humankind in the maintenance, development and improvement of food, feed, fiber and industrial crops. There is a need to share genetic resources and to assume responsibility to conserve and preserve plant germplasm for the future.

The most widely used method for conserving plant genetic resources depends upon seeds which can be dried to low moisture content and stored at low temperature. It is usually safer, cheaper and more convenient to conserve genetic resources as seeds than by any other method. Currently, more than 80% of the genetic resources of plants are conserved as seeds in gene banks around the world. However, several plant species do not produce seed or produce unorthodox seeds, so their germplasm cannot be stored conveniently in seed banks. In many cases it is convenient to conserve the germplasm by other methods. Each method has advantages and limitations. Therefore, there is a need to develop a complementary conservation strategy.

The most common method used to preserve the genetic resources of these problem species is as whole specimen plants in the field, but there are serious drawbacks with field gene banks. Germplasm conservation in field gene banks involves collecting propagation materials and growing them in orchards or fields in a specific location. Field gene banks traditionally have been used for perennial plants, including species that have a long life cycle, to generate breeding and/or planting material. The main disadvantages of this method are pests, diseases, natural disasters and vandalism. Biotechnology continues to have a key role in the conservation and sustainable utilization of biodiversity. Plant conservation biotechnology is an interdisciplinary subject, to which the tools of modern techniques are applied for plant conservation.

Biotechnology techniques offer an alternative method for conservation of such plant material. The biotechnology methodologies of germplasm conservation and utilization fall into three main categories: (1) the use of novel molecular technologies in the assessment of biodiversity and its monitoring; (2) new tools for conservation and the management of collections and (3) use of genetic manipulation to facilitate the widest use of conserved genes (Barlow and Tzotsos 1995). *In vitro* conservation uses tissue culture techniques to store germplasm involving the use of sterile cultures. It is commonly used for vegetative propagated species, recalcitrant-seeded species and species which produce few or no seeds. The tissues conserved include meristems, shoot tips, axillary buds and zygotic embryos. Basically, two methods are currently used – the slow growth method for short- to medium-term conservation of active collections, and cryopreservation for long-term conservation of base collections.

Date palms (*Phoenix dactylifera* L.) have been cultivated and subjected to selection by man since ancient times, and the distinction between wild and cultivated date palms is blurred. The date palm has retained its value for desert dwellers because of its adaptive characteristics to the environment and the wide range of its benefits, aside from its products. It makes a significant contribution toward the creation of equable microclimates within fragile oasis ecosystems, thus enabling sustainable agricultural development in many drought and saline affected regions (Barreveld 1993).

Cultivation of date palm in different ecosystems around the world has resulted in many local cultivars that represent genetic diversity. Currently grown and obsolete cultivars are important genetic resources, and should be preserved for plant breeders to use in the future. Conservation of this genetic diversity is imperative because it is highly heterozygous and has limited importance for conservation by conventional means. Tissue culture and molecular biology techniques have great potential for collection, multiplication and storage of date palm germplasm. Date palm germplasm can be preserved *in vitro* in the form of shoot tips, somatic embryos, and pollen, callus and cell suspension cultures. Miniaturization of explants allows a reduction in space requirements. Disease-free stock is simplifying quarantine procedures for the regional and international exchange of germplasm. Moreover, molecular and biochemical analysis have been used for the characterization of conserved samples. Molecular biology techniques and their use in the conservation, evaluation and utilization of genetic resources can be summarized as: (1) comparisons between collections or populations and measurement of genetic diversity before and after storage and (2) detection and diagnosis of collections. This chapter discusses the general issues and different aspects of biotechnology used for management and operation of *in vitro* conservation of date palm cultivars.

17.2 Need for *In Vitro* Conservation of Date Palm

17.2.1 Threatening Factors

Date palm is a dioecious, perennial monocotyledonous plant successfully cultivated in arid and semiarid regions. The tree has great socioeconomic importance and the fruits nutritional value in different regions of the world, especially in the Middle

East and North Africa. In addition, the date palm tolerates adverse environmental conditions and is important in reducing desertification. There are many factors threatening the genetic diversity and species of date palm including natural and man-made factors. Environmental changes include drought and floods, as well as seasonal fluctuations in temperature and rainfall. Such changes could result in genetic erosion due to crop failure and loss of varieties. Land use changes which include construction and building of roads, factories, canals, dams and new residential areas are other factors. In this respect, in Egypt, date palm tree numbers decreased from 2.5 million to just over 1 million in the Aswan area due to the building of the High Dam (Hussein et al. 1993).

Modern agriculture characterized by the use of improved cultivars in monocropping systems is now taking place, wherein many improved high-yielding varieties are introduced. Expansion in rainfed agriculture and irrigation schemes has been accompanied by changes in the vegetation complexes in rangeland and forest areas and by changing from small-scale agriculture to large-scale modern agriculture. Moreover, biotic factors such as pests and diseases can attack date palms resulting in negative impacts on the genetic variability within species (Bendiab et al. 1993).

In this regard, the principal constraints limiting date palm culture in the Maghreb countries (Morocco, Algeria, Tunisia) are drought, salinity, deserts, development, senescent date palm trees and bayoud disease. The old age of the palm groves is a considerable constraint, since 30% of the date palms in Algeria exceed the most productive age (5–50 years) (Messar 1996). About 45% of the date palm trees in Tunisia are more than 50 years old (Rhouma 1996). An urgent renovation of the plantations is necessary. The loss of date palm diversity was significant due to either wars or indirectly by habitat destruction. Before 1991 Iraq was the largest producer of dates in the world (FAO 2004) and had the most extensive date 'forest' in the world, on the Al-Faw Peninsula (MacFarquhar 2003). However, during the Gulf and Iran-Iraq wars, many palm trees were destroyed and more died when the southern marshes of Iraq were drained. For example, these wars destroyed most of the date palm plantations and by 2003 only 3 million date trees survived out of 16 million trees around Basra before the wars started (MacFarquhar 2003) and may well have resulted in genetic erosion of the natural populations.

17.2.2 *Germplasm Conservation*

Like many other plants, the date palm has been threatened by human intervention and exploitation. Date palm growing countries have a great responsibility to conserve and safe guard date palm genetic diversity to utilize for genetic improvement and development of crop cultivars for domestic and foreign markets. The collected and conserved date palm genetic resources should be readily available to researchers and other interested parties; with detailed characterization, evaluation, and utilization of conserved genetic material as well as proper documentation, including setting up a website for public access.

There are two approaches to conservation of plant genetic resources – *in vivo* and *in vitro*. The *in vivo* conservation approach generally comprises the following methods: seed storage, field gene banks and botanical gardens. Conservation of plant diversity using reserves/protected areas, on-farm and home gardens is considered an *in vivo* conservation approach. Many important varieties of field, horticultural and forestry species are either difficult or impossible to conserve as seeds (i.e. no seeds are formed or if formed, the seeds are recalcitrant) or reproduce vegetatively.

Several techniques to conserve vegetatively propagated species have recently been developed and some are undergoing rigorous testing. For certain species, *in vitro* conservation is the only option available. Although tissue culture offers a great potential for germplasm conservation of vegetatively-propagated material, two factors have been of major technical hindrance to it. First is the genetic instability of the material conserved as tissue culture due to somaclonal variation (Harding 1999). Second, the storage duration of tissue is limited. It is important to emphasize that the two approaches of conservation (*in vitro* and *in vivo*) are complementary.

Conservation of plant gene pool should employ a combination of methods including nature reserves, gene banks, and others, as no single method can conserve all the diversity. The appropriate balance between different methods employed depends on factors such as the biological characteristics of the gene pool, infrastructure and human resources, number of accessions in a given collection and its geographic site and the intended use of the conserved germplasm. The costs of individual conservation methods could be another important factor in developing the complementary conservation strategy.

17.2.3 Constraints of Traditional Preservation

Since date palm is a dioecious and heterozygous fruit tree, and for commercial purposes most often vegetatively propagated through offshoots, its germplasm cannot be stored or handled easily by conventional means. At present, the most common method used to preserve the genetic resources of date palm is *in vivo* as whole plants in the field. There are, however, several problems facing the field gene banks. Collections are exposed to natural disasters and attacks by pests and pathogens and to intrusion by neighboring rural dwellers; moreover, labor cost and requirements for technical personal are very high. In addition, distribution and exchange from a field gene bank is difficult because of the vegetative nature of the material and the greater risk of disease transfer. Vegetatively propagated material, either as whole plants or offshoots, represents the highest risk category involving potential spread of all stages of all types of pests such as insects, mites, fungi, bacteria, nematodes and viruses. Biotechnology can be integrated into all aspects of date palm germplasm conservation such as characterization, acquisition, germplasm exchange and genetic resource management. Future prospects are highly encouraging in terms of the development and application of new techniques and protocols within the context of germplasm conservation. Micropropagation has been successfully achieved either through direct or indirect morphogenesis for different date palm cultivars

(Mater 1986; Taha et al. 2001; Tisserat 1984; Zaid and Tisserat 1983). The procedures developed can also be utilized for conservation by inducing medium- and long-term preservation (Bekheet et al. 2001, 2007).

17.3 *In Vitro* Conservation of Date Palm

Conservation of date palm genetic resources is particularly important to ensure future access to valuable genes for plant improvement programs. The date palm is a usually vegetatively propagated plant from offshoots because seeds do not ensure true-to-type palms. Tissue culture offers a great potential for the conservation of germplasm of vegetatively-propagated crops. Regeneration and successful propagation of genetically-stable seedlings from cultures are prerequisites for *in vitro* conservation efforts. Significant work is being done on both aspects and for some species (Blake 1983; Khalil et al. 2002) tissue culture maintenance is relevant due to improved techniques resulting in low levels of somaclonal variation. Cryopreservation technique is used for long-term storage of *in vitro* cultures, and is being widely used. *In vitro* methods to propagate the date palm have been developed which may be employed to preserve germplasm (Tisserat 1981).

Tissue culture in combination with molecular biology techniques are of great interest for collecting, multiplication and storage of date palm germplasm. Miniaturization of explants allows reduction in space requirements and consequently labor cost for the maintenance of germplasm collections. The potential of *in vitro* culture methods for the conservation of genetic resources of vegetatively propagated crop species was recently recognized. Two types of *in vitro* gene banks for conservation have been reported: (1) slow growth in cold storage, and (2) cryopreservation (Withers and Williams 1985). In cold storage, temperature varies depending on the origin of stored species. Temperate species may be stored at 4°C, whereas tropical plants are require temperatures in the range of 15–20°C. This technique enables extending the subculture periods from 12 month up to 4 years for many species (Ashmore 1997). Most cold preservation protocols were performed under either low light intensity or complete darkness (Wang and Charles 1991; Zandvoort et al. 1994), yet several plant species were cold preserved under light conditions (Hvoslef-Eide 1992). Otherwise, several types of plant materials including shoot tips, nodal segments and rooted shoots are used for low temperature preservation (Orlikowska 1992). Cryopreservation generally refers to storage between −79°C and −196°C in liquid nitrogen. The major advantage of cryo-storage is that both metabolic processes and biological deterioration are considerably slowed or even halted (Kartha 1981). In addition, it is believed that cryopreserved material remains genetically stable, thus affording an advantage over conventional conservation methods (Withers 1980, 1983). So far, cryopreservation protocols for plant tissues have mainly been based on slow freezing in the presence of cryoprotective mixtures containing DMSO (dimethyl sulphoxide), sugars, glycerol and/or proline. Slow freezing results in freeze-dehydration, leaving less water in the cells to form lethal ice crystals upon exposure to extreme low temperatures. However, several new cryopreservation procedures such as vitrification, encapsulation-dehydration, preculture-dehydration, and encapsulation/vitrification have been established, which are all based on vitrification (Sakai and Engelmann 2007).

The crucial factor for the successful use of cryo-storage of *in vitro* cultures is plant regeneration, and maintaining their genetic fidelity. Many publications indicate that characteristics such as growth rate, regeneration capability, ploidy levels and RAPD markers are retained after cryogenic storage (Kobayashi et al. 1990; Wang et al. 1994; Ward et al. 1993). Solid regeneration medium is usually used for regrowth of cryopreserved explants (Hirai and Sakki 1999); nevertheless, some results suggest that liquid medium may be better (Pains et al. 2002). Although plant regeneration can be accomplished from adventitious buds and somatic embryos derived from leaf, stem, root or callus, such undifferentiated tissues should be conserved with caution owing to their potential for somaclonal variation. Tissues for *in vitro* conservation should be pre-indexed for pathogens.

Pollen storage has also been considered as an emerging technology for genetic conservation of date palm. Pollen storage was mainly developed as a tool for controlled pollination of asynchronous flowering genotypes, especially in fruit tree species (Alexander and Ganeshan 1993). Cryopreservation has more potential to succeed in pollen storage as compared to other storage techniques routinely employed for pollen. Pollen can be collected easily and cryopreserved in large quantities in a relatively small space. In addition, exchange of germplasm through pollen poses fewer quarantine problems as compared to seed or other propagules.

Cryopreservation techniques have been developed for pollen of an increasing number of species and pollen cryo-banks of fruit tree species have been established in several countries i.e., India, Indonesia, Italy and South Africa (Barnabas and Kovacs 1997; Bhat and Seetharam 1993). Appropriate molecular markers have been identified to verify true-to-type date palm nursery material; and have great potential of identifying unique value-added individuals for the long-term conservation. The six major steps defined in the conservation are: collection; quarantine; disease indexing and eradication; propagation; characterization; and storage and distribution (Withers 1983). Some problems may arise with the maintenance and repeated transfer of *in vitro* stored date palm germplasm such as danger of contamination during each transfer, which may result in loss of elite plant material and genetic fidelity. Loss of morphogenic potential, and other problems associated with tissue culturing, would be minimized by cryo-storage of tissues in liquid nitrogen (−196°C). Successful cryogenic methods for doing this are now being developed. A brief review of such efforts and their use are described here.

17.3.1 Cold Storage

The main objective of slow growth *in vitro* preservation of germplasm is to limit the number of subcultures and to maintain the genetic diversity of a species under sterile conditions without endangering plant genetic stability. The slow growth method for an active germplasm collection aims to minimize cell division and growth to increase longevity without genetic changes. Slow growth procedures allow clonal plant material to be held for 1–15 years under tissue culture conditions with periodic sub-culturing, depending on species. Standard culture conditions can be used for short-term storage only with slow growing plant species (coffee and oil palm) However, for most plant species, modified conditions such as environmental conditions and/or the culture medium are necessary for short and medium-term storage; subculturing for 6–12 months, depending on species. The advantage is that the time between transfers is lengthened thereby prolonging storage and reducing maintenance. This is accomplished by using media containing growth retardants and osmotic agents, or low temperature and light intensity (Withers 1991).

Medium selection may be an important factor as not all species can survive in the same medium (Dussert et al. 1997). It is also possible to limit growth by modifying the culture medium, mainly by reducing the sugar and/or mineral element concentrations. Addition of osmotic growth inhibitors (e.g. mannitol) or growth retardants (e.g. absisic acid) is also employed successfully to reduce growth. Osmotic agents are materials that reduce the water potential of cells. Adding osmotic agents to the culture has proven efficient in reducing growth and increasing the storage life of many *in vitro*-grown tissues of different plant species. According to the hypothesis for turgor-driven growth and cell expansion, high levels of osmotic agents in the medium would act against the creation of a critical turgor pressure, which must be established before cell expansion can occur (Zimmermann 1978). This stress condition will inhibit both callus growth and shoot formation (Brown et al. 1979). In this regard, mannitol, sucrose, sorbitol, tributyl-2,4 dichlorbenzylphosphonium chloride (phosphon D), malic hydrazide, succinic acid-2, 2-dimethyl hydrazide (B-995) and ancymidole were reported ideal materials to lengthen the storage life of *in vitro* grown tissues.

The role of sucrose is two-fold, a carbon/energy source and an osmotic agent. High sucrose concentration can be used to reduce plant growth *in vitro*. Mannitol can also be used as an osmotic agent; it is a sugar alcohol produced as a primary photosynthetic product by some plants and can be metabolized by them. Sorbitol is another sugar alcohol that inhibits shoot growth of *in vitro* grown tissue cultures. Shibli et al. (1999) reported that elevated sucrose, sorbitol or mannitol reduced growth of bitter almond micro-shoots significantly and extended the subculture interval to 4 months at room temperature. The type of explant as well as its physiological state at the time of storage can influence the duration of storage. For example, the presence of a root system generally increases the storage capacities. Moreover, organized cultures such as shoots are used for slow growth storage since undifferentiated tissues such as callus are more vulnerable to somaclonal variation. Also, the type of culture vessel, its volume as well as the type of closure of the culture vessel can greatly influence the survival of stored cultures.

Plant material can be stored for longer periods by increasing the size of the storage containers. Moreover, replacing cotton plugs with polypropylene caps, thus reducing the evaporation of the culture medium, increased the survival rate during storage (Sharma and Chandel 1992; Withers 1992). Alternative techniques include modifications of the gaseous environment of cultures, and desiccation and/or encapsulation of explants. Growth reduction can be achieved by reducing the quantity of oxygen available to cultures. The simplest method consists of covering the explants with parafilm, mineral oil or liquid medium. Reduction of the quantity of oxygen can also be achieved by decreasing the atmospheric pressure within the culture chamber or by using a controlled atmosphere. In this respect, oil palm polyembrogenic cultures were conserved for 4 months at room temperature in a controlled atmosphere with 1% oxygen (Engelmann 1990).

In most cases, a low temperature, often in combination with low light intensity or even darkness, is used to limit the growth. Temperatures in the range of 0–5°C are employed with cold tolerant species (Monette 1986), but for certain tropical species, which are generally cold sensitive, temperatures of 15–20°C, are used (Banerjee and De Langhe 1985; Corbineau et al. 1990). Storage at reduced temperatures, for example, has been used with strawberry (Reed 1992), grape vines (Galzy and Compan 1988), potato (Sarkar and Naik 1998), grasses (Dale 1980), legumes (Mandal 1995). However, despite the apparent success in the application of slow growth techniques, a number of concerns remain. A main concern is the problem of genetic stability (Harding 1991, 1994). In this respect, *in vitro* conservation by slow growth has thus been developed to resolve some of these problems by prolonging the subculture interval (Yu-Jin and Xiu-Xin 2003).

17.3.2 Short and Mid-Term Storage

The objectives of slow growth in date palm are: (1) to devise an effective and reproducible protocol for minimizing growth and development of plant material, (2) to avoid induction of abnormal growth and development during storage and (3) to preserve viability of stored plant population as *in vitro* cultures. Slow growth is achieved in date palm for short (3–6 months) and mid-term conservation (9–12 months). In this respect, Bekheet et al. (2001) described a method for preservation of Egyptian date palm (Zagloul cv.) tissue cultures by slow growth. Shoot buds and callus cultures were successfully stored for 12 months at 5°C in the dark. Uniform proliferated shoot buds and equal pieces of callus cultures were cultured on standard growth media [MS-hormone-free for shoot buds and MS supplemented with 10 mg/l 2,4-dichlorophenoxyacetic acid (2,4-D) and 3 mg/l dimethylaminopurine (2ip) for callus] and incubated at 5°C and complete dark conditions. One hundred percent of survival and healthy shoot bud cultures with no vitrification were obtained after 3 months of incubation. However, 70% of bud cultures survived for 12 months of storage (Fig. 17.1a). However, callus cultures remained fully viable with slight browning after 12 months of storage at 5°C in the dark. The role of sorbitol as an osmotic stress agent in the *in vitro* storage of date palm cultures was examined. Under normal growth conditions, healthy shoot bud cultures were obtained after 6 months of storage in a medium containing 40 mg/l sorbitol. However, the period was extended to 9 months in the case of callus cultures (Fig. 17.1b). In this regard, 18 date palm cultivars produced through tissue culture from different laboratories around the world were transplanted into the date palm germplasm bank at King Faisal University, Saudi Arabia (Al-Ghamdi 2001).

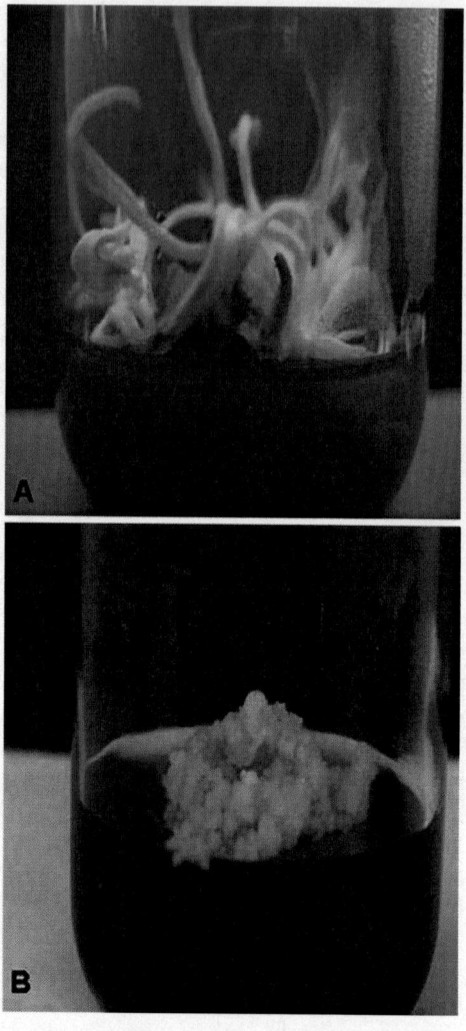

Fig. 17.1 (**a**) Shoot buds of date palm stored for 12 months at 5°C in the dark. (**b**) Callus culture of date palm stored for 9 months on medium contained 40 mg/L sorbitol at normal growth conditions

17.3.3 Cryopreservation

17.3.3.1 Advantages of Cryopreservation

Cryo-storage is a long-term *in vitro* conservation method, which is carried at between –79°C and –196°C. At these temperatures, cell division and metabolic and biochemical processes are arrested and thus the cells retain their properties unchanged for an indefinite period of time (Nimo et al. 1992; Paul et al. 2000). Also, cryopreservation offers maximum stability of phenotypic and genotypic characteristics of stored germplasm. Ultra-low temperature effectively suspends or reduces metabolic activities of tissues to near zero thereby allowing, in theory, near infinite conservation without genetic alteration. The main advantage of cryopreservation is the limited need for viability indexing or subculturing, besides being relatively cheap and manageable. Moreover, cryopreservation is the only available method for long-term conservation of vegetatively-propagated plant germplasm.

In vitro conservation of a wide array of tissues ranging from shoot tips, buds, embryos to protoplast has been successfully cryopreserved in liquid nitrogen (LN). However, desiccation of sensitive tissue such as vegetative plant parts including meristems, shoot tips, and axillary buds, are easily injured even with slight desiccation. Long-term storage using cryopreservation offers an alternative to slow growth. It is attractive because of the potential reduction in maintenance costs and reduced exposure of germplasm to contamination and genetic change. In addition, it is not demanding of space in that a very large collection can occupy a relatively small area. The technology for cryopreservation has advanced significantly in the past decade, especially for tropical crops. Techniques have now been developed for more than 200 different plant species cultured in various forms, including cell suspensions, calli, apices, somatic and zygotic embryos (Reed 2008).

17.3.3.2 Cryopreservation Procedures

In vitro collections of date palm plantlets have been established by the Groupe de Recherche Français sur le Palmier Dattier, France. Management of large-scale *in vitro* collections poses numerous practical problems. Moreover, risks of contamination and of somaclonal variation increase over time. It is therefore essential to develop long-term conservation techniques to conserve elite date palm cultivars. This technique is being optimized by several research groups in order to apply to a wide range of genotypes, and maintain genetic integrity. For successful cryopreservation, it is essential to avoid lethal intracellular freezing which occurs during rapid cooling in LN. Recently, new approaches have been developed for cryopreservation of plant apices by directly plunging them into LN such as is used in vitrification. This is a physical process that can be defined as the transition of aqueous solution from liquid into amorphous glass at the glass transition temperature, while avoiding ice crystallization. The vitrification procedure requires a highly concentrated solution, which sufficiently dehydrates cytosols without causing injury so that they turn into a stable glass when plunged into LN. Vitrification-based procedures involve removal of most or all freezable water by physical or osmotic dehydration of explants, followed by ultra-rapid freezing which results in vitrification of intracellular solutes, i.e. formation of an amorphous glassy structure without occurrence of ice crystals which are detrimental to cellular structural integrity. These techniques are more appropriate for complex organs like embryos and shoot apices; they are also less complex and do not require a programmable freezer, hence are suited for use in any laboratory with basic facilities for tissue culture (Charoensub et al. 1999; Niino et al. 1992a; Scocchi et al. 2004).

Engelmann (2000) described seven vitrification-based procedures in use for cryopreservation: (1) encapsulation-dehydration, (2) vitrification, (3) encapsulation-vitrification, (4) desiccation, (5) pregrowth, (6) pregrowth-desiccation and (7) droplet freezing. With the advent of these new cryogenic procedures, especially vitrification, encapsulation-vitrification and encapsulation-dehydration, the number of cryopreserved species has increased significantly in recent years.

Encapsulation and dehydration cryopreservation methods are based on embedding plant material in alginate beads which protect them from dehydration and freezing damage. This method has been successfully applied to coconut plumules. Apical domes excised from embryos were used as plant material; pretreatment duration, sugar concentration, dehydration period and freezing were tested as key factors. A good survival level around 60% was obtained and 20% of leaf shoots were developed from the cryopreserved explants (N'Nan et al. 2008). Even though cryopreservation is suitable for vegetatively-propagated species, it seems unsuitable for seed propagated species, which is due to high sensitivity to desiccation, structural complexity and heterogeneity in terms of developmental stage and water content at maturity.

17.3.3.3 Development of Date Palm Cryopreservation

For long-term conservation of date palm, cryopreservation in LN appears to be a logical choice, since all metabolic process and all cell division are arrested at those temperatures. Cryopreservation is recognized as having the distinct advantage of allowing long-term conservation with minimum space and maintenance. In this regard, Finkle et al. (1979) investigated the possibility of conservation of a tropical palm and the date palm using cryopreservation. Tisserat et al. (1981) reported that embryogenic callus cultures of date palm were subjected to 0°C, –15°C, –23°C, –30°C and –196°C for up 3 months in the presence of protective mixture. Revived cultures developed callus upon recovery and development of thawed callus from all frozen temperatures was rapid after a quiescent period of 2–4 weeks. Thawed callus cultures were nodular and yellow-brown in color during the log phase. The nodules enlarged and differentiated into embryos as new growth. Plantlets were produced from the callus within 3 months.

Ulrich et al. (1982) stated that embryogenic date palm (Medjool cv.) callus cultures were treated with a cryoprotective mixture of polyethylene glycol (Carbowax 6000), glucose, and dimethyl sulfoxide (10%/8%/10%, w/v); treated with the mixture, frozen to –196°C, and then thawed; or left untreated. Growth subsequent to treatment was measured as fresh weight increase and the number of embryos produced during 18 weeks of culture. The addition of a cryoprotective mixture to the unfrozen controls at 0°C did not adversely affect growth and morphogenesis of callus cultures of date palm, but treatment at –196°C decreased the subsequent rate of culture growth, compared to the control treatments. This inhibitory effect was probably directly related to the number of cells injured or killed by the freeze-thawing process. Later, Tisserat et al. (1985) reported that cryopreserved pollen of date palm (Deglet Noor cv.) was dusted on freshly opened spathes of 10-year-old Deglet Noor female trees; fruit yield and developments were similar using either frozen or non-frozen pollen.

Date palm embryos can continue normal growth and development after cryopreservation provided they are pretreated with a cryoprotectant mixture of glycrol and sucrose and then dried to a water content of 0.4–0.7 g/g. In this respect, Mater (1987) reported that date palm callus was treated with a cryoprotective mixture and frozen to –25°C for 4 months. Freezing did not affect the potential of embryogenesis induction from callus although growth during the first 2 months of culture was inhibited. Studies on the cryopreservation of date palm for germplasm collections were initiated by Towill et al. (1989). Shoot-tips were excised from 2-month-old seedlings of Medjool cv., precultured for 2 days and then cooled to LN temperatures. Viability of treated shoot-tips was assessed by growth *in vitro*. Dimethyl sulfoxide (DMSO) in concentrations up to 10% was not toxic, although growth was slower than untreated shoot-tips. Several combinations of DMSO and sucrose were effective in obtaining survival after LN exposure. In most cases, the LN-treated shoot tips developed directly into a shoot without callus formation. In this respect, Bagniol et al. (1992) suggested that gradients may be exhibiting both the outflow of water and the penetration of the cryoprotectants during the process of date palm cryopreservation. MyCock et al. (1995, 1997) developed date palm plantlets using somatic tissue previously frozen for several months in LN. They reported that late globular/early torpedo stage date palm embryos were successfully cryopreserved. Drying on its own resulted in a better percentage survival than when the samples were pre-treated with cryoprotectants only; as before, when the pre-treatments were combined the survival rate was substantially improved.

A number of steps are followed for cryopreservation, which include: preculture in media with osmotically active compounds, treatment with cryoprotective agents, cooling and storage at –196°C, thawing, post-thaw treatments and recovery of growth. Capacity to survive storage in LN is dependent upon many factors including the physiological status of the explant. Type and nature of cells determine the ability of cells to withstand freezing stress. In general it is recommended that explants be taken from rapidly growing cultures since actively dividing cells have dense cytoplasm and little developed vacuolar systems which make them more likely to withstand freezing and remain viable. Three types of meristematic and regenerative *in vitro* tissues can be obtained from date palm: (1) individual meristems isolated from shoot-tip cultures; (2) highly proliferating meristem cultures containing nodular-like meristem clusters; (3) somatic embryos. Cryopreservation methods have been developed for nodular tissue by Bekheet et al. (2007). Undifferentiated tissue cultures (nodular cultures) were successfully cryopreserved by freezing methods; subsequently the plants were regenerated.

The potential for dehydration caused by air drying in the cryopreservation of date palm tissue cultures through direct immersion in LN was studied. Among different types of sugars (e.g. fructose, glucose, sorbitol and sucrose) used as osmotic agents in preculture medium, sucrose proved to be the best for the survival of cryopreserved date palm tissue cultures. To determine the potential of vitrification on freezing tolerance, cultures were exposed to a vitrification solution [22% (w/v) glycerol, 15% (w/v) ethylene glycol, 15% (w/v) propylene glycol and 7% (w/v) dimethyl sulfoxide] for 20–100 min. The highest survival rate was obtained with cultures exposed for 80 min at 0°C followed by 40 min at 25°C. Cultures were kept in LN (–196°C) for 48 h. The storage duration in LN is not a critical factor affecting the recovery of cryopreserved plant material. Once the plant material is plunged into LN, the internal solutes will be vitrified and biochemical or biophysical events will no longer occur. Thawing is done at room temperature using a laminar airflow cabinet. For recovery of cryostored tissue cultures, various techniques have been used to detect viable plant cells, including fluorescein diacetate (FDA) staining (Ashmore 1997) and a triphenyl tetrazolium chloride (TTC) reduction assay (Wang et al. 2002). The TTC test distinguishes between survived and non-survived plant cells, tissues and organs on the basis of their respiration rate. Regrowth of the cryopreserved plant material is another indicator of plant viability, as evidenced by cellular growth and greening of the apical regions, or in the form of callus.

17.4 Synthetic Seeds

Advances in biotechnology, especially with *in vitro* culture techniques and molecular biology, provide important tools for improving conservation and management of plant genetic resources. Somatic embryos are excellent material for cryopreservation of plant germplasm. Somatic embryogenesis is a technique that lends itself to the production of synthetic seeds. Cryopreservation of encapsulated somatic embryos/synthetic seed is yet another viable proposition for long-term conservation and exchange of germplasm. Synthetic or artificial seeds have been defined as somatic embryos engineered for use in the commercial propagation of plants. Through the combination of vegetative propagation, long-term storage, and clonal propagation, synthetic seeds can have many applications in agriculture. Somatic embryogenesis is the production of embryo-like structures from somatic cells. A somatic embryo (SE) is an independent bipolar structure, not physically attached to the original tissue that develops in the same way as a zygotic embryo. Two pathways of development are possible for SEs; they can either develop directly from the tissue being cultured or indirectly from callus. The latter method, indirect embryogenesis from liquid cell suspensions, is desirable for micropropagation, providing that clonal integrity can be guaranteed. Using this system, large numbers of SEs can potentially be produced in small volumes of culture media in a synchronous manner, thereby allowing partial mechanization and reduced labor costs.

Somatic embryos can be easily managed using bioreactors. Previously, synthetic seeds were simply hydrated somatic embryos produced from vegetative cells in plant tissue culture. These had the advantage of rapid multiplication, but the system was labor intensive, costly and the propagules were very delicate. To improve the process a new technique was developed in which a single embryo is enclosed in an alginate capsule. Prolonged storage of these synthetic seeds is now possible as the somatic embryos can be dried to a moisture content of less than 20%. Capsules of synthetic seeds are produced by mixing embryos in sodium alginate followed by dropping them into a calcium salt solution (i.e. calcium chloride) or by inserting them into a drop of sodium alginate just as they are falling into the calcium chloride. Gel capsules can be amended with nutrient, growth regulators and carbohydrates to facilitate the rapid growth and survival of embryos. In this regard, it was reported that the maturation phase and quality of somatic embryos are critical for achieving high conversion frequencies of any encapsulation or coating systems (Redenbaugh et al. 1986).

By combining a vegetative propagation system with the capability of long-term storage and clonal multiplication, synthetic seeds have many diverse applications. Establishment of synthetic seeds has multiple advantages, including ease of handling, potential long-term storage and low cost of production and subsequent propagation. In this respect, natural unipolar propagules like microbulbs, rhizomes, protocorms, nodal cuttings and shoot buds, besides bipolar somatic embryos, have been subjected to encapsulation for storage, easy handling, transport, delivery and their establishment under *in vitro* and/or *extra vitrum* conditions. Although a variety of natural and synthetic polymers are available for encapsulation, sodium alginate is the most commonly used gel-matrix because of its favorable gelling properties, non-toxicity and low cost. Different concentrations (1.5–6%) of sodium alginate have been used for different systems. Higher or lower levels of sodium alginate reduced the conversion frequency.

Synthetic seed research through somatic embryos has been achieved on numerous crop plants including conifers (Fowke et al. 1994), high-value vegetables like celery and lettuce (Sanada et al. 1993) and papaya (Castillo et al. 1998). At present, the alfalfa synthetic seed system is the most advanced (Redenbaugh 1993) and has progressed to direct field planting of synthetic seeds (Fujii et al. 1992). The use of encapsulated shoot-tips as artificial seeds is reported on banana (Ganapathi et al. 1992). A synthetic seed protocol for garlic conservation as an asexually propagated plant was established using *in vitro* regenerated bulblets (Bekheet 2006).

In date palm, a reliable system for preservation of germplasm via artificial seeds was recognized by Bekheet et al. (2005). Somatic embryos proliferated *in vitro* from shoot-tip cultures were encased in sodium alginate capsules and stored for 12 months. Somatic embryos in four maturation stages i.e. globular, torpedo, cotyledon and late cotyledon were taken and dried in a laminar flow bench (Fig. 17.2a) and then mixed with gel of 3% sodium alginate prepared in distilled water (Fig. 17.2b). An antibiotic mixture containing rifampicin (60 mg), cefatoxine (250 mg) and tetracycline HCL was used to avoid contamination. The embryos were placed into calcium chloride solution (2.5%) for 30 min and then stored at 5 and 25°C. After 12 months, the encapsulated embryos were sown in distilled water and then cultured on MS-hormone-free medium (Fig. 17.2c). Using 3% sodium alginate as a gel matrix was the best for viability and conversion to plantlets (Fig. 17.2d). For plantlets recovery and development from encapsulated somatic embryos, 20 g/L of sucrose was added to the culture medium. Observations indicated that the highest percentages of viability were recorded with the cotyledon stage (80% at 5°C and 40% at 25°C). However the highest percentages of conversion to plantlets were registered with the late cotyledon stage (65% at 5°C and 35% at 25°C). Similarly, somatic embryogenesis and synthetic seeds of Egyptian date palm cvs. Gondeila, Sewi and Zagloul were studied by Ibrahim et al. (2006).

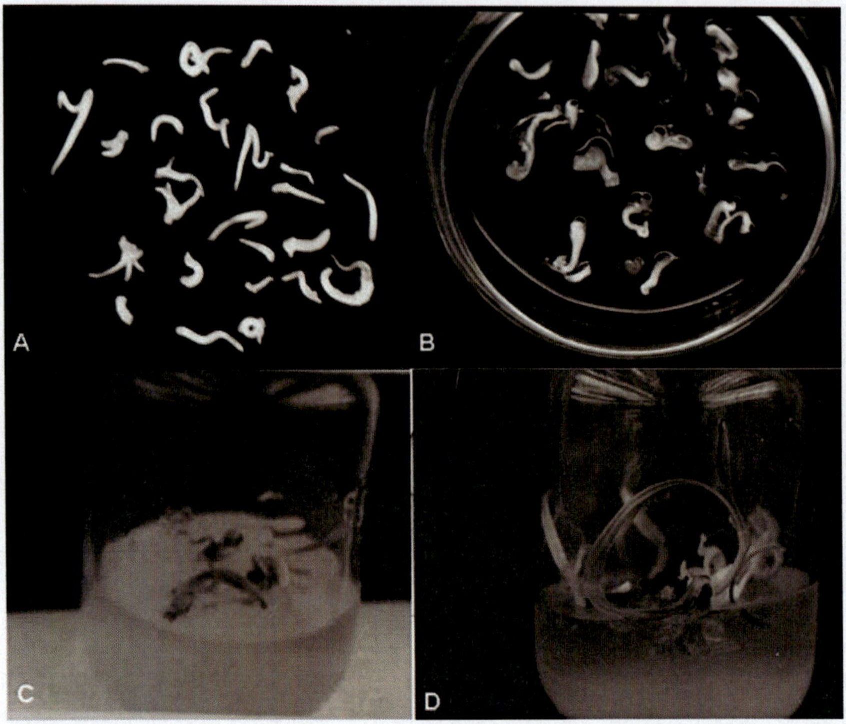

Fig. 17.2 (**a**) Air drying of proliferated somatic embryos of date palm, (**b**) somatic embryos coated with 3% sodium alginate, (**c**) coated somatic embryos recovered on MS-hormone-free medium, and (**d**) plantlets development from encapsulated somatic embryos

17.5 Characterization of Diversity and Genetic Stability

17.5.1 Biochemical and Molecular Characterization

Germplasm characterization is important for varietal identification, genetic conservation and utilization for genetic improvement of date palm (Al-Moshileh et al. 2004). Correct identification of cultivars usually is not possible until fruits are produced. In addition the characterization of cultivars and the evaluation of genetic diversity require a large set of phenotypic data that are often difficult to assess and sometimes variable due to environmental influences. The ability to identify genetic variation is indispensable to effective management and use of genetic resources. Traditionally, diversity is assessed by measuring variation in phenotypic traits such as flower color, growth habit or quantitative agronomic traits like yield potential, stress tolerance, etc., which are of direct interest to users (Rao 2004). The genetic information provided by morphological characters is often limited and the expression of quantitative traits is subjected to strong environmental influence. Earlier biochemical methods based on seed protein and enzyme electrophoresis were introduced. They proved particularly useful in analysis of genetic diversity by revealing differences between seed storage proteins or enzymes encoded by different alleles at one (allozymes) or more gene loci (isozymes). Isozymes represent biochemical markers which are successfully used as a possible alternative or complementary method for characterization of crop plant cultivars (Cardy and Kannenberg 1982; Torres and Tisserat 1980)

The use of biochemical methods eliminates the environmental influence; however, their usefulness is limited due to their inability to detect low levels of variation. Earlier isoenzyme marker analysis proved useful in identification of date palm cultivars (Baaziz and Saaidi 1988; Bendiab et al. 1993). However, they provide limited number of informative markers and give no direct assessment of the genomic variation. By using polymorphic isoenzyme markers, Bendiab et al. (1998) studied genetic diversity of the Moroccan palm groves and estimated variability more than 90% of total diversity of the date palm within-populations, whereas variability between the populations is limited to about 10%. In this regard, a study to assess levels of alloenzym diversity in collection of 29 date palm cultivars belonging to three main date-growing regions in Tunisia was conducted by Salem et al. (2001). A total of 19 different profiles were detected in the 29 cultivars using four enzyme systems i.e., glutamate oxaloacetate transaminase (GOT), phosphoglucomutase (PGM), shikimate dehydrogenase (SDH) and phosphoglucoisomerase (PGI). Results of electrophoretic analysis revealed five polymorphic loci at the four enzymes and 27 (93.3%) of the 29 cultivars were identified uniquely.

Recently introduced DNA-based techniques have the potential to identify polymorphisms represented by differences in DNA sequences. These methods are being used as complementary strategies to traditional approaches for the assessment of genetic diversity. The major advantage of DNA-based techniques is to analyze variation at the DNA level without environmental influences. The analysis can be performed at any plant growth stage using any plant part and requires only small amount of material (Rao 2004). Molecular characterization has a number of applications in the management of germplasm collections. These include elucidating systematic relationships between accessions; assessing gaps and redundancies in the collection; development of core subsets; characterizing newly acquired germplasm; maintaining trueness-to-type; monitoring shifts in population genetic structure in heterogeneous germplasm; monitoring genetic shifts caused by differential viability in storage or *in vitro* culture; exploiting associations among traits of interest and genetic markers; and genetic enhancement (Bretting and Widrlechner 1995). The DNA polymorphisms are exploited by an ever-increasing number of molecular marker techniques for differentiation between individuals, accessions and plant species. Their higher resolution compared with all other markers makes them a valuable tool for varietal and parental identification. Although there has been a small amount of molecular characterization of the date palm holdings, various molecular marker systems, such as isozymes, restriction of fragment length polymorphisms (RFLPs), random amplified polymorphic DNAs (RAPD) and amplified fragment length polymorphism (AFLPs) have been tested for their application in date palms (Ben-Abdalla et al. 2000; Bennaceur et al. 1991; Corniquel and Mercier 1994; Sedra et al. 1998).

17.5.2 Genetic Stability of Conserved Collections

Cryopreservation and cold storage are the most appropriate techniques to conserve large collections of plants (Ashmore 1997). Both techniques involves *in vitro* culture manipulation of the stored tissues that could be considered a potential risk for the regeneration of genetic instability (Brar and Jain 1998). Number of studies indicate that plants recovered from cold storage or slow growth have no genetic alteration (Hao et al. 2004; Renau-Morata et al. 2006). In this respect, Scocchi et al. (2004) tested the genetic fidelity of Chinaberry tree (*Melia azedarach* L.) plants obtained from cryopreserved apical meristem-tips using isozyme and RAPD markers. The isozyme systems presented identical electrophoretic bands and Rf values. In regard to RAPD analysis, none of the primers generated a unique profile for the control and the plants obtained from cryopreserved apical meristem-tips.

Concerning the stability of date palm tissue cultures, genetic variations of *in vitro* grown date palm Deglet Noor cv., tissue cultures were tested by random amplified polymorphic DNA (RAPD) profiles (Othmani et al. 2009). Nine arbitrary 10-mer primers were used to amplify DNA from 180 plantlets. RAPD patterns of the plantlets were identical with the original mother plant, indicating that no somaclonal variation was detected by the RAPD technique. Genetic stability is the key to reliability in the *frozen storage* of date palm germplasm. In this regard date palm callus previously stored at −196°C for 3 months had no observed changes in the patterns of survival or morphology of resulting plantlets (Tisserat et al. 1981). Ulrich et al. (1982) observed the stability of the date palm plantlets, regenerated from cryopreserved callus. Isozyme analysis of leaves revealed that within the Medjool cv. no enzyme differences were expressed among the leaflets of plantlets from untreated, PGD-treated, and frozen cultures.

Salman et al. (1988) analyzed isozyme polymorphism and chromosome number of tissue culture derived palm plants and found that all regenerated plants had the basic chromosome numbers $2n = 36$ except one $2n = 70$. A histo-cytological study was performed on apices of *in vitro* date palm plantlets cv. Bou Sthammi Noir subjected to cryopreservation (Bagniol et al. 1992). After freezing in LN, the apices showed cellular heterogeneity. Some cells conserved their meristematic characters. This was the case in the cellular layers corresponding to the meristem itself, whereas in the underlying zone, where the cells were more vacuolated, some were damaged, showing broken cell walls in some samples. During culture, prior to the cryoprotective treatment, obvious starch synthesis occurred in some cells of the samples. RAPD analysis was used to examine the genetic stability of cryo-stored date palm tissue cultures (Bekheet et al. 2007). Both treated and non-treated date palm tissue cultures in addition to field grown plants were identical with the primers used (Fig. 17.3).

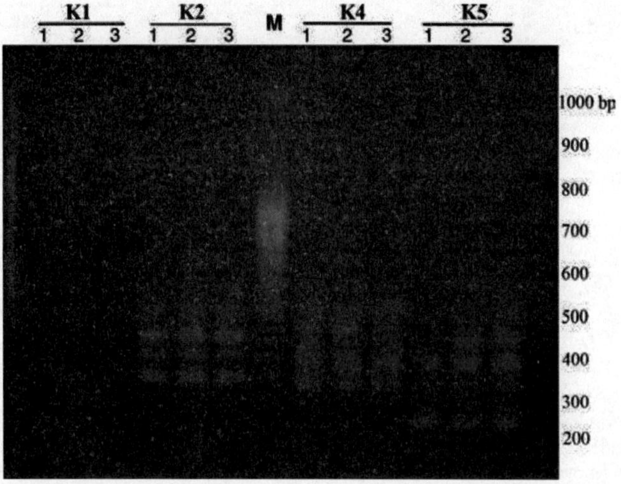

Fig. 17.3 RAPD profile of *in vivo* grown plant (lane 1), non-treated tissue cultures (lane 2), cryopreserved tissue cultures of date palm (lane 3) and the DNA marker (M) using random primers i.e., K1, K2, K4 and K5

17.6 Conclusions and Prospects

Conservation of plant genetic resources via tissue cultures had an immense effort by research work in the last three decades. *In vitro* techniques and storage methods are enabling the establishment of extensive collection using minimum space. Storage under low temperature is considered one of the major tissue culture techniques used for preservation of genetic resources of date palm. Maintaining date palm material *in vitro* under slow growth conditions could reduce cost associated with less frequent subculturing. However, cryopreservation offers long-term storage capability, maximal genetic stability of stored germplasm, minimal storage space and maintenance requirements. Cryopreservation is, so far, the only viable procedure for long-term germplasm conservation of vegetatively-propagated species such as date palm and, therefore, can be used for base collections. Slow-growth techniques are in a more advanced state of development than cryopreservation techniques, which still require improvement before they can be used on a routine basis in date palm. Adjustment of these protocols to the gene bank level would be necessary to actually exploit the advantages of *in vitro* preservation. Otherwise, establishment of synthetic seeds system is of great value in storage and propagation of valuable germplasm. Encapsulated embryos of date palm can be handled and *in vitro* planted providing large numbers of disease-free propagules. The capsules surrounding the somatic embryos would maintain the hydration of embryos, allowing temporary storage. In parallel, molecular and biochemical markers can be used to analyze date palm genetic diversity and detection of *in vitro* preserved derived plants.

References

Abd El-Wahab RH, Zaghloul MS, Moustafa AA (2004) Conservation of medicinal plants in St. Catherine Protectorate, South Sinai. I. Evaluation of ecological status and human impact. Proceedings of first international conference on strategy of Egyptian herbaria, Giza, pp 231–251

Alexander MP, Ganeshan S (1993) Pollen storage. In: Chadha KL, Pareek OP (eds.) Advances in horticulture. Vol. I – Fruit crops. Malhotra Publishing House, New Delhi, pp 481–496

Al-Ghamdi AS (2001) Date palm (*Phoenix dactylifera* L.) germplasm bank in King Faisal University, Saudi Arabia. Survival and adaptability of tissue cultured plantlets. Acta Hort 560:241–244

Al-Moshileh AM, Motawei MI, Al-Wasel A, Abdel-Latif T (2004) Identification of some date palm (*Phoenix dactylifera* L.) cultivars in Saudi Arabia using RAPD fingerprints. Agric Mar Sci 9:1–3

Arora RK (1997) Biodiversity convention, global plan of action and the national programmes. In: Hossain MG, Arora RK, Mathur PN (eds.) Plant genetic resources, Bangladesh perspective, proceedings of a national workshop on plant genetic resources. Bangladesh Agricultural Research Council. BARC-IPGRI, Dhaka, pp 28–35

Ashmore SE (1997) Status report on the development and application of *in vitro* techniques for the conservation of plant genetic resources. International Plant Genetic Resources Institute, Rome

Baaziz M, Saaidi M (1988) Preliminary identification of date palm cultivars by esterase isoenzymes and peroxidase activities. Can J Bot 66:89–93

Bagniol S, Engelmann F, Michaux FN (1992) Histocytolgical study of apices from *in vitro* plantlets of date palm (*Phoenix dactylifera* L.) during cryopreservation process. Cryo Lett 13:405–412

Banerjee N, De Langhe E (1985) A tissue culture technique for rapid clonal propagation and storage under minimal growth conditions of *Musa* (banana and plantain). Plant Cell Rep 4:351–354

Barlow B, Tzotsos GT (1995) Biotechnology. In: Heywood VH, Gardner K (eds.) Global biodiversity assessment. Cambridge University Press, Cambridge, pp 671–710

Barnabas B, Kovacs G (1997) Storage of pollen. In: Shivanna KR, Sawhney VK (eds.) Pollen biotechnology for crop production and improvement. Cambridge University Press, Cambridge, pp 293–314

Barreveld WH (1993) Date palm products. FAO Agricultural Services Bulletin No. 101. Food and Agriculture Organization of the United Nations, Rome

Bekheet SA (2006) A synthetic seed method through encapsulation of *in vitro* proliferated bulblets of garlic (*Allium sativum* L). Arab J Biotechnol 9:415–426

Bekheet SA, Taha HS, Saker MM (2001) *In vitro* long-term storage of date palm. Biol Plant 45:121–124

Bekheet SA, Taha HS, El-Bahr MK (2005) Preservation of date palm cultures using encapsulated somatic embryos. Arab J Biotechnol 8:319–328

Bekheet SA, Taha HS, Solliman ME, Hassan NA (2007) Cryopreservation of date palm (*Phoenix dactylifera* L.) cultured *in vitro*. Acta Hort 736:283–291

Ben-Abdalla A, Stiti K, Lepoivre P, Jardin PD (2000) Date palm (*Phoenix dactylifera* L.) cultivar identification using random amplified polymorphic DNA (RAPD). Cah Agric 9:103–107

Bendiab K, Baaziz M, Brakez Z, Sedra H (1993) Correlation of isoenzyme polymorphism and Bayoud-diseases resistance in date palm cultivars and progeny. Euphy 65:23–32

Bendiab K, Baaziz M, Majourhat K (1998) Preliminary date palm cultivar composition of Moroccan palm groves as revealed by leaf isoenzyme phenotypes. Biochem Syst Ecol 26:71–82

Bennaceur M, Lanaud C, Chevalier MH, Bounagua N (1991) Genetic diversity of the date palm (*Phoenix dactylifera* L) from Algeria revealed by enzyme markers. Plant Breed 107:56–69

Bhat BV, Seetharam A (1993) Pollen storage and viability – an Indian perspective. In: Veeresh GK, Umashankar R, Ganeshaiah KN (eds.) Pollination in the tropics. Proceedings of the international symposium on pollination in tropics. IUSSI, GKVK, Bangalore, pp 335–355

Blake J (1983) Tissue culture propagation of coconut, date, oil palm. In: Dodds JH (ed.) Tissue culture of trees. Croom-Helm, Ltd., Beckenham, pp 29–50

Brar DS, Jain SM (1998) Somaclonal variation: mechanism and applications in crop improvement. In: Jain SM, Brar DS, Ahloowalia BS (eds.) Somaclonal variation and induced mutations in crop improvement. Kluwer, Dordrecht, pp 15–37

Bretting PK, Widrlechner MP (1995) Genetic markers and plant genetic resource management. Plant Breed Rev 13:11–86

Brown DCW, Leung DWM, Thorpe TA (1979) Osmotic requirements for shoot formation in tobacco callus. Physiol Plant 46:36–41

Cardy BJ, Kannenberg LW (1982) Allozymic variability among maize inbred l'mes and hybrid, application for cultivars identification. Crop Sci 22:1016–1020

Castillo B, Smith MA, Yadava UL (1998) Plant regeneration from encapsulated somatic embryos of *Carica papaya* L. Plant Cell Rep 17:172–176

Charoensub R, Phansiri S, Sakai A, Yongma-Nitchai W (1999) Cryopreservation of cassava *in vitro*-grown shoot tips cooled to –196° by vitrification. Cryo Lett 20:89–94

Corbineau F, Engelmann F, Côme D (1990) Ethylene production as an indicator of chilling injury in oil palm (*Elaeis guineensis* Jacq.) somatic embryos. Plant Sci 71:29–34

Corniquel B, Mercier L (1994) Date palm (*Phoenix dactylifera* L.) cultivar identification by RFLP and RAPD. Plant Sci 101:163–172

Dale PJ (1980) A method for *in vitro* storage *Lolium multiflorum* Lam. Ann Bot 45:497–502

Dussert S, Chabrillange N, Antony F et al (1997) Variability in storage response within a coffee (*Coffea* spp.) core collection under slow growth conditions. Plant Cell Rep 16:344–348

Engelmann F (1990) Use of atmospheres with low oxygen contents for the storage of oil palm (*Elaeis guineensis* Jacq) somatic embryos cultures. C R Acad Sci Paris III 310:679–684

Engelmann F (2000) Importance of cryopreservation for the conservation of plant genetic resources. In: Engelmann F, Takagi H (eds.) Cryopreservation of tropical plant germplasm. Current research progress and application. IPGRI, Rome, pp 8–20

FAO (2004) Food and Agriculture Organization statistical databases (FAOSTAT), 24 May, 2004. http://faostat.fao.org/

Finkle BJ, Ulrich JM, Rains DW et al (1979) Survival of alfalfa (*Medicago sativa*), rice (*Oryza sativa*) and date palm (*Phoenix dactylifera* L.) callus after liquid nitrogen freezing. Cryobiology 16:583

Fowke LC, Attree SM, Pomeroy MK (1994) Production of vigorous desiccation-tolerant white spruce (*Picea glauca* [Moench] Voss.) synthetic seeds in a bioreactor. Plant Cell Rep 13:601–606

Fujii J, Slade D, Aguirre-Rascon J, Redenbaugh K (1992) Field planting of alfalfa artificial seeds. In Vitro Cell Dev Biol Plant 28:73–80

Galzy R, Compan D (1988) Growth and nutrition of grapevine during *in vitro* long-term storage. Plant Cell Tissue Org Cult 13:229–237

Ganapathi TR, Suprasanna P, Bapat VA, Rao PS (1992) Propagation of banana through encapsulated shoot tips. Plant Cell Rep 11:571–575

Hao YJ, Wen XP, Deng XX (2004) Genetic and epigenetic evaluation of citrus calluses recovered from slow growth cultures. J Plant Physiol 161:479–484

Harding K (1991) Molecular stability of the ribosomal RNA genes in *Solanum tuberosum* plants recovered from slow growth and cryopreservation. Euphy 55:141–146

Harding K (1994) The methylation status of DNA derived from potato plants recovered from slow growth. Plant Cell Tissue Org Cult 37:31–38

Harding K (1999) Stability assessments of conserved plant germplasm. In: Benson EE (ed.) Plant conservation biotechnology. Taylor & Francis, London, pp 97–107

Hirai D, Sakki A (1999) Cryopreservation of *in vitro*-grown meristems of potato (*Solanum tuberosum* L.) encapsulation-vitrification. Pot Res 42:153–160

Hussein F, El-Kholy MH, Abou-Sayed Ahmed TA (1993) Organic-chemical constituents of some Egyptian dry-date cultivars grown at Aswan. Zagazig J Agric Res 20:1313–1321

Hvoslef-Eide AK (1992) Effect of pre-storage conditions on storage of *in vitro* cultures of *Nephrolepis Exallata* L. and *Cordyline fruticosa* L. Plant Cell Tissue Org Cult 28:167–174

Ibrahim IA, Gargas AA, Nasr ME et al (2006) Biotechnological studies on the production of somatic embryogenesis and synthetic seeds of date palm. Part I: Features leaves ultrastructural surface variations. Bull Fac Agric Cairo Univ 57:97–105

Kartha KK (1981) Meristem culture and cryopreservation methods and applications. In: Thorpe TA (ed.) Plant tissue culture, methods and applications in agriculture. Academic, New York, pp 181–212

Khalil SM, Cheah KT, Perez EA et al (2002) Regeneration of banana (*Musa* spp. AAB cv. Dwarf Brazilian) via secondary somatic embryogenesis. Plant Cell Rep 20:1128–1134

Kobayashi S, Sakai A, Oiyama I (1990) Cryopreservation in liquid nitrogen of cultured navel orange (*Citrus sinensis* Osb.) nucellar cells and subsequent plant regeneration. Plant Cell Tissue Org Cult 23:15–20

MacFarquhar N (2003) Forbidden fruit: Iraq dates hit by war and sanctions. Intl. Herald Trib 1 July, 2004. http: www.iht.com/article/83184.html

Mandal BB (1995) Methods of *in vitro* conservation: principles, prospects and constraints. In: Rana RS, Chandel KPS, Bhat SR et al (eds.) Plant germplasm conservation: biotechnological approaches. NBPGR, New Delhi, pp 83–87

Mater AA (1986) *In vitro* propagation of *Phoenix dactylifera* L. Date Palm J 4:137–152

Mater AA (1987) Production of cryogenic freezing of date palm germplasm and regeneration of plantlets from frozen material. Iraq J Agric Sci Zanco 5:35–49

Messar EM (1996) Le secteur phoenicicole algérien: situation et perspectives à l'horizon 2010. Opt Médit A28:23–44

Monette PL (1986) Cold storage of kiwi shoot tips *in vitro*. HortSci 21:1203–1205

MyCock DJ, Wesley-Smith J, Berjak P (1995) Cryopreservation of somatic embryos of four species with and without cryoprotectant pre-treatment. Ann Bot 75:331–336

MyCock DJ, Berjak P, Pammenter NW, Vertucci CW (1997) Cryopreservation of somatic embryos of *Phoenix dactylifera* L. In: Ellis RH, Black M, Murdoch AL, Hong TD (eds.) Basic applied aspects of seed biology. Kluwer, Dordrecht, pp 75–82

N'Nan O, Hocher V, Verdeil JL et al (2008) Cryopreservation by encapsulation-dehydration of plumules of coconut (*Cocos nucifera* L.). Cryo Lett 29:339–350

Niino T, Sakai A, Enomoto S et al (1992) Cryopreservation of *in vitro*-grown shoot tips of mulberry by vitrification. Cryo Lett 13:303–312

Nimo T, Sakai A, Yakuwa H, Nojin K (1992) Cryopreservation of *in vitro*-grown shoot tips of apple and pear by vitrification. Plant Cell Tissue Org Cult 28:261–266

Orlikowska T (1992) Effect of *in vitro* storage at 4°C on survival and proliferation of two apple rootstocks. Plant Cell Tissue Org Cult 31:1–7

Othmani A, Bayoudh C, Drira N et al (2009) Regeneration and molecular analysis of date palm (*Phœnix dactylifera* L.) plantlets using RAPD markers. Afr J Biotechnol 8:813–820

Pains B, Strosse H, Van den Hende S, Swennen R (2002) Sucrose preculture to simplify cryopreservation of banana meristem cultures. Cryo Lett 23:375–384

Paul H, Daigny G, Sangwan-Norreel BS (2000) Cryopreservation of apple (*Malus domestica* Borkh.) shoot tips following encapsulation-vitrification. Plant Cell Rep 19:768–774

Rao NK (2004) Plant genetic resources: advancing conservation and use through biotechnology. Afr J Biotechnol 3:136–145

Redenbaugh K (1993) Introduction. In: Redenbaugh K (ed.) Synthetic seeds: application of synthetic seeds to crop improvement. CRC Press, Boca Raton, pp 3–10

Redenbaugh K, Paasch B, Nichol J et al (1986) Somatic seeds: encapsulation of asexual plant embryos. Biotechnology 4:797–801

Reed BM (1992) Cold storage of strawberries *in vitro*: a comparison of three storage systems. Fruit Var J 46:98–102

Reed BM (2008) Plant cryopreservation: a practical guide. Springer, New York

Renau-Morata B, Arrillaga I, Segura J (2006) *In vitro* storage of cedar shoot culture under minimal growth conditions. Plant Cell Rep 25:636–642

Rhouma A (1996) Le palmier dattier en Tunisie: un secteur en pleine expansion. Opt Médit A28:85–104

Sakai A, Engelmann F (2007) Vitrification, encapsulation-vitrification and droplet-vitrification. Cryo Lett 28:151–172

Salem A, Trifi M, Salhi-Hannachi A et al (2001) Genetic variability analysis of Tunisian date-palm (*Phoenix dactylifera* L.) cultivars. J Genet Breed 55:269–278

Salman M, Al-Jibouri M, Al-Quadhy K, Omar S (1988) Isozyme and chromosomal analysis of tissue culture derived date palm. Date Palm J 6:401–411

Sanada M, Sakamoto Y, Hayashi M et al (1993) Celery and lettuce. In: Redenbaugh K (ed.) Synseeds: applications of synthetic seeds to crop improvement. CRC Press, Boca Raton, pp 305–327

Sarkar D, Naik PS (1998) Factors affecting minimal growth conservation potato microplants vitro. Euphy 102:275–280

Scocchi A, Faloci M, Medina R et al (2004) Plant recovery of cryopreserved apical meristem-tips of *Melia azedarach* L. using encapsulation/dehydration and assessment of their genetic stability. Euphy 135:29–38

Sedra H, Lashermes P, Trouslot P, Combes C et al (1998) Identification and genetic analysis of date palm (*Phoenix dactylifera* L.) varieties from Morocco using RAPD markers. Euphy 103:75–82

Sharma N, Chandel K (1992) Low temperature storage of *Rauvolfia serpentine* Benth ex Kurz: an endangered, endemic medicinal plant. Plant Cell Rep 11:200–203

Shibli RA, Shatnawi MA, Ajlouni MM et al (1999) Slow growth *in vitro* conservation of bitter almond (*Amygdalus communis* L.). Adv Hort Sci 13:133–134

Taha HS, Bekheet SA, Saker MM (2001) Factors affecting *in vitro* multiplication of date palm. Biol Plant 44:431–433

Tisserat BH (1981) Date palm tissue culture. USDA, Oakland

Tisserat BH (1984) Propagation of date palm by shoot tip cultures. HortSci 19:230–231

Tisserat BH, Ulrich JM, Finkle BJ (1981) Cryogenic preservation and regeneration of date palm tissue. HortSci 16:47–48

Tisserat BH, Gabr MF, Sabour MT (1985) Viability of cryogenically treated date palm pollen. Date Palm J 4:25–32

Torres AM, Tisserat BH (1980) Leaf isozymes as genetic markers in date-palms. Am J Bot 67:162–167

Towill LE, Zaid A, Hughes HG (1989) Cryopreservation of date palm shoot tips. Abstract ISHS meeting (PS II), p 91

Ulrich JM, Finkle BJ, Tisserat BH (1982) Effects of cryogenic treatment on plantlet production from frozen and unfrozen date palm callus. Plant Physiol 69:624–627

Wang PJ, Charles A (1991) Micropropagation through meristem culture. In: Bajaj YPS (ed.) The biotechnology in agriculture and forestry: high-tech and propagation, vol 17. Springer-Verlag, Berlin/Heidelberg, pp 32–52

Wang ZY, Legris G, Nagel J et al (1994) Cryopreservation of embryogenic cells suspensions in *Festuca* and *Lolium* species. Plant Sci 103:93–106

Wang Q, Gafny R, Sahar N et al (2002) Cryopreservation of grapevine (*Vitis vinifera* L.) embryogenic cell suspension by encapsulation-dehydration and subsequent plant regeneration. Plant Sci 162:551–558

Ward ACW, Benson EE, Blackhall NW et al (1993) Flow-cytometric assessment of ploidy stability in cryopreserved dihaploid *Solanum tuberosum* and wild *Solanum* species. Cryo Lett 14:145–152

Withers LA (1980) The cryopreservation of higher plant tissue and cell cultures – an overview with some current observation and future thoughts. Cryo Lett 1:239–250

Withers LA (1983) Germplasm storage in plant biotechnology. In: Mantell SH, Smith H (eds.) Plant biotechnology. Cambridge University Press, Cambridge, pp 187–218

Withers LA (1991) Biotechnology and plant genetic resources conservation. In: Paroda RS, Arora RK (eds.) Plant genetic resources, conservation and management concepts and application. International Board for Plant Genetic Resources, New Delhi, pp 273–297

Withers LA (1992) *In vitro* conservation. In: Hammerschlag F, Litz RE (eds.) Biotechnology of perennial fruit crops. CABI, Wallingford, pp 269–200

Withers LA, Williams JT (1985) Research on the long-term storage and exchange of *in vitro* plant germplasm. In: Biotechnology in international agricultural research, proceeding inter-center seminar on international agricultural research centers (IARCs) and biotechnology. International Rice research Institute, Los Baños, 11–24 pp

Yu-Jin H, Xiu-Xin D (2003) Genet stable regeneration apple plants slow growth. Plant Cell Tissue Org Cult 72:253–260

Zaid A, Tisserat B (1983) *In vitro* shoot tip differentiation in *Phoenix dactylifera* L. Date Palm J 2:163–183

Zandvoort EA, Hulshof MJH, Staritsky G (1994) In vitro storage of Xanthosoma spp. Under minimal growth conditions. Plant Cell Tissue Org Cult 36:309–316

Zimmermann U (1978) Physics of turgor and osmo-regulation. Ann Rev Plant Physiol 29:121–148

Chapter 18
Molecular Markers in Date Palm

C. Cullis

Abstract Molecular markers are an increasingly important resource for all crops. DNA markers, especially those based on simple sequence repeats and single nucleotide polymorphisms, are playing an increasingly important role in plant variety identification, germplasm resource collection and breeding activities. The major types of DNA markers are described and the resources available to the date palm community are identified. In general, the molecular marker resources for date palm are somewhat limited. However, most of the available DNA marker types have been used on some material, mostly to cluster date palm varieties into related groups. The most profound effect on the development of the DNA marker resources for date palm is the newly available shotgun sequence. Mining this sequence database and the steady lowering of the costs of high throughput sequencing will increase rapidly the molecular marker resources and their application to date palm over the next few years.

Keywords DNA markers • Simple sequence repeats • Single nucleotide polymorphisms • Variety identification

18.1 Introduction

The development of molecular tools has changed the way in which individual varieties can be identified and useful information concerning the genetic control of many agronomic characteristics can be analyzed. The ability to apply these molecular tools depends to some extent on the amount of other genomic information available for the specific plant species. The overall molecular toolbox for date palm is

C. Cullis (✉)
Department of Biology, Case Western Reserve University, 10900 Euclid Avenue, Cleveland, OH 44106-7080, USA
e-mail: cac5@case.edu

S.M. Jain et al. (eds.), *Date Palm Biotechnology*,
DOI 10.1007/978-94-007-1318-5_18, © Springer Science+Business Media B.V. 2011

limited, although the resources including a draft shotgun genome sequence that recently has been developed, will allow a rapid expansion of the types of questions that can be asked. The molecular marker technologies reach back to the use of isozymes followed by a series of DNA marker technologies and most recently by possibilities to compare complete genomes. The focus here will be on the use of DNA fragments as markers for varietal identification, for the elimination of off-types from *in vitro* propagated date palms and for possible marker-assisted breeding for disease resistance. An increasing number of publications are appearing on the subject, including a recent review specifically focused on Tunisian date palm germplasm (Rhouma et al. 2010).

18.2 Evolution of Marker Resources: RFLPS to Complete Genome Comparisons

18.2.1 Restriction Fragment Length Polymorphisms

The difference in DNA fragment lengths for homologous sequences is the basis for most DNA marker technologies. The first of the DNA markers to be developed was based on restriction fragment length polymorphisms (RFLPs). As the name implies, these markers are based on assaying variations in the DNA sequence identified through the use of restriction enzymes. When a restriction enzyme is used to fragment genomic DNA then characteristic lengths of DNA between sites are the result. These fragments are resolved through the use of gel electrophoresis and the DNA fragments are transferred to nylon filters and then identified by hybridization to specific probes. The polymorphisms can be the result of various alterations in the genomic DNA sequence. The most usual cause is the loss of a recognition site for that restriction enzyme, thus preventing the enzyme from cutting the DNA, although in rare cases a new restriction site can be produced.

An alternative cause of fragment size variation is by the insertion or deletion of a DNA sequence between adjacent restriction sites. The most useful RFLPs have been in regions of single (low) copy number components of the genome, as these regions can also be useful genetic markers. If the polymorphism was generated by the loss of a restriction site, then two outcomes are possible following hybridization. If the probe is confined to a region of the genomic DNA completely within one of the restriction fragments then a larger band would be observed (Fig. 18.1, probe b). However, if the probe spanned the altered restriction site (Fig. 18.1, probe a) then a new band the size of the sum of the other two bands would be observed. When an insertion or deletion is the cause of the RFLP then no matter what probe is used only a single band is identified with different sizes in various individuals (assuming they are homozygous). Another indication that the RFLP is a result of an insertion or deletion is that an RFLP is identified with a single probe following digestion with a number of different restriction enzymes.

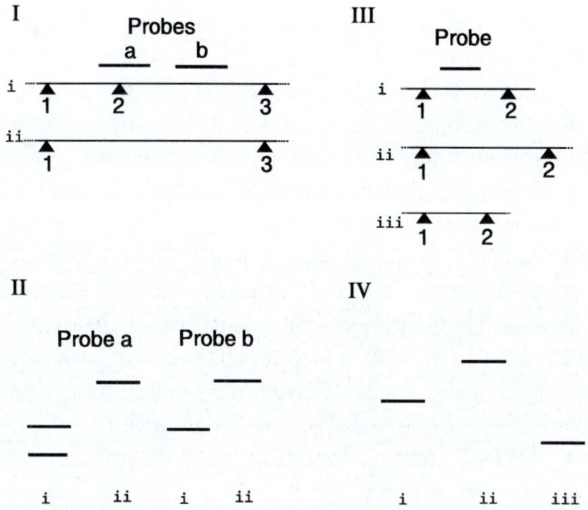

Fig. 18.1 The various outcomes of hybridization with different probes to identify the basis of various RFLPs. *I* – An RFLP caused by a loss of a restrictions site. *II* – The banding patterns after southern blotting and hybridization with probes a and b. *III* – RFLPs caused by an insertion (*ii*) or a deletion (*iii*) in the original allele (*i*). *IV* – The banding patterns after southern blotting and hybridization with the probe for these three alleles

Traditional RFLPs are identified through Southern blots and hybridization such that they are time-consuming and require large amounts of starting DNA. Therefore they are now rarely used in the standard form but have essentially been superseded by polymerase chain reaction (PCR)-based methods such as random amplified polymorphic DNAs (RAPDs), amplified fragment length polymorphisms (AFLPs) and simple sequence repeats (SSRs). These methods, particularly AFLPs, are essentially RFLPs assayed using PCR methods.

18.2.2 Random Amplified Polymorphic DNA

The polymerase chain reaction (PCR) opened the door to many applications and has revolutionized DNA marker technology. However, the major limitation to applying PCR-based methods is that sequence information is required to design the primers to be used in the PCR. To overcome the need for sequence information it was determined from statistical calculations that a single ten base primer could be used in the PCR reaction and result in a small number of amplified bands that would be useful as molecular markers. Experimentally this theoretical analysis was confirmed to be very successful in identifying large numbers of polymorphisms (Williams et al. 1990) in plant and other genomes. Therefore, since hundreds of random decamers are commercially available, the method can be applied when no other genomic information is known. The technique does have some drawbacks.

First, the technique has been reported to suffer from irreproducibility between laboratories and sources of thermostable enzyme, although, within a laboratory, reproducible results can be achieved (Jones et al. 1997). Newer versions of the Taq polymerases have been developed which do not function as well at low temperatures, and therefore are not useful for producing RAPD patterns since the short length of the primers requires an annealing temperature of about 40°C, rather than the more normal 60°C for most PCR reactions.

Second, the assignment of RAPD bands to a molecular map is genome specific, thus for every individual, a unique RAPD map needs to be produced since amplified fragments cannot be correlated across genomes.

Finally, RAPDs are dominant markers – a band is produced whether the fragment is homozygous or heterozygous so larger segregating populations are needed to produce accurate molecular maps. The basis for RAPDs can be a single nucleotide change in the primer sequence (since 1 base change is equivalent to a 10% mismatch in the primer sequence which will cause a 10°C change in the stability of the annealed primer) or insertion or deletion between the primer sites. Thus, overall, RAPDs are useful for DNA fingerprinting and identifying relationships between varieties, but less useful for genetic mapping.

In RFLP analysis, band size is the determining factor for identifying polymorphisms, but with RAPDs it is the presence or absence of a band. Since only one copy of an amplifiable sequence is necessary to result in an amplified band, RAPDs are dominant markers while RFLPs are generally codominant markers. For RFLPs the genomic change needs to be in the restriction site or as an insertion deletion. In the case of RAPDs single base changes anywhere in the primer will result in the loss of an amplified fragment as well as fragment size variation caused by insertions and deletions. In cases where the insertion is large, a loss of band will occur as the size of the amplifiable fragment exceeds that possible under the amplification conditions.

RAPDs have been applied to date palm variety identification (Adawy et al. 2002; Ahmed et al. 2006; Hussein et al. 2002; Sedra et al. 1998; Saker and Moursy 1998) and to the identification of somaclonal variation (Saker et al. 2006). These data were able to distinguish among varieties and place them in related groups, but no useful specific markers were found which could distinguish among somaclonal variants.

18.2.3 Amplified Fragment Length Polymorphisms

These markers are essentially RFLPs detected by PCR amplification (Vos et al. 1995). The primers used for the amplification are attached to the ends of the restriction fragments and then the complete genomic set is amplified. The method involves the digestion of the genomic DNA with two different restriction enzymes. Adaptors are then ligated to the ends of restriction fragments, with different adaptors added to each end. The two adaptors are then used as primers in a PCR reaction. The result of the amplification is a very large number of fragments. Any RFLP will result in a change in fragment size of a specific band, but this will be present against a background of all of the possible sized restriction fragments that can be amplified. The complex mixture of amplified bands needs to be separated on gels or by using automated DNA sequencers. Any polymorphic bands that are observed can be excised from the gel, cloned and sequenced to generate sequence-tagged sites (STSs).

These STS can be amplified with primers designed from the genomic sequence resulting in a single band which becomes a molecular marker. AFLPs are dominant markers since generally the presence/absence of a band is scored. However, the development of an STS will convert these regions into codomiant markers. The use of methylation-sensitive and -insensitive restriction enzyme isoschizomers allows for the analysis of potential epigenetic effects resulting from hyper- or hypo-methylated regions of the genome using AFLPs.

AFLPs have been used for date palm germplasm characterization (El-Assar et al. 2003, 2005). Forty-seven samples of date palm (*Phoenix dactylifera* L.) from Egypt were studied using four sets of amplified fragment length polymorphism (AFLP) markers (El-Khishin et al. 2003). A total of 350 bands were scored and of these 233 (66.6%) were found to be polymorphic. In another study AFLPs were used to survey genetic diversity in 40 ecotypes of date palm from oases in Tunisia (Rhouma et al. 2007). Six primer pairs resulted in a total of 428 AFLPs. The analysis of the data revealed that Tunisian date-palm germplasm has a large range of genetic diversity characteristics. The ecotypes were clustered into two groups that were independent of their geographic origin or the sex of the trees. This work has also been extended with the use of ISSRs (Zehdi-Azouzi et al. 2009). The study of regenerated plants to identify markers for somaclonal variation was carried out using AFLPs as well as RAPDS (Saker et al. 2006).

18.2.3.1 Microsatellites and SSRS

Microsatellites or SSRs are genetic markers that are derived from short (usually <6 bp) tandemly repeated sequences such as $(CT)_n$, $(AAT)_n$, $(GT)_n$. The terms microsatellite and SSR are often used interchangeably, although microsatellites are generally longer than the 2- to 3-bp unit of the SSRs. These SSR regions are widely dispersed through most animal and plant genomes and are also frequently polymorphic. The polymorphisms are due to the variability in the number of repeats that are present at a given site (Fig. 18.2). The length polymorphisms are produced by the changes in the number of the repeats such as that for a CT repeating unit as shown in Fig. 18.2. The three repeat classes shown in Fig. 18.2 are with 4 repeats, b with 2 repeats and c with 7 repeats. Since this SSR is a dinucleotide repeat, the actual size difference between the bands in a and b is 4 bp, that between a and c is 6 bp and that between b and c is 10 bp. These are small differences in length and therefore the separation of the amplified products needs to be done with high resolving power, frequently beyond the capability of agarose gels. Therefore the early ABI automatic sequencers are finding a renewed lease on life in the characterization of SSR length polymorphisms.

As with RFLPs, the primers for each SSR need to be isolated for each species. The current technology using the second-generation sequencing has made their isolation more rapid from genomic libraries or enriched genomic libraries (Panaud et al. 1996; Santana et al. 2009) or generated from an analysis of cDNA sequences. The enrichment techniques involve the use of biotinylated SSR fragments that are

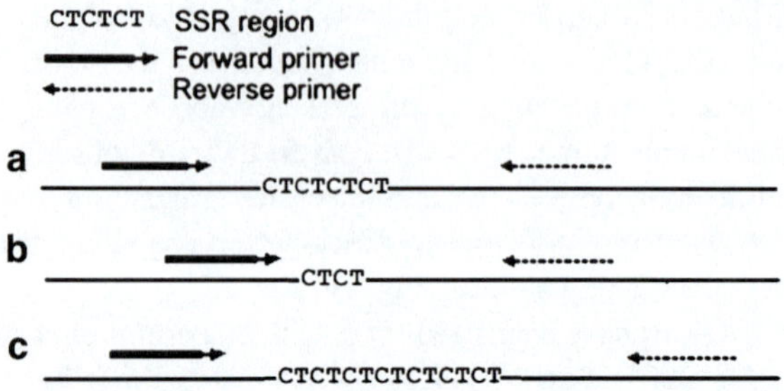

Fig. 18.2 Length polymorphisms resulting from differences in the number of SSR repeats at a locus

used to physically remove complementary sequences from a PCR amplified set of genomic fragments. The early iterations of the technique used the AFLP primers to provide the genomic representation. Initially all of the recovered fragments needed to be cloned and sequenced. However, using the 454 next generation sequencing methodology the complete set of fragments can be sequenced and assembled. The contigs are then analyzed using the program SSRIT (http://www.gramene.org/db/markers/ssrtool) to identify the contigs containing SSRs.

Where there is sufficient sequence surrounding the SSR region, primers are designed and the germplasm of interest is screened to find those which are polymorphic. Thus, although the identification of the SSRs regions is now much more rapid than before, each of them needs to be screened to find those that are polymorphic, which can be time consuming.

A second source of SSRs is from EST sequencing data (Ellis and Burke 2007). Frequently triplet repeats can occur in mRNAs since an additional amino acid will be inserted into the protein with no other changes in the structure and sequence of the protein. They can also be present in the 5′ and 3′ untranslated regions of the mRNA. The identification of these SSRs obviously needs a large EST sequence database and so is restricted to those species where a substantial sequence database is available. It is also possible that the level of polymorphisms present in EST-derived SSRs is lower than those identified from the genomic enrichment methods.

A modification of SSR markers is the inter-simple sequence repeat (ISSR) polymorphisms. ISSR is a general term for a region of the genome between microsatellite loci. They are assayed by using the complementary sequences to two neighboring microsatellites as the PCR primers. The variable region between these two regions is amplified, but by limiting the amplification parameters in the PCR reaction, the result is a mix of a variety of amplified DNA strands which can vary in length. The banding patterns frequently appear very similar to RAPD amplifications, and can therefore be used for DNA fingerprinting.

Data for date palm SSR and ISSR characterizations are available (Adawy et al. 2002; Zehdi et al. 2004; Zehdi-Azouzi et al. 2009). The genetic markers generated from seven selected ISSR primers were used to assess genetic diversity among a set of twelve Tunisian date palm varieties yielding 77 polymorphic markers which were sufficient to identify all of the varieties (Zehdi et al. 2004; Rhouana et al. 2009). The data for all three marker types (RAPDS, AFLPs and ISSRs) gave similar phylogenies within the tested date palm varieties. Each of these methods can be used separately or together for developing relationships between date palm varieties.

18.3 Single-Nucleotide Polymorphisms

Single-nucleotide polymorphisms (SNPs) are DNA sequence variations that occur when a single nucleotide (A, T, C, or G) in the genome sequence differs between two individual DNA samples. For example, a SNP might change the DNA sequence AGGATTCA to AGGATTTA. SNPs can occur in both coding (gene) and noncoding regions of the genome. As detailed above, SNPs are responsible for that class of RFLPs that result from the loss or gain of a restriction site. Many SNPs have no effect on cell function because they may not change protein structure (in fact, any SNP that occurs at the third position in the amino acid codon will have no effect if it does not change the amino acid sequence of the resulting protein). Their high frequency (perhaps as high as 2–3% in plant DNAs) means that they can be particularly useful in linkage mapping (Kristensen et al. 2001; Lai 2001). Again, the identification of SNPs depends on having sequence information available. Additionally, the sequence information needs to be obtained either from different varieties or from outbred individuals so that heterozygosity will be present.

Informatics tools can be used to compare the sequences and identify variations, but the raw data in the form of trace files may be important in deciding which polymorphisms may be real. When ESTs, for example, are built into unigene sets, any differences that may be present are eliminated in the formation of the consensus sequence and would need to be retrieved. Because there is no *a priori* way of differentiating between a true SNP and sequencing errors, each potential SNP must be validated. Even at a frequency of 1% these polymorphisms would generate an exceptionally large number of haplotypes if every polymorphism could be inherited independently. However, relatively few haplotypes are observed, indicating that perhaps the rate of SNP production is similar to the rate at which recombination occurs across the regions of the genome making up the haplotype blocks. Therefore, SNPs are most likely to be useful for defining haplotypes, rather than for their information individually, and so the use of SNPs is likely to involve linkage disequilibrium studies using the haplotype rather than the use of specific SNPs as individual molecular markers.

18.4 Genome Sequencing

The draft genome of date palm using second-generation sequencing has been reported for the Khalas cultivar. The draft genome, which constitutes approximately 20× coverage of the approximately 550 million base pairs genome, will contribute to a more complete understanding of date palm genetics as well as insights into improving yield, quality and disease resistance. Availability of the draft genome sequence will also be exceptionally useful for identifying molecular markers for date palm. The complete sequence can be screened for SSR regions that can then be used to develop a full suite for both genetic mapping and varietal identification. Since date palms are outcrossing they will be heterozygous at most loci. Therefore inspection of the sequence can reveal SNPs that are apparent in the assembly of the sequence information. In the announcement of the draft genome about 850,000 new SNPs have been identified from the comparisons between parental alleles present in the cultivar Khalas. In the reporting of the data, the manual inspection of the assembled contigs appear to be consistently correct for scaffolds that are 12,000 bases or less while the longer range assembly which is developed for spanning gaps is less certain. This is likely due to the interspersion of retrotransposable elements within the date palm genome that makes the assembly of short read shotgun sequencing problematical in the absence of any additional genomic resources such as BAC sequences. The date palm draft genome is available online at: http://qatar-weill.cornell.edu/research/datepalmGenome/download.html. This resource is freely available and should therefore provide for a rapid increase in the availability and use of date palm markers.

18.5 Conclusion and Prospective

It is clear that the date palm genome is structured similarly to that of other characterized plants. Therefore all the tools that have been developed for using DNA markers are available. Preliminary studies have demonstrated that population structures and lineage relationships can be identified with the current crop of DNA markers. As noted, the availability of the complete genome sequence will facilitate the development of a suite of different marker types to be applied appropriately. The development of a series of sequenced tagged sites (probably based in SSRs) will supply resources needed for the screening of collections to reduce the number of samples kept in germplasm banks. They will also add impetus to identifying markers linked to the various disease-resistant genes. With the steady increase in the sequencing resources, SNPs will also become more useful but the relative costs of SNP and SSR analyses may well determine which of the two-marker systems becomes most widely used. Although few publications using molecular markers are currently available, it is expected that this literature will substantially increase over the next few years.

Prospects for the application of molecular markers to date palms are still very minimal. The availability of the shotgun sequence and the steady lowering of the costs of high throughput sequencing will increase the resources and their application rapidly over the next few years. As with many plant species, decisions will need to be made concerning the level of whole genome sequencing compared to targeted re-sequencing as the most efficient method for useful applications. It is undoubted that the collection of many high polymorphism information content SSR primer pairs and validated SNPs will provide the tools for phylogenetic analyses as well as germplasm conservation. However, once genomic regions associated with important characteristics such as disease resistance, taste and post-harvest stability, the sequencing of these regions and the identification of the actual bases for these characteristics can be incorporated into the breeding and improvement programs. The identification of off-types arising in tissue culture propagation and the complete genome sequencing of normal and off-type individuals will lead to the identification of both markers for assessing off-type individuals in the regenerated plants as well as the 'mutations' responsible for these off phenotypes. Therefore these molecular markers and the tools developed through their use will facilitate the improvements in available germplasm for increasing the area under date palm cultivation as well as for the overall improvement of the plant material available to growers.

References

Adawy SS, Hussein EHA, El-Khishin D, Saker MM, El-Itriby HA (2002) Genetic variability studies and molecular fingerprinting of some Egyptian date palm (*Phoenix dactylifera* L.) cultivars II-RAPD and ISSR profiling. Arab J Biotechnol 5:225–236

Ahmed MMM, Soliman SS, Elsayed EH (2006) Molecular identification of some Egyptian date palm males by females varieties (*Phoenix dactylifera L.*) using DNA markers. J Appl Sci Res 2:270–275

El-Assar AM, Krueger R, Devanand PS, Chao CT (2003) Genetic analyses of date palms (*Phoenix dactylifera* L.) from Egypt using fluorescent – AFLP markers. HortScience 38:734

El-Assar AM, Krueger R, Devanand PS, Chao CT (2005) Genetic analysis of Egyptian date (*Phoenix dactylifera* L.) accessions using AFLP markers. Genet Resour Crop Evol 52:601–607

El-Khishin DA, Adawy SS, Hussein EHA, El Itriby HA (2003) AFLP fingerprinting of some Egyptian date palm (*Phoenix dactylifera* L.) cultivars. Arab J Biotechnol 6(2):223–234

Ellis JR, Burke JM (2007) EST-SSRs as a resource for population genetic analyses. Heredity 99:125–132

Hussein EHA, Adawy SS, El Khishin D, Moharam H, El-Itriby HA (2002) Genetic variability studies and molecular fingerprinting of some Egyptian date palm (*Phoenix dactylifera* L.) cultivars. 1-A preliminary study using RAPD markers. Arab J Biotechnol 5(2):217–224

Jones CJ, Edwards KJ, Castaglione S et al (1997) Reproducibility testing of RAPD, AFLP and SSR markers in plants by a network of European laboratories. Mol Breed 3:381–390

Kristensen VN, Kelefiotis D, Kristensen T, Borresen-Dale AL (2001) High-throughput methods for detection of genetic variation. Biotechniques 30:318–321

Lai E (2001) Application of SNP technologies in medicine: Lessons learned and future challenges. Genome Res 11:927–929

Panaud O, Chen X, McCouch S (1996) Development of microsatellite markers and characterization of simple sequence length polymorphism (SSR) in rice (*Oryza sativa* L.). Mol Gen Genet 252:597–607

Rhouma S, Zehdi-Azouzi S, Salem AOM et al (2007) Genetic diversity in ecotypes of Tunisian date-palm (*Phoenix dactylifera* L.) assessed by AFLP markers. J Hort Sci Biotechnol 82:929–933

Rhouma S, Ould Mohamed Salem A, Zehdi-Azouzi S et al (2009) Comparative analysis of genetic diversity in Tunisian date palm (*Phoenix dactylifera* L.) as revealed by RAPDs and AFLPs. Acta Hort 814:125–130

Rhouma S, Zehdi-Azouzi S, Dakhlaoui-Dkhil S et al (2010) Genetic variation in the Tunisian date palm (*Phoenix dactylifera* L). In: Ramawat KG (ed.) Desert plants. Springer, Berlin, pp 355–370

Saker MM, Moursy HA (1998) Molecular characterization of Egyptian date palm: 11 RAPD fingerprints. Proceeding first international conference on date palms, Al-Ain, pp 173–182

Saker MM, Adawy SS, Mohamed AA, El-Itriby HA (2006) Monitoring of cultivar identity in tissue culture-derived date palms using RAPD and AFLP analysis. Biol Plant 50:198–204

Santana QC, Coetze MPA, Steenkamp ET et al (2009) Microsatellite discovery by deep sequencing of enriched genomic libraries. BioTechniques 46:217–223

Sedra MH, Lashermes P, Trouslot P et al (1998) Identification and genetic diversity analysis of date palm (*Phoenix dactylifera* L.) varieties from Morocco using RAPD markers. Euphy 103:75–82

Vos P, Hogers R, Bleeker M et al (1995) AFLP: a new technique for DNA fingerprinting. Nucleic Acids Res 23:4407–4414

Williams KJ, Kubelik A, Livak K et al (1990) DNA polymorphism amplified by arbitrary primers are useful as genetic markers. Nucleic Acids Res 18:6531–6535

Zehdi S, Sakka H, Rhouma S et al (2004) Analysis of Tunisian date palm germplasm using simple sequence repeat primers. Afr J Biotechnol 3:215–219

Zehdi-Azouzi S, Rhouma S, Ould Mohamed Salem A et al (2009) Comparative analysis of genetic diversity in Tunisian collections of date palm cultivars based on random amplified polymorphic DNA and inter simple sequence repeats fingerprints. Acta Hort 814:149–156

Chapter 19
Biodiversity in Date Palm: Molecular Markers as Indicators

S. Elshibli and H. Korpelainen

Abstract Date palm, a tree known only as cultivated and under domestication for thousands of years with a wide range of distinct morphological diversity, has been a target of considerable research work, including, e.g. studies on both phenotypic and genetic diversity. Most research has focused on cultivar identification, where both types of markers – phenotypic and genetic – have been utilized. A wide range of molecular markers have been applied, yet, a relatively modest amount of results has been produced. In this chapter, existing knowledge of date palm biodiversity is reviewed. The usability and potential applications of molecular markers to detect biodiversity in plant species are discussed with special emphasis on date palm. Some powerful applications of molecular markers are not utilized or are underutilized, such as DNA sequencing approaches, which will provide a complete view of different events behind the evolution of the date palm genome, and positively enhance its utilization and conservation, especially under the circumstances of climate change and the scenarios expected to follow. Applications, such as germplasm conservation and management – apart from *in situ* conservation and seed bank strategies – will follow successful germplasm identification and development of efficient *in vitro* cloning and cryopreservation protocols. A detailed assessment of the date palm biodiversity and availability of information and research results are key factors for efficient utilization of date palm germplasm.

Keywords DNA sequencing • Genetic diversity • Germplasm • Molecular markers

S. Elshibli (✉) • H. Korpelainen
Department of Agricultural Sciences, University of Helsinki, FI-00014 Helsinki, Finland
e-mail: sakina.elshibli@helsinki.fi; helena.korpelainen@helsinki.fi

S.M. Jain et al. (eds.), *Date Palm Biotechnology*,
DOI 10.1007/978-94-007-1318-5_19, © Springer Science+Business Media B.V. 2011

19.1 Introduction

Locating and describing the existing diversity, identifying the materials to conserve, and developing conservation methods constitute the foundation for the conservation and utilization of date palm genetic resources. Genetic resources of plant species are direct consequences of genetic diversity and they constitute a major element of the global biological diversity. Phenotypic diversity, the apparent indicator of organismal diversity, is a consequence of the interaction effect between genetic and ecological diversities. Genetic diversity of plant species is also affected by human activities, such as domestication and cultivation processes. Genetic diversity among and within plant species can be detected directly by identifying variation in its components or indirectly by quantifying variation in phenotypic features that are assumed to have a genetic basis. Application of molecular markers in studying biodiversity of plant species has provided information on organismal and ecological aspects that have influenced the extent and distribution of genetic diversity, and on the processes that have given rise to observed patterns of variation. Date palm as an important multipurpose tree species of dry lands has been a target of considerable research work, including studies on both phenotypic and genetic diversity, where a wide range of molecular markers have been applied. The aim of writing this chapter is to review the existing knowledge of date palm biodiversity and to focus on the usability and potential applications of molecular markers to detect and assess biodiversity in plant species with special emphasis on date palm.

19.2 Biodiversity

Biodiversity or *biological diversity* refers to the variation of life at different levels of biological organization (Gaston and Spicer 2004). It comprises all three main elements of biological diversity, namely genetic diversity (from nucleotides, genes, chromosomes, individuals to populations), organismal diversity (from individuals to kingdoms) and ecological diversity (from populations, through habitat and ecosystem to biome) (Frankel et al. 1995; Gaston and Spicer 2004). Concerning genetics, biodiversity is the diversity of genes and organisms, where processes, such as mutations, gene exchanges and genome dynamics that occur at the DNA level and generate evolution can be detected among and within species (Frankel et al. 1995). Genetic diversity represents heritable variation within and between populations of organisms, and in the context of this chapter, within plant species. Such diversity includes genetic variation between individuals within a population, as well as variation between different populations of the same species. This pool of genetic variation present in populations is the basis for selection as well as for plant improvement. Thus, conservation of plant genetic diversity is essential for present and future human well-being (Ford-Lloyd and Jackson 1986). During recent years, there has been increasing awareness of the importance of adopting a holistic view of biodiversity, including agricultural biodiversity and its conservation for sustainable development and utilization. These principles have been incorporated also in the Convention on Biological Diversity and the Global Plan of Action of the Food and Agriculture Organization of the United Nations.The emphasis is now to understand the distribution and extent of genetic diversity available in plant species. Proper conservation and utilization of plant genetic resources is of prime importance for the whole of human society.

19.2.1 The Role of Domestication

Domestication of a plant species has important consequences on its genetic composition and characteristics, and it indeed has evolutionary significance that affects the genetic composition of the species. Based on studies in cultivated crop plants, like cereals, legumes, maize, wild yam, and vegetable and fruit species, three main pathways of domestication have been suggested, and they are distinguished by the prevailing breeding system (Frankel et al. 1995). Those pathways include the following: very rapid genotypic change for inbreeding species, gradual genotypic change for outbreeders, and plastic phenotypic change without immediate genotypic change for vegetatively-propagated species. Considering domestication without immediate genetic modification, two types can be described: (1) Phenotypic change without directional genotypic change. In this case, domestication amounts to the manipulation of pre-existing responses of a plant to specific environmental conditions. (2) Selective clonal domestication, where domestication amounts to a shift from sexual to vegetative reproduction. In the latter kind of domestication, important characteristics, such as distinctive characters of fruit, can become fixed and adopted as a standard for domesticated races. The domestication of ancient fruit species of Near Eastern origin, including date palm, belongs to this category of domestication, where plastic phenotypic changes take place without immediate genetic modifications. Such species are also characterized by high heterozygosity, as they are cross-pollinated. However, vegetative reproduction provides a chance for outstanding individuals to be cloned (Zohary and Spiegel-Roy 1975).

In all domestication pathways, two genetic events are of particular significance in the domestication process and subsequent evolution of a quantitative character of crop plants. These events are gene mutation and polyploidy; the importance of mutations is to provide an umbrella for selection of genes fitting a plant to the principal elements of cultivation as well as many environmental adaptations (Frankel et al. 1995). Different theories and models have been developed of the role of each major gene and polygenic variation to achieve adaptation, basically related to selection strength as well as the amount of phenotypic change required to achieve specific adaptation (Hoffmann and Parsons 1997; Lande 1983; Macnair 1991; Orr and Coyne 1991). Whenever a species has been taken into domestication that means it has been modified by the environment and by the cultural methods adopted by the domesticator. Evolution under natural conditions results in a slow accumulation of adaptations favoring survival and reproduction, while domestication may lead to rapid changes, which are due to the loss of effective products of particular genes (Lester 1989). For example, under domestication, many morphogenetic characters have evolved based on recessive mutations, while characters of a survival value based on dominant alleles may have been lost. Although polyploidy is not addressed in date palm, combining genes of several species was a radical evolutionary event that greatly expanded and affected the diversity of many crops like wheat (Frankel et al. 1995). It is now recognized that the evolutionary process through domestication from wild ancestors to advanced cultivars along with natural selection has resulted in great changes in the genetic diversity of crops. The entire range of domestic crops used in world agriculture is derived from wild species that have been modified through domestication, selective breeding and hybridization. Most remaining world centers of diversity contain populations of variable and adaptable landraces as well as wild and weedy relatives of crops, all of which provide valuable genetic resources for crop improvement (Harlan 1992).

19.2.2 Causes of Biodiversity Loss

The biodiversity that exists on earth today is a result of long-term evolution of different species, while some other species have become extinct at some point due to human induced as well as ecological and genetic factors (Lande 1999). Generally, there are two main factors under which other factors that threaten biodiversity can be explained (Gepts 2006). These factors are the increasing population pressure and the increased globalization of communications, travel and trade. Activities to support the increasing needs of humankind are considered as the major factor in the loss of biodiversity; for example, deforestation, the replacement of traditional agricultural systems by modern industrial methods, introduction of new species to non-original habitats, introduction of pesticides and insecticides, and also the homogenization of agricultural species. Thus, increased globalization leads to a loss of cultural diversity and, hence, to a reduction in biodiversity.

New varieties and cultivars have been developed due to domestication, selection and cultivation of wild plant species, as well as modern large-scale mechanized farming. However, diversity within crop species has declined and landraces have been displaced by bred cultivars, while the wild types of some plant species have disappeared (Gepts 2006). Although it is expected that changes in the diversity of tree species may count less than that of field crops, due to the long life cycles and slowness of selection and breeding, utilization of forest wood timber products as well as tree selection for fruit quality constitute a high pressure that reshapes diversity in tree species.

Loss of biodiversity can be detected as a decline in any of the different elements of biodiversity, such as a decrease in genetic diversity, species extinction, increasingly fragmented habitats and unbalanced ecosystems. Genetic erosion, or the reduction in genetic diversity in crop plants, takes place in various ways, including a reduced number of different plant species being grown and a decrease in genetic diversity within a species. For example, of the nearly 8,000 varieties of apples that grew in the United States a hundred years ago, more than 95% no longer exist. In Mexico, only 20% of the maize types recorded in 1930 can now be found, and only 10% of the 10,000 wheat varieties grown in China in 1949 remain in use (reviewed by Gepts 2006).

Within species, specifically in small populations, four main direct factors are responsible for the loss of genetic diversity, namely genetic drift, founder effect, bottleneck effect and inbreeding (Conner and Hartl 2004; Pannell and Charlesworth 2000; Slatkin 1977). A genetic bottleneck means a loss of genetic diversity after an event that severely reduces population size. A founder effect refers to a particular type of genetic bottleneck that occurs when a new population is started by a small number of individuals. Genetic drift causes a shift in the genetic composition of a population based on random factors rather than natural selection. These events affect the equilibrium of allele frequencies proposed based on the Hardy-Weinberg equilibrium (Hardy 1908; Weinberg 1908) and result in population restructuring, can be traced and detected using different molecular markers and analysis methods (Conner and Hartl 2004).

19.2.3 Measurement of Biodiversity

Given its great complexity, there is no single measure of biodiversity. Instead, it is important to concentrate on a specific element of biodiversity and address some specific question. However, two things are essential to consider, the number of entities and the degree of differences or similarities between those entities, described, for example, as species richness and species evenness (Purvis and Hector 2000). When discriminating between two groups of plant species, you may consider aspects such as morphology, genetics, biochemistry, physiology, biogeography, evolutionary history and ecological role (Frankel et al. 1995; Gaston and Spicer 2004). Biodiversity within a species can be also detected as differences or similarities in morphological and apparent characters as well as differences in responses to biotic and abiotic stresses. The major characteristic component behind these responses is the genetic makeup of the species, as well as its interaction with the environment. Many tools are now available to assess intra- and inter-specific biodiversity, and different methodologies have been developed for different systems and purposes.

19.3 Molecular Markers and Diversity Assessment

Molecular marker technology that relies on DNA or protein analysis provides the preferred tools to assess biodiversity at different levels of biological diversity among and within species, species-environment relationships, as well as biodiversity within an ecosystem, and molecular genetics and adaptation. Many molecular markers are highly polymorphic, they are not influenced by the environment or management practices, and usually they can be assessed using fast, non-destructive assays. Protein markers and enzyme electrophoresis have played an important role in exploring basic diversity information mostly related to the life cycle and behavior of plant species and to discriminate among accessions within plant species (Arulsekar and Parfitt 1986; Godt and Hamrick 1998; González-Pérez et al. 2004; Hamrick and Godt 1990, 1996; Hamrick et al. 1992; Vezvaei 2003; Wilson et al. 2001). However, protein markers have been largely replaced by DNA-based approaches, mainly due to the fact that protein markers are limited in number and are influenced by different environments as well as the developmental stage of the plant (Winter and Kahl 1995).

A range of different methods have been developed and extensively applied for detecting variability at the DNA level. Molecular markers are good indicators of genetic distances among accessions, because the commonly used markers are neutral in the face of selection. However, each of these markers exhibits different properties, for example, with regard to the type of data generated; there are two types: codominant markers and dominant markers, where the information that may be extracted from them and the numerical tools employed in data analysis are different. Allozymes and SSRs are codominant markers, which means that it is possible to identify the two alleles present at a particular locus of a diploid organism and that heterozygotes can be distinguished from homozygotes, enabling the estimation of allele frequencies and inbreeding levels in population genetic studies. Marker data can be considered as allele frequencies or as discrete data in a binary form, i.e., presence or absence of a DNA fragment with a specific length, to generate genetic distance matrices and to conduct other analyses. RAPD, ISSR and AFLP markers segregate as dominant markers and must be always treated as binary characters (presence/absence data). These data are usually converted into similarity matrices for the calculation of genetic distances. However, indirect estimation of allele frequencies for further analyses is possible, although conditioned by several assumptions (reviewed by Escudero et al. 2003).

More recently, the use of functional molecular markers, principally gene-targeting approaches to study biodiversity and genetic variation in plant species, have increased researchers' attention to carry out phylogenetic studies using DNA sequencing and targeting variation at single nucleotides (Single Neucleotide Polymorphisms, SNPs) (Barcaccia 2010; Gepts 2006; van Tienderen et al. 2002). Large-scale DNA sequencing facilities enable the assessment of genetic diversity using markers directly targeted at specific genes or gene families, a method applicable to *ex situ* management of genetic resources, ecological studies of diversity and conservation of endangered species (van Tienderen et al. 2002). Assessing biodiversity for ecologically important traits using DNA sequence approaches is more indicative for variation at loci that determine the ecological and functional distribution of the species and explain the ecological-genetic pattern of a species, such as adaptation to dry or saline environment.

Different molecular markers have different ways to support studies on the conservation and utilization of plant genetic resources as direct consequences of molecular diversity information; they have proved to be useful for investigating germplasm resources and in providing estimates of the amount of genetic diversity, geographical distribution and differentiation of populations within species. Each marker type has specific advantages and disadvantages and their applications vary depending on the nature and objective of the investigation and the properties of the species; some reviews provide thorough information on different DNA-based markers, their specific features, different applications and some hints on data management approaches (Karp et al., 1998; Kumar and Hirochika 2001; Muchugi et al. 2008; Newton et al. 1998; Savolainen and Karhu 2000).

Within-plant cell organization, the three genomes – nuclear, chloroplast and mitochondrial – offer useful models for molecular diversity assessments of plant species (Newton et al. 1998; Petit et al. 2005). Mitochondrial DNA exists basically in all eukaryotes while chloroplast DNA is specific to plants. Nuclear DNA, the biparentally inherited genome, provides markers that reflect gene dispersal by parental dispersing agents, seeds and pollen (Prentice et al. 2008). Chloroplast and mitochondrial DNA are usually maternally inherited but there are exceptions in some plant species concerning the inheritance pattern (Neale and Sederoff 1989; Petit et al. 2005; Reboud and Zeyl 1994). Mitochondrial DNA is generally the most slowly evolving genome of the three plant cell genomes on the sequence level with a tendency for gross structural rearrangements, while the more variable chloroplast DNA, also presumed to be inherited as a single unit without recombination (Knoop 2004; Newton et al. 1998), is considered useful in providing an attractive reservoir of phylogenetic information to trace the phylogeny of plant species. Maternally inherited genomes have been found to have a major effect on genetic differentiation with more subdivisions than paternally and biparentally inherited genomes (Petit et al. 2005). However, differences in the mode of inheritance, geographic-phylogenetic relationships as well as genome association have been reported in different species (Desplanque et al. 2000; Mohanty et al. 2000, 2003).

19.3.1 Extent and Pattern of Genetic Diversity

Genetic diversity is a critical component of biodiversity, and it can be measured directly by identifying variation in nucleotides, genes and chromosomes, or indirectly by quantifying phenotypic variation assumed to have a genetic basis. In a population, genetic diversity can be measured at the gene level by the total number of different genotypes present (richness) and by the equality or the evenness of genotype frequencies. In the case of alleles at a single genetic locus, these measures are: first, allelic richness (the total number of distinct alleles at that locus in the population or a sample) and the second measure is allelic evenness (Frankel et al. 1995).

It is generally recognized that plant genetic diversity changes in time and space. The extent and pattern of genetic variation in a plant species depend on its evolutionary history and breeding system, ecological and geographical factors as well as human factors (Lande 1999). Genetic variation at the species level generally shows the same trend as variation at population levels. However, some categorization has been noticed (Frankel et al. 1995; Hamrick and Godt 1996; Hamrick et al. 1992); for example, for taxonomic status, gymnosperms have been found more variable than monocots, while monocots appear more variable than do dicots. Categorization among species also includes latitudinal distribution as well as successional stages.

Obtaining sufficient information on the extent and pattern of genetic diversity within a species and differentiation occurring over geographical ranges, and understanding the ecological and genetic relationships among individuals and populations are essential for establishing guidelines for the utilization and conservation of the genetic resources of that species, especially when biotic and abiotic stresses are considered (Bates 1985; Bradshaw 1975; Frankel et al. 1995; Namkoong 1989; Virchow 1999). Such knowledge is essential for collecting and use of any plant species and its wild relatives. In order to manage conserved germplasm better, there is also a need to understand the genetic diversity that is present in collections. This will help us to rationalize collections and develop and adopt better protocols for the regeneration of germplasm, whether seeds or other propagules. Through improved characterization and development of germplasm collections based on genetic diversity information, it will be possible to exploit the available resources in more valuable ways. Yet, the optimal strategies for sampling individuals and populations for the purpose of conserving or using their genes differ among species (Ford-Lloyd and Jackson 1986; Frankel et al. 1995).

Measuring the extent of genetic diversity of a species is critical for the correct assessment of the performance of populations and possible threats (Frankham 1995). On the other hand, knowing the pattern and genetic structure of this diversity is also critical to understand the importance of the interaction between specific genome and environment in time and space and to trace different forces behind genome evolution (Berry 1989; Falk and Holsinger 1991). Such kinds of studies are also applicable for detecting genetic variability in a rare or endemic species for conservation genetic purposes, endangered diversity within a specific ecosystem, such as variation in natural populations of different species within a forest ecosystem, as well as understanding the biological nature of invasive plant species (Conner and Hartl 2004). The spread of invasive species constitutes one of the major threats to global environment and native species and is found to have a negative impact on about half of all threatened and endangered species in the U.S. (Conner and Hartl 2004). Studying the extend and pattern of genetic variation helps understanding of the causes, progression and consequences of biological invasions, which help in mitigating the considerable economic and biological costs that results from invasion (Dlugosch and Parker 2008).

The degree of richness and evenness of alleleic variation can be indexed by the level of heterozygosity at three different hierarchical levels of population structure: individuals, subpopulations and populations (Sunnucks 2000). Exact tests of differentiation for diploid populations, such as the F_{ST} estimator and goodness of fit tests are common when examining genetic differentiation and discovering whether genotype frequencies deviate from the Hardy-Weinberg equilibrium. In the absence of evolutionary forces, migration, mutation, natural selection and random genetic drift, and under random mating, allele frequencies do not vary through generations and follow a pattern known as the Hardy-Weinberg equilibrium (HWE) (Hardy 1908; Weinberg 1908). Exact test computation is based on the observed heterozygosity obtained from the data and the expected heterozygosity obtained using HWE principles. A reduction in observed heterozygosity below equilibrium levels is expected to have detrimental effects which are accompanied by inbreeding, meaning an increase in the average relatedness of breeding individuals as a result of genetic drift in a small population. It is hypothesized that increased inbreeding lowers individual fitness and lowers a population's adaptability (Conner and Hartl 2004; Lande 1988). Low genetic diversity indeed needs attention to be paid on: what are the reasons of this low diversity and what kind of evolutionary forces are there? For example, genetic information contributes to the identification of endangered plant species, as inbreeding is one of the factors implicated in species extinctions (Lande 1988). On the other hand, an excess of heterozygosity beyond the HWE may indicate the occurrence of processes, such as outcrossing and admixture effect, recent bottleneck or founder effect which have resulted in an excess of heterozygotes relative to the number of alleles present in the population; also, excess heterozygosity can be detected when a specific locus is linked to a gene that shows overdominance.

Generally, molecular data analysis follows approaches closely related to basic statistics familiar to researchers, including analyses of molecular variance, spatial autocorrelation, Mantel testing and diverse multivariate analyses (Sunnucks 2000). Depending on the marker system, molecular data based on allele frequencies can provide valuable information focusing on evolutionary relationships among individual alleles sampled from populations within species. For example, by analyzing allele frequencies of microsatellite data, genetic processes, such as effective population size, inbreeding, dispersal and migration as well as relatedness and parentage, can be estimated using different analyses and indices to explore information. For instance, recent changes in the effective population size, dispersal rates, geographic patterns or directions of a population colonization/expansion, current gene flow, rates of outcrossing and pollen immigration as well as associations between relatedness and isolation by distance can be estimated (Loikart and England 1999). Sunnucks (2000) provided a detailed review of the basic properties of molecular markers, recent developments in data analysis as well as of the markers of choice when dealing with population biological questions.

A series of computer packages have been developed, continuously updated and available to assist researchers in performing molecular diversity analyses like FSTAT (Goudet 1995), GENEPOP (Raymond and Rousset 1995), Structure 2.3.3. (Falush et al. 2003; Pritchard, et al. 2000), GENETIX (Belkhir et al. 2004), GenAlex 6 (Peakall and Smouse 2006), PHYLIP (Felsentein 2005), Arlequin 3.5 (Excoffier and Lischer 2010) and others. These programs have been widely used in molecular ecology and conservation genetics. They complement each other and can handle several data types like RFLPs, DNA sequences and microsatellite data, both allele frequencies and multi-locus genotypes, while allowing the user to carry out the same types of analyses irrespective of the data type (Excoffier et al. 2005).

19.3.2 Germplasm Identification

Within-species genetic variation is important for cultivar identification, and knowledge of it is indeed required for breeding, distribution and protection of proprietary rights of germplasm collections. For example, with respect to other markers, AFLP were confirmed to be an effective tool in detecting variation within several plant species and identifying accessions from different geographic sources (Polegri and Negri 2010). Studies on genetic diversity and phylogenetic relationship using AFLP markers have provided strong evidence that Italian landraces of cowpea are distinguishable from other landraces from abroad and from commercial varieties that can be found in the market. This information was found helpful when planning on-farm conservation of cowpea landraces in Italy and to secure the farmers' rights in keeping and protecting genotypes (Polegri and Negri 2010). DNA markers targeted at highly variable regions of the genome, such as microsatellites, are considered more suitable for varietal identification and have been extensively employed in DNA fingerprinting for the detection of genetic variation among cultivars (Morgante and Olivieri 1993). As microsatellite markers are species-specific, they need to be developed and selected for each economically-important plant species. Microsatellites have been developed and used for the assessment of genetic diversity and phylogenetic relationships of germplasm collections for a wide range of plant species, such as cereals (Song et al. 2005; Stephenson et al. 1998), vegetables (Spooner et al. 2007), forest trees (Barkley et al. 2005) and fruit tree species (Barkley et al. 2006).

Traditional varieties are synonymous to landraces when describing species selected for fruit quality and subjected, in their centers of origin or centers of diversity, to high pressure of selection. Due to the continued cycles of local planting, harvesting and farmer selection over time, landraces will be selected for local environmental and agroecosystem conditions and, hence, are adapted to their growing conditions (Villa et al. 2006). Important characteristics when identifying a traditional variety are historical origin, local adaptation and identity (Villa et al. 2006), which can be confirmed by using a molecular marker system of a powerful discrimination ability. Generally, most conservation efforts have focused on agriculturally important crops. For example, species used in plant genetic resources (PGR) research published in four specialized scientific journals in 1996, categorized by major crops, indicated that cereals, vegetables and legumes constituted 28%, 18% and 17%, respectively, while forestry and fruits constituted 6% and 3%, respectively, of research in PGR (Dudnik et al. 2001). On the other hand, one third of all *ex situ* accessions in gene banks represent just five plants, wheat, barley, rice, maize and beans (Lanteri and Barcaccia 2005). The success of crop germplasm conservation strategies, including on-farm ones, is facilitated by the short life cycle and the annularity of crop species. However, the same approaches are applicable to fruit tree species, although the processes are more complicated due to the biological nature of tree species. An example is the identification of traditional mango varieties among cultivated fruit trees using AFLP and SSR markers, which were used to discriminate landraces native to Mexico from some cultivars (Gálvez-López et al. 2009).

19.3.3 DNA Barcoding

DNA barcoding projects aim to provide an efficient method for species-level identification, which, in turn, contribute powerfully to taxonomic and biodiversity research. DNA barcoding proposes to use sequence information within a standardized genomic region common across taxa (Hebert et al. 2003a). Barcoding was applied first to animal species (Hebert et al. 2003b) and is now being established also for fungi and plants, including trees (Erickson et al. 2008; Gonzalez et al. 2009). The DNA barcoding approaches of plants may facilitate conventional taxonomic work, simplify species identification and support detection of species for specific applications, such as the detection of undesirable toxic species in food and migration of invasive species. The increasing interest in the application of DNA barcoding and the success in identifying species with universal barcodes may lead to applications that broaden our understanding of phylogenetics, and the extent and nature of population divergence, and facilitate comparative studies of population diversity in plant species (Hajibabaei et al. 2007).

19.3.4 Origin, Domestication and Distribution of Plant Species

The geographic distribution of the wild and ancestral relative of a plant species is among different types of evidence that is utilized to determine the origin of domestication in a plant species (de Candolle 1882; Zohary 1970). Such knowledge contributes to our knowledge of the pattern of diversity found in crop plants and past processes that have led to the present patterns. Molecular markers have been used to identify centers of domestication in crops, by identifying wild populations that are most closely related to the domesticated gene pool and the way they differentiate. Examples of these studies include tracing the origin and domestication of maize (Doebley 1990; Matsuoka et al. 2002), common bean (Gepts 1988), einkorn wheat (Heun et al. 1997), cacao (Motamayor et al. 2002) sunflower (Bruke and Wills 2006) and apple (Gharghani et al. 2010); while, some genes associated with domestication have been detected (Simons et al. 2006). As a result of domestication, some cultivated tree species shift from sexually reproducing wild populations to vegetatively propagated cultivated populations; e.g., the Neotropical fruit tree, *Spondias purpurea* (Anacardiaceae). Miller and Schaal (2006) employed AFLP data to explore the amount, structure, and distribution of variation in clonally propagated, domesticated populations and sexually reproducing wild populations of *S. purpurea*. The extent and pattern of genetic diversity revealed in this study suggested that *S. purpurea* was domesticated in two distinct regions within Mesoamerica.

The differences in the mode of inheritance of chloroplast and mitochondrial genomes have provided researchers with opportunities to study the genetic differentiation in plant species considering one parental genome. Uniparentally inherited chloroplast DNA has allowed the discovery of plant migration and spread often more efficiently than biparentally inherited nuclear markers (Prentice et al. 2008). For instance, studies using chloroplast DNA have been very useful in exploring the history of migration routes and recolonization of some European tree species, where the phylogeographical variation of species, such as oak, Saxifragaceae, *Corylus avellana* and silver birch, has been related to postglacial history (Ferris et al. 1993, 1998; Soltis et al. 1991). The same choloplast DNA approaches of tracing the history of distribution and structuring of natural population have been used in some herb species (Prentice et al. 2008). Colonization routes of Scots pine (*Pinus sylvestris*) have been revealed from the distribution and variation of mitochondrial DNA (Pyhäjärvi et al. 2008). Such kinds of investigations also give information on genetic drift and bottleneck events that lead to the restructuring of the genome of tree species and explore specific factors behind their evolution, e.g., studies on *Pinus* species (Bucci et al. 1997; Magri et al. 2007; Vendramin et al. 1998).

19.3.5 Core Collection and Genebank Management

To conserve agricultural biodiversity, genebanks are used to help preserve genetic resources of major crop plants for food security (Gepts 2006). However, models to conserve other species, such as medicinal plants and forest trees, are also demonstrated by Nageswara Rao et al. (2007) and Uma Shaanker et al. (2002). The use of molecular markers for genebank management requires that a decision is made to conserve representative material and core collections are established. The conserved material is managed by continuous testing of the original genetic diversity in a collection and by assessing genetic redundancy, which may take place due to the loss of seed viability and methods used to multiply and increase the collections. Molecular markers are also used to assess the genetic integrity of accessions during long-term storage and possibilities of accumulation of deleterious mutation within accessions (Spooner et al. 2005). The application of molecular markers has become a routine measure in plant germplasm management and conservation projects.

19.3.6 Molecular Diversity Databases

The accumulation of data generated by different types of molecular markers and large-scale DNA sequencing as well as by the use of different types of material and resources has enabled the establishment of databases for protein and DNA sequences, genetic mapping and data resulting from functional analyses and genotype-phenotype polymorphisms of different types of plant species. Most of the data are freely available to the public via the Internet. However, it is especially important for researchers to benefit from the possibilities of technology standardization across laboratories and plant species, as well as from the use of a platform for global exchange of data and efforts to build globally-valuable research projects. The most important sites that provide access to DNA and protein sequence databases and sequence analysis tools include GenBank and EMBL developed, respectively, by The National Center for Biotechnology Information in the USA and European Molecular Biology Laboratory in Europe. For example, these databases enable the identification of matches among sequences via BLAST (Basic Local Alignment Search Tool) programs. Other general database sites include, e.g. Genomic Diversity and Phenotype Connection (GDPC), which enables data to be integrated and reanalyzed in a standard format http://www.maizegenetics.net/gdpc/index.html. Integration of molecular data would enhance, among others, testing patterns of genetic structure and diversity over a wide geographical range and combining efforts of research groups from different locations following the establishment of standardization of methods and techniques within cooperating groups.

More specifically, the TropGene database project manages genetic and genomic information of tropical crops studied by CIRAD (http://tropgenedb.cirad.fr/), while species-specific databases also exist for legumes, beans, grains, Solanaceae, Rosaceae and others. Examples of *Arabidopsis thaliana* databases include The Arabidopsis Gene Regulatory Information Server (AGRIS), http://arabidopsis.med.ohio-state.edu/ and The Arabidopsis Gene Expression Database (AREX), http://www.arexdb.org/database.jsp. *A. thaliana*, was the first sequenced genome of a higher plant (The Arabidopsis Genome Initiative 2000), and it is used as a model plant that provides more detailed comparable information on the genetics of plant species, such as transcription factors and their target genes, gene expression as connected to specific function or different organs of plant species, and genetic bases for adaptation.

For tree species, the Dendrome Project (http://dendrome.ucdavis.edu) and associated TreeGenes database provide genetic information of forest trees for the international forest genetics community (Wegrzyn et al. 2008). Dendrome is a project of the Institute of Forest Genetics, USDA Forest Service. It is part of a larger collaborative effort to construct genome databases for major crop and forest species. It is worth mentioning that researchers at Weill Cornell Medical College (WCMC) in Qatar have completed a draft version of the date palm genome, generated by whole genome shotgun next generation DNA sequencing. This sequence draft was made available to researchers online at: http://qatar-weill.cornell.edu/research/datepalmGenome/download.html. This genome project is expected to enhance date palm research by developing new molecular markers with different applications, including studies on date palm biodiversity, such as detection of unique sequences for specific cultivars or characters.

19.4 Date Palm Germplasm

19.4.1 The Role of Biological Nature

The main biological features of tree species, such as date palm, include the characteristic long life cycle and the lack of mobility, which necessitate a need to withstand exposure to large fluctuations in their environment. Consequently, the need for the adaptability is extremely high in trees compared to other organisms. To fulfill these demands, tree species have to maintain large amounts of genetic variation to preserve adaptation potential and to be able to survive to subsequent generations (Müller-Starck and Gregorius 1986). Thereby, trees have developed different mechanisms to maintain high levels of genetic variation within species, including high rates of outcrossing, and the dispersal of pollen and seeds over wide areas, as well as different modes of reproduction. These mechanisms, combined with changing environments, have contributed to the evolution of forest and tree species, which are genetically diverse organisms compared to herbaceous species (Hamrick and Godt 1996; Hamrick et al. 1992). However, climate change and increasing human needs for food, water and energy constitute big challenges to biodiversity at all its levels.

19.4.2 Domestication and Distribution

As date palm was taken into domestication and cultivation since pre-historic times, its exact origin is unknown and the wild type is indistinct (Krueger 1995; Nixon 1951; Wrigley 1995). An important part of information was lost and is not available when considering the natural diversity of date palm and its wild ancestor. Date palm germplasm is considered to constitute a complex of wild forms, segregating escapees and cultivated clones, which are genetically interconnected by occasional hybridization, however, botanist consider all belonging to *Phoenix dactylifera* (Zohary and Hopf 2000). It is known that in the old days dates saved human life during travels across deserts for trade, pilgrimage and land exploration. The present evaluation of date palm germplasm and biodiversity is exclusively for already selected germplasm material. Generally, it is known that domestication narrows the genetic base of plant species, and this view can be applied to date palm as well.

The date palm was one of the first fruit trees to be domesticated worldwide (Zohary and Hopf 2000). The knowledge and spread of date palm culture developed under the stimulus of the civilizations in the valleys of the Tigris, Euphrates and Nile Rivers spread to the more important desert oases within their sphere of influence or communication as early as 4,000–5,000 BC (Barreveld 1993; Nixon 1951; Zaid and de Wet 2002a). The main factor behind the historic introduction and distribution of the date palm tree was human; humans played a major role in the dissemination of date palm. Dates were distributed and utilized for food, shelter, medicinal purposes, and they also had a lasting influence on religious and social life of the pioneer utilizers. By means of seeds carried in fruit, date palms were probably first disseminated around the Mediterranean coast and brought into new localities. During more recent times, along with seed dissemination, there have been a number of exchanges of offshoots between date- producing countries, including later also the south-eastern coast of Spain, the only locality in Europe, where commercial date culture has developed. Other regions include the desert valleys of south-central Asia, northern Iran and Afghanistan, the desert region of Australia, South Africa, Namibia, and the tropical and subtropical regions of the New World from California to South America (McCubbin 2007; Nixon 1951; Zaid and de Wet 2002a).

The most striking example of the commercial introduction of the date palm into new World is the administrated introductions of date palm offshoots into the United States, started as early as 1890 with the first importation of named cultivars of offshoots from North Africa and the Middle East, including Algeria, Egypt, Iraq and Morocco (Barreveld 1993; Hodel and Johnson 2007; Nixon 1951). Continuous research and breeding programs have resulted in nine American varieties that are currently grown commercially (Hodel and Johnson 2007). During the domestication process of date palm, humans learned to promote date palm productivity by restricting the number of plants per hectare, by assisting in the pollination process, by eliminating a high percentage of non-productive male palms, and by conducting vegetative propagation of selected cultivars. This whole process has existed for thousands of years, making it possible for humans to live, travel and survive in the most remote places of deserts (Barreveld 1993). On the other hand, the nature of the date palm tree enabled its existence and survival in the desert also by natural reproduction and in some cases without receiving either supplementary watering or any care (Pavez Wellmann et al. 2007). Nowadays, date palm is utilized as a unique multipurpose tree species in dry lands, and is very well suited for different agroforestry systems.

Although date palm culture is known to dominate in dry lands, the existing geographic distribution of date palm covers a wide range of environmental conditions. For instance, it grows and flourishes from 392 m below to 1,500 m above sea level, with an altitude range of 1,892 m (Zaid and de Wet 2002a). The tree also grows and tolerates different levels of salinity and is considered as salt tolerant (Pavez Wellmann et al. 2007). It can be successfully cultivated across a wide range of soil types from light loamy to clay soil due to its characteristic deep network of roots. The main features of roots and leaves of date palm classify it as drought tolerant. However, the wide range of environmental conditions where date palm occurs, as well as the wide range of characteristic features of leaves suggests that date palm or its specific phenotypes or genotypes are adapted to different environments (Ramoliya and Pandey 2003; Wickens 1998).

19.4.3 Genetic Erosion

In many date palm production areas in the world, serious stresses resulting in considerable plant losses have been reported. These stresses include, e.g. diseases, drought, desertification and floods (Baaziz et al. 2000; UNEP 2007; Zaid et al. 2002). Date palm is known as a drought-tolerant tree, and its existence is always connected to dry lands. However, with current climate change and increases in temperature and aridity, date palm culture undergoes stresses, and date palm productivity is greatly affected by drought and poor soil conditions. The annual Nile River floods result in destruction and losses of large numbers of date palm trees in Sudan. In North Africa, date palm culture has been seriously threatened for several decades by the vascular fusariosis bayoud disease caused by the fungus (*Fusarium oxysporum* f. sp. *albedinis*). At present, it occurs only in the date-growing areas of Morocco and in certain oases along the western border of Algeria, but the potential to spread to other date-growing areas of the world exists. The disease is characterized by progressive and rapid blanching and wilting of the leaves followed by the death of the palm, sometimes within 2 months. It probably spreads chiefly through the soil, but is believed to have been carried over long distances by man when transporting offshoots and woody parts of the palm, in which the fungus may survive for months. It is estimated that the bayoud disease has killed more than 13 million trees in less than one century in Morocco and Algeria, and some cultivars have disappeared (Baaziz et al. 2000; Oihabi 2000a, b). The use of resistant genotypes is the only method to protect trees from this disease, and tissue culture techniques are used to clonally multiply such genotypes (Baaziz et al. 2000). Examples of date palm culture under biotic and abiotic stresses are shown in Fig. 19.1.

Date palm has enemies, which in certain localities may occasionally be very destructive, especially during fruiting stages. Control measures have been developed for managing pests, like termites, scale insects and moths. For example, the white scale insect causes serious damage in Algeria, Kuwait, Libya, Mauritania, Morocco and Tunisia; while Iraq, Oman, Saudi Arabia and Sudan consider this pest as a moderate one (Zaid et al. 2002). On the other hand, the green scale insect causes serious damage and results in crop failure in Sudan (Ahmed 2007). These insect pests do not affect very much the survival of date palm trees, but they are very important when considering productivity and fruit quality. On the other hand the red palm weevil, *Rhynchophorus ferrugineus* Olivier, the most important pest of the date palm in the world, results in a direct death of the date palm tree. Symptoms of the red palm weevil infestation are notable only at advanced stages. The pest originates from southern Asia, where it is a serious pest of coconuts. It has advanced westwards very rapidly since the mid-1980s: in about 15 years it was reported in the United Arab Emirates, Kingdom of Saudi Arabia, Iran, Egypt and southern Spain (Gomez and Ferry 1998). Indeed, infestation with red palm weevil has followed exchange and distribution of plant material.

The slowness of natural vegetative propagation as well as the scarcity of designed cloning to multiply specific cultivars, such as those with a high yield or good quality (Zaid and de Wet 2002b), necessitates the establishment of improved tissue culture techniques for date palm cloning. Extensive and successful efforts have been made to propagate date palms through tissue culture (Al-Maarri 1995; Beacuchesne et al. 1986; Zaid and de Wet 2002b). There are two genetic diversity concerns when rapid multiplication through tissue culture is considered. First, the propagation of date palms through tissue culture has been accompanied by challenges that affect the level of certainty of true-to-typeness of propagules. However, commercial micropropagation and distribution of genotypically stable cultivars have been ongoing since early 1990s (Zaid and de Wet 2002b). The modified or mutated *in vitro* plants can result in considerable pollution for the traditional, slowly developed cultivars, if not recognized and carefully considered. The second concern is that the success to clone is found to be cultivar-specific (Al-Khayri and Al-Bahrany 2004; Pinker et al. 2009). Similar selectivity among genotypes has been also observed within natural cloning systems, i.e. vegetative propagation, of date palm. For example, in Sudan, it

Fig. 19.1 Date palm culture under biotic and abiotic stresses, examples: (**a**) River Nile floods in Sudan (UNEP 2007), (**b**) drought and desertification in Sudan (Elshibli 2009d), (**c**) Bayoud disease in Morocco (Photo provided by Mohammed Baaziz); the upper and lower arrows indicate bayoud disease external symptoms, whitening and wilting of leaves, respectively

is well-known that the dry type cv. Barakawi produces more offshoots per mother tree than the soft type cv. Mishrig Wad Laggai. The issue of farmers' shift to monoculture is again raised when disease-tolerant cultivars are considered, as resistance to bayoud disease has been found to be cultivar-specific (Baaziz et al. 2000).

Reported constraints suggest that large numbers of traditional cultivars in many countries may be endangered or diminishing, and it has been suggested that urgent conservation efforts are needed worldwide (Ahmed and Al-Qaradawi, 2009; Baaziz et al. 2000; Chao and Krueger 2007). It is worth mentioning that, to the best of our knowledge, there is as yet no network activity directed towards conservation of date palm genetic resources.

19.5 Date Palm Biodiversity

19.5.1 Phenotypic Diversity

Thousands of date palm cultivars exist in different countries covering a large geographic area. These cultivars have been developed by continuous selection performed by date palm growers all over the world, mainly to improve crop yield and quality, while natural selection, where genotypes that tolerate different types of environment, is also taking place. Generally, the selection, cloning and distribution of cultivars are mainly farmers' activities. Yet, date palm specialists have attempted to list and botanically describe those cultivars in their respective countries (Zaid and de Wet 2002c). There are up to 5,000 date palm cultivars all around the world (Jaradat and Zaid 2004). Based on botanical descriptions, there are about 1,000 cultivars in Algeria (Benkhalifa 1999), 400 in Iran, 370 in Iraq, 250 in Tunisia, 244 in Morocco (Zaid and de Wet 2002c) and 400 in Sudan (Osman 1984), as well as many additional cultivars in the other major date palm growing countries. The nomenclature of date palm cultivars is complicated, as it often describes an important trait, for example Hamra or Safra in Sudan and Tunisia (Elshibli and Korpelainen 2008; Zehdi et al. 2004) in north Kordofan in Sudan farmers describe date cultivars as two types depending on the color of fruits at the edible rutab (khalal) stage (*Hamra* means red in color and *Safra* means yellow in color). These two categories of date classification are also documented by the date palm research group in the Plant Production Department in King Saud University http://colleges.ksu.edu.sa/FoodsAndAgriculture/PlantProduction/Pages/Date%20palm%20research%20unit.aspx (2010) and in Hodel and Johnson 2007 (Fig. 19.2). Some introductions keep their original names or have minor changes, for example: Bitamoda and Gondaila, and Bertamoda and Gondila in Sudan and Egypt, respectively (Elshibli and Korpelainen 2008; Saker et al. 2006). Modification in the written names happens especially when translating the names from one language to another (El-Assar et al. 2005; Elshibli and Korpelainen 2008). Also, the same names may describe different genotypes in different locations. However, the names of the economically most

Fig. 19.2 At the edible rutab (khalal) stage, two categories of date classification exist depending on the color of fruits; (*left*) Barhee dates (originally have yellow rutab fruits), the main variety consumed at this firm stage in the U.S., at Flying Disc Ranch in Thermal, California; (*right*) Khisab dates (originally have red rutab fruits) at the USDA date collection at the Coachella Valley Agricultural Research Station, Thermal, California (Photos provided by David Karp)

important cultivars are well recognized and confirmed (Zaid and de Wet 2002c), although these cultivars are few in number when compared to the entire date palm germplasm.

Although yield capacity is among the most important characters that affect selection of date palm cultivar (Zaid and de Wet 2002c), they are commonly identified by a wide range of morphological features describing trees, fruits and fruit stalks (Al-Yahyai and Al-Khanjari 2008; Elhoumaizi et al. 2002; Elshibli and Korpelainen 2009a; Hodel and Johnson 2007; Ould Mohamed Salem et al. 2008). Within a date palm leaf, a very detailed description for the leaflets and spines was demonstrated as characteristic features for cultivar identity (Sedra 2001; Elhoumaizi et al. 2002; IPGRI et al. 2005; Ould Mohamed Salem et al. 2008; Fig. 19.3). Sedra (2001) and IPGRI et al. (2005) published the only descriptor on date palm varieties in the Maghreb countries; Fig. 19.4 shows some examples of tree and fruit characters described. Chemical characters of fruits, including sugars in their different forms, volatile matters, dietary fiber, acidity, as well as the pattern of changes that take place during different stages of ripening are among the wide range of apparent characters of fruits (Ahmed et al. 1995; Elshibli and Korpelainen 2009a, 2010). Elhoumaizi et al. (2002) studied the phenotypic diversity of 26 date palm cultivars from Morocco in an attempt to identify the most relevant characters that can be used as descriptors at early stages of tree growth apart from fruit characterization. In this study, Elhoumaizi et al. (2002) evaluated characters describing leaf, pinnae and spines. High variability was detected between cultivars, and a possible relationship between cultivars and selection for resistance to bayoud disease was suggested (morphologically related cultivars). Jaradat and Zaid (2004) evaluated 203 cultivars

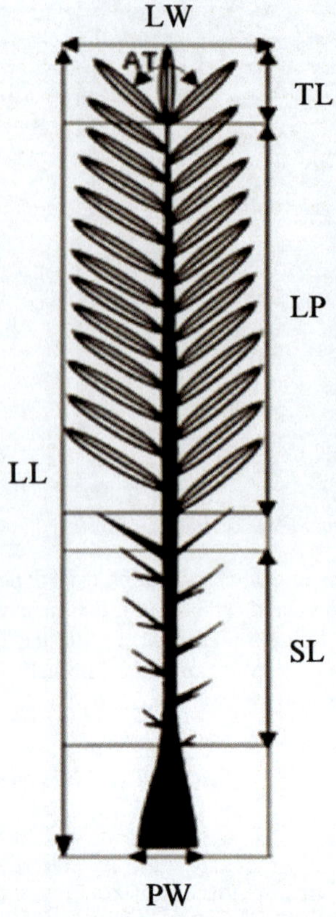

Key to the diagram

Parameter	Label
Leaf length	LL
Leaf width	LW
Leaf angle	LA
Spineted part length	SL
Length of leafleted part	LP
Petiole width	PW
Rachis thickness between the last spine and the first leaflet	RT
Leaf lets number	LN
Terminal leaflet length	TL
Terminal leaflet width	TW
Ventral angle of middle leaflet	VA
Middle leaflet width	WM
Middle leaflet length	LM
Leaflets spacing index at the middle	LI
Angle of leaflets on both sides of the terminal one	AT
Spine number	SN
Middle spine width	WS
Middle spine length	LS

Fig. 19.3 Detailed morphological traits of date palm tree leaf as measured and demonstrated by Ould Mohamed Salem et al. (2008); see also Elhoumaizi et al. (2002) and IPGRI et al. (2005)

from 20 ecogeographical regions in six countries in the Arabian Peninsula; cultivar characterization was more oriented to selection criteria, the quality and economic value of fruits. In this study Jaradat and Zaid (2004) reported high diversity of cultivars among and within countries. However, it was also detected that geographic origin was not a determinant criterion for cultivar grouping. It is worth mentioning that differences in morphological characters of a cultivar are always accompanied by pronounced differences in fruit characters.

Some attempts have been made to characterize date palm germplasm for their physiological and morphological responses to water stress (Elshibli et al. 2009b). The discrimination of cultivars in response to water stress was not found to be cultivar-specific but phenotype-specific, depending on being either a soft or dry type cultivar. This type of discrimination was observed also among date palms that were not subjected to different treatments, in terms of both fruit characters and tree morphology (Elshibli and Korpelainen 2009a). It must also be noted that soft and dry phenotypes were grouped similarly when the cultivars were characterized based on DNA markers. Another type of grouping patterns was observed by Bendiab et al. (1993),

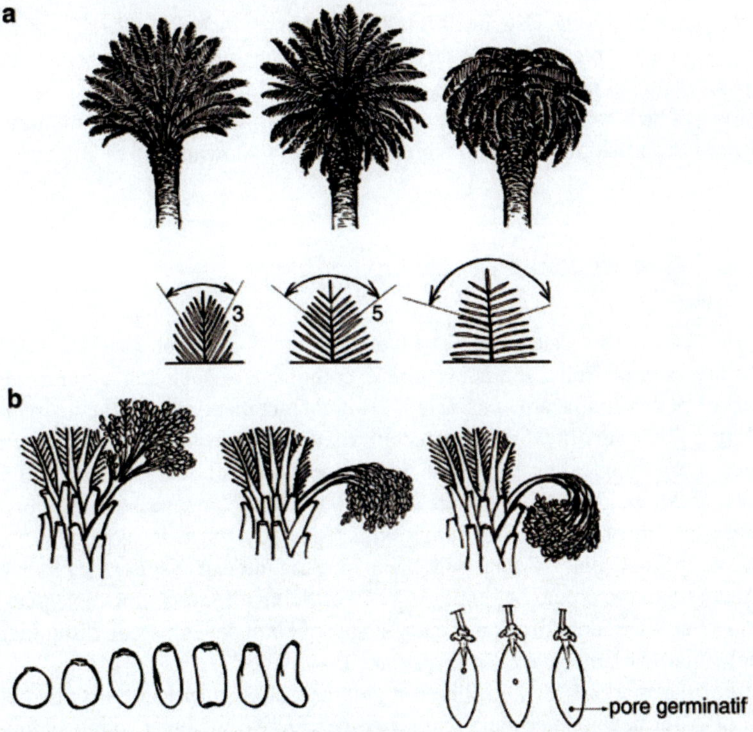

Fig. 19.4 Examples of morphological characters demonstrated in the descriptor and published by Sedra (2001): (**a**) Tree architecture from whole canopy to single leaf, (**b**) fruit stalk position, fruit shape and seed characters

when isozyme markers were applied. In that study, grouping was based on resistance to the bayoud disease. Both findings are very important as cultivar descriptors, yet, these findings need to be validated and studied further across a wide range of cultivars from diverse geographic locations with different ecological characteristics. Physiological studies on the date palm's responses to salinity and different water regimes, and their effects on yield and fruit quality provide implications for selecting for specific-traits, such as tolerance to salinity (Kurup et al. 2009).

The classification of dates into soft, semi-dry and dry types, mainly based on the texture of the ripe fruit, is thought to be associated with the content of particular sugars and water (Barreveld 1993; Cook and Furr 1953). Soft and dry types of dates exist in different producing countries (Zaid and de Wet 2002c), although the ecological distribution of these types has been reported only in Sudan (Osman 1984). Dry types of dates are exceptionally important for the date palm culture of Sudan, as they serve as food for date consumers year-round. Dry types are considerably easier to store than soft types that are consumed fresh during the production season. The introduced Deglet Noor cv. in northern Sudan named *Tunisi* has very dry fruits with an exceptionally high sucrose concentration when left to Tamar stage (Elshibli and Korpelainen 2009a; Osman 2001). We believe that there is wide range of cultivar diversity in each producing country, although such information is either unpublished or included in reports which are not obtainable. A solution could be a public database, which would contribute to the spreading and utilization of date palm information and would also help with maintaining the ownership of cultivars.

19.5.2 Genetic Richness Among Cultivars

It is well known that date palm has a high amount of morphological and compositional diversity of fruits, and is even more complicated when xenia and metaxenia effects of pollen grains are considered. Fruit characters constitute the main criteria used for cultivar identity. Other characters also used to describe date palm cultivars include morphological characters of trees, especially leaf characters as well as the general architecture of the tree and fruit stalks; these characteristics may undergo changes due to environmental conditions, especially when introductions to new environments take place. Along with the increasing interest to study the biodiversity of date palm in the context of exploring and utilizing all levels of its resources, and to resolve cultivar identification at early stages of plant development, different types of molecular markers have been investigated.

Molecular markers are considered highly useful, as they can be studied at any stage of plant development and using non-destructive methods, small samples of leaves being enough. The finding that isozyme polymorphism exists in date palm (Torres and Tisserat 1980) encouraged researchers to conduct further biochemical studies on date palm in Morocco and Tunisia, including isozyme analyses and activity analyses of peroxidases (Baaziz et al. 2000). Considerable diversity was detected among cultivars based on isoenzymes: such diversity has potential for practical applications when the activity of a specific enzyme has a role in plant infection and disease resistance (Baaziz and Saaidi 1988; Bendiab et al. 1993). Random amplified polymorphic DNA (RAPD) fingerprints have been used to study the genetic diversity and identify 13 date palm cultivars in Saudi Arabia (Al-Khalifah and Askari 2003). In this study, 140 RAPD primers were screened, of which 37 primers revealed polymorphism among cultivars. The number of polymorphic bands per primer varied between 2 and 17, with a mean of 4 major bands per primer. In Egypt, Soliman et al. (2003) and Adawy et al. (2006) tested four female cultivars and four unknown males and discovered a high genetic similarity (88–99%) among the tested samples. In addition, polymorphisms of date palm cultivars from California (Cao and Chao 2002), Egypt with a different introduction background (El-Assar et al. 2005), and Iraq (Jubrael et al. 2005) have been independently examined by amplified fragment length polymorphism (AFLP) markers. The Jaccard similarity index (JSI) for 18 Iraqi cultivars, studied by Jubrael et al. (2005), ranged between 11% and 76%.

The attributes of codominance, reproducibility and high resolution have all contributed to the popularity of nuclear microsatellites as powerful tools for population genetic analyses and within-species genotype identification. Recently, a microsatellite-enriched

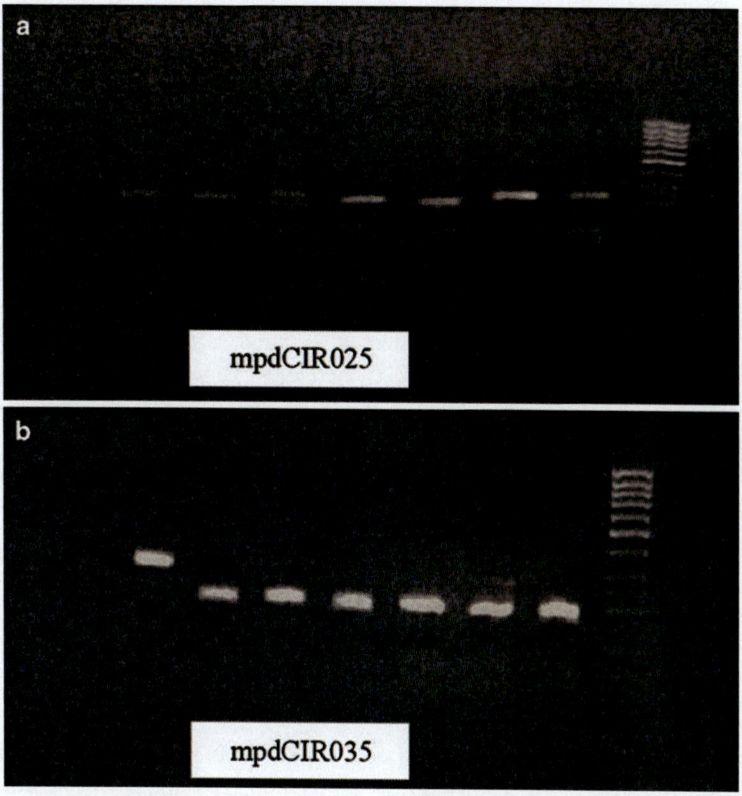

Fig. 19.5 Polymorphism detected at two microsatellite loci developed by Billotte et al. (2004) in seven date palm cultivars from Sudan; cvs. from *left* to *right*: Tonisi, Asada, Sultani, Galisoog, Siringai, Hamra and Safra. (**a**) Shows the codominance of microsatellite markers, heterozygosity present at locus mpdCIR025 in the cv. Sultani, (**b**) Shows differences in the product sizes of the same locus mpdCIR035. Photos provided by Sakina Elshibli

library was constructed and 16 microsatellite loci were characterized in date palm (Billotte et al. 2004). Development of these markers for date palm enabled more detailed genetic diversity assessments of cultivars from different producing countries. Figure 19.5 shows an example of the application of microsatellite markers for the detection of polymorphism in date palm cultivars. Microsatellite markers have been applied to assess the genetic diversity and relationships of 49 accessions of date palm from three oases in Tunisia (Zehdi et al. 2004). Elshibli and Korpelainen (2008) studied the genetic diversity among 45 date palm cultivars as well as 23 male accessions collected from Sudan and Morocco. In both studies, high genetic diversity was revealed in the tested germplasm. The mean expected heterozygosity (H_{exp}) was 0.70 in Tunisian samples, 0.84 in Sudanese cultivars, 0.82 in Moroccan cultivars and 0.80 in Sudanese males. The Moroccan cultivars showed significant differentiation in relation to the Sudanese group, as indicated by F_{ST} values and genetic distances (Elshibli and Korpelainen 2008; Zehdi et al. 2004). In some groups tested by Zehdi et al. (2004), heterozygote deficiency was observed with a significant deviation from the Hardy Weinberg equilibrium (HWE). The proposed reason is strong selection operating in the collection sites in Tunisia. On the other hand, excess heterozygosity was observed among the tested cultivars as well as populations collected from Sudan and Morocco, although deviations from HWE were insignificant.

Excess heterozygosity may result from a combination of sexual and asexual reproduction systems during the selection and multiplication process, as well as from the long-distance exchange of plant material and merger of small populations into one. On the other hand, heterozygote deficiency may be a sign of a fragmented or small population. In such a case, populations may lose allelic richness and genetic diversity, and have increased population differentiation due to genetic drift and inbreeding depression (Buza et al. 2000; Lienert 2004; Tomimatsu and Ohara 2003). Using the same set of microsatellite markers, Ahmed and Al-Qaradawi (2009) analyzed the genetic diversity among 15 cultivars collected from the experimental farm of Qatar University. A total of 40 alleles with an average of 4 alleles per locus were scored. Similarity values computed according to the Jaccard similarity coefficient ranged from 0.000 to 0.750. Within-cultivar genetic diversity has been detected also by Devanand and Chao (2003) in Medjool cv. using AFLP markers.

19.5.3 Pattern of Genetic Diversity

Sedra et al. (1998) investigated the reliability of the RAPD marker system as a tool for the identification of date palm cultivars and examined 43 cultivars from Morocco. In this and other studies, it has become apparent that RAPD markers show considerable difficulties when characterizing cultivars, and these difficulties include mainly low polymorphisms, irreproducibility and lack of evident organization (Benkhalifa 1999; Sedra et al. 1998; Trifi et al. 2000). When Jubrael et al. 2005 conducted a genetic clustering analysis for 11 Iraqi cultivars using AFLP-based genetic markers, the highest shown bootstrap values ranged between 26% and 73%. This kind of weak genetic relationships and inability to discriminate between individual cultivars from different geographic locations has been observed also in other studies with different markers (Adawy et al. 2006; Cao and Chao 2002; El-Assar et al. 2005; Elshibli and Korpelainen 2008, 2010).

Microsatellite markers have been applied to assess the genetic diversity and relationships of 49 accessions of date palm from three oases in Tunisia (Zehdi et al. 2004). Elshibli and Korpelainen (2008) studied the genetic diversity among 45 date palm cultivars and 23 male accessions collected from Sudan and Morocco. In both studies, only little geographic structuring was observed among tested germplasms (Elshibli and Korpelainen 2008; Zehdi et al. 2004). However, Elshibli and Korpelainen (2009b) detected significant structuring of date palm populations according to locations for the production of soft and dry types of dates. This pattern of genetic structuring needs further investigations in relation to the ecological factors that affect the adaptability of each group to specific environment. A similar grouping trend has been reported by Bendiab et al. (1993) in relation to resistance to bayoud disease and isoenzyme polymorphism. Also a potential for using a mitochondrial molecular marker to detect resistance to the bayoud disease was raised in investigations on some date palm cultivars by (Quenzar et al. 2001).

Clonal propagation with tissue culture techniques accompanied by some uncertainties concerning the true type of propagated material (Al Kaabi et al. 2005; Al-Wasel 2005) necessitates the need for methods to confirm the validity of miropropagation techniques in date palm culture. The genetic variability of tissue culture-derived date palms compared to mother plants has been assessed using different genetic markers, and a wide range of variability has been reported (El-Assar et al. 2005; Saker et al. 2000, 2006). However, no links have been identified between genetic variability and the observed morphological and/or physiological disorders. Al-Ruqaishi et al. (2008) used microsatellite markers to screen and analyze the genetic diversity among clonal genotypes of date palm derived by somatic embryogenesis in Oman. A total of 21 palms, representing 14 Omani, 5 Bahraini, 1 Iraqi and 1 Moroccan genotypes, were screened with 10 microsatellite markers. All primer pairs produced an amplification product in the expected size range and detected high levels of polymorphism among the analyzed samples. A correspondence analysis revealed that the genotypes from Bahrain and Iraq showed a close relationship with accessions already grown in Oman, while the genotype from Morocco (Medjool) appeared distinct from the rest of the material.

19.6 Conclusion and Prospective

It is notable that in most of the previously mentioned studies the main goal was to identify cultivars rather than studying the genetic diversity of date palm in general, or genetic diversity within cultivars, and only one of the studies was about the population genetics of date palm. Molecular markers (RAPD, microsatellites and AFLP) were found unable to discriminate between different cultivars: discrimination was impossible between cultivars within a country. In other studies, the discriminating procedure in data analysis missed the bootstrapping of the data, which is expected to give stronger statistical estimates of the degree of genetic differentiation among cultivars and populations. We provide a summary of different genetic diversity estimates in Table 19.1. However, different numbers of samples and different markers used mean that it is difficult to compare results obtained in various studies. It is notable that among all tested markers there was a weak genetic association among most tested samples. The assumption was always that these cultivars are clones, while the results indicated the possibility of an introduction of sexually produced plants at different stages of cloning and distribution, a process that takes place by farmers and date palm growers. It is well documented that the large number of described cultivars are developed by farmers, each in its specific region with very distinct morphological, compositional and agronomic characters, while a wide range of adaptive traits is expected among these traditional cultivars due to the

Table 19.1 Polymorphism in date palm germplasm as revealed by different types of molecular markers

Molecular marker	No. cultivars	Main objective	No. of markers screened	No. of markers used	Total no. of alleles or bands	Range of alleles or bands/marker	Genetic distance and diversity indices	Main result	Reference
RAPD	43	IDC	123	19	37	1.9	0.13–0.77[a]	Weak cluster	Sedra et al. (1998)
RAPD	13	IDC	140	37	149	2–17	19–89[b]	Weak cluster	Al-Khalifah and Askari (2003)
		GD							
RAPD	4 females	IDC	5	5	–	4–12	88–99[b]	–	Soliman et al. (2003)
	4 males								
SSR	49	IDC	16	14	100	4–10	0.71 H_{exp} 0.61 H_{obs}	Some cvs identified	Zehdi et al. (2004)
		GD							
SSR	45 cvs	IDC	16	1s6	343	14–44	0.85H_{exp} 0.53–3.64[a]	Weak cluster	Elshibli and Korpelainen (2008)
	23 males	GD							
ISSR	28 cvs	GD	12	7	67	7–11	0.23–0.98[a]	–	Zehdi et al. (2005)
	6 males								
SSR	21	IDC	16	10	–	–	–	Weak cluster	Al-Ruqaishi et al. (2008)
		In vitro clones							
SSR	15	GD	16	10	40	3–6	–	–	Ahmed and Al-Qaradawi (2009)
AFLP	21	IDC	32	4	328	–	–	Weak cluster	Cao and Chao (2002)
AFLP	18	GD	122	4	17.4	–	0.11–0.76[b]	Weak cluster	Jubrael et al. (2005)

AFLP	47	GD In vitro clones	4 sets	4	350	87.5 avg	–	–	El-Assar et al. (2005)
RDA	2	IDC	–	–	–	–	–	–	Vorster et al. (2002)
Mitocho-ndrial	36	RSB	–	–	–	–	–	–	Quenzar et al. (2001)

RAPD random amplified polymorphic DNA, *ISSR* inter simple sequence repeats, *SSR* microsatellites, *AFLP* amplified fragment length polymorphism, *RDA* representational difference analysis, *IDC* identification of cultivars, *GD* genetic diversity, *RSByoud* R and S plasmid for resistance to Bayoud disease, H_{exp} expected heterozygosity, H_{obs} observed heterozygosity

[a] Genetic distance

[b] Similarity index

distribution under a wide range of environmental conditions. Molecular markers, especially DNA sequencing approaches, provide a potential for better discrimination ability to maximize the utilization of these genetic resources and to secure the property rights of these genotypes for the producers.

In connection with the conservation of plant genetic resources, the information on the extent and pattern of diversity in different locations and its distribution in relation to environmental conditions is a fundamental element that helps the development of cost-effective conservation strategies and allows a choice of how much and of which germplasm should be maintained *ex situ* or managed *in situ* in order to achieve the identified conservation objectives. Yet, with date palm, some aspects of cloning and cryopreservation need to be studied further and settled before it is possible to decide how and what to conserve. Generally, for tree species, both *ex situ* and *in situ* conservation methods are challenging: *ex situ* conservation is challenged by the development of suitable methods of *in vitro* and cryopreservation; while *in situ* conservation is challenged by climate change and natural disasters. However, possibilities of on-farm conservation and development of cryopreservation techniques for date palm germplasm management will follow the successful choice of molecular markers for concrete germplasm identification and inspection of somatic variants.

On the other hand, studying the population genetics of date palm originating from different geographic locations will introduce the genetic richness of date palm germplasm in its widest sense, including the diversity within and among populations, and also the male trees that exist in each of these locations. Moreover, studying the genetic diversity and structure of date palm populations will add knowledge on how date palm genes disperse and the route of distribution, and the existence of rare genes which can be directly linked to specific traits, genetic erosion, fitness, genetic drift and adaptation. The effect of exchanging plant material as well as the effect of the date palm-specific method of pollination may have a great impact on the genetic structure of these populations, although there is significant structuring of population groups according to dry and soft phenotypes, or bayoud-tolerant genotypes. Further studies to trace genetic bases behind this classification may add information on date palm responses to environment and adaptation.

Some powerful applications of molecular markers are not utilized or are underutilized, such as DNA sequencing approaches, compared to DNA separation by gel electrophoresis and genotyping. Although no complete nuclear genome is yet sequenced for a tree species, it is feasible that complete sequencing of tree genomes becomes increasingly common with the use of high throughput DNA sequence facilities. On the other hand, valuable genetic information is expected to result when analyzing date palm's mitochondrial as well as chloroplast genomes. Some well-distinguished apparent characters like soft and dry types, or red and yellow khalal types, type and concentration of sugars at the tamar stage, and other distinct morphological characters can be readily traced by different marker systems to reveal genetic-phenotype relationships.

The nature of date palm culture and utilization of date palm tree practiced by date palm growers is apparent in the high genetic diversity observed in most reported investigations, and in the large numbers of morphologically distinct cultivars and groups of cultivars. However, a detailed assessment of genetic diversity and genetic structure among date palms is very important when aiming to reveal different events behind the evolution of the date palm genome. There is a great need for breeding programs, which will positively enhance the utilization, but especially the conservation, under the circumstances of climate change, natural disasters, pests and diseases, unsolved issues related to tissue culture cloning and distribution, and the tendency to monoculture that affects the genetic diversity of date palm. Availability of information and research results is the key factor for scientists to work effectively. It offers the possibility to compare different germplasm and various types of criteria used for classification, and it enhances selection for future research material that suits different research activities and objectives. Such kinds of services can be obtained through the construction of a proper database system, development of networking activities as well as through research projects conducted on a global basis.

References

(The) Arabidopsis Genome Initiative (2000) Analysis of the genome sequence of the flowering plant *Arabidopsis thaliana*. Nature 408:796–815

Adawy SS, Hussein EHA, Ismail SEME, El-Itriby HA (2006) Genomic diversity in date palm (Phoenix dactylifera L.) as revealed by AFLPs in comparison to RAPDs and ISSRs. In: Abstracts of third international date palm conference, 19–21 Feb, Abu Dhabi

Ahmed MA (2007) The efficacy of four systemic insecticides using two methods of application against the green date palm pit scale insect (*Asterolecanium phoenicis* Rao.) (*Palmapsis phoenicis*) (Homoptera: Asterolecaniidae) in Northern Sudan. Acta Hort 736:369–389

Ahmed TA, Al-Qaradawi AY (2009) Molecular phylogeny of Qatari date palm genotypes using simple sequence repeats markers. Biotechnology 8:126–131

Ahmed IA, Ahmed AWK, Robinson RK (1995) Chemical composition of date varieties as influenced by the stage of ripening. Food Chem 54:305–309

Al Kaabi HH, Zaid A, Ainsworth C (2005) Plant-offtypes in tissue culture-derived date palm (*Phoenix dactylifera* L). Proceedings of the international workshop on true-to-typeness of date palm tissue culture-derived plants, 23–25 May. National Institute of Agronomic Research, Morocco, pp 14–26

Al-Khalifah NS, Askari E (2003) Molecular phylogeny of date palm (*Phoenix dactylifera* L.) cultivars from Saudi Arabia by DNA fingerprinting. Theor Appl Genet 107:1266–1270

Al-Khayri JM, Al-Bahrany AM (2004) Genotype-dependent *in vitro* response of date palm (*Phoenix dactylifera* L.) cultivars to silver nitrate. Sci Hort 99:153–162

Al-Maarri KW (1995) Date palm propagation through tissue culture techniques. Dabs-Dimishg, Syria

Al-Ruqaishi IA, Davey M, Alderson P, Mayes S (2008) Genetic relationships and genotype tracing in date palms (*Phoenix dactylifera* L.) in Oman based on microsatellite markers. Plant Genet Res 61:70–72

Al-Wasel ASAA (2005) Survey study on: somaclonal variations in in vitro-derived date palm trees. Proceedings of the international workshop on true-to-typeness of date palm tissue culture-derived plants, 23–25 May. National Institute of Agronomic Research, Morocco, p 44

Al-Yahyai R, Al-Khanjari S (2008) Biodiversity of date palm in the Sultanate of Oman. Afr J Agric Res 3:389–395

Arulsekar S, Parfitt DE (1986) Isozyme analysis procedures for stone fruits, almond, grape, walnut, pistachio, and fig. HortSci 21:928–933

Baaziz M, Saaidi M (1988) Preliminary identification of date palm cultivars by esterase isoenzymes and peroxidase activities. Can J Bot 66:89–93

Baaziz M, Majourhat K, Bendiab K (2000) Date palm culture in the Maghreb countries: constraints and scientific research. Proceedings of the date palm international symposium, Windhoek, Namibia, 22–25 Feb, pp 306–311

Barcaccia G (2010) Molecular markers for characterizing and conserving crop plant germplasm. In: Jain SM, Brar DS (eds.) Molecular techniques in crop improvement, 2nd edn. Springer, Netherlands, pp 231–256

Barkley NA, Newman M, Wang ML et al (2005) Assessment of the genetic diversity and phylogenetic relationships of a temperate bamboo collection by using transferred EST-SSR markers. Genome 48:731–737

Barkley NA, Roose ML, Krueger RR, Federici CT (2006) Assessing genetic diversity and population structure in a *Citrus* germplasm collection utilizing simple sequence repeat markers (SSRs). Theor Appl Genet 112:1519–1531

Barreveld WH (1993) Date palm products. Agricultural Services Bulletin 101. Food and Agriculture Organisation of the United Nations, Rome

Bates DM (1985) Plant utilization: patterns and prospects. Econ Bot 39:241–265

Beacuchesne G, Zaid A, Rhiss A (1986) Meristematic potentialities of bottom of young leaves to rapidly propagate date palm. Proceedings second symposium on the date palm, 3–6 March, King Faisal University, Al-Hassa, Saudi Arabia, pp 87–93

Belkhir K, Borsa P, Chikhi L et al (2004) Genetix 4.05 software. http://www.univ-montp2.fr/~genetix/genetix/genetix.htm

Bendiab K, Baaziz M, Brakez Z, Sedra MH (1993) Correlation of isoenzyme polymorphism and bayoud-disease resistance in date palm cultivars and progeny. Euphy 65:23–32

Benkhalifa A (1999) Gestion de la diversité génétique du palmier dattier en Algérie. Paper presented at the workshop constitution et organisation d'équipes de recherche scientifique dans les domaines de foresterie et des arbres fruitiers, 13–15 April. Marrakech, Morocco

Berry RJ (1989) Ecology: where genes and geography meet. J Anim Ecol 58:733–759

Billotte N, Marseilla CN, Brottier P et al (2004) Nuclear microsatellite markers for the date palm (*Phoenix dactylifera* L.): characterisation, utility across the genus *Phoenix* and in other palm genera. Mol Ecol Notes 4:256–258

Bradshaw AD (1975) Population structure and the effects of isolation and selection. In: Frankel OH, Hawkes JG (eds.) Crop genetic resources for today and tomorrow. Cambridge University Press, London, pp 37–51

Bruke JM, Wills DM (2006) Chloroplast DNA variation confirms a single origin of domesticated sunflower (*Helianthus annuus* L.). J Hered 97:403–408

Bucci G, Vendramin GG, Lelli L et al (1997) Assessing the genetic divergence of *Pinus leucodermis* Ant. endangered populations: use of molecular markers for conservation purposes. Theor Appl Genet 95:1138–1146

Buza L, Young A, Thrall P (2000) Genetic erosion, inbreeding and reduced fitness in fragmented populations of the endangered tetraploid pea *Swainsonia recta*. Biol Conserv 93:177–186

Cao BR, Chao CT (2002) Identification of date cultivars in California using AFLP markers. HortSci 37:966–968

Chao CT, Krueger RR (2007) The date palm (*Phoenix dactylifera* L.): overview of biology, uses and cultivation. HortSci 42:1077–1082

Conner JK, Hartl DL (2004) A primer of ecological genetics. Sinauer Associates, Sunderland

Cook JA, Furr JR (1953) Kinds and relative amount of sugar and their relation to texture in some American-grown date varieties. Am Soc Hort Sci 61:286–292

de Candolle AP (1882) Origins of cultivated species. Reprint 1967. Hafner, London

Desplanque B, Viard F, Bernard J et al (2000) The linkage disequilibrium between chloroplast DNA and mitochondrial DNA haplotypes in *Beta vulgaris* ssp. *maritima* (L.): the usefulness of both genomes for population genetic studies. Mol Ecol 9:141–154

Devanand PS, Chao CT (2003) Identification of genetic strains of 'Medjool' and 'Deglet Noor' date palm cultivars in California using amplified fragment length polymorphism (AFLP) markers. Acta Hort 623:333–340

Dlugosch KM, Parker IM (2008) Founding events in species invasions: genetic variation, adaptive evolution, and the role of multiple introductions. Mol Ecol 17:431–449

Doebley JF (1990) Molecular evidence and the evolution of maize. Econ Bot 44:6–27

Dudnik NS, Thormann I, Hodgkin T (2001) The extent of use of plant genetic resources in research – a literature survey. Crop Sci 41:6–10

El-Assar AM, Krueger RR, Devanad PS, Chao CT (2005) Genetic analysis of Egyptian date (*Phoenix dactylifera* L.) accessions using AFLP markers. Genet Res Crop Evol 52:601–607

Elhoumaizi MA, Saaidi M, Oihabi A, Cilas C (2002) Phenotypic diversity of date-palm cultivars (*Phoenix dactylifera* L.) from Morocco. Genet Res Crop Evol 49:483–490

Elshibli S (2009d) Genetic diversity and adaptation of date palm (*Phoenix dactylifera* L.). Academic Dissertation, University of Helsinki

Elshibli S, Korpelainen H (2008) Microsatellite markers reveal high genetic diversity in date palm (*Phoenix dactylifera* L.) germplasm from Sudan. Genetica 134:251–260

Elshibli S, Korpelainen H (2009a) Biodiversity of date palms (*Phoenix dactylifera* L.) in Sudan: chemical, morphological and DNA polymorphisms of selected cultivars. Plant Genet Res Charac Util 7:194–203

Elshibli S, Korpelainen H (2009b) Excess heterozygosity and scarce genetic differentiation in the populations of *Phoenix dactylifera* L.: human impact or ecological determinants. Plant Genet Res Charac Util 7:95–104

Elshibli S, Korpelainen H (2010) Identity of date palm (*Phoenix dactylifera* L.) germplasm in Sudan: from the morphology and chemical characters to molecular markers. Acta Hort 859:143–153

Elshibli S, Elshibli E, Korpelainen H (2009) Drought adaptation of date palm (*Phoenix dactylifera* L.): morphological and physiological comparison of selected cultivars. In: Elshibli S. Genetic diversity and adaptation of date palm (*Phoenix dactylifera* L.). Academic Dissertation, University of Helsinki

Erickson DL, Spouge J, Resch A et al (2008) DNA barcoding in land plants: developing standards to quantify and maximize success. Taxon 57:1304–1316

Escudero A, Iriondo JM, Torres ME (2003) Spatial analysis of genetic diversity as a tool for plant conservation. Biol Conserv 113:351–365

Excoffier L, Lischer HEL (2010) Arlequin suite ver 3.5: a new series of programs to perform population genetics analyses under Linux and Windows. Mol Ecol Res 10:564–567

Excoffier L, Laval G, Schneider S (2005) Arlequin (version 3.0): an integrated software package for population genetics data analysis. Evol Bioinformatics Online 1:47–50

Falk DA, Holsinger KE (1991) Genetics and conservation of rare plants. Oxford University Press, Oxford

Falush D, Stephens M, Pritchard JK (2003) Inference of population structure using multilocus genotype data: linked loci and correlated allele frequencies. Genetics 164:1567–1587

Felsentein J (2005) PHYLIP (Phylogeny Interference Package) version. 3.6. Distributed by the author. Department of Genome Sciences, University Washington, Seattle

Ferris C, Oliver RP, Davy AJ, Hewitt GM (1993) Native oak chloroplasts reveal an ancient divide across Europe. Mol Ecol 2:337–344

Ferris C, King RA, Väinölä R, Hewitt GM (1998) Chloroplast DNA recognizes three refugial sources of European oaks and suggests independent eastern and western immigrations to Finland. Heredity 80:584–593

Ford-Lloyd B, Jackson M (1986) Plant genetic resources: an introduction to their conservation and use. Edward Arnold, London

Frankel OH, Brown AHD, Burdon JJ (1995) The conservation of plant biodiversity. Cambridge University Press, Cambridge

Frankham R (1995) Conservation genetics. Ann Rev Genet 29:305–327

Gálvez-López D, Hernández-Delgado S, González-Paz M et al (2009) Genetic analysis of mango landraces from Mexico based on molecular markers. Plant Genet Res Charac Util 7:244–251

Gaston KJ, Spicer JI (2004) Biodiversity: an introduction, 2nd edn. Wiley-Blackwell, New York

Gepts P (1988) Phaseolin as an evolutionary marker. In: Gepts P (ed.) Genetic resources of *Phaseolus* beans. Kluwer, Dordrecht, pp 215–241

Gepts P (2006) Plant genetic resources conservation and utilization: the accomplishments and future of a societal insurance policy. Crop Sci 46:2

Gharghani A, Zamani Z, Talaie A et al (2010) The role of Iran (Persia) in apple (*Malus* × *domestica* Borkh.) domestication, evolution and migration via the Silk Trade Route. Acta Hort 859:229–233

Godt MJW, Hamrick JL (1998) Allozyme diversity in the grasses. In: Cheplick GP (ed.) Population biology of grasses. Cambridge University Press, Cambridge, pp 11–29

Gomez M, Ferry S (1998) The red palm weevil in the Mediterranean Area. http://www.palms.org/palmsjournal/2002/redweevil.htm. Accessed May 2010

Gonzalez MA, Baraloto C, Engel J et al (2009) Identification of Amazonian trees with DNA barcodes. PLoS ONE 4(10):e7483. doi:10.1371/journal.pone.0007483

González-Pérez MA, Caujapé-Castells J, Sosa PA (2004) Allozyme variation and structure of the Canarian endemic palm tree *Phoenix canariensis* (Arecaceae): implications for conservation. Heredity 93:307–315

Goudet J (1995) FSTAT version 1.2: a computer program to calculate *F*-statistics. J Hered 86:485–486. URL. http://www.unil.ch/popgen/softwares/fstat.htm

Hajibabaei M, Singer GAC, Hebert PDN, Hickey DA (2007) DNA barcoding: how it complements taxonomy, molecular phylogenetics and population genetics. Trends Genet 23:167–172

Hamrick JL, Godt MJW (1990) Allozyme diversity in plant species. In: Brown AHD, Clegg MT, Kahler AL, Weir BS (eds.) Plant population genetics, breeding, and genetic resources. Sinauer, Sunderland, pp 43–63

Hamrick JL, Godt MJW (1996) Effects of life history traits on genetic diversity in plant species. Philos Trans R Soc Lond B 351:1291–1298

Hamrick JL, Godt MJW, Sherman-Broyles SL (1992) Factors influencing levels of genetic diversity in woody plant species. New For 6:95–124

Hardy HG (1908) Mendelian proportions in a mixed population. Science 28:49–50

Harlan JR (1992) Crops and man, 2nd edn. American Society of Agronomy-Crop Science Society, Madison

Hebert PDN, Cywinska A, Ball SL, DeWaard JR (2003a) Biological identifications through DNA barcodes. Proc R Soc Lond Biol Sci B 270:313–321

Hebert PDN, Ratnasingham S, DeWaard JR (2003b) Barcoding animal life: cytochrome *c* oxidase subunit 1 divergences among closely related species. Proc R Soc Lond Biol Sci Ser B 270:96–99

Heun M, Schäfer-Pregl R, Klawan D et al (1997) Site of einkorn wheat domestication identified by DNA fingerprinting. Science 278:1312–1314

Hodel DR, Johnson DV (2007) Imported and American varieties of dates (*Phoenix dactylifera*) in the United States. University of California ANR Publication 3498, Oakland, CA

Hoffmann AA, Parsons PA (1997) Extreme environmental change and evolution. Cambridge, Cambridge University Press

IPGRI, INRAA, INRAM, INRAT, FEM, PNUD (2005) Descripteure du palmier dattier (*Phoenix dactylifera* L.). Institut international des recoures phytogenetiqes, Rome, Italie; Fonds pour l' Environnement Mondial, Washington, Etats-Unis; Programme des Nations Unies pour le Developpement, New York, Etats-Unis; Institut National de la rescherche Agronomique, d' Algerie, du Maroc et de Tunsie

Jaradat AA, Zaid A (2004) Quality traits of date palm fruits in a center of origin and center of diversity. Food Agric Environ 2:208–217

Jubrael JMS, Udupa SM, Baum M (2005) Assessment of AFLP-based genetic relationships among date palm (*Phoenix dactylifera* L.) varieties of Iraq. J Am Soc Hort Sci 130(3):442–447

Karp A, Isaac PG, Ingram DS (1998) Molecular tools for screening biodiversity-plants and animals, 1st edn. Chapman & Hall GmbH, Weinheim

Knoop V (2004) The mitochondrial DNA of land plants: peculiarities in phylogenetic perspective. Curr Genet 46:123–139

Krueger R (1995) Mystique of the date palms links old and new worlds. Diversity 11:128–129

Kumar A, Hirochika H (2001) Applications of retrotransposons as genetic tools in plant biology. Trends Plant Sci 6:127–134

Kurup SS, Hedar YS, Al Dhaheri MA et al (2009) Morpho-physiological evaluation and RAPD markers-assisted characterization of date palm (*Phoenix dactylifera* L.) varieties for salinity tolerance. J Food Agric Environ 7:503–507

Lande R (1983) The response to selection on major and minor mutations affecting a metrical trait. Heredity 50:47–65

Lande R (1988) Genetics and demography in biological conservation. Science 241:1455–1460

Lande R (1999) Extinction risks from anthropogenic, ecological and genetic factors. In: Landweber LF, Dobson AP (eds.) Genetics and the extinction of species. Princeton University Press, Chichester West Sussex, pp 1–22

Lanteri S, Barcaccia G (2005) Molecular markers based analysis for crop germplasm preservation. In: Proceedings international workshop on the role of biotechnology for characterisation and conservation of crop, forestry, animal and fishery genetic resources (available at www.fao.org/biotech/docs/lanteri.pdf)

Lester RN (1989) Evolution under domestication involving disturbance of genic balance. Euphy 44:125–132

Lienert J (2004) Habitat fragmentation effects on fitness of plant populations – a review. J Nat Conserv 12:53–72

Loikart G, England PR (1999) Statistical analysis of microsatellite DNA data. Trends Ecol Evol 14:253–256

Macnair MR (1991) Why the evolution of resistance to anthropogenic toxins normally involves major gene changes: the limits to natural selection. Genetica 84:213–219

Magri D, Fineschi S, Bellarosa R et al (2007) The distribution of *Quercus suber* chloroplast haplotypes matches the palaeogeographical history of the western Mediterranean. Mol Ecol 16:5259–5266

Matsuoka Y, Vigouroux Y, Goodman MM et al (2002) A single domestication for maize shown by multilocus microsatellite genotyping. Proc Natl Acad Sci USA 99:6080–6084

McCubbin MJ (2007) The South African date palm industry - strengths and weaknesses. Acta Hort 736:59–63

Miller AJ, Schaal BA (2006) Domestication and the distribution of genetic variation in wild and cultivated populations of the Mesoamerican fruit tree *Spondias purpurea* L. (Anacardiaceae). Mol Ecol 15:1467–1480

Mohanty A, Martín JP, Aguinagalde I (2000) Chloroplast DNA diversity within and among populations of the allotetraploid *Prunus spinosa* L. Theor Appl Genet 100:1304–1310

Mohanty A, Martin JP, Gonzales LM et al (2003) Association between chloroplast DNA and mitochondrial DNA haplotypes in *Prunus spinosa* L. (Rosaceae) populations across Europe. Ann Bot 92:749–755

Morgante MN, Olivieri AM (1993) PCR-amplified microsatellites as markers in plant genetics. Plant J 3:175–182

Motamayor JC, Risterucci AM, Lopez PA et al (2002) Cacao domestication I: the origin of the cacao cultivated by the Mayas. Heredity 89:380–386. doi:10.1038/sj.hdy.6800156

Muchugi A, Kadu C, Kindt R et al (2008) Molecular markers for tropical trees, a practical guide to principles and procedures. In: Dawson I, Jamnadass R (eds.) ICRAF Technical Manual 9. World Agroforestry Centre, Nairobi

Müller-Starck G, Gregorius H-R (1986) Monitoring genetic variation in forest tree populations. Proceedings of the 18th IUFRO World Congress, division 2, vol. II. Forest plants and forest protection. IUFRO World Congress, Vienna, Yugoslavia, pp 589–599

Nageswara Rao M, Ganeshaiah KN, Uma Shaanker R (2007) Assessing threats and mapping sandal (*Santalum album* L.) resources in peninsular India: Identification of genetic hot-spot for *in-situ* conservation. Conserv Genet 8:925–935

Namkoong G (1989) System of gene management. In Gibson GL, Griffin AR, Matherson AC (eds.) Breeding tropical trees: population structure and genetic strategies in clonal and seedling forestry. Proceedings of IUFRO Conference, Pattaya, Thailand, pp 1–8

Neale DB, Sederoff RR (1989) Paternal inheritance of chloroplast DNA and maternal inheritance of mitochondrial DNA in loblolly pine. Theor Appl Genet 77:212–216

Newton AC, Allnut TR, Gillies CCM et al (1998) Molecular phyllogeography, intra-specific variation and the conservation of tree species. Trends Ecol Evol 14:140–145

Nixon RW (1951) The date palm: "tree of life" in the subtropical deserts. Econ Bot 5:274–301

Oihabi A (2000a) Date palm in North Africa and Near East: production and constraints. In: The date palm international symposium, Namibia, 22–25 Feb 2000, pp 312–316

Oihabi A (2000b) Date palm genetic resources in North Africa. Proceedings of date palm international symposium, 22–25 Feb 2000. Namibia, pp 333–335

Orr HA, Coyne JA (1991) The genetics of adaptation: a reassessment. Am Nat 140:725–742

Osman AMA (1984) The performance of date palms in the Sudan. Acta Hort 143:231–237

Osman AMA (2001) Development of date palm culture in Republic of Sudan. Paper presented at a workshop on date palm culture and dates production in Republic of Sudan. Date Palm Research and Development Network, Khartoum, Sudan, 22–17 August

Ould Mohamed Salem A, Rhouma S, Zehdi S et al (2008) Morphological variability of Mauritanian date palm (*Phoenix dactylifera* L.) cultivars as revealed by vegetative traits. Acta Bot Croatica 67:81–90

Pannell JR, Charlesworth B (2000) Effects of metapopulation processes on measures of genetic diversity. Philos Trans R Soc Lond B 355:1851–1864

Pavez Wellmann A, Escobar Araya H, Johnson DV (2007) Date palm cultivation in Chile and Peru (South America): current status and future prospects for development. Acta Hort 736:71–85

Peakall R, Smouse PE (2006) GENALEX 6: genetic analysis in Excel. Population genetic software for teaching and research. Mol Ecol Notes 6:288–295

Petit RJ, Duminil J, Fineschi S et al (2005) Comparative organisation of chloroplast, mitochondrial and nuclear diversity in plant populations. Mol Ecol 14:689–711

Pinker I, Ibraheem YM, Böhme MH (2009) Propagation of some date palm cultivars by using tissue culture methods. Acta Hort 839:71–78

Polegri L, Negri V (2010) Molecular markers for promoting agro-biodiversity conservation: a case study from Italy. How cowpea landraces were saved from extinction. Genet Res Crop Evol, DOI 10.1007/s10722-009-9526-z

Prentice HC, Malm JU, Hathaway L (2008) Chloroplast DNA variation in the European herb *Silene dioica* (red campion): postglacial migration and interspecific introgression. Plant Syst Evol 272:23–37

Pritchard JK, Stephens M, Donnelly P (2000) Inference of population structure using multilocus genotype data. Genetics 155:945–959

Purvis A, Hector A (2000) Getting the measure of biodiversity. Nature 405:212–219

Pyhäjärvi T, Salmela MJ, Savolainen O (2008) Colonization routes of *Pinus sylvestris* inferred from distribution of mitochondrial DNA variation. Tree Genet Genom 4:247–254

Quenzar B, Bouachrine B, Hartmann C et al (2001) A mitochondrial molecular marker of resistance to bayoud disease in date palm. Theor Appl Genet 103:366–370

Ramoliya PJ, Pandey AN (2003) Soil salinity and water status affect growth of *Phoenix dactylifera* seedlings. N Z J Crop Hort Sci 31:345–353

Raymond M, Rousset F (1995) GENEPOP (version 1.2): population genetics software for exact tests and ecumenicism. Heredity 86:248–249

Reboud X, Zeyl C (1994) Organelle inheritance in plants. Heredity 72:137–140

Saker M, Bekheet S, Taha H et al (2000) Detection of somaclonal variation in tissue culture-derived date palm plants using isozyme analysis and RAPD fingerprints. Biol Plan 43:347–351

Saker MM, Adawy SS, Mohamed AA, El-Itriby HA (2006) Monitering of cultivar identity in tissue culture-derived date palms using RAPD and AFLP analysis. Biol Plan 50:198–204

Savolainen O, Karhu A (2000) Assessment of biodiversity with molecular tools in forest trees. In: Jain SM, Minocha SC (eds.) Molecular biology of woody plants. Springer, Netherlands, pp 395–406

Sedra, MH (2001) *Descripteur du Palmier dattier (Phoenix dactylifera* L.). INRA, Rabat, Morocco

Sedra MH, Lashermes P, Trouslot P et al (1998) Identification and genetic diversity analysis of date palm (*Phoenix dactylifera* L.) varieties from Morocco using RAPD markers. Euphy 103:75–82

Simons KJ, Fellers JP, Trick HN et al (2006) Molecular characterization of the major wheat domestication gene *q*. Genetics 172:547–555

Slatkin M (1977) Gene flow and genetic drift in a species subject to frequent local extinction. Theor Popul Biol 12:253–262

Soliman SS, Ali BA, Ahmed MMM (2003) Genetic comparisons of Egyptian date palm cultivars (*Phoenix dactylifera* L.) by RAPD-PCR. Afr J Biotechnol 2:86–87

Soltis DE, Mayer MS, Soltis PS, Edgerton M (1991) Chloroplast DNA variation in *Tellima grandiflora* (Saxifragaceae). Am J Bot 78:1379–1390

Song QJ, Shi JR, Singh S et al (2005) Development and mapping of microsatellite (SSR) markers in wheat. Theor Appl Genet 110:550–560

Spooner D, van Treuren R, de Vicente MC (2005) Molecular markers for genebank management. IPGRI Technical Bulletin No. 10 [Online]. www.ipgri.cgiar.org/publications/pdf/1082.pdf

Spooner DM, Nunez J, Trujillo G et al (2007) Extensive simple sequence repeat genotyping of potato landraces supports a major re-evaluation of their gene pool structure and classification. Proc Natl Acad Sci USA 104:19398–19403

Stephenson P, Bryan G, Kirby J et al (1998) Fifty new microsatellite loci for the wheat genetic map. Theor Appl Genet 97:946–949

Sunnucks P (2000) Efficient genetic markers for population biology. Trends Ecol Evol 15:199–203

Tomimatsu H, Ohara M (2003) Genetic diversity and local population structure of fragmentation of *Trillium camschatcense*. Biol Conserv 109:249–258

Torres A, Tisserat B (1980) Leaf isozymes as genetic markers in date palm. Am J Bot 67:162–167

Trifi M, Rhouma A, Marrakchi M (2000) Phylogenetic relationships in Tunisian date-palm (*Phoenix dactylifera* L.) germplasm collection using DNA amplification fingerprinting. Agron 20:665–671

Uma Shaanker R, Ganeshaiah KN, Nageswara Rao M, Ravikanth G (2002) Forest gene banks – a new integrated approach for the conservation of forest tree genetic resources. In: Engels JMM, Brown AHD, Jackson MT (eds.) Managing plant genetic resources. CABI, UK, pp 229–235

UNEP (2007) Natural disasters and desertification In: Sudan – post – conflict environmental assessment. United Nations Environment Programme, Nairobi, Kenya, pp. 56–69

van Tienderen PH, de Haan AA, van der Linden CG, Vosman B (2002) Biodiversity assessment using markers for ecologically important traits. Trends Ecol Evol 17:577–582

Vendramin GG, Anzidei M, Madaghiele A et al (1998) Distribution of genetic diversity in *Pinus pinaster* Ait. as revealed by chloroplast microsatellites. Theor Appl Genet 97:456–463

Vezvaei A (2003) Isozyme diversity in Iranian almond. Acta Hort 622:451–456

Villa TCC, Maxted N, Scholten M, Ford-Lloyd B (2006) Defining and identifying crop landraces. Plant Genet Res 3:373–384

Virchow D (1999) Conservation of genetic resources – cost and implications for a sustainable utilization of plant genetic resources for food and agriculture. Springer Verlag, Berlin

Vorster BJ, Kunert KJ, Cullis CA (2002) Use of representational difference analysis for the characterization of sequence differences between date palm varieties. Plant Cell Rep 21:271–275

Wegrzyn JL, Lee JM, Tearse, BR, Neale DB (2008) TreeGenes: a forest tree genome database. Int J Plant Genom, Article ID 412875, 7 pp, doi:10.1155/2008/412875

Weinberg W (1908) On the demonstration of heredity in man. In: Boyer SH (ed.) (1963) Papers on human genetics, Prentice-Hall, Englewood Cliffs, pp 4–15

Wickens GE (1998) Ecophysiology of economic plants in arid and semi-arid lands: adaptations of desert organisms. Springer-Verlag, Berlin

Wilson BL, Kitzmiller J, Rolle W, Hipkins VD (2001) Isozyme variation and its environmental correlates in *Elymus glaucus* from the California Floristic Province. Can J Bot 79:139–153

Winter P, Kahl G (1995) Molecular marker technologies for plant improvement. World J Microbiol Biotechnol 11:438–448

Wrigley G (1995) Date palm (*Phoenix dactylifera* L.) In: Smartt J, Simmonds NW (eds.) The evolution of crop plants. Longman, UK, pp 399–403

Zaid A, de Wet PF (2002a) Origin, geographical distribution and nutritional values of date palm. In: Zaid A (ed.) Date palm cultivation. Rev. Ed. Plant production and protection paper 156. Food and Agriculture Organisation of the United Nations, Rome, pp 29–44

Zaid A, de Wet PF (2002b) Date palm propagation In: Zaid A (ed.). Date palm cultivation. Rev. Ed. Plant production and protection paper 156. Food and Agriculture Organisation United Nations, Rome, pp 73–105

Zaid A, de Wet PF (2002c) Botanical and systematic description of the date palm. In: Zaid A (ed.) Date palm cultivation. Rev. Ed. Plant production and protection paper 156. Food and Agriculture Organisation United Nations, Rome, pp 1–28

Zaid A, de Wet PF, Djerbi M, Oihabi A (2002) Diseases and pests of date palm. In: Zaid A (ed.) Date palm cultivation. Rev. Ed. Plant production and protection paper 156. Food and Agriculture Organisation United Nations, Rome, pp 227–242

Zehdi S, Trifi M, Billotte N et al (2004) Genetic diversity of Tunisian date palms (*Phoenix dactylifera* L.) revealed by nuclear microsatellite polymorphism. Heredity 141:278–287

Zehdi S, Sakka H, Ould Mohamed Salem A et al (2005) Molecular polymorphism and genetic relationships in a Tunisian date-palm (*Phoenix dactylifera* L.) collection using ISSR amplification fingerprinting. PGR Newslett 144:39–44

Zohary D (1970) Centers of diversity and centers of origin. In: Frankel OH, Bennett E (eds.) Genetic resources of plants – their exploration and conservation. pp 33.42. Blackwell, Oxford.

Zohary D, Hopf M (2000) Domestication of plants in the Old World, 3rd edn. Oxford University Press, Oxford

Zohary D, Spiegel-Roy P (1975) Beginnings of fruit growing in the Old World. Science 187:319–327

Chapter 20
Polymorphism and Genetic Relationship in Date Palm Using Molecular Markers

S. Zehdi-Azouzi, S. Rhouma, S. Dkkhil-Dakhlaoui, A.O.M. Salem, E. Cherif, A. Othmani, M. Marrakchi, and M. Trifi

Abstract Over recent decades date palm (*Phoenix dactylifera* L.) polymorphism and genetic diversity have been studied extensively, bringing about a revolution in this area of research. Molecular markers have been successfully designed and data have proved their efficiency in the genome assessment of this palm. In fact, these now constitute an important tool for many date palm investigations. Tunisian date palm germplasm is characterized by high genetic diversity because more than 250 cultivars have been identified. This local germplasm is, however, seriously threatened by severe genetic erosion due to diverse biotic and abiotic stresses. Evaluation of polymorphism and genetic diversity has become a prerequisite to establishment of a research program aimed at rational germplasm conservation. For that objective, research work has focused on the development of phenotypic and biochemical characterization of Tunisian date palms. Moreover, DNA-based methods have been successfully realized which permit establishment of cultivar identification keys, as well as to determine the genetic relationships among them. Data are discussed in relation to the opportunities presented by designed markers in the improvement and breeding of date palm.

Keywords Biotechnology • DNA • Germplasm • Molecular markers • Polymorphism

---

S. Zehdi-Azouzi (✉) • S. Rhouma • S. Dkkhil-Dakhlaoui • E. Cherif • A. Othmani • M. Marrakchi • M. Trifi
Laboratoire de génétique moléculaire, immunologie et biotechnologie, Faculté des Sciences de Tunis, Université de Tunis El Manar, Campus Universitaire, 2092 El Manar Tunis, Tunisia
e-mail: mokhtar.t@fst.rnu.tn

A.O.M. Salem
Département de Biologie, Faculté des Sciences et Techniques, Université de Nouakchott, B.P 5026 Nouakchott, Mauritania

S.M. Jain et al. (eds.), *Date Palm Biotechnology*,
DOI 10.1007/978-94-007-1318-5_20, © Springer Science+Business Media B.V. 2011

20.1 Introduction

Date palm (*Phoenix dactylifera* L.), (2n = 36), a dioecious perennial crop within the Arecaceae family, is remarkable among desert plants, as well as one of the many wonders in the biological structure of oases. It also provides food, farm income and other products for local desert dwellers. Date palm is probably the oldest fruit crop grown in tropical and subtropical areas, having been cultivated since ancient times for fruit production and all parts of the tree are used for many subsistence and industrial purposes (Hodel and Johnson 2007; Munier 1973). It constitutes, therefore, the main factor for establishment of favorable conditions in the oasian agroecosystem. There exist hundreds of date palm cultivars around the world and these have been selected mainly for their fruit qualities, and propagated by farmers via offshoots. For example, more than 250 date palm cultivars have been identified in Tunisian groves (Rhouma 1994, 2005).

It is worth noting that in the North African and Middle Eastern countries, many oases consist of date palm groves where the palms constitute the main component in the social, environmental and economic stability of these fragile ecosystems. Tunisia is among the major commercial date palm producing areas of the world and oasis farmers depend upon date palms as a principal financial resource and food source; moreover, the palms contribute to the development of subjacent cultures of fruit trees, saffron, forage and vegetables. More than 10% of the North African people depend upon date palm cultivation (Haddouch 1996). It should be emphasized that the date palm is propagated clonally via offshoots removed from the mother tree. This method of propagation, however, is relatively circumscribed since only a limited number of offshoots are produced by each mother tree and this falls short of the needs to renovate old and establish new plantings. To overcome this difficulty, large quantities of plantlets have been generated by *in vitro* culture methods and utilized to improve modern date palms plantations.

20.2 Local Date Palm Germplasm

Among the countries known for date palm fruit production (i.e. North Africa, Middle East and South Asian countries), a large number of cultivars are grown. According to Munier (1973) more than 600 cultivars grown worldwide have been reported. However, the precise number of cultivars per date-producing country is not clearly established. For instance more than 250 cultivars have been identified in Tunisian date palm groves (Rhouma 1994, 2005). At a commercial scale, Tunisia is included in the main date-palm producing area of the world. It should be stressed that date palm is clonally propagated via offshoots produced by the mother tree. However this method of propagation is relatively slow since a very limited number of offshoots are produced by the mother tree and uncover the needs for renovation and establishment of new plantations. To overcome this difficulty, large numbers of plantlets have been generated via *in vitro* culture methods and used to enhance modern date-palms plantations.

In regions known for date palm fruit production (i.e. North Africa, Middle East as well as Asian countries), a large number of cultivars are grown and produce dates with different colors, flavors, sweetness, acidity and textures. However, the mentioned regions have their favorites but different countries have their elite cultivars: In UAE and Eastern Saudi Arabia the preferred fruit is cv. Khalas but others such as Zaghloul, Khuneizi, Hilali, Howaiz, Naghal and Jaberi Fardh are also favored (Wrigley 1995). Other cvs. popular in different countries include Medjool in Morocco; in Iraq Amir Hajj and Ashrashi (Al-Bekr 1972); Saidy and Hayany in Egypt, Deglet Noor and Thoory in Algeria, Ruzeiz, Bukeira, Nebut, Seif, Barhee and Abou Meaan in Oman; Deglet Noor in Tunisia and Ahmar in Mauritania.

As of 2008, the world's largest producer countries were: Egypt (1,326,133 mt), Iran (1,006,406 mt), Saudi Arabia (986,000 mt), United Arab Emirates (755,000 mt), Pakistan (680,107 mt), Algeria (552,765 mt), Iraq (476,318 mt), Sudan (336,000 mt), Oman (255,871 mt), Libya (150,000 mt), Tunisia (127,000 mt).

20.2.1 Phenotypic and Biochemical Variation in Date Palms

20.2.1.1 Phenotypic Variation

For several decades, characterization of genetic variation has been an imperative in order to establish conformity of date palms from offshoots and *in vitro* derived progeny. Among the methods used, those based on morphological traits are of some benefit in the evaluation of date palm genetic resources. Numerous studies have addressed this issue, and described the use of either morphological traits or other analytical parameters to identify date palm varieties (Belguedj 2002; Bouabidi et al. 1996; El Houmaizi et al. 2006; Mohamed et al. 1983; Ould Mohamed Salem et al. 2001; Reynes et al. 1994; Rhouma 1994, 2005). Criteria related to either vegetative or fruit parameters are useful for cultivar characterization, phenotypic diversity analysis and the exploration of phylogenic relationships among date palm genotypes. Such studies constitute a logical first step to provide reliable evidence of phenotypic diversity in the design of programs to improve germplasm management and utilization of the crop.

20.2.1.2 Genetic Diversity as Inferred from Isozymes

Isozymes have also been reported as a means to characterize date palms (Baaziz and Saaidi 1988; Bendiab et al. 1993; Bennaceur et al. 1991; Booij et al. 1995; Ould Mohamed Salem et al. 2001). Such biochemical markers are of some benefit to precisely characterize the diversity in this crop. However, most morphological traits, including isozymes, are highly influenced by environmental conditions and/or vary depending upon the developmental stage of the plant. It should be stressed that such studies are limiting due to low levels of polymorphism, taking into account that the traits mentioned are very few in number, and the techniques time- and cost-consuming.

20.2.2 Molecular Polymorphisms and Genetic Relationships in Date Palm Ecotypes

In order to overcome the inadequacies reported above, numerous studies have addressed this issue and have described the use DNA-based techniques. As a consequence many molecular markers have been developed, and have proved effective in surveying genetic diversity because they access an almost unlimited source of potential markers to uncover differences at the molecular level. Consequently, data based on molecular markers such as: cleaved amplified polymorphic sequences, (CAPSs) – inter-simple sequence repeats (ISSRs); random amplified polymorphic DNAs (RAPD) PCR – polymerase chain reaction; RAMPO – markers; RFLP – restriction fragment length polymorphism; SSR – (microsatellites), simple sequence repeats restriction fragment length polymorphisms (RFLPs), (RAPDs) and cleaved amplified polymorphic sequences (CAPSs) have been used to characterize date palm genotypes (Ben Abdallah et al. 2000; Corniquel and Mercier 1994; Sakka et al. 2004a; Sedra et al. 1998; Trifi 2001; Trifi et al. 2000). Moreover, additional molecular markers have been investigated for a deeper knowledge of the genetic organization in date palm germplasm. For this purpose, inter simple sequence repeats (ISSRs) (SSRs), random amplified microsatellite polymorphisms (RAMPOs) and amplified fragment length polymorphisms (AFLPs) have been investigated and provided precise data on genetic diversity of local date palm germplasm (El Assar et al. 2005; Rhouma et al. 2007, 2008; Zehdi-Azouzi 2004; Zehdi et al. 2004).

20.2.2.1 Genetic Diversity as Inferred from Restriction Fragment Length Polymorphisms (RFLPs)

The development of RFLPs has been reported by Corniquel and Mercier (1994, 1996). Trifi et al. (2000) also has designed these molecular markers in Tunisian date palms. In fact, that research group has described the use of homologous as well as heterologous labeled probes to discriminate among date palm cultivars. However, the number of genotypes studied was limited to a small set of cultivars that are widely cultivated in oases and/or characterized by fleshy dates, and they are not representative of the germplasm. Moreover, taking into account the use of radio-labeled probes, the large amounts of DNA used in these studies, the development of additional molecular markers has become imperative to overcome these inconveniences.

20.2.2.2 Genetic Diversity as Inferred from Random Amplified Polymorphic DNAs (RAPDs) and Inter Simple Sequence Repeats Markers (ISSRs)

The search for faster techniques has become a compelling objective to characterize date palm germplasm and to survey relationships among cultivars. Taking into account the advantages of the RAPD and ISSR techniques (i.e. small amounts of DNA, fast, no knowledge of the genome is needed and low costs), work has focused on development of the mentioned PCR DNA-based methods to examine genetic diversity as well to characterize genotypes. Studies have described the use of universal random primers to generate molecular markers suitable to more precise genetic diversity at the DNA level. In addition, starting from a group of Moroccan date palm cultivars, Sedra et al. (1998) reported evidence of RAPDs correlated with the resistance to bayoud disease. Moreover, both Trifi et al. (2000) and Zehdi et al. (2004) reported the use of RAPD and ISSR to generate discrete markers correlated with fruit parameters such as fruit size and color of Tunisian cultivars. On the other hand, starting from a set of local and introduced cultivars, together with male genotypes, either RAPDs or ISSRs derived genotypic clustering was achieved, independent of the gender of the trees and/or the geographical origin of the cultivars (Fig. 20.1). This suggests a common genetic basis which characterizes date palm cultivars and concurs with the Mesopotamian domestication origin (i.e. the Fertile Crescent) of this fruit crop as proposed by Munier (1973) and Wrigley (1995).

20.2.2.3 Genetic Diversity as Inferred from Amplified Fragment Length Polymorphisms Markers (AFLPs)

Vos et al. (1995) designed an AFLP method for either Tunisian or Egyptian date palm germplasm. It was employed by Rhouma et al. (2007) and El Assar et al. (2005), in those two countries, respectively. In their studies, different sets of primer pairs were tested for their ability to generate AFLP banding patterns using total cellular DNAs as templates. As a result large numbers of AFLPs were easily produced suggesting that the primers tested in the studies are a powerful means to provide evidence of DNA polymorphisms in this crop. This assumption is strongly supported as regards the high percentage values of polymorphic bands (ppb) scored, using each primer set (Table 20.1). Moreover, estimation of the resolution power (Rp) exhibited high rates of collective Rp of 155.3, with a mean of 25.88 (Table 20.1). In addition, the majority of PIC values obtained was in the 0.8–1 range. Therefore, these data suggest that AFLP constitutes a very attractive and informative procedure for providing evidence of the genetic diversity among date palm ecotypes (Fig. 20.2).

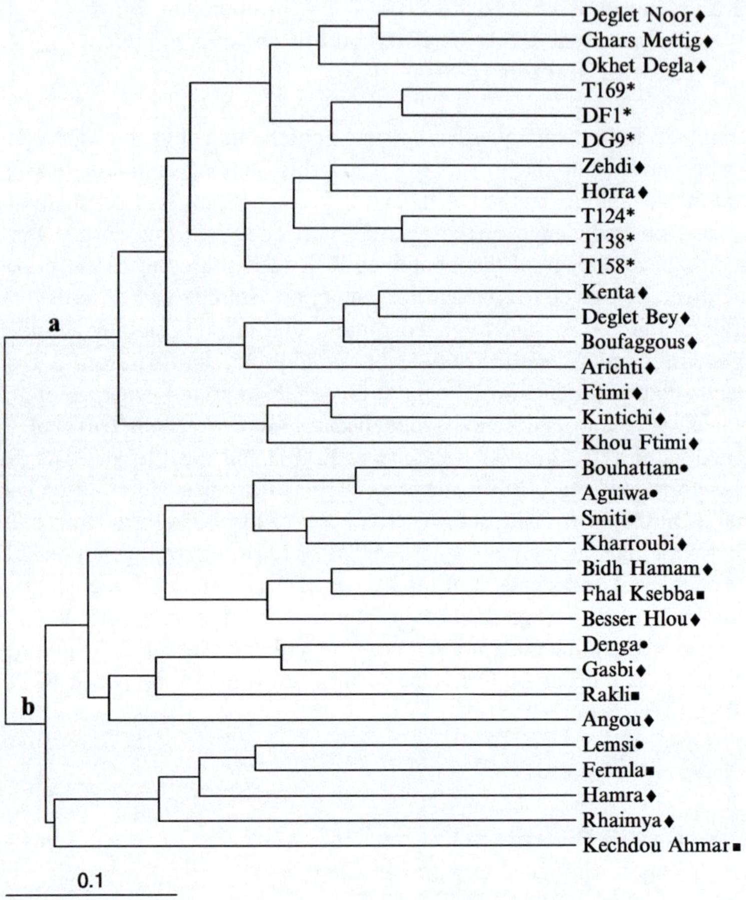

Fig. 20.1 UPGMA Dendrogram of 34 of Tunisian date-palm accessions constructed from Nei & Li genetic distance based on 67 ISSR markers. The scale indicates the relative genetic distance. Accessions originated from Tozeur, Gabès and Kébili oases are labelled (♦), (•) and (■) respectively. Males are labelled with an *asterisk* (From Zehdi-Azouzi (2004))

Table 20.1 Results obtained with primer combinations used for AFLP analysis in a 40 Tunisian date palm cultivars

Primer set	Band numbers		PPB	Rp
	Total	Polymorphic		
E_{AAC}/M_{CAA}	50	50	100	25.60
E_{AGC}/M_{CAA}	56	56	100	23.40
E_{AAC}/M_{CAG}	65	60	92.3	21.55
E_{ACA}/M_{CAG}	67	67	100	23.80
E_{ACC}/M_{CTA}	100	95	95	27.00
E_{AAC}/M_{CAT}	104	100	96.15	33.95
Total	442	428	–	155.3

From Rhouma et al. (2007)

PPB Percentage of polymophic bands, *Rp* Resolving power, E_{NNN}/M_{NNN} primer pairs, *E* and *M* for *Eco* RI and *Mse* I respectively

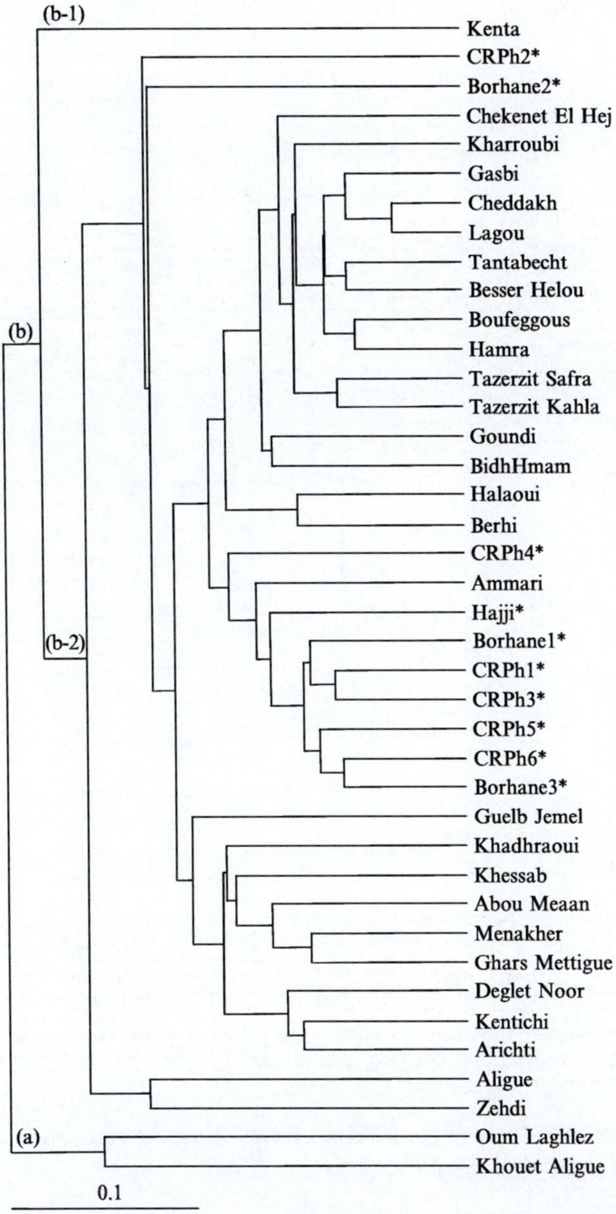

Fig. 20.2 Dendrogram of 40 Tunisian date palm ecotypes constructed by UPGMA method and based on 428 AFLPs (From Rhouma et al. (2007))

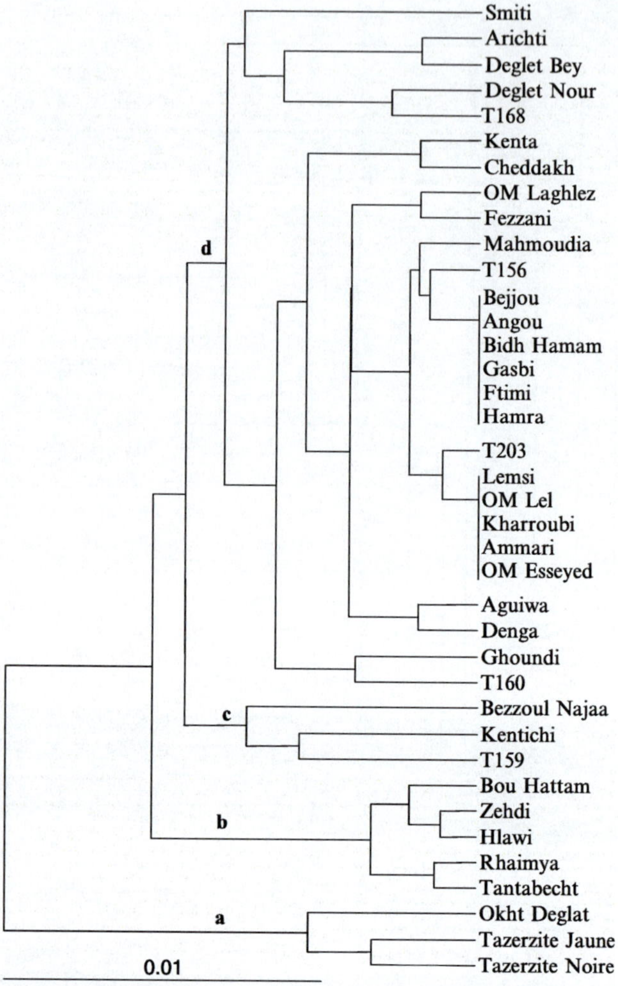

Fig. 20.3 Dendrogram of 38 Tunisian date palm ecotypes constructed by UPGMA method and based on CAPS (From Sakka et al. (2004b))

20.2.2.4 Genetic Diversity as Inferred from Cleaved Amplified Polymorphic Sequence Markers (CAPSs)

Sakka et al. (2004b) reported that 15 plastid fragments were amplified from a set of Tunisian date palm accessions by PCR with consensus primers and analyzed by RFLP. The ctDNA PCR-RFLP method permitted identification of two haplotypes that differ based on the presence or absence of the *Hinf*I restriction site. Phenetic groups were composed of cultivars clustered together, but they do not constitute monophyletic groups. This typology agrees with the organization of the haplotypes and provides a common genetic background within the implied cultivars (Fig. 20.3), which suggests a narrow genetic diversity characteristic of the cultivars studied.

Table 20.2 Results obtained with the primers combinations used for RAMPO analysis

Primer combination	Band numbers Total	Polymorphic	PPB	Rp
OPB04×02	11	10	90.9	3.55
OPB04×06	10	7	70	2.9
OPB04×07	13	11	84.61	6.85
OPB04×09	14	12	85.71	4.2
OPB04×10	16	16	100	8.25
OPB04×14	8	7	87.5	2.15
OPA19×02	9	8	88.88	3.5
OPA19×06	11	9	81.81	4.25
OPA19×07	11	11	100	3.8
OPA19×09	13	12	92.3	3.2
OPA19×10	12	12	100	4.05
OPA19×14	10	10	100	2.6
OPA12×02	12	10	83.3	5.1
OPA12×06	13	10	76.9	2.45
OPA12×07	13	12	92.3	3.45
OPA12×09	13	11	84.6	5.3
OPA12×10	14	12	85.7	6.2
OPA12×14	7	6	85.71	1.1
Total	210	186	88.57	72.9

From Rhouma et al. (2008)

PPB Percentage of polymophic bands, *Rp* Resolving power

20.2.2.5 Polymorphisms as Revealed by Random Amplified Microsatellite Polymorphic Markers (RAMPOs)

The RAMPO procedure was designed by Richardson et al. (1995) as a new source of data on polymorphism in higher plants. The method is based on a random amplification followed by a southern hybridization using microsatellite oligonucleotides as radio-labeled probes. To avoid the use of radio-labeled probes, Chatti et al. (2007) modified the Richardson method to consist of the combination of two PCR-DNA based techniques, namely RAPD and ISSR. Therefore two combinations of primers were tested: universal random decamer oligonucleotides, purchased from Operon Technology Inc. (Alameda CA, USA) to perform RAPD assays; and oligonucleotides that are complementary to simple sequence repeats to perform ISSR assays (Table 20.2). In date palm, RAMPO has been successfully achieved (Rhouma et al. 2008). Therefore, these data suggest that RAMPO constitutes an efficient and informative procedure for providing evidence of the genetic diversity between date palm ecotypes, as well as in the discrimination of date palm genotypes (Fig. 20.4).

20.2.2.6 Genetic Diversity as Inferred from Microsatellites (SSRs)

To expand the number of molecular markers, taking into account the dominant inheritance of the markers reported above, nuclear microsatellite loci were identified by Billotte et al. (2004) and successfully applied to Tunisian date

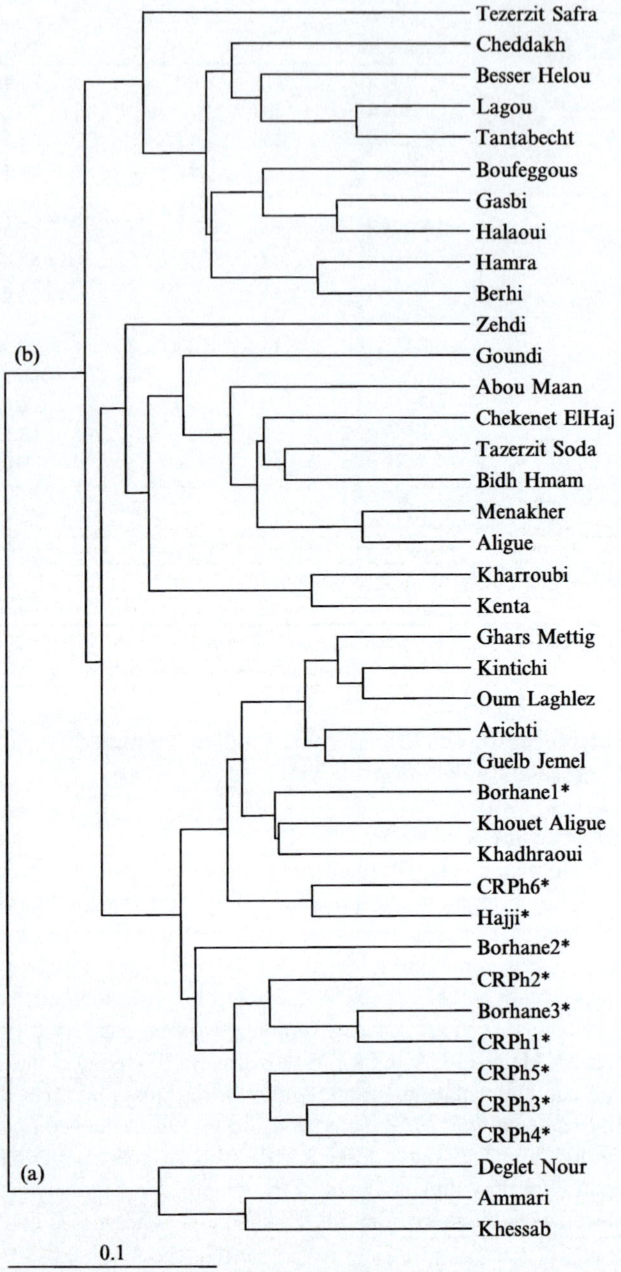

Fig. 20.4 Dendrogram of 40 Tunisian date palm ecotypes constructed by UPGMA method and based on 186 RAMPOs (From Rhouma et al. (2008))

palm cultivars (Zehdi-Azouzi 2004). This author also designed the use of chloroplasic minisatellites in order to examine genetic organization among Tunisian date palm oases (Zehdi et al. 2005). As indicated in Fig. 20.5, the observed

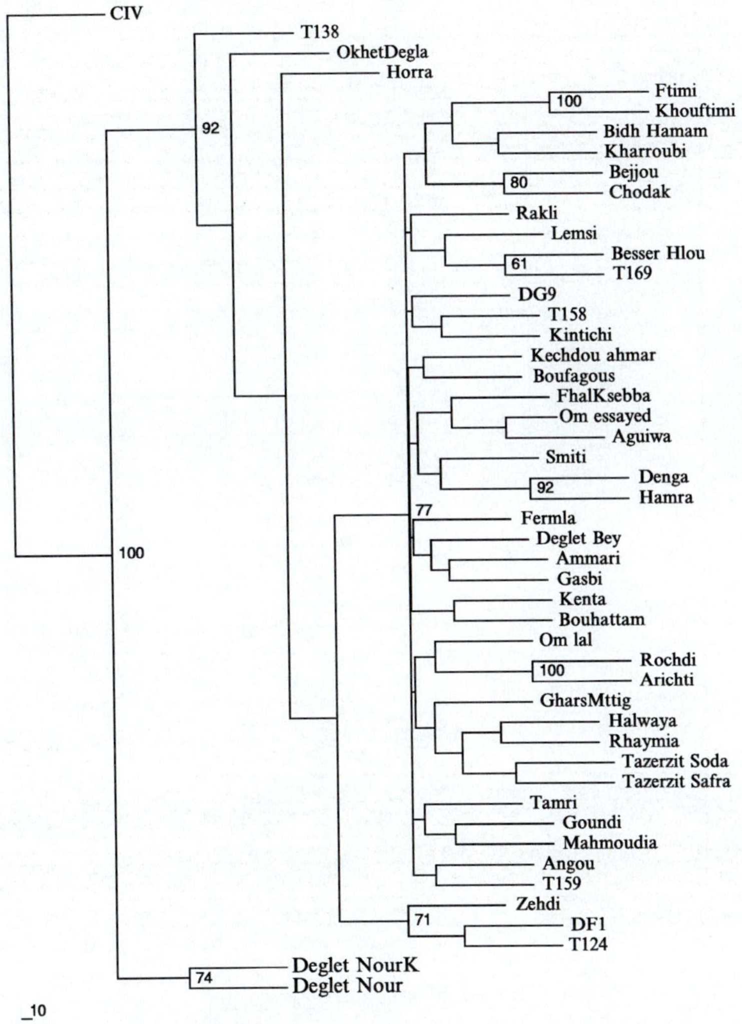

Fig. 20.5 NJ dendrogram of 49 Tunisian date palm accessions constructed with Das genetic distance based on 100 microsatellite alleles. Bootstrap values indicated on branches have been computed over 100 replications (From Zehdi-Azouzi (2004))

clustering topology showed that groupings of accessions are independent of either their geographic origin or their gender. In fact, genetic diversity structure has exhibited spatial organization of accessions which are typically continuous and independent of geographic origin or gender. To explain this organization, Zehdi et al. (2005) suggested a common genetic basis among date palms genotypes, in spite of their phenotypic distinctiveness by morphometric traits, mainly those related to the fruit traits. This result supports the theory of one ancestral date palm population.

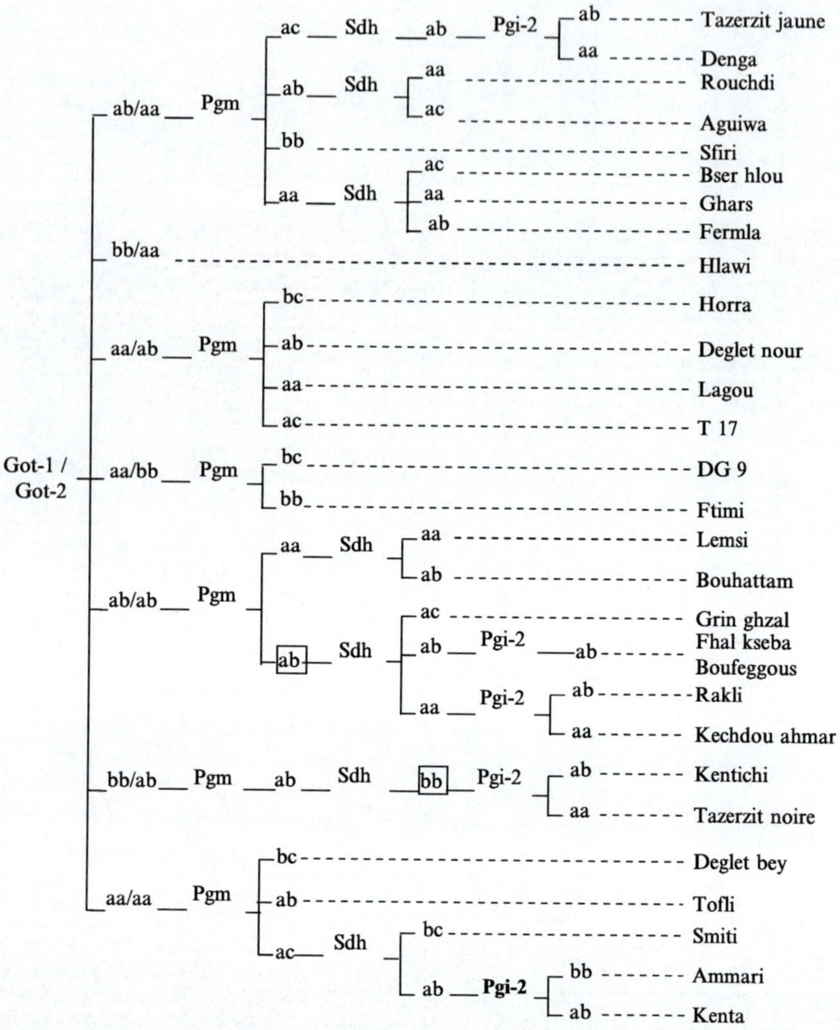

Fig. 20.6 Diagram illustrating the discrimination of 29 Tunisian date palm ecotypes based on isozyme markers (From Ould Mohamed Salem et al. (2001))

20.2.3 *Date Palm Cultivar Genotyping and Identification Keys*

An identification key of date palm cultivars is of great interest because it would permit labelling of offshoots, as well as any other plant material at an early stage, including in *in vitro* plantlets. To this end, studies have been reported utilizing isozymes (Bennaceur et al. 1991; Booij et al. 1995; Ould Mohamed Salem et al. 2001); chloroplast haplotypes (Sakka et al. 2004a); and nuclear microsatellites (Zehdi et al. 2004). Data have proved that the designed markers are suitable to

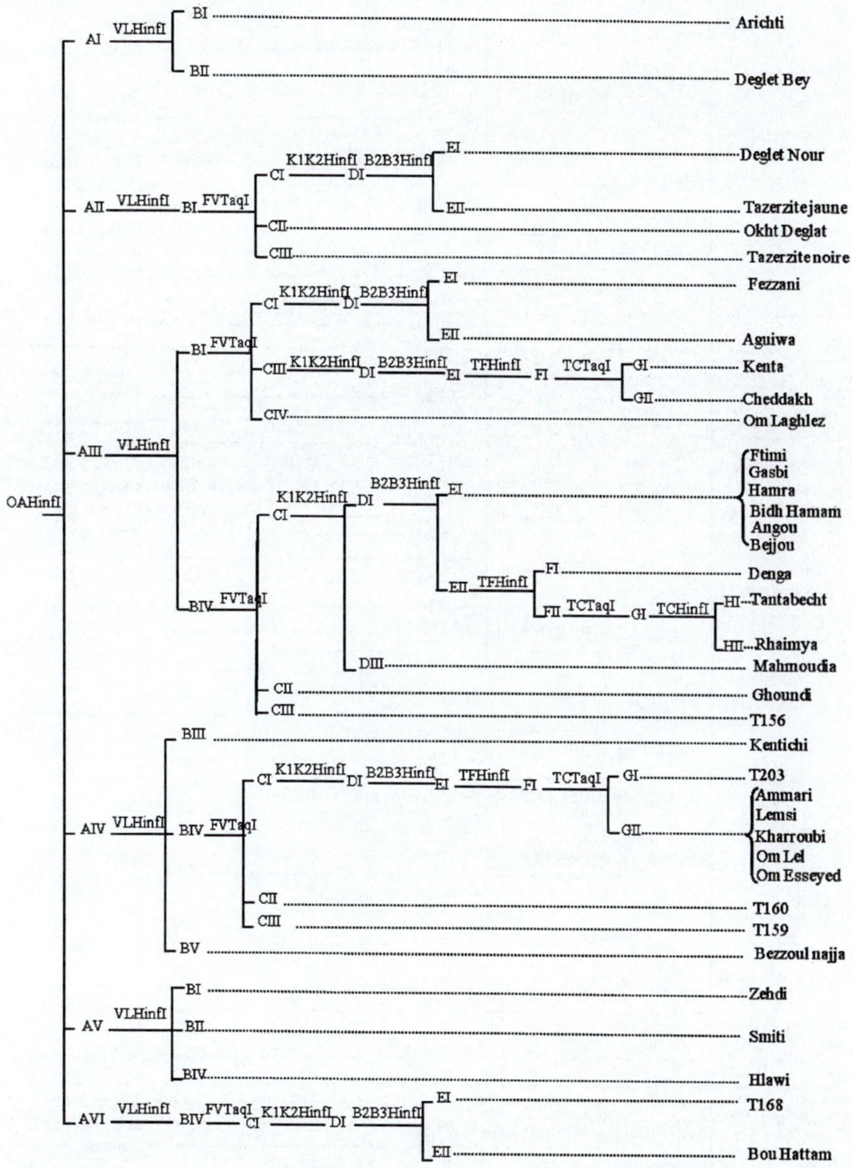

Fig. 20.7 Diagram illustrating the discrimination of 38 Tunisian date palm ecotypes based on CAPS ctDNA (From Sakka et al. (2004a))

successfully achieve establishment of identification keys of date palm cultivars (Figs. 20.6–20.8). Fortunately, these data provide evidence of the possibility of using these powerful markers as descriptors, as well as in the certification and control of origin labels.

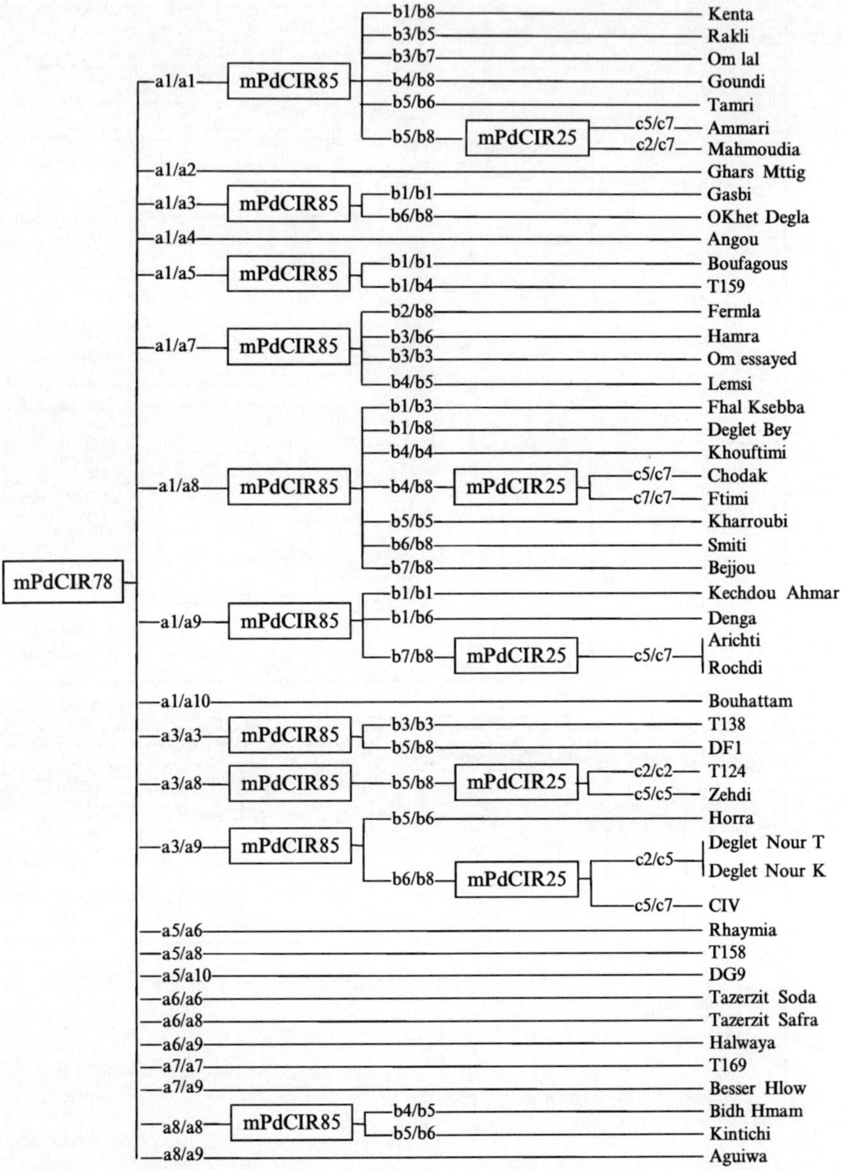

Fig. 20.8 Diagram illustrating the discrimination of 49 Tunisian date palm ecotypes based on three microsatellite loci. Alleles significance (size in pb): a1: 138; a2: 142; a3: 144; a4: 148; a5: 153; a6: 157; a7: 159; a8: 165; a9: 171; a10: 173; b1: 175; b2: 181; b3: 183; b4: 185; b5: 187; b6: 189; b7: 195; b8: 197; c1: 219; c2: 231; c3: 233; c4: 236; c5: 246; c6: 248 and c7: 250 (From Zehdi-Azouzi (2004))

20.3 Molecular Markers and Genomic Stability Among *In Vitro*-Derived Progenies

In date palm, propagation is conventionally via juvenile offshoots removed from mother plants. This clonal propagation mode has several disadvantages: it is time consuming and produces only a small number of shoots from an individual tree. Moreover, the rooting percentage in such shoots is low and a number of cultivars do not produce offshoots. Also, a period of 5–7 years is required to confirm the true-to-typeness of sucker-derived plants. Therefore, cultivar confirmation of offshoot and tissue culture-derived progeny is an imperative to conserve the plant mother characteristics.

Othmani et al. (2009) and Rhouma et al. (2008) recently analyzed polymorphism among and between plantlets produced by *in vitro* culture methods (organogenesis and/or somatic embryogenesis). Those studies reported using the available molecular biology methods of RAPD and AFLP (Vos et al. 1995). Our investigation showed them to be reliable, fast and inexpensive procedures with which to identify clones and to assess somaclonal variation in this crop. As illustrated in Figs. 20.9 and 20.10, the RAPD

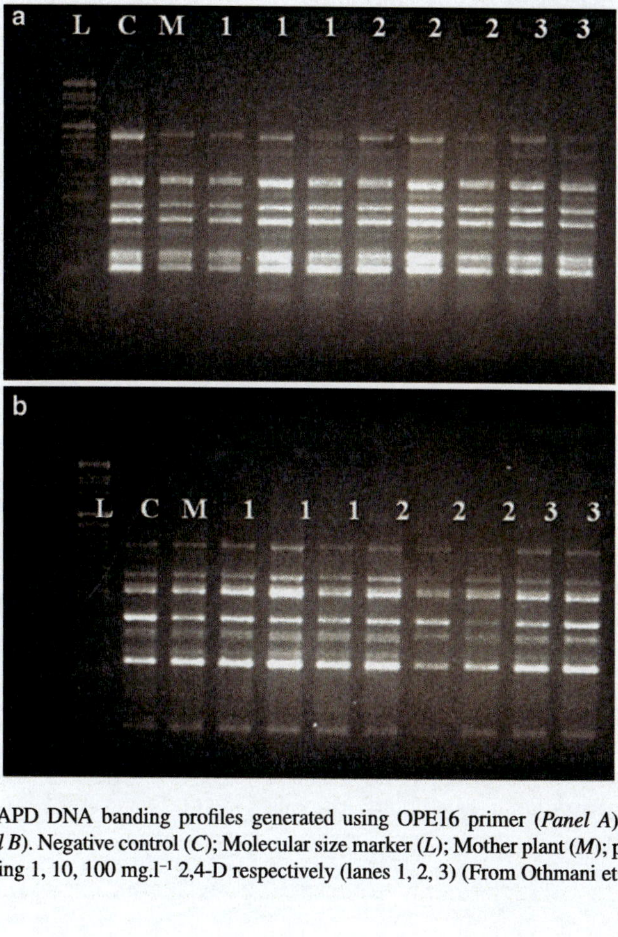

Fig. 20.9 RAPD DNA banding profiles generated using OPE16 primer (*Panel A*) and OPD05 primer (*Panel B*). Negative control (*C*); Molecular size marker (*L*); Mother plant (*M*); plantlets from media including 1, 10, 100 mg.l^{-1} 2,4-D respectively (lanes 1, 2, 3) (From Othmani et al. (2009))

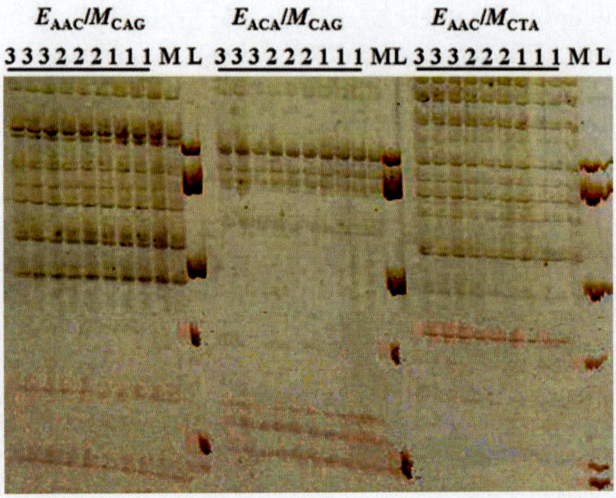

Fig. 20.10 Typical examples of AFLP banding profiles generated by E_{ACA}/M_{CTA} primers' combination. Standard molecular size marker (*L*); Mother plant (*M*); plants from media including 1, 10, 100 mg.l^{-1} 2,4-D (lanes 1, 2, 3 respectively). E_{NNN}/M_{NNN}, primer pairs, *E* and *M* for *Eco*R1 and *Mse*1 restriction enzymes respectively (From Rhouma et al. (2008))

and AFLP banding profiles of the derived progenies were very similar to those of the mother plant, suggesting that no variation had occurred. In fact, the number as well the sizes of the generated bands are similar in these profiles. This result strongly supports the true-to-type nature of the *in vitro*-derived date palm plantlets. Moreover, 2,4-D did not induce somaclonal variation in this crop under the experimental conditions studied.

20.4 Conclusion and Prospects

Date palm has always been looked upon as a key resource of life for survival and evolution of the desert lands, thus forming a base for every ecological pyramid in these areas. Over the long term this important fruit crop has been a very attractive candidate on which to conduct research in many fields such as physiology, micropropagation and agronomy. However, little is known about the date palm genomics, especially its traditional phenotypic markers. It should be stressed that with the advent of molecular markers, a new generation of markers has been introduced over recent decades. In fact, such markers have revolutionized the entire knowledge in this crop: DNA-based molecular markers, therefore, have been designed and function as versatile tools bringing a deeper insight about the genetic organization in date palms. Moreover, the discovery of PCR (polymerase chain reaction) was a landmark in this effort and proved to be an attractive approach in diverse biological fields like genetic variability studies, phylogenetic analysis and DNA fingerprinting. Thus, taking into account the advantages of such markers, we have successfully achieved their development and have established their place in various applications. Fortunately, the present data have proved objectively that the use of molecular markers in date palms is effective for the assessment of genetic polymorphism at the DNA level, the surveying of phylogenic relationships among cultivars and the establishment of cultivars identification keys. As mentioned above, a high level of polymorphism characterizes Tunisian date-palm germplasm. Moreover, DNA fingerprinting of a relatively large number of cultivars has been successfully achieved. On the whole, the discrimination of date-palm genotypes would be of great interest either in the improvement and the breeding objectives or in the rational management of germplasm conservation.

During the early period of date palm cultivation, estimation of genetic diversity in this crop's germplasm was looked upon as an important task to determine inter- and intra-cultivars variability. For this purpose, classical strategies including phonologic, agronomic and analytic parameters were used. Many studies have, therefore, described the use of such traits in date palm and have attested to their suitability for cultivars characterization, phenotypic diversity analysis and the exploration of phylogenic relationships among genotypes. However, with the advent of molecular markers, a new generation of markers has been introduced. In date palm these markers were successfully developed over the last two decades and have revolutionized the entire scenario of polymorphism surveying. In fact, DNA-based molecular markers have served as versatile tools and have found opportunities to more and more characterize local germplasm. In addition, such markers successfully permitted establishment of cultivar identification keys as well as determine precise genetic relationships among them. It is worth noting that the recently online published full sequence of the plastid genome which can be accessed through the URLs:

http://qatar-weill.cornell.edu/research/datepalmGenome/ http://news.med.cornell.edu/wcmc/wcmc_2009/05_01b_09.shtml

This constitutes an attractive scenario for development of additional molecular markers in this crop's improvement programs and will be of great interest to more precise molecular polymorphism in date palm. Such data would enhance date palm genome sequencing initiated by research teams around the world. Such sequences would be of great interest to enlarge the date palm's molecular data bases that are easily exchangeable between laboratories in order to shed light on polymorphism and the domestication process in this important fruit crop.

Acknowledgements The original research was partially supported by grants from the IPGRI CRRAO Degache, and the Tunisian Ministry of Higher Education and Scientific Research project Lab B02. This chapter is dedicated to the late Professor Mohamed Marrakchi (1942–2008), Director of the Lab LRB02 and principal investigator of the original work.

References

Al-Bekr AJ (1972) The date palm. Al-Ani Press, Baghdad

Baaziz M, Saaidi M (1988) Preliminary identification of date palm cultivars by esterase isoenzymes and peroxidase activities. Can J Bot 66:89–93

Belguedj M (2002) Les ressources génétiques du palmier dattier, caractéristiques des cultivars de dattiers dans les palmeraies du Sud-est Algérien. Institut National de la Recherche Agronomique d'Algérie, Algérie

Ben Abdallah A, Stiti K, Lepoivre P, Du Jardin P (2000) Identification de cultivars de palmier dattier (*Phoenix dactylifera* L.) par l'amplification aléatoire d'ADN (RAPD). Cah Agric 9:103–107

Bendiab K, Baaziz M, Brakez Z et al (1993) Correlation of isoenzyme polymorphism and Bayoud-disease resistance in date palm cultivars and progeny. Euphytica 65:23–32

Bennaceur M, Lanaud C, Chevalier MH et al (1991) Genetic diversity of the date palm (*Phoenix dactylifera* L) from Algeria revealed by enzyme markers. Plant Breed 107:56–69

Billotte N, Couvreur T, Marseillac N et al (2004) Nuclear microsatellite markers for the date palm (*Phoenix dactylifera* L.): characterization and utility across the genus *Phoenix* and in other palm genera. Mol Ecol Notes 4:256–258

Booij I, Monfort S, Ferry M (1995) Characterization of thirteen date palm (*Phoenix dactylifera* L.) cultivars by enzyme electrophoresis using the phastsystem. J Plant Physiol 145:62–66

Bouabidi H, Reynes M, Roussi MB (1996) Critères de caractérisation des fruits de quelques cultivars de palmiers dattier (*Phoenix dactylifera* L.) du sud Tunisien. Ann l'INRAT 69:73–87

Chatti K, Saddoud O, Salhi-Hannachi A et al (2007) Analysis of genetic diversity and relationships in a Tunisian fig (*Ficus carica*) germplasm collection by random amplified microsatellite polymorphisms. J Integr Plant Biol 49(3):386–391

Corniquel B, Mercier L (1994) Date palm (*Phoenix dactylifera* L.) cultivar identification by RFLP and RAPD. Plant Sci 101:163–172

Corniquel B, Mercier L (1996) Identification of date palm (*Phoenix dactylifera* L.) cultivars by RFPL: partial characterization of the cDNA probe that contains a sequence encoding a zinc finger motif. Int J Plant Sci 158:152–156

El Assar A, Kruger R, Devanand PS et al (2005) Genetic analysis of Egyptian date (*Phoenix dactylifera* L.) accessions using AFLP markers. Genet Resour Crop Evol 52:601–607

El Houmaizi MA, Devanand PS, Fang J et al (2006) Confirmation of 'Medjool' date as a landrace variety through genetic analysis of 'Medjool' accessions in Morocco. J Am Soc Hortic Sci 131:403–407

Haddouch M (1996) Situation actuelle et perspectives de développement du palmier dattier au Maroc. Cah Options Méditerr 28:63–79

Hodel DR, Johnson DV (2007) Imported and American varieties of dates in the United States. University of California, Agriculture and Natural Resources, Oakland

Mohamed S, Shabana HR, Mawlod BA (1983) Evaluation and identification of Iraqi date cultivars: fruit characteristics of fifty cultivars. Date Palm J 2:27–55

Munier P (ed.) (1973) Le palmier dattier. Maisonneuve et Larose, Paris

Othmani A, Bayoudh C, Drira N et al (2009) Regeneration and molecular analysis of date palm (*Phoenix dactylifera* L.) plantlets using RAPD markers. Afr J Biotechnol 8:813–820

Ould Mohamed Salem A, Trifi M, Salhi-Hannachi A et al (2001) Genetic variability analysis of Tunisian date palm (*Phoenix dactylifera* L.) cultivars. Genet Resour Crop Evol 55:269–278

Reynes M, Bouabidi H, Piombo G et al (1994) Caractérisation des principales variétés des dattes cultivées dans la région de Tozeur. Fruits 49:289–298

Rhouma A (1994) Le palmier dattier en Tunisie: Le patrimoine génétique Vol I. Arabesques Editions et Créations, Tunis

Rhouma A (2005) Le palmier dattier en Tunisie: Le patrimoine génétique Vol II. IPGRI, Rome

Rhouma S, Zehdi-Azouzi S, Ould Mohamed Salem A et al (2007) AFLP markers for genetic diversity analysis in Tunisian date palm (*Phoenix dactylifera* L.) accessions. J Hortic Sci Biotechnol 82(6):929–933

Rhouma S, Dakhlaoui-Dkhil S, Ould Mohamed Salem A et al (2008) Genetic diversity and phylogenic relationships in date-palms (*Phoenix dactylifera* L.) as assessed by random amplified microsatellite polymorphism markers (RAMPOs). Sci Hortic 117:53–57

Richardson T, Cato S, Ramser J et al (1995) Hybridization of microsatellites to RAPD: a new source of polymorphic markers. Nucleic Acids Res 23:3798–3799

Sakka H, Zehdi S, Ould Mohamed Salem A et al (2004a) Tunisian date-palm (*Phoenix dactylifera* L.) genotypes identification mediated by plastid PCR/RFLP based DNA. J Genet Breed 57:259–264

Sakka H, Zehdi S, Ould Mohamed Salem A et al (2004b) Genetic polymorphism of plastid DNA in Tunisian date-palm germplasm (*Phoenix dactylifera* L.) detected with PCR-RFLP. Genet Resour Crop Evol 51:479–487

Sedra My H, Lashermes P, Trouslot P et al (1998) Identification and genetic diversity analysis of date-palm (*Phoenix dactylifera* L.) varieties from Morocco using RAPD markers. Euphytica 103:75–82

Trifi M (2001) Polymorphisme et typage moléculaire de variétés tunisiennes de palmier dattier (*Phoenix dactylifera* L.): relation avec la résistance au bayoud. Thèse Doctorat d'Etat, Faculté des Sciences de Tunis, Université Tunis-El Manar, Tunis

Trifi M, Rhouma A, Marrakchi M (2000) Phylogenetic relationships in Tunisian date palm (*Phoenix dactylifera* L.) germplasm collection using DNA amplification fingerprinting. Agronomie 20:655–671

Vos P, Hoger R, Bleeker R et al (1995) AFLP: a new technique for DNA fingerprinting. Nucleic Acids Res 23:4407–4417

Wrigley G (1995) Date palm (*Phoenix dactylifera* L.). In: Smartt J, Simmonds NW (eds.) The evolution of crop plants, 2nd edn. Longmans, Essex, pp 399–403

Zehdi S, Trifi M, Billotte N et al (2004) Genetic diversity of Tunisian date palms (*Phoenix dactylifera* L) revealed by nuclear microsatellite polymorphism. Hereditas 141:278–287

Zehdi S, Trifi M, Sakka H et al (2005) Plastid minisatellite in date-palms (*Phoenix dactylifera* L): evidencing and use for haplotypic diversity surveying in Tunisian oases. Biol Tunisie 1(3):1–6

Zehdi-Azouzi S (2004) Analyse moléculaire de la diversité génétique chez le palmier dattier (*Phoenix dactylifera* L.). Thèse de Doctorat de biologie, Faculté des Sciences de Tunis, Université Tunis-El Manar, Tunis

Chapter 21
Date Palm Genome Project at the Kingdom of Saudi Arabia

X. Zhang, J. Tan, M. Yang, Y. Yin, I.S. Al-Mssallem, and J. Yu

Abstract The Date Palm Genome Project of the Kingdom of Saudi Arabia is a comprehensive genome research project aimed at sequencing the date palm genome to completion, deciphering the transcriptomes and understanding the biology of date palm for improved cultivation and pest prevention. We introduce plant genomics and its technological advancement, tools and resources used for plant genomics and the scope, goals and recent progress of the Project. Up to date, we generated about

X. Zhang • M. Yang • Y. Yin
Date Palm Genome Project, King Abdulaziz City for Science and Technology, Riyadh, Kingdom of Saudi Arabia

Key Laboratory of Genome Science and Information, Beijing Institute of Genomics, Chinese Academy of Science, Beijing, China
e-mail: zhangxw@big.ac.cn; yangmeng@big.ac.cn; yinyx.1@gmail.com

J. Tan
Date Palm Genome Project, King Abdulaziz City for Science and Technology, Riyadh, Kingdom of Saudi Arabia

Institute of Bioinformatics, Chongqing University of Posts and Telecommunications, Chongqing, China
e-mail: tanjun0528@gmail.com

I.S. Al-Mssallem
Date Palm Genome Project, King Abdulaziz City for Science and Technology, Riyadh, Kingdom of Saudi Arabia
e-mail: imssallem@kacst.edu.sa

J. Yu (✉)
Date Palm Genome Project, King Abdulaziz City for Science and Technology, Riyadh, Kingdom of Saudi Arabia

Key Laboratory of Genome Science and Information, Beijing Institute of Genomics, Chinese Academy of Science, Beijing, China

Plant Stress Genomics Research Center, King Abdullah University of Science and Technology, Thuwal, Kingdom of Saudi Arabia
e-mail: junyu@big.ac.cn; jun.yu@kaust.edu.sa

S.M. Jain et al. (eds.), *Date Palm Biotechnology*,
DOI 10.1007/978-94-007-1318-5_21, © Springer Science+Business Media B.V. 2011

30 M 454 reads (~15× coverage) and have assembly it into 226,501 contigs, with a total length of 416,498,895 bps. In addition, we achieved the whole 158,462 bp double-stranded circular plastid genome, which with a typical quadripartite structure of the large (LSC, 86,198 bp) and small single copy (SSC, 17,712 bp) regions separated by a pair of inverted repeats (IRs, 27,276 bp). Moreover, in excess of 5,000,000 date palm cDNAs have been sequenced by a 454 sequencer, and these EST sequences will play an important role for date palm genome assembly and gene annotation. In future, more than 100× coverage of SOLiD long mate pair reads will be used to increase the quality of genome assembly, and used to construct scaffolds. Further analysis of date palm genome will be performed in the coming months.

Keywords Sequencing technology • Transcriptome • Plastid and mitochondrial genome

21.1 Introduction

Genomics studies the genome of an organism. A genome is the sum total of DNA molecules that harbor all genes of an organism. Thus, genomics is the study of all the genes of a given cell, tissue and organism; it studies not only the DNA (genome) but also RNA (transcriptome), and protein (proteome) in the context of a regulatory network as well across taxa (evolution). The field includes intensive efforts to determine the entire DNA sequence of various organisms and to construct a genetic map, using large-scale sequencing technology, to generate massive, adequate and high-quality data, by using bioinformatic tools for assembly, annotation and in-depth analysis. Genomics was initiated in principle by Sanger et al. (1977) when he first sequenced the complete genomes of a virus and a mitochondrion. A major branch of genomics is still focused on sequencing the genomes of various species, but the knowledge of full genomes has created the possibility for the field of transciptomics, proteomics, bioinformatics, function genomics, metagenomics and system biology. Following is a brief history of genomics:

- In 1972, Walter Fiers and his team first determined the sequence of a coat protein gene of bacteriophage MS2; then in 1976 they sequenced the complete genome of bacteriophage MS2-RNA (Fiers et al. 1976; Min Jou et al. 1972).
- In 1977, Frederick Sanger sequenced the first DNA-based genome phage Φ-X174 (Sanger et al. 1977).
- In 1995, the first free-living organism, *Haemophilus influenzae,* was sequenced (Fleischmann et al. 1995).
- In 2001, the draft human genome was completed and the final map finished in 2007 (Venter et al. 2001).
- In 2002, the draft map of the rice genome was sequenced by Chinese scientists (Yu et al. 2002).
- In 2005, the first genome and metagenomics sequenced using the next-generation sequencer (Margulies et al. 2005; Poinar et al. 2005).

Table 21.1 List of completed plant genomes

Genus/species	Chromosomes	Genome size (Mb)	References
Arabidopsis thaliana (thale cress)	5	157	Mozo et al. (1999)
Cucumis sativus (cucumber)	7	350	Huang et al. (2009)
Brachypodium distachyon (purple false brome)	5	355	IntBrachypodium Initiative (2010)
Carica papaya (papaya)	9	372	Ming et al. (2008)
Oryza sativa (rice)	12	450	Yu et al. (2005)
Lotus japonicus (lotus)	6	470	Sato et al. (2008)
Vitis vinifera (wine grape)	19	500	Zharkikh et al. (2008)
Populus trichocarpa (black cottonwood)	19	550	Tuskan et al. (2006)
Sorghum bicolor (sorghum)	10	760	Paterson et al. (2009)
Glycine max (soybean)	20	1115	Schmutz et al. (2010)

Table 21.2 List of in-progress plant genomes

Species	Chromosomes	Genome size (Mb)	Available resource
Medicago truncatula (barrel medic)	8	500	Transcriptome, plastid genome and genetic maps
Manihot esculenta (cassava)	20	770	Transcriptome, genetic map
Phoenix dactylifera (date palm)	18	800	Transcriptome, plastid genome, sequence map
Solanum tuberosum (potato)	12	840	Transcriptome, plastid genome, genetic map
Solanum lycopersicum (tomato)	12	950	Plastid genome, genetic map
Zea mays (maize)	10	2,400	Transcriptome, plastid genome, mitochondrial genome, genetic map

- Up to June of 2010, there are 6,414 complete genomic sequences, organelle genomes, plasmids as well as draft genomes, including 2,385 viruses, 41 viroids, 95 archaea, 39 plasmids, 1,482 bacteria and 2,372 eukaryota.

To our knowledge, 10 plant genomes have been sequenced and published; most of them are less than 1 Gb in genome size (Table 21.1). With the development of genome sequencing technology, more and more plant genomes will be sequenced in the next few years. Table 21.2 lists the major plant genome sequencing projects, in progress. It is assured that most of important economic plant species will be sequenced in time and with high quality.

21.1.1 DNA Sequencing Technology

DNA sequencing refers to determining the order of the nucleotide (adenine, guanine, cytosine and thymine) in a DNA molecule. The advancement of DNA sequencing technology has significantly accelerated biological research and discovery. The first DNA sequencing work was performed using manual experimental methods based on two-dimensional chromatography. Subsequently, Frederick Sanger used dideoxynucleotide triphosphates (ddNTPs) as DNA chain terminators (Sanger et al. 1977). The dye-based sequencing method was automated and with high-throughput; since then the DNA sequencing has become one of the essential laboratory methods for molecular biology, together with DNA cloning, DNA microarray, polymerase chain reaction (PCR), electrophoresis, chromatography and centrifugation. It has become not only a fundamental method for determining genome sequences but also a basic tool for identifying DNA methylation sites and sequence polymorphisms. In both scale and scope, DNA sequencing dwarfs all other molecular technologies in the acquisition of basic genomic information, paving the way for other genome-scale, technology-oriented disciplines or, in contemporary terminology, the numerous *-omics* (transcriptomics and proteomics being well-accepted examples) leading biologists into future decades, providing new data, ample information, fresh knowledge and innovative insights.

In the middle 1980s, the first generation sequencer was developed in Leroy Hood's laboratory at the California Institute of Technology. Subsequently ABI introduced its first semi-automated DNA sequencer ABI 370. Toward the end of the last century, the second version of the first-generation technology appeared (3730xl, MegaBACE3000 and so on) which can sequence up to 384 DNA samples in a single load and do up to 24 runs a day. It carries out capillary electrophoresis for size separation, detection and recording of dye fluorescence, and data output as fluorescent peak trace chromatograms. Through decades of gradual improvement, the first-generation sequencer has been applied to achieve sequencing length up to 1,000 bp, with raw accuracy higher than 99.999%, at a cost as little as $0.50/kilobase and throughput close to 1.2 mbp/day (Table 21.3). However impressive those numbers may seem, the first-generation technology has reached its peak both in terms of speed and cost.

With the rapid development of genomics research, demands for low-cost and high-throughput DNA sequencing have driven the invention of new sequencing technologies. Second-generation sequencers, which paralleled the sequence process, producing millions or billions of reads at once, were produced by Roche, Illumina and Applied Biosystem (Shendure et al. 2004). We would like to briefly describe the technical parameters of the three major sequencers:

- Genome Sequencer FLX (Roche 454), based on sequencing-by-synthesis, is the initial second-generation sequencer. It offers more than 1 million reads per run and read lengths of 400 bp and the system is ideally suitable for *de novo* sequencing of whole genomes and transcriptomes.
- Hiseq 2000 (Illunina) relies on DNA fragments extended and bridge amplified on a slide to create an ultra-high density sequencing flow cell with hundreds of millions of clusters; it can generate up to 200 Gb per run in a read-length of 2 × 100 bp, equivalent to up to 25 Gb per day.
- SOLiD4 system is a product of Applied Biosystem (also known as Life Technologies, part of Invitrogen). It employs sequencing by the ligation method and delivers 100 Gb of mappable reads per run; accuracy is about 99.94%.

Table 21.3 Specifications and cost of first, second and third-generation sequencers

Sequencer	3730Xl	GS FLX	SOLiD 4	Hiseq 2000
Company	AB	Roche	AB	Illumina
Method	Sanger	Pyro sequence	Sequence by ligation	Sequence by synthesis
Output per day	1.3 Mb	1 Gb	7 Gb	25 Gb
Run time	1 h	8 h	14 days	8 days
Rd length (bp)	1,000	500	50	100
Cost per run ($)	50	7,000	6,000	10,000
Human genome coverage per run	5× (300,000 run)	5× (30 run)	30× (1 run)	30× (0.5 run)
Cost per genome ($)	15,000,000	200,000	6,000	5,000

Sequencer	Hilicos	Complete Genomics#	PacBio
Company	BioScience	Complete genomics	Pacific biotechnology
Method	tSMS technology	cPAL DNA sequencing	SMRT DNA sequencing
Output per day	4 Gb	6 Gb	96 Gb
Run time	8 days	–	15 min
Rd length (bp)	32	35	1,250
Cost per run ($)	12,000	10,000/200 Gbp	–
Human genome coverage per run	40× (4 run)	45×	–
Cost per genome ($)	48,000	1,726	1,000

The second generation sequencing platforms based on sequencing by synthesis or ligation have dramatically increased the speed of data production and reduced sequencing cost per base over the first-generation machines (Kikuya 2009). It is common for these platforms to generate a Giga-base level of sequence data per day at a cost only $500 per Gbp (Table 21.3). However, their short sequence reads are a serious limitation, which is mainly attributable to the dephasing problem of the optical signal in the DNA sequencing cluster. One solution to remedy this would be to eliminate the ensemble effect all together: sequencing on a single molecule (Zhou et al. 2010).

To overcome one of the major drawbacks of second-generation sequencing technology, the relatively short read, efforts have been made to develop single molecule sequencing platforms: a third-generation sequencing technology. A single molecule also helps to increase the number of DNA fragments that can be independently analyzed in a given surface area and, therefore, achieves a much higher level of throughput (Tanaka and Kawai 2009).

Pacific Bioscience Inc. is developing a third-generation sequencer based on a new technology: the SMRT (Single Molecule Real Time) technology. This DNA sequencer uses single DNA molecules as templates. The main characteristic of this sequencer is real-time monitoring of nucleotide incorporation with DNA polymerase.

Unlike the second-generation sequencer, the sequencer can obtain reads of more than 1kb from a single template, and generate about 1 mbp sequence data per second. Other than PacBio, Oxford Nanopore, VisiGen Biotechnologies, Mobious Biosystems and Complete Genomics (Drmanac et al. 2009) will or have already presented their third-generation sequencers separately. To date, there are already two types of third generation sequencing platforms available on the market, from Complete Genomics and Helicos. The third-generation sequencers promise researchers the ability to finish the human genome within a day and for less than USD 1,000. We summarize specification and cost of selected sequencers from the first-generation to the third-generation in Table 21.3.

21.1.2 Date Palm Genome Project

Date palm cultivars are large woody plants in the genus *Phoenix*. There are about 450 domesticated *Phoenix dactylifera* cultivars found in Saudi Arabia alone (Al-Khalifah and Askari 2003) and their genetic diversity is relatively high. It is believed that the date palm originated around the Persian Gulf, and has been cultivated since ancient times from Mesopotamia to prehistoric Egypt, possibly as early as 3,000 BCE (Mahmoudi et al. 2008). The total number of date palm trees in the world is over 100 million with production of 6.9 million mt, and Saudi Arabia possess 8% of the world date palms with 14% of total world production (FAOSTAT). However, there are major challenges encountered in date palm production and marketing, from virulent diseases, dangerous pests, abiotic stresses and new demands for sustainability, quality, transport and yield.

Date Palm trees are mostly diploid (2n = 36) and propagated sexually by seeds and asexually by offshoots (Al-Bakr 1972; Al-Salih et al. 1987). The genome size is estimated to be 1.2 billion base pairs, based on flow cytometry (Zonneveld et al. 2005). Although it is an economically-important plant species, there are very few DNA sequences of date palm found in the GenBank databases. With the development of genome sequencing technology, date palm genome sequencing became the target of two research teams.

A team from Weil Cornell Medical College in Qatar tried to sequence the entire date palm genome using Solexa (illumine) sequencer based on a shotgun method. They announced the finished draft map in 2009 and released the sequence data subsequently: (http://qatar-weill.cornell.edu/research/datepalmGenome/index.html). According to their analyses, the genome assembly has a predicted genome size of ~550Mbp. The following are genome parameters of their draft sequence assembly:

- 45,000 scaffolds greater than 2 kb
- Scaffold N50 is 4,250 bp
- 850,000 novel high quality SNPs between parental alleles
- GC content of the nuclear genome is 37%

- 302 Mb of assembled sequence with 18.5 Mb of ordered gaps
- Unique sequence is 292 Mb at the 24-mer level

As a collaborative effort between the King Abdulaziz City for Science and Technology (KACST) and Beijing Institute of Genomics, Chinese Academy of Science (BIG/CAS), the Date Palm Genome Project (DPGP) was initiated. The DPGP's goals are multifold and composed of several components at the genomics, bioinformatics, genetics and biochemistry levels. The first phase of the DPGP is focused on genomics and bioinformatics that pave the way for genetic and biochemical studies.

The specific aims of the DPGP are: a working draft with sequence coverage; 10× from 454 and 50× from SOLiD; a complete map will be built with end-sequences from BACs and Fosmids; a genome diversity map built with shotgun sequencing of 30 cultivars; each with 30× of SOLiD reads; the date palm transcriptomes: full-length cDNA, over 30,000 unigenes; and expression profiles for leaves, roots, and flowers (~50 tissue samples).

In the following sections, we present some of our preliminary data in genome sequencing and assembly, chloroplast genome sequencing and transcriptomics.

21.2 Date Palm Genome Assembly

21.2.1 Assembly Software Package

We use both Roche 454 and ABI SOLiD for generating sequencing data. Roche 454 produces sequences in a length of 400bp usually; sometimes, we generate paired-end information for a better assembly. SOLiD produces short reads less than 50 bp in length and paired-end information in a range of 3–8 kb.

We carried out *de novo* assembly by using a hybrid strategy where we build contigs with unique reads and add paired-end information to the contigs to obtain supercontigs. To deal with short reads in tens of millions, we use three approaches – De Bruijn graph, traditional overlap-layout-consensus and Prefix tree-based (Trie) – by using several assembly tools. These tools are readily available and have been used for similar data types from different or similar sequencing strategies. We will build scaffolds and try to finish the sequence assembly by using various tools for large plant genome assemblies, such as those used for the date palm and tomato genomes. We are also optimistic about new algorithms and tools being developed by research groups around the world to better deal with a large quantity of data and repeat-rich genomes, such as the large plant genomes (Li et al. 2008; Zerbino and Birney 2008; Simpson et al. 2009). In scaffold building, we will also use paired-end information from large-insert clones such as BACs (120 kb in size) and fosmids (30 kb in size). Some of the major software packages for next-generation sequence assembly are listed in Table 21.4.

Table 21.4 Examples of sequence assembly software packages for data produced by next generation sequencers

Algorithm	Software package	Reference	Institution
De Bruijn graph	ABySS	Simpson et al. 2009	British Columbia Cancer Agency
	ALLPATHS	Butler et al. 2008	Broad Institute of MIT and Harvard
	Contrail	Schatz et al. 2010	University of Maryland
	Euler-SR	Chaisson et al. 2009	University of California San Diego
	Velvet	Zerbino and Birney 2008	Welcome Trust Genome Campus
Traditional overlap-layout- consensus	Celera/WGS	Denisov et al. 2008	J. Craig Venter Institute
	Edena	Hernandez et al. 2008	Geneva University Hospitals and the University of Geneva
	Newbler		F. Hoffmann-La Roche Ltd.
	SeqCon	Rausch et al. 2008	International Max Planck Research School for Computational Biology and Scientific Computing
	SHORTY	Hossain et al. 2009	Stony Brook University
Prefix tree-based (Trie) approach	QSRA	Bryant et al. 2009	Oregon State University
	SHARCGS	Dohm et al. 2007	Max-Planck-Institute for Molecular Genetics
	SSAKE	Warren et al. 2007	Genome Sciences Centre
	VCAKE	Jeck et al. 2007	University of Carolina – Chapel Hill
Smith-Waterman	MOSAIK	Strömberg 2010	Boston College – Higgins Hall
Simple Greedy	MIRA	Chevreux et al. 2004	German Cancer Research Centre Heidelberg
Burrow Wheeler Transformation	SOAP	Li et al. 2008	Beijing Genomics Institute at Shenzhen

21.2.2 Date Palm Genome Assembly

The date palm genome is composed of 18 chromosome pairs, and the estimated size is 250 Mb (Barakat et al. 1999) 350 Mb (Al-Dous et al. 2011). At the same time, oil palm and coconut each have 16 chromosome pairs and around 2G genome sizes. Based on 30M 454 reads (average length is 383 bp, near 15× coverage) produced by KACST, 20-mer filtering was done to pick out unique reads, resulted in about half of them being qualified to be assembled by the Newbler run assembly tool. A total of 226,501 contigs were obtained, with a total length of 416,498,895 bps. That is to say, the estimated genome size of date palm is more than 800 M considering 454 sequencing missed regions.

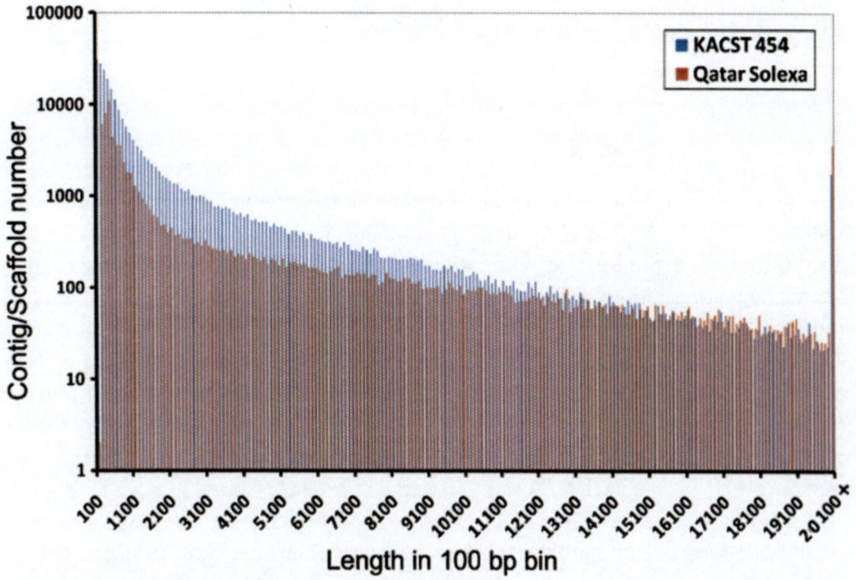

Fig. 21.1 Comparison of contig/scaffold length distribution of KACST 454 and Qatar Solexa assemblies. A cut-off is shown at 500 bp in the Qatar assembly, whereas more than 100,000 contigs of KACST 454 are less than 500 bps. So, the numbers and total lengths are similar for contigs/scaffolds of both researches. But there are more than 30 M 'N/Ns' in Qatar scaffold sequences

Table 21.5 Comparison of assembly results of KACST 454 contigs and Qatar Solexa scaffolds

	KACST 454	Qatar Solexa
Total length (bps)	416,498,895	302,873,018
Contig/Scaffold number	226,501	73,747
Average length (bps)	1,839	4,107
Max length (bps)	89,547	169,087
Min length (bps)	100	222
GC content (%)	39	35
N50 length (bps)	6,075	14,840

21.2.3 Comparison with the Qatar Draft Map

Since we have acquired the majority of the data for our Project, we have compared them to the Qatar scaffolding result that is acquired by using Solexa alone. The Qatar assembly has 18,024 scaffolds with a length more than 4,000 bp (the estimated average gene length of date palm is about 4,000 bp) (Fig. 21.1); it adds up to ~252 Mb in total length. This assembly is similar to our assembled length of 258 Mb in 27,389 contigs (Table 21.5). The next step is to add more than 50X SOLiD data in mate-pairs to obtain supercontigs as well as to improve sequence quality.

21.3 Date Palm Chloroplast Genome

Chloroplasts are one of the many different types of organelles in the cell. In general, they are considered to have originated from cyanobacteria through endosymbiosis. They are found in plant cells and other eukaryotic organisms that conduct photosynthesis. All chloroplasts are thought to derive directly or indirectly from a single endosymbiotic event. The chloroplast genome has long been a focus of research in plant molecular evolution and systematics because of its small size, high copy number, conservation and extensive characterization at the molecular level (Raubeson and Jansen 2005). In angiosperms, most cp genomes are circular DNAs with the length from 120 to 160 kb and have a quadripartite organization consisting of two copies of an inverted repeats (IRs) of ~20–28 kb in size, dividing the rest into an 80–90 kb large-single-copy region (LSC) and a 16–27 kb small-single-copy (SSC) region. Usually, the gene content of an angiosperm cp genome is rather conserved, encoding 4 rRNAs, ~30 tRNAs and ~80 unique proteins (Chumley et al. 2006).

Since the publication of the first angiosperm chloroplast genome sequence of the genus *Nicotiana*, which includes tobacco, in 1986 (Shinozaki et al. 1986), researchers have been keen to use complete chloroplast genome sequences to infer a wide range of information about plants, in which the phylogenetic relationships is most expressed in three forms: restriction fragment/site comparisons, structural rearrangements and DNA sequencing. In the past 10 years, as there was widespread application of new DNA sequencing technology, the complete chloroplast genomes available in public databases rapidly increased. Currently, there are 132 complete chloroplast genomes of land plants deposited in the GenBank Organelle Genome Resources; most of them are angiosperms (Table 21.6).

The dissimilar variations of land plants brought challenges to biologists interested in the origin and evolution of these species, and resolving these issues critically depends on having a well resolved and strongly supported phylogenetic framework. Over the past 20 years, numerous phylogenetic studies have used both morphological and molecular data to assess relationships among the major clades. For nearly two decades, most phylogenetic analyses of land plants have relied on DNA sequences of one to several genes from the plastid, mitochondrial and nuclear genomes; however, the support is often low due to insufficient data. The coding sequences from the completely sequenced genomes provide more facilities to classify the relationship of land plants and find the earliest lineages of angiosperms. A recent phylogenetic study has used 61 protein-coding genes from completely sequenced plastid genomes to estimate the relationships among the major clades of angiosperms (Leebens-Mack et al. 2005). Another famous study used 81 genes from 64 plastid genome to identify genome-scale evolutionary patterns to provide strong support for the position of the genus *Amborella* as the earliest diverging lineage of flowering plants (Jansen et al. 2007).

As a subject of the DPGP, the study of its chloroplast (cp) genome will provide great facility to interpret the evolutional changes among angiosperms. We have finished the sequencing and assembling of the date palm chloroplast

Table 21.6 Completely sequenced land plant chloroplast genomes (as of 1 July 2010)

Higher classification	Family	Genus/species	Accession number	Length
Mosses	Pottiaceae	*Syntrichia ruralis*	NC_012052	122,630
	Funariaceae	*Physcomitrella patens*	NC_005087	122,890
Liverworts	Marchantiaceae	*Marchantia polymorpha*	NC_001319	121,024
	Aneuraceae	*Aneura mirabilis*	NC_010359	108,007
Ferns	Selaginellaceae	*Selaginella uncinata*	AB197035	144,170
	Selaginellaceae	*S. moellendorffii*	NC_013086	143,780
	Lycopodiaceae	*Huperzia lucidula*	NC_006861	154,373
	Pteridaceae	*Adiantum capillus-veneris*	NC_004766	150,568
	Psilotaceae	*Psilotum nudum*	NC_003386	138,829
	Marattiaceae	*Angiopteris evecta*	NC_008829	153,901
	Cyatheaceae	*Alsophila spinulosa*	NC_012818	156,661
Gymnosperms	Cupressaceae	*Cryptomeria japonica*	NC_010548	131,810
	Cycadaceae	*Cycas taitungensis*	NC_009618	163,403
	Ephedraceae	*Ephedra equisetina*	NC_011954	109,518
	Gnetaceae	*Gnetum parvifolium*	NC_011942	114,914
	Pinaceae	*Keteleeria davidiana*	NC_011930	117,720
	Pinaceae	*Picea sitchensis*	NC_011152	109,798
	Pinaceae	*Pinus contorta*	NC_011153	115,615
	Pinaceae	*P. gerardiana*	NC_011154	116,997
	Pinaceae	*P. koraiensis*	NC_004677	117,190
	Pinaceae	*P. krempfii*	NC_011155	115,555
	Pinaceae	*P. lambertiana*	NC_011156	112,521
	Pinaceae	*P. longaeva*	NC_011157	115,896
	Pinaceae	*P. monophylla*	NC_011158	114,607
	Pinaceae	*P. nelsonii*	NC_011159	110,575
	Pinaceae	*P. thunbergii*	NC_001631	119,707
	Welwitschiaceae	*Welwitschia mirabilis*	NC_010654	119,726
Angiosperms				
Monocots	Acoraceae	*Acorus americanus*	NC_010093	153,819
	Acoraceae	*A. calamus*	NC_007407	153,821
	Araceae	*Lemna minor*	NC_010109	165,955
	Arecaceae	*Phoenix dactylifera L.*	GU811709	158,462
	Dioscoreaceae	*Dioscorea elephantipes*	NC_009601	152,609
	Orchidaceae	*Phalaenopsis aphrodite*	NC_007499	148,964
	Orchidaceae	*Oncidium Gower Ramsey*	NC_014056	146,484
	Poaceae	*Agrostis stolonifera*	NC_008591	136,584
	Poaceae	*Anomochloa marantoidea*	NC_014062	138,412
	Poaceae	*Bambusa oldhamii*	NC_012927	139,350
	Poaceae	*Brachypodium distachyon*	NC_011032	135,199
	Poaceae	*Dendrocalamus latiflorus*	NC_013088	139,394
	Poaceae	*Festuca arundinacea*	NC_011713	136,040
	Poaceae	*Hordeum vulgare subsp. vulgare*	NC_008590	136,462
	Poaceae	*Lolium perenne*	NC_009950	135,282
	Poaceae	*Oryza nivara*	NC_005973	134,494

(continued)

Table 21.6 (continued)

Higher classification	Family	Genus/species	Accession number	Length
	Poaceae	*O. sativa Indica Group*	NC_008155	134,496
	Poaceae	*O. sativa Japonica Group*	NC_001320	134,525
	Poaceae	*Saccharum hybrid cultivar SP-80-3280*	NC_005878	141,182
	Poaceae	*S. officinarum*	NC_006084	141,182
	Poaceae	*Sorghum bicolor*	NC_008602	140,754
	Poaceae	*Triticum aestivum*	NC_002762	134,545
	Poaceae	*Coix lacryma-jobi*	NC_013273	140,745
	Poaceae	*Zea mays*	NC_001666	140,384
	Typhaceae	*Typha latifolia*	NC_013823	161,572
Dicots	Amaranthaceae	*Spinacia oleracea*	NC_002202	150,725
	Amborellaceae	*Amborella trichopoda*	NC_005086	162,686
	Apiaceae	*Daucus carota*	NC_008325	155,911
	Araliaceae	*Panax ginseng*	NC_006290	156,318
	Asteraceae	*Helianthus annuus*	NC_007977	151,104
	Asteraceae	*Guizotia abyssinica*	NC_010601	151,762
	Asteraceae	*Lactuca sativa*	NC_007578	152,765
	Asteraceae	*Parthenium argentatum*	NC_013553	152,803
	Berberidaceae	*Nandina domestica*	NC_008336	156,599
	Brassicaceae	*Lobularia maritima*	NC_009274	152,659
	Brassicaceae	*Draba nemorosa*	NC_009272	153,289
	Brassicaceae	*Arabis hirsuta*	NC_009268	153,689
	Brassicaceae	*Aethionema cordifolium*	NC_009265	154,168
	Brassicaceae	*A. grandiflorum*	NC_009266	154,243
	Brassicaceae	*Arabidopsis thaliana*	NC_000932	154,478
	Brassicaceae	*Capsella bursa-pastoris*	NC_009270	154,490
	Brassicaceae	*Barbarea verna*	NC_009269	154,532
	Brassicaceae	*Olimarabidopsis pumila*	NC_009267	154,737
	Brassicaceae	*Lepidium virginicum*	NC_009273	154,743
	Brassicaceae	*Nasturtium officinale*	NC_009275	155,105
	Brassicaceae	*Crucihimalaya wallichii*	NC_009271	155,199
	Buxaceae	*Buxus microphylla*	NC_009599	159,010
	Calycanthaceae	*Calycanthus floridus*	NC_004993	153,337
	Campanulaceae	*Trachelium caeruleum*	NC_010442	162,321
	Caricaceae	*Carica papaya*	NC_010323	160,100
	Ceratophyllaceae	*Ceratophyllum demersum*	NC_009962	156,252
	Chaetosphaeridiaceae	*Chaetosphaeridium globosum*	NC_004115	131,183
	Chloranthaceae	*Chloranthus spicatus*	NC_009598	157,772
	Convolvulaceae	*Cuscuta obtusiflora*	NC_009949	85,286
	Convolvulaceae	*C. gronovii*	NC_009765	86,744
	Convolvulaceae	*C. reflexa*	NC_009766	121,521
	Convolvulaceae	*C. exaltata*	NC_009963	125,373
	Convolvulaceae	*Ipomoea purpurea*	NC_009808	162,046
	Cucurbitaceae	*Cucumis sativus*	NC_007144	155,293
	Euphorbiaceae	*Manihot esculenta*	NC_010433	161,453

(continued)

Table 21.6 (continued)

Higher classification	Family	Genus/species	Accession number	Length
	Euphorbiaceae	*Jatropha curcas*	NC_012224	163,856
	Fabaceae	*Lathyrus sativus*	NC_014063	121,020
	Fabaceae	*Pisum sativum*	NC_014057	122,169
	Fabaceae	*Medicago truncatula*	NC_003119	124,033
	Fabaceae	*Cicer arietinum*	NC_011163	125,319
	Fabaceae	*Trifolium subterraneum*	NC_011828	144,763
	Fabaceae	*Phaseolus vulgaris*	NC_009259	150,285
	Fabaceae	*Lotus japonicus*	NC_002694	150,519
	Fabaceae	*Vigna radiata*	NC_013843	151,271
	Fabaceae	*Glycine max*	NC_007942	152,218
	Geraniaceae	*Pelargonium x hortorum*	NC_008454	217,942
	Magnoliaceae	*Liriodendron tulipifera*	NC_008326	159,886
	Malvaceae	*Gossypium hirsutum*	NC_007944	160,301
	Malvaceae	*G. barbadense*	NC_008641	160,317
	Moraceae	*Morus indica*	NC_008359	158,484
	Myrtaceae	*Eucalyptus globulus*	NC_008115	160,286
	Nymphaeaceae	*Nymphaea alba*	NC_006050	159,930
	Nymphaeaceae	*Nuphar advena*	NC_008788	160,866
	Oleaceae	*Olea europaea*	NC_013707	155,872
	Oleaceae	*Jasminum nudiflorum*	NC_008407	165,121
	Onagraceae	*Oenothera parviflora*	NC_010362	163,365
	Onagraceae	*O. biennis*	NC_010361	164,807
	Onagraceae	*O. argillicola*	NC_010358	165,055
	Onagraceae	*O. glazioviana*	NC_010360	165,225
	Onagraceae	*O. elata*	NC_002693	165,728
	Orobanchaceae	*Epifagus virginiana*	NC_001568	70,028
	Piperaceae	*Piper cenocladum*	NC_008457	160,624
	Platanaceae	*Platanus occidentalis*	NC_008335	161,791
	Polygonaceae	*Fagopyrum esculentum*	NC_010776	159,599
	Ranunculaceae	*Ranunculus macranthus*	NC_008796	155,129
	Ranunculaceae	*Megaleranthis saniculifolia*	NC_012615	159,924
	Rubiaceae	*Coffea arabica*	NC_008535	155,189
	Rutaceae	*Citrus sinensis*	NC_008334	160,129
	Salicaceae	*Populus alba*	NC_008235	156,505
	Salicaceae	*Populus trichocarpa*	NC_009143	157,033
	Schisandraceae	*Illicium oligandrum*	NC_009600	148,553
	Solanaceae	*Solanum tuberosum*	NC_008096	155,298
	Solanaceae	*S. bulbocastanum*	NC_007943	155,371
	Solanaceae	*Solanum lycopersicum*	NC_007898	155,461
	Solanaceae	*Nicotiana tomentosiformis*	NC_007602	155,745
	Solanaceae	*N. sylvestris*	NC_007500	155,941
	Solanaceae	*N. tabacum*	NC_001879	155,943
	Solanaceae	*Atropa belladonna*	NC_004561	156,687
	Vitaceae	*Vitis vinifera*	NC_007957	160,928
	Winteraceae	*Drimys granadensis*	NC_008456	160,604

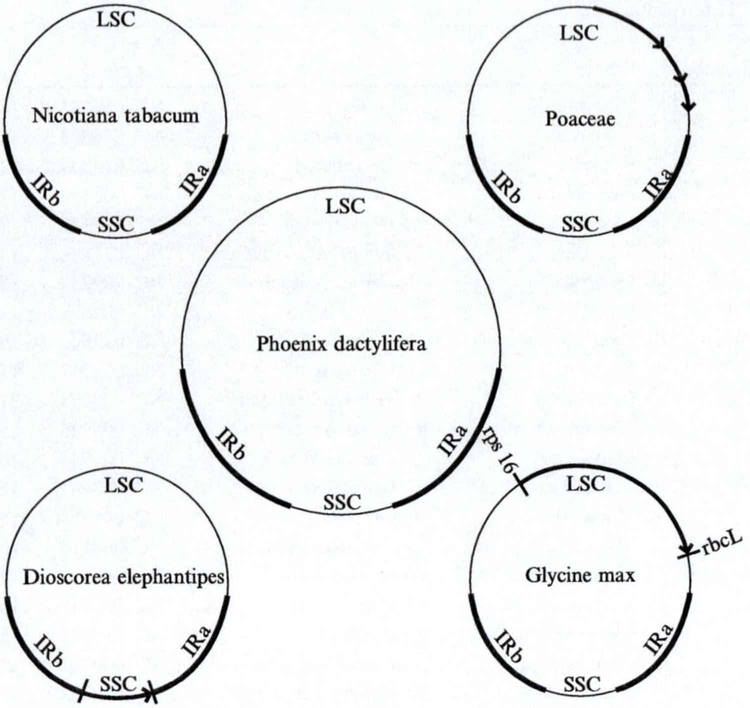

Fig. 21.2 General description of chloroplast genome organization in representative Angiosperms. The arcs with arrows indicate the inversion fragments

genome using the pyrosequencing technology (Yang et al. 2010). The genome is a 158,462 bp double-stranded circular DNA molecule, with a typical quadripartite structure of the large (LSC, 86,198 bp) and small single copy (SSC, 17,712 bp) regions separated by a pair of inverted repeats (IRs, 27,276 bp). Similar to other cp genomes (Gao et al. 2009; Raubeson et al. 2007) the date palm cp genome is also rich in A and T (62.77%). The value is decreasing in regions of non-coding, protein-coding, tRNA and rRNA, which is 66.60%, 61.03%, 57.94% and 52.19%, respectively. As a regular cp genome, the gene content and structure are more like that in tobacco without rearrangements and gene gain or loss (Fig. 21.2). Inverted repeats have expanded into the LSC, but not the SSC region. The number of repeat sequence (>=30bp with 90% identity but not include the IRs) is lower than that in the grass family Poaceae and some dicots. Intravarietal SNP also exist not only in intergenic region, but also in coding regions of genes with essential function, and some SNPs can also be detected in oil palm. Transcriptome analyses showed three genes (atpF, trnA-UGC, and rrn23) have dominant number of transcripts and most protein-coding genes belong to polycistronic transcripts. We built a rectangular phylogenetic tree to estimate the

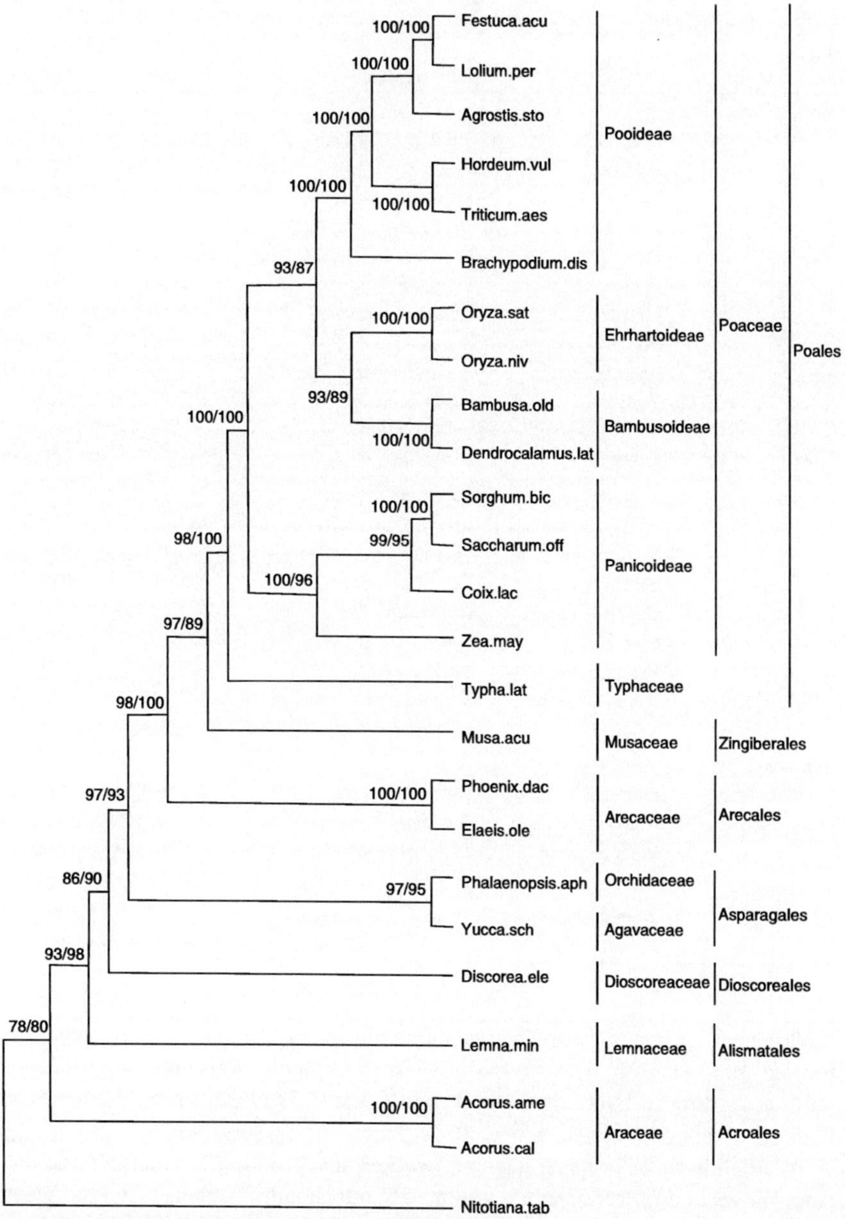

Fig. 21.3 A phylogenetic tree of 76 concatenated cp genes from 25 taxa, derived from both MP and ML methods. Branch-swapping algorithm of both methods is based on TBR. Numbers at nodes indicate the bootstrap support

Table 21.7 Completely sequenced land plant mitochondrial genomes (as of 1 July 2010)

Higher classification	Family	Genus/species	Accession number	Length
Angiosperms				
Monocots	Poaceae	*Bambusa oldhamii*	EU365401	509,941
	Poaceae	*Oryza rufipogon*	NC_013816	559,045
	Poaceae	*Tripsacum dactyloides*	NC_008362	704,100
	Poaceae	*Oryza sativa* Indica Group	NC_007886	491,515
	Poaceae	*O. sativa* Japonica Group	NC_011033	490,520
	Poaceae	*Sorghum bicolor*	NC_008360	468,628
	Poaceae	*Triticum aestivum*	NC_007579	452,528
	Poaceae	*Zea luxurians*	NC_008333	539,368
	Poaceae	*Z. mays* subsp. *mays*	NC_007982	569,630
	Poaceae	*Z. mays* subsp. *parviglumis*	NC_008332	680,603
	Poaceae	*Z. perennis*	NC_008331	570,354
Dicots	Solanaceae	*Nicotiana tabacum*	NC_006581	430,597
	Brassicaceae	*Arabidopsis thaliana*	NC_001284	366,924
	Vitaceae	*Vitis vinifera*	NC_012119	773,279
	Cucurbitaceae	*Citrullus lanatus*	NC_014043	379,236
	Cucurbitaceae	*Cucurbita pepo*	NC_014050	982,833
	Caricaceae	*Carica papaya*	NC_012116	476,890
	Brassicaceae	*Brassica napus*	NC_008285	221,853
	Amaranthaceae	*Beta vulgaris* subsp. *vulgaris*	NC_002511	368,801
Gymnosperms				
Liverworts	Cycadaceae	*Cycas taitungensis*	NC_010303	414,903
	Marchantiaceae	*Marchantia polymorpha*	NC_001660	186,609
	Pleuroziaceae	*Pleurozia purpurea*	NC_013444	168,526
Hornworts	Dendrocerotaceae	*Megaceros aenigmaticus*	NC_012651	184,908
	Anthocerotaceae	*Phaeoceros laevis*	NC_013765	209,482
Mosses	Funariaceae	*Physcomitrella patens*	NC_007945	105,340

phylogenetic position using 76 concatenated genes from 26 taxa of *Nicotiana* as the outgroup (Fig. 21.3). Date palm shares extremely high homology with oil palm according to bootstrap sampling test, and it is closer to the Poales order than other monocot families.

In addition to chloroplast genome, we are now working to assemble the date palm mitochondrial genome. Up to now, 25 mitochondrial genomes in land plants (different species or cultivars) have been completely sequenced and the genome sizes range from 200 to 1,000 kb (Table 21.7). We have assembled the major contigs of date palm mitochondrial genome and the genome size is estimated to be 600 kb. The essential genes accepted in the mitochondrial genome of other land plants have been predicted in date palm. The assembling process is also based on pyrosequencing technology, and mitochondrion-related reads were first extracted from the whole genome sequencing data to assemble the preliminary contigs. The following work describes efforts to extend these contigs, filter the contamination from nuclear and chloroplast reads, and organize the order of all contigs until the final genome sequence are obtained.

21.4 Date Palm Transcriptome Analysis

Recently, a next-generation sequencer has developed a scalable, highly parallel DNA sequencing system that is faster than standard sequencing methods. This increase in throughput comes at the expense of read length. In contrast, the read-length limitation associated with next generation sequencers is less of a concern for transcriptome sequencing and analysis. And they avoid expensive cloning-based library construction, and it is feasible to sequence a wide variety of cDNA samples, thereby increasing the recovery of highly specialized transcripts (Jarvie and Harkins 2008). Moreover, the new technologies are particularly well suited for EST-based gene discovery because it generates hundreds of thousands of tags per run, greatly increasing the chances of capturing rare transcripts. Table 21.8 summarizes the top 20 plant Unigene entry collections from the public databases.

Table 21.8 A list of top 20 plant Unigene entry collections from selected species[a]

Species	No. of unigenes entries	No. of ESTs
Zea mays	97,486	2,021,364
Artemisia annua	88,127	94,734
Oryza sativa	40,978	1,272,399
Triticum aestivum	40,870	1,099,713
Glycine max	33,001	1,465,708
Arabidopsis thaliana	30,579	1,788,991
Brassica napus	27,139	656,152
Theobroma cacao	24,958	164,435
Nicotiana tabacum	24,069	326,135
Malus x domestica	23,731	329,920
Hordeum vulgare	23,595	533,726
Picea glauca	22,472	341,325
Vitis vinifera	22,083	364,469
Panicum virgatum	20,973	442,269
Gossypium hirsutum	20,671	270,813
Picea sitchensis	19,828	290,364
Solanum tuberosum	18,825	245,308
Raphanus raphanistrum	18,794	164,120
Medicago truncatula	18,785	285,262
Solanum lycopersicum	18,346	300,688

[a] This list is ordered by the total number of Unigene entries at http://www.ncbi.nlm.nih.gov/, as of 2010-7-10

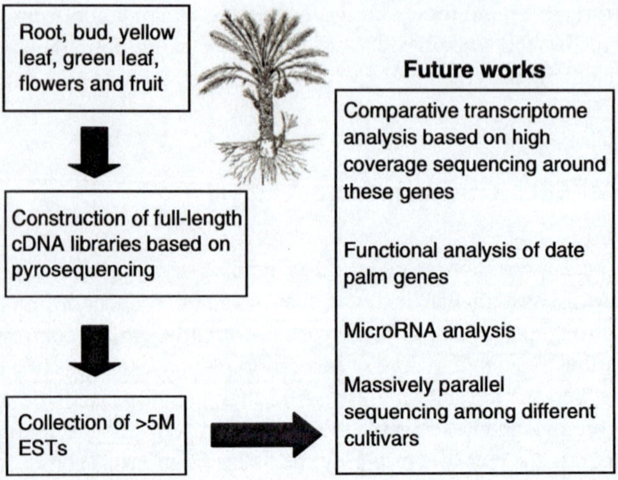

Fig. 21.4 Schematic diagram of large-scale cDNA collection, sequencing and future works

Full-length cDNA are necessary to identify exon-intron boundaries and gene-coding regions within genomic sequences and for comprehensive gene function analysis at the transcriptional and translational levels. More than 5,000,000 date palm cDNAs have been sequenced, and these sequences have not only advanced the functional analysis of the date palm genes, but have also played an important role in date palm genome annotation (Fig. 21.4). To identify and characterize all the transcribed regions in a genome is one of the greatest challenges. In order to provide genomic resources for the analysis of date palm, and for the analysis of genes differentially expressed among different tissues, the objectives of this study were to create cDNA libraries (Beldade et al. 2006) suitable for the analysis of expressed sequence tags (ESTs) and to generate an EST resource for date palm to annotate the entire genome, and also this technique of sequencing cost just a fraction of the entire genome.

As transcriptome sequencing (mRNA transcript-expression analysis, novel gene discovery, gene space identification in novel genomes and assembly of full-length genes, etc.) become more encompassing, it is now possible to detect and identify nearly every class of molecules. Naturally, it is one of the most efficient ways for the identification of differentially expressed genes. Furthermore, a certain number of scientific propositions are being solved and future perspectives were revealed just as illustrated in Fig. 21.5. Transcriptomes are very important for both understanding the physiology of date palm and annotation of the genome. To the present, significant progress has been made in transcriptome analysis of the genomes based on new sequencers. We review the recent advances in new methods (SOLiD total RNA-Seq; cDNA rapid library preparation method – Roche 454) for a deep and comprehensive transcriptome analysis of the date palm genome.

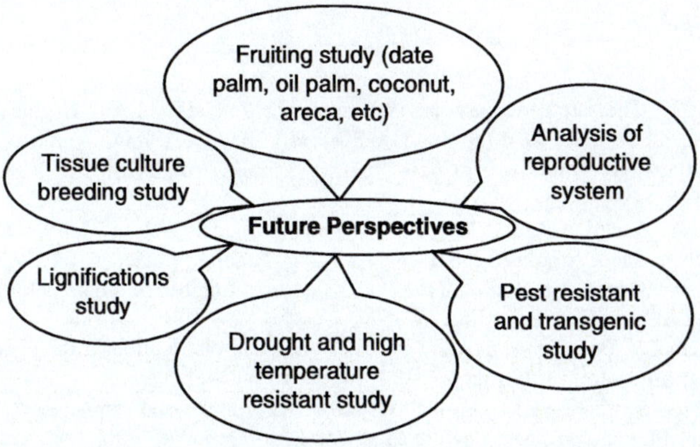

Fig. 21.5 Schematic diagram of future perspectives

The alignment of full-length cDNA sequences to the genome sequence has provided direct experimental evidence for many of the gene models, which were predicted by computer programs. But, a large number of the gene models should be confirmed by even more experimental data (Kikuchi et al. 2007). Current computer programs still are not enough for gene structural annotation, and the problem of exon-intron structure is still a major challenge for date palm genome biologists. In addition, we need further collection and complete sequencing of cDNA, and these data will improve current date palm genome annotation.

21.5 Conclusion and Prospect

As shown above, the Date Palm Genome Project is multifold and composed of several components at the genomic, bioinformatic, transcriptomic and post-genomic levels. Data have been generated by using second-generation sequencers and sequence assembling has been working on most likely in a complex process where different types of data are integrated to ensure both quality and contiguity. This ongoing work will be published in the near future. The sequence of the date palm genome can provide insights into the mechanisms underlying fruit development, sex determination, insect resistance, drought resistance and other important biological processes at the phenotypic level. The genome can also advance our knowledge of the evolution of the palm family (Arecaceae).

Acknowledgement The Date Palm Genome Project is supported by a research grant from KACST and institutional support from BIG and CAS. It has also received support from the Chinese Ministry of Sciences and Technology for administration and travel

References

Al-Bakr A (1972) The date palm: past, present and future, 2nd edn. Al-Ani, Baghdad

Al-Dous EK, George B, Al-Mahmoud ME et al (2011) De novo genome sequencing and comparative genomics of date palm (*Phoenix dactylifera*). Nature Biotech 29:521–528

Al-Khalifah NS, Askari E (2003) Molecular phylogeny of date palm (*Phoenix dactylifera L.*) cultivars from Saudi Arabia by DNA fingerprinting. Theor Appl Genet 107:1266–1270

Al-Salih AA, Hussain N, Al-Jarrah A (1987) Chromosome number of a date palm male cultivar Ghanname Akhdar. Date Palm J 5:128–133

Barakat A, Han DT, Benslimane AA et al (1999) The gene distribution in the genomes of pea, tomato and date palm. FEBS Lett 463:139–142

Beldade P, Rudd S, Gruber JD, Long AD (2006) A wing expressed sequence tag resource for *Bicyclus anynana* butterflies, an evo-devo model. BMC Genome 7:130

Bryant DW Jr, Wong WK, Mockler TC (2009) QSRA – a quality-value guided de novo short read assembler. BMC Bioinform 10:69

Butler J, MacCallum I, Kleber M et al (2008) ALLPATHS: de novo assembly of whole-genome shotgun. Genome Res 18:810–820

Chaisson MJ, Brinza D, Pevzner PA (2009) De novo fragment assembly with short mate-paired reads: does the read length matter? Genome Res 19(2):336–346

Chevreux B, Pfisterer T, Drescher B et al (2004) Using the miraEST assembler for reliable and automated mRNA transcript assembly and SNP detection in sequenced ESTs. Genome Res 14:1147–1159

Chumley TW, Palmer JD, Mower JP et al (2006) The complete chloroplast genome sequence of *Pelargonium* x *hortorum*: organization and evolution of the largest and most highly rearranged chloroplast genome of land plants. Mol Biol Evol 23:2175–2190

Denisov G, Walenz B, Halpern AL et al (2008) Consensus generation and variant detection by Celera Assembler. Bioinformatics 24:1035–1040

Dohm JC, Lottaz C, Borodina T et al (2007) SHARCGS, a fast and highly accurate short-read assembly algorithm for de novo genomic sequencing. Genome Res 17:1697–1706

Drmanac R, Sparks AB, Callow MJ et al (2009) Human genome sequencing using unchained base reads on self-assembling DNA nanoarrays. Science 327:78–81

Fiers W, Contreras R, Duerinck F et al (1976) Complete nucleotide sequence of bacteriophage MS2 RNA: primary and secondary structure of the replicase gene. Nature 260:500–507

Fleischmann RD, Adams MD, White O et al (1995) Whole-genome random sequencing and assembly of *Haemophilus influenzae* Rd. Science 269:496–512

Gao L, Yi X, Yang YX et al (2009) Complete chloroplast genome sequence of a tree fern *Alsophila spinulosa*: insights into evolutionary changes in fern chloroplast genomes. BMC Evol Biol 9:130

Hernandez D, François P, Farinelli L et al (2008) De novo bacterial genome sequencing: millions of very short reads assembled on a desktop computer. Genome Res 18:802–809

Hossain MS, Azimi N, Skiena S (2009) Crystallizing short-read assemblies around seeds. BMC Bioinform 10(Suppl 1):S16

Huang S, Li R, Zhang Z et al (2009) The genome of the cucumber, *Cucumis sativus* L. Nat Genet 41:1275–1283

Jansen RK, Cai Z, Raubeson LA et al (2007) Analysis of 81 genes from 64 plastid genomes resolves relationships in angiosperms and identifies genome-scale evolutionary patterns. Proc Natl Acad Sci USA 104:9369–9374

Jarvie T, Harkins T (2008) Transcriptome sequencing with the genome sequencer FLX system. Nature Methods 5 http://www.nature.com/nmeth/journal/v5/n9/full/nmeth.f.220.html

Jeck WR, Reinhardt JA, Baltrus DA et al (2007) Extending assembly of short DNA sequences to handle error. Bioinformatics 23:2942–2944

Kikuchi S, Wang GL, Li L (2007) Genome-wide RNA expression profiling in rice. In: Upadhyaya NM (ed.) Rice functional genomics: challenges, progress and prospects. Springer, New York, pp 31–59

Kikuya K (2009) Impact of the next generation DNA sequencers. Int J Clin Exp Med 2:193–202

Leebens-Mack J, Raubeson LA, Cui L et al (2005) Identifying the basal angiosperm node in chloroplast genome phylogenies: sampling one's way out of the Felsenstein Zone. Mol Biol Evol 22:1948–1963

Li R, Li Y, Kristiansen K, Wang J (2008) SOAP: short oligonucleotide alignment program. Bioinformatics 24:713–714

Mahmoudi H, Hosseininia G, Azadi H et al (2008) Enhancing date palm processing, marketing and pest control through organic culture. J Organ Sys 3:29–39

Margulies M, Egholm M, Altman WE et al (2005) Genome sequencing in microfabricated high-density picolitre reactors. Nature 437:326–327

Min Jou W, Haegeman G, Ysebaert M, Fiers W (1972) Nucleotide sequence of the gene coding for the bacteriophage MS2 coat protein. Nature 237:82–88

Ming R, Hou S, Feng Y et al (2008) The draft genome of the transgenic tropical fruit tree papaya (*Carica papaya* Linnaeus). Nature 452:991–996

Mozo T, Dewar K, Dunn P et al (1999) A complete BAC-based physical map of the *Arabidopsis thaliana* genome. Nat Genet 22:271–275

Paterson AH, Bowers JE, Bruggmann R et al (2009) The Sorghum bicolor genome and the diversification of grasses. Nature 457:551–556

Poinar HN, Schwarz C, Qi J et al (2005) Metagenomics to paleogenomics: large-scale sequencing of mammoth DNA. Science 311:392–394

Raubeson LA, Jansen RK (2005) Chloroplast genomes of plants. In: Plant diversity and evolution: genotypic and phenotypic variation in higher plants. CABI, London, pp 45–68

Raubeson LA, Peery R, Chumley TW et al (2007) Comparative chloroplast genomics: analyses including new sequences from the angiosperms *Nuphar advena* and *Ranunculus macranthus*. BMC Genome 8:174

Rausch T, Emde AE, Reinert K (2008) Robust consensus computation. BMC Bioinform 9 (Suppl 10):P4

Sanger F, Air GM, Barrell BG et al (1977) Nucleotide sequence of bacteriophage phi X174 DNA. Nature 265:687–695

Sato S, Nakamura Y, Kaneko T et al (2008) Genome structure of the legume, *Lotus japonicus*. DNA Res 15:227–239

Schatz MC, Delcher AL, Salzberg SL (2010) Assembly of large genomes using second-generation sequencing. Genome Res 20:1165–1173

Schmutz J, Cannon SB, Schlueter J et al (2010) Genome sequence of the palaeopolyploid soybean. Nature 463:178–183

Shendure J, Mitra R, Varma C et al (2004) Advanced sequencing technologies: methods and goals. Nat Rev Genet 5:335–344

Shinozaki K, Ohme M, Tanaka M et al (1986) The complete nucleotide sequence of the tobacco chloroplast genome: its gene organization and expression. EMBO J 5:2043–2049

Simpson JT, Wong K, Jackman SD et al (2009) ABySS: a parallel assembler for short read sequence data. Genome Res 19:1117–1123

Strömberg M (2010) http://bioinformatics.bc.edu/marthlab/Mosaik

Tanaka H, Kawai T (2009) Partial sequencing of a single DNA molecule with a scanning tunneling microscope. Nat Nanotechnol 4:518–522

The International Brachypodium Initiative (2010) Genome sequencing and analysis of the model grass *Brachypodium distachyon*. Nature 463:763–768

Tuskan GA, Difazio S, Jansson S et al (2006) The genome of black cottonwood, *Populus trichocarpa* (Torr. & Gray). Science 313:1596–1604

Venter JC, Adams MD, Myers EW et al (2001) The sequence of human genome. Science 291:1304–1351

Warren RL, Suttonl GG, Jones SJM et al (2007) Assembling millions of short DNA sequences using SSAKE. Bioinformatics 23:500–501

Yang M, Zhang X, Liu G et al (2010) The complete chloroplast genome sequence of date palm (*Phoenix dactylifera* L.). PLoS ONE 5:99–106

Yu J, Hu S, Wang J et al (2002) A draft sequence of the rice genome (*Oryza sativa* L. ssp. *indica*). Science 296:79–92

Yu J, Wang J, Lin W et al (2005) The Genomes of *Oryza sativa*: a history of duplications. PLoS Biol 3:38

Zerbino DR, Birney E (2008) Velvet: algorithms for de novo short read assembly using de Bruijn graphs. Genome Res 18:821–829

Zharkikh A, Troggio M, Pruss D et al (2008) Sequencing and assembly of highly heterozygous genome of *Vitis vinifera* L. cv Pinot Noir: problems and solutions. J Biotechnol 136: 38–43

Zhou X, Ren L, Li Y et al (2010) The next-generation sequencing technology: a technology review and future perspective. China Life Sci 53:44–57

Zonneveld BJM, Leitch IJ, Bennett MD (2005) First nuclear DNA amounts in more than 300 Angiosperms. Ann Bot 96:229–244

Chapter 22
Potential of Arbuscular Mycorrhizal Technology in Date Palm Production

G. Shabbir, A.J. Dakheel, G.M. Brown, and M.C. Rillig

Abstract Arbuscular mycorrhizal symbiosis is a mutualistic relationship between the roots of higher plants and a wide variety of fungi belonging to the phylum Glomeromycota. Arbuscular mycorrhizal fungi are a ubiquitous and widespread component of agro-ecosystems. They have the potential to promote plant growth as well as to increase salinity and drought tolerance of plants. Date palm is one of the most important fruit crops in the world, which is usually grown under harsh climatic conditions, often on marginal soils, restricted water-availability and high salinity. Mycorrhizae occur naturally under saline and water-deficient environments, and they can help date palms alleviate stress associated with salinity and drought through various physiological and biochemical mechanisms. There is, therefore, a clear rationale for utilizing mycorrhizal technology in date palm production systems, particularly at the nursery stage. Inoculated palms grow better under low fertility and saline field conditions, clearly demonstrating that mycorrhizae can enhance the fertilizer and water-use efficiency of the palms, which ultimately leads to a more sustainable production.

Keywords Arbuscular mycorrhizae • Drought • Mutualistic • Salinity • Symbiosis

---

G. Shabbir (✉)
Department of Field and Forage Crops, International Center for Biosaline Agriculture (ICBA), P.O. Box 14460 Dubai, UAE

Institut für Biologie, Plant Ecology, Freie Universität, Berlin, Germany
e-mail: g.shabbir@biosaline.org.ae

A.J. Dakheel
International Center for Biosaline Agriculture, Dubai, United Arab Emirates
e-mail: a.dakheel@biosaline.org.ae

G.M. Brown
Department of Biology, Sultan Qaboos University, Al-Khod, Sultanate of Oman
e-mail: gmarbrown@yahoo.co.uk

M.C. Rillig
Institut für Biologie, Plant Ecology, Freie Universität, Berlin, Germany
e-mail: rillig@zedat.fu-berlin.de

S.M. Jain et al. (eds.), *Date Palm Biotechnology*,
DOI 10.1007/978-94-007-1318-5_22, © Springer Science+Business Media B.V. 2011

22.1 Introduction

Soil salinization is a steadily increasing global problem, particularly in the arid and semiarid areas (Al-Karaki 2006; Giri et al. 2003). Saline soils occupy 7% of the earth's land surface (Ruiz-Lozano et al. 2001) and increased salinization will result in a 50% loss of arable area by 2050 (Wang et al. 2003). At present, of the 1.5 billion hectares of cultivated land around the world, about 77 million hectares (5%) are affected by salinity (Sheng et al. 2008), mainly due to inappropriate irrigation and the use of fertilizers (Abrol 1986; Al-Karaki 2000; Copeman et al. 1996), as well as low precipitation and high temperatures (Cantrell and Linderman 2001; Mouk and Ishii 2006). Salinity has a major impact on agriculture (Tester and Davenport 2003) by reducing establishment and growth of plants, leading to large production losses (Giri et al. 2003; Mathur et al. 2007). The direct effects of salinity on plant growth may involve reduced osmotic potential of the soil solution, toxicity of excessive Na^+ and Cl^- ions and nutrient imbalance in the plant. Reduction in osmotic potential causes physiological drought and plants must maintain lower internal osmotic potentials to allow water uptake (Feng et al. 2002; Jahromi et al. 2008). Ion toxicity damages cell organelles and plasma membranes and disrupts the structure of enzymes, macromolecules, photosynthesis, respiration and protein synthesis (Feng et al. 2002; Juniper and Abbott 1993). Nutrient imbalance is caused by nutrient uptake and/or transport to the shoot, leading to ion deficiencies (Adiku et al. 2001; Marschner 1995).

The date palm (*Phoenix dactylifera* L.), grown primarily for its edible fruit, is one of the most important socioeconomic crops in the arid and semiarid environments of the Middle East and North Africa. Worldwide date production has increased by 6.8% over the past decades (Chao and Krueger 2007). Much of the world's date production is concentrated in the Middle East, North Africa and Pakistan and this area provides almost 90% of the total global date consumption (FAO 2002, 2003, 2007). Due to its shallow and coarse root system, date palms are highly mycotrophic. They benefit enormously from mycorrhizal associations, especially under the saline and water-deficient conditions that often prevail in date palm production systems. Date palms are now often propagated by tissue culture (Awad 2008), which in spite of its great potential, is often problematic due to the limited outplanting success of the plants in the field. For some cultivars, the survival rate can be as low as 40–50% (Zaid and de Wet 1999). However, inoculation with AMF can enhance the establishment and survival of the transplanted seedlings.

Arbuscular mycorrhizal fungi (AMF) are ubiquitous soil microbes forming symbiotic associations with the root systems of 70–90% of terrestrial plant species (Smith and Read 2008). AM can significantly increase plant growth under saline conditions by forming extensive hyphal networks (Daei et al. 2009; Hildebrandt et al. 2007; Miransari et al. 2007, 2008, 2009a, b; Rillig 2004) which can enhance nutrient uptake (Al-Karaki and Al-Raddad 1997), produce plant growth hormones (Lindermann 1994), alter physiological and biochemical properties of the host (Smith and Read 1995) and defend roots against soil-borne pathogens (Dehne 1982; Elsen et al. 2008). In addition, AMF can improve water-absorption capacity, osmotic balance and composition of carbohydrates of the host plants

(Al-Karaki and Clark 1998; Rosendahl and Rosendahl 1991; Ruiz-Lozano 2003; Ruiz-Lozano and Azcón 2000). In general, the symbionts trade nutrients; the AMF obtains carbon from the plant while providing the plant with an additional supply of water and otherwise hard-to-absorb nutrients. As a consequence, the AMF symbiosis is of tremendous significance to life on this planet in both natural and agricultural ecosystems. Due to these benefits, the application of mycorrhizal technologies can be viewed as an inexpensive option to alleviate salt stress compared to the more expensive approach of physical salt leaching or bio-amelioration of saline soils. This review covers the occurrence of AMF in date palms and saline soils, as well as the interaction between salinity and AMF. It also examines the literature relating to the alleviation of salt stress by AMF, and in this context the beneficial effects of AMF on growth, changes in biochemical, physiological and molecular mechanisms. However, high levels of a particular stress may turn the symbiosis into parasitism, as unfavorable conditions may adversely influence AM performance (Hildebrandt et al. 2007; Miransari et al. 2007, 2008, 2009a, b; Rillig 2004), which in turn may affect plant growth due to decreased colonization rate, spore germination and hyphal growth (Evelin et al. 2009; Jahromi et al. 2008).

22.2 Arbuscular Mycorrhizal Fungi (AMF)

The word mycorrhiza is derived from the Greek *mukes/mykes* meaning *mushroom/fungus,* and *rhiza* meaning *root*. Mycorrhiza is a specialized, the most common and possibly the most important mutualistic symbiotic relationship formed between the roots of the higher (vascular) plants and fungi (Frank 1885). This association occurs in the vast majority of both wild and cultivated plants. The arbuscular mycorrhizal (AM) symbiosis is mostly a mutualistic association formed between the plants and a wide variety of the fungi in the phylum Glomeromycota (Schüßler 2002; Schüßler et al. 2001).

22.2.1 Types of Mycorrhizal Fungi

Mycorrhizae are commonly divided into two main types: ectomycorrhizae and endomycorrhizae. An overview of these two types is given in the following sections.

22.2.1.1 Ectomycorrhizae

Ectomycorrhizae (EM) are typically associated with the roots of around 10% of plant families, mostly woody plants, ranging from shrubs to forest trees. Typical examples include birch, dipterocarp, eucalyptus, oak, pine and rose, which belong to the families Betulaceae, Dipterocarpaceae, Myrtaceae, Fagaceae, Pinaceae and Rosaceae, respectively (Wang and Qiu 2006). The ectomycorrhizal fungi belong primarily to the phyllum Basidiomycota, with Ascomycota and Zygomycota less commonly represented.

The diagnostic feature of ectomycorrhizae is the presence of hyphae outside (between) the root cortical cells, producing a netlike structure called the *Hartig net*, named after Robert Hartig. In some cases the hyphae may also penetrate the plant cells. This is then referred to as ectendomycorrhiza. Many EM also have a sheath or mantle of fungal tissue that may completely cover the root tips (usually the fine feeder roots). The mantle can vary widely in thickness, color and texture depending on the particular plant-fungus association. The mantle increases the surface area of absorbing roots and often affects fine-root morphology, resulting in root bifurcation and clustering. Contiguous with the mantle, are hyphal strands that extend into the soil and often aggregate to form rhizomorphs that may be visible to the naked eye. The internal portion of rhizomorphs can differentiate into tube-like structures specialized for long-distance transport of nutrients and water. Nutrients can move between different plants through the fungal network (sometimes called the *wood wide web*). For instance, carbon has been shown to move from trees of paper birch to Douglas fir, thus associated with vegetation succession in ecosystems (Suzanne et al. 1997). The ectomycorrhizal fungus *Laccaria bicolor* has been found to lure and kill tiny insect-like springtails to obtain nitrogen, some of which may then be transferred to its host plant. Eastern white pine inoculated with *L. bicolor* was able to derive up to 25% of its nitrogen from springtails (Klironomos and Hart 2001; Wang 2001).

Many EM fungi produce mushrooms and puffballs on the forest floor. Some fungi, such as *Boletus betulicola* on *Betula* spp., have a narrow host range while in others, such as *Pisolithus arhizus* (= *P. tinctorius*), it is very broad, forming ectomycorrhizae with more than 46 tree species belonging to at least eight genera.

22.2.1.2 Endomycorrhizae

The term *endomycorrhiza* is generally used for all mycorrhizal types in which the fungus grows inside (within) root cortical cells, but neither the fungal cell wall nor the host cell membrane are breached. As the fungus grows, the host cell membrane invaginates and envelops the fungus, creating a new compartment where highly complex molecular material is deposited. This apoplastic space prevents direct contact between the plant and fungal cytoplasm, and facilitates efficient transfer of nutrients between the symbionts. Endomycorrhizae are variable and have been further classified as arbuscular, ericoid, arbutoid, monotropoid and orchid mycorrhizae (Peterson et al. 2004).

Arbuscular Mycorrhizae Fungi (AMF)

With arbuscular mycorrhizae (AM), formerly known as vesicular-arbuscular mycorrhizae (VAM), the hyphae enter the plant cells and produce either balloon-like

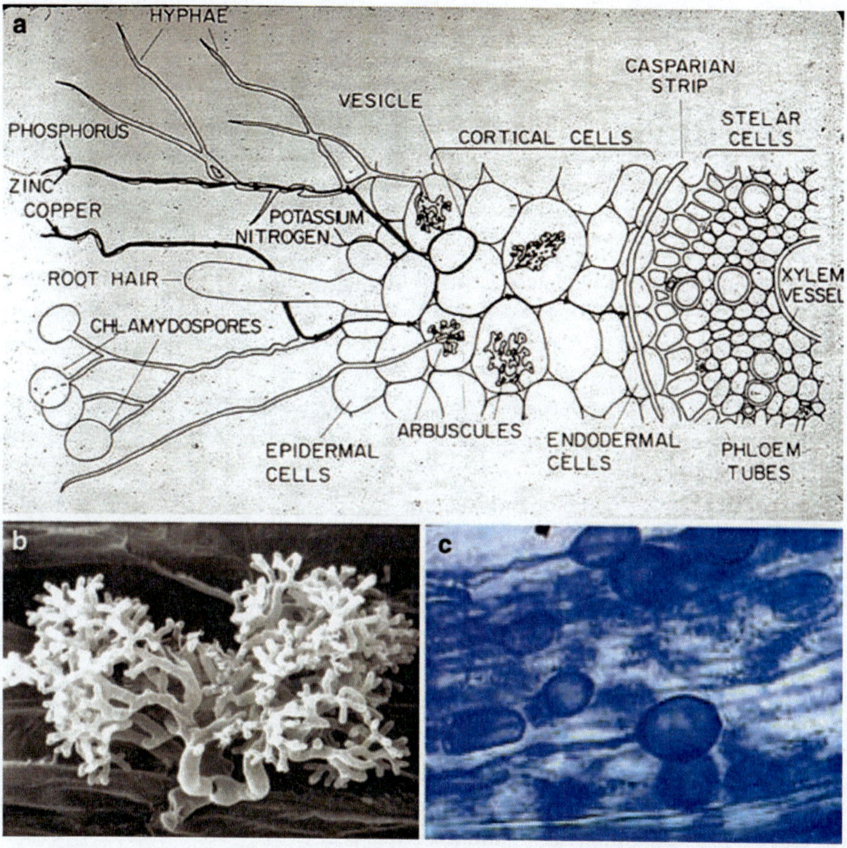

Fig. 22.1 Illustrations showing different structures of AM. (**a**) AM hyphae, arbuscules and vesicles growing inside (within) root cortical tissue, (**b**) AM arbuscule, (**c**) AM vesicles

structures called vesicles, or dichotomously-branching invaginations known as arbuscules (Fig. 22.1). The vesicles are thin-walled, lipid-filled structures thought to function as the storage compartments of the fungi. The arbuscules are relatively short lived (<15 days) and are often difficult to see in field samples. The fungal hyphae do not in fact penetrate the protoplast, but invaginate the cell membrane. The structure of the arbuscules greatly increases the contact surface area between the hyphae and the cytoplasm, facilitating the transfer of nutrients between them. Arbuscular mycorrhizae are formed only by fungi of the phyllum Glomeromycota. Fossil evidence (Remy et al. 1994) and DNA sequence analyses (Simon et al. 1993) suggest that this mutualism appeared 400–460 million years ago when first terrestrial plant started colonizing land. Arbuscular mycorrhizae are found in 85% of all plant families and occur in many crop species (Wang and Qiu 2006).

Ericaceous/Ericoid Mycorrhizae

The term *ericaceous* is applied to mycorrhizal associations found in plants belonging to the order Ericales. Ericoid mycorrhizae are the third of the three more ecologically important types. Many plants in the order Ericales form ericoid mycorrhizae, while some members form arbutoid and monotropoid mycorrhizae. Ericoid mycorrhizae have a simple intraradical (growing in cells) phase consisting of dense coils of hyphae in the outermost layer of root cells. There is no periradical phase and the extraradical phase consists of sparse hyphae that do not extend very far into the surrounding soil. Ericoid mycorrhizae have also been shown to have considerable saprotrophic capabilities that enable plants to receive nutrients from not-yet-decomposed materials via the decomposing actions of their ericoid partners (Read and Perez-Moreno 2003). The hyphae in the roots can penetrate cortical cells (the endomycorrhizal habit); however, they do not form arbuscules. Ericaceous mycorrhizae are subdivided into three major forms:

- Ericoid mycorrhizae are formed by ascomycete fungi of the genus *Hymenoscyphus*, and the inner cortex root cells become packed with fungal hyphae. A loose welt of hyphae grows over the root surface, but a true mantle is not formed. Ericoid mycorrhizae are found in plant species of various genera, such as *Calluna* (heather), *Rhododendron* (azaleas and rhododendrons) and *Vaccinium* (blueberries). They are characterized by very fine root systems and typically grow in acid, peaty soils.
- Arbutoid mycorrhizae possess characteristics both of ecto- and endomycorrhizae and are formed by members of the basidiomycetes. Intracellular penetration can occur, a mantle forms and a *Hartig net* is present. These associations are found in the genera, *Arbutus* (e.g. Pacific madrone), *Arctostaphylos* (e.g. bearberry) and several species belonging to the Pyrolaceae.
- Monotropoid mycorrhizae colonize achlorophyllous (chlorophyll-lacking) plants in the family Monotropaceae (e.g. Indian pipe). They typically produce a *Hartig net* and mantle. The same fungi also form EM associations with trees and thereby develop channels through which carbon and other nutrients can flow from the autotrophic host plant to the heterotrophic parasitic plant.

Orchidaceous Mycorrhizae

Orchidaceous mycorrhizal fungi have a unique role in the life cycle of orchids. All orchids are mycoheterotrophic at some stage during their life cycle, and they form associations with a range of basidiomycete fungi. Orchids typically have very small seeds with little nutrient reserve. The seedlings become colonized shortly after germination and the mycorrhizal fungus supplies carbon and vitamins to the developing embryo. Achlorophyllous species depend throughout their life on the carbon supply of the fungal partner. The fungus grows into the plant cell, invaginates the cell membrane and forms hyphal coils. These coils are active for a short time, after which they degenerate and release the nutrient contents that are subsequently absorbed by the

Table 22.1 Taxonomy and classification of arbuscular mycorrhizal fungi

Phylum Glomeromycota			
Class Glomeromycetes			
Order	Family	Genus	Number of Species
Glomerales	Glomeraceae	*Glomus*	120
Diversisporales	Gigasporaceae	*Gigaspora*	9
		Scutellospora	29
		Racocetra	10
	Acaulosporaceae	*Acaulospora*	34
		Kuklospora	2
	Entrophosporaceae	*Entrophospora*	2
	Pacisporaceae	*Pacispora*	7
	Diversisporaceae	*Diversispora*	3
		Otospora	1
Paraglomerales	Paraglomeraceae	*Paraglomus*	2
Archaeosporales	Geosiphonaceae	*Geosiphon*	1
	Ambisporaceae	*Ambispora*	8
	Archaeosporaceae	*Archaeospora*	2
		Intraspora	1

developing orchid. Orchidaceous mycorrhizae belong to the phylum Basidomycota, similar to those involved in decaying wood (e.g., the genera *Coriolus*, *Fomes*, *Marasmius*) and pathogenesis (e.g., the genera *Armillaria* and *Rhizoctonia*). In mature orchids, mycorrhizae also help in nutrient uptake and translocation.

22.2.2 *Taxonomy and Classification*

The taxonomy of AM fungi is in a state of flux. According to the latest taxonomy and classification system, arbuscular mycorrhizal fungi are placed in four orders, i.e., Glomerales, Diversisporales, Paraglomerales and Archaeosporales. These orders belong to the class Glomeromycetes of the phylum Glomeromycota, and comprise 10 families and 15 genera with around 231 known species (Oehl and Sieverding 2004; Palenzuela et al. 2008; Schüßler et al. 2001; Sieverding and Oehl 2006; Spain et al. 2006; Walker and Schüßler 2004; Walker et al. 2007a, b). They are shown in Table 22.1.

22.3 Salinity, Arbuscular Mycorrhizae and Plant Growth

Soil stress caused by salinity and drought can decrease plant growth and production. High levels of salts (>4 dSm^{-1}) in soils is caused mainly by secondary salinization (Juniper and Abbott 1993, 2006; Richards 1954) as a result of the accumulation of soluble salts present in the irrigation water, and fertilizers used in agricultural production systems (Abrol 1986; Al-Karaki 2000; Copeman et al. 1996), low precipitation, high temperature and over-exploitation of available water resources (Al-Karaki 2006; Cantrell and Linderman 2001; Mouk and Ishii 2006). Plants are usually colonized by AM fungi in their natural habitats. This association can improve plant performance under conditions of stress, and as a consequence, enhance yield (Brown 1974; Creus et al. 1998; Levy et al. 1983).

22.3.1 Mycorrhizae in Saline Environments

AMF occur naturally in saltmarsh plants (Carvalho et al. 2001; Hilderbrandt et al. 2001; Khan 1974; Rozema et al. 1986; Sengupta and Chaudhuri 1990) and colonize halophytes under saline environments (Allen and Cunningham 1983; Brown and Bledsoe 1996; Brundrett 1991; Carvalho et al. 2001; Harisnaut et al. 2003; Hilderbrandt et al. 2001; Hoefnagels et al. 1993; Khan 1974; Mason 1928; Pond et al. 1984; Rozema et al. 1986; Sengupta and Chaudhuri 1990; Yamato et al. 2008), with varying effects on plant performance. For instance, the mycorrhized halophytic grass *Distichlis spicata* has similar or even lower biomass than non-mycorrhized plants in saline soils (Allen and Cunningham 1983). In contrast, a beneficial effect of AMF symbiosis was observed on the water status, accumulation of osmolytes and growth of *Phragmites australis* under salt-stress conditions (Al-Garni 2006). An improved water relationship was also reported in mycorrhized *Aster tripolium* plants in saline soils (Rozema et al. 1986). AMF can tolerate 50 mg total salts ml^{-1} in soil water suggesting that the mycorrhizal benefits to halophytes might occur primarily through improved mineral nutrition rather than through increased biomass (Johnson-Green et al. 2001).

A total of 33 species representing three AMF genera (*Archaeospora*, *Acaulospora* and *Glomus*) were found at ECe 40.2 dSm^{-1}, mostly at a depth of 0–40 cm (Wang et al. 2004). AMF species distribution was unlikely to follow the salinity gradient: spores showed higher species richness while the overall mycorrhizal diversity in the roots was comparably low under natural field conditions (Wilde et al. 2009).

Mean AM spore density was high in saline areas (Aliasgharzadeh et al. 2001; Bhaskaran and Selvaraj 1997; Khan 1974; Landwehr et al. 2002). However, it was not correlated with the salinity gradient (Aliasgharzadeh et al. 2001). AMF spore density showed a positive correlation with soil pH and organic carbon, but a negative one with available soil P and Na (Ho 1987; Mathur et al. 2007; Mohammad et al. 2003), available soil Mg, Ca, Cl, clay, soil salinity, SO_4 and the sodium absorption ratio (Aliasgharzadeh et al. 2001; Saint-Etienne et al. 2006). The higher fungal spore density in saline soils may not necessarily reflect a greater abundance of mycorrhizae. An alternative explanation could be that in some species, salt stress stimulates sporulation on one hand, but on the other, inhibits spore germination leading to an accumulation of spores in the soil at higher salinity levels (Aliasgharzadeh et al. 2001; McMillen et al. 1998; Tressner and Hayes 1971). This, however, contradicts reports that have recorded low or even no spore populations at very high salinity levels (ECe of the saturation extract >45 dS m^{-1}) (Barrow et al. 1997; Carvalho et al. 2001; Hirrel et al. 1978; Kim and Weber 1985).

Salinity can negatively affect colonization capacity, inhibit spore germination and hyphal growth of the AMF (Estaun 1989; Hirrel 1981; Jahromi et al. 2008; McMillen et al. 1998). The growth rate of hyphae was reduced with increasing concentrations of NaCl, suggesting that it is more sensitive to NaCl than is the case with spore germination, which is delayed, but not necessarily prevented (Cantrell and Linderman 2001; Juniper and Abbott 2006). Growing pre-inoculated seedlings can avoid the inhibitory effect of salts on spore germination (Al-Karaki 2006; Cantrell and Linderman 2001). Germination of AMF spores may also depend on salt type because different salts impart differential effects on the rate and maximum germination of spores (Juniper and Abbott 1993). Propagules and spores of the same fungal genus can vary in their relative tolerance to salinity. Propagules of *Glomus* sp. in the colonized root pieces grew in 300 mM NaCl, but not the spores. This may be due to the differences in the energy status of the spores, or the amount of water and energy required to initiate germination (Juniper and Abbott 2006). Further studies are required on this aspect. Hyphal length and spore production of *G. intraradices* was significantly decreased at 100 mM NaCl, suggesting that persistent salt stress can reduce or inhibit colonization and symbiotic capability of AMF and reduce its inoculum potential (Jahromi et al. 2008). Colonization by endomycorrhizae was found to be negatively correlated with ECe, concentrations of the specific saline ion and cation exchange capacity. This suggests that AM fungi are sensitive to elevated salt concentrations, independent of the type of salt-specific anion (Fuzy et al. 2010). Colonization by some AMF was reduced in the presence of NaCl (Duke et al. 1986; Giri et al. 2007; Hirrel and Gerdemann 1980; Juniper and Abbott 2006; Menconi et al. 1995; Ojala et al. 1983; Poss et al. 1985; Rozema et al. 1986; Sheng et al. 2008), probably due to direct effect of NaCl on the fungus (Juniper and Abbott 2006), and thus indicating that salinity suppresses the formation of AM (Sheng et al. 2008; Tian et al. 2004). In contrast to these findings, several studies have reported that sporulation and colonization of AM colonization was not reduced under saline (Aliasgharzadeh et al. 2001; Hartmond et al. 1987; Levy et al. 1983) and highly saline conditions (Yamato et al. 2008). These results suggest that salt-tolerant species of AMF can be isolated and potentially be used to alleviate of salt stress in vascular plants (Fig. 22.2).

22.3.2 *Salinity Amelioration by Mycorrhizae*

The ability to help host plants counter the adverse effects of salinity may depend on the behavior of the particular AM species. Various species of *Glomus*, including *G. geosporum*, *G. intraradices*, *G. versiforme* and *G. etunicatum*, are the most prominent AMF species found in highly-saline soils (Allen and Cunningham 1983; Ho 1987; Wang et al . 2004), where an average of 80% of the spores belong to a single species, *G. geosporum*. However, this does not necessarily indicate that this species will necessarily alleviate salt stress in the plants concerned (Wilde et al. 2009). For

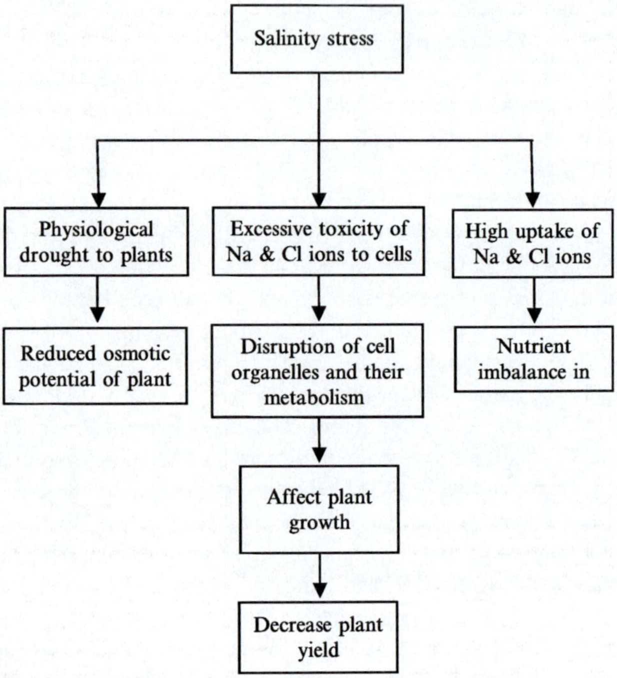

Fig. 22.2 Flow diagram showing how salinity can affect the plant growth by causing physiological drought, imbalance in nutrient composition and excessive toxicity due to Na and Cl ions which leads to reduced osmotic potential of plants, disruption of cell organelles and their metabolism which ultimately affect plant growth and reduce the yield

instance, *G. geosporum* isolated from an inland salt marsh consistently failed to show positive effects on plant growth (Fuzy et al. 2008). *Glomus mosseae* isolated from saline soil had a lower capacity to alleviate salinity stress in cotton than the same species isolated from non-saline soil (Tian et al. 2004). *Glomus mosseae* was the most efficient of the three species *G. mosseae*, *G. intraradices* and *G. claroideum*, in alleviating salt stress in olive trees under nursery conditions (Porras-Soriano et al. 2009).

AM-symbiosis often helps plants to grow better under saline conditions by a combination of several nutritional, biochemical and physiological processes, such as increased nutrient uptake, accumulation of osmoregulators, increased photosynthetic rate and water-use efficiency.

22.3.2.1 Enhanced Nutrient Uptake and Fertilizer-Use Efficiency

Mycorrhizal associations have been shown to enhance plant growth in a number of higher plant taxa under saline conditions. Various species inoculated with AM grew better than non-inoculated plants under salt stress (Al-Karaki 2000; Cantrell and Linderman 2001; Giri et al. 2003; Sannazzaro et al. 2007; Zuccarini and Okurowska 2008).

AM-inoculated *Acacia nilotica* seedlings had higher root and shoot dry matter than the non-mycorrhizal seedlings (Giri et al. 2007). Shoot and root dry matter, fresh fruit yield and number of fruits of AM-inoculated tomato plants were higher than in non-inoculated plants (Al-Karaki 2000). *Cucurbita pepo* plants inoculated with *Glomus intraradices* had improved growth, yield, water status, nutrient content and fruit quality under saline conditions (Colla et al. 2008). Improved nutrient absorption mediated by mycorrhizae, especially P, generally increases plant growth under normal (Plenchette and Duponnis 2005; Sharifi et al. 2007) and saline conditions (Al-Karaki and Clark 1998). This is primarily regulated by the nutrient supply to the root system (Giri and Mukerji 2004) and increased transport (absorption and/or translocation) by mycorrhizae (Al-Karaki 2000; Sharifi et al. 2007). Mycorrhizal dependency increases with increasing salt concentrations (Giri and Mukerji 2004). Absorption of mineral nutrients, especially P, is significantly reduced under saline conditions because of its low solubility. P concentrations in *Trifolium alexandrium* plants were found to decline with increasing salinity levels (Shokri and Maadi 2009). P solubilization or fertilization is therefore necessary to mitigate salt stress (Cantrell and Linderman 2001).

An extensive system of extraradical mycorrhizal mycelia (ERM) helps plants to explore larger soil volumes and absorb higher amounts of P (Ruiz-Lozano and Azcón 2000). It is estimated that ERM deliver up to 80% of the total P required by plants (Matamoros et al. 1999). Better P nutrition in AM-inoculated plants may improve their growth rate, increase antioxidant production and enhance nodulation and nitrogen fixation in legumes (Alguacil et al. 2003; Feng et al. 2002; Garg and Manchanda 2008). Mycorrhizal symbiosis can also enhance N assimilation in the host plants. For instance, N accumulation in the shoots of *Sesbania grandiflora* and *S. aegyptiaca* was increased by mycorrhizae (Giri and Mukerji 2004). A higher Na/K ratio in the plant cells, caused by salinity, disrupts various plant metabolic pathways (Giri et al. 2007). This can be alleviated by a higher uptake of K, and it is known that mycorrhizae are able to selectively enhance the absorption of this ion (Alguacil et al. 2003; Giri et al. 2007; Sharifi et al. 2007; Zuccarini and Okurowska 2008) and prevent Na translocation to the shoot tissues (Dixon et al. 1993; Sharifi et al. 2007; Zuccarini and Okurowska 2008) under saline conditions thus increasing the K/Na ratio in roots and shoots of mycorrhizal plants (Giri et al. 2007). Application of arbuscular mycorrhiza can reduce the uptake of Cl ions (Zuccarini and Okurowska 2008). However, decreased P uptake at higher salinity level (9.5 dSm^{-1}) by mycorrhizal plants suggests that mycorrhizae can respond only up to a certain level of salinity (4.7 dSm^{-1}) (Al-Karaki 2000; Giri et al. 2007). Mycorrhizae can also strongly affect the Ca content in the plants. Mycorrhizae increased Ca uptake in lettuce (Cantrell and Linderman 2001) and banana (Yano-Melo et al. 2003). Higher Ca alleviates the toxic effects of NaCl by facilitating higher K/Na selectivity leading to plant adaptation to salinity (Cramer et al. 1985; Rabie and Almadini 2005). However, there was no difference in the Ca concentration in the shoots of mycorrhizal and non-mycorrhizal *Acacia auriculiformis* plants (Giri et al. 2003). Mycorrhizae help plants in the synthesis of chlorophyll by enhancing uptake of Mg (Giri et al. 2003). This reduces salt-mediated reduction of chlorophyll synthesis (Giri and Mukerji 2004), thus improving photosynthetic efficiency and plant growth.

22.3.2.2 Biochemical Changes

Higher soil salinity can result in physiological drought, but plants adapted to growing under saline conditions are able to maintain uptake of water by means of osmotic adjustment or osmoregulation. This mechanism involves the active accumulation of organic ions and solutes (Hoekstra et al. 2001; Morgan 1984), including proline (Ashraf and Foolad 2007; Jain et al. 2001; Parida et al. 2002; Sannazzaro et al. 2007; Stewart and Lee 1974), betaines and polyamines (Rabie and Almadini 2005). Mycorrhiza-mediated increases in proline content have been recorded in *Vigna radiata* at 12.5 and 25 mM NaCl (Jindal et al. 1993) and soybean (*Glycine max*) at 0, 50, 100, 150 and 200 mM NaCl salinity levels (Sharifi et al. 2007). However, Rabie and Almadini (2005) reported that non-mycorrhized *Vicia faba* plants accumulated more proline than mycorrhized plants at 0–6 dSm^{-1}. Accumulation of proline could therefore be a response to salinity and not necessarily associated with mycorrhizal colonization (Sannazzaro et al. 2006). Similarly, accumulation of glycine betaines in mycorrhized plants was higher than non-mycorrhized plants at higher salinity levels (Al-Garni 2006).

AM-inoculation increases free polyamine concentrations under saline conditions. Total free polyamine pools in *Lotus glaber* plants colonized by *Glomus intraradices* were found to be increased, possibly as one of the myccorhizal-induced mechanisms that enable plants to adapt to saline soils (Sannazzaro et al. 2007). Colonization of common reed (*Phragmites australis*) by *Glomus fasciculatum* and soybean roots by *G. intraradices* increased soluble sugar levels under elevated levels of salinity (Al-Garni 2006; Porcel and Ruiz-Lozano 2004). Conversely, Pearson and Schweiger (1993) reported a negative correlation between carbohydrate concentration and percent of root colonization. No relationship was found between soluble sugar content and substrate salinity in soybean plants colonized by *G. etunicatum* (Sharifi et al. 2007). In other cases, it has been observed that AM symbiosis enables plants to alleviate salt stress by enhancing the activities of antioxidant enzymes (Alguacil et al. 2003; Harisnaut et al. 2003; Zhong Qun et al. 2007). The plants possess higher antioxidant enzyme activities as a result of mycorrhizal colonization, but the response of the individual enzymes varies with respect to the host plant and the fungal species. This variation may also depend on the micronutrients available to some of the enzymes (Alguacil et al. 2003).

22.3.2.3 Physiological Changes

Salt stress can affect plant growth by decreasing photosynthetic efficiency, gas exchange, lowering of internal water status and cause membrane disruption. Chlorophyll content has been shown to decrease under conditions of higher salinity (Sheng et al. 2008) due to the suppression of specific enzymes responsible for the synthesis of photosynthetic pigments (Murkute et al. 2006). However, mycorrhizal plants had a higher chlorophyll content under saline conditions (Colla et al. 2008; Giri and Mukerji 2004; Sannazzaro et al. 2006; Sheng et al. 2008; Zuccarini 2007). It can therefore be assumed that mycorrhization can counteract the effects of salinity stress (Zuccarini 2007). This may be due to the fact that the antagonistic effect of Na on Mg uptake is diminished or even suppressed in the presence of mycorrhiza (Giri et al. 2003). Chlorophyll a fluorescence studies have shown that the Fv/Fm ratio, as a measure of the efficiency of PS II, is significantly higher in the leaves of mycorrhizal plants under saline conditions than in non-mycorrhized plants (Sheng et al. 2008; Zuccarini and Okurowska 2008). The toxic influence of salinity on the PS II reaction center could therefore possibly be mitigated by AM symbiosis, which also regulates the energy bifurcation between photochemical and non-photochemical events (Sheng et al. 2008).

AM-inoculation enables plants to maintain a higher electrolyte concentration by improved integrity and stability of the membrane (Feng et al. 2002; Garg and Manchanda 2008; Kaya et al. 2009). Mycorrhizal *Cajanus cajan* roots showed a higher relative permeability at ECe 4, 6 and 8 dSm^{-1} (Garg and Manchanda 2008). AM-inoculated *Capsicum annum* plants had a relatively lower electrolyte leakage in the leaves at 50 and 100 mM NaCl (Kaya et al. 2009). Mycorrhiza-mediated enhanced P uptake and elevated production of antioxidants increase membrane stability of maize plants (Feng et al. 2002). It appears as if mycorrhization can alter abscisic acid (ABA) levels in the host plants of some species (Estrada-Luna and Davies 2003; Ludwig-Muller 2000). For instance, *Lotus glaber* plants colonized by *Glomus intraradices* had higher ABA levels than non-colonized plants, under saline conditions (Sannazzaro et al. 2007). In contrast, Jahromi et al. (2008) reported that lettuce plants colonized by *G. intraradices* accumulated lower ABA, which indicates that AM plants were less stressed by salinity than non-AM plants. This may also suggest that the effect of AM species on ABA content is host-specific.

Salinity decreases symbiotic nodule formation (Garg and Manchanda 2008; Harisnaut et al. 2003; Rabie and Almadini 2005), resulting in a decrease in N-fixation (Delgado et al. 1994). Application of mycorrhizae can counteract the effects of salinity on nodulation and nitrogen fixation in legumes. Mycorrhizal inoculation had a strong effect on nodule formation under salt stress (Giri and Mukerji 2004), and the number of nodules was increased in the presence of mycorrhizae (Garg and Manchanda 2008; Giri and Mukerji 2004; Rabie and Almadini 2005). This may point to a positive influence of AM on legume-nitrogen-fixing bacteria symbiosis. Mycorrhizal plants possess a higher nitrogenase activity and hence higher N fixation rate due to enhanced P uptake which is vital for the functioning of the nitrogenase (Founoune et al. 2002). It may be suggested that mycorrhizal and nodule symbioses often act synergistically on infection rate, mineral nutrition and plant growth of leguminous species (Patreze & Cordeiro 2004; Rabie 2005).

22.3.2.4 Mycorrhizae and Water-Use Efficiency (WUE)

Drought is a common stress factor affecting plant growth and yield (Feng et al. 2002; Kramer and Boyer 1997). Under saline conditions, Na and Cl ions bind the water molecules and cause physiological drought (Fuzy et al. 2008). AM-symbiosis can help plants to alleviate drought stress (Bray 1997). Mycorrhizae improve the hydraulic conductivity of the root at low water potentials (Kapoor et al. 2008) Mycorrhizal plants are therefore able to maintain a relatively higher water content than non-mycorrhizal plants (Auge 2001; Colla et al. 2008; Jahromi et al. 2008; Sheng et al. 2008; Subramanian et al. 2006). Mycorrhizal plants exhibit higher stomatal conductance, which in turn results in enhanced transpiration rates (Jahromi et al. 2008; Sheng et al. 2008). Higher turgor potential and lower water saturation deficits in mycorrhized plants also improve the water status of the plant (Al-Garni 2006; Sheng et al. 2008). Moreover, mycorrhizal plants can absorb forms of N that are unavailable to non-mycorrhizal plants under conditions of low water availability, resulting in higher growth (Subramanian et al. 2006). Mycorrhizal symbiosis enhances plant drought tolerance by altering its physiology and gene expression (Aroca et al. 2008; Boomsma and Vyn 2008; Ruiz-Lozano et al. 2006). It is also noteworthy that AM can indirectly promote plant growth under water stress by modifying soil structure. This is because AM hyphae produce glomalin, a substance that binds soil particles and is important in soil aggregate formation, thus serving to increase soil moisture retention capacity (Auge 2001; Auge et al. 2004; Rillig 2004; Ruiz-Lozano 2003).

22.4 The Date Palm

Date palm (*Phoenix dactylifera* L.) belongs to the family Arecaceae (Dransfield et al. 2008; Munier 1973). The scientific name is presumably derived from the words *phoenix*, a Phoenician name meaning date palm, and *dactylifera*, which is Greek *daktulos* for finger (Linnee 1734). The date palm is one of the oldest crops known to man, and it is grown primarily for its edible fruit throughout the Middle East and North Africa. It thrives in areas with long dry summers and mild winters, and grows best in light, deep soils. It is able to tolerate prolonged drought and fairly high levels of salinity. The date palm has high socioeconomic and environmental importance in these regions. It is grown on a domestic to large commercial scales primarily for its fruit production. However, it is also extensively used in landscaping, and frequently planted as windbreak along roadsides. Worldwide date production has increased rapidly over the last three decades, from 1.8 mt in 1962 to 6.9 mt in 2005 (FAO 2002, 2003, 2007). Fruit production will continue to increase, especially in the Middle East, despite a number of production challenges (Chao and Krueger 2007). The date industry is mostly concentrated in the Middle East, North Africa and Pakistan, which account for almost 90% of total world production (FAO 2002, 2003, 2007).

22.4.1 *Botanical Description and Root System*

The date palm is a perennial, diploid, dioecious monocot characteristic of harsh arid environments. It has unique biological and developmental features that necessitate specialized propagation, cultural and management techniques. Being a monocot,

Table 22.2 Date palm root morphology and distribution

Order	Origin	Form	Length (m)	Diameter (mm)	Characteristics
Primary	Trunk base	Cylinder	4–10	7–12	Vertical, adventitious, no root hair, conic tip, called auxirhyzes
Secondary	Primary roots	Cylinder	0.20–0.25	3.5	Called mesorhyzes
Tertiary	Secondary roots	Thin cylinder	0.02–0.1	0.3–.5	Low growth, short, abundant, called brachyrhizes

date palm has a fasciculate system of fibrous roots in which secondary roots appear on the primary root, which develops directly from the seed. The secondary roots then produce lateral roots (tertiary roots and so on) with approximately the same diameter throughout their length. Date palm root morphology and distribution are illustrated in Table 22.2.

All date palm roots bear respiratory organs called pneumatics. A substantial portion of the root system, almost 85%, is located in the upper 2 m of the soil surface and extends only about 2 m laterally (Munier 1973). The entire root system can be divided into four zones (Oihabi 1991).

- *Zone I* (respiratory zone): It consists, mainly of primary and secondary roots, most of which show a negative geotropism and play a respiratory role. This zone is localized near the palm's trunk base up to 25 cm depth and a lateral distribution of up to 0.5 m away from the stipe.
- *Zone II* (nutritional zone): It is a large zone and contains the highest proportion of primary and secondary roots (1,000 roots m^{-2} and >1.60 gm of roots/100 gm soil). These roots develop at 0.90–1.50 m depth and could laterally be found beyond the projection of the tree's canopy.
- *Zone III* (absorbing zone): This zone is usually found at a depth of 1.5–1.8 m and consists mostly of primary roots whose density decreases from top to bottom. Zone III has lower root density as compare to Zone II (about 200 roots m^{-2}).
- *Zone IV:* The largest portion of this zone is dependent on the depth of ground water table. At a shallower depth, it becomes difficult to distinguish between Zone III and Zone IV. Roots in zone IV show a positive geotropism and could reach to a greater depth when the ground water table is deeper.

22.5 Occurrence of Mycorrhizal Fungi in Date Palm

Many reports indicate the presence of AM associations and their beneficial role on date palm growth. One of the earliest observations on the presence of AMF in date palms was reported by Khudairi (1969) from the Fertile Crescent in Iraq, where mycorrhizae were believed to benefit mineral nutrition and water supply in the plants. Similar observations were made in date palms growing in the Qassim oasis region of Saudi Arabia (Khaliel and Abou-Hailah 1985). More recent studies have shown that well-established date palms are usually colonized by AMF. Bouamri et al. (2006) found that the frequency and mean intensity of root colonization reached 72% and 43%, respectively, and that the spore population varied from 238 to 1,840 spores/10 g of soil. A total of ten AMF species, including *Glomus mosseae*, *G. fasciculatum*, *G. constrictum*, *G. aggregatum*, *G. macrocarpum*, three undescribed species of *Acaulospora* and two of *Scutellospora* genera were isolated from ten palm groves in the arid part of southwestern Morocco (Bouamri et al. 2006). Nine phylogenetic taxa of AMF were found to be associated with date palm roots, eight of which could be attributed to the *Glomus* group A and one to the *Scutellospora* group (Al-Yahya 2008). These observations could be a general case of occurrence of mycorrhizal associations with date palms in other areas as well.

22.6 Benefit of Mycorrhizal Fungi to Date Palms

Plant species vary greatly in their dependency on mycorrhizal associations (Janos 2007). Date palms are commonly grown in desert oases, which are characterized by high temperature, low rainfall, poor soil, low organic matter and nutrients, salinity and drought (Zaid et al. 2002). Moreover, palms are susceptible to nutrient deficiencies (Downer 2004) and large amounts of mineral fertilizers are added annually to improve plant growth to achieve high yields and superior fruit quality (Barreveld 1993). However, it is unlikely that much of the N and P that is applied is taken up by the plants and large amounts probably therefore leach down into the ground water. Although date palms can withstand long periods of drought, large amounts of water are required for vigorous growth and high yields (Chao and Krueger 2007). Good quality water is not available for irrigation purposes in many areas, and so brackish groundwater is used, which frequently results in the accumulation of salts in the root zone. AMF could have the potential to improve sustainability of date palm production systems in various ways, for instance by enhancing soil aggregate stability, fixing mobile sand and improving physical and chemical soil conditions (Bearden and Petersen 2000), as well as by enhancing water and nutrient uptake under saline conditions. Inoculation of date palm seedlings with AMF has been reported to increase the absorption of K and P and enhance growth over the non-inoculated seedlings. The degree of colonization of date palm seedlings at the nursery stage after inoculation with AMF were 66.0% and 55.5% respectively, in the absence and presence of Mg in the nutrient solution (Al-Whaibi and Khaliel 1994; Bouhired et al. 1992). The availability of mineral nutrients in the desert soils is enhanced in the presence of mycorrhiza (Al-Karaki et al. 2007). Inoculation with AM increased the growth of tissue culture date palm seedlings at low fertilizer rate and irrigation with highly saline water (Shabbir et al. unpublished data) (Fig. 22.3).

Blal et al. (1990) reported a 4–5 fold increase in the coefficient of P utilization in micropropagated oil palms after inoculation with mycorrhizae. Micropropagated

Fig. 22.3 Growth of date palm seedlings, 300 days after transplanting at low and high salinity water; *Left* – Low fertilizer rate, non-inoculated with mycorrhizae, *Right* – Low fertilizer rate, inoculated with mycorrhizae

oil palms cannot efficiently absorb P and generally grow well without the presence of mycorrhizae. Inoculation of the palms resulted in a 2.7–5.6 fold increase in fertilizer use efficiency (Blal and Gianinazzi-Pearson 1990). Date palms are usually given an annual treatment of organic and inorganic fertilizers. Although organic fertilizers are generally compatible with mycorrhizae, organic P is not available to the plants unless hydrolyzed into inorganic form. Mycorhizae enables date palms to absorb more P by enhancing the enzymatic activity of the phosphatase in the roots (Fries et al. 1998; Tawaraya and Saito 1994). The enzyme phosphatase, catalyze the conversion of organic P to inorganic form, which then becomes available to the plants. Although the addition of inorganic fertilizers leads to an increase in nutrient uptake, as well as to higher fruit yields in date palms (Bacha and Abo-Hassan 1983), phosphorus-rich inorganic fertilizers are known to inhibit AM fungi (Amaya-Carpio et al. 2009; Ezawa et al. 2000). AMF colonization levels of field date palm roots were found to be negatively correlated ($r^2 = 0.64$) with soil phosphorus content (Bouamri et al. 2006). AMF colonization can improve water-use efficiency of date palms (Sieverding 2008).

Date palms suffer from many diseases and pests. The majority of diseases are associated with fungi (Zaid et al. 2002) and nematodes (Eissa et al. 1998). One of the most serious fungal diseases in North Africa is bayoud disease, which is caused by *Fusarium oxysporuym*. This fungal infection results in huge losses to the trees (Jaiti et al. 2007). The red palm weevil (*Rhynchophorus ferrugineus*) is considered one the most destructive insect pests in date palms, especially in the Middle East (Abraham et al. 1998). It is clear that chemical treatments alone are not effective in controlling such diseases and pests. AMF has been recognized as a highly potential means in plant protection and pest management (Quarles 1999). In several cases, AMF has demonstrated direct biocontrol potential especially with plant diseases caused by *Phytophtora* and *Fusarium* fungi (Vigo et al. 2000). These interactions suggest that AMF might affect plant and soil microbial activity by stimulating the production of root exudates, phytoalexins and phenolic compounds (Jaiti et al. 2007; Morandi 1996). In some cases it appears that disease is prevented because the fungus is physically unable to penetrate the space where mycorrhiza exists. Some AMF produce antibiotics that attack pathogens, or produce structures preventing diseases from entering the roots (Harrison 1997). In addition, AMF use active mechanisms to trap and strangle root-feeding nematodes. The AMF *Glomus intraradices* induced systemic resistance against two parasitic nematodes, *Radopholus similis* and *Pratylenchus coffeae*, in banana plants, and reduced their infestation by more than 50%, even when the AMF and the nematodes were spatially separated (Elsen et al. 2008).

The presence and high colonization of date palms roots by AMF indicates that they significantly benefit from AM associations (Bouamri et al. 2006). Mycorrhizal symbiosis can therefore be regarded as a significant adaptation and survival mechanism to harsh and unfavorable growing conditions.

22.7 Utilization of Mycorrrhizal Fungi for Date Palm Propagation

Date palms are commonly grown in deserts, which tend to be unstable ecosystems and subject to frequent disturbance (Herrera et al. 1993). For example, topsoil erosion generally results in the loss or reduction of indigenous mycorrhizae present there (Brundrett 1991; Herrera et al. 1993; Jasper et al. 1991; Jeffries and Barea 1994). This can be detrimental because mycorrhizal symbioses are regarded as key components of desert ecosystems (Brundrett 1991; Carpenter and Allen 1988). The low density of mycorrhizal propagules in semiarid ecosystems (McGee 1989) may limit the successful establishment of native plants, including date palms (Sylvia 1990). Low or ineffective mycorrhizal inoculum potential needs to be reconstituted for date palm cultivation by adding an appropriate mycosymbiont population under field conditions (Allen et al. 1992; Jasper 1994). Alternatively, date palm seedlings can be inoculated at the propagation stage and then transplanted to the field, where they are grown using low levels of chemical fertilizers. Under normal nursery conditions, date palm seedlings are grown in artificial growth media, which lack mycorrhizal inoculum, such as peat moss and compost. When nursery-grown plants or seedlings are transplanted to the field, especially into disturbed soils, their roots are not colonized by mycorrhizal fungi and they grow more slowly and have lower survival rates. The innovative technique of artificial inoculation of date palms during propagation at the nursery stage and field planting might reverse these effects and result in improved plant establishment, growth rates and yields, with complementary reductions in water and fertilizer use (Sieverding 2008).

When mature seedlings are transplanted to the field, the benefits of mycorrhizal inoculation become much less apparent, because planting holes are usually heavily fertilized, and this may inhibit mycorrhizal colonization of the roots. However, as soil-P levels drop with increasing plant growth, AMF from the surrounding soil can start to colonize and develop in the growing root system, thus negating the need for inoculum.

22.8 Economic Considerations of Using AMF for Date Palms

The demand for high quality date palm cultivars has increased during the last three decades, which has been encouraged by the use of micropropagation and tissue culture techniques for mass production of the transplants (Awad 2008). In spite of this great potential, transplantation success to the field has been rather low, and in some cases, cultivars have shown only a 40–50% survival rate (Zaid and de Wet 1999). Under such conditions, inoculation with AMF could be an important step to enhance the establishment and survival of the transplanted seedlings. However, to date most work involving mycorrhizae and date palms has concentrated on the potential of AMF to improve date palm growth. Schultz (2001) reported that *post vitro* survival rate and growth of oil palm ranged between 83% and 100% after 3 months of *in vitro* inoculation with AMF, compared to only 55% of the non-inoculated plants. From an economic perspective, enhancing growth characteristics may reduce costs, but the potential saving by improving the survival rate is greater, a fact that has largely been overlooked by researchers. These findings could lead to enormous savings for nurseries. Given the high costs associated with the loss of nursery plants as well as high mortality and slow growth rate at the outplanting stage, experiments are needed to assess the potential savings that could be made through mycorrhizal inoculation of date palms.

22.9 Conclusions and Prospective

There are some doubts as to whether artificial mycorrhizal inoculation can improve date palm productivity in established plantations, especially if growers continue in their use of high levels of fertilizers and pesticides. Proper soil and nutrient management in date palm fields can enhance native AM levels in the soil which can help plants to tolerate stresses and grow better under harsh climatic conditions.

Nevertheless, the fact that date palms are very responsive to mycorrhizal inoculation provides a clear indication that mycorrhizae could have a substantial role to play in production systems. Inoculation with AM-fungi can brings enormous benefit to the establishment of the date palm seedlings under saline conditions. Although, AM are naturally present in the field soil, they are not sufficiently present in nursery substrates, therefore artificial inoculation is essential. It is advisable to inoculate date palm seedlings, especially those from tissue culture, with high quality artificial mycorrhizal inoculum.

It appears that biological production systems are superior to chemical systems. Utilization of innovative mycorrhizal technology during propagation and transplanting phase in date palm production may alleviate problems associated with high chemical inputs, and lead the way to more natural and sustainable production systems characterized by improved growth rates, better plant establishment and more consistent and stable yields. There is enormous potential to reduce overall production costs by reducing mortality of tissue-cultured plants in addition to complementary reductions in fertilizer and water use.

References

Abraham VA, Al-Shuaibi MA, Aleiro JRF et al (1998) An integrated management approach for red date palm weevil, *Rhynchophorus ferrugineus* Oliv., a key pest of date palm in the Middle East. Sult Qaboos Univ J Sci Res Agric Sci 3:77–84

Abrol JP (1986) Fuel and forage production from salt affected wasteland in India. Recl Reveg Res 5:65–74

Adiku G, Renger M, Wessolek G et al (2001) Simulation of dry matter production and seed yield of common beans under varying soil water and salinity conditions. Agric Water Manage 47:55–68

Al-Garni SMS (2006) Increasing NaCl-salt tolerance of a halophytic plant *Phragmites australis* by mycorrhizal symbiosis. Am Euras J Agric Environ Sci 1:119–126

Alguacil MM, Hernandez JA, Caravaca F et al (2003) Antioxidant enzyme activities in shoots from three mycorrhizal shrub species afforested in a degraded semi-arid soil. Physiol Plant 118:562–570

Aliasgharzadeh N, Saleh RN, Towfighi H et al (2001) Occurrence of arbuscular mycorrhizal fungi in saline soils of the Tabriz Plain of Iran in relation to some physical and chemical properties of soil. Mycorrhiza 11:119–122

Al-Karaki GN (2000) Growth of mycorrhizal tomato and mineral acquisition under salt stress. Mycorrhiza 10:51–54

Al-Karaki GN (2006) Nursery inoculation of tomato with arbuscular mycorrhizal fungi and subsequent performance under irrigation with saline water. Sci Hort 109:1–7

Al-Karaki GN, Al-Raddad A (1997) Effect of arbuscular mycorrhizal fungi and drought stress on growth and nutrient uptake of two wheat genotypes differing in drought resistance. Mycorrhiza 7:83–88

Al-Karaki GN, Clark RB (1998) Growth, mineral acquisition and water use by mycorrhizal wheat grown under water stress. J Plant Nutr 21:263–276

Al-Karaki GN, Othman Y, Al-Ajmi A (2007) Effects of mycorrhizal fungi inoculation on landscape turf establishment under Arabian Gulf region conditions. Arab Gulf J Sci Res 25(3):147–152

Allen EB, Cunningham GL (1983) Effects of vesicular-arbuscular mycorrhizae on *Distichlis spicata* under three salinity levels. New Phytol 93:227–236

Allen MF, Clouse SD, Weinbaum BS et al (1992) Mycorrhiza and the integration of scales: from molecules to ecosystems. In: Allen MF (ed) Mycorrhizal functioning, 1st edn. Chapman & Hall, New York, pp 488–515

Al-Whaibi MH, Khaliel AS (1994) The effect of Mg on Ca, K and P content of date palm seedlings under mycorrhizal and non-mycorrhizal conditions. Mycoscience 35:213–217

Al-Yahya M (2008) Arbuscular mycorrhizal fungal communities associated with date palms in a traditional and a modern experimental plantation and with desert plants in the adjacent natural habitats in Southern Arabia. PhD thesis, University of Basel, Switzerland

Amaya-Carpio L, Davies FT Jr, Fox T et al (2009) Arbuscular mycorrhizal fungi and organic fertilizer influence photosynthesis, root phosphatase activity, nutrition, and growth of *Ipomoea carnea* subsp. *fistulosa*. Photosynthetica 47:1–10

Aroca R, Vernieri P, Ruiz-Lozano JM (2008) Mycorrhizal and non-mycorrhizal *Lactuca sativa* plants exhibit contrasting responses to exogenous ABA during drought stress and recovery. J Exp Bot 8:2029–2041

Ashraf M, Foolad MR (2007) Roles of glycine betaine and proline in improving plant abiotic stress resistance. Environ Exp Bot 59:207–216

Auge RM (2001) Water relations, drought and vesicular arbuscular mycorrhizal symbiosis. Mycorrhiza 11:3–42

Auge RM, Sylvia DM, Park SJ et al (2004) Partitioning mycorrhizal influence on water relations of *Phaseolus vulgaris* into soil and root components. Can J Bot 82:503–514

Awad MA (2008) Promotive effects of a 5-aminolevulinic acid-based fertilizer on growth of tissue culture-derived date palm plants (*Phoenix dactylifera* L.) during acclimatization. Sci Hort 118:48–52

Bacha M, Abo-Hassan AA (1983) Effects of soil fertilization on yield, fruit quality and mineral content of Khudairi date palm variety. In: Proceedings first symposium on the date palm in Saudi Arabia, pp 174–179

Barreveld WH (1993) Date palm products. Agricultural services Bull 101. FAO, Rome

Barrow JR, Havstad KM, McCaslin BD (1997) Fungal root endophytes in four wing saltbrush, *Atriplex canescens* on arid rangeland of southwestern USA. Arid Soil Res Rehabil 11:177–185

Bearden BN, Petersen L (2000) Influence of arbuscular mycorrhizal fungi on soil structure and aggregate stability of a vertisol. Plant Soil 218:173–183

Bhaskaran C, Selvaraj T (1997) Seasonal incidence and distribution of VAM fungi in native saline soils. J Environ Biol 18:209–212

Blal B, Gianinazzi-Pearson V (1990) Interest of endomycorrhizae for the production of micropropagated oil palm clones. Agric Ecosyst Environ 29:39–43

Blal B, Morel C, Gianinazzi-Pearson V et al (1990) Influence of vesicular-arbuscular mycorrhizae on phosphate fertilizer efficiency in two tropical acid soils planted with micropropagated oil palm (*Elaeis guineensis* Jacq.). Biol Fert Soil 9:43–48

Boomsma C, Vyn T (2008) Maize drought tolerance: potential improvements through arbuscular mycorrhizal symbiosis? Field Crop Res 108:14–31

Bouamri R, Dalpe Y, Serrhini MN et al (2006) Arbuscular mycorrhizal fungi species associated with rhizosphere of *Phoenix dactylifera* L. (date palm) in Morocco. Afr J Biotechnol 5:510–516

Bouhired L, Gianinazzi S, Gianinazzi-Pearson V (1992) Influence of endomycorrhizal inoculation on the growth of *Phoenix dactylifera*. In: Micropropagation, root regeneration and mycorrhizae. Joint meeting between Cost 87 and Cost 8.10. Dijon, France 53

Bray EA (1997) Plant responses to water deficit. Trends Plant Sci 2:48–54

Brown ME (1974) Seed and root bacterization. Ann Rev Phytopathol 12:181–197

Brown AM, Bledsoe C (1996) Spatial and temporal dynamics of mycorrhizas in *Jaumea carnosa*, a tidal saltmarsh halophyte. J Ecol 84:703–715

Brundrett M (1991) Mycorrhizas in natural ecosystems. In: Begon M, Fitter AH, Macfadyen A (eds.) Advances in ecological research, vol 21. Academic Press Limited, London, pp 171–313

Cantrell IC, Linderman RG (2001) Preinoculation of lettuce and onion with VA mycorrhizal fungi reduces deleterious effects of soil salinity. Plant Soil 233:269–281

Carpenter AT, Allen MF (1988) Responses of Hedysarum boreale Nutt. to mycorrhizas and Rhizobium: plant and soil nutrient changes in a disturbed shrub-steppe. New Phytol 109: 125–132

Carvalho LM, Correia PH, Martins-Loucao A (2001) Arbuscular mycorrhizal fungal propagules in a salt marsh. Mycorrhiza 14:165–170

Chao CT, Krueger RR (2007) The date palm (*Phoenix dactylifera* L.): overview of biology, uses, and cultivation. Hort Sci 42:1077–1082

Colla G, Rouphael Y, Cardarelli M et al (2008) Alleviation of salt stress by arbuscular mycorrhizal in zucchini plants grown at low and high phosphorus concentration. Biol Fert Soil 44: 501–509

Copeman RH, Martin CA, Stutz JC (1996) Tomato growth in response to salinity and mycorrhizal fungi from saline or nonsaline soils. Hort Sci 31:341–344

Cramer GR, Lauchli A, Polito VS (1985) Displacement of Ca^{2+} by Na^{+} from the plasmalemma of root cells: a primary response to salt stress? Plant Physiol 79:207–277

Creus CM, Sueldo RJ, Barassi CA (1998) Water relations in *Azospirillum* inoculated wheat seedlings under osmotic stress. Can J Bot 76:238–244

Daei G, Ardekani M, Rejali F et al (2009) Alleviation of salinity stress on wheat yield, yield components and nutrient uptake using arbuscular mycorrhizal fungi under field conditions. J Plant Physiol 166:217–225

Dehne HW (1982) Interaction between vesicular-arbuscular mycorrhizal fungi and plant pathogens. Phytopathology 72:1115–1119

Delgado MJ, Ligero F, Lluch C (1994) Effects of salt stress on growth and nitrogen fixation by pea, faba-bean, common bean, and soybean plants. Soil Biol Biochem 26:371–376

Dixon RK, Garg VK, Rao MV (1993) Inoculation of *Leucaena* and *Prosopis* seedlings with *Glomus* and *Rhizobium* species in saline soil: rhizosphere relations and seedlings growth. Arid Soil Res Rehabil 7:133–144

Downer J (2004) Palm tree management. Part I: Selecting the right palms. Landsc Note 18(4):4

Dransfield J, Uhl NW, Asmussen CB et al (2008) Genera palmarum: the evolution and classification of palms. Kew Publishing, Kew

Duke ER, Johnson CR, Koch KE (1986) Accumulation of phosphorus, dry matter and betaine during NaCl stress of split-root citrus seedlings colonized with vesicular-arbuscular mycorrhizal fungi on zero, one or two halves. New Phytol 104:583–590

Eissa MFM, El-Sharief MA, Abdel-Gawad MM et al (1998) Histological response of susceptible and resistant date-palm cultivars to *Meloidogyne incognita* infection. Pak J Nematol 16:103–109

Elsen AD, Gervacio R, Swennen D et al (2008) AMF-induced biocontrol against plant parasitic nematodes in *Musa* sp.: a systemic effect. Mycorrhiza 18:251–256

Estrada-Luna AA, Davies FT (2003) Arbuscular mycorrhizal fungi influence water relations, gas exchange, abscissic acid and growth of micropropagated chile ancho pepper (*Capsicum annuum*) plantlets during acclimatization and post-acclimatization. J Plant Physiol 160: 1073–1083

Estaun MV (1989) Effect of sodium chloride and mannitol on germination and hyphal growth of the vesicular-arbuscular mycorrhizal fungus *Glomus mosseae*. Agric Ecosyst Environ 29:123–129

Evelin H, Kapoor R, Giri B (2009) Arbuscular mycorrhizal fungi in alleviation of salt stress: a review. Ann Bot 104:1263–1280

Ezawa T, Yamamoto K, Yoshida S (2000) Species composition and spore density of indigenous vesicular-arbuscular mycorrhizal fungi under different conditions of P-fertility as revealed by soybean trap culture. Soil Sci Plant Nutr 46:291–297

FAO (2002, 2003, 2007) Agro-Statistics Database

Feng G, Zhang FS, XI L et al (2002) Improved tolerance of maize plants to salt stress by arbuscular mycorrhiza is related to higher accumulation of soluble sugars in roots. Mycorrhiza 12:185–190

Fernandez-Munoz CJ (1999) Effects of salinity on tomato. Sci Hort 78:83–125

Founoune H, Duponnis R, Ba AM et al (2002) Influence of the dual arbuscular endomycorrhizal/ ectomycorrhizal symbiosis on the growth of *Acacia holosericea* (*A. Cunn.* ex G. Don) in glasshouse conditions. Annu For Sci 59:93–98

Frank AB (1885) Über die auf Würzelsymbiose beruhende Ehrnährung gewisser Bäum durch unterirdische Pilze. Ber Deut Botan Ges 3:128–145

Fries LLM, Pacovsky RS, Safir GR et al (1998) Phosphorus effect on phosphatase activity in endomycorrhizal maize. Physiol Plant 103:162–171

Fuzy A, Biro B, Toth T et al (2008) Drought, but not salinity, determines the apparent effectiveness of halophytes colonized by arbuscular mycorrhizal fungi. J Plant Physiol 165:1181–1192

Fuzy A, Biro B, Toth T (2010). Short communication. Effect of saline soil parameters on endomycorrhizal colonisation of dominant halophytes in four Hungarian sites. Span J Agr Res 8: (Sp. Iss. 1) S144–S148

Gallagher JK (1985) Halophytic crops for cultivation at seawater salinity. Plant Soil 89:323–336

Garg N, Manchanda G (2008) Effect of arbuscular mycorrhizal inoculation of salt-induced nodule senescence in *Cajanus cajan* (pigeon pea). J Plant Growth Regul 27:115–124

Gavito ME, Miller M (1998) Changes in mycorrhiza development in maize induced by crop management practices. Plant Soil 198:185–192

Giri B, Mukerji KG (2004) Mycorrhizal inoculant alleviates salt stress in *Sesbania aegyptiaca* and *Sesbania grandiflora* under field conditions: evidence for reduced sodium and improved magnesium uptake. Mycorrhiza 14:307–312

Giri B, Kapoor R, Mukerji KG (2003) Influence of arbuscular mycorrhizal fungi and salinity on growth, biomass and mineral nutrition of *Acacia auriculiformis*. Biol Fert Soils 38: 170–175

Giri B, Kapoor R, Mukerji KG (2007) Improved tolerance of *Acacia nilotica* to salt stress by arbuscular mycorrhiza, *Glomus fasciculatum* may be partly related to elevated K^+/Na^+ ratios in root and shoot tissues. Microb Ecol 54:753–760

Glenn EP, O'Leary JW (1985) Productivity and irrigation requirements of halophytes grown with seawater in the Sonoran desert. J Arid Environ 9:81–91

Graham JH, Syversten JP (1984) Influence of vesicular arbuscular mycorrhiza on the hydraulic conductivity of roots of two *Citrus* rootstocks. New Phytol 97:277–284

Harisnaut P, Poonsopa D, Roengmongkol K et al (2003) Salinity effects on antioxidant enzymes in mulberry cultivar. Sci Asia 29:109–113

Harrison MJ (1997) The arbuscular mycorrhizal symbiosis. In: Stancey G, Keen NT (eds.) Plant-microbe interactions Vol 3. Chapman & Hall, New York, pp 1–34

Hartmond U, Schaesberg NV, Graham JH, Syversten JP (1987) Salinity and flooding stress effects on mycorrhizal and non-mycorrhizal citrus rootstock seedlings. Plant Soil 104:37–43

Hejiden JN, Klironomos M, Ursic P et al (1998) Mycorrhizal fungal diversity determines plant biodiversity, ecosystem variability and productivity. Nature 396:69–72

Herrera MA, Salamanca CP, Barea JM (1993) Inoculation of woody legumes with selected arbuscular mycorrhizal fungi and rhizobia to recover desertified Mediterranean ecosystems. Appl Environ Microbiol 59:129–133

Hildebrandt U, Regvar M, Bothe H (2007) Arbuscular mycorrhiza and heavy metal tolerance. Phytochemistry 68:139–146

Hilderbrandt U, Janetta K, Ouziad F et al (2001) Arbuscular mycorrhizal colonization of halophytes in Central European salt marshes. Mycorrhiza 10:175–183

Hirrel MC, Mehravaran H, Gerdemann JW (1978) Vesicular-arbuscular mycorrhiza in the Chenopodiaceae and Cruciferae: do they occur? Can J Bot 56:2813–2817

Hirrel MC, Gerdemann JW (1980) Improved growth of onion and bell pepper in saline soils by two vesicular-arbuscular mycorrhizal fungi. Soil Sci Soc Am J 44:654–665

Hirrel MC (1981) The effect of sodium and chloride salts on the germination of *Gigaspora margarita*. Mycologia 73:610–617

Ho I (1987) Vesicular-arbuscular mycorrhizae of halophytic grasses in the Alvord desert of Oregon. Northwest Sci 61:148–151

Hodge A, Campbell CD, Fitter AH (2001) An arbuscular mycorrhizal fungus accelerates decomposition and acquires nitrogen directly from organic material. Nature 413:297–299

Hoefnagels MH, Broome SW, Shafer SR (1993) Vesicular-arbuscular mycorrhizae in salt marshes in North Carolina. Estua 16:851–858

Hoekstra FA, Golovina EA, Buitink J (2001) Mechanisms of plant desiccation tolerance. Trends Plant Sci 6:431–438

Jahromi F, Aroca R, Porcel R et al (2008) Influence of salinity on the *in vitro* development of *Glomus intraradices* and on the *in vivo* physiological and molecular responses of mycorrhizal lettuce plants. Microb Ecol 55:45–53

Jain M, Mathur G, Koul S et al (2001) Ameliorative effects of proline on salt stress-induced lipid peroxidation in cell lines of groundnut (*Arachis hypogaea* L.). Plant Cell Rep 20:463–468

Jaiti FA, Meddich A, El Hadrami I (2007) Effectiveness of arbuscular mycorrhizal fungi in the protection of date palm (*Phoenix dactylifera*) against bayoud disease. Physiol Mol Plant Pathol 71:166–173

Janos DP (2007) Plant responsiveness to mycorrhizas differs from dependence upon mycorrhizas. Mycorrhiza 17:75–91

Jasper DA (1994) Management of mycorrhiza in revegetation. In: Robson AD, Abbot LK, Malajczuk N (eds) Management of mycorrhizas in agriculture, horticulture and forestry. Kluwer Academic, Dordrecht, pp 211–219
Jasper DA, Abbot LK, Robson AD (1991) The effect of soil disturbance on vesicular-arbuscular mycorrhizal fungi in soils from different vegetation types. New Phytol 118:471–476
Jeffries P, Barea JM (1994) Biogeochemical cycling and arbuscular mycorrhizas in the sustainability of plant-soil systems. In: Gianinazzi S, Schüpp SH (eds.) Impact of arbuscular mycorrhizas on sustainable agriculture and natural ecosystems. Birkha¨user Verlag, Basel, pp 101–115
Jindal V, Atwal A, Sekhon BS et al (1993) Effect of vesicular-arbuscular mycorrhizae on metabolism of moong plants under NaCl salinity. Plant Physiol Biochem 3:475–481
Johnson-Green P, Kenkel NC, Booth T (2001) Soil salinity and arbuscular mycorrhizal colonization of *Puccinellia nuttalliana*. Mycol Res 105:1094–1110
Juniper S, Abbott LK (1993) Vesicular-arbuscular mycorrhizas and soil salinity. Mycorrhiza 4:45–57
Juniper S, Abbott LK (2006) Soil salinity delays germination and limits growth of hyphae from propagules of arbuscular mycorrhizal fungi. Mycorrhiza 16:371–379
Kapoor R, Sharma D, Bhatnagar AK (2008) Arbuscular mycorrhizae in micropropagation systems and their potential applications. Sci Hort 116:227–239
Kaya C, Ashraf M, Sonmez O et al (2009) The influence of arbuscular mycorrhizal colonization on key growth parameters and fruit yield of pepper plants grown at high salinity. Sci Hort 121:1–6
Khaliel AS, Abou-Hailah AN (1985) Formation of vesicular-arbuscular mycorrhiza in *Phoenix dactylifera* L. cultivated in Qassim region, Saudi Arabia. Pak J Bot 17:267–270
Khan AG (1974) The occurrence of mycorrhizas in halophytes, hydrophytes and xerophytes of endogone spores in adjacent soils. J Gen Microbiol 81:7–14
Khudairi AK (1969) Mycorrhiza in desert soils. BioScience 19:598–599
Kim CK, Weber DJ (1985) Distribution of VA mycorrhiza on halophytes on inland sea playas. Plant Soil 83:207–214
Klironomos JN, Hart MM (2001) Animal nitrogen swap for plant carbon. Nature 410:651–652
Kramer PJ, Boyer JS (1997) Water relations of plants and soils. Academic, San Diego
Landwehr M, Hilderbrandt U, Wilde P et al (2002) The arbuscular mycorrhizal fungus *Glomus geosporum* in European saline, sodic and gypsum soils. Mycorrhiza 12:199–211
Levy Y, Dodd J, Krikun J (1983) Effect of irrigation water salinity and rootstock on the vertical distribution of vesicular-arbuscular mycorrhiza in citrus roots. New Phytol 95:397–403
Lindermann RG (1994) Role of VAM in biocontrol. In: Pfleger FL, Linderman RG (eds.) Mycorrhizae and plant health. American Phytopathological Society, St. Paul, pp 1–26
Linnee (1734) cited in Keaney TH (1906) Date varieties and date culture in Tunis. Bureau of Plant Industry, Bulletin No. 92, Washington, USDA.
Ludwig-Muller J (2000) Indole-3-butyric acid in plant growth and development. Plant Growth Regul 32:219–230
Marschner H (1995) Mineral nutrition of higher plants, 2nd edn. Academic, London
Mason E (1928) Note on the presence of mycorrhizae in the roots of saltmarsh plants. New Phytol 27:193–195
Matamoros MA, Baird LM, Escuredo PR et al (1999) Stress-induced legume root nodule senescence: physiological, biochemical and structural alterations. Plant Physiol 121:97–111
Mathur N, Singh J, Bohra S et al (2007) Arbuscular mycorrhizal status of medicinal halophytes in saline areas of Indian Thar Desert. Int J Soil Sci 2:119–127
McGee P (1989) Variation in propagule numbers of vesicular-arbuscular mycorrhizal fungi in a semi-arid soil. New Phytol 92:28–33
McMillen B, Juniper S, Abbott LK (1998) Inhibition of hyphal growth of a vesicular-arbuscular mycorrhizal fungus in soil containing sodium chloride limits the spread of infection from spores. Soil Biol Biochem 30:1639–1646
Menconi M, Sgherri CLM, Pinzino C, Navari-izzo F (1995) Activated oxygen species production and detoxification in wheat plants subjected to a water deficit programme. J Exp Bot 46: 1123–1130

Miransari M, Bahrami HA, Rejali F et al (2007) Using arbuscular mycorrhiza to reduce the stressful effects of soil compaction on corn (*Zea mays* L.) growth. Soil Biol Biochem 39:2014–2026

Miransari M, Bahrami HA, Rejali F et al (2008) Using arbuscular mycorrhiza to reduce the stressful effects of soil compaction on wheat (*Triticum aestivum* L.) growth. Soil Biol Biochem 40:1197–1206

Miransari M, Rejali F, Bahrami HA et al (2009a) Effects of soil compaction and arbuscular mycorrhiza on corn (*Zea mays* L.) nutrient uptake. Soil Till Res 103:282–290

Miransari M, Rejali F, Bahrami HA et al (2009b) Effects of arbuscular mycorrhiza, soil sterilization and soil compaction on wheat (*Triticum aestivum* L.) nutrient uptake. Soil Till Res 104:48–55

Mohammad MJ, Hamad SR, Malkani HI (2003) Population of arbuscular mycorrhizal fungi in semi-arid environment of Jordan as influenced by biotic and abiotic factors. J Arid Environ 53:409–417

Morandi D (1996) Occurrence of phytoalexins and phenolic compounds in endomycorrhizal interactions, and their potential role in biological control. Plant Soil 185:241–251

Morgan JM (1984) Osmoregulation and water stress in higher plants. Ann Rev Plant Physiol 33:299–319

Mouk BO, Ishii T (2006) Effect of arbuscular mycorrhizal fungi on tree growth and nutrient uptake of *Sclerocarya birrea* under water stress, salt stress and flooding. J Jpn Soc Hort Sci 75:26–31

Mozafar A, Anken T, Ruh R et al (2000) Tillage intensity, mycorrhizal and nonmycorrhizal fungi, and nutrient concentrations in maize, wheat and canola. Agron J 92:1117–1124

Munier P (1973) Le palmier-dattier – techniques agricoles et productions tropicales. Maison Neuve et Larose, Paris

Muralev E, Nazarenko PI, Poplavskij VM et al (1997) Seawater desalination. In: Nuclear desalinization of seawater. Proceedings of a symposium in Taejon, Republic of Korea. International Atomic Energy Agency, Vienna, Austria, pp 355–366

Murkute AA, Sharma S, Singh SK (2006) Studies on salt stress tolerance of citrus rootstock genotypes with arbuscular mycorrhizal fungi. Hort Sci 33:70–76

Oehl F, Sieverding E (2004) *Pacispora,* a new vesicular arbuscular mycorrhizal fungal genus in the Glomeromycetes. J Appl Bot 78:72–82

Oihabi A (1991) Effect of vesicular arbuscular mycorrhizae on bayoud disease and date palm nutrition. PhD thesis, University Marrakech

Ojala JC, Jarrell WM, Menge JA, Johnson ELV (1983) Influence of mycorrhizal fungi on the mineral nutrition and yield of onion in saline soil. Agron J 75:255–259

Palenzuela J, Ferrol N, Boller T et al (2008) *Otospora bareai,* a new fungal species in the Glomeromycetes from a dolomitic shrubland in Sierra de Baza National Park (Granada, Spain). Mycologia 100:296–305

Parida A, Das AB, Das P (2002) NaCl stress causes changes in photosynthetic pigments, proteins and other metabolic components in the leaves of a tree mangrove, *Bruguiera parviflora,* in hydroponic cultures. J Plant Biol 45:28–36

Patreze CM, Cordeiro L (2004) Nitrogen fixing and vesicular arbuscular mycorrhizal symbioses in some tropical legume trees of tribe Mimosaceae. For Ecol Manage 196:275–285

Pearson JN, Schweiger P (1993) *Scutellospora calospora* (Nicol. and Gerd.) Walker and Sanders associated with subterranean clover: dynamics of colonization, sporulation and soluble carbohydrates. New Phytol 124:215–219

Peterson RL, Massicotte HB, Melville LH (2004) Mycorrhizas: anatomy and cell biology. National Research Council Research Press, Washington DC

Plenchette C, Duponnis R (2005) Growth response of the salt brush *Atriplex numularia* L. to inoculation with the arbuscular mycorrhizal fungus *Glomus intraradices*. J Arid Environ 61:535–540

Plenchette C, Fortin JA, Furlan V (1983) Growth responses of several plant species to mycorrhizae in a soil of moderate P-fertility. I Mycorrhizal dependency field conditions. Plant Soil 110:199–209

Pond EC, Menge JA, Jarrell WM (1984) Improved growth of tomato in salinized soil by vesicular arbuscular mycorrhizal fungi collected from saline sites. Mycologia 76:74–84

Porcel R, Ruiz-Lozano JM (2004) Arbuscular mycorrhizal influence on leaf water potential, solute accumulation and oxidative stress in soybean plants subjected to drought stress. J Exp Bot 55:1743–1750

Porras-Soriano A, Soriano-Martin ML, Porras-Piedra A et al (2009) Arbuscular mycorrhizal fungi increased growth, nutrient uptake and tolerance to salinity in olive trees under nursery conditions. J Plant Physiol. doi:10.1016/j.jplph.2009.02.010

Poss JA, Pond EC, Menge JA, Jarrell WM (1985) Effect of salinity on mycorrhizal onion and tomato in soil with and without additional phosphate. Plant Soil 88:307–319

Quarles W (1999) Plant disease biocontrol and VAM fungi. IPM Pract 21:1–9

Rabie GH (2005) Influence of VA-mycorrhizal fungi and kinetin on the response of mung bean plants to irrigation with seawater. Mycorrhiza 15:225–230

Rabie GH, Almadini AM (2005) Role of bioinoculants in development of salt-tolerance of *Vicia faba* plants under salinity stress. Afr J Biotechnol 4:210–222

Read DJ, Perez-Moreno J (2003) Mycorrhizas and nutrient cycling in ecosystems – a journey towards relevance? New Phytol 157:475–492

Remy W, Taylor TN, Hass H et al (1994) 4 hundred million year old vesicular-arbuscular mycorrhizae. Proc Natl Acad Sci USA 91:11841–11843

Richards LA (1954) Diagnosis and improvement of saline and alkali soils. US Department of Agriculture, Handbook 60. Washington, DC

Rillig MC (2004) Arbuscular mycorrhizae and terrestrial ecosystem processes. Ecol Lett 7:740–754

Rosendahl CN, Rosendahl S (1991) Influence of vesicular-arbuscular mycorrhizal fungi (*Glomus* spp.) on the response of cucumber (*Cucumis sativus* L.) to salt stress. Env Exp Bot 31:313–318

Rozema J, Arp W, Van Diggelen J et al (1986) Occurrence and ecological significance of vesicular-arbuscular mycorrhiza in the salt marsh environment. Acta Bot Neerl 35:457–467

Ruiz-Lozano JM (2003) Arbuscular mycorrhizal symbiosis and alleviation of osmotic stress: new perspectives for molecular studies. Mycorrhiza 13:309–317

Ruiz-Lozano JM, Azcón R (2000) Symbiotic efficiency and infectivity of an autochthonous arbuscular mycorrhizal *Glomus* sp from saline soils and *Glomus deserticola* under salinity. Mycorrhiza 10:137–143

Ruiz-Lozano JM, Collados C, Barea JM et al (2001) Arbuscular mycorrhizal symbiosis can alleviate drought induced nodule senescence in soybean plants. Plant Physiol 82:346–350

Ruiz-Lozano JM, Porcel R, Aroca R (2006) Does the enhanced tolerance of arbuscular mycorrhizal plants to water deficit involve modulation of drought-induced plant genes? New Phytol 171:693–698

Saint-Etienne L, Paul S, Imbert D et al (2006) Arbuscular mycorrhizal soil infectivity in a stand of the wetland tree *Pterocarpus officinalis* along a salinity gradient. For Ecol Manage 232:86–89

Sanan-Mishra N, Pham XH, Sopori SK et al (2005) Pea DNA helicase 45 overexpression in tobacco confers high salinity tolerance without affecting yield. Proc Natl Acad Sci USA 102:509–514

Sannazzaro AI, Ruiz OA, Albetro EO et al (2006) Alleviation of salt stress in *Lotus glaber* by *Glomus intraradies*. Plant Soil 285:279–287

Sannazzaro AI, Echeverria M, Albertó EO et al (2007) Modulation of polyamine balance in *Lotus glaber* by salinity and arbuscular mycorrhiza. Plant Physiol Biochem 45:39–46

Schultz C (2001) Effect of (vesicular) arbuscular mycorrhiza on survival and *post vitro* development of micropropagated oil palms (*Elaeis guineensis* Jacq.). PhD thesis. Universität Göttingen, Germany

Schüßler A (2002) Molecular phylogeny, taxonomy and evolution of *Geosiphon pyriformis* and arbuscular mycorrhizal fungi. Plant Soil 244:75–83

Schüßler A, Schwarzott D, Walker C (2001) A new fungal phylum, the Glomeromycota: phylogeny and evolution. Mycol Res 105:1413–1421

Sengupta A, Chaudhuri S (1990) Vesicular arbuscular mycorrhiza (VAM) in pioneer salt marsh plants of the Ganges river delta in West Bengal (India). Plant Soil 122:111–113

Sharifi M, Ghorbanli M, Ebrahimzadeh H (2007) Improved growth of salinity-stressed soybean after inoculation with pre-treated mycorrhizal fungi. J Plant Physiol 164:1144–1151

Sheng M, Tang M, Chan H et al (2008) Influence of arbuscular mycorrhizae on photosynthesis and water status of maize plants under salt stress. Mycorrhiza 18:287–296

Shokri S, Maadi B (2009) Effects of arbuscular mycorrhizal fungus on the mineral nutrition and yield of *Trifolium alexandrium* plants under salinity stress. J Agron 8:79–83

Sieverding E (2008) Mycorrhizae and water management in crops, lawn and turf. Lands 8:14–16

Sieverding E, Oehl F (2006) Revision of *Entrophospora* and description of *Kuklospora* and *Intraspora*, two new genera in the arbuscular mycorrhizal Glomeromycetes. J Appl Bot Food Qual 80:69–81

Simon L, Bousquet J, Lévesque RC et al (1993) Origin and diversification of endomycorrhizal fungi and coincidence with vascular land plants. Nature 363:67–69

Smith SE, Read DJ (1995) Mycorrhizal symbiosis. Academic, New York

Smith SE, Read DJ (1997) Mycorrhizal symbiosis, 2nd edn. Academic, San Diego

Smith SE, Read DJ (2008) Mycorrhizal symbiosis, 3rd edn. Academic, Amsterdam

Spain JL, Sieverding E, Oehl F (2006) *Appendicispora*: a new genus in the arbuscular mycorrhiza-forming Glomeromycetes with a discussion of the genus *Archaeospora*. Mycotaxon 97:163–182

Stewart CR, Lee JA (1974) The rate of proline accumulation in halophytes. Planta 120:279–289

Subramanian K, Santhanakrishnan P, Balasubramanian P (2006) Responses of field grown tomato plants to arbuscular mycorrhizal fungal colonization under varying intensities of drought stress. Sci Hort 107:245–253

Suzanne WS, David AP, Melanie DJ et al (1997) Net transfer of carbon between ectomycorrhizal tree species in the field. Nature 388:579–582

Sylvia DM (1990) Inoculation of native woody plants with vesicular-arbuscular mycorrhizal fungi for phosphate mine land reclamation. Agric Ecosyst Environ 31:253–261

Tang W, Peng X, Newton RJ (2005) Enhanced salt tolerance in transgenic loblolly pine simultaneously expressing two genes encoding mannitol-1- phosphate dehydrogenase and glucitol-6-phosphate dehydrogenase. Plant Physiol Biochem 43:139–146

Tawaraya L, Saito M (1994) Effect of vesicular-arbuscular mycorrhizal infection on amino acid composition in roots of onion and white clover. Soil Sci Plant Nutr 40:339–343

Tester M, Davenport R (2003) Na^+ tolerance and Na^+ transport in higher plants. Ann Bot 91:503–527

Tian CY, Feng G, Li XL et al (2004) Different effects of arbuscular mycorrhizal fungal isolates from saline or non-saline soil on salinity tolerance of plants. Appl Soil Ecol 26:143–148

Tressner HD, Hayes JA (1971) Sodium chloride tolerance of terrestrial fungi. Appl Microbiol 22:210–213

Vigo C, Norman JR, Hooker JE (2000) Biocontrol of the pathogen *Phytophthora parasitica* by arbuscular mycorrhizal fungi is a consequence of effects on infection loci. Plant Pathol 49:509–514

Walker C, Schüßler A (2004) Nomenclatural clarifications and new taxa in the Glomeromycota. Mycol Res 108:979–982

Walker C, Vestberg M, Demircik F et al (2007a) Molecular phylogeny and new taxa in the Archaeosporales (Glomeromycota): *Ambispora fennica* gen. sp. nov., Ambisporaceae fam. nov., and emendation of Archaeospora and Archaeosporaceae. Mycol Res 111:137–13

Walker C, Vestberg M, Schüßler A (2007b) Nomenclatural clarifications in Glomeromycota. Mycol Res 111:253–255

Wang L (2001) Fungi kill insects and feed host plants. Sci News Apr 17, 2002. BNET.com

Wang B, Qiu YL (2006) Phylogenetic distribution and evolution of mycorrhizas in land plants. Mycorrhiza 16:299–363

Wang W, Vinocur B, Altman A (2003) Plant responses to drought, salinity and extreme temperatures: toward genetic engineering for stress tolerance. Planta 218:1–14

Wang FY, Liu RJ, Lin XG et al (2004) Arbuscular mycorrhizal status of wild plants in saline-alkaline soils of the Yellow River Delta. Mycorrhiza 14:133–137

Wei-Feng XU, Wei-Ming SHI, Ueda A et al (2008) Mechanisms of salt tolerance in transgenic *Arabidopsis thaliana* carrying a peroxisomal ascorbate peroxidase gene from barley. Pedosph 4:486–495

Wilde P, Manal A, Stodden M et al (2009) Biodiversity of arbuscular mycorrhizal fungi in roots and soils of two salt marshes. Environ Microbiol 11:1548–1561

Wu YY, Chen QJ, Chen M et al (2005) Salt-tolerant transgenic perennial ryegrass (*Lolium perenne* L.) obtained by *Agrobacterium tumefasciens*-mediated transformation of the vacuolar Na^+/H^+ antiporter gene. Plant Sci 169:65–73

Yamato M, Ikeda S, Iwase K (2008) Community of arbuscular mycorrhizal fungi in coastal vegetation on Okinawa Island and effect of the isolated fungi on growth of sorghum under salt-treated conditions. Mycorrhiza 18:241–249

Yano-Melo AM, Saggin OJ, Maia LC (2003) Tolerance of mycorrhized banana (*Musa* sp. cv. Pacovan) plantlets to saline stress. Agric Ecosyst Environ 95:343–348

Zaid A, de Wet PF (1999) Date palm propagation. In: Zaid A (ed.) Date palm cultivation. FAO Rome, pp 74–106

Zaid A, de Wet PF, Djerbi M et al (2002) Diseases and pests of date palms. In: Zaid A (ed) Date palm cultivation, rev. edn. FAO, Rome, pp 227–281

Zhang HX, Blumwald E (2001) Transgenic salt-tolerant tomato plants accumulate salt in foliage but not in fruit. Nat Biotechnol 19:765–768

Zhong Qun H, Chao Xing H, Zhibin Z et al (2007) Changes in antioxidative enzymes and cell membrane osmosis in tomato colonized by arbuscular mycorrhizae under NaCl stress. Coll Surf B Biointerfaces 59:128–133

Zuccarini P (2007) Mycorrhizal infection ameliorates chlorophyll content and nutrient uptake of lettuce exposed to saline irrigation. Plant Soil Environ 53:283–289

Zuccarini P, Okurowska P (2008) Effects of mycorrhizal colonization and fertilization on growth and photosynthesis of sweet basil under salt stress. J Plant Nutr 31:497–513

Part IV
Genetics and Genetic Improvement

Chapter 23
Date Palm Genetics and Breeding

A. El Hadrami, F. Daayf, and I. El Hadrami

Abstract Date palm, *Phoenix dactylifera* L., is one of the oldest domesticated plants, but research to understand its diversity, evolution and adaptability is still lagging behind. One of the major limitations has been the lengthy life-cycle of this plant. Meanwhile, breeding challenges and opportunities are enormous, which in turn highlights the importance of how unraveling the evolutionary driving forces of this desert crop could help preserve its socio-economical, botanical and nutritional values. This review represents an up-to-date account of date palm progress in terms of genetics and breeding and underlines the challenges facing these two fields of research in this special crop. It also highlights some of the opportunities on the rise for breeding in this crop.

Keywords Abiotic stress • Biotechnology • Biotic stress • Conventional breeding • Genetics • Micropropagation • Somatic embryogenesis • Tissue culture

---

A. El Hadrami (✉)
OMEX Agriculture Inc. Canada, P.O. Box 301, 290 Agri Park Road, Oak Bluff, Manitoba R0G 1N0, Canada
e-mail: abdele@omexcanada.com

F. Daayf
Department of Plant Science, University of Manitoba, 222 Agriculture Building, Winnipeg, MB R3T 2 N2, Canada

I. El Hadrami
Faculty of Sciences Semlalia, Department of Biology Laboratoire de Biotechnologies, Protection et Valorisation des Ressources Végétales (Biotec-VRV), Faculté des Sciences Semlalia, University Cadi Ayyad, B.P. 2390, 40 000 Marrakech, Morocco

S.M. Jain et al. (eds.), *Date Palm Biotechnology*,
DOI 10.1007/978-94-007-1318-5_23, © Springer Science+Business Media B.V. 2011

23.1 Introduction

Date palm, *Phoenix dactylifera* L., is a perennial long-lived monocotyledonous species that is grown over a million hectares worldwide (Fig. 23.1). It is an economically important crop in the sub-Saharan countries as well as for other countries that import its fruits (dates) or other by-products. The world annual production of dates had peaked at 6.4 million mt 2003 (El Hadrami and El Hadrami 2009). Currently, this production is estimated at over 6.8 million mt (FAO Statistics 2010). According to this database, leading producer countries are Egypt, Iran, Saudi Arabia and United Arab Emirates, with almost two thirds of the world production (4.1 million mt/year). The annual trade export value of dates is estimated at almost USD 2 billion. The world's largest importer of dates currently is India, with over 250,000 mt and an expenditure of USD 70 million per year. On the other hand, the leading exporter country is Iran with over 240,000 mt, generating annual revenues of over USD 80 million. The most popular and appreciated cultivar is Medjool (originating from Morocco) and the most exported cultivar is Deglet Noor, from Algeria and Tunisia.

This desert species has also ecological value, because other crops can grow under its canopy and around the oases. This species provides local populations with raw material for many of their daily-basis uses and needs. Besides, this species represents an important botanical evolutionary link for monocotyledonous species.

Given the harsh conditions in the habitat of date palm trees, the crop faces many challenges with respect to soil, salinity, heat and water scarcity. Also, the plant is often grown in large plantations that require very little maintenance and inputs. Therefore, such an important botanical, ecological and economical species deserves a breeding

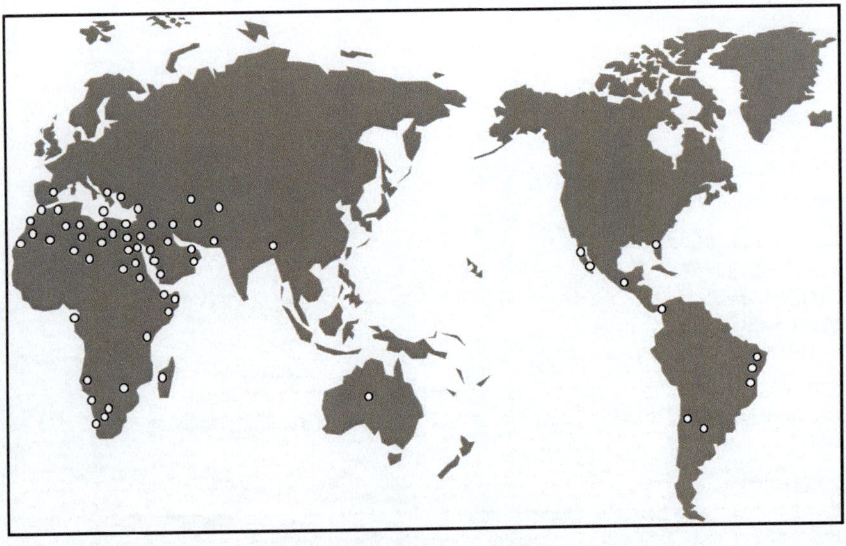

Fig. 23.1 World distribution of date palm

program in accordance with the needs of its market growth and development and of its productivity and date quality.

Date palm is traditionally propagated vegetatively through the separation of basal offshoots (suckers) from the mother trees followed by their planting in soil (El Hadrami and El Hadrami 2009; Zaid 2002). The species can also be propagated through seeds, which are not dormant and can immediately germinate after dissemination without requiring any scarification or treatment. Development of adult plants from seeds usually takes 5–7 years (Baaziz 2000; Zaid 2002).

Because date palm is a dioecious species, 50% of the produced trees could be either males or females. The distinction between the two genders is impossible before the adult stage. Only female trees are able to produce fruit after pollination. This process relies essentially on wind. However, to guarantee high productivity, artificial pollination is commonly practiced in commercial plantations. This consists of placing a portion of a male flower spike on the female inflorescence.

Vegetative propagation of date palm through offshoots preserves the genetic integrity of the cultivars. However, this method does not meet the needs of large industrial plantations because only a few shoots are produced by a tree over its lifetime (El Hadrami and El Hadrami 2009; Zaid 2002). Seed propagation is possible but faces difficulties in distinguishing female bearing trees among the seed-propagated material before they mature. To alleviate these challenges, an extensive effort has been given to rapid propagation of date palm through tissue culture (Beauchesne 1983; Beauchesne et al. 1986; El Hadrami et al. 1995, 1998; Omer et al. 1992; Tisserat 1979; Zaid 2002; Zouine and El Hadrami 2007). This clonal propagation of the species accelerated the establishment of large populations suited to commercial plantations using several high-producing cultivars. However, there are still problems arising from the multiplication of off-types due to variants observed during somatic embryogenesis or calli establishment (Al-Wasel 2005; Azequour et al. 2002; Djerbi 2000; McCubbin et al. 2004). In the last few years, a number of molecular markers have been developed to detect variants in tissue cultured-plants (Al Khalifah and Askari 2003; Azequour et al. 2002; El-Assar et al. 2005; Saker et al. 2006).

Despite the socio-economical and botanical values of date palm, and the extensive use of micropropagation techniques, its breeding is still lagging behind. This is mainly due to the slow growth of the plant and to the lengthy time needed to produce progeny. A number of advances have been made in terms of developing and controlling *in vitro* propagation techniques to reduce the generation time. An extensive body of knowledge has also been gathered regarding the characterization of local germplasm and the study of interactions with some of the crop's notorious pathogens. Recently, a sequencing project has led to the development of the first draft of the entire genome of this plant species. Progress has also been made using proteomic approaches to unveil differences between the zygotic and somatic embryos.

The current contribution reviews the challenges and opportunities facing date palm breeding and reports on the latest advances made towards the improvement of yield and quality as well as resistance/tolerance to abiotic and biotic stresses.

23.2 Origin and Diversity of Date Palm

23.2.1 Center of Origin and Domestication

According to numerous references, including Biblical texts, date palm is believed to be the oldest fruit-bearing tree domesticated. Representations of the tree appear in the hieroglyphics from ancient Egypt and the Neolithic civilizations from Mesopotamia. The earliest records are dated to 5,000–6,000 B.C. in Iran, Egypt and Pakistan, while early cultivation was found around 4,000 B.C. in Eridu and lower Mesopotamia, nowadays the Persian Gulf by the Euphrates River. Date palm is also mentioned in Akkadian and Sumerian cuneiform sources dated as early as 2,500 B.C. Historically, date palm covered areas stretching from the Indus valley (now Pakistan) to Mesopotamia in the Tigris/Euphrates valleys (now Iraq), to the Nile valley, Southern Persia (Iran) and the Eastern Mediterranean.

The center of origin of date palm is still debatable although claims mention that it originated from Babel in Iraq, Dareen or Hofuf in Saudi Arabia or Harqan and other islands of the Arabian Gulf in Bahrain. The oldest ^{14}C-dated seeds were found in Dalma, one of the Abu Dhabi Islands. Two seeds were found in 1998, the oldest was dated at 5,110 B.C. and the other at 4,670 B.C.

Beccari (1890) considered the center of origin of this species to be in the Persian Gulf, stating that "It is only in the orient, that the true home of the date palm is to be found; in the orient where the true center of formation of the genus *Phoenix* is located and where the wild species most closely related to the domestic one -namely, *Phoenix sylvestris* – exists." In more detail Beccari elaborates that the center of origin of date palm should be established once the distribution of the whole genus *Phoenix* is determined. This genus is believed to have originated from India but the morphological and physiological characteristics of date palm do not correspond to a species that has originated from this region. Date palm thrives in alkaline soils from sub-Saharan and sub-tropical regions where rainfall is scarce and an excess of soil moisture still exists around the roots. These specifics allowed Beccari to formulate his hypothesis that date palm originates from a region with a subtropical climate and scarcity of rainfall, close to the sea or some other salty water bodies that provide the required moisture and saline subsoil for which the palm is manifestly adapted: "This region can only be sought to the west of India, in Southern Persia or on the Arabian coast of the Persian Gulf". Although based on some interesting botanical, physiological and meteorological facts, Beccari's point of view is not definite because the current abundant cultivation of date palm in these regions does not necessarily mean that it started there thousands of years ago.

To animate the debate further on the center of origin of this species, Barrow (1998) examined the properties of the vegetative and reproductive systems of the *Phoenix* genus and conducted a phylogenetic study using molecular markers for the 5S intergenic spacer of the ribosomal DNA region, and published a monograph revising this botanical clade. These investigations allowed the author to establish a relationship between morphological and molecular data and links among the tested *Phoenix* species. She provided new data for the discussion of date palm origin and new taxonomic criteria that have never been considered in previous taxonomical or botanical studies. Barrow (1998) validated certain species described after Beccari's earlier monograph (1890). These include *P. caespitosa* Chiov. from Somalia, *P. atlantica* A. Chev. from the Atlantic Islands and *P. theophrasti* Greuter from Crete.

Similarly, Baker et al. (1999) established phylogenetic relationships among species belonging to the Palmae family using molecular markers from the chloroplast genome. The authors used a highly conserved DNA region, namely *trnL – trnF*, and focused on the few parsimonies present in the aligned data matrices to establish a number of relationships between species and genera belonging to this family. Their analyses highlighted congruencies as well as discrepancies with the current adopted classification of palms.

23.2.2 Evolution

The earliest documented fossils of palms are leaves of *Sabal magothiensis* and stems of *Palmoxylon cliffwoodensis* from the late Cretaceous, some 80 million years ago. A cross-section of a palm branch (*Palmae*) is the first, from a modern family of monocots, to be clearly represented in the fossil record. During the Eocene, nearly 60 million years ago, palms were widespread and abundant. A diversity of what is now present among modern genera was represented, including *Phoenix, Sabal, Serenoa, Livistona, Trachycarpus* and *Oncosperma*. These fossils have been found in locations where palms do not now occur. Today, species of *Sabal* are associated with hot climates and poor soils. Some of these genera are considered primitive while others seem to be from the origin of diversification of this monocotyledonous family. Because palms separated from the monocots earlier than other families, they developed more intra-familial specialization and diversity (Wrigley 1995). By tracing back these diverse palm characteristics to the basic structures of monocots, palms are valuable in studying the evolution of this botanical group.

The geographical distribution of palms, including *Phoenix*, is tied to many factors such as bird and other animal movements. Seeds carried in the stomachs of animals led to the dispersal of many palm species from their original zone to other locations where they now can be found. Date palm, on the other hand, is believed to have been dispersed through human activity. Successful establishment of the culture depended on temperature and moisture, as well as the presence of alkaline soils.

Wild stock of the earliest domesticated cultivars is believed to exist in the southern Near East (Zohary and Spiegel-Roy 1975). Other *Phoenix* species may have contributed to the development of current cultivars through natural hybridization. The earliest cultivation of date palm is recorded in 3,700 B.C. (Munier 1973) in the area between the Euphrates and the Nile rivers. From there it expanded to other areas where climatic conditions are suitable for the plant, mainly between the parallels 9–39° North latitude (Munier 1973) (Fig. 23.1) especially in dry and semi-arid regions.

The Moors brought date palm to Iberia, and the Spanish during their conquest and colonization introduced it into the Americas, especially Mexico, around Sonora,

Sinaloa, and Baja California. Later, date palm was introduced to drier parts of southern California and Arizona. Currently, date palm is found in temperate and arid regions as well as in subtropical and tropical countries in the Middle East, Africa, southern Spain and the Mediterranean coast of Africa, west Asia, USA, Australia and several countries in Latin America such as Jamaica, Brazil, Argentina, Peru and Venezuela, where the plant thrives in drier zones (Fig. 23.1). Hundreds of date palm cultivars are grown worldwide. Their fruits have different colors, flavors, sweetness, acidity and textures given the diversity of their growth conditions.

23.2.3 Diversity and Adaptation

The date palm growing belt stretches from the Indus Valley in the east to the Atlantic Ocean in the west (Zaid 2002). This broad geographic distribution of the species from the Near East and North Africa expanded to the Americas and Australia, affording opportunities for adaptation and diversification. Zaid (2002) examined the species distribution according to latitude, elevation and world abundance of date palms. On the basis of latitude, it was noted that the northern limit of date palm cultivation in Asia is at 32°N, in the Indus Valley. This northern limit continues along the edge of the Perso-Afghan Mountains, until it reaches 35°N in Iraq and then turns south-west to the Mediterranean Sea at the Gulf of Gaza. Along the North African Mediterranean coast, the date palm growing area expands from Tunisia to the southern edge of Morocco and on to the Atlantic Ocean. Date palm culture also occurs in Elche, southern Spain and in Turkmenistan at 39°N. The 17° north parallel represents the southern limit of date palm in the Sahara. From 15°N in Sudan, it follows the coast of the Red Sea and the Gulf of Aden, till it drops to 10°N to include the northern part of Somalia. This southern line continues until it reaches the coasts of Arabia and Pakistan to the limit of Indus River. Favorable areas are located between 24° and 34°N (Morocco, Algeria, Tunisia, Libya, Egypt, Iraq and Iran).

On the North American continent, date palms were introduced in Southern California (33–35°N) and some 7° farther south in Baja California and the mainland of Mexico. Other introductions took place into Jamaica, Brazil, Argentina, Venezuela and Peru. Date palm culture occurs also in Australia. Beyond these geographic limits, date palm will still grow, but does not bear normal fruit. As far as elevation is concerned, date palms are grown from 392 m below to 1,500 m above sea level (Zaid 2002).

Sedra et al. (1998) studied the genetic diversity among several date palm cultivars including 37 accessions from Morocco and 6 cultivars from Iraq and Tunisia. The study revealed a certain level of polymorphism using random amplified polymorphic DNA (RAPD) and demonstrated that morphologically related genotypes were clustered together. A lack of genetic relatedness was also detected among the tested Moroccan cultivars.

Recently, Elshibli and Korpelainen (2009b) investigated the genetic diversity and differentiation of date palm and used 200 individuals from 19 geographically-distinct populations in Sudan. Ten simple sequence repeat (SSR) markers were used to investigate the genetic diversity within and among the populations. Correlations were established between the genetic diversity level and the geographic distance. A high level of polymorphism was detected with an average of 26.1 alleles per locus (261 alleles for 10 tested loci). Although the Hardy-Weinberg test revealed that the populations were panmictic, fixation indices were quite high (–0.163), suggesting a high level of heterozygosity maintained among them. Ninety-five percent of the genetic variation was observed within the populations and a significant differentiation was shown between groups (i.e. soft vs. dry dates). Clustering of the populations based on geographical distances, although initially stated as an objective for the study, was revealed to be difficult due to the ways plantations are maintained (e.g. uncontrolled introductions of diverse material by growers and traders, heterogeneous seed dispersal, etc.).

23.2.4 Genetic Structure and Cellular Organization

Date palm is a diploid species with 2n = 2x = 28 chromosomes as first reported by Nemec (1910). Other studies reported numbers varying from 26 to 36 according to the cultivar and its multiplication, or the way the material was cytogenetically prepared (Al-Salih and Al-Jarrah 1987; Al-Salih and Al Rawi 1987; Al-Salih et al. 1987; Beal 1937; Ibrahim et al. 1998; Loutfi 1999; Loutfi and El Hadrami 2004). Sexual chromosomes carrying distinctive nucleolar heterochromatin were also revealed using chromocyanin staining (Siljak-Yakovlev et al. 1996). Characterization of the date palm genome size, using flow cytometry (Ouenzar 2003; Ouenzar et al. 2003) and the thale cress (*Arabidopsis thaliana* L.) genetic material for calibration, estimated the 2X DNA of the species to 0.51 pg (~490 Mpb). The recently drafted sequence of the entire genome estimates this size to be around 550 Mbp (Al-Dous et al. 2009). Either way, date palm genome size appears to be relatively small in comparison to other monocotyledons and perennial species.

In recent years, genomic approaches allowed the gathering of new data related to variation among sexual chromosomes in many eukaryotes. Currently, two main theories have been formulated to explain the antagonistic variation among these chromosomes. According to the first theory (Rice 1984) antagonistic polymorphism, in which an advantageous allele from one sex has a deleterious effect in the opposite sex, should be common on the female chromosome (X or W in female-heterogametic species) but rare on the autosomes. The second theory (Fry 2010) stipulates that sexually antagonistic variation exists also on the autosomes and the situation of the X chromosome is only a *hot spot* of such a variation. In date palm, the latest evidence revealed the existence of equal sex chromosomes in this species (Gorelick 2005). Sex is believed to be determined by an extra-chromatin on one of the male chromosomes (Siljak-Yakovlev et al. 1996). In addition, epigenetic cytosine methylation controls the evolution of such chromosomes along with the dioecy of the species (Gorelick 2003; Gorelick and Osborne 2002).

23.3 Date Palm Breeding

23.3.1 Date Palm Germplasm

The date palm genetic pool is rich with over 5,000 named cultivars. Depending on the country, these resources are more or less characterized. Several cultivars are no longer grown due to susceptibility to diseases or pests, or to low yield as compared to some of the elite cultivars with high productivity and/or tasty appreciated dates. This has raised concerns about the impoverishment and narrowing of date palm genetic diversity to guarantee homogeneous production. Worldwide, wherever date palms thrive, new commercial plantations are becoming monocultures of one or a few elite cultivars. In Tunisia, for instance, cultivation of the most marketable date cultivar, Deglet Noor, represents over 65% of the date palms on plantations. A similar trend is observed in Algeria and other countries. In southern Morocco, BouFeggous-ou-Moussa and Jihel are two appreciated cvs. in the Draa Valley, while Medjool is the cv. best adapted to conditions of the Tafilalet region.

Over recent decades, many studies have reported on the use of molecular markers to genetically characterize or fingerprint the local date germplasm of many countries (Al Khalifah and Askari 2003; Al-Ruqaishi et al. 2008; Elshibli and Korpelainen 2008, 2009a; Saker et al. 2006; Sedra et al. 1998; Soliman et al. 2003; Talaat and Al-Qaradawi 2009; Trifi et al. 2000; Zehdi et al. 2004a,b; Zivdar et al. 2008).

23.3.2 Concepts of Phenotype and Genotype

The genotype represents the genetic make-up and identity of a date palm tree that is specific to the cultivar. It is determined by the nuclear and extra-nuclear DNA that the cultivar carries. The phenotype, on the other hand, is the manifestation of the cultivar characteristics, depending on the genotype and its interaction with local environmental conditions. In addition, age can play an important role in this interaction, because all of the genes are not expressed at the same time. The phenotype evident at a younger age may not necessarily remain the same at a later stage of growth and development due, among other factors, to the dioecy of the species. Since date palm is a diploid species, three genotypes (AA, Aa, aa) and two phenotypes (A or a) are to be expected for any given locus. If the two alleles are alike, the genotype is considered homozygotic, when they are not alike, the genotype is heterozygotic. The levels of heterozygosity in date palm are estimated to be high, which leads to off-types, especially during tissue culture propagation. Other genetic phenomena such as dominance, co-dominance, additive/multiplicative effects, pleiotropy and epistasis may also occur during cultivar development and propagation.

Date palm is a dioecious species having male and female individuals on separate trees. During fertilization, recombination and other linkage disequilibrium events occur, leading to coupling and repulsion arrangements between loci that can be either synthenic or non-synthenic. Syntheny represents the way loci physically co-localize on the same chromosome of a given cultivar. It is related to the genetic linkage between loci, representing lower-than-expected recombination frequencies in panmictic populations. Theoretically, all linked loci are inevitably synthenic, but not all synthenic loci are linked. In genomics for instance, the genetic loci on a given chromosome are synthenic regardless of how they were brought together (i.e. sequencing and assembly, physical localization or mapping).

Syntheny is an important notion in breeding because it allows prediction of the co-localization of loci on chromosomes from related species used in a cross. Genetic events such as translocation that occurs during genome re-arrangements may result in a loss of syntheny among synthenic loci or in a gain of syntheny between non-synthenic ones. When syntheny exceeds expectancy, a selection is possible for functional relationships between synthenic genes. This is often explored to increase inheritance of alleles that would provide a substantial advantage when inherited in association rather than individually (i.e. improving the yield) or when the two alleles share the same regulatory mechanisms.

23.3.3 Inheritance and Genetic Linkage

To the present time, inheritance in date palm in not fully understood due to the unavailability of enough segregating populations with sufficient time-depth following their establishment by a series of crosses and backcrosses. No physical or linkage maps have yet been constructed for this crop. Physical mapping of plant chromosomes usually requires either classical or molecular physical mapping techniques or a combination of both. Classical physical mapping techniques rely on strategies such as the knobb mapping practiced in maize, deletion/substitution/aneuploid mapping or chromosome banding used in many other species. However, molecular physical mapping techniques use contour-clamped homogeneous electric field (CHEF) gel mapping, radiation hybrid mapping, large-insert clone libraries [Yeast Artificial Chromosomes (YACs), Bacterial Artificial Chromosomes (BACs)], Cosmids, Fosmids, fluorescent hybridization or mapping gene space. The latter was used recently to draft the genetic map of date palm that is still under construction as we were writing this contribution (Al-Dous et al. 2009).

The continuous phenotypic variation observed in a segregating population that is attributable to a genetic variation is called heritability. In breeding, this term is used *senso stricto* to qualify the proportion of the phenotypic variation attributable to genes with additive effects, and, in a more general sense, it refers to the proportion of the phenotypic variation linked to all types of genetic variation. Heritability is the summation of all variances due to additive effects of the genes, their dominance or epistasis as well as the variances due to the genotype x environment (G×E) interactions. These notions are important in conventional breeding to select better performing genotypes with desirable characters. For example, Salem et al. (2001) reported on the genetic inheritance of a selected set of isozymes in four date palms, and described five polymorphic loci involved with 12 alleles.

The genetic differentiation and structure of date palm is governed by many factors. Three of these appear to greatly affect the genetic differentiation of date palm populations and/or cultivar groups. These factors include (i) the geographic isolation of the species, (ii) the long biological life-cycle of the trees and (iii) the impact of the regional environment, where the plant thrives, especially with respect to aridity that often affects the organoleptic, textural and nutritional characteristics of the dates. Cultural practices are a very difficult parameter to trace and judge but have certainly played an important role in the making of the current genetic structure of date palm. This also includes offshoots trade or exchange that occurred among regions or countries. In addition, it is noteworthy to point out that growers' practices regarding the choice of well appreciated and/or adapted cultivars, along with male pollinators for their date palm trees, may have contributed to the genetic structure of date palms on a global scale. A few studies have already shown, using highly repetitive microsatellite markers, that date palm populations are structured into groups on the basis of their geographic location and the type of fruits they produce (Elshibli and Korpelainen 2008, 2009a; Zehdi et al. 2002).

The date palm draft genome released recently (Al-Dous et al. 2009) will soon lead to the establishment of a genetic linkage map for the species. Assembly of the scaffolds is still ongoing and physical and linkage maps are being built. Hopefully, this will stimulate and reactivate some of the breeding activities being abandoned in this crop.

23.3.4 Conventional Breeding

Conventional breeding activities in date palm are very scarce, especially over the past three to four decades. A substantial effort was carried out in the 1960s and 1970s in terms of assessing cultivars' susceptibility to bayoud (a vascular wilt –see below) and establishing targeted and non-targeted crosses to combine resistance to this disease with maintaining the quality of the produced dates. This effort has slowly vanished facing the long life-cycle of the plant that makes the establishment of segregating populations, backcrosses and recurrent selection an enormous burden to the whole breeding program. On the other hand, much more attention was given to vegetative clonal propagation and tissue culture to reduce the time required for improving traits such as fruit size and quality or tree height. Sudhersan et al. (2009) reported on the use of such a technique to assist conventional breeding in Kuwait, intending to reduce the height of the trees. In crosses involving tall female cvs. Barhee, Medjool and Sultana pollinated with dwarf *Phoenix pusilla* L., all fertilized female cultivars have set fruit, but the development of the embryos was stopped in the seed, leading to seedless fruits. The embryos were further rescued and used to regenerate adult plants *in vitro*. According to the authors, the produced plants were acclimatized and transferred into the field for more follow-up on their growth and development as well as their flowering capabilities.

23.3.4.1 Statement of Breeding Objectives

A successful breeding program usually relies on three pillars consisting of (i) identifying objectives for the breeding; (ii) assessing the current genetic variability among the germplasm and creating resources; and (iii) identifying and selecting new genotypes with traits that correspond to the initial breeding objectives. In date palm, the absence of a worldwide consortium makes breeding objectives regionally defined, depending on the locally-faced biotic or abiotic stress. Most breeding programs often concur in terms of their general objectives, given the arid and/or semi-arid habitat and the growth habits of the species. The selection of parents, creation of genetic variability, and evaluation and selection of progenies are continuously evolving and may vary from one program to another but the traditional breeding objectives always include yield enhancement, disease and pest resistance and tolerance to salt and drought. For instance, a tremendous effort has been made to assess the available variability among the local germplasm in many countries (Al Khalifah and Askari 2003; Al-Ruqaishi et al. 2008; Elshibli and Korpelainen 2009b; Saker et al. 2006; Sedra et al. 1998; Trifi et al. 2000; Zivdar et al. 2008). However, very little has been achieved in trait improvement due to the lengthy life-cycle of the plant. Given the recent intentions to use date fruits as a source for dietary fibers or as a functional food and nutraceutical, several considerations have emerged in terms of restating new objectives for the breeding programs. In the near future, it is going to be possible to breed date palm for an enhanced physico-chemical composition to provide dates on the basis of processing needs and health benefits, while improving storability and flavor and extending shelf-life.

Numerous constraints and challenges face date palm breeding programs. Some of these challenges are of worldwide interest while others are regional and country-specific. To respond to these constraints the objectives of the breeding programs are set according to short-, mid- and long-term goals. The short-term objective is to replace and provide date palm groves with juvenile material in order to sustain the perennial aspect of the culture. This goes along with the diversification of the genetic basis that has been narrowed in recent years due to the expansion of monoculture. The mid-term challenge represents the core of the breeding program wherein resistance to different biotic and abiotic stresses is pursued and the *date palm complex system* is studied. As for the long-term objective, the creation of new cultivars using conventional and/or non-conventional approaches would lead hopefully to resolve some of the date palm constraints, especially bayoud disease and red palm weevil, and respond to growing opportunities in terms of nutraceutical uses and understanding the evolution of monocotyledonous families.

Date palm breeding objectives depend also on producers needing to grow high-yielding and healthy palms for a long period of time with minimum input, and on market acceptability of dates. In this fruit crop, consumer preference in terms of color, taste, and texture of the fruit are a defining criterion for any successful improvement and selection of new cultivars. Acceptance of new products is always a challenging task and to succeed date palm breeding programs need to include an extension service that would assess the producers and public acceptance as well as the marketability of new cultivars, hence redirecting breeders toward new targets. The third party in guaranteeing a successful breeding program is the processors and their ability to efficiently develop products and uses from the raw materials whether it is the fruit or other plant parts. Having a variety of products and a growing market share usually enhances in return the initial breeding program. In date palm, this step is being reached and some attention has recently been drawn to the plant products outside of the producing areas. In this perspective, there is an urgent need to assess what is in the traits pipeline that is available to growers and commercially, to improve the value-added of date palm and attract substantial funding and investments that will give a boost to the breeding of this species.

23.3.4.2 Creation and Introduction of the Genetic Variation

To be able to improve a crop in general it is important to gather sufficient knowledge to characterize the germplasm and to identify genetic variations. Depending on the breeding objective, this genetic variation is explored to improve elite cultivars or create new ones. Traditional and artificial hybridizations along with induced mutagenesis are among the techniques used by breeders to generate new material. Recent technological advances have increased our knowledge about the palms genome, proteome and metabolome, and opened new opportunities to further expand the identification and creation of genetic variation. In date palm, conventional breeding, relying on crosses and backcrosses, has been almost futile due to the lengthy life-cycle of the species (over 30 years back to back). Recourse to biotechnologies, including various techniques of tissue culture (El Hadrami and El Hadrami 2009), have somewhat accelerated certain processes but there are still many challenges to overcome.

Genetic Variation Through Artificial Hybridization

Available genetic variation to date palm breeders is either intra-specific within the species *dactylifera* or inter-specific across species within the genus *Phoenix*. For instance, seeking dwarfing in date palm, Sudhersan et al. (2009) pollinated female cvs. Barhee, Medjool and Sultana with pollen grains originating from the dwarf congeneric species *P. pusilla*. The resulting progenies are currently undergoing selection and testing for the inheritance of dwarfism. Another source of genetic variation relies on the use of artificial hybridization across genera belonging to the same botanical family. Species in palm genera such as *Sabal, Serenoa, Livistona, Trachycarpus* and *Oncosperma* could also be included in future crosses if desirable traits are found in them and compatibility with *P. dactylifera* permits. Very little has been achieved so far in date palm using this approach. This is probably due to the fact that after artificial hybridization, there is a need for one or more reproductive cycles to allow the recombination between chromosomes from crossed genotypes, which in date palm will dramatically increase the length of the breeding process. However, this approach could be useful to create or improve parents to be used in crosses rather than in creating segregating populations.

As in many other breeding programs, the major focus in date palm is on creating and selecting segregating populations. However, creating and increasing variation (i.e. by artificial hybridization) among the parents is as important as selecting variants among the progenies. Often ignored or set aside, since the recognition of the use of elite cultivars in breeding, this technique could lead to enrichment of the germplasm and to the creation of extensive genetic variation. This along with the recurrent selection applied to segregating populations would ultimately increase the chances of selecting individuals that surpass the original parents and the currently deployed cultivars.

Genetic variation created through hybridization is a criterion that is often used along with agronomic performance to select parents and establish crosses. Although an appreciation of the agronomic performance could be measured and quantified, genetic variation remains, until now, at the discretion of the scientists's intuition in most breeding programs including the one of date palm. Given the fact that all loci are not equal, breeders often have recourse to a series of crosses in unlimited numbers to hopefully find interesting lines to be further developed into commercial cultivars. A concerted effort has been made during the last few decades in developing molecular tools that allow the identification of parents with the desirable traits within the germplasm. Polymorphic molecular markers such as restriction fragment length polymorphism (RFLP), amplified fragment length polymorphism (AFLP), simple sequence repeat (SSR) and single nucleotide polymorphism (SNP) have been developed and screened in recent years by many research teams. The advancement of new sequencing technologies had also made it possible to highlight SNPs that could differentially distinguish parents with distinct traits. For instance, with the newly drafted genome using Illumina™ technology (Al-Dous et al. 2009), 850,000 SNPs were shown to be potentially useful in breeding. Future data mining across the entire genome will yield the discovery of functions for various genes and establish strategies to manipulate them in order to improve cultivars of interest and create new ones.

Genetic Variation Through Biotechnologies

Creation of variation can also be achieved through artificial hybridization combined with other methods to include inter-specific and intra-generic variations. These comprise induced mutation using physical or chemical mutagens (Ahloowalia et al. 2004; Jain 2006), somaclonal variation and embryo rescue as well as other new mutant screening technologies not yet adopted to date palm (El Hadrami and El Hadrami 2009; McCallum et al. 2000) and cytogenetic chromosome manipulation and transformation (Fedak 1999; Jain 2006; Siljak-Yakovlev et al. 1996). Transformation offers endless possibilities for creating genetic variation and manipulating genes in order to clone them into cultivars of interest. However, public acceptance and the current debate on such technologies are still constraining these tools from achieving the expected results. Also, all these methods are far from being applied on a regular basis in date palm breeding due to other enormous challenges that still have to be overcome in terms of controlling the micropropagation of the plant and guaranteeing the fidelity of genotypes.

Genetic Variation Through Apomixis

Apomixis represents a cloning procedure of plants through seeds and can be a great supplement to creating variation within date palm breeding program. This approach, when successful, opens doors for improving the crop with new genetic variations from the natural large reservoir of wild relative species or after chemical treatment (i.e. with gibberellins). The approach has been explored through the application of GA_3 on non-pollinated female inflorescence (Ben Abdallah et al. 2001; Snoussi et al. 2001). This has led to the development of high fruit set. Seeds, obtained through GA_3 treatment, developed normally and contained viable embryos. Histological and isozyme investigations revealed the presence of homozygotes and heterozygotes among the apomictic plants, ruling out any potential effect of self-pollination. The authors reported also that additional RAPD data showed the existence of doubled haploids in the apomictic progenies (Ben Abdallah and Lepoivre 2000; Ben Abdallah et al. 2000; Ben Abdallah et al. 2001). However, the true-to-typeness has never been assessed. Using AFLP markers, Snoussi et al. (2001) showed that the plants derived from GA_3-induced apomixis were actually off-types and exhibited a distinct AFLP profile when compared to the mother plant. Although the authors never undertook further studies to characterize this material at the developmental and inheritance levels, one could hypothesize that these apomictic seeds had suffered from an epigenetic load. This often results from seed tissues of both the endosperm and embryo, exhibiting frequent defects due to unnecessary parental genomic contributions and/or atypical methylation patterns, as shown in other species. For instance in maize, where a good understanding of the genetic basis of apomixis has been made, the mapping of *Tripsacum* hybrids revealed that this genetic variation results from a single chromosome, able to reverse the phenotype if deleted. Inheritance studies showed, on the other hand, that female gametes inheriting this chromosomal segment were unequivalent in carrying the genetic information according to their origin; that is unreduced gametes transmit a functional segment and reduced ones reproduced sexually. The observation on apomixis in date palm could also be explained based on what was demonstrated in barley where chromosomal or genomic dosage showed variation, hence affecting apomictic phenotypes and suggesting independence from ploidy. Regardless, the plausible explanation clearly is that epigenetic information in apomictic seeds restricts their development and inflicts constraints against their propagation. The nature of the mechanisms and pathways involved has still to be determined. However, this explains why such an approach has not found its way into many breeding programs, including those on date palm.

23.3.4.3 Assessment and Use of Genetic Variation

Genetic variation of traits is either qualitative or quantitative. Breeding for Mendelian qualitative, non-discrete and discontinuous traits is easier in terms of selection while that of quantitative discrete and continuous traits remains a daunting task. A number of statistical criteria and parameters (i.e. gene frequency, gene action, heritability, recombination) are often calculated by breeders to estimate the genetic variation in the established crosses based on various morphological traits. These were used to develop the so-called chromosome linkage maps (Allard 1956). The arrival of polymorphic molecular markers [SCAR (Sequenced Characterized Amplified Regions), RFLP, AFLP, SSR and lately SNP] took the approach to a new level. The development of the so-called QTL (Quantitative Trait Loci) has made it possible to correlate trait variability with the genetic variability among the molecular markers segregating within a progeny (Tanksley 1993). Denser linkage maps have been published for many crops in recent years allowing for a more precise determination of QTLs associated with either agronomic traits or resistance/tolerance to stress. A date palm linkage map was far from being developed due to the time required for establishing progenies and conducting backcrosses and recurrent selection. The recent sequencing of the entire genome (Al-Dous et al. 2009), currently undergoing scaffolds assembly, will in the near future provide such a map and make it possible to calculate genetic and statistical parameters and determine QTLs linked to various stress factors and agronomical traits, including yield and fruit quality. According to the initial data arising from the draft sequencing, the date palm genome appears to be small in comparison with other monocotyledonous and perennial species. This should be an advantage in order to make up ground in understanding the inheritability and the genetics of this species.

Recent developments in cytogenetics and genomics have allowed for the acquisition of new evidence that explain the evolutionary features for many plant species including date palm. For instance, it has been determined that epigenetic changes, such as embryonic cytosine methylation, have contributed to the development and evolution in currently grown species of characters such as dioecy and sex chromosomes (Gorelick 2003; Gorelick and Osborne 2002). In date palm, a dioecious species, sex chromosomes appear to be homomorphic. The presence of an extra-heterochromatin on one or both arms of the male chromosomes is thought to be a sex determinant (Siljak-Yakovlev et al. 1996). This could be used to manipulate chromosomes to guarantee the creation of female-producing trees.

Similarly, Swingle (1928) reported on an interesting genetic phenomenon that occurs in date palm at the time of maturity of the fruit. The author noticed the so-called metaxenia, where the pollen laying outside of the embryo sac influences the development of the maternal structures, including the seed coat and pericarp. The time of maturity and the size of the fruit were interrelated with the type of pollen used in the fertilization. Having no clear explanation, the author hypothesized that metaxenia results from the fact that the embryo and/or endosperm release hormones or hormone-like substances that diffuse into the wall of the seeds or the fruits, exerting specific effects according to the particular male parent used in the cross. Knowing that differences in size and shape of the fruit can be due to factors such as the number, state of maturity and the genetic make of the seeds they contain, Swingle's explanation has yet to be experimentally tested but seems to have some merit. Unlike other species, where the fruit results from the transformation of more than one carpel, hence leading to variation in shape, flavor, color, etc, date results from the transformation of a single carpel. In addition, endosperm tissues often lack uniformity and can conceivably lead to observed variations. Whether it is due to hormonal or non-hormonal effects, metaxenia represents a good source of genetic variation to be explored in date-palm breeding.

In plant breeding, maximizing the genetic variability is the key to successful development of cultivars with new agronomic or resistance/tolerance traits. The difficulty is to dissociate advantageous variability from the disadvantageous one, generated by mechanisms such as peliotropy, epistasis and G×E interactions. Such a difficulty led many breeding programs to subdivide their activities into an aspect of cultivar development, with plants having mainly desirable characters, and another aspect with an objective of enriching the germplasm with genotypes still having a large number of undesirable and wild traits. These two categories usually have their importance in the short versus long-term objective of breeding including in date palm. Although thus far there is no official consortium or organization to manage the preservation of the worldwide germplasm of date palm, several regional programs are active in assuming their share of the responsibility in rescuing cultivars under threat of extinction and preserving their propagation, through either vegetative or biotechnological methods.

Unlike crops intended for intensive cropping systems, for which the breeding programs are mainly private and benefit from substantial funding opportunities and technological advances, horticultural crops such as date palm predominantly rely for their breeding efforts on the public sector and often universities. Intermittently, cooperative collaborations occur but have never been sustained for a long period of time. Date palm breeding over the last few decades has lacked sufficient funding and private investments, which have been triggered, after the establishment of the first crosses, by the lengthy life-cycle of the plant, and the time necessary for establishing progenies and conducting backcrosses and recurrent selection. Most of the recent published data on the species have focused on advancements in tissue culture (El Hadrami and El Hadrami 2009; El Hadrami et al. 1995, 1998; Zouine and El Hadrami 2007) and the development of various molecular markers to either study the diversity within local germplasm (Al Khalifah and Askari 2003; Al-Ruqaishi et al. 2008; Elshibli and Korpelainen 2009b; Majourhat et al. 2002; Saker et al. 2006; Sedra et al. 1998; Trifi et al. 2000; Zivdar et al. 2008) or screen for somaclone variants (Al-Wasel 2005; Saker et al. 2000). The influx of genetic data arising from these studies, along with the newly available drafted genome (Al-Dous et al. 2009), will make it possible to re-elicit investments in the breeding of this species, and to botanically and evolutionary investigate its genetic constitution and understand its interaction with the environment and pathogens and pests.

23.3.4.4 Identification and Selection of Progenies

After defining breeding objectives and targeting the genetic variation to be included in the bred material, a cross can be made between parents harboring desirable traits. A selection is then effected from among the progenies on the basis of the so-called selection units, which for date palm correspond to the individual seedlings in the case of conventional breeding and calli or cell suspensions or seedling lots in the case of biotechnology. The level of inbreeding, representing the generation of selection (i.e. the number of backcrosses or re-culturing cycles *in vitro*), as well as the selection criteria to be included in the process have to be determined according to the stated objectives and rigorously applied, without high pressure at the early stages of selection, especially for traits with low inheritance. Selection criteria may include the assessment of selection efficiency and genetic gain; establishing selection indices (i.e. Smith-Hazel index, Bernardo 2002), conducting an indirect selection or assessing the genotypic stability. Baenziger and Peterson (1992) determined the concepts of genetic gain and heritability to improve selection efficiency. Given that the expression of a particular genotype (G) into a phenotype (P) is dictated by the genetic constitution of the genotype, the environment (E) in which the expression and interaction (G×E) occur, additive effects are observed as follows: $P+G+E+(G\times E)+\text{Error}$. Heritability in its broad sense is defined as the genotypic variance divided by the phenotypic variance. *Sensu stricto*, heritability represents the additive effect that results from genotypic variance, due to additive effects of dominance, pleiotropy and epistasis, divided by the phenotypic variance. In either case, protocols that allow for minimizing the error due to environmental or other external conditions are often adopted in breeding to better appreciate the variation due to the genotypic expression.

Increasing Selection Efficiency

Increasing the selection efficiency can be achieved through the modification or deliberate choice of a selection environment that would allow certain selection units to stand out from the rest of the population. This is often practiced under controlled conditions to screen for resistance to pathogens and pests. However, results from such selections must be validated using field experiments. Most date palm screenings for resistance to disease and pests as well as to tolerance to drought and salinity have so far been conducted using *in vitro* material (El Hadrami and El Hadrami 2009; El Hadrami et al. 2005) or seedling grown under controlled or nursery conditions. This often allows for a better control of the environment and maximizes the efficiency of the selection process. Few reports exist on the use of field conditions to increase selection efficiency.

Indices-Based Selection

Selection based on indices is often used to select for multiple traits and/or operate multiple types of selection within a selection unit. The Smith-Hazel index is the most commonly referred to in plant breeding (Bernardo 2002); however, other simplified indices have been developed according to the specifics of each crop undergoing breeding and selection. In date palm, only a few crosses have been made and maintained over the years to allow any development or comparison of selection indices. However, in many countries, surveys of local germplasm have used various criteria such as the morphology of vegetative and reproductive parts of the plant (i.e. El Houmaizi et al. 2002) to make associations between cultivars with closely-related traits or establish indices that could be used in selection.

Other Methods of Selection

Recourse to indirect selection is practiced when the main interest trait is not the selection criterion. Instead, another criterion tied to the primary one is selected for. This strategy is often used in breeding programs intended to enhance yield; yield being a complex and multi-factorial trait to select for. Among the indirect selection techniques, the marker-assisted selection (MAS) is one that combines the principle of indirect selection with that of the selection based on indices. This method requires close linkage between the molecular marker used and the gene/QTL controlling the inheritance of the desirable trait. In successful cases, the marker has been the gene itself or a DNA sequence within the gene (Ellis et al. 2002). Since its introduction, the notion of MAS has not yet yielded any immense success despite the increasing number of high-resolution QTL maps, higher throughput marker technology and an integration of functional genomics with QTL mapping (Collard et al. 2005). The usefulness of such a technique to improve polygenic traits in the long term has even been questioned in many instances (i.e. Bernardo 2001). On the other hand, it remains a useful tool to routinely fingerprint commercial cultivars and even accelerate gene transfers from or into them during backcrosses. This technique represents also an important advantage for breeding programs focusing on gene pyramiding to develop polygenic and durable resistance to pathogen and pests or improve yield.

An interesting cost-to-benefit comparison has been recently conducted within the breeding program of a sister plant to date palm, the African oil palm (Wong and Bernardo 2008), between the use of marker-assisted recurrent selection (MARS) and an approach calling for a genome-wide selection. In this study, the authors revealed that for a segregating population of as small as 50 trees, genome-wide selection reduced the time necessary for screening and was cost effective.

Whether it is a direct or an indirect selection operated against a single trait or an index of complex traits, breeders in general and for date palm in particular, intend to select for an overall genotype mean performance. Consistency of this performance across environments, i.e. genotypic stability, represents one of the time-consuming tasks in breeding, and in date palm this represents a lengthy process.

Fig. 23.2 Various steps of production of date palm through somatic embryogenesis micropropagation. (**a**) date palm basal offshoots *a.k.a.* suckers; (**b**) Ex-plants used to initiate calli; (**c**) Regenerative calli on culture media; (**d**) Close shot on embryogenic calli; (**e**) Vitroplants regenerated by somatic embryogenesis; (**f**) Shoot induction of vitroplants on culture media; (**g**) Root induction; (**h**) Potted young seedling regenerated from tissue culture; (**i**) Adult date palm trees

23.3.5 Biotechnologies

Propagation of date palm by tissue culture (Fig. 23.2) began three decades ago (Reuveni 1979; Reynolds and Murshige 1979; Tisserat 1979). Due to the inherent habit of the species to grow slowly and the inadequate resources available in places where the date palm culture was given attention, micropropagation progressed at a relative snail's pace (Al-Khayri 2007). It took nearly a decade from initial investigation of date palm tissue culture to the publication of reports describing the regeneration of seedlings and plants through indirect somatic embryogenesis or organogenesis (Benbadis 1992; Omer et al. 1992; Tisserat 1984). The ability to regenerate date palm directly without calli was also investigated by Sudhersan et al. (1993), but has never been taken to a mass-production level.

Somatic embryogenesis exploration began to micropropagate date palm in the early 1990s (Al-Khayri 2005; Bhaskaran and Smith 1992; El Hadrami 1995; El Hadrami and Baaziz 1995; El Hadrami and Coumans 1994; El Hadrami et al. 1995).

Studies are still ongoing to examine the impact of various components used in the culture media (Al-Khayri 2001, 2003; Al-Khayri and Al-Bahrany 2001, 2004a; Bekheet et al. 2001; El Bellaj et al. 2000; Veramendi and Navarro 1996). Progress on generation of cell suspension, somatic embryogenesis and embryo rescue have also been made in recent years (Al-Khayri 2002; Al-Khayri and Al-Bahrany 2004b; Fki et al. 2003; Zouine et al. 2005). Difficulties are still being encountered in protoplast research. Two fields are currently gaining heightened interest: (i) the production of synthetic seeds (Bekheet et al. 2002) and (ii) the use of tissue culture to induce mutation and increase variability of the germplasm (El Hadrami et al., Chap. 9 in this book; El Hadrami and El Hadrami 2009; El Hadrami et al. 2005; Jain 2006, 2007, 2009; Jain and Gupta 2005). However, progress thus far in this regard has been modest.

Biotechnologies represent an important component of date palm breeding and provide a major support to the conventional breeding in terms of creating and introducing genetic variations and the selection of fast-propagating new genotypes with traits of interest.

23.4 Breeding Challenges and Opportunities

A numbers of challenges face date palm breeding with respect to agronomic traits, cultivar replacement, enhancing yield, quality and resistance/tolerance to abiotic and biotic stresses. On the other hand, many opportunities are also open to the breeding of this species to provide nutraceutical and health-beneficial substances. Recent genomic and proteomic studies have also expanded the body of knowledge regarding the genetic constitution of this species and its evolutionary progress from domestication to the present.

23.4.1 Breeding for Yield and Quality

Providing enhanced production and well adapted trees to growers remains the most important focus for date palm breeding, as in many other crops. Often this should be supplemented with specific recommendations that allow farmers to grow the new material under the best conditions, which adds a dimension to the breeding program; that is farm management. In date palm, this is still lacking and apart from some scarce extension services provided in certain countries where the crop is grown, management practices have not yet been standardized. Cultural practices are currently available (i.e. FAO) but often not properly used in groves and plantations. Extension programs to provide training and to raise farmer awareness about the issues and challenges facing their crop are matters that need to be addressed in a near future.

In all likelihood, it is the highest-yielding and best-adapted date palm germplasm that has a chance to be deployed among producers. This elite material has had consumer acceptance and remains also the target for date palm breeders to introduce any changes in agronomic or resistance/tolerance traits. Yield and fruit quality then represent the utmost priorities. Several elite cultivars are currently grown in large plantations worldwide because of their high-yield potential and fruit quality. A single mature fully productive tree from these cultivars can bear up to ten clusters, carrying over 100–200 kg of dates. Medjool cv. is susceptible to devastating bayoud disease and can no longer be grown in zones where the disease exists (e.g. Morocco, Algeria). Replacement of this cultivar with others that show resistance has been hampered by the poor quality of their fruit. Breeding initiatives began several decades ago to incorporate these resistances into cultivars with a better fruit quality. Crossing and backcrossing to stabilize resistance and fruit quality traits are lengthy and often delay the breeding progress. Nevertheless, in Morocco several interesting cultivars are being selected and propagated to replace the declining Medjool cv.

23.4.2 Breeding for Dwarfism

Reducing plant height is one of the major targets of date palm breeding. Conventional breeding techniques combined with the use of tissue culture techniques were recently implemented to study the outcome of crosses between tall date palm cultivars and a dwarf palm belonging to a congeneric species in Kuwait (Sudhersan et al. 2009). Female date palm trees from the cvs. Barhee, Medjool and Sultana were manually pollinated using male pollen from the dwarf date palm (*Phoenix pusilla* L.). All fertilized trees from the three tested cultivars achieved fruit set. The fruits were seedless and arrested in development leading to embryo abortion. Immature hybrid embryos were tentatively isolated and sub-cultured *in vitro*. Plantlets were regenerated from these embryos and rooted to be acclimatized and planted in the field. Once the progenies of these crosses are evaluated the impact of dwarfism on the various date palm criteria will be assessed. An establishment of the inheritance of dwarfism trait will be also examined.

23.4.3 Breeding for Tolerance to Abiotic Stress

Date palm growth and development, yield and quality are affected by the adverse effect of many stress conditions such as scarcity of water and drought, erosion, salinity and excessive heat among other abiotic constraints. These unfavorable conditions usually alter many physiological and metabolic processes such as the reduction of the net photosynthetic activity, the disequilibrium of the C/N ratio, and the drop in pigment contents. A decline in photosynthesis often leads to an increase in the dark reaction at the level of photosystem II, hence reducing the overall biomass and yield.

Some questions arise from breeding date palm cultivars for abiotic stress tolerance to the satisfaction of growers and consumers, which may ultimately have an impact on commercial aspects. Will stress-tolerant cultivars achieve a better yield in comparison to the traditional? Does the phenotype or the genotype of the stress-resistant/tolerant cultivar differ too much from the local one? Will these cultivars be able to withstand biotic stresses for the benefit of farmers without having any adverse impact on yield and quality? Will farmers need to adopt specific agronomic practices to grow abiotic stress tolerant date palm cultivars without loss of stress tolerance and popularity among growers?

In an era where powerful tools and modern approaches in molecular biology are flourishing to functionally study the expression of genes involved in stress (Cushman and Bohnert 2000), it has become conceivable to precisely map and manipulate genes that are implicated in stress-responses within breeding programs and eventually transfer them into crops of interest. In model plant species such as the thale cress *Arabidopsis thaliana*, hundreds of genes have been associated with resistance to drought and water stress and have been annotated and sub-cellulary localized under various environments (Seki et al. 2005). A large gap remains, though, between translating these molecular findings into an improved crop ready for use by farmers, especially date palm producers. In addition to the time spent over the past two decades acquiring the basic knowledge using transgenic approaches, and understanding plant responses to drought and salt stresses, another lengthy time period is still required to apply these findings in the field.

23.4.3.1 Drought

The date palm is a desert plant and most cultivars grown exhibit some level of tolerance to drought and water stress (Al-Khayri and Al-Bahrany 2004b). New cultivars released from conventional and non-conventional breeding programs are usually screened for drought tolerance (Al-Khayri and Al-Bahrany 2004b; El Hadrami and El Hadrami 2009). Elshibli and Korpelainen (2009b) examined the response of several date palm cultivars, under greenhouse conditions, to water stress in terms of morphological and physiological changes, alteration of gas exchange and photosynthetic activity. Dry and soft types of date palm were subjected to four regimes of water stress including 10%, 25%, 50% and 100% of field capacity, under a stepwise varying concentration of CO_2. Their results showed that water stress reduced the photosynthetic capacity and led to an alteration of the growth and morphology. Interaction between water levels and the elevated CO_2 concentrations was also detected.

23.4.3.2 Salinity

El-Sharabasy et al. (2008) evaluated salt tolerance of three *in vitro*-grown date palm cvs. Samani, Sewy and Bartamuda at different concentrations of NaCl (0; 4,000; 8,000 and 12,000 ppm). Differences in the responses of the cultivars were revealed, and most of the tested cultivars were tolerant up to 4,000–8,000 ppm. At 12,000 ppm, signs of toxicity were detected, resulting in leaf necrosis.

Al Mansoori et al. (2007) screened four date palm cultivars, derived from calli generated from immature embryos, against NaCl during dedifferentiation and the exponential growth phase when cell division is at its maximum. The results showed calli more efficient in screening salt tolerant lines. A complete inhibition of embryonic activity accompanied by a slow growth of calli was detected on media containing 3% NaCl. Both stages exhibited excessive dehydration with a dramatic increase in Na^+ and proline content and a reduction of K^+ levels. Furthermore, free proline content was higher in micropropagated cultivars with a higher rate of calli induction.

23.4.4 Breeding for Resistance to Biotic Stress

Date palm is host to a number of diseases and pests, some of which have a devastating economic impact and threaten the sustainability of the crop. Others are still not well characterized and little is known about their interaction with date palm. Breeding date palm for resistance to diseases and pests is still at a juvenile stage due to the lack of funding and investments to sustain long-lasting phases of establishing progenies and making backcrosses and recurrent selection within conventional breeding programs. Creation of resistant cultivars through biotechnology is also still challenging due to the peculiarities of the species. However, efforts have been made, over the years, in terms of studying plant *x* pathogen interactions and preparing the ground for future breeding for resistance.

According to published data, and our own research experience with date palm, several clones with interesting traits have been obtained through targeted and non-targeted genetic crosses (El Hadrami and El Hadrami 2009; El Hadrami et al. 1998). However, most of the attempts to establish date palm crosses and progenies with a good level of resistance to pathogens and pests have failed or been abandoned for several reasons. First, often introducing resistance is dependant on maintaining date fruit quality. Second, a high number of backcrosses to stabilize the selected resistance traits and to recurrently restore fruit quality is needed, which is incompatible with the lengthy life-cycle of the species. Third, sex determination is not possible until the plant reaches maturity (5–8 years), reducing the final number of producing trees at the end of a selection program. Last, continuity of research programs on this horticultural crop is poorly funded and often pursued on restricted budgets.

23.4.4.1 Breeding for Resistance to Bayoud Disease

Bayoud is a devastating vascular disease in date palm that originated in Morocco back in the 1870s. It is a fusariosis caused by *Fusarium oxysporum* f. sp. *albedinis* that led to the near extinction of the highly-valued cv. Medjool inNorth Africa. Most breeding for resistance against this disease has been conducted in the country of origin, Morocco, and to some extent in Algeria and Tunisia.

Carpenter and Ream (1976) reported on a breeding program active until the 1980s with three main objectives: (i) to develop stable genotypes able to propagate through seed and produce true-to-type plants, (ii) to create male trees genetically not differentiated and closely related to female trees showing promising traits and (iii) to make inter-varietal crosses among the progenies of the most interesting clones in order to advance the technological and commercial values of the culture. From this program only a limited number of genotypes survived three to five backcrosses and retained the mother plant traits. A high level of heterozygosity and lack of genetic stability and fidelity were also observed, making it difficult and almost impossible to pursue breeding through this method.

In Morocco, initial selection programs were carried out in 1949–1956. The main objective was to select date palm clones with a certain degree of resistance to bayoud disease through targeted crosses. However, all recorded results were lost during the war period 1956–1962. After Moroccan independence, date palm selection was back on track and benefitted from the international research subsidies and collaboration. The program followed a number of developmental steps. First, a large survey was conducted to assess the local germplasm. Second, resistant cultivars with good fruit quality were selected from the natural date palm plantations and used for controlled crossings.

In parallel, bayoud resistance testing was conducted in an experimental naturally-infested station in Zagora, Morocco from 1964 to 1972. These field trials led to the selection of date palm cultivars possessing various levels of resistance including nearly complete resistance in six cvs. originally found in the Draa Valley and in the region of Bani. Namely these cvs. were: Bousthami Noire, Bousthami Blanche, Iklane, Tadment, BouFeggous-ou-Moussa and Sair Layalet. None of these cultivars was recorded in either Tafilalet or in eastern Morocco. Other cultivars were shown to be partially resistant or completely susceptible (Jihel, Medjool) (El Hadrami and El Hadrami 2009; El Hadrami et al. 1998; Toutain and Louvet 1972). It is also noteworthy that among the cultivars tested for resistance at the same experimental station were two Allegs and two Deglet Noor cvs. from Tunisian germplasm, which were all scored as highly susceptible. Unfortunately, no association between good fruit quality and a substantial level of resistance was recorded in these screening trials (El Hadrami et al. 1998).

The next step was research conducted in the 1970–1980s with a focus on multiplying the resistant cultivars and establishing targeted and non-targeted crosses to re-establish the quality of dates in the resistant clones. Extensive testing under controlled conditions and in the field was conducted until the 1990s, where most of the focus was redirected toward accelerating multiplication processes of the clones through *in vitro* techniques (El Hadrami 1995; El Hadrami et al. 1995).

Similarly, several other investigations to develop commercial cultivars of date palm with various traits, including resistance to bayoud, are tentatively being carried out in Algeria and Tunisia.

23.4.4.2 Breeding for Resistance to the Red Palm Weevil

The red palm weevil, *Rhynchophorus ferrugineus* Oliv., is one of the most devastating Coleoptera in date palm growing regions (Al-Ayedh 2008). Yield losses to this pest can be dramatic in severe cases and the tree killed. Larvae are the most damage-inflecting stage of the coleopterum because of their feeding habit. The quality and health state of date palm trees also play an important role in the establishment, survival and development and reproduction of this pest. Very little is currently known about the host *x* pest interaction, and most of the research effort has been made to eradicate or establish control measures to minimize the red palm weevil impact on date palm yield and productivity.

Recently, Al-Ayedh (2008) investigated the capabilities of four date palm cultivars in nurturing the red weevil for two consecutive generations. The cvs. were Khalas, Sukkary, Khasab and Sillaj. Overall, the study revealed differential responses among the tested cultivars in terms of the pest adult lifespan and the production of cocoons, and showed no effect on the ratio male to female among the population of the pest. Of interest, the author reported that on Sukkary, the pest established and developed well and this was translated by the increase in the measured parameters in the study (i.e., length, width and weight at larval, pupal and adult stages). In addition, *R. ferrugineus* was able to lay eggs in numbers that were significantly higher than the other tested cultivars, most likely, according to the author, to the higher sugar content of the cultivar.

Breeding for resistance to the red palm weevil will definitely be promoted in the upcoming years due to the recent spread of the pest to non-infested areas, and to its development on other palm species. This will also benefit from the recent development in terms of genomic and sequencing of the entire genome of date palm (Al-Dous et al. 2009).

23.4.5 Challenges and Opportunities on the Rise

Although there are many challenges facing date palm breeding, new opportunities have been emerging in recent years. These are in relation with the high flux of genetic data acquired or to be gathered in the upcoming years regarding the genome, proteome and metabolome of this plant species. This will open doors to new selection techniques (i.e. molecular markers-assisted approaches; genome-wide screening approaches) to be adopted in the breeding program as well as in terms of innovative development of cultivars through strategies such as synthetic seeds, harboring somatic embryos, to fulfill the demand of large date palm plantations. Recent claims about the health benefits of dates consumption have also drawn some attention to the species as a source for functional food and nutraceuticals.

23.4.5.1 Physical and Genomic Mapping

A draft version of the date palm genome was recently made available (Al-Dous et al. 2009). This effort benefitted from the use of a whole genome shotgun second generation of sequencing Illumina's genome Analyzer II with paired-end reads. The cv. Khalas was used for sequencing and that should lead to the understanding of date palm genetics, reproduction and yield as well as fruit quality and resistance to biotic and abiotic stress. The approach that was taken during the sequencing took advantage of the lower repetitive DNA in the date palm genome and allowed for a 1,000X increase in publically available knowledge about date palm genes space. Thus far, a 20X coverage of the genome has been made. Early estimates place the date palm genome at around 550 millions bp with the availability of 850,000 new SNP markers to differentiate parental alleles. Manual inspection of the contigs suggested the correctness of the assemblies involving scaffolds of more or less 12,000 bp. Over 19,000 genes have so far been fully or partially predicted from the functional annotated of mRNA or proteins. Al-Dous et al. (2009) reported also on the similarity of the draft version of the genome with those of monocotyledonous species such rice and papaya. Genome assembly and annotation is still ongoing to validate longer range contiguity and gaps that may be uncertain in this first sequencing. The sequencing and annotation information are available online (Al-Dous et al. 2009). Efforts are continuing to produce draft linkage and physical maps taking advantage of the Expressed Sequence Tags (ESTs) and the traditional whole-genome mapping methods.

23.4.5.2 Marker-Assisted Selection

Although some still question the long term usefulness of marker assisted selection as a technique to improve polygenic traits (Bernardo 2001), a few studies have been published on its implementation in date palm breeding. El-Kharbotly et al. (1998) reported the use of AFLP markers to study two F_1 and F_4 back-crossed populations to tentatively map traits related to the fruit and seed weights, seed volume and the total soluble solids (TSS), as well as other measured morphological characters. The authors first used the AFLP markers on parental clones and tested 35 primer combinations to detect polymorphic markers to further construct a linkage map. *In vitro* regeneration of their material seemed to have been a limiting factor to achieve this objective. Hela et al. (2000), on the other hand, constructed a random genomic library of Tunisian date palm varieties using RAPD sequences. The recombinant DNA included fragments ranging from 200 to 1,600 bp inserts. According to the authors, this library could potentially be used to anonymously probe southern hybridizations and in the development of markers associated with bayoud resistance and sex determination of the phylogenetic trees.

For future use in date palm breeding, it is worth mentioning the findings of Wong and Bernardo (2008), who compared the costs and benefits of using marker-assisted recurrent selection to another selection technique, based on the genome-wide screening, in an African oil palm breeding program. The authors found that in selection units as small as 50 segregating progenies, marker-assisted recurrent selection was more costly and time-consuming as compared to the other considered selection technique. Further considerations should be given to the advantages and disadvantages of using such a technique in date palm breeding, once large segregating populations are established and maintained over time.

23.4.5.3 Genome-Wide Selection Approaches

Now that the draft genome of date palm has been made publically available (Al-Dous et al. 2009), an extensive effort is yet to be made in order to functionally characterize and annotate a wide array of gene networks and understand their interplays and regulatory mechanisms. A massive flow of data is to be expected, which will consequently serve the objectives of the breeding program in the short- and the long-terms. A genome-based strategy for screening date palm progenies would be advantageous in order to reduce generation time, the high cost of maintaining breeding populations and to minimize the size of the selection units. Wong and Bernardo (2008) compared such a method to others, used in an African oil palm breeding program, and found that the genome-wide selection procedure was able to increase the selection gain per unit time *x* cost, using small populations.

23.4.5.4 Proteomic Approaches

Somatic embryogenesis represents an alternative for date palm micropropagation (El Hadrami and El Hadrami 2009; El Hadrami et al. 1995, 1998; Zouine and El Hadrami 2007). Somatic embryos lack dormancy as compared to their zygotic counterparts and have been used to study many aspects of plant growth and development. However, their deficiency in seed integuments and an endosperm, which are both required for seed survival and germination, often lead to a diminished vigor. A recent study in date palm compared the proteome map (pH 5.0–8.0) of somatic and zygotic embryos derived from the commercially grown cv. Deglet Noor (Sghaier-Hammami et al. 2009). Using earlier knowledge acquired through SDS-PAGE and EDMAN sequencing (Sghaier et al. 2008) and 2-D combined with mass spectrometry analysis, they revealed qualitative and quantitative differences among the two tested embryos. Total protein content was higher in zygotic than in somatic embryos. Among the differential proteins identified in somatic embryos there was a series of enzymes involved in the glycolysis pathways. On the other hand, in zygotic embryos, storage and stress-related proteins were most differentially represented.

Development of proteomic approaches will yield an understanding of the differences between somatic and zygotic embryos and open the door for possible creation of synthetic seeds, harboring somatic embryos and multiplied vegetatively.

23.4.5.5 Readiness to Nutraceutical and Health Benefit Demands

Date fruit consumption has been practiced for centuries by local populations where the culture thrives. Exports have also made the fruit popular among people of importing countries. Recent advances in determining the fruit physico-chemical composition and nutritional value have yielded many reports, assigning to dates antibacterial, antifungal, antitumor, antiulcer and immuno-modulatory properties (El Hadrami et al. 2010). These health benefit claims, along with the extensive attention drawn by private and public sectors to natural products in order to develop functional food and nutraceutical uses, should be seen in the near future as increasing opportunities, for which a response from the breeding program has to be formulated. One would expect to see a definition of the breeding objectives specifically for this new expanding sector of research and development. For instance, dates are very rich in secondary metabolites such as phenolics that are widely known for their antioxidant and active oxygen and radicals-scavenging properties. The metabolic pathways leading to these compounds are well studied in many model and cultivated plant species such as the thale cress, legumes, wheat and potatoes and could be targeted for future modifications through breeding to increase the content of specific substances with a medicinal, antioxidant or nutritional value. The available sequence of the genome will definitely accelerate the cloning of homologous genes involved in the biosynthesis pathways of these metabolites. If successful, this strategy will also have a substantial impact on other activities of a breeding program such as incorporating resistance to bayoud disease, since we have shown in recent years the involvement of some specific constitutive and induced hydroxycinnamates in the resistance to the disease (Daayf et al. 2003; El Hadrami 2002; El Hadrami et al. 1998; El Hassni et al. 2004a, b, 2007; J'Aiti et al. 2009).

23.5 Concluding Remarks and Future Prospects

It is quite evident that the conventional date palm breeding is a time-consuming process and constitutes a laborious task. The biotechnologies and biochemical tools, developed over the last few decades, have allowed date palm breeders to overcome some of the hurdles encountered and to evaluate new clones with interesting agronomic and resistance traits. The gender of produced trees was and still is one of the most sought characters, to be determined as early as seedlings start their growth and development. Knowing gender early is of a great importance in order to shorten the selection process of date-producing palms. Given the lack of an international consortium to manage date palm germplasm, international cooperation should be promoted among research groups and institutions. Cooperation with the private sector needs attention in order to strengthen date palm breeding programs. This will certainly help meet the increasing demand for dates worldwide and promote their industrial food uses such as in health products and as a potential source for dietary fibers, and functional food and nutraceuticals.

Biotechnology applications such as genetic transformation, together with conventional breeding programs will form a basis for development of new date palm cultivars with high yield and superior fruit quality, and resistance to biotic and abiotic stress. Genetically-modified date palm fruit may be unacceptable to consumers and exporters initially, but in the meantime will still be a useful technological tool to understand the genetic complexity of this species. Induced mutagenesis could be used to isolate useful mutants and develop molecular programs based on functional genomics, metabolomics and proteomics technologies to improve date palm. Now that the entire genome of date palm is fully sequenced and publically available, a massive flow of data is to be expected, which will ultimately take the breeders, pathologists, physiologist and biochemists to another level of understanding of the genetic bases of this crop and its interaction with the environment, as well as in discovering new uses and traits of interest.

Finally, the conservation of genetic resources by *in vivo* and *in vitro* approaches are ideal for the establishment of germplasm banks to supply national and international researchers and promote sustainable date palm production as a fruit tree and as an industrial crop. These contributions must be promoted with synergistic actions to allow modernization of the date palm sector.

Acknowledgments This review has been supported by equal contribution from several research grants to Professor I. El Hadrami (IFS/AUF/PRAD/CNRST/AI Morocco-Tunisia/AI Morocco-France). The authors would like to thank Drs S. M. Jain and J. M. Al-Khayri for their critical review of the first draft of this contribution.

References

Ahloowalia BS, Maluszynski M, Nichterlein K (2004) Global impact of mutation-derived varieties. Euphytica 135:187–204

Al Khalifah NS, Askari E (2003) Molecular phylogeny of date palm (*Phoenix dactylifera* L.) cultivars from Saudi Arabia by DNA fingerprinting. Theor Appl Genet 107:1266–1270

Al Mansoori TA, Alaa El-Deen MN, Caligari PDS (2007) Evaluation of *in vitro* screening techniques for salt tolerance in date palm. Acta Hortic 736:301–307

Al-Ayedh H (2008) Evaluation of date palm cultivars for rearing the red date palm weevil, *Rhynchophorus ferrugineus* (Coleoptera: Curculionidae). Fla Entomol 91:353–358

Al-Dous EK, George B, Salameh YM et al (2009) Qatar researchers sequence draft version of date palm genome. URL: http://qatar-weill.cornell.edu/research/datepalmGenome/download.html

Al-Khayri JM (2001) Optimization of biotin and thiamine requirements for somatic embryogenesis of date palm (*Phoenix dactylifera* L.). In Vitr Cell Dev Biol Plant 37:453–456

Al-Khayri JM (2002) Growth, praline accumulation, and ion content in NaCl-stressed callus cultures of date palm (*Phoenix dactylifera* L.). In Vitr Cell Dev Biol Plant 38:79–82
Al-Khayri JM (2003) *In vitro* germination of somatic embryos in date palm: effect of auxin concentration and strength of MS salts. Curr Sci 84:101–104
Al-Khayri JM (2005) Date palm *Phoenix dactylifera L.*. In: Jain SM, Gupta PK (eds) Protocols of somatic embryogenesis in woody plants. Springer, Berlin, pp 309–320
Al-Khayri JM (2007) Date palm *Phoenix dactylifera* L. micropropagation. In: Jain SM, Häggman H (eds) Protocols for micropropagation of woody trees and fruits. Springer, Dordrecht, pp 509–526
Al-Khayri JM, Al-Bahrany AM (2001) Silver nitrate and 2-isopentyladenine promote somatic embryogenesis in date palm (*Phoenix dactylifera* L.). Sci Hortic 89:291–298
Al-Khayri JM, Al-Bahrany AM (2004a) Genotype-dependent *in vitro* response of date palm (*Phoenix dactylifera* L.) cultivars to silver nitrate. Sci Hortic 99:153–162
Al-Khayri JM, Al-Bahrany AM (2004b) Growth, water content, and proline accumulation in drought-stressed callus of date palm. Biol Plant 48:105–108
Allard RW (1956) Formulas and tables to facilitate the calculation of recombination values in heredity. Hilgard 24:235–278
Al-Ruqaishi IA, Davey M, Alderson P, Mayes S (2008) Genetic relationships and genotype tracing in date palms (*Phoenix dactylifera* L.) in Oman based on microsatellite markers. Plant Genet Res 61:70–72
Al-Salih AA, Al Rawi A (1987) A study of the cytology of two female cultivars of date palm. Date Palm J 5:123–124
Al-Salih AA, Al-Jarrah A (1987) Chromosome number of a date palm male: cultivar Ghannami Akhdar. Date Palm J 5:128–133
Al-Salih AA, Al-Najjar NR, Al-Mashhadani AN (1987) A study on the chromosome number of two specific female date palm cultivars. Date Palm J 5:134–143
Al-Wasel ASAA (2005) Survey study on somaclonal variations in *in vitro*-derived date palm trees. In: Proceeding of the international workshop on true-to-typeness of date palm tissue culture-derived plants, Institut National de Recherche Agronomique, Morocco, 23–25 May 2005. http://www.inra.org.ma/ist/publications/ouvrages/truetype.pdf
Azequour M, Majourhat K, Baaziz M (2002) Morphological variations and isoenzyme polymorphism of date palm clones from *in vitro* culture acclimatized and established on soil in South Morocco. Euphytica 123:57–66
Baaziz M (2000) Date palm culture in the Maghreb countries: constraints and scientific research. In: Proceedings of the date palm international symposium. Namibia, 22–25 Feb 2000, pp 306–311
Baenziger PS, Peterson CJ (1992) Genetic variation: its origin and use for breeding self-pollinated species. In: Stalker TM, Murphy JP (eds) Plant breeding in the 1990s. Proceedings symposium on plant breeding in the 1990s, Raleigh, March 1991. CAB International, Wallingford, pp 69–92
Baker WJ, Asmussen CB, Barrow SC et al (1999) A phylogenetic study of the palm family (Palmae) based on chloroplast DNA sequences from the *trnL - trnF* region. Plant Syst Evol 219:111–126
Barrow S (1998) A monograph of *Phoenix* L. (Palmae: Coryphoideae). Kew Bull 53:513–575
Beal JM (1937) Cytological studies in the genus *Phoenix*. Bot Gaz 99:400–407
Beauchesne G (1983) Vegetative propagation of date palm (*Phoenix dactylifera*) by *in vitro* culture. In: Proceedings of the first symposium of date palm. King Faisal University, Al-Hassa, Saudi Arabia, 23–25 March 1982, pp 698–699
Beauchesne G, Zaid A, Rhiss A (1986) Meristematic potentialities of bottom of young leaves to rapidly propagate date palm. In: Proceeding of second symposium of date palm vol 1. King Faisal University, Al-Hassa, 3–6 March 1986, pp 87–93
Beccari O (1890) Revista monografica delle species del genera *Phoenix* L. Malesia 3:345–416
Bekheet SA, Saker MM, Taha HS, Moursy HA (2001) Plant regeneration *via* somatic embryogenesis of date palm (*Phoenix dactylifera* L.). Arab J Biotechnol 4:111–118
Bekheet SA, Taha HS, Saker MM, Moursy HA (2002) A synthetic seed system of date palm through somatic embryos encapsulation. Annu Agric Sci Ain Shams Univ Cairo 47:325–337
Ben Abdallah A, Lepoivre P (2000) Production of date palm (*Phoenix dactylifera* L.) fruit and diploid embryos following gibberillic acid treatment of unpollinated female influorescences. Cah Agric 9:467–473

Ben Abdallah A, Stiti K, Lepoivre P, du Jardin P (2000) Date palm cultivar identification using random polymorphic DNA RAPD markers identification. Cah Agric 9:103–107

Ben Abdallah A, Lepoivre P, du Jardin P (2001) Apomixis induction possibility explored in date palm (*Phoenix dactylifrea* L.), In: Proceedings second international conference on date palms. Al-Ain, UAE, 25–27 March 2001, pp 164. URL: http://www.pubhort.org/datepalm/index2.htm

Benbadis BK (1992) Coconut and date palm. In: Hammerschlag FA, Litz RW (eds) Biotechnology of perennial fruit crops. CAB International, Wallingford, pp 383–400

Bernardo R (2001) What if we knew all the genes for a quantitative trait in hybrid crops? Crop Sci 41:1–4

Bernardo R (2002) Breeding for quantitative traits in plants. Stemma Press, Woodbury

Bhaskaran S, Smith RH (1992) Somatic embryogenesis from shoot tip and immature inflorescence of *Phoenix dactylifera* cv. Barhee. Plant Cell Rep 12:22–25

Carpenter JB, Ream CL (1976) Date palm breeding, a review. Date Grow Inst 53:25–33

Collard BCY, Jahufer MZZ, Brouwer JB, Pang ECK (2005) An introduction to markers, quantitative trait loci (QTL) mapping and marker-assisted selection for crop improvement: the basic concepts. Euphytica 142:169–196

Cushman JC, Bohnert HJ (2000) Genomic approaches to plant stress tolerance. Curr Opin Plant Biol 3:117–124

Daayf F, El Bellaj M, El Hassni M et al (2003) Elicitation of soluble phenolics in date palm (*Phoenix dactylifera*) callus by *Fusarium oxysporum* f. sp. *albedinis* culture medium. Environ Exp Bot 49:41–47

Djerbi M (2000) Abnormal fruiting of the date palm trees derived from tissue culture. In: Proceedings of date palm international symposium. Windhoek, Namibia. 22–25 Feb 2000, p 73

El Bellaj M, El Jaafari S, El Hadrami I (2000) L'AIA-oxydase marqueur et régulateur potentiel de l'embryogenèse somatique chez *Phoenix dactylifera* L. Cah Agri 9:193–195

El Hadrami I (1995) L'embryogenèse somatique chez *Phoenix dactylifera* L. quelques facteurs limitants et marqueurs biochimiques. Thesis Doctorat d'Etat, Faculté des Sciences Semlalia, Univ Cadi Ayyad, Marrakech, Morocco

El Hadrami I (2002) Infections racinaires localisées et rôle des dérivés hydroxycinnamiques dans la résistance du palmier dattier (*Phoenix dactylifera* L.) au *Fusarium oxysporum* f. sp. *albedinis*, agent causal du Bayoud. Polyphen Actual 22:19–27

El Hadrami I, Baaziz M (1995) Somatic embryogenesis and analysis of peroxidases in *Phoenix dactylifera* L. Biol Plant 37:197–203

El Hadrami I, Coumans M (1994) Somatic embryogenesis in *Phoenix dactylifera L*: relation with phenolic compounds. In: de Kouchkovsky Y, Larher F (eds) Plant sciences. SFPV-Université de Rennes I, France, p 155

El Hadrami I, El Hadrami A (2009) Breeding date palm. In: Jain SM, Priyadarshan PM (eds) Breeding plantation tree crops. Springer, New York, pp 191–216

El Hadrami I, Cheikh R, Baaziz M (1995) Somatic embryogenesis and plant regeneration from shoot-tip explants in *Phoenix dactylifera* L. Biol Plant 37:205–211

El Hadrami I, El Bellaj M, El Idrissi A et al (1998) Biotechnologies végétales et amelioration du Palmier dattier (*Phoenix dactylifera* L.), pivot de l'agriculture oasienne marocaine. Cah Agric 7:463–468

El Hadrami A, El Idrissi-Tourane A, El Hassni M et al (2005) Toxin based in vitro selection and its potential application to date palm for resistance to the bayoud fusarium wilt. A review. C R Biol 328:732–744

El Hadrami I, Daayf F, El Hadrami A (2010) On the socio-economical importance and traditional, health benefits, functional food and nutraceutical uses of dates. In: Manickavasagan A et al (eds) Dates – production, processing, marketing, food and medicine

El Hassni M, El Hadrami A, Daayf F, El Hadrami I (2004a) Chitosan, antifungal product against *Fusarium oxysporum* f. sp. *albedinis* and elicitor of defence reactions in date palm roots. Phytopathol Mediterr 43:195–204

El Hassni M, J'Aiti F, Dihazi A et al (2004b) Enhancement of defense responses against Bayoud disease by treatment of date palm seedlings with an hypoaggressive *Fusarium oxysporum* isolate. J Phytopathol 152:1–8

El Hassni M, El Hadrami A, Daayf F et al (2007) Biological control of Bayoud disease in date palm: selection of microorganisms inhibiting the causal agent and inducing defense reactions. Environ Exp Bot 59:224–234

El Houmaizi MA, Saaidi M, Oihabi A, Cilas C (2002) Phenotypic diversity of date-palm cultivars (*Phoenix dactylifera* L.) from Morocco. Genet Res Crop Evol 49:483–490

El-Assar AM, Krueger RR, Devanad PS, Chao CT (2005) Genetic analysis of Egyptian date (*Phoenix dactylifera* L.) accessions using AFLP markers. Genet Res Crop Evol 52:601–607

El-Kharbotly A, El-Mardi M, Al-Saadi N, Al-Mahruki Y (1998) Towards the construction of a genetic map of date palms using the amplified fragment length polymorphism technique (AFLP). In: Proceedings of the first international conference on date palms. Al-Ain, United Arab Emirates, pp 194–207

Ellis MH, Spielmeyer W, Gale KR et al (2002) "Perfect" markers for the *Rht-B1b* and *Rht-D1b* dwarfing genes in wheat. Theor Appl Genet 105:1038–1042

El-Sharabasy SF, Wanas WH, Al-Kerdany AY (2008) Effect of salinity stress on some date palm cultivars during proliferation stage *in vitro*. Arab J Biotechnol 11:273–280

Elshibli S, Korpelainen H (2008) Microsatellite markers reveal high genetic diversity in date palm (*Phoenix dactylifera* L.) germplasm from Sudan. Genetica 134:251–260

Elshibli S, Korpelainen H (2009a) Excess heterozygosity and scarce genetic differentiation in the populations of *Phoenix dactylifera* L.: human impact or ecological determinants. Plant Genet Res 7:95–104

Elshibli S, Korpelainen H (2009b) Biodiversity of date palms (*Phoenix dactylifera* L.) in Sudan: chemical, morphological and DNA polymorphisms of selected cultivars. Plant Genet Res 7:194–203

Fedak G (1999) Molecular aids for integration of alien chromatin through wide crosses. Genome 42:584–591

Fki L, Masmoudi R, Drira N, Rival A (2003) An optimized protocol for plant regeneration from embryogenic suspension cultures of date palm (*Phoenix dactylifera* L.) cv. Deglet Nour. Plant Cell Rep 21:517–524

Fry JD (2010) The genomic location of sexual antagonistic variation: some cautionary comments. Evolution 64:1510–1516

Gorelick R (2003) Evolution of dioecy and sex chromosomes *via* methylation driving Muller's ratchet. Biol J Linn Soc 80:353–368

Gorelick R (2005) Theory for why dioecious plants have equal length sex chromosomes. Am J Bot 92:979–984

Gorelick R, Osborne R (2002) Inducing sex change and organogenesis from tissue culture in the endangered African cycad *Encephalartos woodii* (Cycadales, Zamiaceae). S Afr J Sci 98:114–117

Hela S, Trifi M, Oould Mohamed Salem A et al (2000) Rapid construction of a random genomic library from date palm (*Phoenix dactylifera L.*). Plant Mol Biol Rep 17:1–7

Ibrahim AMF, El Kobbia AM, Kitat FM, Abd El Kawy MM (1998) Cytological studies on date palm. I. Chromosomal behavior during meiosis of two date palm (*Phoenix dactylifera* L.) male types. Alex J Agric Res 43:237–246

J'Aiti F, Verdeil JL, El Hadrami I (2009) Effect of jasmonic acid on the induction of polyphenoloxidase and peroxidase activities in relation to date palm resistance against *Fusarium oxysporum* f. sp. *albedinis*. Physiol Mol Plant Pathol 74:84–90

Jain SM (2006) Radiation-induced mutations for developing Bayoud disease resistant date palm in North Africa. In: Proceedings of the international workshop on true-to-typeness of date palm tissue culture-derived plants, Morocco, 23–25 May 2005. UAE University, Date Palm Global Network, Al Ain, United Arab Emirates, pp 31–41

Jain SM (2007) Recent advances in plant tissue culture and mutagenesis. Acta Hortic 736:205–211

Jain SM (2009) In vitro culture and mutagenesis in genetic improvement of date palm. In: Proceedings of 4th Date palm conference, Saudi Arabia, 5–8 May 2007 (in press)

Jain SM, Gupta P (eds) (2005) Protocols of somatic embryogenesis in woody plants. Springer, Dordrecht

Loutfi K (1999) Organogenèse et embryogenèse somatique à partir des tissus floraux du palmier dattier (*Phoenix dactylifera* L.) cultivés *in vitro*. Aspects histologiques et caryologie des vitro-plants. Thesis Doctorat d'Etat, Faculté des Sciences Semlalia, Marrakech, Morocco

Loutfi K, El Hadrami I (2004) *Phoenix dactylifera* L.: date palm. In: Litz RE (ed) Biotechnology of fruit and nut crops, vol 29. CAB International, Wallingford, pp 144–156

Majourhat K, Bendiab K, Medraoui L, Baaziz M (2002) Diversity of leaf peroxidases in date palm (*Phoenix dactylifera* L) as revealed in an example of marginal (seedling-derived) palm groves. Sci Hortic 95:31–38

McCallum CM, Comai L, Greene EA, Henikoff S (2000) Targeting induced local lesions in genomes (TILLING) for plant functional genomics. Plant Physiol 123:439–442

McCubbin MJ, Zaid A, van Staden J (2004) A southern African survey conducted for off-types on date palms produced using somatic embryogenesis. Emir J Agric Sci 16:8–14

Munier P (1973) Le palmier dattier. Coll. Techniques Agricoles et Productions Tropicales, XXIV. Maisonneuve and Larose, Paris

Nemec B (1910) Das problem der befruchtungsvorgänge und andere zytologische Fragen. Borntraeger, Berlin

Omer MS, Hameed MK, Al-Rawi MS (1992) Micropropagation of date palm (*Phoenix dactylifera* L.). In: Bajaj YPS (ed) Biotechnology in agriculture and forestry, vol 18, High-tech and micropropagation II. Springer-Verlag, Berlin, pp 471–490

Ouenzar B (2003) Contribution à l'amélioration du Palmier dattier: recherche de marqueurs moléculaires précoces de la détermination du sexe et de la résistance au Bayoud. Thesis Doctorat d'Etat, Faculté des Sciences Semlalia, Marrakech, Morocco

Ouenzar B, Hartmann C, Rode A et al (2003) Détermination de la taille génomique du palmier dattier par cytométrie en flux. In: El Hadrami I, Daayf F (eds) Biotechnologies végétales: de la structure des génomes à l'amélioration des plantes. El Watanya, Marrakech, pp 70–77

Reuveni O (1979) Embryogenesis and plantlets growth of date palm (*P. dactylifera* L.) derived from callus tissues. Plant Physiol 63:138 (Abstract)

Reynolds JF, Murshige T (1979) Asexual embryogenesis in callus cultures of palms. In Vitro 15:383–387

Rice WR (1984) Sex-chromosomes and the evolution of sexual dimorphism. Evolution 38:735–742

Saker M, Bekheet S, Taha H et al (2000) Detection of somaclonal variation in tissue culture-derived date palm plants using isozyme analysis and RAPD fingerprints. Biol Plant 43:347–351

Saker MM, Adawy SS, Mohamed AA, El-Itriby HA (2006) Monitoring of cultivar identity in tissue culture-derived date palms using RAPD and AFLP analysis. Biol Plant 50:198–204

Salem AOM, Trifi M, Sakka H, Rhouma A, Marrakchi M (2001) Genetic inheritance analysis of four enzymes in date palm (*Phoenix dactylifera* L.). Genet Res Crop Evol 48:361–368

Sedra MH, Lashermes P, Trouslot P et al (1998) Identification and genetic diversity analysis of date-palm (*Phoenix dactylifera* L.) varieties from Morocco using RAPD markers. Euphytica 103:75–82

Seki M, Narusaka M, Abe H et al (2005) Monitoring the expression pattern of 1300 *Arabidopsis* genes under drought and cold stresses by using a full-length cDNA microarray. Plant Cell 13:61–72

Sghaier B, Bahloul M, Bouzid GR, Drira N (2008) Development of zygotic and somatic embryos of *Phoenix dactylifera* L. cv. Deglet Nour: comparative study. Sci Hortic 116:169–175

Sghaier-Hammami B, Drira N, Jorrín-Novo JV (2009) Comparative 2-DE proteomic analysis of date palm (*Phoenix dactylifera* L.) somatic and zygotic embryos. J Proteomics 73:161–177

Siljak-Yakovlev S, Benmalek S, Cerbah M et al (1996) Chromosomal sex determination and heterochromatin structure in date palm. Sex Plant Reprod 9:127–132

Snoussi H, Du Jardin P, Ben Abdallah A, Lepoivre P (2001) Assessment of genetic variation within date palm (*Phoenix dactylifera* L.) using amplified fragment length polymorphism (AFLP) – genotyping of apomictic seedlings as a case study. In: Proceedings second international conference on date palms. Al-Ain, UAE, 25–27 Mar 2001, pp 678–683. *URL:* http://www.pubhort.org/datepalm/index2.htm

Soliman SS, Ali BA, Ahmed MMM (2003) Genetic comparisons of Egyptian date palm cultivars (*Phoenix dactylifera* L.) by RAPD-PCR. Afr J Biotechnol 2:86–87

FAO Statistics (2010) www.fao.org

Sudhersan C, Abo El-Nil M, Al-Baiz A (1993) Occurrence of direct somatic embryogenesis on the sword leaf of *in vitro* plantlets of *Phoenix dactylifera* L. cultivar Barhee. Curr Sci 65:887–888

Sudhersan C, Al-Shayji Y, Jibi Manuel S (2009) Date palm crop improvement *via* TxD hybridization integrated with *in vitro* culture technique. Acta Hortic 829:219–224

Swingle WT (1928) Metaxenia in the date palm. J Hered 19:257–268

Talaat AA, Al-Qaradawi A (2009) Molecular phylogeny of Qatari date palm genotypes using simple sequence repeats markers. Biotechnology 8:126–131

Tanksley SD (1993) Mapping polygenes. Annu Rev Genet 27:205–233

Tisserat B (1979) Propagation of date palm (*Phoenix dactylifera* L.) *in vitro*. J Exp Bot 30:1275–1283

Tisserat B (1984) Propagation of date palms by shoot tip cultures. Hort Sci 19:230–231

Toutain G, Louvet J (1972) Resistance to Bayoud in varieties of date palm. In: First international seminar and workshop on bayoud, Oct 1972, ITAS, Alger, pp 208–210

Trifi M, Rhouma A, Marrakchi M (2000) Phylogenetic relationships in Tunisian date- palms (*Phoenix dactylifera* L.) germplasm collection using DNA amplification fingerprinting. Agronomie 20:665–671

Veramendi J, Navarro L (1996) Influence of physical conditions of nutrient medium and sucrose on somatic embryogenesis of date palm (*Phoenix dactylifera* L.). Plant Cell Tissue Org Cult 45:159–164

Wong CK, Bernardo R (2008) Genomewide selection in oil palm: increasing selection gain per unit time and cost with small populations. Theor Appl Genet 116:815–826

Wrigley G (1995) Date palm (*Phoenix dactylifera* L.). In: Smartt J, Simmonds NW (eds) The evolution of crop plants. Longman, London, pp 399–403

Zaid A (2002) Date palm cultivation. Plant Production and Protection Paper 156. FAO, Rome

Zehdi S, Trifi M, Salem AOM (2002) Survey of inter simple sequence repeat polymorphisms in Tunisian date palms (*Phoenix dactylifera* L.). J Genet Breed 56:77–83

Zehdi S, Sakka H, Rhouma A et al (2004a) Analysis of Tunisian date palm germplasm using simple sequence repeat primers. Afr J Biotechnol 3:215–219

Zehdi S, Trifi M, Billotte N et al (2004b) Genetic diversity of Tunisian date palms (*Phoenix dactylifera* L.) revealed by nuclear microsatellite polymorphism. Hereditas 141:278–287

Zivdar S, Mousawi M, Alemzadeh Ansari N (2008) Genetic stability in date palm micropropagation. Asian J Plant Sci 7:775–778

Zohary D, Spiegel-Roy P (1975) Beginnings of fruit growing in the old world. Science 187:319–327

Zouine J, El Hadrami I (2007) Effect of 2,4-D, glutamine and BAP on embryogenic suspension culture of date palm (*Phoenix dactylifera* L.). Sci Hortic 112:221–226

Zouine J, El Bellaj M, Meddich A et al (2005) Proliferation and germination of somatic embryos from embryogenic suspension cultures in *Phoenix dactylifera* L. Plant Cell Tissue Org Cult 82:83–92

Chapter 24
Development of New Moroccan Selected Date Palm Varieties Resistant to Bayoud and of Good Fruit Quality

MyH. Sedra

Abstract Bayoud disease, caused by *Fusarium oxysporum* f. sp *albedinis*, is one of the date palm diseases of the world which is difficult to control. The elite commercial cultivars have shown high susceptibility to the disease. The genetic strategy to use resistant cultivars is the most predominant method, up to now, to control this disease. The diffusion of existing resistant cultivars has been limited due to the mediocre quality of the fruits that do not possess the commercial quality wanted by producers. Research undertaken by INRA on genetic improvement of date palm has permitted the selection of cultivars combining good fruit quality and resistance to bayoud. Studies of the performance of these cultivars have shown that they possess several agro-morphological characters that are more desirable than those of the major common and commercial cultivars in Morocco. Some of these new cultivars have already been multiplied on a large scale by tissue culture and distributed to farmers, not only to rejuvenate palm plantations ravaged by bayoud but also to enhance productivity of traditional low-yielding palm groves. Given their diversity in agronomic characters, the selected cultivars can be of value to other countries contaminated with and threatened by bayoud. Research continues, to identify and multiply other new more effective cultivars.

Keywords Cultivar characterization • Micropropagation • New cultivars • Selection

MyH. Sedra (✉)
Laboratoire de phytopathologie, Génétique et de Lutte Intégrée, Centre Régional de Marrakech, Institut National de Recherche Agronomique (INRA), BP. 533, Ménara 40000, Marrakech, Morocco
e-mail: sedramh@hotmail.com; mhsedra@yahoo.fr

S.M. Jain et al. (eds.), *Date Palm Biotechnology*,
DOI 10.1007/978-94-007-1318-5_24, © Springer Science+Business Media B.V. 2011

24.1 Introduction

The date palm tree, *Phoenix dactylifera* L., is a tree of major interest because of its high productivity, the high nutrient value of its very desirable fruits and its adaptation to conditions of the Saharan regions. In fact, it constitutes the keystone species of the oasis ecosystem because of the favorable microclimate it creates for the growth of annual crops and other fruit trees in the shade created by the palms.

The exact origin of the date palm is not known with certainty. According to ancient writings (3,000 years BC), the geographical origin of date palm tree is probably the region of Mesopotamia and the valley of the Nile in Egypt (Al-Bakr 1972). The species was likely spread in association with human migrations. The culture of date palm tree extended eastward (Iran and the Indus River valley) and westward, from Egypt, then to Libya as well as the other countries of North Africa, the countries of the Sahel; also to Iberia, during the period of the Arabian presence. The introduction of date palm into a given region typically is made by offshoots that generate progeny identical to the mother palm, maintaining a high level of conformity in contrast to seed multiplication, used in older times, which produces heterogeneous progeny (i.e. both genders and different agro-morphological criteria).

Nevertheless, the selection process followed by farmers in their palm groves was focused for centuries not only on certain fruit criteria (taste, quality, appearance, color, use of fruit, etc.) but also on the local adaptation of cultivars; all the factors which help in understanding the makeup of the cultivars existing today. The selection criteria farmers employed were subjective and took into consideration their own preferences and those of tradesmen, for large brown dates which can be stored under ambient conditions. In Morocco, as in some other countries of North Africa, sexually-propagated palms descended from natural crosses were important and given names (*khalts*, progeny of spontaneous seedlings) which exhibit variability according to the regions and the technological level of the farmer. Cultivars of agronomic interest were multiplied and then designated, deriving common names from individual people, a location, agronomic character, shape of the fruit or other distinctive feature. These palms were propagated by offshoots in surrounding lands, then in other more distant localities. The individual tree types retained the names they were given and become cultivars. Currently, more than 4,000 named cultivars exist under different local and regional names in the more of 30 countries that produce dates. However, in general for every country, only about ten cultivars are most important some recognized for their good commercial fruit quality and these have come to dominate the national markets (Table 24.1).

Moroccan date palm groves occupy a surface area of more than 48,000 ha with a population of 4.8 million trees; they are composed of about 55.6% seed-derived *khalts* representing about 2.5 million trees, each of which is genetically different. The rest of the material is composed of known cultivars (more than 223) represented by about 2.3 millions trees (Anonymous 2009). The number of cultivars recently has been more than doubled to some 450 cultivars by Sedra (2010, unpublished results) and can be further enriched by new cloned cultivars in future. The average annual per tree fruit productivity (18–20 kg/tree) is relatively low and national

Table 24.1 List of important date palm cultivars in several countries (cultivars in alphabetical order)

Countries	Cultivars
Algeria	Deglet Noor (60%), Ghars, Hartan, Hmira, Takerboucht, Tazerzayte,Tinnnacer
Saudi Arabia	Al Chalabi, Anasefri, Khalas, Nabout Saif, Rhziz, Rothona
Egypt	Amhat, Amri, Apremi, Bentaîcha, Bertoma, Eglani, Gendila, Hayani, Saâidi, Semani, Sioui
United Arab Emirates	Achahl, Anouane, Chechi, Debbassi, Hassab, Jabri, Kenzi, Khadraoui, Khalas, Loulou, Mektoumi, Naghel, Raziz
Iraq	Aghrass, Barhee, Borbon, Braîme, Hallaoui, Khadraoui, Khastaoui, Maktoume, Texbarzal, Zahdi
Jordan	Ahmartalal, Aqabaoui, Barhee, Hayani, Khadraoui, Zahdi
Kuwait	Barhee, Hallaoui, Khadraoui, Khalas, Sammaran
Libya	Abel, Adoui, Broumi, Deglet Noor, Halima, Hallaoui, Jedoug, Khadri, Mersim, Saâidi, Soufir, Tabouni, Tafsirt, Tagyiat, Talis, Tedis
Morocco	Aguelid, Ahardane, Assiane, Azigzao, Aziza bouzid, Boucerdoune, Boufeggous, Bouittob, Bouskri, Bouslikéne, Bousthammi Noire, Bourar, Jihel, Medjool, Taâbdounte
Mauritania	Ahmar, Oum-aârich. Soukkani,Tiguedert, Tinterguel, Tijib
Qatar	Arzaizi, Barhee, Chechi, Hilali, Khalas, Khassab, Khénazi, Nebtat Sif
Sudan	Abdelkrim, Barakaoui, Deglet Noor, Gondalla, Koulma, Medino, Michreg/oued Khatib, Michreg/oued Laggai, Tamoud
Oman	Berni, Bounerinja, Chahla, Fard, Hilali, Jabri, Khalas, Khassab, Khenazi, Mabsali, Madlouki, Naghel, Oach, Unsilla, Salani
Tunisia	Ain Hnach, Alig, Bouhatime, Deglet Noor, Grin lghzal, Igoua, Kenta, Lemsi, Rochdi, Smiti
Yemen	Ajoua, Arkdi, Azar, Bakiah, Baomairah, Batahi, Bator, Boram, Dabach, Dalil, Deniari, Dkhara, Dhkour, Foufal, Hachadi, Hajoui, Hamraâ, Hokabi, Jabri, Jahara, Jazaz, Khabri, Khafouch, Khoudari, Koui, Loban, Machtoum, Madyani, Majabrah, Makari, Maksab, Mechar, Mokbaran, Mokhalis, Omani, Oraegi, Saâfani, Sabai, Sareh Maktari, Tael, Tobaki, Zaneki, Wak

Some information derived from a meeting of the national experts held to the United Arab Emirates (FAO 1995)

annual production of dates rarely reaches 100,000 mt. Otherwise, the palm groves are characterized by the advantage of a notably long production season which begins in June and extends until the end of November, and by very important exceptional plant biologic diversity (Fig. 24.1). However, the development and the modernization of the date palm sector are hampered by several important constraints.

Indeed, deterioration of the palm groves has continued in various localities because of constraints which can be summarized as follows:

- Permanent stress from the impact of the destructive bayoud disease caused by the soil fungus (*Fusarium* f. sp *oxysporum albedinis*) (Fig. 24.2); one of the known diseases which is difficult to control in the world and which has decimated more than 10 million trees in Morocco over a century (Djerbi 1988; Sedra 2003a, b, c, 2007b), contributing to ecosystem degradation of oases, desertification and accentuating the rural-to-urban exodus. Moreover, the steady loss of date palm genotypes also constitutes a perilous threat to phytogenetic resources of the oases.

Fig. 24.1 Some date cultivars showing the diversity of the Moroccan date patrimony

Fig. 24.2 Example of bayoud disease effects on Jihel cv. in Draa Valley, southern Morocco

- Prolonged drought and traditional water management.
- Inefficient management of the array of cultivars.
- Insufficiency of suitable palm practices in relation to the physiognomy of palm groves.
- Insufficient valorization of predominant cultivars having low commercial value.

Many efforts are being extended to overcome these constraints and results of these actions are positive and promising for a satisfactory future of the oases and their date palms.

The spread of bayoud disease in North Africa has destroyed 3 million palms in Algeria (Djerbi 1988) and some thousands of palms in Mauritania since its discovery there in 1999 (Sedra 1999, 2002, 2003a, b, c, 2007a, b). The best Moroccan commercial cvs. (Medjool, Boufeggous, Jihel, Bouskri) and Algerian cvs. (Deglet Noor, Ghras) are all very susceptible to the disease. Bayoud has certainly killed several thousands of individuals of natural-occurring *khalts* (Sedra 2003a), as well as two Moroccan cvs. Idrar and Berni (Pereau-Leroy 1958). The disease has also reduced significantly the populations of extensively-cultivated cultivars like Medjool, Boufeggous and Jihel.

24.2 The Genetic Method to Combat Bayoud Using Resistant Cultivars

In order to combat bayoud disease, the use of resistant cultivars appears, up to now, the only practical means to protect and cultivate date palm trees, even in the presence of the disease. Although complex and requiring much time, it is promising and the genetic approach was adopted by INRA (Institut National de la Recherche Agronomique) in Morocco in the 1960s and also in Algeria in the 1980s. The selection of productive cultivars having good fruit quality and with resistant to the disease requires a rigorous methodology, especially at the stage of the resistance assessment. To that end, reliable and rapid methods of cultivar screening have been developed in the field using artificial inoculation of palm trees with the pathogen, and in the laboratory on young plants (Sedra 1993, 1994a, b; Sedra and Besri 1994), or by the use of pathogen toxins (El Fakhouri et al. 1996, 1997; Sedra et al. 1993, 1998b, 2008)

Among 32 Moroccan cultivars tested at the Zagora Experiment Station of INRA, six (Boufeggous ou Moussa, Black Bousthammi, White Bousthammi, Iklane, Sair-Layalate and Tadmainte) were found to have bayoud resistance, the research dating back to 1973 (Louvet and Toutain 1973; Saaidi 1992) and a seventh cv. (Boukhanni) was selected 20 years later (Sedra 1992, 1993, 1995). In Algeria, only the Takerbouchte cv. is recognized as resistant (Bulit et al. 1967). Tirichine (1991) added another resistant cv., Akerbouch, in the M'zab region of Algeria. Among these resistant cvs. only Boukhanni, Sair-Layalate and Takerbouchte were relatively acceptable but not as good as elite cvs. like Medjool and Deglet Noor. As for Moroccan-grown foreign cvs., notably six from Iraq (Barhee, Hallawy, Khastawy, Khadrawy, Sair and Zahdi) and six Tunisian cvs. (Boufeggous, Besser Lahlou, Gondi, Horra, Kenka and Kentichi) have all shown susceptibility to bayoud disease (Djerbi and Sedra 1982; Sedra 1992, 1993, 1995). The last synthesis concerning the selection of new genotypes for their resistance and fruit quality from different genetic sources was made by Sedra (1990, 1995, 1997, 2003a, b), Sedra et al. (1996)

Fig. 24.3 Plantation and development of vitroplants derived from selection using an irradiation and screening test with pathogen toxins. These selected mutants, and other plants as controls, are under field study at the Zagora Experiment Station, INRA

and Zaher and Sedra (1998). Several clones have been identified from mass selection among a population of *khalts*, derived from controlled crosses. In addition, several hundred crosses (F_1, backcross, sib) have been achieved in order to explore their variability, to obtain new effective hybrids and to study heredity of the characters (Sedra 2003b). A first series of a quarantined individuals has been selected, some of which have been identified and characterized (Sedra 1990, 1995, 2003a, b; Sedra et al. 1996; Zaher and Sedra 1998). Other genotypes are under assessment and characterization. A process of micropropagation by tissue culture has been developed by INRA and adapted to more than 30 cvs. (Abahmane 2010; Anjarne et al. 2010) and this technology transferred to private companies that can multiply and market more than 700,000 acclimatized vitroplants to farmers for their date palm groves, not only to rejuvenate plantations ravaged by bayoud disease but also to improve traditional date groves with low productivity. Research led by INRA, in partnership with the International Atomic Energy Agency (IAEA), permitted *in vitro* propagation of some resistant mutants of the commercial susceptible cv. Boufeggous employing gamma irradiation and *in vitro* selection using pathogen toxins (Bougerfaoui et al. 2006; Sedra et al. 2008). The selected mutants are under study in the field at the Zagora Experiment Station (Fig. 24.3). In order to preserve genetic date palm resources, cultivar collections have been established at five different INRA Experiment Stations, representing more than 5,000 genotypes and in excess of 8,000 specimen plants (Sedra 2007b).

Table 24.2 Different steps of selection and evaluation of new selected cultivars of date palm

Sequence	Procedures
Step 1	Selection of genotypes (a single tree) among sexually propagated palms descended from seeds on the basis of fruit quality in comparison with dates of good commercial cultivars but susceptible to bayoud
Step 2	Planting of offshoots at the Zagora Experiment Station within the framework of a larger experiment concerning a number of raised of genotypes tested in comparison with susceptible and resistant cultivars as a control. The trees have been inoculated in the field with the pathogen, the causal agent of bayoud, using the technique of inoculation and production of contamination on the roots according to methodology developed by (Sedra 1994a), then monitored for several years. Afterwards, preliminary selection is done in comparison with control cultivars
Step 3	Proposal of selected individuals for rapid multiplication by tissue culture. After production of rooted vitroplants in sufficient number by genotype, evaluation of the levels of resistance or susceptibility to bayoud using a statistical approach by testing a sufficient number of vitroplants inoculated by the pathogen in the greenhouse under controlled conditions according to the methods developed by (Sedra 1994b; Sedra and Besri 1994) and using the pathogen toxins according to the technique of (Sedra et al. 1993, 1998b)
Step 4	Confirmation in the field of resistance of selected cultivars by planting of vitroplants in sufficient numbers for statistical analysis in experimental plots contaminated with the pathogen in comparison with susceptible and resistant cultivars as a control. Moreover, testing the true-to-type of tissue culture-derived plants using organogenesis of some Moroccan cultivars determined by using phenological descriptors and molecular markers (Sedra 2005)
Step 5	Identification and characterization of selected cultivars. This step begins before field planting while evaluating some important criteria of vegetative organs and fruits. Then overall description of the trees and characterization of cultivars based on morphological and agronomic characters, as well as molecular markers
Step 6	To name and propose selected cultivars for inscription into an official catalogue then proposing them for mass multiplication carried out by private laboratories within the framework of agreements and conventions. Thereafter, multiplied cultivars are handled by the state (Ministry of Agriculture), and distributed to farmers whose groves have been destroyed by bayoud in order to rejuvenate those ravaged areas and to investors to encourage establishment of new groves

This paper presents initial results of the characterization of new selected cultivars having good fruit quality and exhibiting resistance to bayoud disease, as well as certain other selected cultivars with high fruit quality but susceptible to the disease that need to be preserved. A population of common cultivars is being grown as controls for comparison.

Promising clones of cultivars derived from mass selection among several hundred genotypes are under field cultivation at the Zagora Experiment Station, and are composed of 15 female and 2 male cvs. Selection has been achieved through several steps and time periods: assessment in the laboratory and the field, mass multiplication of selected cultivars, and farm assessment by producers and at the market level. The different steps of evaluation and selection are presented in Table 24.2. Since

Fig. 24.4 Development of a screening resistance test on date palm vitroplants inoculated by the pathogen causal agent of bayoud (test duration 2–3 months)

being planted, these date palms have been monitored over 25 years in the field for their susceptibility to bayoud and for their productivity. They have been inoculated experimentally with the pathogen according to a screening method developed by Sedra (1994a). Most of these cultivars have been proposed for *in vitro* multiplication but only some have been multiplied and distributed to farmers. The behavior of vitroplants at the young stage of some of these new cvs., and their reactions to the pathogen and its toxins has been evaluated under greenhouse and growth-room conditions, following techniques developed by (El Fakhouri et al. 1996, 1997; Sedra 1994b; Sedra and Besri 1994) (Fig. 24.4) and Sedra et al. (1993) (Fig. 24.5). The new cvs. studied and their INRA references numbers, are: Najda (3014), Ayour (3415), Hiba (3419), Tanourte (3414), Al-Baraka (3417), Tafoukte (3416), Mabrouk (1394) and Khair (3300). According to the availability of material, cultivars used as resistant controls (Black Bousthammi and/or Sairlayalate) and susceptibility control (Medjool and/or Boufeggous and/or Jihel and/or Bouskri*)* have always been part of the trials. The agronomic, morphological and physical characteristics of trees and fruits have been recorded in accordance with the descriptors developed by Sedra (2001). The results presented, concerning only some chosen characteristics, correspond to calculated global averages based at least on about 15 years of observations, according to the cultivar.

Table 24.3 presents a listing of new selected cultivars nominated and proposed for mass micropropagation in order to rejuvenate areas in Morocco ravaged by bayoud and to renovate underproductive palm groves to improve date production quality and quantity. The Najda cv. (INRA-3014) (Fig. 24.6) was the first new

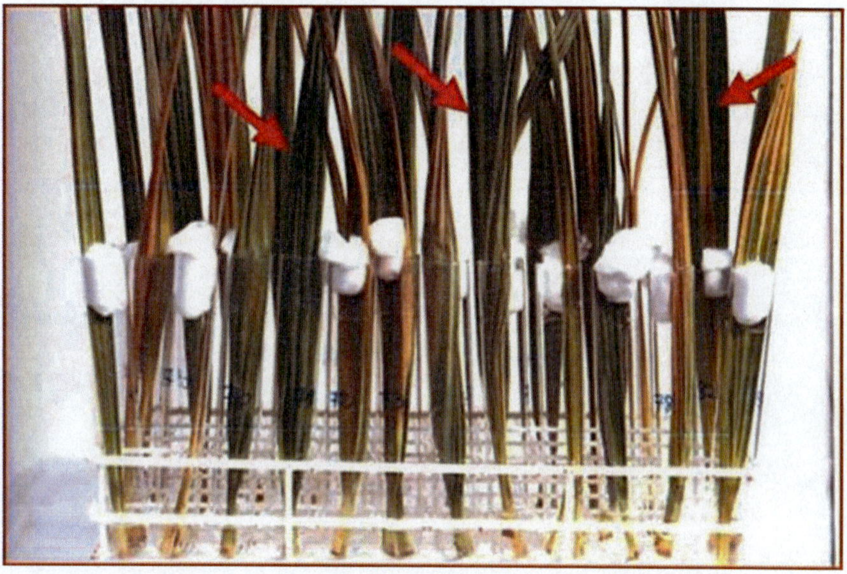

Fig. 24.5 Development of a screening resistance test on leaves of date palm using pathogen toxins. The green leaves without symptoms correspond to resistant cultivars (test duration 7–10 days)

Table 24.3 New selected date palm cultivars proposed for mass micropropagation

Series	Cultivar	INRA reference no.	Name significance
1	Najda	3014	First-help
	Mabrouk	1394	Newborn
2	Ayour	3415	Moon
	Hiba	3419	Grant
	Tanourte	3414	Big round bread
	Tafoukte	3416	Sun
	Al-Baraka	3417	Sufficiency
	Khair	3300	The good
	Al-Fayda	1447	Benefit
	Bourrihane	1414	Father of challenge
	Al-Amal	1443	Hope
	Nebch-Bouskri	NP3	From locality Nebch-Parcel Bouskri cultivar
	Nebch-Boufeggous	NP4	From locality Nebch-Parcel Boufegous cultivar
3	Sedrat	3003	For Mr Sedra
	Darâaouia	1445	From the Draa valley
	Not yet named	3010	–

selected cultivar released to farmers, especially for a plantation in Foci infected with bayoud. More than 60,000 plants have been distributed over the past 20 years. The dates produced from the Najda cv. are well accepted and are beginning to

Fig. 24.6 (**A**) Date production of selected cv. Najda (INRA-3014); (**B**) Area ravaged by bayoud then repopulated by the planting in rows of the vitroplants of the cv. Majda; (**C**) Example of a tree in production of selected cv. Al-Amal (INRA-1443)

Fig. 24.7 Examples of new selected cultivars effectively resistant to bayoud, except for cv. Ayour

appear in the markets. This has precipitated an increased demand for plants by the farmers. The second and third series of new cultivars are enriching the national date patrimony of cultivars in order to improve date production and/or to contribute

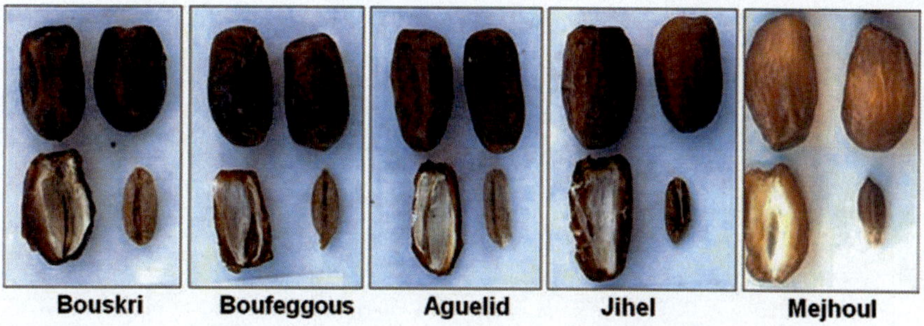

Fig. 24.8 Examples of the major common commercial cultivars susceptible to bayoud

Fig. 24.9 Examples of the major common commercial and resistant cultivars to bayoud

to the control of bayoud; for example: Ayour (3415), Hiba (3419), Tanourte (3414), Tafoukte (3416), Al-Baraka (3417), Khair (3300), Al-Amal (1443), Al-Fayda (1447), Bourrihane (1414), Mabrouk (1394) and INRA-3010 (unnamed). Most of these cultivars are undergoing mass multiplication. The last series of new selected cultivars proposed in 2010 were: Darâaouia (1445) and Sedrat (3003). Figures 24.7–24.9 illustrate, respectively, some examples of selected cultivars, both commonly susceptible and resistant ones. Other new selected cultivars released from different crosses (F_1, backcross) are under evaluation and characterization. Additional new cultivars will be proposed at the beginning of 2011.

Table 24.4 presents important quantitative and qualitative agro-morphological characters of the first selected cultivars and the major commercial cultivars in Morocco that, in addition to high productivity of the trees, should be of interest to date producers.

24.3 Behavior of Selected Cultivars to Bayoud

Field evaluations showed that the susceptible control (cv. Jihel) and selected cultivars for their fruit quality, notably Ayour (3415), Hiba (3419), Tanourte (3414), Tafoukte (3416), Al Baraka (3417) and Khair (3300), are attacked by bayoud

Table 24.4 Agro-morphological characteristics of new selected female cultivars compared to major common Moroccan cultivars

Cultivars (INRA reference numbers)	Behavior towards bayoud	Fruit appearance	Fruit color	Fruit consistency	Fruit weight[b]	Fruit pulp (%)	Heat requirements for fruit maturity (°C)[a]	Fruit stalk length (cm) and characterization	Leaf length (cm) and characterization	Thorns, total number	Thorn angle at middle of rachis
Selected cultivars											
Daraouia[c] (1445)	Resistant	Excellent	Clear brown	Semi-soft	21.9 VH	93 VH	4,000–4,500	121 (very long)	502 (very long)	68 VH	37 M
Sedrat[c] (3003)	Resistant	Excellent	Clear brown	Dry	21.2 VH	91.5 VH	<3,500	96.5 (long)	463 (very long)	30 L	27 L
Al-Amal (1443)	Resistant	Excellent	Clear brown	Dry	21 VH	95.2 VH	3,501–4,000	65 (moderate)	426.5 (long)	40 M	32 L
INRA-3010[c]	Resistant	Excellent	Clear brown	Semi-dry	18 H	93.9 VH	4,000–4,500	93 (long)	443 (long)	36 M	30 L
Al-Fayda (1447)	Resistant	Excellent	Clear brown	Semi-soft	14.2 M	92.4 VH	3,501–4,000	170 (very long)	442 (long)	24 H	52 H
Bourihane (1414)	Resistant	Excellent	Clear brown	Semi-soft	14.1 M	92.9 VH	3,501–4,000	60.6 (moderate)	329.2 (moderate)	43.5 H	30 H
Mabrouk (1394)	Resistant	Very good	Clear brown	Semi-soft	14.3 M	90.9 VH	4,501–5,000	63 (short)	440.5 (long)	43.5 H	57 H
Najda (3,014)	Resistant	Good	Clear brown	Semi-soft	17.7 H	92.1 VH	4,000–4,500	77.2 (moderate)	417 (long)	41.5 H	20 L
Khair (3,300)	Susceptible	Good	Clear brown	Semi-soft	12.9 M	86 M	4,501–5,000	65 (moderate)	360 (moderate)	36.5 M	32 L
Tanourte (3414)	Susceptible	Good	Brown	Semi-soft	13.5 M	90.3 VH	3,501–4,000	75 (moderate)	305 (short)	34 M	40 M
Ayour (3,415)	Susceptible	Very good	Clear brown	Semi-soft	24 VH	92.5 VH	4,501–5,000	97 (long)	472 (very long)	40.5 H	30 L
Tafoukte (3416)	Susceptible	Fair	Brown	Semi-dry	10 L	89 H	<3,500	74.3 (moderate)	555 (very long)	28.5 L	22 L
Hiba (3419)	Susceptible	Good	Clear brown	Dry	18 H	91.4 VH	3,501–4,000	65.3 (moderate)	439.5 (long)	44 H	38 M
Common cultivars											
Medjool	Very susceptible	Excellent	Dark brown	Semi-soft	25 VH	93.6 VH	>5,000	143 (very long)	308 (short)	34 M	20 L
Boufeggouss	Very susceptible	Good	Dark brown	Soft	19 H	92 VH	4,501–5,000	115 (very long)	270 (very short)	29.5 L	22 L
Jihel	Very susceptible	Good	Clear brown	Dry	12.2 M	90 H	>5,000	77 (moderate)	363 (moderate)	38 M	19 VL

Bouskri	Very susceptible	Fair	Dark brown	Dry	9,8 L	84 M	4,501–5,000	66.2 (moderate)	305 (short)	27.5 L	20 L
Aguélid	Fairly susceptible	Poor	Clear brown	Semi-dry	8,8 L	86 M	<3,500	107.2 (long)	380 (moderate)	32 M	40 M
Iklane	Resistant	Poor	Black	Soft	9.55 L	93 VH	>5,000	71.4 (moderate)	398 (moderate)	22 VL	41 M
Saîrlayalate	Resistant	Fair	Clear brown	Semi-dry	10.9 M	90 H	4,501–5,000	105.5 (long)	369 (moderate)	30.5 M	15 VL
Bousthammi Noire	Resistant	Poor	Black	Soft	6 L	90 H	4,501–5,000	120 (very long)	498 (very long)	22.5 VL	41 M

VH very high, *H* high, *M* medium, *L* low, *VL* very low

[a]The sum of the daily average of temperature above 18°C since flowering until the date maturity that influences the precocity or lateness of fruit maturity: precocious cultivars (<3,500°C), fairly precocious (3,501–4,000°C), in season (4,000–4,500°C), fairly late (4,501–5,000°C) and late (>5,000°C)

[b]The presented value represents the average of values of 12 samples. The calculated median weight of date fruits represents the average of the weight of 100 sampled dates at random from normal productions based on at least 15 years of production. These descriptors of date palm are defined by Sedra (2001)

[c] New cultivars proposed in early 2010: Daraouia (INRA-1445) and Sedrat (INRA-3003); the clone-variety INRA-3010 is not yet named)

(Table 24.4). Some of these cultivars have been lost and to preserve them, genetic material was produced by tissue culture and preserved in the collection of the INRA Menara Experiment Station, in the Marrakech region, in an area free of bayoud . All of the susceptible cultivars have been distributed to farmers for their fruit quality and not for their bayoud resistance. It is advisable to plant these cultivars, as is the case of the susceptible cvs. Medjool and Boufeggous, in areas free of bayoud in order to better valorize them without risk. On the other hand, the following cultivars grown as a resistant control group (cv. Black Bousthammi) (Table 24.4) as well as the male clones Nebch-Bouskri (NP3) and Nebch-Boufeggous (NP4) (Table 24.5), showed no symptoms of the disease in spite of numerous artificial inoculations with the pathogen in the field. After cv. Najda (3014) is added, the seven new selected and promising cultivars to control bayoud are the following: Drâaouia (1445), Sedrat (3003), Al-Amal (1443), Al-Fayda (1447), Bourihane (1414), Mabrouk (1394) and 3010 (unnamed) as well as the selected males Nebch-Bouskri (NP3) and Nebch-Boufeggous (NP4). Some of these cvs. such as Sedrat (3003), Al-Amal (1443) and Daraâouia (1445), exhibit high performance and are resistant to bayoud.

24.3.1 *Fruit Characters*

Most of the selected cultivars are attractive, bearing fruits of a clear brown color and with flesh of moderate consistency (semi-soft), comparable to Medjool. Others are dry dates such as the Jihel cv. (Table 24.4 and Fig. 24.7). Among 66 genotypes composed of 32 cultivars and a sample of 34 selected clones (Sedra 2003a), a ranking of these genotypes based on the fruit weight (tamar stage) revealed that of the first 20 common and selected cultivars, Medjool and Boufeggous occupy first and sixth places, respectively. Indeed, the fruit of Medjool cv. is the largest with only 40 fruits to the kg. Cultivars that occupy the last five positions, not in descending order, are namely: Oum'hal, Bouittob, Black Bousthammi, Azigzao and Hafs; these bear fruits weighing on average 3.4 g, requiring about 294 dates to the kg. On the other hand, the majority of selected cultivars are regrouped in the first half of the ranks of the set of genotypes. The selected cultivars with fruit weight exceeding 20 g, and the same approach to the Medjool cv. are, in descending order: Daraâouia (1445), Sedrat (3003) and Al-Amal (1443) (Table 24.4). Ayour cv. (3415) which has the highest fruit weight among the first selected cultivars, is unfortunately susceptible to bayoud. Others such as the clone-cvs. 3010 and Najda (3014) are classified in the group of Boufeggous cvs. The Najda cv. (3014) was selected about 15 years ago (Sedra 1993, 1995, 2003a, b) and has been important for distribution to farmers up to the present. The selected cvs. Al-Fayda (1447), Bourrihane (1414) and Mabrouk (1394) are classified in the group as being similar to the Jihel cv. Besides, most cultivars selected for their bayoud resistance and fruit quality possess a percentage of fruit flesh above 92%, with a maximum of 95.2% in the Al-Amal cv. (1443) and 93.9% for cultivar-clone (3010), comparable to that of Medjool (93.6%) (Table 24.4).

Table 24.5 Some agro-morphological and biologic characters of selected male cultivars to control bayoud

Males cvs. (INRA Ref No.)	Leaf length (cm)	Density of leaves in crown	Pollen production	Germination rate of pollen	Pollen fertility	Distorted or aborted pollen	Resistance to bayoud
Nebch-Bouskri (NP3)	383 moderate	Slightly open	Moderate	Very high	Very high	Low	Resistant
Nebch-Boufeggous (NP4)	438 long	Open	High	Very high	Very high	Low	Resistant

24.3.2 Heat Requirements, Precocity and Late Fruit Ripening

The heat requirement is a biologic and ecological character determining possible cultivation areas and localities, as well as the period of date maturity. For the selected and promising cultivars to control bayoud, and under the conditions at Zagora, the Sedrat cv. (3003) is precocious; Al-Amal (1443), Bourrihane (1414) and Al-Fayda (1447) are fairly precocious; cvs. Darâaoui (INRA-1445), Najda (3014) and the unnamed variety (3010) produce fruits in season, whereas the Mabrouk cv. (1394) is late (Table 24.4). Except for the precocious cv. Aguelid, as a reference, the other commercial cvs. such as Medjool and Jihel are relatively late. Boufeggous and Bouskri are somewhat late and resistant common cvs., notably Iklane, Black Bousthammi and Sairlayalate, are late ripening.

24.3.3 Length of Leaves and Fruit Stalks

Date palms selected as common cultivars have long to very long leaves requiring tree spacing of at least 10 × 10 m (100 trees per ha), whereas in the case of those having short to medium length leaves, the planting field can be more intensively utilized by adopting a higher a density of up to 8 × 8 m (150 palm trees per ha). On the other hand, cultivars with long stalks are most desired by farmers because this character facilitates certain agricultural practices such as bending, reducing length and number of inflorescences, fruit covering for protection during ripening and during harvest. Such is the case with the commercial cvs. Deglet Noor, Medjool and Boufeggous and some selected cvs. like Darâouia (1445), Al-Fayda (1447) and Sedrat (3003). It appears that the best cvs. are Medjool, Boufeggous, Jihel and Bouskri; some common bayoud-resistant cvs. (e.g. Black Bousthammi, Iklane and Slairlayalate) present favorable agro-morphological characters (Table 24.4). Figure 24.6 shows an example of a tree of selected cv. Al-Amal (1443).

24.3.4 Thorns – Number and Angle of Insertion

Those cultivars with thorns borne at a low angle to the rachis are very desirable because this character facilitates entering into the crown for pollination, bunch thinning, harvesting, etc. It is no surprise that the major and most cultivated cvs., such as Bouskri, Boufeggous and Medjool, possess these characters (Table 24.4). The eight selected resistant cultivars showed a diversity of these characters, notably for the number of thorns (Table 24.4) but only four among them present a low thorn insertion angle at the leaf midpoint. The Sedrat cv. (3003) is the only resistant and effective variety that possesses a low number of thorns along with a low insertion angle.

24.3.5 Selected Male Cultivars

The selected male cvs, Nebch-Bouskri (NP3) and Nebch-Boufeggous (NP4) possess, in addition to their resistance to bayoud, an important character of notably good quality pollen (Table 24.5), according to the descriptors developed by Sedra (2001). These males deserve to be exploited to the scale of palm groves within the framework of the national program of rejuvenation of palm groves ravaged by bayoud.

24.4 Conclusion and Prospective

Based on morphological and agronomic characters, results have shown a diversity of selected clones and cultivars with regard to some important characters in date palm cultivation. The characterization of some cultivars by using molecular approaches has been made (Sedra 2001; Sedra et al. 1998a) and continues, currently using new molecular biology techniques. The results of molecular characterization and identification are not presented here. Except for the Najda cv. (3014), multiplied and distributed in relatively significant quantities, the other affected cultivars did not exhibit satisfactory results. In order to control bayoud disease and to preserve genetic diversity within the date palm groves, and to rebuilt the framework of the national plan of reconstitution of oases set forth in the *Green Plan Morocco*, it is urgent to focus efforts on mass multiplication of these new selected cultivars that have been developed; their performances are reported in this paper. In fact, several of the new cultivars are currently being multiplied on a large scale and they will contribute to improve the quantity and quality of date production, leading to increased farmer income and the promotion of human development in the Saharan areas.

Acknowledgments Thanks to all the people who facilitated and assisted in field and laboratory (technicians, students, training persons) in the achievement of this work over 25 years. I also thank the researchers of the tissue-culture laboratory (INRA-Marrakech) who have supplied me with vitroplants to confirm the bayoud resistance of certain cultivars.

References

Abahmane L (2010) Micropropagation of date palm (*Phoenix dactylifera* L.) selected genotypes from inflorescence tissues by using somatic embryogenesis technique. Fourth International Conference on date Palm, Abu Dhabi, United Arab Emirates; 15–17 March 2010, Abstracts – Poster Session, p 49

Al-Bakr A (1972) The date palm: a review of its past and present status and its culture, industry and trade. Edit Alaïn Press, Iraq

Anjarne M, Bougerfaoui M, Abahmane L (2010) *In vitro* propagation as tools for the development of Morroccan date palm groves. Fourth International Conference on date Palm, Abu Dhabi, UAE; 15–17 March 2010, Abstracts – Poster Session, p 50

Anonymous (2009) Situation de l'Agriculture agricole. N°7. Ministère de l'Agriculture et de la Pêche Maritime, Avril 2009, Royaume du Maroc

Bougerfaoui M, Abahmane L, Anjarne M, Sedra MyH (2006) Utilisation de mutagenèse induite pour l'amélioration de la résistance aux toxines du bayoud chez le palmier dattier (*Phœnix dactylifera* L.). Conférence Régionale sur Mutagenèse Induite et Biotechnologies d'Appui pour la Protection du Palmier Dattier Contre le Bayoud projet AIEA – RAF/5/049. Alger, 17–18

Bulit J, Louvet J, Bouhot D, Toutain G (1967) Recherches sur les fusarioses. I. Travaux sur le bayoud, fusariose vasculaire du palmier dattier en Afrique du Nord. Ann Epiphytes 18:231–239

Djerbi M (1988) Les maladies du palmier dattier. Projet régional de lutte contre le Bayoud. FAO (Rab/84/0/8), Alger

Djerbi M, Sedra MyH (1982) Screening commercial Iraqi date varieties to bayoud. Nenadates News 2. PNUD/FAO

El Fakhouri R, Lazrek HB, Bahraoui E, Sedra MyH, Rochat H (1996) Preliminary investigation on a phytotoxic peptide produced *in vitro* by *Fusarium oxysporum* f.sp. *albedinis*. Phytopathol Mediterr 35:121–123

El-Fakhouri R, Lotfi F, Sedra MyH, Lazrek HB (1997) Production et caractérisation chimique des toxines sécrétées par *Fusarium oxysporum* f. sp. *albedinis*, agent causal du bayoud. Al Awamia 93:83–93

FAO (1995) Report of the expert consultation on date palm pest problems and their control in the Near East. Al-Ain, UAE 22–26 April 1995

Louvet J, Toutain G (1973) Recherches sur les fusarioses. VII. Nouvelles observations sur la fusariose du palmier dattier et précisions concernant la lutte. Annu Phytopathol 5:35–52

Pereau-Leroy P (1958) Le palmier dattier au Maroc. Ministère de l'Agriculture. Instit Franc. Rech, Outre-mer

Saaidi M (1992) Comportement au champ de 32 cultivars de palmier dattier vis-à-vis du bayoud: 25 ans d'observations. Agron 12:259–370

Sedra MyH (1990) Preliminary results on the evaluation of the resistance to the bayoud of the clones (Khalts), cultivars and some hybrids of the date palm trees selected on the fruit quality criterium. Eighth congress of Mediterranean Phytopathology Union, Agadir-Morocco

Sedra MyH (1992) Evaluation and selection of the resistant good cultivars and clones of date palm to the bayoud disease. Arab Soc Plant Prot 10:155–160

Sedra MyH (1993) Lutte contre le bayoud, fusariose vasculaire du palmier dattier: sélection des cultivars et clones de qualité résistants, et réceptivité des sols de palmeraies à la maladie. Thèse de Doctorat d'Etat es-Sciences PhD. Université Cadi Ayyad Marrakech, Maroc

Sedra MyH (1994a) Evaluation de la résistance à la maladie du bayoud causé par *Fusarium oxysporum* f. sp. *albedinis* chez le palmier dattier: recherche d'une méthode fiable d'inoculation expérimentale en pépinière et en plantation. Agron 14:445–452

Sedra MyH (1994b) Mise au point d'une méthode pour l'évaluation rapide de la resistance au bayoud de plantules du palmier dattier issus de semis. Al Awamia 86:21–41

Sedra MyH (1995) Triage d'une collection de génotypes de palmier dattier pour la résistance au bayoud causé par *Fusarium oxysporum* f. sp. *albedinis*. Al Awamia 90:9–18

Sedra MyH (1997) Diversité et amélioration génétique du patrimoine phénicicole marocain. In: Proceeding du séminaire national sur "Ressources Phytogénétiques et Développement Durable" Actes Editions IVA Hassan II, Rabat. pp 283–308

Sedra MyH (1999) Prospections et importance du bayoud en Mauritanie et actions urgentes à prendre pour lutter contre la maladie. Rapport de mission de consultation FAO effectuée du 19/10/99 au 18/11/1999 en République Islamique de Mauritanie et proposition de projet de lutte contre le bayoud dans ce pays. Mission financée par le projet Développement des Oasis, phase II, UFT/MAU/020//MAU

Sedra MyH (2001) Descripteurs du palmier dattier (*Phoenix dacttylifera* L.). Editions-INRA, Maroc, Imprimerie Al-Watania, Marrakech

Sedra MyH (2002) Rapport de mission de consultation d'expert international FAO effectuée du 06 Novembre 2001 – 04 Janvier 2002 en République Islamique de Mauritanie. Projet: UFT/MAU//025//MAU, Assistance technique de protection des palmeraies d'Adrar, FADES/FAO

Sedra MyH (2003a) Le bayoud du palmier dattier en Afrique du Nord, FAO, RNE/SNEA-Tunis. Editions FAO sur la protection des plantes. Imprimerie Signes, Tunis, Tunisie

Sedra MyH (2003b) Le palmier dattier base de la mise en valeur des oasis au Maroc. Techniques phoénicicoles et Création d'oasis. Edit INRA Maroc, Imprimerie Al Watania-Marrakech Maroc

Sedra MyH (2003c) Date palm cultivation, characterization and classification of main Mauritanian varieties. Edition AOAD, Al-Khartoume

Sedra MyH (2005) Phenological descriptors and molecular markers for the determination of true-to-type of tissue culture-derived plants using organogenesis of some Moroccan date palm (*Phoenix dactilyfera* L.) varieties. Al Awamia 113:87–101

Sedra MyH (2007a) Discovery of bayoud disease on date palm tree in Mauritania: its importance, kind and control strategies. Proceedings Fourth Symposium on Date Palm in Saudi Arabia. King Faisal University, Hofuf, 5–8 May 2007, Kingdom of Saudi Arabia

Sedra MyH (2007b) Bayoud disease of date palm in North Africa: recent distribution and remarks about its characterization, diagnosis and origin. Proceedings Fourth Symposium on Date Palm in Saudi Arabia. King Faisal University, Hofuf, 5–8 May 2007, Kingdom of Saudi Arabia

Sedra MyH, Besri M (1994) Evaluation de la résistance au bayoud du palmier dattier causé par *Fusarium oxysporum* f. sp. *albedinis*: Recherche d'une méthode de discrimination des vitroplants acclimatés en serre. Agron 14:467–472

Sedra MyH, El Fakhouri R, Lazrek HB (1993) Recherche d'une méthode fiable pour l'évaluation de l'effet des toxines secrétées par *Fusarium oxysporum* f. sp. *albedinis* sur le palmier dattier. Al Awamia 82:89–104

Sedra MyH, El Filali H, Benzine A et al (1996) La palmeraie marocaine: évaluation du patrimoine phénicicole. Fruits 51:247–259

Sedra MyH, Lasherme P, Trouslot P et al (1998a) Identification and genetic diversity analysis of date palm (*Phoenix dactylifera* L.) varieties from Morocco using RAPD markers. Euphy 103:75–82

Sedra MyH, Lazrek HB, Amraoui H, Nour S (2008) Pathogen toxins of bayoud disease on date palm: *in vitro* selection and biological and specificity activities. J Plant Pathol 90 Suppl: S3.57

Sedra MyH, Lazrek H, Lotfi F, Rochat H (1998b) *Fusarium oxysporum* f. sp. *albedinis* toxin isolation and use for screening of date palm plants for resistance to the bayoud disesase. Proceedings of XXV International Horticultural Congress (IHC), 2–7 August 1998 Brussels, Belgium

Tirichine M (1991) Caractérisatiques des palmeraies du M'zab et de Metlili. Ressources génétiques du palmier dattier – Comportement vis-à-vis du bayoud. Communication présentée au séminaire sur la génétique du palmier dattier tenu à Adrar, Algérie en Décembre 1990. Reprise au Bulletin du Réseau Maghrébin de Recherche sur la Phéniciculture et la Protection du palmier dattier, PNUD/FAO, Vol.1 N°3, 1991, pp 7–10

Zaher H, Sedra MyH (1998) Evaluation of some date palm hybrids for fruit quality and bayoud disease resistance. In: Proceedings of date palm research symposium, réseau de recherches et du développement du palmier dattier. Marrakech, 16–18 février 1998, Maroc, pp 181–192

Chapter 25
Molecular Markers for Genetic Diversity and Bayoud Disease Resistance in Date Palm

MyH. Sedra

Abstract The date palm (*Phoenix dactylifera L*) is a monocotyledoneus woody perennial and dioecious plant with a long generation life time. Traditional and modern genetic improvement in date palm need long time and considerable funds. The molecular markers can assist the selection and give better and efficient research strategies. Several researches cited in this overview paper showed the use molecular markers as tools to evaluate genetic diversity and genotyping of date palm cultivars. Based on statistical analysis, some informative molecular markers which are associated to some phenological characters in date palm. Previous study of the date-palm mitochondrial DNA had evidenced two plasmid-like DNAs that seem to be linked to resistance to Bayoud disease but these markers cannot distinguish both studied cultivars. The research using several hundred RAPD and ISSR primers allowed identifying several markers candidates which can distinguish partially or totally between resistant and susceptible cultivars of date palm. The difficulty and relatively weak efficiency were probably due to the nature of genetic status of resistance. These preliminary researches open new doors to explore in the use of molecular technologies in the development of programme breeding of date palm in order to select rapidly new varieties desired by farmers and more demanded by different markets. They also may give area in research and construction programme of date palm genetic map.

Keywords Date palm • Bayoud • Molecular techniques • Selection • Resistance • Fingerprinting

MyH. Sedra (✉)
Laboratoire de Phytopathologie, Génétique et de Lutte Intégrée, Centre Régional de Marrakech, Institut National de Recherche Agronomique (INRA),
BP. 533, Ménara 40000, Marrakech, Morocco
e-mail: sedramh@hotmail.com; mhsedra@yahoo.fr

S.M. Jain et al. (eds.), *Date Palm Biotechnology*,
DOI 10.1007/978-94-007-1318-5_25, © Springer Science+Business Media B.V. 2011

25.1 Introduction

The date palm (*Phoenix dactylifera L*) is a monocotyledonous woody perennial belonging to the Arecaceae family, which comprises 183 genera and more than 2,300 species (Moore 1973). Date palm is a dioecious plant with a long generation life time (a period of 4–5 years is necessary to reach the first flowering). One of the earliest cultivated tree crops it is believed to be native to the Arabian Gulf region, possibly southern Iraq. In very early times, the date palm was introduced by man into northern India, North Africa and southern Spain, where it plays a major role in arid zones. The date palm has traditionally been propagated vegetatively from offshoots produced by elite individual trees and more recently by plants derived from tissue culture. Although the number of cultivars in the world is about 5,000, the offshoot mode of distribution has probably limited the genetic diversity. In Morocco, the date palm is one of the most important traditional crops of the oases. In addition to its important ecological and social roles, this tree plays a significant role in human nutrition and animal feed, and is used to produce a wide range of end-products. In Morocco, around 4.8 million date palms are cultivated over an area of approximately 48,000 ha. To the present, more than 453 known cultivars have been identified. These are represented by some 2.1 million trees; the remaining 2.7 million trees originate from natural seed propagation and are commonly known as *khalts* (Sedra et al. 1996). The major constraints of date palm production are drought, low productivity and post-harvest techniques on traditional farms, as well as pests and diseases. Bayoud is the most serious fungal disease of the date palm caused by *Fusarium oxysporum* f. sp. *albedinis* which occurs in the major date palm-growing areas of Morocco, in a large portion of western and southern Algeria and in some areas of Mauritania (Sedra 2003a, c, 2007a, b). It is a serious disease in these North African countries and represents a serious threat to those countries which are still bayoud-free. Since 1963, the Moroccan National Agricultural Research Institute (INRA) has carried out, in collaboration with national and international partners, several scientific and applied investigations in order to serve date palm farmers and preserve the ecosystem of oases.

At present, the most promising means to control bayoud disease is the genetic approach using resistant cultivars. Several clones resistant to bayoud have been selected (Sedra 1990a, 1995, 1997, 2001, 2003a,b, 2005a), but they are each represented by only one to a few trees. In order to produce sufficient numbers of nursery plants for the reconstitution of date palm groves destroyed by bayoud, mass micropropagation of selected resistant date palms clones is essential. Among Moroccan cultivars, seven are resistant to bayoud and the behavior towards this disease of several hundred other cultivars among the majority of *khalt* trees is still unknown. Field observations alone of this behavior are not sufficient and the evaluation of this material by artificial inoculation using pathogens needs considerable funding and relatively lengthy time periods (Sedra 1993). Phenological and agro-morphological characters as quantitative or qualitative descriptors have been determined (Sedra 2001), but they cannot distinguish between resistant and susceptible cultivars, except for the importance of a black color at the leaf bases, which is indicative (Sedra 1990b). This approach, based on phenological characters, does not easily allow early detection of resistant lines for young plants. To the contrary, molecular markers may identify the change in behavior of palm trees, which is expressed from certain regions of the DNA, or the total composition of DNA.

Pathologists and breeders cannot ignore the progress made in plant biotechnology, including the application of DNA-based markers for quality assurance. DNA-based tests for date palm identification and detection of some characters include techniques such as: RAPD (Random Amplified Polymorphic DNA); RFLP (Restriction Fragment Length Polymorphism) (Botstein et al. 1980); random-amplified polymorphism DNA markers (RAPD) (Williams et al. 1990), AFLP (Amplification Fragment Length Polymorphism) (Blears et al. 1998; Vos et al. 1995); RDA (Representational Difference Analysis) (Cullis et al. 1999; Powell et al. 1996) simple sequence repeat polymorphism or microsatellites (SSR) (Akkaya et al. 1992; Gupta et al. 1994; Morgante and Oliviery 1993) and inter simple sequence repeat (ISSR) (Ajibade et al. 2000; Fang and Roose 1999; Fang et al. 1997; Stepansky et al. 1999).

25.2 Genetic Diversity and Characterization of Cultivars

Sedra (2001) enumerated 342 date-palm descriptors of which 105 are descriptive characters (tree, inflorescence, fruit); 132 agronomic characters (maturity, pollination, resistance, etc.), 62 chemical characters (chemical structure, etc.); 6 biochemical scorers (enzymes, etc.) and 37 RAPD molecular markers. The statistical analyses using the Static and Systat programs on a sample of more than 90 date palm cultivars grown in North Africa permitted an appreciation of the interrelationships observed between all quantitative and qualitative characters and to determine those that proved to be discriminative (Sedra 2001, 2003c; Sedra et al. 1996). Therefore, 25 quantitative characters have been recognized as discriminative, of which 9 are very highly discriminative (0–1%), 16 highly discriminative (1–5%) and 11 correlated to highly discriminative quantitative characters (Sedra 2001).

RAPD technology appears very effective for identifying accessions of date palm, although the overall polymorphism exhibited is rather low (Sedra et al. 1998) in comparison with results reported for other cultivated species (Hu and Quiros 1991; Koller et al. 1993; Mossler et al. 1992; Wolff and Van Run 1993; Yang and Quiros 1993). Previous molecular markers studies (Aît Chitt et al. 1995; Corniquel and Mercier 1994) involved a restricted set of date palm cultivars and were less rewarding. The relatively low polymorphism and lack of evident organization observed among the date palm cultivars grown in Morocco could be related to the mode of introduction and maintenance of germplasm (Sedra et al. 1998). The foundation of the germplasm is somewhat limited. The fact that the cultivars from Tunisia and Iraq did not markedly diverge from the genetic diversity present in Morocco suggests a narrow genetic diversity of populations from which the present cultivars have been derived and maintained over several centuries (Sedra et al. 1998). Exchange of cultivars between plantations and periodic development of new recombinant cultivars through sexual reproduction and seedling selection also may have played an important role. In addition, selection by farmers concerns mainly end-use quality-related genes which may represent only a small fraction of the date palm genome.

Cultivars showing resistance to bayoud disease in three different genetic groups may indicate the presence of several genetic resistance sources (Sedra et al. 1998). A combination of such potentially different sources of resistance could therefore be of interest in the framework of breeding programs. Date palm breeding and genetic improvements have relied, and continue to rely, on traditional methods. Advances in selection for agronomically-important traits, such as fruit quality or disease resistance, are difficult due to the species' long generation time. The Moroccan experience in date palm genetic improvement by crosses and mass selection reveals this reality (Djerbi et al. 1986; Louvet and Toutain 1973; Saaidi et al. 1981; Sedra 1995, 1997, 2003a, 2005a). The resistance character to bayoud disease seems the least complicated characters for date palm tree. Relatively fast methods were developed for the assessment of resistance to bayoud in different stages of plant growth (seedlings, young plants and adult trees) in the field and in laboratory (*in vitro* screening and in the greenhouse) (Sedra 1994a,b; Sedra and Besri 1994; Sedra et al. 1993). These methods of conventional breeding and hybridization programs are limited practically and financially when it is necessary to evaluate populations with a very large number of individuals.

Genetic engineering and molecular markers have not thus far been used for the improvement of date palm tree, but they are likely to play an important role in the future development of this crop. Until recently, there was little study of crops at the genome level; however, the next few years should bring molecular breeding technology to this plant group. Molecular breeding can be defined as the application of DNA-based analysis of genome polymorphism to breeding programs. Molecular markers can be linked to genes of interest like disease resistance, allowing indirect selection of the desired genotypes. Technologies for genome fingerprinting (molecular marker detection) include RFLP, RAPD, AFLP, SSR and ISSR. The use of amplified fragment length polymorphism (AFLP) has some advantages in terms of use in the identification of diagnostic or specific markers. Although these markers are generally dominants, the AFLP technique does not require previous knowledge of the DNA sequence, generates reproducible fingerprinting profiles and allows the amplification of a high number of DNA fragments per reaction, enabling the detection of specific amplified fragments (Vos et al. 1995). All of the molecular techniques (RAPD, AFLP, SSR and ISSR) have been applied to evaluate the genetic diversity and identification of date palm cultivars (Adawy et al. 2002, 2005; Cao and Chao 2002; Lacaze and Brackpool 2000; Saker and Moursy 1999; Sedra 2007c; Sedra et al. 1998) and for genetic comparison and the identification of vitroplants obtained by tissue-culture techniques from mother adult palms (Diaz et al. 2003; Saker et al. 2006; Sedra 2005b). Detection of somaclonal variations in tissue culture-derived date palm plants of some Egyptian cultivars has been accomplished using isozyme analysis and RAPD fingerprints (Saker et al. 2000). The genetic variations occurred in approximately 4% of the analyzed plants representing 70 regenerated plants. On the other hand, Rival et al. (1998) identified a molecular marker linked to somaclonal variations in African oil palm.

Molecular markers associated with bayoud resistance would be suitable for rapid and efficient screening of field grown palms produced by mass selection, hybrid plants from crosses in breeding programs, as well as plantlets from *in vitro* tissue culture.

25.3 Resistance in Date Palm

25.3.1 Biochemical and Plasmid Mitochondrial DNA Markers

In this approach, many markers that are correlated to resistant palms have been reported, such as isozymes (Baaziz 1990; Bendiab et al. 1993; Bennaceur et al. 1991), polyphenolics (El Hadrami et al. 1996; El Idrissi-Tourane et al. 1996), and mitochondrial plasmid-like DNAs (Benslimane et al. 1994). However, the correlation between the date-palm phenotype and the described marker has not been clearly established. In fact, previous studies of date-palm mitochondrial DNA provided evidence of two plasmid-like DNAs, called the S and R plasmids that are of 1,454 and 1,345 bp respectively (Benslimane 1995). These plasmids are of about 99% sequence similarity. A 109 bp sequence is only present in the S plasmid (Benslimane et al. 1996). The S and R plasmids were found in the mitochondria of two Moroccan cultivars: the first one is bayoud susceptible and contained the S plasmid, and the second containing the R plasmid is bayoud resistant. This suggested that S and R DNAs could be correlated to date-palm susceptibility to and resistance against bayoud. In Tunisia, Trifi (2001) has extended a similar study to nine Tunisian date-palm cultivars in order to obtain a deeper insight of the relationship that exists between these plasmids and the bayoud-tree phenotype (susceptibility/resistance). The analysis of results based on the detection of the mitochondrial plasmids did not agree with the date-palm phenotype against the bayoud, because seven of the nine cultivars tested showed the S plasmid. Surprisingly, the S plasmid and both S and R plasmids have been detected respectively in cvs. Deglet Noor and Horra. However, both of the tested cultivars have been shown susceptible to bayoud according to Saaidi (1992) and Sedra (1992). This finding showed that the correlation between the date-palm phenotype and these markers has not been clearly established. Trifi (2001) suggested that this feature could be justified by interrelations involving nuclear and mitochondrial genomes. This is supported by Flamand et al. (1993) who reported that the mitochondrial plasmids that arise by recombination events are controlled by the nuclear genome and by Sedra et al. (1998) who suggested a multi-gene control of date-palm bayoud-resistance/susceptibility. Otherwise, the genetic resistance can be controlled by dominant genes and additive genes (Djerbi and Sedra 1986; Sedra 2003a). This has been shown when the percentage of resistant individuals of progenies from S (susceptible) × R (resistant) or R × S parents and S × S parents reaches, respectively, more than 50% and 5%.

The abovementioned situation encouraged researchers to develop other efficient and potential molecular markers associated with bayoud resistance in date palm and how the polymerase chain reaction (PCR) technology allowed a powerful approach suitable in the rapid screening of selected bayoud-resistant individuals.

25.3.2 *RAPD and ISSR Markers for Characterization and Relationship with Phenological and Agronomic Traits Including Resistance to Bayoud Disease*

Data based on molecular markers such RAPDs, have been developed to molecularly characterize date-palm genotypes of cultivars and to examine their phylogenetic relationships (Sedra 2000; Sedra et al. 1998; Soliman et al. 2003; Trifi et al. 2000). Therefore, the search of many other markers is required to obtain a deeper understanding of the genetic organization in date-palm cultivars. Among the markers that can be investigated, microsatellites may be the efficient. In fact, microsatellites are interspersed in the genomes (Gupta et al. 1994; Sanchez et al. 1996) and they constitute discrete markers suitable in DNA fingerprinting. In addition, microsatellites are also informative about many loci and are suitable to discriminate closely-related genotype variants (Fang and Roose 1997). Microsatellites are small arrays (typically <100 bp) of simple di- and tri-nucleotide repeats (Scribner and Pearce 2000). The inter simple sequence repeat (ISSR) technique permits the detection of polymorphism in microsatellites and inter-microsatellites loci without previous knowledge of the DNA sequence. The sequences of repeats and anchored nucleotides are randomly selected (Fang et al. 1997). ISSR strategy was therefore performed to access the DNA diversity among crop genotypes. The ISSR markers were tested to examine the genetic variability and molecular fingerprinting of Egyptian date palm cultivars (Adawy et al. 2002, 2005; Ben Saleh and El-Helaly 2003). A similar strategy has been employed to distinguish ecotypes in closely-related groups such as vigna bean (Ajibade et al. 2000), citrus (Fang and Roose 1999; Fang et al. 1997) and melon (Stepansky et al. 1999).

The ISSRs and RAPDs markers were tested as informative markers to identify some other markers eventually related to date palm cultivar phenotype responses to bayoud disease. For this molecular approach, at least seven resistant and seven susceptible cultivars of date palm tree (Table 25.1) and numerous susceptible and resistant young plant-hybrids derived from controlled crosses Black Bousthammi (resistant female) × INRA-A18 (susceptible male) and Jihel (susceptible female) x male INRA-NP4-Boufegopus (resistant male) were used. The DNA extraction method was based on CTAB protocol. Yield of genomic DNA varied from 250 to 450 μg from 4 g of non-dried samples or from 0.2 to 0.3 g of dried-lyophilized leaflets samples.

The RAPD analysis and PCR were achieved according to the protocol used by Sedra et al. (1998) with some slight modifications. Oligonucleotide primers (10 mers)

Table 25.1 Moroccan date palm cultivars tested in a study concerning the research on molecular markers associated with resistance to bayoud disease

Cultivars	Main geographical area	Fruit quality	Phenotype to bayoud disease
Boufeggous ou Moussa	Bani	Fair	Resistant
Black Bousthammi	Bani, Draa valley,Tafilalet, Saghro, Anti-Atlas	Fair	Resistant
Boukhanni	Draa valley	Moderate	Resistant
Iklane	Anti-Atlas, Bani, Draa valley, Saghro	Fair	Resistant
Najda	Draa valley	Good	Resistant
Sairlayalate	Bani	Moderate	Resistant
Tademainte	Anti-Atlas, Bani, Draa valley, Oriental, Saghro	Fair	Resistant
White Bousthammi	Anti-Atlas, Bani	Fair	Resistant
Ahardane	Draa valley, Anti-Atlas, Saghro, Bani, Oriental		Susceptible
Boufeggous	Draa valley, Tafilalet, Ziz, Anti-Atlas, Saghro, Bani, Ferkala, Gheris, Guir, Todra, Oriental, between Saghro and High-Atlas	Good	Susceptible
Bourar	Draa valley, Bani, Saghro, Tafilalet	Good	Susceptible
Bouskri	Bani, Draa valley, Saghro,Todra, Oriental, Tafilalet, between Saghro and High-Atlas, Anti-Atlas	Moderate	Susceptible
Deglet Noor	Oriental	Good	Susceptible
Jihel	Draa valley, Bani, Anti-Atlas, Tafilalet, Saghro, between Saghro and High-Atlas	Good	Susceptible
Medjool	Tafilalet, Ziz valley, Oriental, Draa valley, Saghro, between Saghro and High-Atlas	Excellent	Susceptible

were purchased from Operon Technologies Inc., Alameda CA (OP) and The University of British Columbia (UBC) Vancouver. A total of 550 primers were used; 13 (OP) were already selected by Sedra et al. (1998) and 100 (UBC) interesting primers which were selected among more than 400 primers (Sedra 2007c) as primers generating polymorphism. Afterward, other RAPD primers were selected. For ISSR analysis, PCR was realized using 46 oligonucleotides composed of defined, short tandem repeat sequences representing different ISSR microsatellites used as genetic primers in PCR amplifications in Moroccan date palm cultivars. PCR was performed and adapted according to the protocol by Sedra and Zhar (2010). Only distinct reproducible, well-resolved fragments were scored as present (1) or absent (0) for both date cultivars. PCR generated band profiles obtained in seven resistant and seven susceptible cultivars cited in the Table 25.1 were analyzed to produce a genetic distance matrix using the formula of Nei and Li (1979). The genetic distance matrix was then computed using the UPGMA cluster analysis. A dendrogram was constructed using the average linkage between groups in order to evaluate the relation between resistant cultivars.

Table 25.2 Examples of selected RAPD primers that revealed interesting molecular markers which allowed good genetic diversity of date palm (bases of nucleotides between 3′ and 5′)

Primer sequences			
CACAGACACC	CCTGGGCTTA	AAGCTGCGAG	GATCCATTGC
CCGCTACCGA	CCTGGGTGGA	GGTCTCTCCC	GCTGGGCCGA
GAGGGAAGAG	TCCGGGTTTG	GCTTCCCCTT	AGGGAGTTCC
GGGAACGTGT	CCGGCCCCAA	TGTCGGTTGC	TCCACGGACG
GTCCACTGTG	GCGGCTGGAG	ATGTGTTGCG	CTGAGGCAAA
CCGAACACGG	GGGCAATGAT	CTGGGGATTT	CTGTCCAGCA
CTCCATGGGG	CTCGGGTGGG	GAGCACTTAC	CTGAAGGGGA
CCACACTACC	CGTCTGCCCG	CACGGCGAGT	ATGTTCCAGG
CACCCGGATG	AGTAGACGGG	CGGTTTGGAA	GCCCGACGCG
TGGTCGCAGA	ATCCCAAGAG	TGCACTGGAG	GGGTGAACCG
GGACACCACT	TGACCGAGAC	GAAGCGCGAT	TACCGACGGA
CTGGGGACTT	TTAGCGGTCT	CAGCGAACTA	GAGTAAGCGG
GTCGCCGTCA	ATTGGGCGAT	CATGTGCTTG	GCATCTACCG
TCTGGTGAGG	CGGTTACTAG	TCACACGTGC	CGACAGTCCC
GGTCTACACC	GTAGACGAGC	ACAGGTAGAC	CGGTGGCGAA
CACCGTATCC	GCGGTTGAGG	CCCGTCAATA	CGGCCCACGT
CATCCGTGCT	ATCTGGCAGC	AAGCCTCCCC	GCTGGTACCC
AGGGCGTAAG	GAGCCAGAAG	CTAGGGGCTG	ATCTAGGGAC
CTCACGTTGG	ACAGGGAACG	GTCGCCGTCA	TCTGGTGAGG
CTCCATGGGG	CTGGGGACTT	TCGTCTAGCT	GGTCTACACC
CCACACTACC	CCGAACACGG	AGGGCGTAAG	CACCGTATCC
CTCACGTTGG	GGGAACGTGT	GGACACCACT	CACCCGGATG
CACAGACACC	ACAGGCAGAC	GAGGGAAGAG	TGGTCGCAGA
ATACAGGGAG	CATCCGTGCT	GTCCACTGTG	CCGCTACCGA

25.3.2.1 Selection of RAPD and ISSR Molecular Markers Allowing Genetic Diversity and Cultivar Genotyping

In order to select interesting primers, a total of 550 RAPD primers 10-Decamer were tested on date palm DNA and 170 of them were selected permitting identification of more than 300 polymorphic markers which are able to detect polymorphism and genetic diversity and to identify date-palm cultivars. The percentage of polymorphism may reach 70% and 1–5 polymorphic bands per primer were generally generated. Tables 25.2 and 25.3 and Fig. 25.1 present examples of selected primers (Sedra 2007c). For ISSR primers, the PCR analysis applied to 46 primers (short sequences) tested, allowed the selection of 21 primers that showed a high rate of polymorphism among 45 date palm cultivars studied. These primers have revealed more than 80 polymorphic markers. The percentage of polymorphism varies from 43% to 100% (average 80%) according to primers and the average of 6.1 polymorphic bands per primer was generated. Table 25.4 and Fig. 25.2 present some examples of selected ISSR primers. For both RAPD and ISSR markers, the results were reproducible.

Table 25.3 Examples of selected RAPD molecular markers and their sizes that revealed good genetic diversity of date palm and cultivar identification

RAPD primers	Primer sequences	Molecular marker weight (kb)
OP-D3	5′-GTCGCCGTCA-3′	0.15
OP-D4	5′-TCTGGTGAGG-3′	0.7
OP-D10	5′-GGTCTACACC-3′	0.1
OP-D12	5′-CACCGTATCC-3′	1.1
OP-D12	5′-CACCGTATCC-3′	1.9
OP-D15	5′-CATCCGTGCT-3′	1
OP-D15	5′-CATCCGTGCT-3′	1.06
OP-D16	5′-AGGGCGTAAG-3′	0.83
OP-D16	5′-AGGGCGTAAG-3′	1.06
OP-J4	5′-CCGAACACGG-3′	2.5
OP-J5	5′-CTCCATGGGG-3′	1.12
OP-J13	5′-CCACACTACC-3′	0.9
OP-J14	5′-CACCCGGATG-3′	1.1
OP-J18	5′-TGGTCGCAGA-3′	1.47
OP-M5	5′-GGGAACGTGT-3′	1.16
OP-M11	5′- GTCCACTGTG-3′	1.16
OP-M11	5′- GTCCACTGTG-3′	0.94
OP-M11	5′- GTCCACTGTG-3′	0.67
OP-N1	5′-CTCACGTTGG-3′	0.5
OP-L6	5′-GAGGGAAGAG-3′	0.86
OP-X4	5′-CCGCTACCGA-3′	2.29
UBC-3	5′-CCTGGGCTTA-3′	0.38
UBC-3	5′-CCTGGGCTTA-3′	0.46
UBC-3	5′-CCTGGGCTTA-3′	0.7
UBC-13	5′-CCTGGGTGGA-3′	0.76
UBC-173	5′-CAGGCGGCGT-3′	0.75
UBC-201	5′-CTGGGGATTT-3′	1.47
UBC-209	5′-TGCACTGGAG-3′	1.4
UBC-213	5′-CAGCGAACTA-3′	0.63
UBC-256	5′-TGCAGTCGAA-3′	1.42
UBC-276	5′-AGGATCAAGC-3′	0.35
UBC-302	5′-CGGCCCACGT-3′	2.52
UBC-304	5′-AGTCCTCGCC-3′	0.75
UBC-310	5′-GAGCCAGAAG-3′	1.02

25.3.2.2 Molecular Markers and Quantitative and Qualitative Descriptors of Date Palm

Preliminary studies based on statistical analysis revealed the relationship between 31 quantitative and qualitative descriptors of date palm tree and 34 selected RAPD molecular markers (from m1 to m34) (Sedra 2007c). As cited above, phenological and agro-morphological characters cannot distinguish between resistant and susceptible

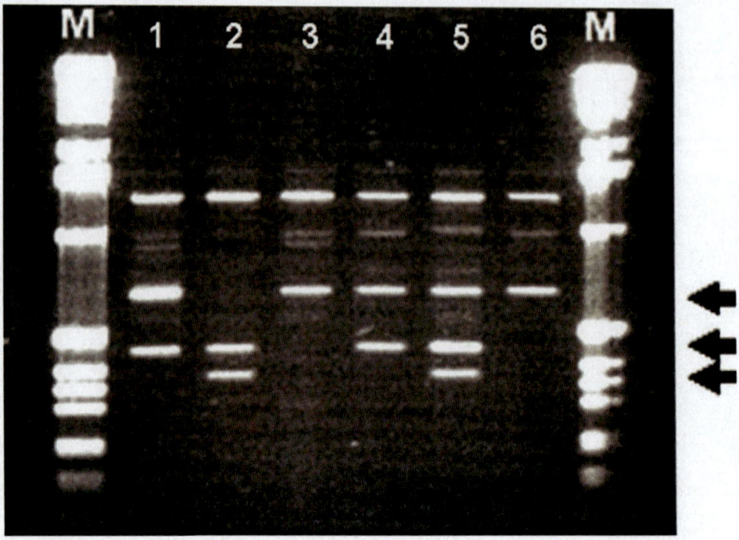

Fig. 25.1 Examples of date palm DNA fragments amplified using RAPD primer UBC-3. Ethidium bromide-stained agarose gel of amplification fragments. From lane *1* to lane *6*: Sampled Moroccan cvs. respectively, Boukhani, Bouzeggar, Boufegous ou Moussa, Outokdime, Iklane and Ahardane. M: Standard molecular weight size (λ/EcoR1/Hind III.BAP). The *arrows* indicate the discriminative markers

Table 25.4 Examples of selected ISSR primers that revealed interesting molecular markers which allowed good genetic diversity of date palm (bases of nucleotides between 3′ and 5′)

ISSR primers	Primer sequences	ISSR primers	Primer sequences
Mic 3	$(AC)_{10}$	Mic 15	$(TC)_{10}$
Mic 4	$(AGT)_5$	Mic 16	$(AG)_{12}$
Mic 5	$(ATC)_5$	Mic 17	$(GA)_{13}$
Mic 6	$(GATA)_4$	Mic 18	$(GGGA)_4$
Mic 7	$(GACA)_4$	Mic 19	$(TA)_{14}$
Mic 9	$(TGTC)_4$	Mic 21	$(CTCACA)_4$
Mic 10	$(AAC)_8$	Mic 50	$(AGG)_8$ TC
Mic 11	$(TATG)_4$	Mic 51	$(TCC)_5$ AG
Mic 12	$(AAG)_8$	Mic 52	$(ATG)_5$ AG
Mic 13	$(TTC)_8$	Mic 54	$(TCC)_5$ G
Mic 14	$(CT)_8$		

cultivars, except the importance of the black color in palm leaf bases which showed an indication of a resistant cultivar (Sedra 1990b). The results based on molecular analysis permitted the identification of some informative molecular markers that have a relationship with some descriptors of date palm; for example, width of the stalk carrying the fruits; width of the spathe; length of the thorns (spines) situated in the middle of the leaf; number of inflorescences (tiny flowers) on the spike situated in the

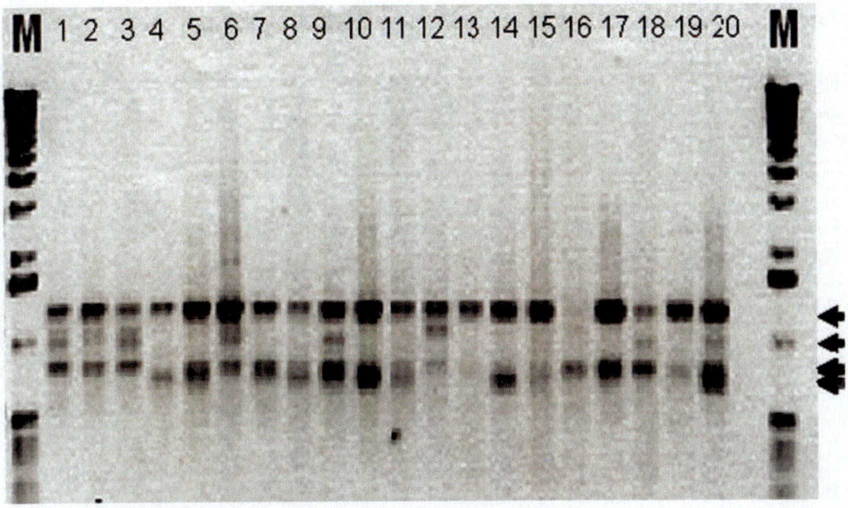

Fig. 25.2 Examples of DNA fragments amplified using ISSR primer $(CTCACA)_4$ in 20 Moroccan date palm cultivars. Ethidium bromide-stained agarose gel of amplification fragments. Lanes *1* and *20* contain fragments of sampled cultivars. M: Standard molecular weight size (λ/EcoR1/Hind III. BAP). The *arrows* indicate the discriminative markers

middle; angle made between the apical leaflets of the palm leave; date-fruit shape; color of the bases of the palm leaves (Table 25.5) (Sedra 2007c). This last case is interesting for our present studied topic. The informative molecular markers grouped with this character are m11, m21, m33 and m25 of which primer name and marker size are cited in Table 25.5.

25.3.2.3 Molecular Markers and Resistance to Bayoud Disease in Date Palm

Table 25.6 indicates the examples of molecular markers revealed by RAPD and ISSR techniques which are candidates to be associated with resistance to bayoud disease in the date palm. It appears that the markers cited have been detected in different resistant cultivars. The RAPD marker UBC-145-1.22 is present in five resistant cultivars among seven studied (Table 25.7 and Fig. 25.3). The RAPD-UBC-578-1.50 is present in 5/6 resistant cultivars. The ISSR marker Mic19-1.37 is detected in both six resistant cultivars studied (Table 25.7 and Fig. 25.4). Other markers are only detected in one or a few resistant cultivars. In results not presented here, these markers have been revealed in the majority of resistant hybrids (young plantlets) derived from crossing a resistant parent with the other susceptible parent. These markers therefore can be transmitted to the progeny. These results suggest that the resistance could be encoded by different genes.

Table 25.5 Relationship between certain informative molecular markers and certain quantitative and qualitative descriptors of date palm

Descriptors	Informative molecular markers
Quantitative descriptors	
Percentage of length of the part of the thorns (spines)	m1
Width of the spathe	m10, m25, m30
Width of the stalk carrying the dates	m35
Length of the thorn at the middle	m14, m2, m4, m14
Total number of the inflorescences or tiny flowers on middle spike	m30
Total number of the thorns	m10, m30, m25
Angle made between the apical leaflets of the palm leaf	m30, m20, m25, m10
Angle made between the thorn at the middle and the rachis	m27, m3
Qualitative descriptors	
Date shape	m24, m9, m22, m5, m12, m27, m32
Density of leave grouping	m22, m1, m15, m11, m9, m17, m21
Consistency of the leaflets of the palm	m14, m17, m21
Consistency of thorns	m11
Color of the bases of the palm leaves	m11, m21, m33, m25
The arching of the palm leaf in relation to the trunk of tree	m13, m31, m34, m11, m30, m37

Table 25.6 Examples of the sizes of some informative RAPD molecular markers

Informative molecular markers	Name and size (kb)
m11	OP-J14-0.12
m19	OP-D12-1.9
m21	OP-D12-1.58
m22	OP-J4-2.5
m24	OP-D16-0.83
m25	OP-N1-0.5
m32	OP-L6-0.86
m33	OP-M5-1.16
m34	OP-M11-1.16

The dendrogram of 14 Moroccan date-palm cultivars (seven susceptible and seven resistant cited in Table 25.1) constructed by genetic distance using 79 ISSR markers showed two groups which each contains two sub-groups (Fig. 25.5). Each sub-group comprises at least one or two resistant cultivars. This supposes that the resistance may have several sources localized in different regions. These data agree with those describing the application of RAPD molecular tools in date-palm variability analysis and previously reported (Sedra et al. 1998).

Table 25.7 Examples of molecular markers revealed by RAPD and ISSR techniques and that are candidates to be associated with resistance to bayoud disease in the date palm

Primer	Total number of revealed bands	Size of markers candidates (Kb)	Number and origin (cultivars): No. markers detected/studied resistant cultivars	Number and origin (cultivars): Name of cultivars
RAPD-OP-D16	6	1.06	3/7	Black Bousthammi, Iklane, Tadmainte
RAPD-OP-D19	6	0.10	4/7	Black Bousthammi, White Bousthammi, Boukhani, Tadmainte
RAPD-UBC-145	5	1.22	5/7	Black Bousthammi, White Bousthammi, Boufeggous ou Moussa, Boukhani, Tadmainte
RAPD-UBC-578	4	1.50	5/6	Black Bousthammi, White Bousthammi, Boufeggous ou Moussa, Tadmainte
RAPD-UBC-594	14	0.64	3/6	Iklane, Sairlayalate, Tadmainte
ISSR-Mic 19	11	1.37	6/6	Black Bousthammi, White Bousthammi, Boufeggous ou Moussa, Iklane, Sairlayalate, Tadmainte
		1.01	2/6	Black Bousthammi, Iklane

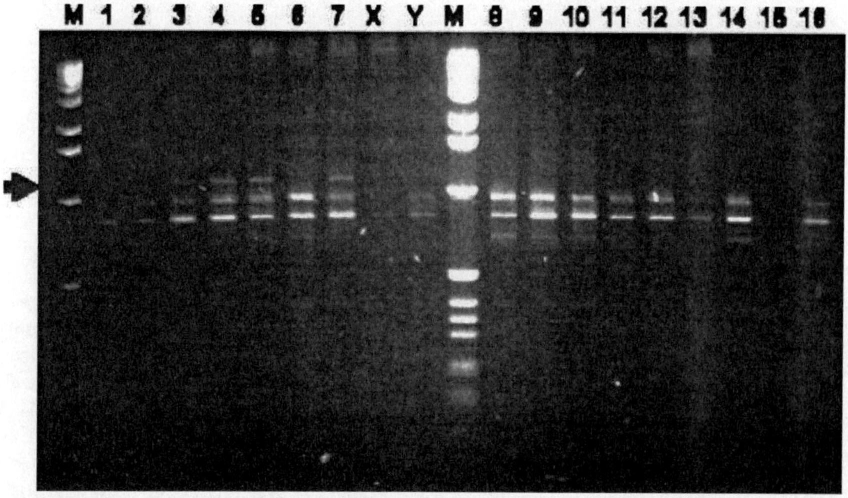

Fig. 25.3 DNA fragments amplified using RAPD-UBC-145 primer. Ethidium bromide-stained agarose gel of amplification fragments. Resistant cvs. *1* Black Bousthammi, *2* White Bousthammi, *3* Tadmainte, *4* Iklane, *5* Sairlayalate, *6* Boufegous ou Moussa, *7* Boukhani, X and Y males. Susceptible cvs. *8* Ahardane, *9* Boufegous, *10* Bourar, *11* Bouskri, *12* Deglet Noor, *13* Jihel, *14* Bouittob, *15* Mejhoul, *16* Outokdime. M: fragments of molecular weight markers (λ/EcoR1/Hind III.BAP). The *arrow* indicates the discriminative marker

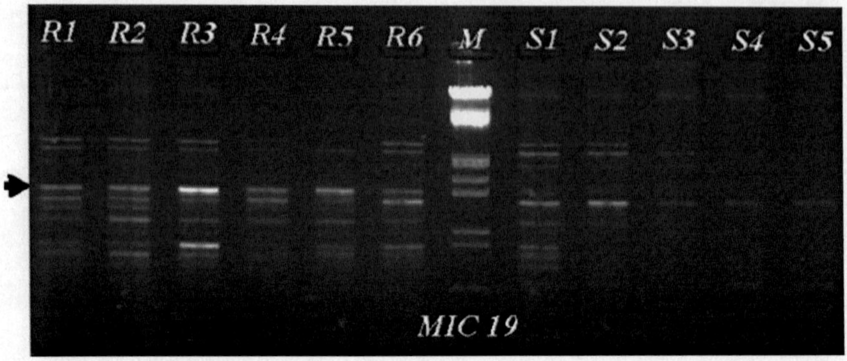

Fig. 25.4 DNA fragments amplified using ISSR Mic19. Ethidium bromide-stained agarose gel of amplification fragments. Resistant cvs.: *R1* Black Bousthammi, *R2* Iklane, *R3* Tadmainte, *R4* Sairlayalate, *R5* Boufegous ou Moussa, *R6* White Bousthammi; Susceptible cvs.: *S1* Boufegous, *S2* Oum N'hale, *S3* Jihel, *S4* Hafs, *S5* Ademou, *S6* Belhazit (non visualized). *M* fragments of molecular weight markers (λ/EcoR1/Hind III.BAP). The *arrow* indicates the discriminative marker

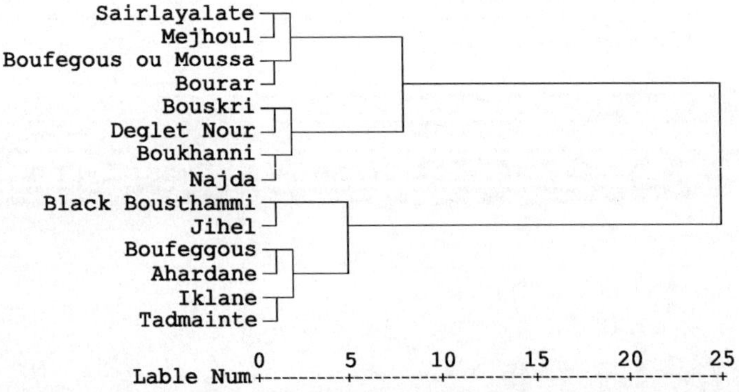

Fig. 25.5 Dendrogram of 14 Moroccan date-palm cultivars constructed by genetic distance using 79 ISSR markers. Clustering was with the UPGMA method. Resistant cvs.: Black Bousthammi, Boufegous ou Moussa, Boukhani, Iklane, Najda, Sairlayalate and Tadmainte. Susceptible cvs.: Ahardane, Boufegous, Bourar, Bouskri, Deglet Nour (Deglet Noor), Jihel and Mejhoul (Medjool)

25.4 Conclusion and Prospective

Traditional and modern genetic improvement in date palm need extended time periods and considerable funds. Therefore, they can be assisted by molecular markers that give better and more efficient research strategies. Several research results cited in this chapter show the use of molecular markers as tools to evaluate genetic diversity and genotyping of date-palm cultivars. Based on statistical analysis, Sedra (2007c) reported certain informative molecular markers which are associated with specific phonological characters in date palm. Previous study of date-palm mitochondrial DNA gave evidence of two plasmid-like DNAs that seem to be linked to bayoud-disease resistance (Benslimane et al. 1996) but these markers cannot distinguish both cultivars studied (Trifi 2001). Each marker corresponds to one part of date palm DNA and the genome has the size estimated to 1.7 pg and it is constituted of more than 10^{12} nucleic bases. These data seem to suggest that the higher the number of markers used the greater the probability to achieve more precise results. Our research using several hundred RAPD and ISSR primers allowed identifying several markers as candidates which can distinguish partially or totally between resistant and susceptible cultivars of date palm. The difficulty and relatively weak efficiency were probably due to the nature of the genetic status of resistance. These preliminary research results open new doors to explore the use of molecular technologies in the development of a breeding program of date palm in order to rapidly select new cultivars desired by farmers, and fruit more in demand by different markets. They also may provide an area of research and a construction program of the date palm genetic map.

Acknowledgements Thanks are due to my research institution (Moroccan INRA) and other international institutions and organizations (IAEA, AOAD, ICGEB, FAO, IFS, IFAD, IBD, ACSAD, IRD and other partners) for help and funding support and to all the individuals who helped from near and far: laboratory technicians, training people and students and the INRA-Zagora experimental station staff.

References

Aît Chitt M, Mantell SH, Thangavelu M, Ainsworth CC (1995) Cloning date palm (*Phoenix dactylifera* L.) DNA and characterization of low, medium and high copy DNA sequences. Elaeis 1:57–63

Adawy S, Hussein E, El-Khishin D et al (2002) Genetic variability studies and molecular fingerprinting of some Egyptian date palm (*Phoenix dactylîfera* L.) cultivars. II-RAPD and ISSR profiling. Arab J Biotechnol 5:225–236

Adawy SS, Hussein E, Ebtissam HA et al (2005) Genomic diversity in date palm (*Phoenix dactylifera* L.) as revealed by AFLP in comparison to RAPDs and ISSRs. Arab J Biotechnol 8:99–114

Ajibade SR, Weeden NF, Chite SM (2000) Inter simple sequence repeat analysis of genetic relationships in the genus Vigna. Euphytica 111:47–55

Akkaya M, Bhagwat AA, Cregan PB (1992) Length polymorphisms of simple séquence repeat DNA in soybean. Genetics 132:1131–1139

Baaziz M (1990) Contribution à l'étude des peroxydases du palmier dattier (*Phoenix dactylifera* L.). Relations avec la résistance de la plante à la maladie du bayoud, fusariose vasculaire à *Fusarium oxysporum* f.sp. *albedinis*. Thèse de Doctorat d'Etat, Université Cadi Ayed, Faculté des Sciences Sémlalia, Marrakech, Maroc, 173 p

Bendiab K, Baaziz M, Brakez Z, Sedra MyH (1993) Correlation of isoenzyme polymorphism and bayoud-disease resistance in date palm cultivars and progeny. Euphytica 65:23–32

Ben Saleh M, El-Helaly R (2003) Using ISSR to investigate the similarities and dissimilarities between some Tunisian date palm (*Phoenix dactylifera* L.) cultivars. (in Arabic and abstract in English). International Conference on Date Palm, Kingdom of Saudia Arabia, 16–19 Sept 2003

Bennaceur M, Lanaud C, Chevalier MH, Bounaga N (1991) Genetic diversity of date-palm (*Phoenix dactylifera* L.) from Algeria detected by enzyme markers. Plant Breed 107:56–69

Benslimane AA (1995) Plasmides mitochondriaux du palmier dattier (*Phoenix dactylifera* L.). Mise en évidence et caractérisation moléculaire. Thèse de Doctorat d'Etat, Université Cadi Ayed, Faculté des Sciences Semlalia, Marrakech, Maroc, 219 p

Benslimane AA, Rode A, Quetier F, Hartmann C (1994) Characterization of two minicircular plasmide-like DNAs isolated from date-palm mitochondria. Curr Genet 26:535–541
Benslimane AA, Hartmann C, Ouenzar B, Rode A (1996) Intramolecular recombination of a mitochondrial minicircular plasmide-like DNA of a date-palm mediated by a set of short direct repeat sequences. Curr Genet 29:591–593
Blears MJ, De Grandis SA, Lee H, Trevors JT (1998) Amplified fragment length polymorphism (AFLP): a review of the procedure and its applications. J Ind Microbiol Biotechnol 21:99–114
Botstein D, White RL, Skolnick MH, Davis RW (1980) Construction of a genetic map in man using restriction fragment length polymorphisms. Am J Hum Genet 32:314–331
Cao BR, Chao CCT (2002) Identification of date cultivars in California using AFLP markers. Hort Sci 37:966–968
Djerbi M, Sedra MyH (1986) Résistance to *Fusarium oxysporum* f.sp. *albedinis* in *Phoenix dactylifera*: Evaluation of new screening method and performance of american high quality backcrossed males. Proceeding of the Second Symposium on the Date Palm. March 3-6/1986, Date Palm Research Center, King Faisal University, El-Hassa, Saoudi Arabia, pp 400–412
Djerbi M, Aouad L, El Filali H et al. (1986) Preliminary results on selection of high quality Bayoud resistant lines among natural date palm population in Morocco. Proceeding of the second symposium on the date palm. Mar 3-6/1986, Date Palm Research Center, King Faisal University, El-Hassa, Saoudi Arabia, pp 386–399
Corniquel B, Mercier L (1994) Date palm (*Phoenix dactylifera* L.) cultivar identification by RFLP and RAPD. Plant Sci 101:163–172
Cullis C, Rademan S, Kunert KJ (1999) Method for finding genetic markers of somaclonal variation. International publication number WO 99/53100
Diaz S, Pire C, Ferrer J, Bonete MJ (2003) Identification of *Phoenix dactylifera* L. varieties based on amplified fragment length polymorphism (AFLP) markers. Cell Mol Biol Lett 8:891–899
El Hadrami I, Ramos T, Macheix JJ (1996) Caractérisation de nouveaux dérivés hydroxycinnamiques amines chez *Phoenix dactylifera* L.: relation avec le brunissement des tissus et la résistance des cultivars au bayoud. Polyphénols Communications 2:341–342
El Idrissi-Tourane A, Ramos T, El Bellaj M, El Hadrami I (1996) Importance des phénolamides et des polyamines des racines du palmier dattier dans la résistance/sensibilité au Bayoud. Polyphénols Communications 2:371–372
Fang DQ, Roose ML (1997) Identification of closely related *Citrus* cultivars with inter-simple sequence repeat markers. Theor Appl Genet 95:8–417
Fang DQ, Roose ML, Krueger RR, Federici CT (1997) Fingerprinting trifoliate orange germplasm accessions with isozymes, RFLPs, and inter-simple sequence repeat markers. Theor Appl Genet 95:211–219
Fang DQ, Roose ML (1999) Inheritance of inter simple sequence repeat markers in *Citrus*. J Hered 90:247–249
Flamand MC, Duc G, Goblet JP et al (1993) Variant mitochondrial plasmids of broad bean arose by recombination and are controlled by the nuclear genome. Nucl Acids Res 21:5468–5473
Gupta M, Chyi Y-S, Romeo-Severson J, Owen JL (1994) Amplification of DNA markers from evolutionarily diverse genomes using single primers of simple-sequence repeats. Theor Appl Genet 89:998–1006
Hu J, Quiros CF (1991) Identification of broccoli and cauliflower cultivars with RAPD markers. Plant Cell Rep 10:505–511
Koller B, Lehmann A, McDermott JM, Gessler C (1993) Identification of apple cultivars using RAPD markers. Theor Appl Genet 85:901–904
Lacaze P, Brackpool A (2000) Molecular fingerprinting of date palm cultivars using AFLP. Date Palm International Symposium. Windhoek, Namibia, 22–25 Feb 2000
Louvet J, Toutain G (1973) Recherches sur les fusarioses VII. Nouvelles observations sur la fusariose du palmier dattier et précisions concernant la lutte. Ann Phytopathol 5:35–52
Moore HE (1973) The major groups of palms and their distributions. Ferites herbarium 1:27–141
Morgante M, Oliviery AM (1993) PCR-amplified microsatellites as markers in plant genetics. Plant J 1:175–182

Mossler A, Egger KN, Hughes GA (1992) Low levels of genetic diversity in red pive confirmed by random amplified polymorphic DNA markers. Can J For Res 22:1332–1337

Nei M, Li WH (1979) Mathematical models for studying genetic variation in ternis of restriction endonucleases. Proc Natl Acad Sci USA 76:5269–5273

Powell W, Morgante M, Andre C, Hanafey M, Vogel J, Tingey S, Rafalski A (1996) The comparison of RFLP, RAPD, AFLP, and SSR microsatellite markers for germplasm analysis. Mol Breed 2:225–238

Rival A, Tregear J, Verdeil JL et al (1998) Molecular search for mRNA and genomic markers of the oil palm "mantled" somaclonal variations. Acta Hort 461:165–171

Saaidi M (1992) Comportement au champ de 32 cultivars de palmier dattier vis à vis du bayoud: 25 années d'observations. Agronomie 12:359–370

Saaidi M, Bannerot TG, Louvet J (1981) La sélection du Palmier-dattier (*Phoenix dactylifera* L.) pour la résistance au Bayoud. Fruits 36:241–249

Saker MM, Moursy HA (1999) Molecular characterization of Egyptian date palm: II – RAPD Fingerprints. Arab J Biotechnol 2:71–78

Saker MM, Bekheet SA, Taha HS et al. (2000) Detection of somaclonal variations in tissue culture-derived date palm plants using isozyme analysis and RAPD fingerprints. Date palm international symposium. Windhoek, Namibia. 22–25 Feb 2000

Saker MM, Adawy SS, Mohamed AA, El-Itriby HA (2006) Monitoring of cultivar identity in tissue culture-derived date palms using RAPD and AFLP analysis. Biol Plant 50:198–204

Sanchez De La Hoz MP, Davalia JA, Loarce Y, Ferrer E (1996) Simple sequence repeat primers used in polymerase chain reaction amplification to study genetic diversity in barley. Genome 39:112–117

Sedra MyH (1990a) Preliminary results on the evaluation of the resistance of the Bayoud of the clones (Khalts), cultivars and some hybrids of the date palm trees selected on the fruit quality *criterium*. 8th congress or Mediterranean Phytopathological Union, Agadir Morocco, pp 529

Sedra MyH (1990b) Observations on some phenotypical criteria that are eventually in relation with the date palm tree resistance to Bayoud. 8th congress of the Mediterranean Phytopathological Union, 28 Oct 3 Nov 1990, Agadir, Maroc

Sedra MyH (1992) Evaluation and selection of the resistant good cultivars and clones of date palm to the Bayoud disease. Arab Society for Plant Protection 10(2):155–160

Sedra MyH (1993) Lutte contre le Bayoud, fusariose vasculaire du palmier dattier causée par *Fusarium oxysporum* f.sp. *albedinis :* sélection des cultivars et clones de qualité résistants et réceptivité des sols de palmeraies à la maladie. Thèse de Doctorat d'Etat es-Sciences, Fac. Sc., Sémlalia, Marrakech, Maroc, 142 p

Sedra MyH (1994a) Mise au point d'une méthode pour l'évaluation rapide de la resistance au Bayoud de plantules du palmier dattier issues de semis. INRA-Rabat Maroc. Al Awamia 86:21–41

Sedra MyH (1994b) Evaluation de la résistance à la maladie du Bayoud causé par *Fusarium oxysporum* f.sp. *albedinis* chez le palmier dattier: Recherche d'une méthode fiable d'inoculation expérimentale en pépinière et en plantation. Agronomie 14:445–452

Sedra MyH (1995) Triage d'une collection de génotypes de palmier dattier pour la résistance au Bayoud causé par *Fusarium oxysporum* f.sp. *albedinis.* INRA-Rabat Maroc. Al Awamia 90:9–18

Sedra MyH (1997) Diversité et amélioration génétique du patrimoine phénicicole marocain. Proceeding du Séminaire National sur "Ressources Phytogénétiques et Développement Durable" Actes Editions IVA Hassan II, Rabat, pp 283–308

Sedra MyH (2000) Biological and genetic characteristics of Bayoud resistant Moroccan date palm cultivars and strains. Agric Water 20:55–66

Sedra MyH (2001) Descripteurs du palmier dattier (*Phoenix dactilyfera* L.). Edit. INRA Maroc, Imprimerie Al Watania-Marrakech Maroc), 135 p

Sedra MyH (2003a) Le Bayoud du palmier dattier en Afrique du Nord, FAO, RNE/SNEA-Tunis. Editions FAO sur la protection des plantes. Imprimerie Signes,Tunis, Tunisie, 125 p

Sedra MyH (2003b) Le Palmier Dattier base de la mise en valeur des oasis au Maroc. Techniques phoénicicoles et Création d'oasis. Edit. INRA Maroc, Imprimerie Al Watania-Marrakech Maroc, 276 p

Sedra MyH (2003c) Date palm cultivation, characterization and classification of main Mauritanian varieties. Edition AOAD, Al-Khartoum, Sudan, 261 p

Sedra MyH (2005a) Caractérisation des clones sélectionnés du palmier dattier et prometteurs pour combattre la maladie du Bayoud. The international symposium: sustainable agricultural development of oasian systems. March 07–10 2005, Erfoud Morocco

Sedra MyH (2005b) Phenological descriptors and molecular markers for the determination of true-to-type of tissue culture-derived plants using organogenesis of some Moroccan date palm (*Phoenix dactilyfera* L.) Varieties. INRA-Rabat Maroc. Al Awamia 113:87–101

Sedra MyH (2007a) Discovery of Bayoud disease on date palm tree in Mauritania: its importance, kind and control strategies. Proceeding of the fourth symposium on Date Palm King Faisal University, Hofuf, 5–8 May 2007, Kingdom of Saudi Arabia

Sedra MyH (2007b) Bayoud disease of date palm in North Africa: recent distribution and remarks about its characterization, diagnosis and origin. Proceeding of the fourth symposium on Date Palm King Faisal University, Hofuf, 5–8 May 2007, Kingdom of Saudi Arabia

Sedra MyH (2007c) Selection of morphological characteristics and molecular markers and their use for identification and distinguishing between date palm varieties and the plants issued from tissue culture. Proceeding of the fourth symposium on Date Palm King Faisal University, Hofuf, 5–8 May 2007, Kingdom of Saudi Arabia

Sedra MyH, Besri M (1994) Evaluation de la résistance au Bayoud du palmier dattier causé par *Fusarium oxysporum* f.sp. *albedinis*: Recherche d'une méthode de discrimination des vitroplants acclimatés en serre. Agronomie 14:467–472

Sedra MyH and Zhar N (2010) Genetic variability analysis of populations of *Fusarium oxysporum* f. sp. *albedinis*, causal agent of Bayoud disease of date palm and other *Fusarium oxysporum* using molecular techniques. (Proc. 4th Int Date Palm conference United Arab Emirates, Abu Dhabi). ACTA Hort 882:491–504

Sedra MyH, El Fakhouri R, Lazrek HB (1993) Recherche d'une méthode fiable pour l'évaluation de l'effet des toxines secrétées par *Fusarium oxysporum* f.sp.*albedinis* sur le palmier dattier. INRA-Rabat Maroc. Al Awamia 82:89–104

Sedra MyH, El Filali H, Benzine M et al (1996) La palmeraie dattière marocaine: Evaluation du patrimoine phoénicicole. Fruits 51:247–259

Sedra MyH, Lashermes P, Trouslot P et al (1998) Identification and genetic diversity analysis of date palm (*Phoenix dactylifera* L.) varieties from Morocco using RAPD markers. Euphyt 103:75–82

Scribner KT, Pearce JM (2000) Microsatellites: evolutionary and methodological background and empirical applications at individual, population, and phylogenetic levels. In: Baker A (ed) Molecular methods in ecology. Blackwell Science, London, pp 235–271

Soliman SS, Ali BA, Ahmed MMM (2003) Genetic comparisons of Egyptian date palm cultivars (*Phoenix dactylifera L.*) by RAPD-PCR. Afr J Biotechnol 2:86–87

Stepansky A, Kovalski I, Perl-Treves R (1999) Intraspecific classification of melons (*Cucumis melo* L.) in view of their phenotypic and molecular variation. Plant Syst Evol 217:313–332

Trifi M (2001) Polymorphisme et typage moléculaire de variétés tunisiennes de palmier dattier (*Phoenix dactylifera* L.): relation avec la résistance au bayoud. Thèse de Doctorat d'Etat, Université de Tunis El Manar, Faculté des Sciences de Tunis, Tunisie, 141 p

Trifi M, Rhouma A, Marrakchi M (2000) Phylogenetic relationships in Tunisian date-palm (*Phoenix dactylifera* L.) germplasm collection using DNA amplification fingerprinting. Agronomie 20:665–671

Vos P, Hogers R, Bleeker M, Reijans M et al (1995) AFLP: a new technique for DNA fingerprinting. Nucleic Acids Res 23:4407–4414

Williams JGK, Kubelik AR, Livak KJ et al (1990) DNA polymorphïsm amplified primers are useful as genetic markers. Nucleic Acids Res 18:6531–6535

Wolff K, Van Run JP (1993) Rapid detection of genetic variability in chrysanthemum (*Dendranthema grandiflora* Tzvelev) using random primers. Heredity 71:335–341

Yang X, Quiros C (1993) Identification and classification of celery cultivars with RAPD markers. Theor Appl Genet 86:205–212

Chapter 26
Towards Sex Determination of Date Palm

S.A. Bekheet and M.S. Hanafy

Abstract In the date palm (*Phoenix dactylifera*), a dioecious mode and late initial reproductive age of 5–10 years are major practical constraints to genetic improvement. Improvement of the existing palm cultivars or selection of new ones with superior characters is a tedious endeavor due to the long life cycle of the date palm tree and its heterozygous nature. Sexual propagation method cannot be used commercially for propagating the cultivars of interest in a true-to-type manner. Currently there is no reliable method to identify sex at the early seedling stage. Early sex identification of young seedlings could enhance breeding programs and generate experimental male and female genetic stocks that will help the genetic improvement of the date palm. Moreover, the selection and identification of superior seedling characters for yield enhancement and to improve the physical and chemical properties of fruits is of great commercial interest. There has been significant progress in our understanding of sex-determining mechanisms in date palm using traditional means. But physiological and cytological methods do not give obvious differences between male and female date palms. Biotechnology, as a new tool in date palm breeding, can be useful to improve the qualities of palm trees through early sex identification. Although molecular markers have been introduced in date palm programs, few research efforts have been geared toward studying the early sex determination in the plant. This chapter will focus on genetic and molecular basis of sex determination in date palm in attempting to develop reliable methods to identify sex at an early stage of seedlings.

Keywords Biotechnology • Breeding • Sex determination • Molecular markers

---

S.A. Bekheet • M.S. Hanafy (✉)
Plant Biotechnology Department, National Research Center, El-Tahrir Str.,
Dokki, 12622, Cairo, Egypt
e-mail: shawky005@yahoo.com; mshanafy@yahoo.com

S.M. Jain et al. (eds.), *Date Palm Biotechnology*,
DOI 10.1007/978-94-007-1318-5_26, © Springer Science+Business Media B.V. 2011

26.1 Introduction

The genetic control of sex determination is well understood in several animal systems, particularly *Drosophila melanogaster*, *Caenorhabditis elegans* and mammals. In plants, understanding of the sex determination system is closely connected with an understanding how separate sexes evolved, and current theoretical ideas about this also illuminate the evolution of sex chromosomes. Angiosperms are of particular interest for empirical studies of sex chromosome evolution because they probably evolved separate sexes repeatedly and relatively recently. Other plants, particularly bryophytes also have interesting independently evolved sex chromosomes (Okada et al. 2001). In many sexually reproducing plant species all individuals are essentially alike in their gender condition. Many such *sexually monomorphic* species are hermaphroditic. The term *cosexual* is used when individual plants have both sex functions, whether present within each flower (hermaphrodite) or in separate male and female flowers (monoecious) (Lloyd and Bawa 1984). A minority of plant species are *sexually polymorphic*, including dioecious species, with separate males and females. Many dioecious species with hermaphrodite relatives have evident rudiments of opposite sex structures in flowers of plants of each sex, suggesting recent evolution of unisexual flowers. The low frequency and scattered taxonomic distribution of dioecy and sex chromosomes suggest that cosexuality is the ancestral angiosperm state (Charlesworth 1985; Renner and Ricklefs 1995). Sex chromosomes therefore probably evolved repeatedly and quite recently.

Sex inheritance and sex chromosomes in plants are strikingly similar to those in animals. Sex-determination systems in plants have evolved independently many times, and are just one of the strategies that promote outcrossing and thus help avoid inbreeding. The majority of plants studied have heterozygous males, or, when the chromosomes are visibly different, perhaps half of plants that have separate sexes, male heterogamety (XY males, XX females). In many dioecious plants, males are *inconstant*, i.e. produce occasional fruits (Lloyd 1975). Self fertilization of such plants in several species has provided genetic evidence that males are heterozygous. The presence or absence of the X chromosome in male gametes provides an efficient mechanism for sex determination.

Knowledge of sex determination in plants indicates that only about 5% of flowering plants, such as hops, date palms and spinach form individuals with separate sexes. Papaya plants can turn out male, female or hermaphroditic. Not all plants with separate sexes have sex chromosomes that look different from their partner. Although economically important, palms are a much neglected plant group in terms of understanding genetics and development potential; therefore much effort must be expended to resolve this problem.

In dioecious plants cultivated for fruit or seed, it is often difficult to identify females at an early stage of growth. Research comparing allocation patterns between genders in dioecious species indicates that female plants usually have higher resource requirements and/or reproduction imposes a greater drain on resources. Perhaps for this reason, a recurring pattern observed among dioecious plant species is an increasing proportion of males within populations along an axis of decreasing site productivity. Thus early sex identification and genetic characterization of the unknown scattered genotypes of dioecious tree such as date palm resulting from seeds, represent a very important item.

A major problem for farmers is to identify gender at an early stage so that they can cultivate in their orchards a sufficiently large number of productive female trees with only a minimal number of male trees. Moreover, it is important to select and identify superior males in terms of fertilization. This direct influence of the male parent on the development of the date fruit is precise and definite and varies with the particular male used to fertilize the female flowers. Date palm pollen has been found to exert a direct influence on the size, shape and color of the seed, and also on the size of the fruit, on the speed of development of the fruit and on the time of ripening of the fruit. This direct effect of the pollen on the parts of the seed and fruit lying outside the embryo and endosperm is called *metaxenia*. Date palm breeding is a long-term endeavor (Carpenter 1979). The genetics, morphology, morphogenesis and physiology of date palms are somewhat less understood than other fruit-tree crops. It has been difficult to study because it is native to the subtropics, has a long life cycle and has diverse and unique growth habits compared to other fruit-producing trees. This species is slow flowering and fruiting and it is difficult to determine the sex of the trees before the first flowering, when they are about 5 years of age. Propagation of date palm through seeds or zygotic embryos is desirable for improvement of the cultivars and for selection of diseases resistance, fruit quality and high yield. Also, the early determination of sex type is very important for speeding up breeding programs. Breeding programs to maintain genetic diversity have not been employed because the sex of a date palm cannot be known until it reaches reproductive age (5–10 years) (Bendiab et al. 1993).

Biotechnology, as one of the newest tools in plant breeding, can be helpful for breeders and producers of date palm to improve the qualities through early sex identification. The data suggest that *in vitro* tissue culture conditions can modulate sex modification (Komai et al. 2003), particularly by the reactivation of the arrested organs. Moreover, molecular biology techniques such as Restriction Fragment Length Polymorphism (RFLP), Random Amplification of Polymorphic DNA (RAPD) and Inter Simple Sequence Repeat (ISSR) can be used to identify sex-specific DNA markers of date palm. With the availability of molecular techniques, we may now hope to understand more about how sex chromosomes evolve. Mapping data, even with anonymous markers, should give estimates of the fraction of X-linked loci that are located in the pairing and differential regions. Once genes have been identified and sequenced, we will be able to estimate how long sex chromosome evolution takes. This should help us to evaluate the plausibility of the proposed mechanisms for the process. In recent years, there have been serious efforts to understand the genetic basis of sex determination in plants and to develop methods to identify sex at an early stage by using molecular marker tools (Biffi et al. 1995; Hormaza et al. 1994; Mulcahy et al. 1992). To achieve progress in understanding sex chromosome evolution and organization in plants, sex-linked genetic markers are required.

26.2 Sex Type of Date Palm

Plant sexuality covers the wide variety of systems found across the plant kingdom. The complexity of the systems and devices used by plants to achieve sexual reproduction has resulted in botanists and evolutionary biologists using numerous terms to describe physical structures and functional strategies. Dellaporta and Calderon-Urrea (1993) list and define a variety of terms used to describe the modes of sexuality at different levels in flowering plants. This list is generalized to fit more than just plants that have flowers, and expanded to include other terms and more complete definitions. Bisexual or perfect flowers have both male (androecium) and female (gynoecium) reproductive structures, including stamens, carpels and an ovary. Flowers that contain both androecium and gynoecium are called *androgynous* or *hermaphroditic*. Other terms widely used are *hermaphrodite*, *monoclinous* and *synoecious*. A complete flower is a perfect flower with petals and sepals. A unisexual-reproductive structure is either functionally male or functionally female. In angiosperms this condition is also called *diclinous*, *imperfect* or *incomplete*. Many plants have complete flowers that have both male and female parts, others only have male or female parts and still other plants have flowers on the same plant that are a mix of male and female flowers. Certain plants even have mixes that include all three types of flowers, where some flowers are only male, some female and some both male and female. A few plants also undergo what is called sex-switching, like *Arisaema triphyllum*, expressing sexual differences at different stages of growth. In some arums smaller plants produce all or mostly male flowers and as plants grow larger over the years the male flowers are replaced by more female flowers on the same plant. Other species have plants that produce more male flowers early in the year and, as plants bloom later in the growing season, they produce more female flowers.

Dioecious refers to a species having separate male and female plants. That is, no individual plant of the species produces both microspores and megaspores; individual plants are either male (producing microspores) or female (producing megaspores). Dioecy is a rare sexual system in flowering plants, occurring in only 4–6% of species (Guttman and Charlesworth 1998; Renner and Ricklefs 1995). In addition, only a minority of these dioecious species have heteromorphic sex chromosomes. Their sex determination system is based on the X/ autosome ratio or on the X/Y ratio (Ainsworth 2000); the Y chromosome is dominant (Charlesworth et al. 2005; Liu et al. 2004). Male genomes consist of an association of three nuclear subgenomes: the autosomes, the X and the Y. The male flower phenotype is not dependent on the presence of the Y chromosomes, but they are necessary for the production of fertile pollen (Negrutiu et al. 2001). The best known cultivated dioecious species for determination of sex expression are: hops (*Humulus lupulus*), spinach (*Spinacia oleracea*), asparagus (*Asparagus officinalis*), carob (*Ceratonia siliqua*), date palm (*Phoenix dactylifera*), fig (*Ficus carica*) and papaya (*Carica papaya*).

The date palm is an important horticultural crop grown mainly in the Middle East and Arabian region (FAO 1984). It is an outcrossed, perennial monocotyledon which is very heterozygous. Researchers have reported that the chromosome number of date palm is 2n=2x=36 (Al-Salih et al. 1987; Beal 1937; Ibrahim et al. 1998). Sexual phenotype is a particularly important problem in dioecious plants that are cultivated for agricultural purposes, as illustrated in studies of the date palm. Date palm is a dioecious species and consequently half of the progeny will be male and half female, with no certain way to determine at an early stage the gender of the progeny nor fruit or pollen quality prior to flowering. As a dioecious species, date palm has male and female flowers being produced in clusters on separate palms. These flowering clusters are produced with axils of leaves of the previous year's growth. In rare cases both pistillate and staminate flowers are produced on the same spike while the presence of hermaphrodite flowers in the inflorescence has also been reported (Mason 1915). Palms which carry both unisexual and hermaphrodite flowers are known as *polygamous*.

At the maturation stage of the date palm tree, it is easy to identify the male and female flowers. The unisexual flowers are pistillate (female) and staminate (male) in character; they are borne in a large cluster (inflorescence) called a *spadix* or *spike,* which consists of a central stem called a *rachis* and several strands or *spikelets.* Male spathes are shorter and wider than the female ones. Each spikelet carries a large number of tiny flowers: as many as 8,000–10,000 in the female and more in the male inflorescence (Chandler 1958). The male inflorescence is crowded at the end of the rachis, while branches of the inflorescence of the female cluster are less densely crowded at the end of the rachis. These characteristics allow the recognition of the inflorescence's sex before it opens. The male flower is sweetly scented and normally has 6 stamens, surrounded by waxy scale-like petals and sepals. Each stamen is composed of two small yellowish pollen sacs. Farmers determine the quality of male date flowers by their smell and other features.

In the wild, the date palm tree pollinates naturally, but fruit production is low. If one were to depend on insects or wind-aided natural pollination, 50% of the trees should be male, which would make date farming by that method uneconomical. Artificial pollination increases production substantially. The dioecious nature of the date palm necessitates the transfer of pollen from a staminate palm to the pistillate in order to obtain an economically feasible yield. Therefore, this necessitates adoption of a manual pollination process to ensure a rich crop. Traditionally to pollinate a spathe on a female palm, a piece of mature male spikelet is inserted into the female spathe as it splits open and is loosely bound around it. The most important benefit of the manual pollination process is that male flowers from a single tree can be used to pollinate 40–50 female date palms.

Sexual propagation is the most convenient method by which to propagate date palm: seeds can be stored for years, they germinate easily and are available in large numbers. The most obvious drawback is the heterozygous characteristics of seedlings which are related to the dioecious nature of the date palm: half of the progeny are generally male, which produce no fruits, and large variations in phenotype can occur in progeny. Furthermore, no method is known at the present for sexing date palm at an early stage of tree development. It is therefore not possible to eliminate non-productive male trees in the nursery before planting on a field scale.

26.3 Sex Identification by Traditional Means

Sex determination is an important developmental event in the life cycle of all sexually reproducing plants and can be controlled genetically by mechanisms also found in the animal kingdom. Dioecious plants provide a particularly interesting system in which to study the genetics and evolution of sex chromosomes. In dioecious species, sex determination is a complex developmental process ending in the differential expression of stamen and carpel genes in male and female individuals. In date palm, the most significant limiting factor in breeding is the time required for maturation of plants, since the palm tree becomes sex identifiable after 5–7 years (Shaheen 1990). Both male and female palms bear spathes; they appear from the bases of the palm leaves towards the top of the palm. The male blossom is fluffy white and star-shaped. Female blossoms resemble beads on a string (Fig. 26.1).

A rapid and easy method for determining the sex of seedlings is considered to be very important for seed cultivation as well as for speeding up breeding programs. In a dioecious plant such as date palm, the genetics of sex determination are far less understood. However, sex expression in many cases is strongly affected by environmental conditions (light, temperature, nutrients) as well as by applied chemical agents (hormones).

The importance of hormones in sex expression has been studied using two different approaches: exogenous administration of growth regulators to developing plants, and quantitative analysis of endogenous hormones. All plant hormones appear to influence sex expression, but unequivocal roles cannot be attributed to a single hormone due to the disparity of the results in different species. Generally it was found that cytokinin; indoylacetic acid and abscisic acid are higher in male than in female plants (Bracale et al. 1990). Accordingly, the induction of the male organ can be attributed to a disturbance in the regulation of gene expression caused by a hormonal balance variation that can lead to an unusual activation of the inhibited male-related loci in the female flower. The success of this derepression or reactivation event is tightly dependent on a distinct stage in the floral developmental process. Similarly, DeMason and Tisserat (1980) reported a female organ induction by a 2,4-dichlorophenoxyacetic acid (2,4-D) treatment of date palm male flowers. This female organ activation might have resulted from a modified gene expression regulation, due to the hormonal treatment of male flowers. However, the authors reported that no ovules were observed in the carpels of the apparent bisexual flowers obtained. They added that, in some cases, apparently bisexual flowers can occur naturally within female date palm trees.

Normal development of male flowers was described by DeMason et al. (1982), who reported that the early development of staminate flowers is identical through carpel initiation. Then, the pistillate and staminate flowers diverge. Recently, a study to investigate the main factors involved in the process of sex modification of female date palm through *in vitro* hermaphrodism induction was reported (Masmoudi-Allouche et al. 2009). That study demonstrated that female date palm flowers can acquire new physiological characteristics and capacities that are quite rarely observed during wild flower development. The vestigial stamens of female date palm flowers display a new and higher capacity to proliferate under particular

Fig. 26.1 (**a**) Male and (**b**) female inflorescences of date palm (Source: Zaid and de Wet 2002)

in vitro conditions, without blocking carpel development, leading to morphologically typical hermaphrodite flowers.

Recent studies of sex determination in many plant species have been useful in identifying the diversity of genetic and epigenetic factors that are involved in determining the sex of the flower or individual. Amenable to genetic analysis, significant progress has been made toward identifying mutations that affect sex expression. In the genes, an angiosperm tree flower, for example, consists of four stacked components. The components include the sepal area, petal area, male area and female area. The sepal and petal areas are not part of the functional reproductive parts but influence air flow and animal pollinator effectiveness. Genetically, the four flower components are controlled by separate gene sets in three zones within a tree flower: zone 1 = female and male part development controlled by specific gene sets, zone 2 = male part and petal area development controlled by gene sets which assure pollen distribution and pollinator attraction and zone 3 = vegetative zone development controlled by specific gene sets which display, support and protect reproductive parts. In this case, instead of one set of genes conveying gender, trees have developed three unique gene sets which control flower formation and gender expression. Each of these gene sets has been separately driven by many agents to affect efficient reproduction.

Papaya researchers report that the plant's sex-determining region is starting to lose genes for nonsexual traits and to accumulate anomalous DNA. The region has only 62% of the gene density of the rest of the papaya chromosomes. It also shows 28% more rogue genetic elements and nearly triple the amount of DNA with a reversed orientation. There is now no major obstacle preventing isolation of a large set of sex-linked genes from other plants such as those of papaya (Ma et al. 2004). Recent investigations of genes that control the process of sex development reveal that involvement of microRNA in both the sex determination of the male inflorescence and its growth pattern (Banks 2008).

The genetic and physiologic bases (DNA content, frequency of repetitive sequences, genes associated with sex chromosome) of sex expression in *Asparagus officinalis* were studied by Bracale et al. (1990). In addition, stamen and carpel specific messages were searched together with hormonal content of male and female flowers. Major results indicated that gene expression in young male and female flowers is very similar to each other (in agreement with their morphological appearance) and only later on differential expression takes place. On the contrary, important differences in hormonal content between sexes (mainly auxin levels) were observed in meiotic flowers, suggesting that this stage is critical step in the pathway of sex differentiation. In some species, sex expression is under epigenetic control mediated by chromatin modifications of the sex determining regions, including DNA methylation and nucleosomal histone acetylation (Vyskot 1999). There is a growing body of evidence suggesting that epigenetic changes, such as nascent cytosine methylation, are responsible for early stages in the evolution of dioecy and sex chromosomes in all eukaryotes (Gorelick 2003; Gorelick and Osborne 2002). There is only one report suggesting that the date palm is one of the rare monocotyledonous dioecious species that possess heteromorphic X/Y sex chromosomes. In date palms, which like papayas are dioecious with homomorphic sex chromosomes, extra heterochromatin on one of the male chromosomes is believed to determine sex (Siljak-Yakovlev et al. 1996).

Among the many dioecious plant species, only a few have evolved sex chromosomes. Sex determination systems based on heteromorphic X and Y sex chromosomes are particularly interesting to study from both a developmental and an evolutionary perspective. A cytological method based on chromomycin staining which demonstrates the occurrence of sexual chromosomes carrying distinctive nucleolar heterochromatin is described in the date palm which offers, for the first time, the possibility of identifying male and female individuals by simple analysis of root meristems (Siljak-Yakovlev et al. 1996); chromomycin A3 was used to stain root chromosomes, thus identifying subtle differences between the heterochromatin of chromosomes isolated from male and female cells. While useful for sex-typing date palm seedlings, this study also illustrates two other important points in understanding sex determination in dioecious species of plants. First, there are often no obvious cytological or genetic differences between male and female plants; second, it is often difficult to study the genetic or molecular basis of sex determination in many species of monoecious or dioecious agronomically-important plants simply because of their longevity. Otherwise, microsporocytes were examined at late diakinesis and metaphase I of meiosis (Ibrahim et al. 1998). Results showed that in both Samany and Zaghloul date cv. male types, chromosomes tended to pair as bivalents. However, loose bivalents and/or univalents were observed in some pollen mother cells (PMCs) during diakinesis, and cases where more than one bivalent associated with the nucleolus were observed in a few PMCs. Eighteen bivalents at diakinesis indicated that both types were diploid with $2n=2x=36$ chromosomes. Otherwise, mitosis was studied in root tip cells of female date cvs. Barhee, Nebut Seif and Succary. Results showed that all three cultivars had a somatic chromosome number of $2n=36$, and chromosomal behavior was normal (Aly and Bacha 2000).

26.4 Sex Determination Using Biotechnology

Sex determination is a process that leads to the physical separation of male and female gamete-producing structures to different individuals of a species. In the past, various morphological, histological and genetic approaches have been used to determine the gender of higher plants. A genetic test to distinguish between male and female plants would prove useful because it is impossible to tell the sexes apart by looking at the chromosomes under the microscope, unlike the case with many other species. Most the studies on palm characterization, detection of genetic variation and gene mutation have concentrated on the variation in chromosome number and biochemical diversity. Despite increasing research efforts on number of different plant species, there is relatively little information available on the molecular basis of sex determination and it is even difficult to estimate the numbers of genes involved, particularly as the genes which result in organ suppression are unlikely to be the primary sex-determining genes.

The development of molecular markers holds many promises to plant breeders and geneticists in different areas; such as in varietal identification or fingerprinting, estimation of relatedness between different genotypes, discernment of evolutionary relationships and introgression of Mendelian traits into a population. Marker-based selection; however, is the area where molecular markers could have the greatest impact in plant breeding. Dellaporta and Calderon-Urrea (1993) mentioned that plants offer unique systems through which to study sex determination. Because the production of unisexual flowers has evolved independently in many plant species, different and novel mechanisms may be operational. Hence there is probably not one unifying mechanism that explains sex determination in plants.

Our understanding of the evolution of plant sex chromosomes and sex determination should be advanced by the use of molecular markers and several groups of researchers are searching for them. Advances in our understanding of sex determination will come from analysis of molecular biology genetics and the biochemistry of genes controlling sexual determination in plants. To date, there is increasing effort to develop molecular markers tightly linked to the sex determining locus in the plant genome and to isolate the corresponding genes (Caporali et al. 1996). For any program of sex determination in dioecious species, several points should be studied: (a) identification of sex type at the molecular level of the DNA fragments bearing the sex controlling genes which will be achieved by searching genetic markers linked to the sex genes, (b) identification of stamen and carpel specific genes, (c) understanding the process of sex differentiation at the biochemical and physiological level.

Early selection of young seedlings could enhance breeding programs and generate experimental male and female genetic stocks, but no cytogenetic protocol exists for sex determination in an immature date palm. Molecular markers can be effectively utilized to diagnose and select a genotype based on linked DNA markers, long before the phenotype is apparent. This is particularly important in date palm given its long juvenile phase. Despite its major status as a cultivated tree crop, little information is available on the genetics and molecular genetics of date palm. Until now, there has not been identified a single gene for ergonomically important traits. Isolation of male-specific cDNAs from developing flower buds or reproductive organs has not yet led to discovery of sex determining genes, probably because sex determination occurs very early in flower development, so the genes identified are controlled in response to sex, rather than the controlling loci.

The region containing the sex determining loci must initially have been fully homologous between the two alternative chromosomes. One goal of plant sex chromosomes studies is therefore to test for homology. Both X- and Y-linked markers are now being discovered in plants, with and without heteromorphic sex chromosomes (Harvey et al. 1997; Mandolino et al. 1999; Polley et al. 1997; Testolin et al. 1995; Zhang et al. 1998). Most markers are, however, anonymous, and cannot tell us which X-linked loci have homologues on the Y chromosomes and which do not. The introduction DNA-based markers in date palm will help in developing a sex specific marker. Developing such a marker would allow early determination of sex in palm offshoots at the seedling stage which can help to speed up breeding programs. The cosegregation of the molecular marker and a trait of interest, in progeny segregating for this trait, is an indication of linkage between them. This marker could then be used for selection instead of morphological characters.

Molecular analysis can predict the superior heterotic combination within a set of genotypes. This is helpful in accelerating a breeding program in date palm, which requires many years before flowering. Molecular techniques are extensively used in constructing genetic maps, marker-based selection, cloning useful genes and fingerprinting in several plant species such as tomato and potato, as well as in weed-like *Arabidopsis.* In palms, it could be possible to screen for a desirable genotype at the seedling stage by using marker-based selection strategy. The same method can be used to distinguish between male and female date palms before flowering. The most general approach to achieve this objective is the initial construction of a complete linkage map (Gebhardt and Salamini 1992).

Selection of desirable palms (male or female) can be carried out based on biochemical and molecular markers. Although molecular cloning approaches have not yet identified primary sex determining genes in any dioecious plant species, a range of molecular markers linked to sex have been generated. These markers have either arisen from genetic mapping programs or from research aimed at finding sex-linked markers for agronomically-important dioecious species. In some plants such as *Silene latifolia*, *Cannabis sativa*, *Phoenix dactylifera* and *Rumex acetosa*, therefore, it is not surprising that male-associated markers are relatively abundant.

In dioecious plants where sex chromosomes have not been identified, markers for maleness indicate either the presence of sex chromosomes which have not been distinguished by cytological methods or that the marker is tightly linked to a gene involved in sex determination. Female-associated molecular markers have been described in *Actinidia* (Harvey et al. 1997) and *Salix viminalis* (Alstrom-Rapaport et al. 1998). These may arise as a consequence of close linkage with a female sex determining gene or may indicate a sequence on the X chromosome inherited from the male parent. *Salix viminalis* is unlikely to have sex chromosomes and probably has a two-locus epistatic system. Moreover, a number of research groups have used subtraction techniques of either cDNA or genomic DNA in attempts to isolate sex determining genes from *Silene latifolia*. Differential screening of a subtracted cDNA library enriched for male-specific sequences enabled nine male enhanced cDNA sequences (Men-1 to -10) to be isolated (Scutt and Gilmartin 1998; Scutt et al. 1997). To date, two groups of researchers have reported randomly amplified polymorphic DNA markers that are highly specific for males and hermaphrodites but absent in females of papaya (Deputy et al. 2002; Urasaki et al. 2002).

The various molecular markers linked to sex include Randomly Amplified Polymorohic DNA (RAPDs), Restriction fragment Length Polymorphisms (RFLPs), Amplified Fragment Length Polymorphism (AFLPs) and microsatellites; all are powerful techniques, which have been developed for molecular analysis of plant genome. In this respect, ALFP has been used to initiate a genetic map of date palm (El-Kharbotly et al. 1998). Five cultivars, i.e. Bunarenga, Damous, Fardh, Khalas and Khenizi and pollen of male palm coded BN-96, DN-96, Fr-96, KL-96, and KN-96 amended by the Rumais Agriculture Research Station, Sultanate of Oman, were used to establish F_1 populations. Data taken from 3-week old seedlings showed that the populations were segregating with 1:0, 1:1 or 3:1 ratios for erect and slanting, respectively. The erect and slanting leaf was controlled with a simple genetic factor following Mendelian inheritance. The AFLP fingerprinting for both parental clones was obtained with 32 primer combinations. Three primer combinations (E-AAC, M-CAG; E-AAG, M-CAT and E-ACG, M-CAA) showed few bands in the male parents while few bands were observed in the female parent using the combinations of E-ACG, M-CAA. The polymorphism between the two parents ranged from 4% to 55% depending on the primer combination. This study showed the possibility of using the AFLP technique to characterize the date palm genome. A map can be constructed based on the pattern of the segregation of ALFP generated bands. The authors mentioned that linkage groups can be constructed based on the cosegregation of different bands.

RFLP have several advantages over morphological and isozyme markers and are currently contributing greatly to the construction of detailed genetic maps. The level of allelic variation of RFLP markers in plant population is much greater than that of morphological or isozyme markers. Furthermore, RFLP markers usually behave in a codominant manner, are apparently free of epistatic effects and are developmentally stable. A detailed RFLP map could lead to the identification of new genetic markers that are tightly linked to sex-determining genes. In this respect, a preliminary genetic map of the dioecious species of *Asparagus officinalis* has been constructed on the basis of Restriction fragment Length Polymorphisms (RFLP) and isozyme markers (Restivo et al. 1995). One isozyme and three RFLP markers were assigned to the sex chromosome. Otherwise, RFLP markers have been used to distinguish between the sexes in *Asparagus* (Biffi et al. 1995). Microsatellite banding patterns have been shown to be sex-specific in *Carica papaya*, (GATA) (Parasnis et al. 1999). In *Phoenix dactylifera*, the sexes can readily be distinguished by cytological examination of interphase nuclei in root tip cells. Cells from male plants carry two fluorescent blocks of unequal intensity while female cells carry two equal blocks (Siljak-Yakovlev et al. 1996).

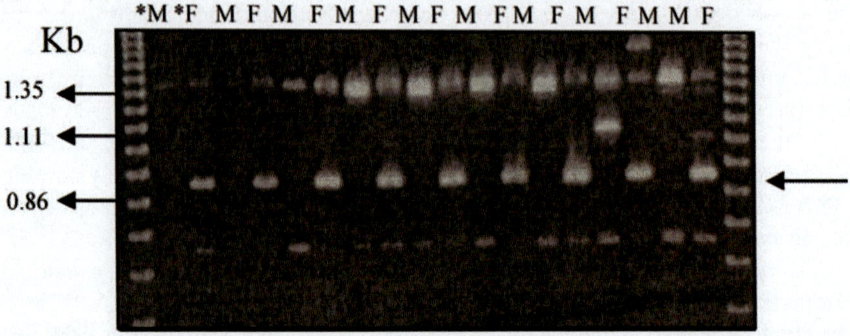

Fig. 26.2 RAPD banding patterns from pooled DNA from male and female 'Kerman' progeny (*M, *F, respectively) and DNA extracted from 7 each of the male (M) and female (F) individuals comprising the pools using primer OPO08. OPO08945 is indicated with an *arrow* (Source: Hormaza et al. 1994)

RAPD technique has been used extensively in plants for various purposes such as genetic diversity, DNA fingerprinting, classification and phylogenetic studies. A glance at the potential applications of RAPD marker technology reveals its importance as a powerful tool in plant molecular biology which can play an important role in genetics and breeding. RAPD markers have already been used for determining sex by bulk segregant analysis in *Pistacia vera*, *Atriplex garretti*, *Trichosanthes dioica* and *Salix viminalis*. Also, RAPD has been used successfully to create a linkage map (Soundur et al. 1996) and sex determination (Saker and Kuehne 1998) in papaya and (Hormaza et al. 1994) in *Pistacia vera* (Fig. 26.2). In this respect, Saker and Rady (2003) analyzed male and female papaya plants using different classes of molecular markers i.e., isozyme and RAPD in order to identify new polymorphisms as a first step towards developing of a universal molecular marker linked to sex in papaya. Analysis of peroxidase isozyme banding patterns indicated that one peroxidase isomer is peculiar only to male trees. RAPD indicated that one polymorphism DNA fragment detected in banding patterns of male clones was absent in female.

In an attempt to determine the genetic difference between male and female date palms, genome DNA was extracted from leaves of four female cvs. (Deglet Noor, Allig, Kentich, Menakher), a male genotype pollinator T23 and F_1 hybrid. The results of RAPD gave reproducible polymorphic bands with 11 primers from 53 primers used. The RAPD thus was successfully used to differentiate between female cultivars, male and F_1 hybrid (Ben-Abdallah et al. 2000). Soliman et al. (2003) used RAPD technique to compare genetic material from four female date palms and four unknown male Egyptian trees. The banding profiles obtained suggested that two male clones are genetically related to the four female date palm cvs. (Zaghloul, Amhat, Samany and Siwi) ranged from 87.5% to 98.9%. In this respect, identification of some Egyptian date palm males from females varieties using molecular markers was reported (Ahmed et al. 2006). Genomic DNA and RNA were extracted, measured and used as a template to detect genetic relationship and similarities among four known females (Sakkoty, Malkabi, Bartamoda and Dagana cvs.) and three unknown males of Egyptian date palm based on DNA and RNA technology. Results showed that differential display and RAPD analysis provided a rapid and effective method to detect the genetic relationship and similarities between four males and females of Egyptian date palms.

For early identification of cultivars and tracing genetic diversity among date palm genotypes of different origin, offshoot-derived, male and female plants of cvs. Barhee and Sukkary, seed-derived plants, and two *in vitro* cultures of both of these cultivars were subjected to RAPD analysis (Al-Khalifa et al. 2006). Similarity matrixes show that offshoot-derived male plant of Barhee was 73.6% genetically similar to its female counterpart, while similarity between male and female plants of Sukkary was 43.1%. In the case of seedlings, male and female plants of Barhee were 87.2% similar and those of Sukkary were 62.3% genetically alike.

Recently, sexual embryos of date palm were *in vitro* cultured and molecular analysis was used for early identification of sex type (Bekheet et al. 2008). In that study, the potential of isozymes and RAPD markers in sex identification of *in vivo* grown and *in vitro* differentiated cultures of date palm was investigated. *In vitro* zygotic lines were proliferated from mature and immature zygotic embryos of date palm. Early estimation of sex type of *in vitro* differentiated lines has been realized via the activity levels of two enzymes. A high level of peroxidase activity has been observed in adult and offshoot females. Acid phosphatase and glutamate oxaloacetate enzymes gave a strong difference between male and female date palms. Otherwise, the RAPD technique was used to compare genetic material from male, female and unknown lines of date palm. RAPD analysis showed a relatively close relation between the two females (adult and offshoot) cultures, since they have large number of homologous bands. Although, there was a low relationship between male and female, results of similarity could not confirm a link to sex or estimate the sex type of unknown clones. Moreover, an attempt to identify sex-specific DNA markers for date palm using molecular technique (RAPD and ISSR) has been achieved by Younis et al. (2008).

Four dry date cvs. (Sakoty, Bertmoda, Malkabi, Dagana) and three males (Dagana, Malkabi, Sakoty) recognized as superior date pollinators were used in the study. RAPD analyses gave three positive specific markers for females and two for males in addition to five positive specific markers for males in ISSR analysis. RAPD markers have also been used for identify the desirable traits in palmyra palm (*Borassus flabellifer*) to identify sex and high sap-yielding types (Ponnuswami et al. 2008).

26.5 Conclusion and Prospective

Sex determination in date palm continues to be problematic even with the efforts of researchers in the field. As a consequence, until now there is no reliable molecular method for distinguishing the date palm producing female trees from the male trees before the first flowering, which can occur >5 years after planting. In the date palm, a dioecious mode (separate male and female individuals) and the late initial reproductive age are major practical constraints for its genetic improvement. Furthermore, it is difficult to identify female cultivars according to their morphological characteristics outside fruiting time. Date palm is not of major economic importance to the most technologically advanced countries in North America and Europe. The major producing area of this crop is located in the Near East and North Africa where technical expertise and infrastructure for advanced molecular genetics research are poor. Moreover, there are few research groups which have been involved in the date palm breeding programs. Therefore, comparatively little work has been done on early sex determination of date palm using molecular genetics approach. However, the available data open a new window for identification of a molecular marker linked to sex in date palm. Therefore, in order to solve this problem, a lot of information still needs to be collected and the genome of the date palm must be studied in more detail.

References

Ahmed M, Soliman S, Elsayed E (2006) Molecular identification of some Egyptian date palm males by females varieties (*Phoenix dactylifera* L.) using DNA markers. J Appl Sci Res 2:270–275

Ainsworth C (2000) Boys and girls come out to play: the molecular biology of dioecious plants. Ann Bot 86:211–221

Al-Khalifa NS, Khan FA, Askari E, Hadi S (2006) *In vitro* culture and genetic analysis of male and female date palm. Acta Hort 725:653–661

Al-Salih A, Al-Najjar NR, Al-Mashdani AN (1987) A study on the chromosome number of two specific female date palm cultivars. Date Palm J 5:134–143

Alstrom-Rapaport C, Lascoux M, Wang YC et al (1998) Identification of a RAPD marker linked to sex determination in the basket willow (*Salix viminalis* L.). J Hered 89:44–49

Aly M, Bacha M (2000) Chromosome characteristics of three date palm cultivars (*Phoenix dactylifera* L.). Sultan Qaboos Univ J Sci Res Agric Sci 5:11–14

Banks JA (2008) MicroRNA, sex determination and floral meristem determinacy in maize. Genome Biol 9:204.1–204.3

Beal J (1937) Cytological studies in the genus *Phoenix*. Bot Gaz 99:400–407

Bekheet S, Taha H, Hanafy M, Solliman ME (2008) Morphogenesis of sexual embryos of date palm cultured *in vitro* and early identification of sex type. J Appl Sci Res 4:345–352

Ben-Abdallah A, Stiti K, Lepoivre P, du Jardim P (2000) Date palm (*Phoenix dactylifera* L.) cultivar identification using random amplified polymorphic DNA (RAPD). Cah Agric 9:103–107

Bendiab K, Baaziz M, Brakez Z, Sedra MH (1993) Correlation of isoenzyme polymorphism and Bayoud-disease resistance in date palm cultivars and progeny. Euphy 65:23–32

Biffi R, Restivo F, Tassi F et al (1995) A restriction fragment length polymorphism probe for early diagnosis of gender in *Asparagus officinalis* L. HortScience 30:1463–1464

Bracale M, Galli MG, Longo C et al (1990) Understanding sex differentiation in *Asparagus officinalis*: genetical and physiological approaches. Acta Hort 271:423–431

Caporali E, Carboni P, Nicoloso L et al (1996) Sex determination and differentiation in the dioecious species *Asparagus officinalis* L. Plant Biosyst 130:79–81

Carpenter J (1979) Breeding of date palm in California. Date Grow Inst Rep 54:13–16

Chandler W (1958) Evergreen orchards. Lea & Fabiger, Philadelphia

Charlesworth D (1985) Distribution of dioecy and self-incompatibility in angiosperms. In: Greenwood PJ, Slatkin M (eds) Evolution-essays in honour of John Maynard Smith. Cambridge University Press, Cambridge, pp 237–268

Charlesworth D, Charlesworth B, Marais G (2005) Steps in the evolution of heteromorphic sex chromosomes. Heredity 95:118–128

Dellaporta SL, Calderon-Urrea A (1993) Sex determination in flowering plants. Plant Cell 5:1241–1251

DeMason D, Tisserat B (1980) The occurrence and structure of apparently bisexual flowers in the date palm, *Phoenix dactylifera* L. (Arecaceae). Bot J Linn Soc 81:283–292

DeMason D, Stolte K, Tisserat B (1982) Floral development in *Phoenix dactylifera*. Can J Bot 60:1437–1446

Deputy JC, Ming R, Ma H et al (2002) Molecular markers for sex determination in papaya (*Carica papaya* L.). Theor Appl Genet 106:107–111

El-Kharbotly A, El-Mardi M, Al-Saadi N, Al-Maharuki Y (1998) Towards the construction of a genetic map of date palm using the Amplified Fragment Length Polymorphism technique (AFLP). In: The first international conference on date palm. Al-Ain, March 8–10, pp 194–207

FAO (1984) Production yearbook. Food and Agriculture Organization of the United Nations, Rome

Gebhardt C, Salamini F (1992) Restriction fragment length polymorphism analysis of plant genomes and its application to plant breeding. Int Rev Cytol 135:201–237

Gorelick R (2003) Evolution of dioecy and sex chromosomes via methylation driving Muller"s ratchet. Biol J Linn Soc 80:353–368

Gorelick R, Osborne R (2002) Inducing sex change and organogenesis from tissue culture in the endangered African cycad *Encephalartos woodii* (Cycadales, Zamiaceae). S Afr J Sci 98:114–117

Guttman D, Charlesworth D (1998) An X-linked gene with a degenerate Y-linked homologue in a dioecious plant. Nature 393:263–266

Harvey CF, Gill CP, Fraser LG, McNeilage MA (1997) Sex determination in *Actinidia*. 1. Sex-linked markers and progeny sex ratio in diploid *A. chinensis*. Sex Plant Rep 10:149–154

Hormaza J, Dollo L, Polito V (1994) Identification of a RAPD marker linked to sex determination in *Pistacia vera* using bulked segregant analysis. Theor Appl Genet 89:9–13

Ibrahim AMF, El-Kobbia AM, Kitat FM, Abd El-Kawy NM (1998) Cytological studies on date palm. I. Chromosomal behaviour during meiosis of two date palm (*Phoenix dactylifera* L.) male types. Alex J Agric Res 43:237–246

Komai F, Shikazono N, Tanaka A (2003) Sexual modification of female spinach seeds (*Spinacia oleracea* L.) by irradiation with ion particles. Plant Cell Rep 21:713–771

Liu ZY, Moore PH, Ma H et al (2004) A primitive Y chromosome in papaya marks incipient sex chromosome evolution. Nature 427:348–352

Lloyd D (1975) The transmission of genes via pollen and ovules in gynodioecy angiosperms. Theor Popul Biol 9:299–316

Lloyd D, Bawa K (1984) Modification of the gender of seed plants in varying conditions. Evol Biol 17:255–338

Ma H, Moore PH, Liu Z et al (2004) High-density linkage mapping revealed suppression of recombination at the sex determination locus in papaya. Genetics 166:419–436

Mandolino G, Carboni A, Forapani S et al (1999) Identification of DNA markers linked to the male sex in dioecious hemp (*Cannabis sativa* L.). Theor Appl Genet 98:86–92

Masmoudi-Allouche F, Chaari-Rkhis A, Kriaa W et al (2009) *In vitro* hermaphrodism induction in date palm female flower. Plant Cell Rep 28:1–10

Mason S (1915) Botanical characters of the leaves of the date palms used in distinguishing cultivated varieties, Bulletin 223:1–28. USDA, Washington, DC

Mulcahy DV, Weeden NF, Kesseli RV, Carroll SB (1992) DNA probe for Y-chromosome of *Silene latifolia*, a dioecious angiosperm. Sex Plant Rep 3:1–3

Negrutiu I, Vyskot B, Barbacar N, Moneger F (2001) Dioecious plants. A key to the early events of sex chromosome evolution. Plant Physiol 127:1418–1424

Okada S, Sone T, Fujisawa M et al (2001) The Y chromosome in the liverwort *Marchantia polymorpha* has accumulated unique repeat sequences harboring a male-specific gene. Proc Natl Acad Sci USA 98:9454–9459

Parasnis AS, Ramakrishna W, Chowdari KV et al (1999) Microsatellite (GATA)M reveals sex-specific differences in papaya. Theor Appl Genet 99:1047–1052

Polley A, Seigner E, Ganal MW (1997) Identification of sex in hop (*Humulus lupulus*) using molecular markers. Genome 40:357–361

Ponnuswami V, Jagadeesan R, Ramesh Kumar A et al (2008) Genetic relationship and diversity in palmyrah accessions based on rapid markers. Am Euras J Sust Agric 2:165–171

Renner S, Ricklefs R (1995) Dioecy and its correlates in the flowering plants. Am J Bot 82:596–606

Restivo FM, Tassi F, Biffi R et al (1995) Linkage arrangement of RFLP loci in progenies from crosses between doubled haploid *Asparagus officinalis* L. clones. Theor Appl Genet 90:124–128

Saker M, Kuehne T (1998) RAPD marker and RAPD-based marker (SCAR) distinguish male from female papaya plants *in vitro*. Third Arab Conference on modern biotechnology and areas of applications in the Arab world, Cairo, 14–17 Dec 1998

Saker M, Rady M (2003) Employment of molecular markers for identification of male and female papaya plants. J Gen Eng Biotechnol NRC 1:85–97

Scutt CP, Gilmartin P (1998) The men-10 cDNA encodes a novel form of proline-rich protein expressed in the tapetum of dioecious *Silene latifolia*. Sex Plant Rep 11:236–240

Scutt CP, Li Y, Robertson SE, Willis ME, Gilmartin PM (1997) Sex determination in dioecious *Silene latifolia* (effects of the Y chromosome and the parasitic smut fungus (*Ustilago violacea*) on gene expression during flower development). Plant Physiol 114:969–979

Shaheen M (1990) Propagation of date palm through tissue culture: a review and an interpretation. Ann Agric Sci Ain Shams Univ Cairo Egypt 35:895–909

Siljak-Yakovlev S, Cerbah M, Sarr A et al (1996) Chromosomal sex determination and heterochromatin structure in date palm. Sex Plant Rep 9:127–132

Soliman S, Bahy A, Mohamed Morsy M (2003) Genetic comparisons of Egyptian date palm cultivars (*Phoenix dactylifera* L.) by RAPD-PCR. Afr J Biotechnol 2:86–87

Soundur N, Manshardt R, Stiles J (1996) A genetic linkage map of papaya based on randomly amplified polymorphic DNA markers. Theor Appl Genet 93:547–553

Testolin R, Cipriani G, Costa G (1995) Sex segregation ratio and gender expression in the genus *Actinidia*. Sex Plant Rep 8:129–132

Urasaki N, Tokumoto M, Tarora K et al (2002) A male and hermaphrodite specific RAPD marker for papaya (*Carica papaya* L.). Theor Appl Genet 104:281–285

Vyskot B (1999) The role of DNA methylation in plant reproductive development. In: Ainsworth CC (ed) Sex determination in plants. BIOS Scientific Publisher, Oxford, pp 101–120

Younis R, Ismail O, Soliman S (2008) Identification of sex- specific DNA markers for date palm (*Phoenix dactylifera* L.) using RAPD and ISSR techniques. Res J Agric Biol Sci 4:278–284

Zaid A, de Wet PF (2002) Botanical and systematic description of the date palm, In: Zaid A, Arias-Jimenez EJ (eds) Date palm cultivation, Chapter 1. FAO, Rome, pp 1–47

Zhang YH, DiStilio VS, Rehman F et al (1998) Y chromosome specific markers and the evolution of dioecy in the genus *Silene*. Genome 41:141–147

Chapter 27
Interspecific Hybridization and Embryo Rescue in Date Palm

C. Sudhersan and Y. Al-Shayji

Abstract Date palm breeding is a long-term endeavor. Crop improvement in date palm through breeding is very slow due to slow growth and a long life cycle. Thousands of date-palm cultivars were naturally established through crossing within the species. Recently, interspecific hybridization between two distinct palm species, *Phoenix dactylifera* and *P. pusilla*, was successfully carried out in Kuwait for developing dwarf date palm hybrids to overcome the fruit-production barrier due to excessive tree height. In this interspecific hybridization, embryogenesis and seed development occurred at the initial stages of fruit development. At the later stage of fruit growth, seed development and embryo maturation was unsuccessful due to the failure of endosperm development. Zygotic embryo culture techniques have been developed and used for the rescue and multiplication of interspecific date palm hybrid zygotic embryos at different stages of fruit development. Details of the interspecific hybridization and methods of hybrid plant production through embryo culture technique are described in this chapter.

Keywords Interspecific hybridization • Date palm hybrid • Embryo rescue • Somatic embryogenesis • *In Vitro* culture

27.1 Introduction

In spite of the fact that date palm (*Phoenix dactylifera* L.) is one of the oldest crops known to man, new and improved cultivars through breeding have not been released for cultivation in any of the date-growing countries. Date palm crop improvement

C. Sudhersan (✉) • Y. Al-Shayji
Biotechnology Department, Food Resources and Marine Sciences Division,
Kuwait Institute for Scientific Research, P.O. Box 24885, Safat 13109, Kuwait
e-mail: schellan@kisr.edu.kw; yshayji@kisr.edu.kw

S.M. Jain et al. (eds.), *Date Palm Biotechnology*,
DOI 10.1007/978-94-007-1318-5_27, © Springer Science+Business Media B.V. 2011

through breeding is slow when compared to other crops, due to its long life cycle, heterogeneous nature and slow growth habit. It usually takes 30 years to complete three back crosses and to obtain the first offshoots from the inter-varietal crosses. Due to this long time requirement, new improved cultivars of date palm have not been produced through conventional breeding.

During past centuries, many date-palm cultivars were developed through natural open pollination within the species. These cultivars were mainly selected based on physical properties and sugar content of the fruit. In the Middle East and North Africa, superior cultivars have been selected from chance seedling populations since the dawn of history. In 1948 a date breeding program was started at USDA, Indio, California by Nixon and Farr and their results showed only 1% of the female population had the potential for improvement and inclusion in further advance breeding trials (Carpenter 1979; Carpenter and Ream 1976; Nixon and Farr 1965; Ream 1975). That breeding program was terminated in 1978 and some of the plants developed are conserved in the National Date Palm Germplasm Repository in California (Krueger 1998). The earliest breeding attempts conducted in the University of Arizona in 1912 with date palm cv. Deglet Noor failed to give similar or better quality fruit than the parent (Anon 1982). In Algeria, an early flowering male variety was selected through breeding programs (Monciero 1959). Breeding for bayoud disease resistance was carried out in Morocco (Saaidi et al. 1981). Recently, an interspecific hybridization trial on date palm has been successfully achieved in Kuwait (Sudhersan et al. 2009).

Embryo culture involves aseptical removal of an embryo from the seed and *planting* it in a sterile nutrient medium. Embryo culture *in vitro* was first performed with crucifers (Hanning 1904). Embryo culture has several potential applications in agricultural crop improvement research programs. It is used to save embryos that fail to develop naturally in interspecific or intergeneric hybridization where defective endosperms are common (Hodel 1977). Embryo culture may also be used to reduce lengthy dormancy periods or with seeds difficult to germinate due to physical or physiological factors. Excised embryos cultured *in vitro*, under suitable basal nutrient culture media, usually germinate immediately. Embryo culture also can be useful in seedling developmental studies.

Large-scale plantation growth of genetically superior high-yielding date palms for the date industry was difficult some 40 years ago. This barrier has been removed through the development of plant tissue culture technology. Date palm embryo culture was the basis for the recent developments of date palm mass clonal propagation. Date palm zygotic embryo culture studies were initiated in the 1970s and plantlets obtained through *in vitro* germination of zygotic embryos (Sudhersan 1988; Tisserat 1979; Zaid 1987). Reuveni (1979) reported callus and root development from date palm embryo cotyledonary sheath tissue explants in media containing naphthalene acetic acid (NAA). Ammar and Benbadis (1977) established organogenic callus from the date palm cotyledonary sheath of zygotic embryos germinated *in vitro*. Date palm plantlets were obtained through callus regeneration followed by asexual somatic embryogenesis from zygotic embryos using a medium enriched with 2,4-dichlorophenoxy acetic acid and hormone-free medium

(Tisserat 1979). Recently, Sudhersan et al. (2009) developed an embryo culture technique to rescue hybrid date palm embryos at various stages of fruit development. The details of interspecific hybridization in date palm and the free-living plant production via *in vitro* embryo rescue method are discussed in this chapter.

27.2 Interspecific Hybridization

In date palm cultivation, pruning, pollination, fruit thinning, bunch removal and fruit harvesting are highly essential for good quality fruit production. The cost of date production increases when the trees grow taller, due to the high labor cost in many producing countries. Mechanization is also expensive and unjustifiable in the case of small growers. Frequent climbing for fruit harvesting is highly dangerous in the case of taller old trees. Tree height is one of the major constraints to good quality date production. In order to reduce tree height and to develop dwarf date palms, a related dwarf palm species, *Phoenix pusilla*, was selected and crossed with selected female date palm cultivars.

In this hybridization trial, crossing was carried out between female date palm (*Phoenix dactylifera*) and male dwarf date palm (*P. pusilla*); both species were found to be compatible and fertilization occurred. Post-fertilization changes in the interspecific cross were similar to normal date palm flowers pollinated by date palm pollen. Fruit development and seed development were similar during the early stages. However, seed development was totally arrested in the interspecific cross due to poor endosperm formation at the ripening stage. Previous reports of such interspecific crosses revealed that pollen from *Phoenix reclinata, P. canariensis, P. roebelenii* and *P. rupicola* crossed with date palm for fruit-quality improvement, failed to produce better-quality fruit; whereas the cross between the date palm and *P. sylvestris* produced slightly larger fruits than normal (Nixon 1935). In the present study, using *P. pusilla* pollen, fruits were equal or larger in size depending on different date palm cultivars but were seedless (Fig. 27.1).

27.2.1 Methodology

Dwarf date palm pollen was obtained from *Phoenix pusilla* introduced to Kuwait in 2000 (Sudhersan 2004). Female date palm cvs. Barhee, Medjool and Sultana were selected from the tissue culture date palm orchard established in 2000 at the Kuwait Institute for Scientific Research (KISR). Male pollen was collected from *P. pusilla* during the summer months (May–July) when no date palm pollen was present to avoid pollen mixing. Pollen was stored inside sterile glass bottles at 4°C temperature for 8–9 months. Pollination was conducted with late-maturing female date palm inflorescences. The unopened mature female inflorescences (the spathe showing a small crack) of the selected date palm cultivars were opened with a surgical

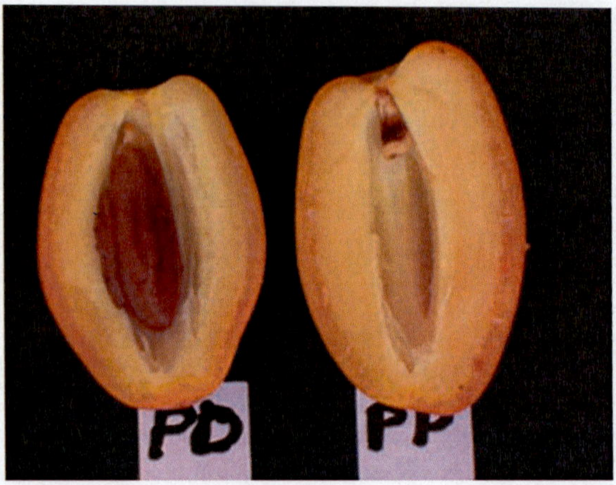

Fig. 27.1 Interspecific hybridization and embryo rescue in date palm. Pollen effect on date fruit (*PD Phoenix dactylifera* pollen, *PP Phoenix pusilla* pollen)

knife and the dwarf date palm pollen dusted over the flowers and covered immediately with paper to avoid date palm pollen mixing. The entire interspecific cross-pollination process was carefully carried out and paper covers were removed only after 15 days. Fruits were carefully observed during their different stages: hababouk (7 days after pollination, DAP) – creamy white in color; kimri (35–120 DAP) – the fruits are green in color; khalal (125–175 DAP) – the fruits are yellow or red in color; rutab (180–200 DAP) – the fruits turn brown color from the base to top; tamar (after 200 DAP) – the fruits turn completely brown or black in color (Sawaya 1987; Zaid and De Wet 2002). Seed development was also observed frequently (every 15 days) by opening the fruit at different stages.

27.2.2 Metaxenia

The effect of pollen on the fruit morphology is termed *metaxenia*. After interspecific cross-pollination, fruit set and development were similar to normal date fruit development up to the kimri stage. Later, the fruit morphological characteristic features changed from khalal stage onwards. At the mature stage, the fruits differed from the normal fruits in shape and size. Normal fruits were oval in shape and in the interspecific cross the fruits were dumbbell shaped due to the seedless cavity. Mature fruits of the cross-pollinated Medjool fruits were larger in size than the control (Fig. 27.1).

27.2.3 Hybrid Seed Development

Seed development occurred in both the treatment and the control at the initial stages of fruit development. Later, at 120 DAP (kimri stage), seed growth and development was arrested. At the late khalal stage (175 DAP), fruits exhibited an

Table 27.1 Date palm cultivar Medjool fruit and seed character of normal and hybrid

	Fruit			Seed			Seed cavity	
Fruit type	Length (cm)	Width (cm)	Weight (g)	Length (cm)	Width (cm)	Weight (g)	Length (cm)	Width (cm)
Normal	4.8	3.1	28.2	2.5	1.2	1.2	4.0	3.1
Hybrid	4.8	3.0	27.1	0.5	0.3	0.04	3.6	1.3

interior cavity and a small rudimentary seed at the anterior end of the fruit. The rudimentary hybrid seeds were small and similar to a grape seed in shape and size. Rudimentary seeds were 5 mm long, 3 mm wide and weighed an average of 40 mg, whereas normal seeds were 25 mm long, 12 mm wide and 1.2 g in weight (Table 27.1). In the early stages, seeds showed embryo development, but the embryos were aborted in the ripened fruits due to the total arrest in endosperm growth and development. Boyes and Thompson (1937) found shrivelled small seeds with floury endosperm in interspecific crosses due to chromosome imbalance in the endosperm. Brink and Cooper (1947) suggested that endosperm breakdown was the main reason for the failure in interspecific and intraspecific crosses in plants. Johnson et al. (1980) suggested that endosperm breakdown was due to post-fertilization incompatibility. The endosperm plays a role in embryo nutrition as it accumulates reserves of starch, proteins and lipids. Genetic analyses suggest that maternal and endosperm tissues may regulate each other's development (Lopes and Larkins 1993). The failure of endosperm development causes embryo abortion after 120 DAP in the hybrid seed. The embryo abortion at late khalal stage onwards was confirmed through *in vitro* seed culture at kimri, khalal, rutab and tamar stages. It was due to arrested endosperm development. The hybrid seed endosperm was very soft and shriveled at maturity.

27.3 Date Palm Zygotic Embryo

Mature seed of *Phoenix dactylifera* varies in length, width and weight depending on the cultivar. The hardy endosperm harbors the embryo in a depression on one side, and on the opposite side a longitudinal groove runs from end to end of the cylindrical seed. The embryo is a small body embedded in the horny endosperm. The mature zygotic embryo is 1 mm long and white in color. The bulk of the embryo is made up of the massive cylindrical cotyledon. The cotyledon sheath encloses a short root-shoot axis. The haustorium is a small buttonlike organ at the apex of the cylindrical cotyledon. The shoot bud of the mature embryo encloses one prophyll promidium (Fig. 27.2).The date palm zygotic embryo has a symmetrical structure. The shoot root axis is straight and vertically disposed. The cotyledon forms an inverted cone over the shoot end. This arrangement imparts a radial symmetry to the embryo. The cotyledon has three distinct regions: haustorium, solid stalk and hollow sheath. The sheathing base of the cotyledon encircles the

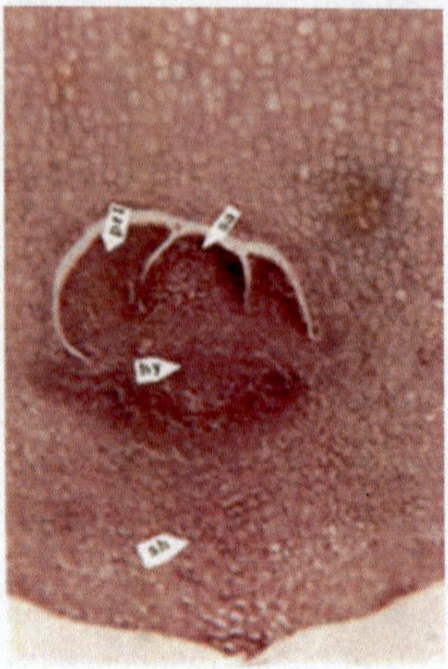

Fig. 27.2 Interspecific hybridization and embryo rescue in date palm. Date palm zygotic embryo vertical section (235 X) (*hy* hypocotyl, *prl* prophyll, *sa* shoot apical meristem, *sh* cotyledon sheath)

axis and encloses the only prophyll. The opening of the sheath closes early in ontogeny and hence the sheath appears to be a closed structure. The vasculature of the embryo is based on a nodal plate complex in the hypocotyl from which bundles are given off to the root, the cotyledon and the prophyll. The haustorium has 12–16 vascular bundles.

27.3.1 *In Vivo Germination*

Date palm seeds are of the orthodox type where the embryo undergoes maturation and desiccation prior to germination (Engelmann 1991). Seed germination in date palm is of the remotive type. During germination, the embryo absorbs water through the hilum and swells. The haustorium expands progressively and digests the endosperm, as the cotyledonary sheath emerges out of the seed carrying the root-shoot axis with it. The growth of the sheath is positively geotropic and the axis is buried deep in the soil before the tap root emerges. The elongating sheath, the apocole, attains a length of about 10 cm. The first prophyll grows in height and pierces through the apocole to emerge out of the soil. The first leaf, eophyll, is produced following the prophyll. The haustorium continues to digest more and more of the horny endosperm as developments in the root and shoot proceed. As the haustorium digests and grows into the endosperm, it develops several folds and pouches, which increases the area of absorption. As the seedling becomes established, the endosperm is totally consumed but for a thin crusty layer which turns very dark. The haustorium, which remained a fleshy soft organ until that point, becomes exhausted and its remains are seen adhering to the crust of the leftover endosperm as a thin, white papery layer. Several adventive roots develop around the cotyledonary node as the tap root establishes itself.

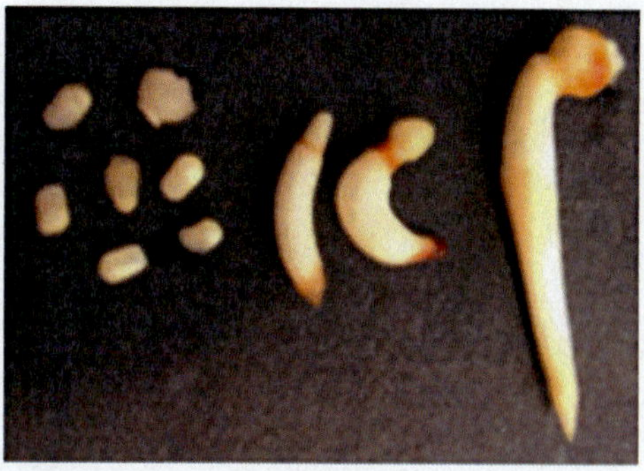

Fig. 27.3 Interspecific hybridization and embryo rescue in date palm. Stages of zygotic embryo germination

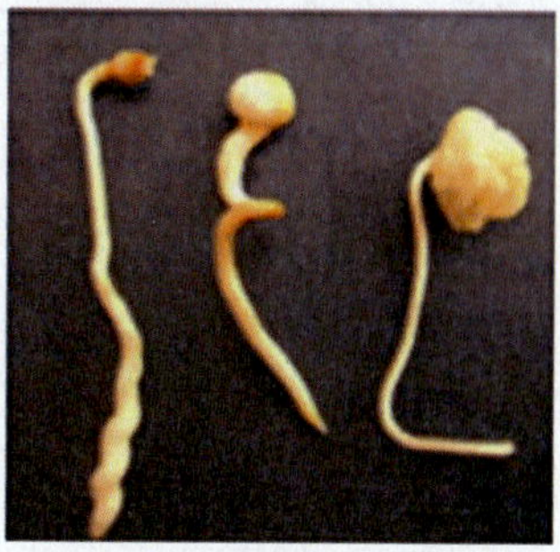

Fig. 27.4 Interspecific hybridization and embryo rescue in date palm. Variation in haustorial development in culture

27.3.2 *In Vitro* Zygotic Embryo Culture

Zygotic embryos were used as explants for the development of a propagation technique beginning in the 1970s (Hodel 1977; Oppenheimer and Reuveni 1972; Reuveni and Lillen-Kipnis 1974; Sharma et al. 1980; Tisserat 1979; Tisserat and De Mason 1980). Mature zygotic embryos undergo germination or callusing depending upon the culture media composition. *In vitro* zygotic embryo germination in normal growth hormone-free MS basal media is similar to the *in vivo* germination but differs in haustorial function. During *in vivo* germination, the cotyledonary sheath of the embryo elongates into the soil taking the root shoot axis 10–12 cm deep. Later, the primary root and prophyll development occurs. Similarly, the cotyledonary sheath of the isolated embryos under *in vitro* culture elongate 5–10 cm deep inside the media carrying the root-shoot axis down, similar to the seed germination. The haustorial part of the embryo enlarges during the cotyledonary sheath elongation, but it is rudimentary in function. After the complete elongation of the cotyledonary sheath (apocole), the tap root and the first prophyll grow simultaneously. The prophyll grows inside the sheath and emerges by piercing it after 15–20 days. Later, the first leaf emerges out through the prophyll, within 20–30 days (Figs. 27.3–27.5).

Fig. 27.5 Interspecific hybridization and embryo rescue in date palm. Stages of embryo to plantlet development

27.4 Embryo Rescue and Multiplication

The term plant embryo rescue refers to *in vitro* techniques that promote the development of an immature or weak embryo into a viable plant. Embryo rescue has been widely used for producing plants from interspecific or intergeneric hybrids in which failure of the endosperm causes embryo abortion resulting in the production of nonviable seeds. Date palm easily hybridizes with other *Phoenix* species (Nixon 1935; Sudhersan et al. 2009). In many cases of interspecific hybridization, seed development is unsuccessful due to the failure of endosperm development resulting in embryo abortion and seed collapse (Ragavan 1977). Similar results were obtained in our hybridization trial on date palm with dwarf date palm. In such a case, *in vitro* embryo culture is the only way to produce the hybrid plants. In embryo rescue procedures, the artificial nutrient medium serves as a substitute for the endosperm thereby allowing the embryo to continue its development. Generally, in embryo culture, excised embryos are placed directly onto suitable culture media. However, since small embryos are very difficult to excise without damage, entire ovules or ovaries are sometimes placed into culture. The success of embryo culture depends on several factors; particularly, the stage of embryo development, culture media components and culture environmental conditions. The type of medium needed for rescuing embryos is strongly dependent on the stage of embryo development. Young embryos require a complex medium with high sucrose concentrations, while more mature embryos can usually develop on a simple medium with low levels of sucrose (Ragavan 2003). In the case of date palm hybrid embryo culture, the mature zygotic embryos germinated in normal MS basal medium with 20–30 g/l sucrose, while the young globular stage embryos required high sucrose concentration and growth hormones for embryo maturation.

27.4.1 Stages of Hybrid Zygotic Embryos

After female date palm flower pollination, fertilization occurs and the zygote develops. During seed development, the zygote develops into an embryo through different stages: proembryo, isobilateral proembryo, globular embryo and mature embryo; this process is called *embryogenesis*. The zygotic embryo stage is very important in successful hybrid plant production (Carmi et al. 1998). Culture media component changes depend upon the stage of embryo. Fully mature zygotic embryos germinate into normal plantlet under any plant growth regulators (PGR)-free culture media containing macro elements, microelements and sugar. The early stages of embryos need a more complicated culture media with high sucrose, vitamins, amino acids and PGRs for embryo maturation prior to germination. In a zygotic embryo rescue study of sour orange (*Citrus aurantium* L.), globular stage embryos required more gibberellic acid than the heart-shaped embryos. Other culture media additives such as yeast extract enhanced the early stage embryos *in vitro*. The ability to form plantlets *in vitro* strongly increased with increasing embryo maturation (Carmi et al. 1998; Ragavan 2003).

27.4.2 Method of Embryo Isolation

Isolation of zygotic embryos at early stages under sterile condition is very difficult. Therefore, seeds at three different stages of fruit development after pollination could be used as explants. In date palms, as mentioned earlier, fruit developmental stages, based on the number of days after pollination, are classified into hababauk (7 DAP), kimri (35–120 DAP), khalal (125–175 DAP), rutab (180–200 DAP) and tamar (>200DAP). Generally, the zygotic embryo matures during the late khalal stage and undergoes dormancy during the tamar stage. Seeds isolated during the last three stages germinate into plantlets under suitable *in vitro* culture media and environment. In our date palm hybridization study, the endosperm and embryo development occurred until the early khalal stage and later the endosperm development became arrested and consequently the embryo aborted at the fruit ripening stage. Therefore, the seeds at kimri, khalal and rutab stages were isolated, sterilized and cultured in three different culture media: (1) MS basal salts with 60 g sucrose/liter; (2) MS basal salt with 40 g sucrose; and (3) MS basal salts + vitamins and PGRs (Kinetin, 2-isopentyl adenine and Napthalene acetic acid). Under total darkness, after 15 days of culture, the area where the embryo was positioned in the seed of the kimri and khalal stages swelled and projected in all the three types of culture media. At this stage the immature embryos were isolated using a sterile scalpel and needle under aseptic conditions, maintained by a laminar hood from the immature seeds.

27.4.3 Hybrid Embryo Culture

Isolated hybrid zygotic embryos cultured on the respective culture media showed different types of morphogenesis depending upon the media composition and embryo stage. Embryo maturation occurred in media with high sucrose and normal embryo germination occurred in media with 30–40 g sucrose/liter. Callusing occurred in media with PGRs. Seeds collected at the rutab stage failed to respond in

Table 27.2 Response of different stages of hybrid date palm somatic embryos to different types of culture media

	EG % of DAP			EC % of DAP		
Media type	100	120	140	100	120	140
EG	0	30	0	0	0	0
EM	0	50	0	0	0	0
SE	0	0	0	60	80	0

100 DAP-kimri stage; 120 DAP-khalal stage; 140 DAP-rutab stage
DAP Days after pollination, *EG* Embryo germination, *EC* Embryo callusing, *EM* Embryo maturation, *SE* Somatic embryogenesis

all the three type culture media (Table 27.2). A majority of the embryos isolated from the early khalal stage seeds germinated into rooted plantlets after 60–90 days of culture in the PGR-free MS culture media with 40 g/l sucrose. Therefore, embryos of the khalal stage seeds are the ideal choice for plantlet production through the embryo-rescue method in date palm interspecific hybrids.

27.4.4 Hybrid Embryo Multiplication and Plant Production

Normally a single embryo produces only one plantlet through embryo culture. In hybrids, a greater number of plantlets are necessary for experimentation. Therefore, embryo multiplication is necessary and can be achieved through somatic embryogenesis. As mentioned earlier, date palm somatic embryogenesis was perfected in the 1970s. Immature or mature somatic embryos cultured on MS culture media containing many organic additives, an auxin 2,4-dichlorophenoxyacetic acid (2,4-D) and a cytokinin 6-β-β- dimethylaminopurine (2-iP) enhanced somatic embryogenesis either directly or indirectly depending upon the culture technique. Immature embryo cotyledons in the presence of 2,4-D produce many globular proembryos after 30 days of culture. On transferring to the embryo maturation media, these globular stage proembryos mature and produce more somatic embryos through secondary somatic embryogenesis. The maintenance of the hybrid zygotic embryo on the same 2,4-D media for 90 days induced somatic embryogenic callus or organogenic callus. Transferring these calli to the maturation media produces somatic embryos or shoot buds followed by secondary somatic embryos which germinate into plantlets in the hormone-free growth media (Figs. 27.6–27.8). The direct and indirect somatic embryogenesis, and organogenesis take place in the same culture medium and the culture time factor controls the two types of regeneration (C. Sudhersan 2010, New method, unpublished). Removal of explants with somatic proembryos from the 2,4-D medium immediately after the initiation and subculture to maturation media avoids somatic embryogenic callus formation. Organogenic callus produces multiple shoots in the germination media (Figs. 27.9–27.10). Secondary somatic embryo formation is a continuous process which occurs in PGR-free MS culture media for several years when the cultures are subcultured every

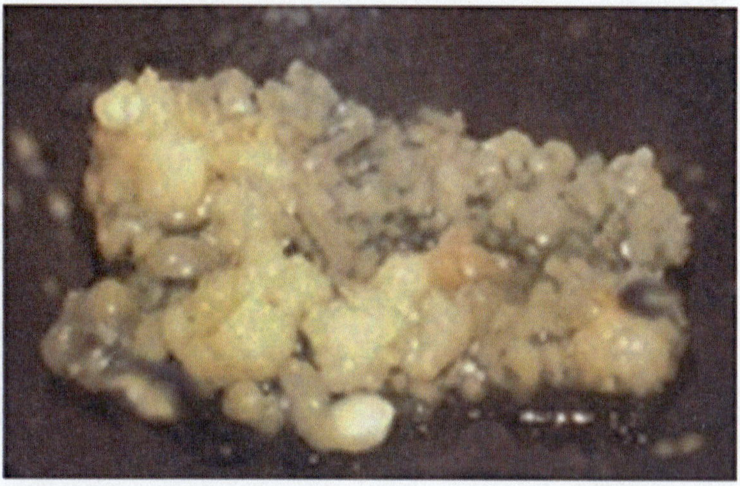

Fig. 27.6 Interspecific hybridization and embryo rescue in date palm. Somatic embryogenic callus

Fig. 27.7 Interspecific hybridization and embryo rescue in date palm. Somatic embryos

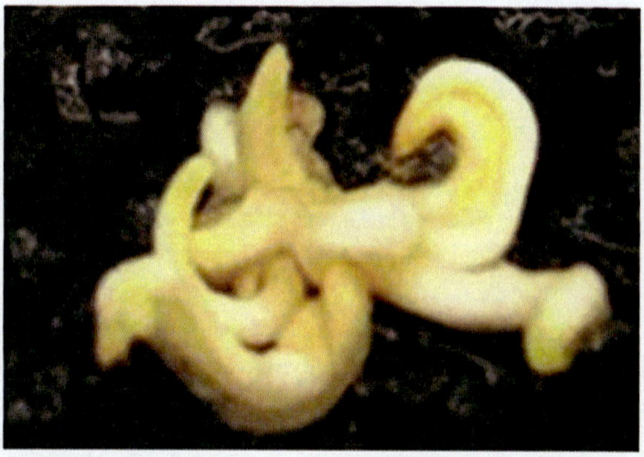

Fig. 27.8 Interspecific hybridization and embryo rescue in date palm. Secondary somatic embryos

Fig. 27.9 Organogenic callus

Fig. 27.10 Plantlet regeneration from organogenic callus

15 days. The mature somatic embryos germinate into plantlets with a single weak tap root (Figs. 27.11–27.12). On trimming the weak taproot and transferring the plantlets to a rooting media containing 0.1 mg/l NAA produces several healthy adventitious roots after 20–30 days in culture.

27.4.5 Comparison of Somatic Embryos and Hybrid Zygotic Embryos

Hybrid somatic embryos are similar to zygotic embryos in structure and behavior. A normal date palm zygotic embryo of the orthodox type undergoes a maturation and desiccation process prior to germination. Similarly in the case of somatic embryos, these two processes are highly necessary for healthy plant production. Even though the somatic embryos germinate before the maturation process, precocious germination, the plantlets obtained will be weak and do not survive during the hardening process. Therefore, the premature somatic embryogenic callus mass

Fig. 27.11 Somatic embryo germination

Fig. 27.12 *In vitro* plantlet from somatic embryo

needs to be maintained in MS culture media containing a high concentration of sucrose for 30–45 days prior to transfer to the normal germination media. At a high level of sucrose in the media, the somatic embryos undergo a maturation process and also desiccation. Mature desiccated somatic embryos germinate into healthy plantlets when transferred to PGR-free MS culture media with normal sucrose level (30 g/l). Somatic embryo vitrification, continuous multiplication by secondary somatic embryogenesis and precocious germination are the three major factors affecting healthy plantlet production. All these problems can be controlled through embryo desiccation prior to embryo multiplication and germination stage by adding more gel strength to the culture media. Desiccated somatic embryogenic callus when transferred into normal hormone-free MS culture media produces more mature somatic embryos that enhance high-quality plant production. Zygotic embryos during germination produce a haustorium which is nonfunctional under *in vitro* condition but, a majority of the somatic embryos never produce a haustorium. However, nonfunctional haustorial development during somatic embryo germination rarely occurs.

Table 27.3 Composition of hybrid embryo rescue and plant production culture media

	Culture media type			
Media components	SE	EM	EG	AR
MS basal salts	+	+	+	+
Activated Charcoal (g/l)	0.5	0.5	0	0
Other additives (mg/l)				
Mayo-Inositol	100	0	0	0
Glutamine	100	0	0	0
Glycine	3	0	0	0
Thiamine-HCl	0.5	0	0	0
Pyredoxine-HCl	0.5	0	0	0
Calcium panthothenate	0.5	0	0	0
Plant growth regulator (mg/l)				
2,4-Dichlorophenoxy acetic acid	100	0	0	0
Kinetin	3	3	0	0
2-Isopentyl adenine	3	3	0	0
Napthalene acetic acid	0	2	0	0.1
Carbohydrate (g/l)				
Sucrose	40	60	30	20
Gelling agent (g/l)				
Agar *or*	8	8	8	8
Phytagel	1.5	1.5	1.5	1.5

SE Somatic embryogenesis, *EM* Embryo maturation, *EG* Embryo germination, *AR* Adventitious rooting

27.4.6 Culture Media and Protocol

A complete protocol for date palm hybrid somatic embryo rescue and multiplication has been established. Four different culture media have been developed by the modifying MS culture media (Table 27.3). Mature hybrid zygotic embryos germinate in the germination media, while immature embryos require maturation and germination medias for plantlet production. Immature and mature somatic embryos need all the four stages of culture media for plant production via somatic embryogenesis or organogenesis.

27.5 Hardening

Hardening is a process through which free-living palms are obtained from embryo culture and it is a highly critical stage in plant production by *in vitro* culture. This process will take about 3–6 months in the case of date palm embryo *in vitro* culture.

27.5.1 Photoautotrophic Culture

Photoautotrophic culture is an intermediate stage between heterotrophic nature of *in vitro* plantlets and complete autotrophic plantlets in the greenhouse. Acclimatization of *in vitro* plantlets is a challenging part of plant micropropagation. Special procedures are necessary for the acclimatization of *in vitro* grown plantlets. Zaid and Hughes (1995) treated *in vitro* grown date palm plantlets with polyethylene glycol during acclimatization. In our laboratory, we developed a photoautotrophic culture for the plantlets produced through embryo rescue. All rooted plantlets were washed in sterile water and planted in media containing sand and peat moss at 1:1 ratio mixed with quarter strength MS nutrient solution along with 0.1 mg/l Benlate® fungicide in large culture vessels under 16 h/day high light intensity (7,000 lux) for 60–90 days. During this culture process, the plantlets produce more cuticle over the leaf epidermis and hardened enough to tolerate the complete autotrophic nature in the greenhouse. This procedure enhanced the percentage of plant survival during the greenhouse nursery establishment. All the plantlets from photoautotrophic culture, when transferred to the temperature, light and humidity-controlled greenhouse, survive better than the plantlets transferred directly from the semi photoautotrophic cultures. Plantlets coming from the photoautotrophic cultures need not be treated with any anti-transpirant chemicals.

27.5.2 Nursery Maintenance

Transfer of plantlets from photoautotrophic cultures to the greenhouse environment is more critical and requires more care. Plantlets cannot be directly exposed to the greenhouse environment; plantlets must be gradually acclimatized from the closed environment with 100% relative humidity (RH) to the open greenhouse environment with 60–70% RH. Keeping the plantlets under mist or fog condition for a few days is the general practice of plant transfer from lab to greenhouse. This procedure failed to produce 100% plant survival in the case of date palm. Mortality was high due to fungal infection. For 100% successful establishment of free-living palms, 0.5 mg/l Benolate fungicide and quarter strength MS macro and micro nutrient solution have to be sprayed regularly over on the plantlets, kept in special nursery beds with sterile sand, every 15 days for 3 months. After 3 months, the plantlets will be hard enough for field experimentations.

27.5.3 DNA Fingerprinting

The clonal nature of the free-living hybrid palms through somatic embryogenesis or organogenesis is achieved through DNA fingerprinting. A new protocol for DNA extraction and purification from date palm leaf samples has been developed at KISR (Al-Shayji et al. 1994) by combining the features from the DNA extraction methods of Murray and Thompson (1980), Dellaporta et al. (1983) and Callahan and Mehta (1991). DNA extracted from leaf samples of randomly selected greenhouse-grown plantlets obtained through somatic embryogenesis subjected to randomly amplified polymorphic DNA (RAPD) and inter simple sequence repeats (ISSR) analysis confirmed true-to-trueness. Plantlets obtained through somatic embryogenesis and organogenesis developed by the authors has not shown any genetic variation at the greenhouse stage.

27.6 Conclusion and Prospective

Interspecific hybridization between the date palm (*Phoenix dactylifera*) and the dwarf date palm (*P. pusilla*) has been successfully carried out, aimed at the development of short hybrid date palms. Culture media and protocols have been developed for the embryo rescue, multiplication, plant production and acclimatization. A few hybrid palms have been planted inside the Kuwait Institute for Scientific Research campus, Kuwait (Figs. 27.13 and 27.14) and maintained for further evaluation. It will take a few more years for the final result of this tall × dwarf (T×D) date-palm hybridization.

Fig. 27.13 Hybrid planted in the field

Fig. 27.14 A well-established T × D hybrid in the field

References

Al-Shayji Y, Saleem M, Al-Amad S et al (1994) Isolation and analysis of the total genomic DNA from the date palm (*P. dactylifera* L.) and related species. Acta Biotechnol 14:163–168

Ammar S, Benbadis A (1977) Multiplication vegetative du palmier dattier (*Phoenix dactylifera* L) par la culture de tissue de jeunes plantes issues de semis. C R Acad Sci 284:1789–1792

Anon (1982) Date production and protection. FAO Plant production and Protection Paper 35

Boyes JW, Thompson WP (1937) The development of the endosperm and embryo in reciprocal interspecific crosses in cereals. J Genet 24:14–227

Brink BA, Cooper DC (1947) The endosperm in seed development. Bot Rev 13:423–541

Callahan FE, Mehta AM (1991) Alternative approach for constant yields of total genomic DNA from Cotton (*Gossypium hirsutem* L.). Plant Mol Biol Rep 9:252–261

Carmi F, De Pasquale F, Puglia AM (1998) *In vitro* rescue of sour orange (*Citrus aurantium* L.) and their detection based on RFLP analysis. Plant Breed 117:261–266

Carpenter JB (1979) Breeding date palms in California. Ann Rep Date Growers Inst 54:13–14

Carpenter JB, Ream CL (1976) Date palm breeding a review. Ann Rep Date Growers Inst 53:25–29

Dellaporta SL, Wood J, Hicks JB (1983) A plant DNA mini preparation: version II. Plant Mol Biol Rep 1:19–21

Engelmann F (1991) *In vitro* conservation of tropical plant germplasm; a review. Euphytica 57:227–243

Hanning E (1904) Uber die kultur von Cruciferne, Embryonen ausenhalb des Embryosachs. Bot. Ztg. 62: 45–80

Hodel D (1977) Notes on embryo culture of palms. Principes 21:103–108

Johnson SA, den Nijs TPN, Peloquin SJ, Hanneman RE Jr (1980) The significance of genic balance to endosperm development in interspecific crosses. Theor Appl Genet 57:5–6

Krueger RR (1998) Date palm germplasm: overview and utilization in the USA. In: Proceedings first international conference date palms, Al-Ain, pp 2–37

Lopes MA, Larkins BA (1993) Endosperm origin, development and function. Plant Cell 5:1383–1399

Monciero A (1959) Le palmier-dattier au Sahara. First FAC international technical meeting on date products and protection, Paper No. 3, Tripoli

Murray MG, Thompson WF (1980) Rapid isolation of high molecular weight plant DNA. Nucleic Acid Res 8:4321–4325

Nixon RW (1935) Metaxenia and interspecific pollination in *Phoenix*. Proc Am Soc Hortic Sci 33:21–26

Nixon RW, Farr JR (1965) Problems and progress in date breeding. Ann Rep Date Growers Inst 42:2–5

Oppenheimer C, Reuveni R (1972) Development of a method for quick propagation of new and superior date varieties. Agricultural Research Organization, Bet Daga

Ragavan V (1977) Applied aspects of embryo culture. In: Reinert J, Bajaj YPS (eds) Plant cell, tissue and organ culture. Springer, Berlin, pp 375–397

Ragavan V (2003) One hundred years of zygotic embryo culture investigations. In Vitro Cell Dev Biol Plant 39:437–442

Ream CL (1975) Date palm breeding – a progress report. Ann Rep Date Growers Inst 52:8–9

Reuveni O (1979) Embryogenesis and plantlet growth of date palm (*Phoenix dactylifera* L.) derived from callus tissues. Plant Physiol 63(Suppl):138

Reuveni O, Lillen-Kipnis H (1974) Studies of the in vitro culture of date palm tissues and organs. Volcani Institute Agricultural Research, Pamphlet 145, Bet Dagan

Saaidi M, Toutain G, Bannerot H, Louvet J (1981) The selection of date palm (*Phoenix dactylifera* L.) for resistance to boyoud disease. Fr d' Outre Mer 35:241–249

Sawaya WN (1987) Dates of Saudi Arabia. In: Hamden I, El-Nawawy Y, Mameesh M (eds.) Advances in food industries development in the Arab World. Kuwait Institute for Scientific Research, Kuwait, pp 109–136

Sharma DR, Kumari R, Chowdhury JB (1980) *In vitro* culture of female date palm (*Phoenix dactylifera* L) tissues. Euphytica 29:169–174

Sudhersan C (1988) Morphological and developmental studies on embryos and seedlings of some palms. Ph.D thesis, Madurai Kamaraj University, Madurai

Sudhersan C (2004) Introduction of a multipurpose palm *Phoenix pusilla* in Kuwait. Palms 48:191–196

Sudhersan C, Al-Shayji Y, Jibi Manuel S (2009) Date palm crop improvement via T x D hybridization integrated with *in vitro* culture technique. Acta Hortic 829:219–224

Tisserat B (1979) Propagation of date palm (*Phoenix dactylifera* L.), *In Vitro*. J Exp Bot 30:1275–1283

Tisserat B, De Mason D (1980) A histological study of the development of adventive embryos in organ cultures of *Phoenix dactylifera* L. Ann Bot 46:465–472

Zaid A (1987) Morphogenetic variation in palm embryos *in vitro*. Date Palm J 5:36–47

Zaid A, De Wet PF (2002) Date palm propagation. In: Zaid A (ed) Date palm cultivation, FAO Plant Production and Protection Paper No. 156, Rome

Zaid A, Hughes H (1995) *In vitro* acclimatization of date palm (*Phoenix dactylifera* L.) plantlets: a quantitative comparison of epicuticular leaf wax as a function of polyethylene glycol treatment. Plant Cell Rep 15:111–114

Chapter 28
In Vitro Flowering of Date Palm

F. Masmoudi-Allouche, B. Meziou, W. Kriaâ, R. Gargouri-Bouzid, and N. Drira

Abstract Flowering is a complex process which is regulated by several intricate external and internal factors that make its induction under *in vitro* culture highly sensitive. Recently, *in vitro* flowering has been extensively investigated for many plant species and significant advances in the understanding of this phenomenon have been made. However, with regard to date palm trees, this research area remains virtually unexplored. Accordingly, the present chapter aims to present some observations concerning *in vitro* flowering of date palm, and discusses the possible factors involved in the *in vitro* flowering induction of these plants. It also describes *in vivo* flowering in date palm and the limitations associated with its requirements. The chapter goes on to present the genes controlling the flowering process and to discuss the *in vitro* induction of bisexual flowers in date palm. The chapter concludes by discussing the importance of *in vitro* flowering to date palm propagation and its implications for future programs of early sex determination and genetic improvement via hybridization practices.

Keywords Biotechnology • Micropropagation • Plant tissue culture

28.1 Introduction

Date palm (*Phoenix dactylifera* L.) is a monocotyledonous, dioecious, heterozygous and perennial tree that belongs to the Arecaceae family. The palm family is composed of various combinations of hermaphrodite species, namely *Nannorrhops ritchiana,*

F. Masmoudi-Allouche (✉) • B. Meziou • W. Kriaâ • N. Drira
Laboratory of Plant Biotechnology, Department of Biology, Faculty of Sciences Sfax, Road Soukra BP. 1171, 3000 Sfax, Tunisia
e-mail: faizamasmoudi@yahoo.fr; besmameziou@yahoo.fr; maswkri@yahoo.fr; drira_riadh@yahoo.fr

R. Gargouri-Bouzid
Department of Biology, Institute for Engineering studies, University of Sfax, Road Soukra BPW, 3038, Sfax, Tunisia
e-mail: gargouriradhia@yahoo.fr

S.M. Jain et al. (eds.), *Date Palm Biotechnology*,
DOI 10.1007/978-94-007-1318-5_28, © Springer Science+Business Media B.V. 2011

Sabal palmetto and *Washingtonia filifera*, and other monoecious species, such as *Elaeis guineensis* and *Cocos nucifera*, as well as dioecious ones, including *Borassus flabellifer* and *Phoenix dactylifera.* At the morphological level, four architectural models have been described for the various structures in palms by Holtum, Corner, Shoute and Tomlinson (Halle et al. 1978). According to the Tomlinson model, the date palm produces offshoots at the base of the tree and lateral inflorescences (Tomlinson 1961).

Date palm constitutes a major income-generating activity and food supplying source to millions of people mostly in the arid and warm regions of the Middle East and North Africa, which are favored by suitable dry subtropical and high temperature. Unfortunately, however, the cultivation of date palm is troubled by several internal and external problems worldwide. Among the most disturbing are the complexities involved in the cultivation process and the exposure to damaging diseases that menace growth, productivity and production yields.

Because date palm is slow flowering and fruiting, a major problem is to identify the sex of the samplings at an early stage, which in turn would allow for the cultivation of adequate number of productive female trees with only a minimal number of male trees.

Apart from the obvious problems emanating from the slow rates of growth and the imperceptibility of sex before flowering, date palm is also prone to severe selective pressures exerted by several pernicious diseases. Prominent among these are the bayoud vascular disease, which is caused by *Fusarium oxysporum* f. sp. *Albedinis* and has prevailed in North Africa for more than a century, and the brittle leaf disease, which emerged in Tunisia in the early 1980s and still is unknown as to its source (Triki et al. 2003). The severity of these diseases and consequent losses in production yields vary from year to year depending on the prevailing conditions.

In order to preserve the date palm and to improve its growth conditions, several propagation techniques have been developed and used with varying degrees of efficiency and success. Conventional propagation methods, to start with, are based on vegetative multiplication using offshoots. Although often leading to seedlings that are closely similar to the mother plant, these techniques have often been reported to be very slow and risky for it can cause the dissemination of infectious diseases. Another popular technique for date palm propagation is through sexual reproduction. This seed-dependant strategy actually generates a very heterogeneous population of male and female individuals in almost equal proportions (50% each). The sex of the date palm remains indiscernible until flowering, which may not occur until the plant reaches the age of 6–8 years.

Since the traditional improvement programs are based on hybridization, the Nixon research group envisaged a selection process for date palm based on repeated pollinations of some elite cultivars, such as Deglet Noor, by males derived from the same cultivars with the hope of adopting multiplication by seedling without losing the essential characteristics during the successive multiplication generations (Nixon 1959; Nixon and Furr 1965). After two generations, however, and due to the slowness of the method and the long vegetative cycle of this species, this strategy was abandoned.

To overcome these limitations and in an attempt to regenerate date palm cultivation in Morocco, as well as to control the damage caused by the bayoud disease, attempts have been envisaged to create new and useful cultivars based on controlled crosses, using resistant male and female plants. The cultivars regenerated were of poor quality. This line of research was also abandoned because of the lack of sufficient information on date palm genetics and difficulties pertaining to its biology (e.g. dioicy, long vegetative cycle) (Saaidi 1979).

Recent research seems to have paid special attention to the development of viable alternative propagation methods that can best surmount the shortcomings and inadequacies of the propagation techniques mentioned above. Of particular interest, *in vitro* culture techniques have been described to offer quite handy opportunities to produce date palm cultivars of high quality and with resistance to biotic stress. Moreover, classic breeding requires sexual hybridization, and the flowering process is crucial for the selection programs. Within this context, *in vitro* flowering seems to present a quite promising candidate that would open new pathways in genetic improvement and selection programs of this species.

In date palm, flowering has long been considered a complex process regulated by intricate internal and external factors and its induction under *in vitro* culture has often been reported to present an extremely sophisticated venture. In fact, only a few studies have so far been carried out to investigate this phenomenon in date palm. In this context, *in vitro* flowering has been successfully induced in vegetatively-propagated plantlets of different date palm cultivars (Masmoudi-Allouche et al. 2009). The highest flower induction rates were obtained through alternating between hormone-free and –rich media under different light/dark conditions. This sex induction constitutes a novel system that may allow for early sex determination and to explore the *in vitro* flowering in relation to the photoperiodic requirements in date palm. The conversion into inflorescences involved the entire apical vegetative meristem of the plantlet without affecting its phyllotaxis. A change in the architectural model of date palm was induced. Such *in vitro* flowering, producing typical female flowers, allows a significant reduction in plant cycle and can, therefore, be considered a valuable tool for future genetic improvement and selection programs in date palm. Moreover, *in vitro* induction of bisexual flowers in date palm has been achieved. The vestigial stamens of female flowers display a new and high capacity to proliferate under particular *in vitro* conditions leading to morphologically typical hermaphrodite flowers. Such hermaphrodism control can provide new prospects for improving the understanding of the genetic mechanism involved in sex organ development in date palm.

28.2 *In Vivo* Flowering

Normal date palm flowering follows an annual cycle consisting of four major phases. Date palm inflorescence development occurs in winter (Northern Hemisphere) while the vegetative growth undergoes a rest phase, during which the growth of all

Table 28.1 Date palm biological cycle (Northern Hemisphere) (Saaidi 1979)

Month	Vegetative development	Reproductive organ development
January	Vegetative growth rest	Inflorescence growth
February		Inflorescence emergence/pollination
March	Vegetative growth departure	Fruit maturation
April		
May		
June	Full vegetative activity	Fruit size increase
July		
August		Maturation
September		
October	Growth slow down	Maturation/harvest
November		
December	Vegetative growth rest	Inflorescence initiation

the organs is blocked. In spring, when the soil temperature exceeds 12°C, the vegetative development takes place again. In summer, the plant undergoes full vegetative activity. In autumn, the date palm shows a slow growth period; it is the season that characterizes fruit maturity and harvest (Table 28.1) (Saaidi 1979).

The date palm inflorescence is composed of an axis, the spadice (or peduncle), which divides into branches (pedicels) that bear the flowers. The latter are unisexual and develop on distinct plants. Each inflorescence is enclosed in a large bract (spathe) whose shape constitutes a sex characteristic. The inflorescences derive from the development of the inflorescential buds located at the leaf axils (Bouguedoura 1991).

Date palm flowers are trimeric and unisexual. In addition to the 3 sepals and 3 petals of the flowers, the male flower is characterized by an androecium composed of 6 stamens formed in 2 whorls of 3 stamens each. The male flower also contains 3 vestigial carpels. The female flower, however, develops a gynoecium composed of 3 free carpels; it also contains 6 vestigial stamens (staminodes) (Masmoudi-Allouche et al. 2009).

In fact, the *in vivo* requirements that govern floral induction, initiation and development are not yet fully known and elucidated. Further research is particularly needed to determine the photoperiodic prerequisites of date palm *in vivo* flowering and floral induction.

28.3 *In Vitro* Flowering

The phenomenon of *in vitro* flowering, though proved as a valuable approach that can be integrated into breeding programs for date palms, has received much less attention than other domains of *in vitro* research. The few reports currently available

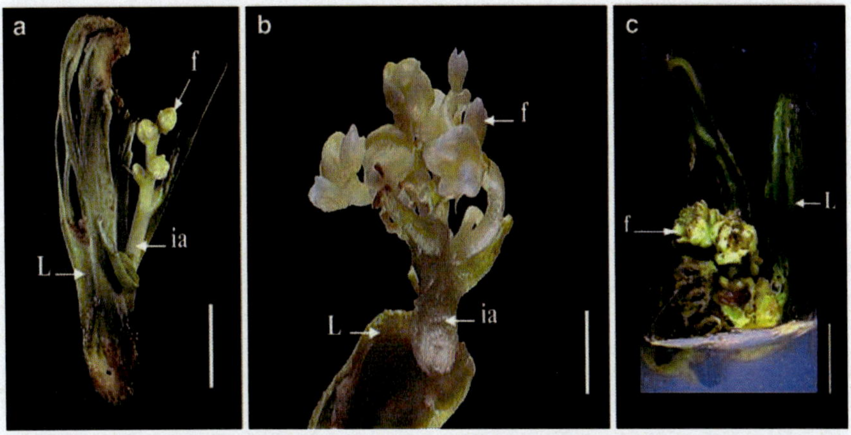

Fig. 28.1 *In vitro* flowering of date palm plantlets. (**a**) Inflorescence reduced to a single floral axis, (**b**) Ramified inflorescence (typical inflorescence), (**c**) Inflorescence reduced to a single flower. The photo shows tow flowers induced on two plantlets cultured together in the same test tube. *L* leaf, *ia* inflorescential axis, *f* flower. Scale bars: 10^4 μm

in the literature include the work conducted by Ammar et al. (1987), investigating the sexual induction of young seedlings obtained from the germination of Deglet Noor seeds. They indicated that both male and female flowers were induced in 5-month-old seedlings on 16 h day at 28°C using BAP (6-benzylaminopurine), IAA (Indole acetic acid) and glucose or sucrose. Masmoudi-Allouche et al. (2010) also achieved *in vitro* flower induction in 1-year-old *in vitro*-grown date palm plantlets (Fig. 28.1), regenerated from bud cultures via organogenesis using juvenile leaves taken from the offshoots (Drira 1983) of several date palm cultivars.

In fact, the scarcity of data in this area of research is attributed to the complex and multifaceted factors governing and affecting *in vitro* date palm flowering.

28.3.1 Factors Involved in In Vitro Flowering Induction

The literature presents a wide array of studies that have been conducted on multiple plant species and that focused on a variety of factors involved in *in vitro* flowering. These factors include light (Heylen and Vendrig 1988), photoperiod (Floh and Handro 2001), pH of the medium (Jumin and Ahmad 1999), lipids (Groenewald and Westhuizen 2004), putrescine and silver nitrate (Bais et al. 2000, 2001), and nutrients (Franklin et al. 2000; Wang et al. 2001; Zhang 2007). Several studies have also focused on the effect of plant growth regulators on the *in vitro* flowering process in some species. The requirement of cytokinin for the growth and development of flower bud has, for instance, been reported in both monocots and dicots (Wang et al. 2001; Zhang 2007; Zhong et al. 1992; Zhou et al. 2004). The effect of BA (6-benzyladenine) on early *in vitro* flowering has also been reported for different plant species (Hee et al. 2007; Jumin and Ahmad 1999; Jumin and Nito 1996; Sim et al. 2008; Wang et al. 2001). Other reports described the combined effect of BA and other hormone species on the induction and stimulation of *in vitro* flowering (Galoch et al. 2002). Similarly, Tang et al. (1983), Das et al. (1996) and Wang et al. (2002) have reported on *in vitro* flowering using BAP (Benzylaminopurine) and GA3 treatments. The influence of cytokinins has also been reported in several studies (Kachonpadungkitti et al. 2001; Taylor et al. 2005; Zhang and Leung 2000). Likewise, the effect of GA3 alone in the induction of *in vitro* flowers in gerbera (*Gerbera jamesonii* Adlam) has been reported (Ranasinghe et al. 2006). Last but not least, the individual and combined effects of abscisic acid (ABA) and proline on *in vitro* flowering in *Vigna aconitifolia* have been described by Saxena et al. (2008).

28.3.1.1 Effect of Plant Growth Regulators and Physico-Chemical Conditions

The significant effect of cytokinins on *in vitro* flowering has been extensively described and documented in the literature (Saritha and Naidu 2007; Wang et al. 2001). The combined effect of auxin and cytokinin on *in vitro* flower induction has also been reported in several studies (Handro 1983; Wang et al. 2002). Likewise, Taylor et al. (2005) reported that phytohormones affected flowering by mediating growth changes within the apical meristem and that cytokinins, in particular, played a key role in the initiation of mitosis and the regulation of cell division and organ formation. Similarly, Galoch et al. (2002) suggested that in the case of morning glory (*Pharbitis nil*), the floral transition involved a multifactorial signaling system, including the photoperiodic conditions, the endogenous phytohormone concentrations and the exogenous phytohormone application, with different phytohormones acting sequentially to trigger various steps in the flowering process.

Furthermore, Ammar et al. (1987) and Masmoudi-Allouche et al. (2010) reported on the combined effect of cytokinin and auxin on date palm flowering. Ammar et al. (1987) described the *in vitro* flowering in date palm seedlings using BAP and IAA treatment. In this study, the cultures were composed either of the cotyledonary petioles enclosing the embryos or the isolated embryos. The experiments were conducted on basal MS (Murashige and Skoog 1962) media supplemented with agar (7 $g \cdot l^{-1}$) and either sucrose (30 $g \cdot l^{-1}$) or glucose (50 $g \cdot l^{-1}$). The phytohormones IAA and BAP were added to the basal medium either alone or in combination at different concentrations. All the cultures were maintained in a growth chamber at 16 h photoperiod at a constant temperature of 28 ± 1°C. The authors reported that both male and female inflorescences were produced in the case of embryos enclosed in the cotyledonary sheath on a medium containing glucose (50 $g \cdot l^{-1}$), IAA (1 $mg \cdot l^{-1}$) and BAP (1 $mg \cdot l^{-1}$). Interestingly, inflorescence production was reported to have always been preceded by leaf initiation and to occur only in the absence of root development. In the case of excised embryo culture, on the other hand, male and female inflorescences were produced axillary to leaves, in branches. Such *in vitro* flower formation was reported to have occurred on a medium supplemented with sucrose

Table 28.2 Composition and concentrations of plant growth regulators of M1, M2, and M3 culture media

Culture medium	M1	M2	M3
Nutrient solution	MS + ½ QL major salts	MS + ½ QL major salts	MS
Plant growth regulators (μM)	NAA: 2.68 BAP: 4.44 Kin: 4.64 IPA: 5.28	No growth hormones	

(30 g·l^{-1}), IAA (0.1 mg·l^{-1}) and BAP (10 mg·l^{-1}) and to be always preceded by the development of very small leaves with reduced limbs.

In a study by Masmoudi-Allouche et al. (2010), *in vitro* flower induction experiments were conducted on basal MS (Murashige and Skoog 1962) media supplemented with sucrose (50 g·l^{-1}), adenine (0.03 g·l^{-1}), L-glutamine (0.1 g·l^{-1}) and agar (8 g·l^{-1}). In some cases, half-strength major salts of Quoirin and Lepoivre's nutrient solution (QL, Quoirin and Lepoivre 1977) were added to the basal MS medium (M1 and M2, Table 28.2). The plantlets were alternatively subcultured every 30 days on M1 (hormone-rich medium) and M2 medium (hormone-free medium) or on M1 and M3 medium (hormone-free medium) (Table 28.2).

The assays performed in Masmoudi-Allouche et al. (2010) involved three sets of experimental protocols. While the first experimental set was exposed to light (15 μM·m^{-2}·s^{-1}) at 16 h photoperiod, the second was exposed to continuous dark and the third to sequential 4-week light/dark treatments. Each treatment was applied to 24 explants of each cultivar (1 explant = 1 replicate) and the number of plantlets that showed initiation and development of inflorescences was recorded on a monthly basis.

Moreover, the experimental battery of tests performed by Masmoudi-Allouche et al. (2010) used a combination of cytokinins and auxin (M1) to evaluate their effect on flowering induction. The findings generated from the data analysis revealed that, in the absence of hormonal alternation (M1/M2 or M1/M3), no floral induction was allowed by all the light/dark conditions tested. However, regardless of the light/dark conditions used, the application of a hormonal treatment corresponding to the monthly subculture alternation of M1/M2 or M1/M3 media allowed the induction of an important flowering capacity for the different varieties tested.

More accurately, when the plants were cultured under light/dark alternations and subjected to alternative subcultures of M1/M2 media, the cumulative 1-year flowering rates reached 48% for the Arichti cv., while they did not exceed 25% for the Deglet Noor and Boufeggous cvs. and 10% for the Bousthammi Noir cv. However, when the light/dark alternation conditions were combined with a hormonal alternation between M1 and M3 media, the best flowering rates were obtained with the Arichti cv. whose cumulative 1-year flowering rate increased to reach 58% (Table 28.3). Furthermore, when the hormonal alternation of M1 and M3 media was coupled with light (16 h photoperiod) and continuous dark conditions, the flowering percentages for the Arichti cv. were 51% and 53%, respectively. These flowering

Table 28.3 The effect of various physico-chemical treatments on the flowering percentage of date palm cultures of Arichti cv. that underwent a monthly alternative subculture on M1/M2 and M1/M3 media

Treatment condition	Cumulative 1-year flowering percentage
M1/M3 alternation	
Light	51
Dark	53
16 h light/dark alternation	58
M1/M2 alternation	
Light	25
Dark	40
16 h light/dark alternation	48

percentages were actually much more important when compared to those produced in light (25%) and dark (40%) conditions in the case of a monthly alternation between M1 and M2 media (Table 28.3).

The authors suggest that the acquisition of an important flowering capacity occurred in plantlets that were subjected to sudden change from hormone-rich to -free media. This flowering ability increased when the plants were subjected to a concentration deviation involving both a medium that was rich in hormones and MS and QL nutrients (M1), and a medium that was hormone-free and that contained only the MS major salts (M3).

28.3.1.2 The Culture Period Effect

In a study by Masmoudi-Allouche et al. (2010) date palm plantlets were subcultured for 45 instead of 30 days and were submitted to a hormone alternation between M1 and M2 or M3 media (Table 28.2). The percentages of flowering explants obtained were even much more significant. For the Arichti cv., the cumulative 1-year flowering rate increased from 25% to 43%. Moreover, although the pH of the culture media was first adjusted to 5.8, it decreased during the culture period. In fact, after 30 and 45 days of culture, this decrease was of a 0.4 and a 1.8 pH unit, respectively. This phenomenon revealed that active exchanges occurred between the culture media and the tissues of the explant during the culture period and, might, therefore, have created a stress condition due to the increase of the subculture period.

The effect of a chemical stress on the enhancement of *in vitro* flowering has previously been suggested by Thorpe (1980) for the common chicory (*Cichorium intybus*) subjected to hydrous stress. Similarly, Neelu (1997) showed that flowering induction can be achieved by chemical stress using a medium containing 100 mM NaCl and an appropriate hormonal composition. These stress conditions favored the induction of *in vitro* flowering on date palm tissues, which consequently allowed the expression of new potentials that were not expressed under natural conditions. In fact, the highest flower induction rates achieved through the joint alternation between hormone-rich and -free media together with the light/dark changes could be attributed to an amplification in terms of stress condition and tissue disturbance, which might have ultimately generated a high ability to undergo a reproductive morphogenesis and flower formation, as previously suggested by Tang et al. (1983).

Table 28.4 The best cumulative means over 1-year flowering percentages of date palm plantlets from the different varieties tested when cultured under light/dark alternations and subjected to alternative subcultures of M1/M3 media

Date palm cultivar	Cumulative 1-year flowering percentage	Time of initial flowering response (month)
Arichti	58	1
Deglet Noor	32	3
Boufeggous	28	3
Bousthammi Noir	13	5

The values represent the cumulative 1-year percentages of monthly means of 24 replicates each

28.3.1.3 The Genotype Effect

Very little data are available in the literature regarding the critical effect of genotype on date palm flowering. To date, and to the authors' knowledge, only one report, by Masmoudi-Allouche et al. (2010), explored the genotype effect on date palm *in vitro* flowering using several date palm cvs., namely Deglet Noor, Arichti, Bousthammi Noir and Boufeggous. The offshoots that were investigated were collected from Deglet Noor and Arichti cvs. in the south of Tunisia and from Bousthammi Noir and Boufeggous cvs. in Morocco. An important flowering capacity was observed for the different cultivars. However, the percentages of the neo-formed flowers and the initial response of the *in vitro*-grown plantlets were tightly dependent on the cultivar of the date palm used (Table 28.4). The *in vitro* flowering efficiency observed in this study varied with the date palm varieties used, which is in agreement with the results described by Kenza and Chlyah (1998) that correlated *in vitro* differences in plant tissue responses with genotype effect.

28.3.2 Histological Analysis of the In Vitro Flowering Process

Histological studies are useful in providing information concerning structural changes during the course of flowering and to determine the optimal timing and conditions for attaining maximal induction outcomes. Such histological examination is very scarce in the literature. In one of the few currently available histological analyses that examined the floral initiation process in palm date, Masmoudi-Allouche et al. (2010) identified the morphological changes that the *in vitro* plantlet apical buds underwent to generate a floral state (Fig. 28.2a, b). In this study, the samples were fixed in Svaloff Navashine solution (chromic acid 0.5%, glacial acetic acid 5%, formaldehyde 15% and ethanol 5%) at room temperature, washed with running water for 24 h, dehydrated in ethanol solution series (50°–100°) and then immersed in xylene-ethanol baths as described by Masmoudi-Allouche et al. (2009). Paraffin inclusion was subsequently performed and 10-μm-thick sections were made using a rotary microtome. The sections obtained were stained with a Regaud ferric hematoxylin solution (Regaud hematoxylin solution 10%, glycerol 10%) (Masmoudi-Allouche et al. 2009) and then observed under a light microscope.

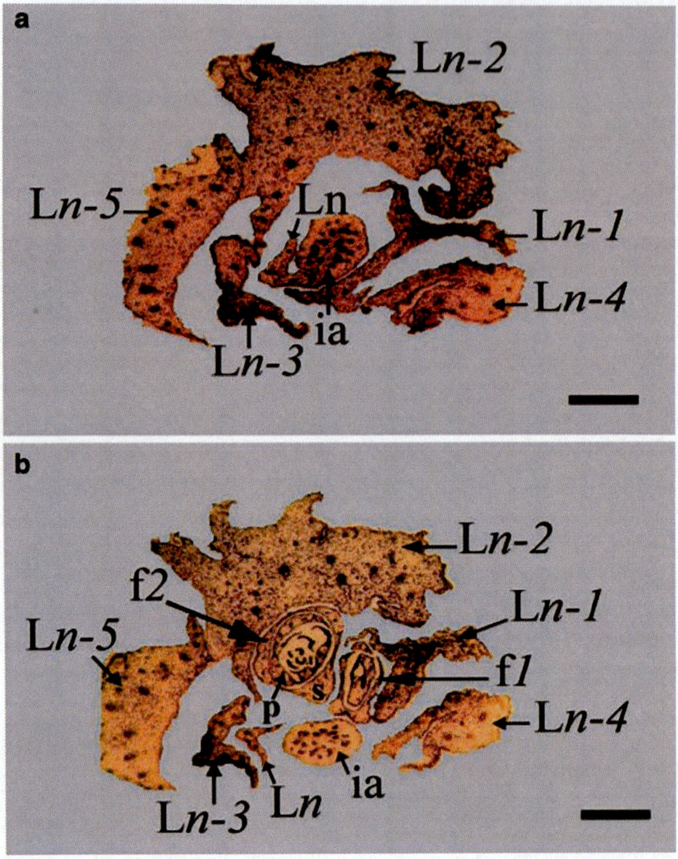

Fig. 28.2 Transverse sections on two different levels of the apical bud of date palm vitroplant cv. Arichti that have reversed to the floral growth. (**a**) Transverse section on the basal level of the apical bud showing the aspect of the last leaf structures formed at the time of the reversion to the floral state. (**b**) Transverse section on the upper level of the apical bud showing the development of the inflorescential axis on which two flowers were initiated. ia: inflorescential axis ; f*1*: flower 1 in longitudinal section ; f2: flower 2 in transverse section showing, from the external to the internal part, the 3 welded sepals (S), the 3 free petals (P) and the primordia of the reproductive organs of the flower. L*n*: leaf of rang n (last leaf that was formed before the reversion process). L*n-1*, L*n-2*, L*n-3*, L*n-4* and L*n-5*: leaves of rang *n-1, n-2, n-3, n-4 and n-5,* respectively. Scale Bars: 1,200 μm

The findings presented in Masmoudi-Allouche et al. (2010) revealed that the apical vegetative meristem of the vitroplants, which were submitted to the flower inductive treatment, continued to form leaf primordia (Fig. 28.2a) according to the usual phyllotaxis of 2/5 index (Bouguedoura 1980). It was then noted to undergo a sudden change, without any transition, into a structure of inflorescential type (Fig. 28.2b). In contrast, Bernier et al. (1981) claimed that the flower evocation in the shoot meristem was accompanied by a number of events that affected the growth habit of the plant including changes in phyllotaxis and leaf shape.

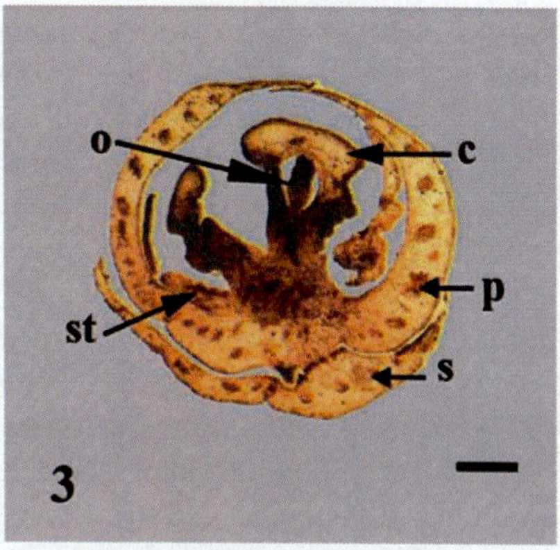

Fig. 28.3 Date palm female flower resulting from *in vitro* culture in longitudinal section. *s* sepal, *p* petal, *c* carpel, *st* staminode, *o* ovule. Scale Bar: 666 μm

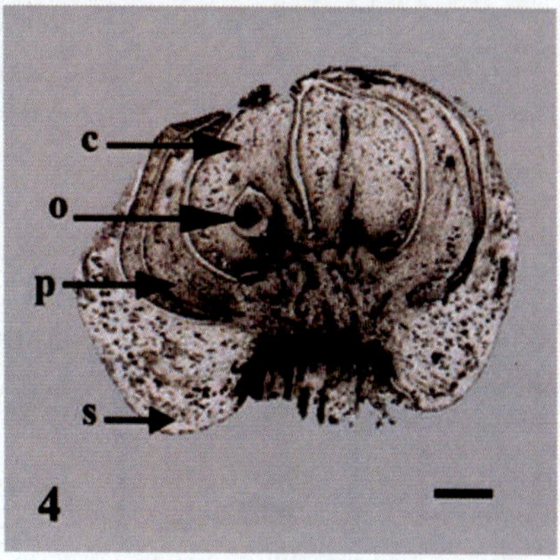

Fig. 28.4 Mature *in vivo* female date palm flower in longitudinal section. *s* sepal, *p* petal, *c* carpel, *o* ovule. Scale bar: 500 μm

Starting from a female plant material, the authors obtained flowers of the same sex. These *in vitro* female flowers were morphologically similar to those formed *in vivo*. The histological examination of longitudinal sections that was performed on *in vitro* and wild type date palm flowers showed that the *in vitro* flower structures (Fig. 28.3) were similar to those of the natural female ones (Fig. 28.4). In fact, both types of flowers were globular and consisted of three fused sepals, three free petals, six staminodes (vestigial stamens) and three carpels that harbor ovules. This confirmed the mature state of both flower types.

28.3.3 Architectural Model in Relation to In Vitro Flowering

Architectural analysis methods have greatly increased our understanding of plant structure and development and have helped construct the architectural models of several plant species. With regards to date palm, four architectural models have been defined (Tomlinson 1961). Of special interest, the results of Masmoudi-Allouche et al. (2010) revealed that, contrary to the natural flowering development, the *in vitro* neo-formed inflorescences were completely uncovered, i.e. lacking a spathe (inflorescence envelope). The authors attributed this distinct morphogenesis to the inflorescential ontogenetic mechanism. In fact, under *in vivo* conditions, date palm flowering is pleonanthic for, according to the Tomlinson model (Tomlinson 1961), the inflorescences evolve from the development of lateral buds located at the leaf axils, and the bract situated at the axillary position of the inflorescential bud develops into a spathe. However, under *in vitro* conditions, the first leaf axils of the plantlets are empty (devoid of lateral buds), as is the case of plants that derive from seedlings (Bouguedoura 1980). In that particular study, the *in vitro* flowering was terminal (hapaxanthic), resulting from the development of the apical bud that was devoid of any bract, which consequently gave rise to uncovered inflorescences.

The results of Masmoudi-Allouche et al. (2010) are in agreement with those of Ammar et al. (1987) who induced sexuality in date palm seedlings and suggested another model corresponding to the neotenous development of this tree as an herb. The latter suggested that the neoteny interpretation cannot be made only when considering a variation in the original architectural model.

Unlike African oil palm (*Elaeis guineensis*) and coconut (*Cocos nucifera*), whose architectural structures conform to the Corner model, the date palm has an architectural structure that conforms to the Tomlinson model (Tomlinson 1961), and this is due to this species' capacity of ramification. Accordingly, only date palm can be vegetatively propagated using its basal axillary buds (Drira 1983). Bouguedoura (1980) and Drira and Benbadis (1985) defined the relationship between the developmental stage of plants issued from offshoots and the nature of the axillary productions that differentiate. The authors indicated that at the first phase of its development, the date palm produces axillary buds that are both vegetative and inflorescential. The latter are, however, abortive. It is the development of the vegetative buds only that permits a basal ramification. The development of the inflorescential buds takes place later at the leaf axils.

Considering the abovementioned data (Bouguedoura 1980; Drira and Benbadis 1985), Ammar (1987) and Benbadis et al. (1985) reported that the multiple expressions of the neoteny make the identification of the architectural model essential. These authors suggest that the neotenic plants were different from the fundamental model since they produce both lateral and terminal inflorescences. In fact, in date palm plants issued from the *in vitro* development of excised embryos, the neoteny was expressed in lateral inflorescences (pleonantic). These inflorescences may be attributed to the developmental stimulation of inflorescential buds, which are normally abortive. Concerning the neoteny that occurred in plants issued from the *in vitro* development of embryos at the onset of germination, the authors indicated that it was terminal, and therefore conforming to the Holtum model (Halle et al. 1978).

Furthermore, and in a previous study by Drira and Benbadis (1985), hormonal treatment performed at precocious stages of floral differentiation was reported to induce other morphogenetic programs into the floral meristem, leading to the emergence of axes from the staminodes and the carpels of the flower. Masmoudi-Allouche et al. (2010), on the other hand, reported that the use of combined hormonal and physico-chemical factors induced the reduction of the inflorescential system, which became limited to the differentiation of either a single flower or a unique axis carrying many flowers.

Taken together, the findings reported by Drira and Benbadis (1985) and Masmoudi-Allouche et al. (2010) confirm the hypothesis of Nozeran (1954) that the trimeric flower of date palm corresponds to one group which is more ramified but highly contracted by the phylogenesis. Its original structure is, therefore, completely masked. This compacted group can appear only accidentally and in particular situations, such as *in vitro* culture conditions.

28.3.4 Photoperiodic Requirements of In Vitro Flowering

In a study by Masmoudi-Allouche et al. (2010), the authors explored the mechanism of *in vitro* flower induction control in relation to the photoperiodic requirements in date palm. Such *in vivo* photoperiodic requirements for date palm flowering are not yet fully known. Nevertheless, under *in vitro* conditions, flowering induction was obtained by varying the nature and concentration of the hormonal factor in both continuous dark and 16 h light photoperiod exposure. These data provide additional evidence confirming that date palm flowering occurred in both long-day (16 h photoperiod) as well as short-day conditions. Similarly, *in vivo* date palm flowering usually lasts (Northern Hemisphere) from March to May but, in some cases, can extend from January to the end of September, including both short and long-day conditions. These observations provide further evidence that date palm is able to flower, whatever the photoperiod is, as long as a sufficient photosynthetic activity is properly available. Under the *in vitro* culture conditions presented in the mentioned study, the availability of organic nutrients in the medium was sufficient to ensure flowering in continuous darkness.

28.4 Genes Controlling the Flowering Process

Bernier et al. (1999) indicated that the flowering process is controlled by several tens of genes. They also suggested the existence of an inhibitory system that is able to block the expression of floral identity meristem genes controlled by a flowering repressor gene (embryonic flower1, EMF1). The activity of this system induces an obligatory vegetative growth in most plants. Nevertheless, when blocked, the plant would be able to flower spontaneously after germination. The authors added that this gene complex is present in all plants. The characterization of *Arabidopsis thaliana* flowering mutants suggested that a very complex system controls the flowering process in plants.

According to Calonje et al. (2004), the genetic and molecular characterization of the flowering process in different species reveals a conservation of the basic genetic mechanisms controlling the early stages of flower formation (Ng and Yanofsky 2001; Theissen and Saedler 1999). In monocots, however, *SQUA*-like genes do not seem to be always functional orthologs of their *Arabidopsis* counterparts, based on their relatively large number and expression patterns (Gocal et al. 2001; Schmitz et al. 2000; Theissen et al. 2000; Yu and Goh 2000). All these data suggest that flower initiation and development may involve common regulatory mechanisms to all the angiosperms as well as species-specific mechanisms whose genetic and molecular bases are yet unknown (Calonje et al. 2004).

In fact future research is needed to understand the genes controlling the flowering process in date palm and to analyze their expression during the reproductive organ development. This knowledge may offer a very important system to identify a sex-marker for this dioecious species, particularly at a time when research studies, namely the one conducted by Masmoudi-Allouche et al. (2009), asserted that in vitro floral hermaphrodism induction could be successfully achieved. Genetic analysis can, therefore, be conducted on the bisexual, as well as on the normal unisexual female and male, date palm flowers.

28.5 *In Vitro* Induction of Bisexual Flowers

The plant reproductive systems that pattern floral and sexual differentiation can be monomorphic, with a single bisexual flower type, or polymorphic, with two or more flower types. The majority of flowering plants are hermaphroditic, developing perfect flowers that contain both pistils and stamens. Hermaphroditic individuals produce both male and female gametes (Irish and Nelson 1989).

In plants, the understanding of the sex determination system is closely connected with the knowledge of how separate sexes evolved. The widespread view that all flowering plants arose from a common hermaphrodite ancestor (Cronquist 1988) suggests that much of the floral developmental program is common to all species. An early theory (Darwin 1877) claims that the first plant species were hermaphroditic (Ainsworth 2000; Charlesworth 2002; Lebel-Hardenack and Grant 1997; Negrutiu et al. 2001). It suggests that, during evolution, 10% or so of these plants have evolved, via different evolutionary routes, to floral unisexuality as the spatial separation of their flowers generate evolution toward monoecy or dioecy (Ainsworth 2000).

Date palm sexuality follows the dioecy system, which is a rare sexual system in flowering plants, occurring only in 4–6% of the species (Ainsworth 2000; Guttman and Charlesworth 1998; Renner and Ricklefs 1995; Tanurdzic and Banks 2004). In some cases, however, apparent bisexual flowers seem to naturally occur within female date palm trees (Demason and Tisserat 1980). In fact, *in vitro* production of bisexual flowers in date palm species has been reported by Demason and Tisserat (1980), who induced the *in vitro* carpel development in male flowers, and by Masmoudi-Allouche et al. (2009), who induced the *in vitro* stamen development in female flowers.

28.5.1 *In Vitro* Carpel Development in Male Flowers

Demason and Tisserat (1980) described the occurrence of apparent bisexual date palm flowers through a 2,4-dichlorophenoxyacetic acid (2,4-D) treatment of male flowers. They postulated, however, that the staminodes in cultured pistillate flowers did not expand under the culture conditions they used. The apparent bisexual flowers harbored carpels without ovules.

28.5.2 *In Vitro* Stamen Development in Female Flowers

Masmoudi-Allouche et al. (2009) investigated the vestigial stamens (staminodes) of female date palm flowers from different Tunisian date palm cvs. (Deglet Noor, Gondi, Boufeggous, Allig, Mattata and Kentichi) and observed that those stamens displayed a new and high capacity to proliferate under particular *in vitro* conditions and hormonal treatment without blocking carpel development, leading to morphologically typical hermaphrodite flowers (Fig. 28.5). The pollen mother cells isolated appeared in the anther locules obtained and underwent an ordinary microsporogenesis process (Masmoudi-Allouche et al. 2009).

Among the different hormonal combinations tested, only the one including IBA and BAP (Table 28.5), which were added at different concentrations in the MS basal medium, showed an efficient reinitiating of anther development within the female flower (Table 28.5). In fact, a remarkable proliferation of stamens (80–90%) in the female flowers was obtained when the supplemented IBA/BAP concentrations corresponded to 4.92/4.44 μM and 9.84/4.44 μM (media 5 and 4, respectively; Table 28.5). By lowering the IBA/ BAP concentrations to 2.46/2.22 μM (medium 3, Table 28.5), an induction of about 50% was obtained within the cultured female flowers. However, higher hormonal concentrations of 9.84/8.88 μM and 19.68/8.88 μM (media 1 and 2, respectively; Table 28.5) generated a low percentage of hermaphrodite flower production.

The findings of Masmoudi-Allouche et al. (2009) confirm the early theory suggesting that dioecious plants derive from a hermaphrodite ancestor. They support the system reported by Lebel-Hardenack and Grant (1997), which postulated that in many dioecious species, unisexual floral meristems are sexually bipotent and that a change in the level or ratio of endogenous hormones can trigger a switch between the alternative developmental programs of the sex-determining genes. As far as the unisexual flowers are concerned, an abortion or arrest of the carpel primordial in the male flower and the stamen primordia in the female flower occur in a later span of time (Dellaporta and Calderon-Urrea 1994; Kater et al. 2001).

Such hermaphrodism control can provide new prospects and opportunities for the investigation of the *in vitro* self-fertilization process. It can also offer useful tools for further understanding the genetic mechanisms involved in the sex organ development of date palm.

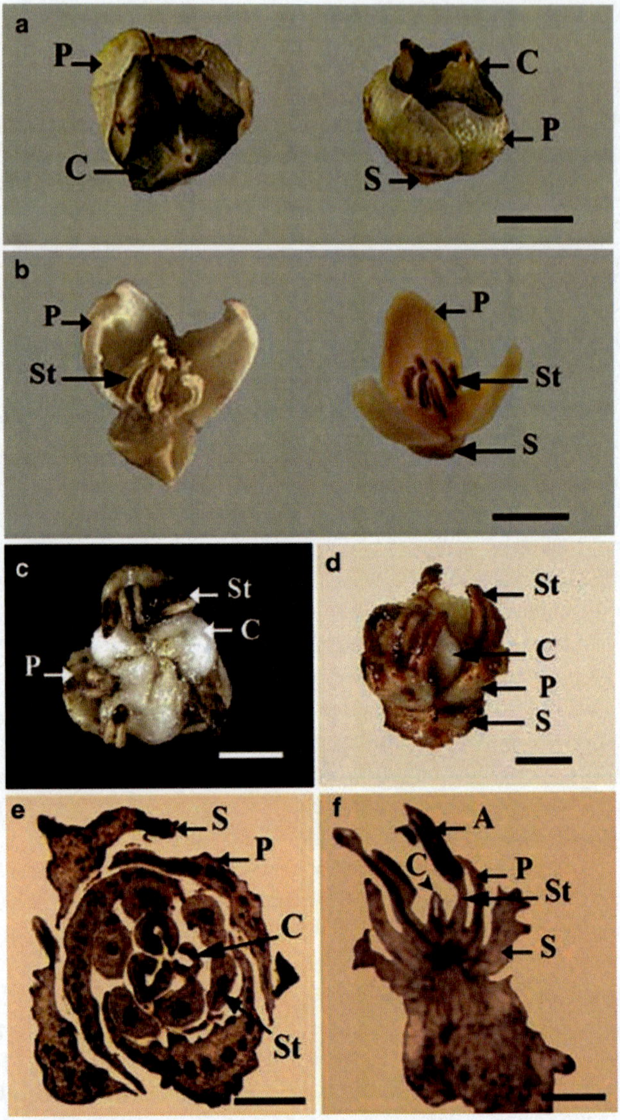

Fig. 28.5 Wild type (**a**, **b**) and hermaphrodite date palm flowers differentiated *in vitro* (**c**–**f**). *A* wild female date palm flowers; frontal (on the *left*) and profile (on the *right*) views. *B* wild male date palm flowers; frontal (on the *left*) and profile (on the *right*) views. *C-F* hermaphrodite flower differentiated under *in vitro* conditions ; frontal (**c**) and profile (**d**) views, in transverse (**e**) and longitudinal (**f**) sections. *S* sepal, *P* petal, *A* anther, *St* stamen, *C* carpel. Scale bars: 1,600 μm (**a**–**d**); 500 μm (e, f)

Table 28.5 The percentage of hermaphrodite flowers induced *in vitro* by culturing Deglet Noor female inflorescences on the different tested media for 15 days. About 500 flowers were used for each assay conditions by several repetitions

Medium	IBA (μM)	BAP (μM)	Hermaphrodite flowers, (%)
1	9.84	8.88	5±0.6
2	19.68	8.88	30±1.15
3	2.46	2.22	50±1
4	9.84	4.44	82±1
5	4.92	4.44	90±1.5

28.6 Conclusion and Prospects

Date palm has a long juvenile phase that delays their reproductive development by between 6 and 8 years. Although *in vitro* flowering induction has been reported for several plant species, only a few studies have so far been carried out to with the aim of accelerating the flowering time of date palm. In fact, the few studies currently available in the literature indicate that *in vitro* flowering induction can be achieved on plantlets regenerated from shoot cultures of different date palm cultivars under particular *in vitro* culture conditions (Masmoudi-Allouche et al. 2010) and also on young seedlings obtained from the germination of Deglet Noor seeds (Ammar et al. 1987). Interestingly, the *in vitro* flowers obtained were histologically and morphologically similar to *in vivo* flowers.

In vitro flowering can be useful to the *in vitro* rejuvenation process which is based on the changes in vegetative characteristics (Hackett 1985; Pierik 1990) and on the flowering ability of cultured shoots (Harada and Murai 1998). These results suggest that the neo-formed flowers have future prospects in developing renovation programs for saving date palm germplasm losing senescence status. *In vitro* flowering will assist to reinitiate the micropropagation process. Furthermore, the significant shortening of the plant cycle through the control of *in vitro* flowering also provides a valuable and promising system for early sex determination. *In vitro* flowering will save time for date palm genetic improvement by using intra- or interspecific crosses. It would help to reveal novel properties and characteristics pertaining to fruit quality and stress resistance.

Moreover, bisexual flowers can be induced *in vitro* through hormonal treatment of female inflorescences leading to high percentages of typical hermaphrodite flowers (Masmoudi-Allouche et al. 2009). Such floral hermaphrodism control will ensure to investigate the *in vitro* self-fertilization process and to identify the sex markers and genes for sex maturation.

Acknowledgements This work was supported by the Ministry of Higher Education, Scientific Research and Technology in Tunisia and the International Atomic Energy Agency, under TC Project RAF/5/049. The authors would like to thank the staff of the Regional Centre of Oasian Agriculture Research in Deguache (southern Tunisia) for their help in providing offshoots from Tunisian cultivars. They would also like to thank Dr. Ismail El Hadrami, Chef of the Plant

Physiology Laboratory in Marrakech for kindly providing plant material from Moroccan cultivars. Thanks are also due to Prof. Anouar Smaoui from the English section at the Sfax Faculty of Science for his constructive language polishing services.

References

Ainsworth C (2000) Boys and girls come out to play: the molecular biology of dioecious plants. Ann Bot 86:211–221

Ammar S (1987) Contribution à l'étude des potentialités morphogénétiques exprimées en culture *in vitro* par de jeunes plantes de palmier dattier (*Phœnix dactylifera* L.) et des tissus qui en dérivent en culture. Doctoral thesis, Faculty of Science, Tunis

Ammar S, Benbadis A, Tripathi BK (1987) Floral induction in date palm seedling (*Phoenix dactylifera* L. var. Deglet Nour) cultured *in vitro*. Can J Bot 65:137–142

Bais HP, Sudha GS, Ravishankar GA (2000) Putrescine and silver nitrate influences shoot multiplication, in vitro flowering and endogenous titers of polyamines in *Cichorium intybus* L. cv. Lucknow local. J Plant Growth Regul 19:238–248

Bais HP, Sudha GS, Ravishankar GA (2001) Influence of putrescine, silver nitrate and polyamine inhibitors on the morphogenetic response in untransformed and transformed tissues of *Cichorium intybus* and their regenerants. Plant Cell Rep 20:547–555

Benbadis A, Ammar S, Drira N et al (1985) Evolution du modèle arcuitectural du palmier dattier (*Phœnix dactylifera* L.) liée aux modalités de la multiplication végétative et à l'expression de la néoténie en culture *in vitro*. International colloque on tree, France, pp 14

Bernier G, Kinet JM, Sachs RM (1981) The physiology of flowering, vol II. Transition to reproductive growth. CRC Press, Boca Raton

Bernier G, Cremer F, Huijser P (1999) Pourquoi les plantes fleurissent. Biofutur 187:22–26

Bouguedoura N (1980) Morphologie et ontogenèse des productions axillaires du palmier dattier (*Phoenix dactylifera* L.). C R Acad Sci Paris 291:857–860

Bouguedoura N (1991) Connaissance de la morphogenèse du palmier dattier *Phœnix dactylifera* L. Étude *in situ* et *in vitro* du développement morphogénétique des appareils végétatif et reproducteur. Doctoral Thesis, USTHB, Alger

Calonje M, Cubas P, Martínez-Zapater JM, Carmona MJ (2004) Floral meristem identity genes are expressed during tendril development in grapevine. Plant Physiol 135:1491–1501

Charlesworth D (2002) Plant sex determination and sex chromosomes. Heredity 88:94–101

Cronquist A (1988) The evolution and classification of flowering plants. New York Botanical Garden, Bronx

Darwin C (1877) The different forms of flowers on plants of the same species. Murray, London

Das P, Samantaray S, Rout GR (1996) Organogenesis and in vitro flowering of *Echinochloa colona*. Effect of growth regulators and explant types. Biol Plant 38:335–342

Dellaporta SL, Calderon-Urrea A (1994) The sex determination process in maize. Sci 266:1501–1505

Demason DA, Tisserat B (1980) The occurrence and structure of apparently bisexual flowers in the date palm, *Phoenix dactylifera* L. (Arecaceae). Bot J Linn Soc 81:283–292

Drira N (1983) Multiplication végétative du palmier dattier (*Phoenix dactylifera* L.) par la culture *in vitro* de bourgeons axillaires et de feuilles qui en dérivent. C R Acad Sci Paris III 296:1077–1082

Drira N, Benbadis A (1985) Multiplication végétative du palmier dattier (*Phoenix dactylifera* L.) par réversion, en culture in vitro, d'ébauches florales de pieds femelles adultes. J Plant Physiol 119:227–235

Floh EIS, Handro W (2001) Effect of photoperiod and chlorogenic acid on morphogenesis in leaf discs of *Streptocarpus nobilis*. Biol Plant 44:615–618

Franklin G, Pius PK, Ignacimuthu S (2000) Factors affecting in vitro flowering and fruiting of green pea (*Pisum sativum* L.). Euphytica 115:65–73

Galoch E, Czaplewska J, Burkacka-Łaukajtys E, Kopcewicz J (2002) Induction and stimulation of *in vitro* flowering of *Pharbitis nil* by cytokinin and gibberellin. Plant Growth Regul 37:199–205

Gocal FW, King RW, Blundell CA et al (2001) Evolution of floral meristem identity genes: analysis of *Lolium temulentum* genes related to *APETALA1* and *LEAFY* of *Arabidopsis*. Plant Physiol 125:1788–1801

Groenewald EG, Van Der Westhuizen AJ (2004) The effect of various lipids on flowering of *Pharbitis nil* in in vitro culture. Biol Plant 48:637–639

Guttman DS, Charlesworth D (1998) An X-linked gene with a degenerate Y-linked homologue in a dioecious plant. Nature 393:263–266

Hackett WP (1985) Juvenility, maturation, and rejuvenation in woody plant. Hortic Rev 7:109–155

Halle F, Oldeman RAA, Tomlinson PB (1978) Tropical trees and forests: an architectural analysis. Springer, Berlin

Handro W (1983) Effects of some growth regulators on *in vitro* flowering of *Streptocarpus nobilis*. Plant Cell Rep 2:133–136

Harada H, Murai Y (1998) *In vitro* flowering on long-term subcultured pear shoots. J Hortic Sci Biotech 73:225–228

Hee KH, Loh CS, Yeoh HH (2007) Early in vitro flowering and seed production in culture in *Dendrobium Chao Praya* Smile (Orchidaceae). Plant Cell Rep 26:2055–2062

Heylen C, Vendrig JC (1988) The influence of different cytokinins and auxins on flower neoformation in thin cell layers of *Nicotiana tabacum* L. Plant Cell Physiol 29:665–671

Irish EE, Nelson T (1989) Sex Determination in monoecious and dioecious plants. Plant Cell 1:737–744

Jumin HB, Ahmad M (1999) High-frequency in vitro flowering of *Murraya paniculata* (L.) Jack. Plant Cell Rep 18:764–768

Jumin HB, Nito N (1996) In vitro flowering of *Fortunella hindsii* (Champ.). Plant Cell Rep 15:484–488

Kachonpadungkitti Y, Romchatngoen S, Hasegawa K, Hisajima S (2001) Efficient flower induction from cultured buckwheat (*Fagopyrum esculentum* L.) node segments in vitro. Plant Growth Regul 35:37–45

Kater MM, Franken J, Carney KJ et al (2001) Sex determination in the monoecious species of cucumber is confined to specific floral whorls. Plant Cell 13:481–493

Kenza L, Chlyah H (1998) Vegetative multiplication of date palms from *in vitro* cultured inflorescences: effect of some growth regulator combinations and organogenetic potential of various cultivars. Agronomy 18:573–580

Lebel-Hardenack S, Grant SR (1997) Genetics of sex determination in flowering plants. Trends Plant Sci 2:130–136

Masmoudi-Allouche F, Chaâri-Rkhis A, Kriaâ W et al (2009) *In vitro* hermaphrodism induction in date palm female flower. Plant Cell Rep 28:1–10

Masmoudi-Allouche F, Meziou B, Kriaâ W et al (2010) *In vitro* flowering induction in date palm (*Phoenix dactylifera* L.). J Plant Growth Regul 29:35–43

Murashige T, Skoog F (1962) A revised medium for rapid growth and bioassays with tobacco tissue cultures. Physiol Plant 15:473–497

Neelu S (1997) *In vitro* completion of vegetative and floral phase of salt - stressed *Brassica juncea* var. BN - 1. Plant Tissue Cult Biotech 3:160

Negrutiu I, Vyskot B, Barbacar N et al (2001) Dioecious plants: a key to the early events of sex chromosome evolution. Plant Physiol 127:1418–1424

Ng M, Yanofsky MF (2001) Function and evolution of the plant MADS-box gene family. Nat Rev Genet 2:186–195

Nixon RW (1959) Pollinization, breeding and selection of date palms. In: FAO International technical meeting date products and processing, pp 22–41

Nixon RW, Furr R (1965) Problems and progress in date breeding. Date Growers Inst Rpt 42:2–5

Nozeran R (1954) Contribution à l'étude de quelques structures florales (essai de morphologie florale comparée). Thèse Doct, Etat Sci Nat Fac Sci, Montpellier

Pierik RLM (1990) Rejuvenation and micropropagation. Newsl Int Assoc Plant Tissue Cult 62:11–21

Quoirin M, Lepoivre P (1977) Etude de milieu adapté aux cultures *in vitro*. C R Acad Sci Paris 281:1309

Ranasinghe RATD, Abayagunawardana AGNI, Hettiarachchi HIDD et al. (2006) In vitro flower induction in Gerbera (*Gerbera jamesonii* Adlam). In: Eighteenth annual congress of the PGIA, pp 16–17

Renner SS, Ricklefs RE (1995) Dioecy and its correlates in the flowering plants. Am J Bot 82:596–606

Saaidi M (1979) Contribution à la lutte contre le Bayoud, fusariose vasculaire du palmier dattier. Thèse Doct, Univ Dijon

Saritha KV, Naidu CV (2007) High frequency plant regeneration and *in vitro* flowering of regenerated plantlets of *Spilanthes acmella* Murr – An important threatened bio-insecticide medicinal plant. Acta Hortic 756:183–198

Saxena SN, Kaushik N, Sharma R (2008) Effect of abscisic acid and proline on in vitro flowering in *Vigna aconitifolia*. Biol Plant 52:181–183

Schmitz J, Franzen R, Ngyuen TH et al (2000) Cloning, mapping and expression analysis of barley MADS-box genes. Plant Mol Biol 42:899–913

Sim GE, Goh CJ, Loh CS (2008) Induction of in vitro flowering in *Dendrobium* Madame Thong-In (Orchidaceae) seedlings is associated with increase in endogenous N^6-(Δ^2-isopentenyl)-adenine (iP) and N^6-(Δ^2-isopentenyl)-adenosine (iPA) levels. Plant Cell Rep 27:1281–1289

Tang AF, Cappadocia M, Byrne D (1983) In vitro flowering in cassava (*Manihot esculenta* Crantz). Plant Cell Tissue Org 2:199–206

Tanurdzic M, Banks JA (2004) Sex determining mechanisms in land plants. Plant Cell 16:61–71

Taylor NJ, Light ME, Staden JV (2005) *In vitro* flowering of *Kniphofia leucocephala*: influence of cytokinins. Plant Cell Tissue Org 83:327–333

Theissen G, Saedler H (1999) The golden decade of molecular floral development (1990–1999): a cheerful obituary. Dev Genet 25:181–193

Theissen G, Becker A, Di Rosa A et al (2000) A short history of MADS-box genes. Plant Mol Biol 42:115–149

Thorpe TA (1980) Organogenesis *in vitro*: structural, physiological and biochemical aspects. In: Vasil IK (ed) Perspectives in plant cell and tissue culture. Academic, New York, pp 71–111

Tomlinson PB (1961) Anatomy of the monocotyledons. II. Palmae. Clarendon, Oxford

Triki MA, Zouba A, Khoualdia O et al (2003) Maladie des feuilles cassantes or brittle leaf disease of date palms in Tunisia: biotic or abiotic disease? J Plant Pathol 85:71–79

Wang S, Tang L, Chen F (2001) *In vitro* flowering of bitter melon. Plant Cell Rep 20:393–397

Wang G, Yuan M, Hong Y (2002) *In vitro* flower induction in roses. In Vitro Cell Dev Biol Plant 38:513–518

Yu H, Goh CJ (2000) Identification and characterization of three orchid MADS-box genes of the *AP1/AGL9* subfamily during floral transition. Plant Physiol 123:1325–1336

Zhang T (2007) In vitro flowering of *Perilla frutescens*. In Vitro Cell Dev Biol Plant 43:91–94

Zhang Z, Leung DWM (2000) A comparison of in vitro with in vivo flowering in gentian. Plant Cell Tissue Org 63:223–226

Zhong H, Srinivasan C, Sticklen MB (1992) In vitro morphogenesis of corn (*Zea mays* L.). Planta 187:490–497

Zhou JH, Zhou JR, Lin BC, Din J (2004) Influences of 6-BA and amino acids on flower formation from cotyledons of cucumber in vitro. Plant Physiol Commun 6:31–33

Chapter 29
Date Palm Cell and Protoplast Culture

A. Assani, D. Chabane, H. Shittu, and N. Bouguedoura

Abstract This chapter describes the current status of cell and protoplast cultures in date palm (*Phoenix dactylifera* L.). Critically important steps toward plant regeneration from recalcitrant date palm protoplasts have been achieved in the recent past. Callus regeneration was achieved in commercial cvs. Deglet Noor, Takerboucht, Barhee and Zaghloul. The use of feeder layer was the main factor for inducing cell divisions as well as subsequent microcallus and callus formation. Presently, a protoplast-to-plant system has been reported for more than 400 species, of which the family Solanaceae is predominantly represented. In crops like potato, tobacco, brassica, citrus and eggplant seed companies are using protoplast fusion based technology to produce new commercial varieties. In the last decade however, little progress has been made regarding protoplast regeneration from the so-called recalcitrant species which includes date palm. There are certain similarities between some protoplast systems and mammalian stem cells. More collaboration between animal and plant scientists is highly encouraged, which would be useful for both parties to help address these challenges.

Keywords Cellulase • Cell suspension • Feeder layer • Microcallus • Somatic embryogenesis • Tissue culture

---

A. Assani (✉)
Department of Plant Agriculture, University of Guelph, Guelph, ON N1G 2W1, Canada
e-mail: aassani@bell.net

D. Chabane • N. Bouguedoura
Laboratoire de Recherche sur les Zones Arides (LRZA), Université des Sciences et de la Technologie Houari Boumediene (USTHB), BP 31 El Alia Bab Ezzouar, Algiers, Algeria
e-mail: chabanedj@yahoo.fr; nadiaboug@voila.fr

H. Shittu
Department of Molecular and Cellular Biology, University of Guelph, Guelph, ON, N1G 2W1, Canada
e-mail: hshittu@uoguelph.ca

S.M. Jain et al. (eds.), *Date Palm Biotechnology*,
DOI 10.1007/978-94-007-1318-5_29, © Springer Science+Business Media B.V. 2011

29.1 Introduction

Date palm (*Phoenix dactylifera* L.) is a dioecious monocotyledonous perennial and diploid (2n = 2x = 36) species of the Arecaceae. It is mainly cultivated in North Africa and the Middle East, and date fruits are consumed as a staple food (Bouguedoura 1991). It also grows in USA, South America, Spain and Italy (Heselmans 1997). Date trees also offer ecological conditions for other cultivated crops in these regions. Date palm trees are propagated sexually by seeds as well as asexually with offshoots; however, seed-derived date palm leads to inconsistent performance due to the genetic heterogeneity. When seed propagation is used, half of the progeny would be male and half female. To maintain the genetic identity of the cultivars, seed-derived tree offshoots have to be separated and established for each individual tree (Saaidi 1979). The traditional vegetative propagation, with a very limited number of offshoots (about 10–15 per tree), is less efficient as compared to *in vitro* micropropagation procedures (Bouguedoura et al. 1990).

Date palm biotechnology is routinely used in tissue, organ and cell culture for large-scale plant production and multiplication. Protoplast technique is yet to reach a stage of being used routinely in date palm genetic improvement, especially for somatic cell hybridization. The protoplast, consisting of the cell cytoplasm bounded by the plasma membrane and free from cell wall, was first initiated by Klercker (1892) in the aquatic plant water soldier (*Stratiotes aloides*). Later Kuster (1909) and Michel (1937) showed that isolated protoplasts could be induced to fuse, allowing coalescence of their cytoplasm. The availability of commercial enzymes enables the production of large numbers of uniform protoplasts. Regeneration of fertile plants from isolated protoplasts was reported in tobacco (*Nicotiana tabacum*) for the first time by Nagata and Takebe (1971) and Takebe et al. (1971). Presently, a protoplast-to-plant system has been reported for more than 400 species, of which the family Solanaceae is predominantly represented, followed by Leguminosae, Gramineae, Compositae, Cruciferae, Umbelliferae and Rosaceae (Davey et al. 2005).

The totipotency of isolated protoplasts is fundamental for genetic manipulation approaches involving somatic hybridization and genetic transformation. Electrofusion and chemical treatments of protoplasts with polyethylene glycol (PEG) techniques are commonly used. Somatic hybrids are routinely produced in citrus, brassica, potato and other members of the Solanaceae (Grosser et al. 2001; Möllers et al. 1994; Sacristan et al. 1989). However, success in monocots, which includes date palm, is very limited. Plant transformation was also successfully performed by uptake of DNA into totipotent protoplasts (Vardi et al. 1990). Chemical agents (PEG), high electrical pulses (electroporation) or a combination of both procedures are commonly used for DNA uptake studies, e.g. in sugar beet (Hall et al. 1996).

To the best of our knowledge there are only two reports on date palm protoplasts. Chabane et al. (2007) reported callus formation from protoplasts in cvs. Deglet Noor and Takerboucht. Similarly, Rizkalla et al. (2007) succeeded in inducing callus from protoplasts in Barhee and Zaghloul cvs. One of the most reliable tools to produce resistant plants with good date fruit quality is genetic improvement.

This approach would enable (1) the selection of resistant cultivars and cultivars with excellent fruit quality through field trials, (2) and then combining both traits in one cultivar through conventional (crossbreeding) or somatic hybridization. Also resistance genes can be taken from a cultivar or species with high resistance level to a particular disease through genetic engineering method. These genes can be inserted into date palm through genetic transformation approach. Once an elite tree is developed, it can be micropropagated through *in vitro* culture. A successful production of a resistant elite cultivar would reduce the use of chemicals for pest and disease control in date palm plantations.

Classical crossbreeding usually takes a long time (15–30 years) for the production of a new cultivar. Somatic hybridization can be a credible alternative. Somatic hybridization has been successfully used in vegetatively propagated banana (*Musa spp.*) (Assani et al. 2005); in Solanaceae, to introduce disease-resistant traits, by transferring resistance from wild relatives to cultivated varieties (Collonnier et al. 2001; Fock et al. 2000); in rice (*Oryza sativa* L.); in barley (*Hordeum vulgare* L.) (Kisaka et al. 1998) and in citrus (*Citrus* spp.) (Grosser et al. 2001). Protoplasts can be isolated from various parts of the plant including leaves, cotyledons, shoots, roots and flowers. In monocot species the ideal material for protoplast isolation is embryogenic cell suspension and prolific growing callus cultures. However, embryogenic callus cultures are most suitable for protoplast isolation in date palm (Chabane et al. 2007). This chapter describes the current status of date palm cell suspension and protoplast culture.

29.2 Cell Suspension Cultures

29.2.1 Source Tissue for Cell Suspension Cultures

There have been several reports on the use of different types of explants in date palm biotechnology for establishing cell suspension cultures including juvenile leaves of offshoots, immature inflorescences (Bhaskaran and Smith 1992; Fki et al. 2003), leaf segments of *in vitro* plantlets (Sudhersan et al. 1993), and shoot tips excised from offshoots (Al-Khayri 2005; Al-Khayri and Al-Bahrany 2004). Seeds of early flowering cultivars also have been used as explants (Sané et al. 2006). Other plant tissues or organs have also been used as explants such as apical zone from apical bud offshoots (Rhiss et al. 1979), zygotic embryos (Tisserat and DeMason 1980) and floral buds (Drira and Benbadis 1985).

29.2.2 Preparation of Explants for Callus Induction

Plant tissues were excised with a sharp knife and then sterilized under aseptic conditions. Several sterilization agents have been employed in date palm. Young leaves and immature inflorescences were sterilized with 0.01% $HgCl_2$ for 1 h and then washed three successive times with sterile distilled water (Fki et al. 2003). Seeds were sterilized with 96% H_2SO_4 for 10 min, rinsed and then soaked in sterile distilled water for 24 h (Sané et al. 2006). After 1 month in the culture room with 12 h/12 h photoperiod and irradiance 80 μE s^{-1}, at 27°C ± 0.2°C, the young white to yellowish leaves were excised from the seedlings and cut into 1 cm long segments. The apices were dissected separately (Sané et al. 2006).

Also, shoot tips were used as explants for callus induction. Shoot tips excised from 3-year-old offshoots were sterilized with 70% ethanol for 1 min followed by immersion in 1.6% w/v sodium hypochlorite for 15 min. This was followed by rinsing with distilled water three times before the tissues were kept in a chilled sterile antioxidant solution of ascorbic acid and citric acid at 150 mg l^{-1} each throughout the manipulation process (Al-Khayri 2005).

29.2.3 Induction of Callus Formation

The explants (young leaves, leaves from seedlings, inflorescences, shoot tips and immature inflorescences) were cut into small segments of about 1 cm long and cultured on callus induction media (medium composition in Table 29.1; Al-Khayri 2005; Chabane 1995; Fki et al. 2003; Sané et al. 2006). After about 2 months, the calli were separated and chopped with a scalpel blade and transferred onto the same fresh callus induction medium. The media used for callus induction vary depending on the type of explants (Al-Khayri 2005; Al-Khayri and Bahrany 2004; Fki et al. 2003; Othmani et al. 2009; Sané et al. 2006). The various media employed for callus induction are summarized in Table 29.1. The callus cultures were maintained in the darkness at 27°C ± 0.2°C for more than 12 months.

29.2.4 Establishment of Cell Suspension Culture

Calli were either directly transferred to the liquid medium or first chopped before transfer. There are two protocols used for chopping calli. In the first procedure, calli were chopped thoroughly with a scalpel blade in a plastic Petri dish containing 5 ml of suspension culture medium as described by Teixeira et al. (1995). The second protocol produces a finer chopped calli. It involves the use of a tissue grinder glass pestle (Sigma) to strain embryogenic callus through a 40 mesh nylon screen (Sigma) with a pore size of 380 μm as described by Othmani et al. (2009). The small pieces obtained were transferred aseptically in suspension culture medium. Chopped calli about 500–1,000 mg were transferred aseptically into 250 ml Erlenmeyer flasks containing 50 ml liquid medium of the same composition as the callus induction medium, but without agar or gelrite (Fki et al. 2003). In some cases, the cell

Table 29.1 Media composition and explants cultured for callus induction of date palm

	Reference:	Fki et al. (2003)		Sané et al. (2006)	Al-Khayri (2005)	Chabane (1995)
Medium composition (mg l^{-1})	Explant:	Inflorescence	Young leaves	Seedling leaves	Shoot tips	Shoot tips
MS salts		4,568	4,568	4,568	4,568	4,568
MS vitamins		1.1	1	–	–	–
Morel vitamins		–	–	–	–	2
Fe-EDTA		65	65	65	–	65
Sucrose		50,000	50,000	30,000	30,000	30,000
Myo-inositol		100	100	100	125	100
Glycine		2	2	0	2	–
Glutamine		100	100	–	200	100
KH_2PO4		120	120	–	–	–
NaH_2PO_4		–	–	–	170	170
Adenine		30	30	–	–	30
Biotine		–	–	0.01	1	0.01
Thiamine-HCl		–	–	–	1	0.1
Nicotinic acid		–	–	–	1	0.1
Pyridoxine-HCl		–	–	–	1	0.1
Ca-pentothenate		–	–	–	1	0.1
Sodium ascorbate		–	–	100	–	–
2,4-D		45 μM	2.2 μM	9 μM	425.5 μM	9 μM
NAA		–	–	–	53.7 μM	–
2iP		–	–	–	14.7 μM	14.7 μM
Agar		8,000	8,000	8,000	7,000	7,000
Activated charcoal		–	–	–	1,500	–

Table 29.2 Media composition for cell suspension culture of date palm

Medium composition (mg l^{-1})	Reference:	Bhaskaran and Smith (1992)	Fki et al. (2003)		Sané et al. (2006)
	Callus source:	Inflorescence shoot tips	Inflorescence	Young leaves	Seedling leaves
MS salts		4,568	2,259	2,259	4,568
MS vitamins		–	1.1	1	–
Fe-EDTA		–	32	32	65
Sucrose		–	30,000	30,000	30,000
Myo-inositol		–	100	100	100
Glycine		–	2	2	–
Glutamine		–	100	100	–
KH_2PO4		70	120	120	–
NaH_2PO_4		–	–	–	–
Adenine		40	30	30	–
Biotine		–	–	–	0.01
Thiamine-HCl		0.4	–	–	–
Sodium ascorbate		–	–	–	100
2,4-D		–	4.5 μM	4.5 μM	9 μM
Activated charcoal		–	300	300	–

suspension cultures were passed through a 500 μm nylon mesh. The cultures were then placed on a rotary shaker 90–120 rpm at 23–28°C under a 16/8 h (light/dark) photoperiod and a 50 μmol m^{-2} s^{-1} photon flux, provided by cool fluorescent lamps. The suspension was subcultured every 15 days on the fresh culture medium (Zouine et al. 2005). About 300 mg fresh weight of cell suspension cultures were transferred into the liquid medium as described in Sané et al. (2006) supplemented with 20 g l^{-1} glucose and 9 μM 2,4-D (2,4-Dichlorophenoxyacetic acid) (Table 29.2). After about 20 days of culture, proembryonic masses (PEMs) were observed. The various media employed for cell suspension cultures in date palm are summarized in Table 29.2.

29.2.5 *Induction of Somatic Embryogenesis*

There are different protocols for induction of somatic embryogenesis from cell suspension cultures. According to Fki et al. (2003) cell suspension cultures were transferred into a new medium weekly (medium composition in Table 29.3, Fki et al. 2003), which enables the proembryonic masses to develop into somatic embryos. The medium was decanted, and the embryos were placed into a new medium (Table 29.3) for embryo maturation and cultured under the same conditions. The proliferation and maturation of somatic embryos occurred simultaneously in the same culture vessels. Similarly, Sané et al. (2006) obtained somatic embryogenesis induction by growing cell suspension cultures in the liquid medium without 2,4-D for 1 month (Table 29.2). The cell suspensions were filtered through a 2,000/1,000 μm filter combination. The sieved cell aggregates (about 50 mg) were transferred into a Petri dish containing basic MS medium

Table 29.3 Media composition for used to produce somatic embryos in date palm

	Author:	Fki et al. (2003)	Sané et al. (2006)
Medium composition (mg l^{-1})	Callus source:	Inflorescence, young leaves	Seedling leaves
MS salts		4,568	4,568
MS vitamins		1.1	–
Fe-EDTA		65	65
Sucrose		50,000	–
Glucose		–	20,000
Myo-inositol		100	100
Glycine		2	0
Glutamine		100	–
KH_2PO4		120	–
Adenine		30	–
BA		–	2 μM
Biotin		–	0.01
Sodium ascorbate		–	100

supplemented with 2 μM BA (6-benzylaminaminopurine) and gelled with 8 g l^{-1} agar. The filter paper with the cell culture was transferred every week to a new medium without plant growth regulator. The somatic embryos were transferred for conversion into plantlets in individual glass tubes on MS medium with or without 5.4 μM NAA (6-naphthaleneacetic acid) for rooting. The various media employed for production of somatic embryos in date palm are summarized in Table 29.3.

29.2.6 *Histological Analysis*

For histological examination, different fixatives and fixation methods have been used for cell suspension cultures in date palm. Rhiss et al. (1979) used a mixture containing 4 ml 25% glutaraldehyde solution, 50 ml phosphate buffer, pH 7.2, 20 ml of 10% paraformaldehyde solution, 1 g caffeine and the total volume was raised to 100 ml by adding 26 ml distilled water. Another fixative, Nawashin fluid has also been used (Sass 1958). After fixation, this was followed by progressive dehydration with ethanol and then embedded in paraffin or epoxy resin. Thin serial sections, 3.5–10 μm thick, were usually made, using a microtome; stained with acetohematoxylin as described by Sass (1958) or Periodic Acid Schiff (PAS) combined with Naphthol Blue Black (NBB), as described by Fisher (1968).

29.2.7 *Determination of Ploidy Level*

In order to determine the ploidy level of the regenerants, the flow cytometry technique was applied. Leaf pieces (1 × 1 cm) from *in vitro* plantlets were chopped with a razor blade in 600 μl buffer solution containing 45 mM $MgCl_2$, and 30 mM trisodium citrate ($Na_3C_6H_5O_7$,$2H_2O$), 20 mM MOPS, 1% triton X-100, 10 mM sodium-metabisulfite ($Na_2S_2O_5$) (pH 7.0) in order to obtain a nuclei suspension. The whole sample was sieved through a 40 μm nylon filter and stained with 6 μg ml^{-1} DAPI (4,6-Diamidino-2-phenylindole) stain. A diploid parental plant was used as an internal standard. Nuclei analysis was done using Partec CA II flow cytometer.

29.2.8 *Growth Responses of Callus and Cell Suspension Culture*

29.2.8.1 Callus Culture

The induction of highly prolific and embryogenic callus in date palm is done to initiate a stable cell suspension culture, which is a very slow process requiring 6–12 months (Bhaskaran and Smith 1992; Chabane et al. 2007). Embryogenic calli are white, display a granular appearance and are friable. A histological study revealed the presence of several globular proembryogenic clusters (Bhaskaran and Smith 1992). In terms of donor material, the ability of inflorescences to form callus is much higher than that of leaves (Fki et al. 2003). Shoot tips have been commonly used for callus induction (Chabane et al. 2007; Zouine et al. 2005). The frequency of explants that produce callus can vary from 20% to 90% depending on genotype.

MS medium (Murashige and Skoog 1962), supplemented with plant growth regulators (PGRs) such as 2,4-D, Picloram, BA and 2iP (6-(γ-γ-dimethylallylamino) purine), is most frequently used for callus induction. In date palm cv. Boufeggous, a high percentage of prolific calli has been accomplished in MS medium containing 45 μM 2,4-D (Othmani et al. 2009); 450 μM 2,4-D in cv. Barhee (Bhaskaran and Smith 1992); 9 μM 2,4-D and 14.8 μM 2iP in cv. Deglet Noor (Chabane 1995); 9 μM 2,4-D in cv. Amsekshi (Sané et al. 2006); a combination of 22.5 μM 2,4-D and 14 μM BA in cvs. Bousthami Noir and Jihel (El Hadrami et al. 1995; Zouine et al. 2005); and a combination of 452 μM 2,4-D and 14.7 μM 2iP in cvs. Barhee, Hillali, Naboot Saif, Ruzaiz and Khusab. Explants on medium added with 4.1, 20.7 and 41.4 μM picloram failed to regenerate embryogenic calli in cv. Boufeggous (Othmani et al. 2009). In terms of vitamin supplements, MS vitamins and Morel and Wetmore vitamins (Morel and Wetmore 1951) are preferred. In some cases, addition of active charcoal in the medium is necessary for callus formation (Othmani et al. 2009). The liquid culture leads to a higher percentage of callus formation than solid culture (Fki et al. 2003).

29.2.8.2 Cell Suspension Culture

The time required to establish a stable cell suspension culture is about 6 months from initiation. A fine chopping of calli and sieving by means of 500-μm filter are

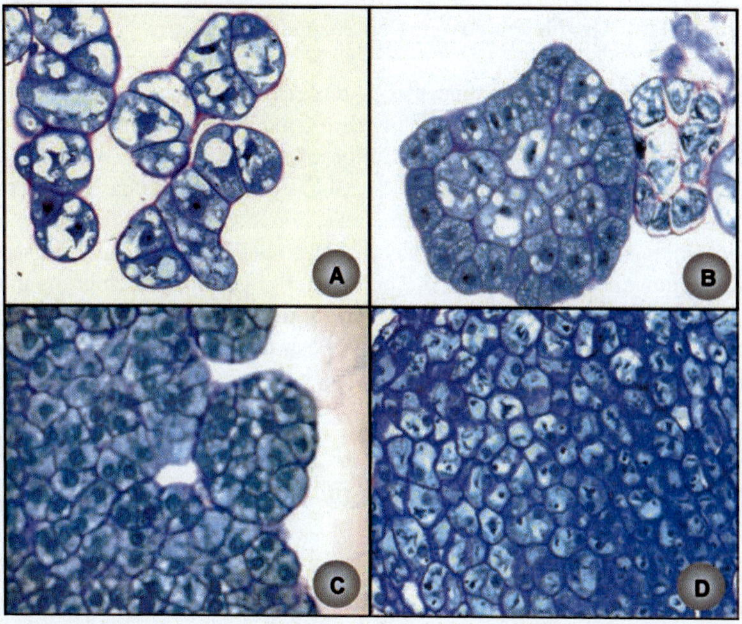

Fig. 29.1 Histocytological analysis of some important structures in date palm suspension cultures. **(a)** Dividing embryogenic cells, **(b)** PEMs, **(c)** Releasing of PEMs from cell, a large aggregate, **(d)** A large cell aggregate

crucial for a successful establishment of embryogenic cell suspension cultures (Fki et al. 2003; Sané et al. 2006). The cell suspensions display a very heterogeneous structure with cells at different developmental and differentiation stages. Depending on the age, the cell suspension consists of (1) single embryogenic cells (15–20 μm diameter) with cytoplasm containing starch and lipoproteic storage grains, with thick polysaccharide walls which stained pink by PAS, (2) nonembryogenic long single cells with less-dense cytoplasm, (3) dividing embryogenic cells, (4) cell aggregates of four to several cells, (5) microcalli whose inner parts are comprised of small meristematic cells with a dense cytoplasm that is rich in soluble protein which stained blue by NBB (Naphthol Blue Black) which could be a credible indicator of the development towards somatic embryogenesis. The outer part of those microcalli displaying cells with less-dense cytoplasm that is rich in starch and lipoprotein storage grains. Fragmentation zones are observed between cells with thickened cell wall and (6) proembryonic clusters which are surrounded by thick polysaccharide outer cells (Bhaskaran and Smith 1992; Sané et al. 2006). The suspension cultures can be maintained for more than 3 years without losing their embryogenic capability (Bhaskaran and Smith 1992). Figure 29.2 displays some important structures in date palm suspension culture, which include dividing embryogenic cells (Fig. 29.1a), PEMs (Fig. 29.1b), release of PEMs from a large aggregate (Fig. 29.1c), a large cell aggregate (Fig. 29.1d) that can evolve into a somatic embryo.

29.2.8.3 Plant Regeneration from Cell Suspension Culture

The chopping of callus into fine pieces increases the number of proembryos, which develop into somatic embryos in about 6 months after establishment of the suspension cultures. Also two-fold dilution of culture medium with MS liquid medium, subsequent addition of 4.5 μM 2,4-D and activated charcoal up to 300 mgl^{-1} positively affects differentiation of the somatic embryos (Fki et al. 2003). The development of somatic embryos in solid medium was slow compared to liquid medium (Zouine et al. 2005). The transfer of germinated embryos in a MS medium supplemented with 4.4 μM BA and 5.4 μM NAA resulted in formation of plantlets. The genomic stability of cell suspension derived plants was confirmed by Fki et al. (2003).

29.2.9 Concluding Remarks on Cell Suspension Culture

Cell suspension cultures in date palm are routinely used in many laboratories for micropropagation purposes. The core factors for establishing stable cell suspensions culture are: (1) starting materials used for callus induction, (2) embryogenic potential of the calli, (3) the fine chopping of the calli, (4) the PGRs employed and (5) the maintenance of the cell suspension cultures.

Different kinds of date palm tissues are suitable for producing embryogenic calli. According to Fki et al. (2003), only calli obtained from floral tissues showed embryogenic potential. Other reports revealed that immature leaves, apical tissues (Sané et al. 2006) and shoot tips (Bhaskaran and Smith 1992) are excellent starting materials for callus cultures. In any case, all those tissues should be tested in preliminary experiments prior to conducting a successful cell suspension procedure.

Only embryogenic calli are able to produce embryogenic cell suspension cultures which can develop into plantlets. Those calli are usually white and friable with a granular appearance (Bhaskaran and Smith 1992). The core element in terms of embryogenic callus is the presence of embryogenic cells in the callus tissue. Those cells display a protein-rich cytoplasm, a small vacuole and a large nucleus. They have intensive mitotic activities, and appear as proembryogenic masses (PEMs) (Chabane et al. 2007). The crucial role of those cells is also reported in other monocot species such as banana (Dhed'a et al. 1991) and coconut (Verdeil et al. 2001).

It has been shown that chopping the calli into fine pieces and sieving them before transferring into liquid medium are critical for cell suspension cultures (Fki et al. 2003; Kreuger et al. 1995; Sané et al. 2006). The effect of embryogenic calli on cell suspension cultures is poorly understood. According to Othmani et al. (2009) some inhibitory substance(s) might be released by surrounding embryogenic cells and a fine chopping might raise this inhibition. Also fine chopping might cause a physical stress which stimulates protein synthesis allowing sustained cell proliferation and differentiation into embryogenic cells. Those embryogenic cells are usually located in the callus tissue interior; a fine chopping may contribute to their isolation, which is essential for the expression of their embryogenic potential (Lowe et al. 1985). Sieving of calli after chopping them might help eliminate the largest cell aggregates which usually result in dead cells and could interact negatively with the living cells. A callus culture without embryogenic cells should not be used for cell suspension experiments.

Basic MS medium is suitable for suspension cultures in date palm; the key PGR for establishment of cell suspension cultures is 2,4-D. A dilution of culture medium with basic MS medium, the subsequent addition of 2,4-D and activated charcoal at a low concentration promoted the differentiation of PEMs into somatic embryos (Fki et al. 2003). In other reports, 2,4-D was not necessary for embryo formation in the suspension culture (Bhaskaran and Smith 1992). According to Zouine et al. (2005), 2,4-D, glutamine and ABA (abscisic acid) together in liquid medium promote substantially somatic embryo formation. The critical challenge is to maintain the embryogenic capability of those suspension cultures over a long period of time. The regular renewal of liquid culture medium and sieving of the cells are also critical to maintain the embryogenic capability of the suspension cultures. Some protocols allow maintaining the suspension culture for 6–36 months without losing their embryogenic potential (Bhaskaran and Smith 1992; Fki et al. 2003).

The conversion of somatic embryo into plantlets is controlled by many parameters such as duration of embryos in liquid medium, reduction of water content through partial desiccation and cutting of cotyledonary leaves. The transfer of germinated embryos into medium containing 4.4 μM BA and 5.4 μM NAA resulted in formation of plantlets (Fki et al. 2003). A similar result was obtained in banana somatic embryos (Assani et al. 2001, 2002).

The commercial relevance of cell suspension culture is that a cell suspension-to-plant-system can be adapted to an automatic bioreactor for large-scale micropropagation in date palm; however field trials are necessary to access the potential effect of somaclonal variation of cell suspension-derived plants. The cell suspension techniques also can be used for transformation experiments and for inducing mutagenesis. Finally, the cell suspension can be employed for production and fusion of protoplasts.

29.3 Protoplast Culture

29.3.1 Source Material for Protoplast Isolation

In dicotyledonous species, leaf tissue has been the most frequently used donor material for protoplast isolation. The growth conditions for the donor plant play a core role for protoplast isolation. In order to ensure the continuous supply of sterile leaf material, *in vitro* shoot cultures are preferentially employed for protoplast isolation in various species (Collonnier et al. 2001; Möllers and Wenzel 1992). For protoplast isolation from callus and cell suspension cultures, the physiological status of the

Table 29.4 Composition of enzyme solutions for date palm protoplast isolation

Components (%, w/v)	Enzyme solutions EC1	EC2	EC3
Cellulase RS (Yakult Pharmaceutical, Japan)	1.5	1.5	1.5
Pectolyase (Kyowa Chem Products, Japan)	0.15	0.2	–
Hemicellulase (Sigma, USA)	0.2	–	–
Macerozyme (Sigma, USA)	–	–	–
Pectinase (Sigma, USA)	–	1	1.5
KCl	3	3	3
$CaCl_2$	0.5	0.5	0.5
pH adjusted with HCl 0.1 N	5.6	5.6	5.6

donor material is also critical. The best response is achieved during the exponential phase of growth of the donor tissues or cells (Ochatt et al. 1989). Nevertheless, in monocots, embryogenic cell suspension cultures are most frequently used as a protoplast source (Assani et al. 2001). In our study on date palm protoplasts, the offshoots grown in open fields (20 cm diameter, 3–6 kg weight) were taken from adult female plants. Young leaves and the apical tip from offshoots were used for protoplast isolation or for callus production (Chabane et al. 2007).

29.3.2 *Isolation of Protoplasts*

The protocols employed for optimizing protoplast production are essentially empirical. In many cases, preincubation treatments are performed before enzyme treatment. Generally, a plasmolysis period (1–2 h) before transfer into enzyme solution, containing the same osmoticum, as to be employed later during enzymatic maceration has improved the performance of protoplast production from leaves, callus and cell suspension (Gamborg et al. 1981; Ochatt et al. 1989). Moreover, calli are chopped into small pieces or sliced thin before enzymatic maceration. For leaves, they can be either digested intact, or have their lower epidermis removed or chopped into small pieces before enzymatic treatment. The most frequently used enzymes are given in Table 29.4. In our investigations on date palm, young leaves of offshoots and callus were used as protoplast source.

29.3.2.1 Protoplasts Isolated from Young Leaves of Offshoots

Protoplasts were isolated from young offshoot leaves around the shoot tip (10–15 cm) from field grown adult female date palm. Leaf pieces of approximately 1 cm in length were excised and surface sterilized with a mercryl lauryle solution (about 5 drops in 100 ml distilled water) for 15 min and followed by three successive washing with a domestic detergent (Domestos, Sunlight, France). This was followed by a repeat surface sterilization with a 30% aqueous solution of 5.4% (v/v) sodium hypochlorite solution for 20 min and rinsed in sterile distilled water three times with a total rinse time of 15 min. Small cuts were made on the lower surfaces of leaf explants (about 1 g fresh weight) and placed in 15 ml enzyme solution with their abaxial surface downward. Three different enzyme solutions EC_1, EC_2 and EC_3 (pH 5.6) (Table 29.4) were tested. The enzyme solutions were sterilized using a 0.2-µm Millipore filter (Millipore, Billerica, MA, USA). The enzyme solution/young leaf mixture was incubated overnight at 27°C for 12–20 h in the dark. Before the purification step, the mixture was observed under the microscope. In case that the number of protoplasts released was not important, the protoplast suspension was transferred to a gyratory shaker (30 rpm) for 15 min.

29.3.2.2 Protoplasts Isolated from Callus

For protoplast isolation, callus cultures were first initiated and maintained. The offshoots (2–4 years old) from field-grown adult female date palm were used for callus initiation. The shoot tip of offshoots were sterilized with 0.3% benlate (methyl [1-(butylcarbamoyl) benzoimidazol-2-yl] aminoformate) (DuPont, France) for 30 min, followed by transferring to a 30% aqueous solution of 5.4% (v/v) sodium hypochlorite in water with two drops of Tween 20 per 500 ml for 45 min. The tip was then rinsed three times with distilled water (duration of each rinse: 10 min). Apical tips (about 5 cm long, 3 cm diameter) from offshoots were excised in small pieces (5 mm length) and cultured on Petri dishes (9.5 cm diameter) on solid medium M_1 (20 ml, 10 pieces per Petri dish) containing MS salts (Murashige and Skoog 1962) and supplemented with 9.0 µM 2,4-D, 14.76 µM 2iP, Morel vitamins (Morel and Wetmore 1951), 87 mM sucrose and 7 g l^{-1} agar (Sigma, USA). The pH was adjusted to 5.7–5.8 before autoclaving (20 min, 120°C, 1 bar). The cultures were kept at 27°C in the dark. The explants were subcultured at 4-week intervals on the same fresh culture medium. After 3–6 months of culture, friable white and yellow nodular calli were formed. Two types of callus derived from shoot tip (3–7 cm) were tested for protoplast isolation: nodular yellow callus and friable white callus. About 800 mg fresh weight of callus was used for each experiment. The callus was cut into small pieces, put into 15 ml enzyme solution in Petri dishes (9.5 cm diameter), and placed in the dark at 27°C for 12–20 h. The same enzyme solutions (Table 29.4) were used for leaf explants.

29.3.3 Purification of Protoplasts

A mixture of protoplasts, cells and cellular debris is generally obtained after enzymatic maceration and needs to be purified in order to obtain debris-free protoplasts. The most frequently employed purification approach comprises filtration and centrifugation (Gamborg et al. 1981). For purification of date palm protoplasts, the following protocol, described in Assani et al. (2006), was employed. After enzymatic maceration, the digested mixture was filtered through 100/25-μm metallic mesh combination to remove debris. Protoplasts were washed three times through centrifugation (65 g for 5 min) with a washing solution that consists 204 *mM* KCl, 67 *mM* $CaCl_2$ with pH 5.7. In order to minimize the salt content in the protoplast suspension, they were rinsed again with 0.5 *M* mannitol and 67 *mM* $CaCl_2$ (centrifugation: 65 g for 5 min).

Usually, for assessment of the cell wall in protoplast suspension, circumstantial evidence is used such as the spherical form of protoplast when observed under a microscope (Evans and Bravo 1983; Fowke and Constabel 1985), or the use of Calcofluor white (Galbraith 1981; Nagata and Takebe 1970). Protoplast viability is determined by fluorescein diacetate (FDA) (Widholm 1972). Protoplast yield is assessed by using a haemocytometer. For manipulation in the laboratory, the density of the protoplast population is expressed in terms of number of protoplasts per ml of suspension. The protoplast yield, in terms of protocol efficiency, is expressed as protoplast number per g fresh weight of source material.

29.3.4 *Transfer of Protoplasts in Culture Medium*

29.3.4.1 Composition of Culture Medium

The media used for protoplast culture is mostly a modification of standard plant cell culture medium. Plant cell culture media generally used as a basis for protoplast media are MS medium (Murashige and Skoog 1962), B_5 medium (Gamborg et al. 1968) and KM medium (Kao and Michayluk 1975). Those media are commercially available. Regarding the changes in the medium composition, reduction or lack of ammonium concentration, addition of calcium, microelements and organic components in the culture medium are beneficial for protoplasts of some species. Ammonium ions are detrimental to protoplast survival; on the other hand calcium ions may play an important role in the membrane stability (Gamborg et al. 1981; Erikson 1977; Evans and Bravo 1983). A carbon supply is needed for cell proliferation because protoplasts are non-photosynthetic. Glucose and sucrose, alone or in combination, are most frequently used as a carbon source, depending on the protoplast systems. Non-metabolizable osmoticum like sorbitol, manitol, inositol and sucrose, alone or as a mixture, have been employed for several plant protoplast systems (Evans and Bravo 1983; Kao and Michayluk 1975). Also solutions of mineral salts, such as KCl and $CaCl_2$ have been used as osmoticum (Assani et al. 2006; Chabane et al. 2007).

Usually, organic compounds such as vitamins (thiamine HCl, pyridoxine HCl, nicotinic acid etc.), organic acids, sugars, sugar alcohols and undefined nutrients such as casamino acids and coconut water are as employed in many species, like cereals (Thomson et al. 1987), legumes (Gilmour et al. 1987), ornamental (Power et al. 1989) and fruit trees (Assani et al. 2001; Patat-Ochatt et al. 1988;). In terms of plant growth regulators (PGRs), both auxins and cytokinins are usually required for protoplast division. Both auxins 2,4-D and NAA are the most frequently used. For some protoplast systems, other auxins such as IBA (Indole-3-butyric acid) are favoured. Both cytokinins BA and zeatin are the most frequently employed in several species. In most protoplast systems, a combination of auxin/cytokinin with preference for auxin is generally used to ensure a sustained protoplast proliferation.

In terms of protoplast density, protoplasts of most species are cultured in a density range of $5 \times 10^2 - 1 \times 10^6$ protoplasts per ml (Assani et al. 2005; Evans and Bravo 1983; Gamborg et al. 1981). Of the numerous culture methods used, liquid culture (Möllers and Wenzel 1992), agarose culture (Shillito et al. 1983), alginate culture (Pan et al. 2004) and feeder layer culture (Assani et al. 2001; Funatsuki et al. 1992) are the most frequently employed. The later system is used for species described as recalcitrant such as banana and date palm. The protocol that has been used for date palm protoplast culture is described in details below:

The protoplasts were cultured at a density of 10^6 per ml. Three culture systems were tested: liquid culture, feeder layer culture and agarose culture.

Liquid culture: Protoplasts were suspended in 4 ml media in small Petri dishes (5.5 cm diameter). The liquid medium (named M_5) containing MS salts, vitamins of Morel, 0.68 *mM* glutamine, 117 *mM* sucrose, 0.4 *mM* glucose, 0.5 *mM* MES (2-(N-morpholino) ethanesulfonic acid), 1.9 mM KH_2PO4, 9.0 µM 2,4-D, 14.76 µM 2iP and 250 mg l^{-1} polyethylene glycol (PEG) 4000. The pH was adjusted to 5.7 with 0.1 N NaOH before filter sterilization (0.22-µm millex GS filters, Millipore, Billerica, MA, USA). The Petri dishes were sealed with parafilm and transferred into the culture room.

Feeder layer culture: Embryogenic offshoot-tip-derived calli of Deglet Noor were used as nurse cells. The nurse culture was prepared the same day when the protoplasts were isolated. The PCM liquid medium (double strength) contained double strength of MS salts, 9.0 *µM* 2,4-D, vitamins of Morel, 2.8 *mM* glucose, 278 *mM* maltose, 170 *mM* sucrose and 2.5 *mM* Myo-inositol (pH 5.7). Callus suspension was made by cutting friable calli into small pieces (0.2 mm) and added to 100 ml PCM culture medium, in order to obtain a final cell concentration of 2% in PCM/agarose mixture. Agarose sea plaque 1.2 g (Sigma, USA) was dissolved separately in 100 ml distilled water (pH was adjusted to 5.7) and then autoclaved. When the temperature of the agarose solution decreased to 30–35°C, it was carefully mixed with a 100-ml PCM medium containing nurse cells. Aliquots of 10–12 ml of this mixture were poured into small Petri dishes (5.5-cm diameter). After solidification, the medium was covered with sterilized nitrocellulose filter (AA type millipores), and 1 ml protoplast suspension in M_5 medium (see above). All cultures were maintained at 27°C in the dark. Cell-wall regeneration was observed with calcofluor white brightener stain (Nagata and Takebe 1970). The microcalli formed were transferred onto a callus induction medium containing MS salts and supplemented with 13.5 *µM* 2,4-D and 14.76 *µM* 2iP, Morel vitamins, and 3 g l^{-1} gelrite. The calli were transferred into the regeneration medium, which consisted of MS salts supplemented with the same level of 2iP of the callus induction medium (see above) and 1.4 μM 2,4-D. The cultures were kept in the dark at 27°C.

Agarose culture: In a study by Rizkalla et al. (2007) to optimize the production of protoplasts from two date palm cvs., Barhee and Zaghloul, cultures containing protoplasts embedded in agarose were also tested. After the purification step, protoplasts were plated at 5×10^5 protoplasts per ml and cultured on 2.5 ml agarose of various concentrations: 0.2%, 0.3%, 0.4% or 0.5%, suspended in 2.5 ml double strength of medium. Protoplast cultures were kept in the dark at 26°C until cell division was evident and the osmotic pressure of the medium was gradually reduced at 14-day intervals. Protoplasts were left undisturbed or 1 ml of culture media was replaced after 7 days to initiate microcalli formation. The microcalli were grown in the dark until they reached a diameter of 0.3–0.4 mm.

29.3.5 Production and Growth Responses of Protoplasts

29.3.5.1 Production of Protoplasts

Highly prolific embryogenic callus (Fig. 29.2a, b) is the material of choice for protoplast isolation in date palm (Chabane et al. 2007). In our study, viable protoplasts were isolated from both cvs. Deglet Noor and Takerboucht (Fig. 29.2b, d). Generally, young leaves gave less viable protoplasts than those from calli. In young leaves, the viability rate was 8% in Deglet Noor and 9% in Takerboucht; in calli it was 65% in Deglet Noor and 57% in Takerboucht. The viability of freshly isolated protoplasts (immediately after isolation) was about 80% in both leaf- and callus derived protoplasts.

29.3.5.2 Factors Affecting Protoplast Isolation

The viability rate of freshly isolated protoplasts was dependent on the composition of the enzyme solution tested. EC_1, the combination of cellulase and pectinase with hemicellulase (Table 29.5) was more efficient for the isolation of protoplasts in date palm. The protoplasts yield was also dependent on donor material, genotype and incubation time. Regarding the donor material, the best response was obtained with nodular callus; the protoplast yields were 5.6×10^5 per g f.wt in Deglet Noor and 4.95×10^5 g f.wt. in Takerboucht (Table 29.5). The protoplast yields were less in young leaves of offshoots (1.97×10^5 per g f.wt. in Deglet Noor and 1×10^5 per g f. wt. in Takerboucht). Regarding the incubation time, the best response was obtained with an incubation period of 12 h. There was a significant difference between the cultivars as shown in Fig. 29.3; protoplast yield were 5.2×10^5 per g f.wt. callus in Deglet Noor and 4.2×10^5 per g f.wt. callus in Takerboucht.

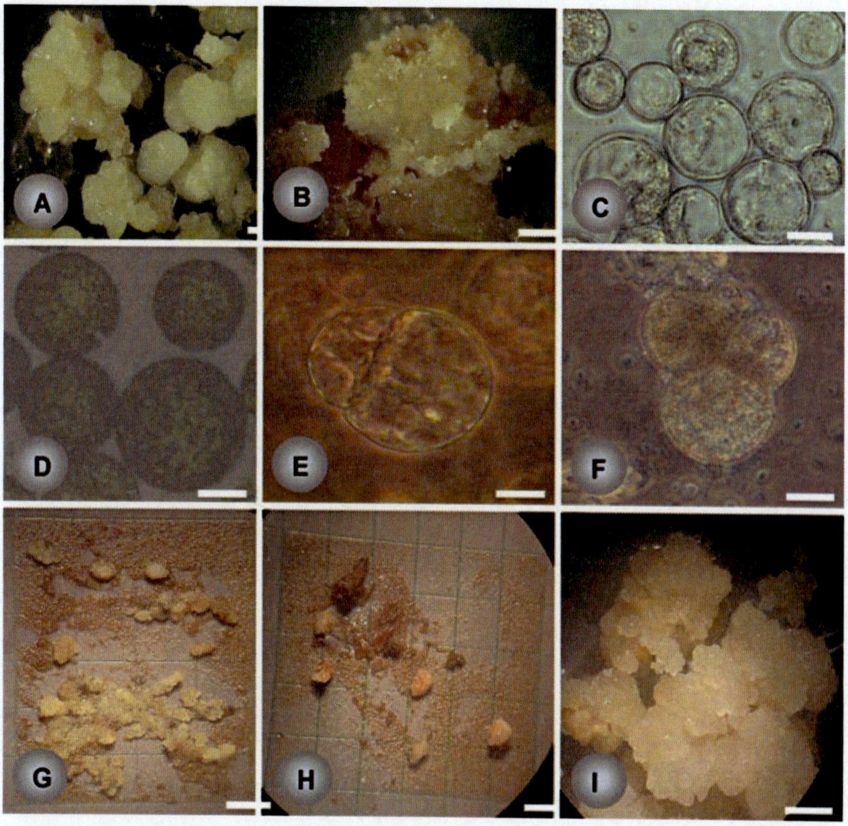

Fig. 29.2 (**a**) Development of nodular callus from meristematic shoots of cv. Deglet Noor after 4 months of culture on induction medium, bar = 6 mm. (**b**) Friable and granular callus development of cv. Deglet Noor after 6 months of culture on induction medium, bar = 0.2 mm. (**c**) Protoplasts isolated from embryogenic callus of cv. Deglet Noor, bar = 30 µm. (**d**) Protoplasts isolated from leaves of cv. Deglet Noor, bar = 30 µm. (**e**) Dividing cell 7 days after protoplast isolation from callus in Deglet Noor, bar = 5 µm. (**f**) Second cell division 10 days after protoplast isolation from callus in Deglet Noor. Bar = 5 µm. (**g**) Microcalli of cv. Deglet Noor on feeder layer 6 weeks after protoplast isolation, bar = 1 cm. (**h**) Microcalli of Takerboucht on feeder layer 8 weeks after protoplast isolation, bar = 1 cm. (**i**) Callus formation from callus-derived protoplasts in Deglet Noor after 4 months of protoplast culture, bar = 6 mm

29.3.5.3 Protoplast Response in Culture

The protoplast viability rate was 80% at day 0 (immediately after isolation) before transferring in culture medium; however, the cell viability diminished with increasing culture duration, depending on the explants, culture system and genotype. With regard to cell viability, callus-derived protoplasts showed the best responsiveness as compared to leaf-derived protoplasts. In Deglet Noor, the cell viability rate of callus-derived protoplasts at day 4 was 65% on the feeder layer and 60% in the liquid

Table 29.5 Influence of genotype, donor material and enzyme solution on protoplast yield ($\times 10^5$)

		Enzyme solutions		
Cultivar	Donor material	EC1	EC2	EC3
		Protoplast yield ($\times 10^5$)		
Deglet Noor	Nodular callus	5.6±0.1 a	2.95±0.1a	0.95±0.09 a
	Friable callus	3.9±0.1 c	2.25±0.2 b	0.68±0.08 cb
	Leaf	1.97±0.05 e	0.72±0.03 c	0.21±0.03 d
Takerboucht	Nodular callus	4.95±0.2 b	2.05±0.2 b	0.78±0.03 b
	Friable callus	3.25±0.3 d	1.85±0.06 b	0.62±0.04 c
	Leaf	1.00±0.06 f	0.53±0.03 c	0.05±0.03 e

Data are means of three independent treatments. Differences between means were assessed using the Student-Newman-Kenls test; means with the same letter are not significantly different at $\alpha = 0.05$

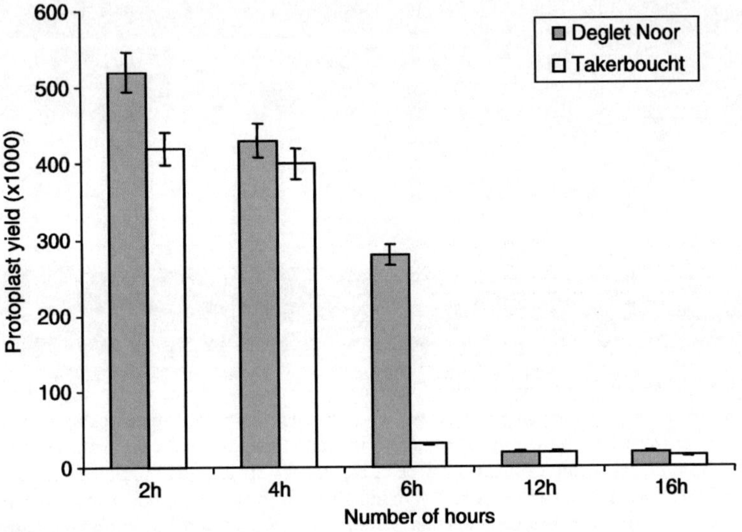

Fig. 29.3 Impact of incubation time on protoplast yield from nodular callus

medium (Table 29.6). In Takerboucht, the frequency of cell viability of callus-derived protoplasts was 57% on the feeder layer and 49% in liquid medium (Table 29.6). Three days after the protoplasts were cultured, the cells began to regenerate the cell wall and became oval. In terms of cell wall regeneration, the best responsiveness was obtained with callus-derived protoplasts, and the best culture system was on the feeder layer. At day 4, the rate of cell-wall regeneration of callus-derived protoplasts in Deglet Noor was 54% on the feeder layer and 25% in liquid medium; in Takerboucht, the frequency of cell-wall regeneration of callus-derived protoplasts was 38% on the feeder layer and 14% in liquid medium (Table 29.6).

Cell division was obtained in both callus and leaf-derived protoplasts; however, the callus-derived protoplasts showed a better response. The highest division rate of callus-derived protoplasts was achieved on the feeder layer; in contrast, leaf-derived

Table 29.6 Influence of genotype, donor material and culture system on frequency of cell viability and cell wall regeneration 4 days after protoplast culture

		% Cell viability		% Cell wall regeneration	
Cultivar	Donor material	Liquid medium	Feeder layer	Liquid medium	Feeder layer
Deglet Noor	Callus	60±2.0 a	65±2.0 a	25±3.0 a	54±3.6 a
	leaf	5±2.0 c	8±2.0 c	3±1.0 c	5±1.0 c
Takerboucht	Callus	49±2.6 b	57±2.6 b	14±1.7 b	38±3.6 b
	leaf	4±1.0 c	9±1.0 c	1±0.3 c	2±1.0 c

Data are means of three independent treatments. Differences between means were assessed using the Student-Newman-Kenls test; means with the same letter are not significantly different at $\alpha = 0.05$

Table 29.7 Influence of genotype, donor material and culture system on cell division rate (%) 10 days after protoplast isolation

Cultivar	Donor material	Liquid medium	Feeder layer
Deglet Noor	Callus	20±2.6 a	30±3.6 c
	leaf	5±2.0 be	6±2.5 e
Takerboucht	Callus	7±1.7 b	15±2.0 d
	leaf	4±1.0 be	5±1.5 e

Data are means of three independent treatments. Differences between means were assessed using the Student-Newman-Kenls test; means with the same letter are not significantly different at $\alpha = 0.05$

protoplasts did not show any significant difference in either culture systems. The cell division rate of callus-derived protoplasts (Fig. 29.2e, f), related to the number of isolated protoplasts at day 0 (immediately after isolation), was 30% on the feeder layer and 20% in liquid medium in Deglet Noor at day 10; in Takerboucht, the division frequency was 15% of feeder layer and 7% in liquid medium (Table 29.7).

29.3.5.4 Regeneration of Protoplasts

The dividing protoplast-derived cells continued to grow and developed into microcalli particularly on the feeder layer 8 weeks after protoplast plating, as shown in Fig. 29.2g, h. The number of microcalli was 14,000 per Petri dish in Deglet Noor and 9,000 per Petri dish in Takerboucht. In both genotypes, about 25% of the microcalli developed to callus (Fig 29.2i) on the callus induction medium. The calli that were transferred onto the regeneration medium failed to regenerate into shoots or roots. The flow scheme of protoplast culture in date palm is given in Fig 29.4.

Critically important steps toward plant regeneration have been achieved during these studies. Plant regeneration from protoplasts is a very complex process involving, composition of the enzyme solution, isolation system, culture system, culture medium in particular the PGRs, genotype, source tissue, environmental condition etc which can be the cause of the lack of plant regeneration in these studies. The combination of cellulase RS-pectolyase-hemicellulase gave a higher number of protoplasts in our investigation. Using the same approach in banana protoplast isolation (Assani et al. 2002), the mixture of cellulase-pectolyase-hemicellulase was more efficient for protoplast isolation from cell suspension after 15–17 h in darkness. The protoplast

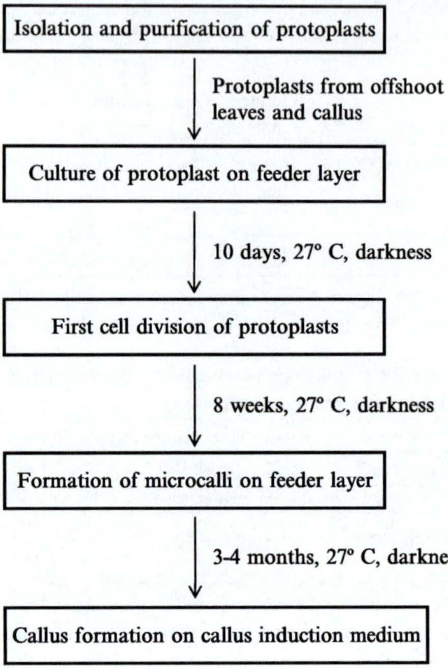

Fig. 29.4 Flow scheme for protoplast culture of date palm

isolation involved the treatment of tobacco leaves with pectinase to separate the cells, followed by cellulase to remove their wall (Takebe et al. 1968). The procedure was simplified by a single treatment with a mixture of enzymes (Power and Cocking 1970). In dicotyledonous plants, it was shown that the combination of cellulase onozuka RS and pectolyase Y-23 significantly improved the yield and the viability of leaf mesophyll protoplasts in the sour cherry, *Prunus cerasus* L., (Mehri-Kamoun 2001). The optimum enzyme mixture for protoplast isolation in callus of the same plant material was cellulase RS and macerozyme R10. It was reported that in the purple coneflower, *Echinacea angustifolia*, the yield of protoplasts released also increased when the cellulase concentration was increased up to 2% (w/v), and the lower cellulase concentration (less than 1%) could not lead to liberation of enough protoplast (Zhu et al. 2005). Cellulase at a concentration higher than 2% may cause overdigestion of plant material (Koster et al. 2003; Monteiro et al. 2003). Several factors influence protoplast release, including the extent of cell-wall thickening, temperature, duration of enzyme incubation, pH of enzyme solution (Sinha et al. 2003), agitation, the nature of osmoticum and plasmolysis prior to enzyme digestion of source tissues, manual or enzymatic removal of the epidermis and conditioning of the donor material or its culture on media containing osmoticum (Davey et al. 2000).

During the present studies, cell-wall regeneration, cell division, and callus formation were obtained. Among the plant growth regulators (PGRs) that have been tested (data not shown), only the combination of 2,4D-2iP induced cell divisions. In earlier studies of rose mesophyll protoplasts, NAA and BA were the most efficient growth regulators for the regeneration of microcalli (Marchant et al. 1997). In lily protoplasts, the addition of picloram to the culture medium was critical of development of microcalli (Horita et al. 2002). Our investigation demonstrates that nurse culture was effective for mitotic activity of date palm protoplasts.

Recently nurse culture technique has been successfully used to address the regeneration issues in recalcitrant protoplasts (Assani et al. 2006; Horita et al. 2002; Sun et al. 2004). The number of microcalli we obtained was close to those obtained in earlier study in banana (Assani et al. 2001).

Previous investigations showed that the impact of genotype on plant regeneration from protoplasts in apple and banana is significant (Assani et al. 2002; Patat-Ochatt and Lespinasse 1993). Currently, protoplasts can be isolated from almost all plant tissues, calli and cell suspensions; however, the type of the donor material is crucial for shoot organogenesis and somatic embryogenesis. Callus tissues contain less embryogenic or dedifferentiated cell aggregates compared to cell suspensions, the most used donor material for protoplast isolation in monocotyledonous plants. The main PGRs, auxin and cytokinin, either sole or in combination, are generally essential for efficient protoplast-to-plant system (Davey et al. 2005). PGR concentrations and combinations need to be optimized for each protoplast developmental stage. The following PGRs were tested in our preliminary experiments: 2,4-D, 2iP, IAA (Indolacetic-3-acid), BA and NAA. Only the combination 2,4-D and 2iP induced sustained cell proliferation and callus formation. None of the PGRs induced plant regeneration, which may be related to the negative interaction between those PGRs and some metabolites produced by callus tissues. The nature of positive or negative factors produced by feeder cells and protoplasts is very time-dependent in banana protoplasts (Assani et al. 2001). During the present studies, liquid, feeder layer and agarose system were used. Alginate technique may be a useful alternative to liquid, agarose or feeder layer method (Pan et al. 2004). The conversion frequency of microcalli into somatic embryos was very low when using a feeder layer compared to when microcalli were transferred onto regeneration medium immediately after their formation (Assani et al. 2001). Each individual factor should be optimized for efficient protocol development.

29.4 Conclusion

Protoplast-to-plant-system remains the limiting factor towards the routine application of somatic hybridization in date palm biotechnology. For optimizing shoot regeneration in date palm protoplasts, further studies should focus on the following: (1) development of a stable embryogenic cell suspension culture as a protoplast source because of their higher frequency of embryogenic cells, (2) optimizing exposition time of microcalli on nurse culture and (3) application of other culture procedures such as alginate culture.

Overall, significant progress has been made in protoplast systems since the totipotence of plant protoplasts was demonstrated about 40 years ago. Many potato, tobacco, brassica, citrus and eggplant seed companies are using protoplast fusion based technology to produce new commercial varieties. However, in the last decade, little progress has been made regarding protoplast regeneration from so-called recalcitrant species, which includes date palm. Basically, the development of regeneration protocol remains empirical. There are some similarities between some protoplast systems and mammalian stem cells regarding their cultures and responses to some growth regulators. More collaboration between animal and plant scientists is highly encouraged, which would be useful for both parties to help address these challenges.

References

Al-Khayri JM (2005) Date palm *Phoenix dactylifera* L. In: Jain SM, Gupta PK (eds) Protocol for somatic embryogenesis in woody plants. Springer, Dordrecht, pp 309–319

Al-Khayri JM, Al-Bahrany AM (2004) Genotype-dependent *in vitro* response of date palm (*Phoenix dactylifera* L.) cultivars to silver nitrate. Sci Hortic 99:153–162

Assani A, Haïcour R, Wenzel G et al (2001) Plant regeneration from protoplasts of dessert banana Grand Naine (*Musa* spp., Cavendish sub-group AAA) via somatic embryogenesis. Plant Cell Rep 20:482–488

Assani A, Haïcour R, Wenzel G et al (2002) Influence of donor material and genotype on protoplast regeneration in banana and plantain cultivars (*Musa* spp.). Plant Sci 162:355–362

Assani A, Chabane D, Haïcour R et al (2005) Protoplast fusion in banana (*Musa* spp.): Comparison of chemical (PEG: Polyethylene Glycol) and electrical procedure. Plant Cell Tissue Org Cult 83:145–151

Assani A, Chabane D, Foroughi-Wehr B, Wenzel G (2006) An improved protocol for microcallus production and whole plant regeneration from recalcitrant banana protoplasts *(Musa* spp.). Plant Cell Tissue Org Cult 85:257–264

Bhaskaran S, Smith R (1992) Somatic embryogenesis from shoot tip and immature inflorescence of *Phoenix dactylifera* L. Cv Barhee. Plant Cell Rep 12:22–25

Bouguedoura N (1991) Connaissance de la morphogenèse du palmier dattier *Phoenix dactylifera* L. Etude in situ et *in vitro* du développement morphogénétique des appareils végétatif et reproducteur. Thèse Doct, Etat. U.S.T.H.B, Alger

Bouguedoura N, Michaux-Ferrière N, Bompar J-L (1990) Comportement *in vitro* de bourgeons axillaires de type indéterminé du palmier dattier (*Phoenix dactylifera* L.). Can J Bot 68:2004–2009

Chabane D (1995) Etude des aptitudes morphogénétiques de divers explants des rejets du palmier dattier *Phoenix dactylifera* L. chez deux variétés Takerboucht et Deglet nour pour induire une embryogenèse somatique. Thèse de Magister U.S.T.H.B, Alger

Chabane D, Assani A, Bouguedoura N et al (2007) Induction of callus formation from difficile date palm protoplasts by means of nurse culture. C R Biol 330:392–401

Collonnier C, Mulya K, Fock I et al (2001) Source of resistance against *Ralstonia solanacearum* in fertile somatic hybrids of eggplant (*Solanum melongena* L.) with *Solanum aethiopicum* L. Plant Sci 160:301–313

Davey MR, Anthony P, Power JB, Lowe KC (2000) Plant protoplasts. In: Spier RE (ed) Encyclopaedia of cell technology. Wiley, New York, pp 1034–1042

Davey MR, Anthony P, Power JB, Lowe KC (2005) Plant protoplast technology: current status. Acta Physiol Plant 27:117–129

Dhed'a D, Dumortier F, Panis B et al (1991) Plant regeneration in cell suspension cultures of cooking banana cv. "Bluggoe" (*Musa* spp. ABB group). Fruits 46:125–135

Drira N, Benbadis A (1985) Multiplication végétative du palmier dattier (*Phoenix dactylifera* L.) par réversion, en culture *in vitro*, d'ébauches florales de pieds femelles adultes. J Plant Physiol 119:227–235

El Hadrami I, Cheikh R, Baaziz M (1995) Somatic embryogenesis and plant regeneration from shoot-tip explants in *Phoenix dactylifera* L. Biol Plant 37:205–211

Erikson T (1977) Technical advances in protoplasts isolation and cultivation. In: Barz W, Reinhard E, Zenk MH (eds) Plant tissue culture and its biotechnological application. Springer, Berlin, pp 313–322

Evans DA, Bravo JE (1983) Protoplast isolation and culture. In: Evans DA, Sharp WR, Ammirato PV, Yamada Y (eds) Handbook of plant cell culture, vol 1. Macmillan, New York, pp 124–177

Fisher DB (1968) Protein staining of ribboned epon sections for light microscopy. Histochemie 16:92–96

Fki L, Masmoudi R, Drira N, Rival A (2003) An optimised protocol for plant regeneration from embryogenic suspension cultures of date palm, Phoenix dactylifera L. cv. Deglet Nour. Plant Cell Rep 21:517–524

Fock I, Collonnier C, Purwito A et al (2000) Resistance to bacterial wilt in somatic hybrids between *Solanum tuberosum* and *Solanum phureja*. Plant Sci 160:165–176

Fowke LC, Constabel F (eds) (1985) Plant protoplasts. CRC Press, Boca Raton

Funatsuki H, Lörz H, Lazzeri PA (1992) Use of feeder cells to improve barley protoplast culture and regeneration. Plant Sci 85:179–187

Galbraith DW (1981) Microfluorimetric quantitation of cellulose biosynthesis by plants protoplasts using Calcofluor white. Physiol Plant 53:111–116

Gamborg OL, Miller RA, Ojima K (1968) Nutrient requirement of suspension cultures of soybean root cells. Exp Cell Res 50:151–158

Gamborg OL, Shyluk JP, Shahin EA (1981) Isolation, fusion and culture of plant protoplasts. In: Thorpe TA (ed) Plant tissue culture methods and application in agriculture. Academic, New York, pp 115–153

Gilmour DM, Davey MR, Cocking EC, Pental D (1987) Culture of low numbers of storage legume protoplasts in membrane chambers. J Plant Physiol 126:457–466

Grosser JW, Ollitrault P, Olivares-Fuster O (2001) Somatic hybridization in citrus: an effective tool to facilitate variety improvement. In Vitro Cell Dev Biol Plant 36:434–449

Hall RD, Riksen-Bruinsma T, Weyens GJ et al (1996) A high efficiency technique of transgenic sugar beets from stomatal guard cells. Nature Biotechnol 14:1133–1138

Heselmans M (1997) Setting research priorities through an international date palm network. Biotechnol Dev Monit 30:18–20

Horita M, Morohashi H, Komai F (2002) Regeneration from flowering plants from difficile lily protoplasts by means of nurse culture. Planta 215:880–884

Kao KN, Michayluk M (1975) Nutritional requirements for growth of *Vicia hajastana* cells and protoplasts at very low population densities in liquid media. Planta 126:1133–1137

Kisaka H, Kisaka M, Kanno A, Kameya T (1998) Intergeneric somatic hybridization of rice (*Oryza sativa* L.) and barley (*Hordeum vulgare* L.) by protoplast fusion. Plant Cell Rep 17:362–367

Klercker JAF (1892) Eine method zur isolierung lebender protoplasten. Ofvers Vetensk Akad Forh Stokh 9:463–475

Koster KL, Reisdorph N, Ramsay JL (2003) Changing desiccation tolerance of pea embryo protoplasts during germination. J Exp Bot 54:1607–1614

Kreuger M, Postma E, Brouwer Y, Holst GJ (1995) Somatic embryogenesis of *Cyclamen persicum* in liquid medium. Plant Physiol 119:227–235

Kuster E (1909) Über die Verschmelzung nakter Protoplasten. Ber Dtsch Bot Ges 27:589–598

Lowe K, Taylor DB, Ryan P, Paterson KE (1985) Plant regeneration via organogenesis and embryogenesis in the maize inbred line B73. Plant Sci 41:125–132

Marchant R, Davey MR, Power JB (1997) Isolation and culture of mesophyll protoplasts from rosa hybrid. Plant Cell Tissue Org Cult 50:131–134

Mehri-Kamoun R (2001) Effet de la pectolyase Y-23et de la cellulase RS sur le rendement en protoplasts viables de *Prunus cerasus* L. "Montmorency". Biotechnol Agron Soc Environ 5:99–104

Michel W (1937) Über die experimentelle Fusion pflanzlicher protoplasten. Arch Exp Zellforsch Bas Gewebe 20:230–252

Möllers C, Wenzel G (1992) Somatic hybridization of dihaploid potato protoplasts as a tool for potato breeding. Bot Acta 105:133–139

Möllers C, Frei U, Wenzel G (1994) Field evaluation of tetraploid somatic potato hybrids. Theor Appl Genet 88:147–152

Monteiro M, Appezzato-da-Glória B, Valarini MJ et al (2003) Plant regeneration from protoplasts of alfalfa *Medicago sativa* via somatic embryogenesis. Sci Agri L 60:683–689

Morel G, Wetmore RM (1951) Fern callus tissue culture. Am J Bot 38:141–143

Murashige T, Skoog F (1962) A revised medium for rapid growth and bioassays with tobacco tissue culture. Physiol Plant 15:473–479

Nagata T, Takebe I (1970) Cell wall regeneration and cell division in isolated tobacco mesophyll protoplasts. Planta 92:301–308

Nagata T, Takebe I (1971) Plating of isolated tobacco mesophyll protoplasts on agar medium. Planta 99:12–20

Ochatt SJ, Davey MR, Power JB (1989) Somatic hybridization of sexually incompatible top-fruit tree rootstocks, wild pear (*Pyrus communis* var *pyraster* L.) and Colt cherry (*Prunus aviumx pseudoceracus*). Theor Appl Genet 78:35–41

Othmani A, Bayoudh C, Drira N et al (2009) Somatic embryogensis and plant regeneration in date palm *Phoenix dactylifera* L., cv Boufeggous is significantly improved by fine chopping and partial desiccation of embryogenic callus. Plant Cell Tissue Org Cult 97:71–79

Pan ZG, Liu CZ, Zobayed SMA, Saxena PK (2004) Plant regeneration from mesophyll protoplasts of *Echinacea*. Plant Cell Tissue Org Cult 77:251–255

Patat-Ochatt EM, Lespinasse Y (1993) Régénération des plantes a partir de protoplastes de pommier (*Malus* x *domestica* Borkh.), études des exigences de culture pour différents génotypes apparentes de ploïdie différente. Thèse, Université de Rennes-1, France

Patat-Ochatt EM, Ochatt SJ, Power JB (1988) Plant regeneration from protoplasts of apple rootstocks and scion varieties (*Malus* x *domestica* Borth.). J Plant Physiol 133:460–465

Power JB, Cocking EC (1970) Isolation of leaf protoplasts: macromolecular uptake and growth substance response. J Exp Bot 21:64–70

Power JB, Davey MR, McLellan MS, Wilson D (1989) Laboratory manual: plant tissue culture. University of Nottingham, Nottingham

Rhiss A, Poulain C, Beauchesne G (1979) *In vitro* culture for the vegetative propagation of date palm *Phoenix dactylifera* L. Fruits 34:551–554

Rizkalla AA, Badr-Elden AM, Nower AA (2007) Protoplast isolation, salt stress and callus formation of two date palm genotypes. J Appl Sci Res 3:1186–1194

Saaidi M (1979) Contribution à la lutte contre le bayoud, fusariose vasculaire du palmier dattier. Thèse de Doctorat, Université de Dijon

Sacristan MD, Gerdemann-Knörck M, Schieder O (1989) Incorporation of hygromycin resistance in *Brassica nigra* and its transfer to *B. napus* through asymmetric protoplast fusion. Theor Appl Genet 78:194–200

Sané D, Aberlenc-Bertossi F, Gassama-Dia YK (2006) Histocytological analysis of callogenesis and somatic embryogenesis from cell suspensions of date palm (*Phoenix dactylifera*). Ann Bot 98:301–308

Sass JE (1958) Botanical microtechnique. Iowa State University Press, Ames

Shillito RD, Paszkowski J, Potrykus I (1983) Agarose plating and a bead type culture technique enable and stimulate development of protoplasts-derived colonies in a number of plant species. Plant Cell Rep 2:244–247

Sinha A, Wetten AC, Caligari PDS (2003) Effect of biotic factors on the isolation of *Lupinus albus* protoplasts. Aust J Bot 51:103–109

Sudhersan C, El-Nil A, Al-Baiz A (1993) Occurrence of direct somatic embryogenesis on leaf of in vitro plantlets of *Phoenix dactylifera* L., cultivar Barhee. Curr Sci 65:887–888

Sun Y, Zhang X, Nie Y et al (2004) Production and characterization of somatic hybrids between upland cotton (*Gossipium hirsutum*) and wild cotton (*G. klotzschianum*) via electrofusion. Theor Appl Genet 109:472–479

Takebe I, Otsuki Y, Aoki S (1968) Isolation of tobacco mesophyll protoplasts in intact and active state. Plant Cell Physiol 9:115–124

Takebe I, Labib G, Melchers G (1971) Regeneration of whole plants from isolated mesophyll protoplast of tobacco. Naturwissenschaften 58:318–320

Teixeira JB, Sondahl MR, Nakamura T, Kirby EG (1995) Establishment of oil palm cell suspensions and plant regeneration. Plant Cell Tissue Org Cult 40:105–111

Thomson JA, Abdullah R, Chen WH, Gartland KMA (1987) Enhanced protoplast division in rice (Oryza sativa L.) following heat shock treatment. J. Plant Physiol 127:367–370

Tisserat B, DeMason D (1980) A histological study of development of adventive embryos in organ cultures of *Phoenix dactylifera* L. Ann Bot 46:465–472

Vardi A, Bleichman S, Aviv D (1990) Genetic transformation of *citrus* protoplasts and regeneration of transgenic plants. Plant Sci 69:199–206

Verdeil JL, Hocher V, Huet C et al (2001) Ultrastructural changes in coconut calli associated with the acquisition of embryogenic competence. Ann Bot 88:9–18

Widholm JM (1972) The use of fluoresceine diacetate and phenosafrnine for determining viability of cultured plant cells. Stain Techhnol 47:189–194

Zhu L, Wang B, Zhou J, Chen L, Dai C, Duan C (2005) Protoplast of callus in *Echinacea augustifolia.* Colloids and Surfaces B. Biointerfaces 44:1–5

Zouine J, El Bellaj M, Meddich A et al (2005) Proliferation and germination of somatic embryos from embryogeneic suspension cultures in *Phoenix dactylifera.* Plant Cell Tissue Org Cult 82:83–92

Chapter 30
Transgenic Date Palm

M.M. Saker

Abstract Modern biotechnology creates unprecedented opportunities to improve agricultural productivity, decrease our dependence on potentially-harmful chemical pesticides and enhance our ability to produce food. The date palm has a considerable socioeconomic value and is mentioned repeatedly with appreciation in the Bible and Quran. The tree plays a key role in the Arabic nations and ought to gain special attention from Arabic governments and scientists. The date palm is the most suitable tree for cultivation in arid and semiarid regions of the world due to its low water demand, tolerance to high temperature, drought and salinity. Unfortunately, there are many biotic stresses that hinder expansion of date palm cultivation, productivity and accordingly date palm revenue. A literature search shows that most publications have been focused on *in vitro* propagation, and molecular characterization of date palm cultivars and tissue culture-derived plants. Research aimed at developing date palm transformation systems is lacking and date palm transformation seems to be in its infancy. The tree is a target host for several pests and diseases, so it is necessary to focus on its *in vitro* propagation and genetic engineering to overcome some of these problems. This review highlights ongoing efforts in date palm transformation and the expected role of genetic transformation in date palm improvement.

Keywords *Agrobacterium* • Particle bombardment • Transformation • Tissue culture

30.1 Introduction

Date palm (*Phoenix dactylifera L.*) is a member of the family Palmae (Arecaceae) and occurs in tropical and subtropical habitats. The Palmae is rich in diversity with 183 genera and more than 2,300 species with the greatest diversity in the Old World

---

M.M. Saker (✉)
Department of Plant Biotechnology, National Research Center, Dokki, Cairo, Egypt
e-mail: msakr@asrt.sci.eg; sakrmahmoud@yahoo.com

S.M. Jain et al. (eds.), *Date Palm Biotechnology*,
DOI 10.1007/978-94-007-1318-5_30, © Springer Science+Business Media B.V. 2011

Fig. 30.1 Date palm photos showed on the walls of Ancient Egyptian temples (Pond in a garden Fragment from the Tomb of Nebamun 1,400 BCE)

(Dransfield et al. 2008). The scientific name for the date palm was derived from *phoenix*, the legendary bird of ancient Greece, and *dactylos* meaning *finger* after the shape of the fruit (El Hadrami and El Hadrami 2009).

Representations of this tree appear in the writings of Neolithic civilizations of Mesopotamia. Probably the earliest wild findings of date palm were recorded around 5,000–6,000 BC from Iran, Egypt and Pakistan while the earliest cultivations were found around 4,000 BC from Eridu and Lower Mesopotamia. Date palm was known in ancient Egypt and representations appear in hieroglyphics on the walls of ancient Egyptian temples which are easily recognized as date palm depictions (Fig. 30.1).

Date palm is mentioned in Akkadian and Sumerian cuneiform sources to 2,500 BC and later. The oldest radiocarbon dated discovery of date seeds was on Dalma island, one of the Abu Dhabi Islands. Two seeds were found in 1998, the oldest from 5,110 BC and the other, 4,670 BC Date palm was introduced into Spain by the Moors and by the Spanish into the Americas. *Phoenix dactylifera* is now found in tropical and subtropical regions all over the world as well as in temperate and arid regions in

USA, Australia, southern Spain and the Mediterranean coast of Africa and West Asia (El Hadrami and El Hadrami 2009).

Date palm is a dioecious species where the male and female organs are borne by separate trees. It is the tallest of the *Phoenix* species and the non-branching trunk can grow, under some conditions, taller than 30 m. The plant has one terminal shoot apex that ensures the growth lengthwise. The root system of a date palm is highly developed. The leaves are large, 4–5 m long, alternate, pinnate, growing upward in a spiral pattern on the trunk and sheathing the top in a dense terminal rosette or crown of 100–120 leaves. Each leaf has an axiliary bud that may be vegetative, floral or intermediate (Bouguedoura 1991; Bouguedoura et al. 1990). A fully productive date palm tree can support up to 10 clusters, which can carry more than 100 kg of fruits. A single date fruit, or *t'mar* in Arabic, is usually cylindrical, occasionally rounded or ovoid, a single-seeded drupe of 2.5–7.5 cm long × 4 cm wide with a fleshy, sugary pericarp, yellowish to reddish brown (Purseglove 1972; Tackholm and Drar 1973). The nutritional quality of date palm fruits is well-known and documented. Date fruits provide more calories than other fruits, cooked rice, bread or meat. Date fruits are also rich in sodium, iron, copper, calcium, zinc, magnesium, phosphorus and sulfur. Fruits also contain vitamins A, B_1, B_2 and B_7, protein (2–3%), fiber (2.5–5%), lipids (0.2%) and are free of cholesterol (Toutain 1967). The wood and leaves of the date palm are as important as the fruit itself. They provide fiber, fuel, clothing, furniture, hats, baskets and housing. The pit of the fruit is used to feed animals and the sap is used to prepare wine. Currently, date palm trees and leaves have become an important component in decoration. In addition to its nutritional value, the date palm tree tolerates adverse environmental conditions (abiotic stresses) and is important in reducing desertification and dune-encroachment. Accordingly it is often called the *tree of life*. Moreover, the date palm is mentioned repeatedly with appreciation in holy books (Qur'an and Bible) and receives special attention from Arabic governments and scientists.

Arab countries grow 62 million of the 105 million trees worldwide on an area of over one million ha and occupy 5% of the total arable land; the date palm is considered one of the most important commercial crops in the Arab world. In Saudi Arabia, for example, date palm trees are grown on about 90% of the cultivated land (Shaheen 1990). According to FAOSTAT, world date fruit production in 2008 reached 7,048,089 mt, from a harvested area of 1,264,611 ha. The world's largest producers are Egypt with 1,313,696 mt followed by Iran (1,000,000 mt), Saudi Arabia (982,546 mt), United Arab Emirates (755,000 mt), Pakistan (557,524 mt), Algeria (526,921 mt) Iraq (440,000 mt), Sudan (332,000 mt) and Oman (255,871 mt). Other significant producing countries are Libya, Tunisia, Morocco, Yemen, Mauritania, Qatar, Chad, Israel, Bahrain and the USA.

Phoenix dactylifera is a widely distributed species covering extensive geographic, soil and climatic areas. Date palm grows in nearly rainless regions between 9° and 39° North Latitude, which is represented by the Sahara and the southern fringe of the Near East (Arabian Peninsula, southern Iraq, Jordan). Both wild and domesticated trees are morphologically and ecologically similar. Wild date fruits are small and nonedible compared to those coming from domesticated trees.

Date palm is a highly salt-tolerant tree and may have good production with water containing under 3,000 ppm salt. Some varieties have even exhibited a high tolerance to total dissolved salt (22,000 ppm) but their productivity was affected (El Hadrami and El Hadrami 2009).

Greater date palm productivity can be achieved through increasing the productivity of the existing trees or expanding of palm cultivated areas, which are unsuitable for growing traditional crops. Both approaches constitute a huge dilemma, because extensive breeding programs for the selection of superior date palm clones through traditional methods are tedious efforts due to the long life cycle and strongly heterozygous nature of the palm tree (Saker and Moursy 2003). Moreover, insufficient and the expensive cost of offshoots is another serious limiting factor, which hinders expansion of palm cultivated area, renewing of existing palm orchards and improving their productivity. Therefore, serious efforts benefitting from the application of gene transfer technology are a must to overcome the date palm dilemma.

30.2 Date Palm Constraints and Biotechnological Challenges

Worldwide gene technology has been almost exclusively focused on crops of high economic importance such as maize, wheat, soybean, sunflower, rice and potatoes. Unfortunately, other species native to developing countries have not attracted the interest of the well-known multinational seed and biotechnology companies due to their low socioeconomic status and minimal contribution to the national treasury. Date palm is one of these neglected crops on a global scale, and efforts at biotechnological improvement of this tree are limited to local researchers within the developing countries.

The date palm faces many constraints, mainly due to its development under harsh desert conditions, where growth takes place without a regular supply of amendments. Date palm also faces many biotic constraints, especially bayoud disease caused by *Fusarium oxysporum* f. sp. *albedinis* (Carpenter and Klotz 1966; Djerbi 1988; Laville 1973; Louvet and Toutain 1973; Malençon 1934). This disease is the most devastating infection of the date palm and was first described in southern Moroccan groves. Currently, it continues to spread across North African countries, especially in Morocco and Algeria where more than 12 million date palm trees have been destroyed so far. No effective means is known to control this disease and only a few cultivars with poor-quality fruits, unfortunately, are known to be resistant to bayoud.

The major date palm problems and the potential contribution of biotechnology to solve date palm problems are summarized (Table 30.1). The problems listed indicate clearly that some of them were solved as a result of adopting simple biotechnology tools such as plant tissue culture techniques. Also, it could be concluded that the remaining date palm problems can be solved using gene transfer and molecular breeding programs, especially since those tools have proved effective in similar crops such as African oil palm (*Elaeis guineensis*) (Saker and Moursy 2003).

Table 30.1 Date palm problem and the role of biotechnology

No	Problem	Contribution of biotechnology	References
1	Shortage of superior disease-free offshoots needed for new cultivation.	Rapid mass micropropagation for the production of cheap, disease-free and superior offshoots needed for new cultivation.	Al-Khayri (2007)
2	Somaclonal variations: Genetic instability of tissue culture-derived plants.	Application of molecular fingerprints for the early detection of genetic variations at early stages of development.	Saker et al. (2006a)
3	Palm breeding program is a long and tedious endeavor due to the long life cycle of the palm tree.	1. Development of molecular marker tightly linked to economically important traits to facilitate marker assisted selection (MAS) program.	Jain (2007)
		2. Application of *in vitro* selection and induced mutation for the selection of new breeds.	El Hadrami and El Hadrami (2009)
4	Palm tree is a target for epidemic diseases such as bayoud disease caused by *Fusarium oxysporium* f. sp. *Albedinis.*	Development of transgenic date palm harboring antifungal proteins auto resist fungal diseases including bayoud.	Aaouine (1998)
5	Palm tree is also target for insects such as red palm weevil, which is considered the most notorious pest in Middle East and Gulf countries.	Productions of transgenic date palm harboring insecticidal proteins (alpha-amylase inhibitor, protease inhibitor, Bt, chitinase, etc.) naturally auto resist insects.	Salama and Saker (2002)
6	Spread of epidemic diseases due to date palm trade (import and export of offshoots and exchange of germplasm) is usually associated with disease transfer.	Artificial seeds and *in vitro* date palm cultures are excellent and biosafe material for exchange of germplasm and offshoots trade.	Bekheet et al. (2001)
7	Due to its very low rate of multiplication using the traditional methods and urbanization, some of rare date palm germplasm will be endangered.	Date palm tissue culture proved to be an excellent technology to preserve rare varieties.	Zaid and Arias-Jiménez (2002)

As a prerequisite to successful genetic transformation, a reproducible and efficient tissue-culture system with minimum somaclonal variation must be available. Although notable progress had been made in date palm tissue culture (Al Khayri 2003; Bekheet et al. 2001, 2002; Bhansali et al. 1988; Saker et al. 2006b; Tisserat 1982; Zaid and Hughes 1995), some evidence points to a high degree of genetic instability of tissue culture-derived date palm clones (Saker et al. 2000, 2006a).

Although genetic improvement of date palm via genetic engineering is the only realistic strategy to rapidly improve this tree, genetic transformation of date palm has not been reported or is still in its infancy (Saker and Ghareeb 2007; Saker et al. 2007, 2009). Therefore, great attention is being given to studies aiming at optimization of factors affecting genetic transformation of date palm.

30.3 Current Status of Date Palm Biotechnology

Analysis of published data on date palm biotechnology indicates that events related to it can be divided into three main eras. The first is the plant tissue culture era, dating back to 1978 when the first article on date palm tissue culture was published (El-Hennawy and Wally 1978). The second era started in the early 1990s and dealt with the deployment of molecular markers and DNA fingerprinting in genetic analysis of tissue culture-derived date palm offshoots, cultivar identification, development of molecular markers for the diagnosis and characterization of bayoud disease and sex determination (Al-Khalifah and Askari 2003; Elshibli and Korpelainen 2008; Saker et al. 2000, 2006a). The current (third era) of date palm biotechnology is focusing on date palm transformation and genome analysis. Recently, a few reports have investigated different factors affecting either *Agrobacterium* or biolistic-mediated transformation (Saker et al. 2007, 2009).

30.4 Date Palm Tissue Culture

The global need to expand date palm groves has led to the development of methods for tissue culture propagation. Most of the plants generated from tissue culture are normal and true-to-type. *In vitro* techniques have completely or partially replaced traditional vegetative propagation practices of date-palm growing in many countries. Many different date palm tissues such as leaves, apical dome, shoot tips, lateral buds and roots have proven to be useful as explants for initiating tissue culture (Zaid and Tisserat 1983). Tissue culture such as callus, cell-suspension cultures, root cultures, shoot tip and meristem, embryo or micro-spore cultures have been developed for date palm with more or less success in generating plantlets (El Bellaj et al. 2000; Fki 2005; Fki et al. 2003).

Analysis of published transformation protocols of date palm shows that somatic embryos can often start from a single cell, which makes them the ultimate candidate for applying transformation in date palm. However, somatic embryos germinate at a very low rate that makes them insignificant for date palm transformation. During the last three decades, several protocols for *in vitro* propagation of date palm via somatic embryogenesis have been developed (Aaouine 1998; Al-Khayri 2007; Shaheen 1990). Furthermore, tremendous progress has been made during the last few years regarding *in vitro* regeneration of date palm from zygotic or somatic explants through either organogenesis or somatic embryogenesis. This progress has made possible large-scale production of *in vitro* plants.

The regeneration of date palm plantlets through organogenesis involves accomplishing several steps that are more or less critical: (i) meristem induction; (ii) shoot multiplication; (iii) shoot elongation and (iv) acclimatization. Approximately 1–2 years after induction, date palm plantlets can be regenerated and transferred to the greenhouse for evaluation/screening. Root formation occurs easily during the final stages of *in vitro* culture (El Hadrami and El Hadrami 2009).

Both organogenesis and somatic embryogenesis are being used to produce large numbers of date palm plants on a commercial scale. Embryo-like structures of date palm derived from somatic (asexual) embryogenesis are able to develop into a whole plant in a similar way as zygotic embryos. Two consecutive steps have been known to generate date palm plantlets through somatic embryogenesis (Poulain et al. 1979; Reynolds and Murashige 1979; Tisserat 1979). Poulain et al. (1979) initially described the initiation of vegetative buds from the date palm offshoot heart (apical meristem). In the same year, Reynolds and Murashige (1979) and then Tisserat (1979) described the induction and regeneration of somatic embryos from these cultures and utilized various other tissues as explants. Critical factors controlling the establishment of date palm embryogenic cultures include explant type, genotype and plant-growth regulators. Both offshoots and flower buds were successfully used as explants to regenerate embryogenic cultures (El Hadrami and El Hadrami 2009). Other protocols have been recently described concerning the establishment of the embryogenic suspension cultures in date palm (Fki 2005). However, studies dedicated to anther and ovule cultures recovery are scarce for date palm (Bouguedoura 1991; Chaibi et al. 2002).

30.5 Breeding and Genetic Manipulation of Date Palm

Both conventional and non-conventional date palm-breeding programs rely at certain time points on biotechnologies and specifically the use of tissue culture for regeneration of plants and/or transformation. Many advances have been made in this field of investigation regarding the determination of condition factors and stimuli that control the date palm tissue plasticity and totipotency. These two concepts, as in many other systems, are still empirical making the identification of culture conditions and stimuli extremely difficult to gather.

Only a few reports describe the use of cell suspensions as a tool of genetic manipulation of date palm (Fki 2005). Until now, no study or report has been published regarding date palm protoplast isolation and culture. Serious difficulties were encountered to overcome browning and rapid death of protoplasts.

Induced mutagenesis uses either ionizing radiation or chemical mutagens combined with various *in vitro* cultures to induce mutagenesis and they represent the simplest, fastest and most highly efficient method for improving crops. It can result in the development of mutant cultivars able to exhibit resistances to biotic or abiotic stresses, to produce a desirable quality and/or an improved quantity of fruits as well as specific morphological features (Jain 2002). Moreover, *in vitro* techniques currently allow the induction of mutations in a large number of propagules within a reduced working space. Many breeding programs around the world use this technology to induce variation within the stock. The improvement of *in vitro* techniques for date palm has made it possible to irradiate these cultures on a large scale and to maintain them within the same collection (Jain 2006). Moreover, recombinant DNA research and the use of labeled probes such as RFLP, microsatellite based DNA fingerprinting, developed for cloning and mapping plant genes or transgenesis, are able to trace such modifications within the genome. Currently, it is routine to identify and analyze mutants using DNA fingerprinting or mapping genetic alterations using PCR (polymerase chain reaction) based markers, such as RAPD, AFLP, SSR, SNP and consequently tagging mutants (Beetham et al. 1999; Zhu et al. 1999). Stable mutations then can be linked to noticeable changes in the DNA sequence of specific plant traits and subsequently mapped and located on the chromosomes before being analyzed by functional genomics and transferred into a desirable background variety. In turn, breeding processes can be accelerated to lead to new varieties of crops enhanced for their yield, quality or resistance to biotic/abiotic stresses. The most suitable markers, those discriminating sex within progeny, would be markers for the productivity of date palm trees (yield and quality of dates) and for the resistance/tolerance to bayoud and other biotic and abiotic constraints.

30.6 Date Palm Transformation

Today's global human population is about 6.5 billion and is expected to reach 8.5 billion by the year 2025. This situation simply means that our present food crop production level must be increased by 30–40%. Therefore, it is expected that genetically engineered date palm will naturally overexpress biopesticide and antifungal proteins automatically to resist serious pests and epidemic diseases; this will be the main focus of date palm biotechnology research in the third millennium. Growing of such palms will significantly reduce the hundreds of tons of pesticide applied yearly with their associated health risks and degradation of the ecosystem. Only through gene transfer technology can outstanding date palm cultivars, often the results of time-consuming selection programs, be improved further, if only one or a few characters are changed without altering the remaining genetic makeup.

Transformation is most simply defined as a *change*. In the plant biotechnology community, transformation can be more precisely defined as the process of DNA introduction into a plant cell, leading to a permanent change in the genetic makeup of the target cell and its derivatives. Although the transformation process itself was initially limiting, all crops of major interest have been successfully transformed and many if not most transformation technologies are considered routine. Some crops do remain a bit recalcitrant to transformation and improvements in the methods for production of stably transformed plants are still needed (Finer 2010). The current limitations in the production of transgenic plants for both basic research and commercial application include more efficient production of transformed plants and obtaining more predictable insertion and expression of the introduced DNA. The ability to produce whole plants from transformed plant cells, first reported by Horsch et al. (1985), has revolutionized the plant sciences through the success and implementation of genetically modified plants.

A large number of plant species have subsequently been genetically transformed, primarily using two different strategies for DNA delivery into totipotent cells. The first strategy employed T-DNA delivery mediated by disarmed strains of the soil bacterium *Agrobacterium tumefaciens* (De Block et al. 1984; Horsch et al. 1984). *Agrobacterium tumefaciens* was initially used in transformation of a broad range of dicotyledonous crops (Horsch et al. 1987). However, in the beginning, it appeared that *Agrobacterium* would not be useful for transformation of monocotyledonous plants (Potrykus 1990; Vain et al. 1995). This led to the contemporary development of direct DNA delivery methods as the second strategy for transformation of intractable monocot species. Fertile, transgenic plants produced via direct DNA delivery methods were reported soon after the success with *A. tumefaciens* (Paszkowski et al. 1984).

All published data on palm transformation over the past 10 years focused on African oil palm (Abdullah et al. 2003, 2005). Special emphasis was given to develop techniques for gene transfer and production of transgenic oil palms to express *Bacillus thuringiensis* (*Bt*) crystal insecticidal protein genes, to address problems related to insect pests (Sharma et al. 2000, 2002) and chitinase (to address problems related to basal stem rot). The first successful transfer, insertion and expression of economically important genes in palms, was reported by Lee et al. (2006) in African oil palm. They developed a rapid and efficient method for the transformation and evaluation of *CryIA (b)* expression in oil palm. More than 700 putative transformants were generated. It was observed that pre-treatment of target tissues with phytohormones is essential for increasing the transformation efficiency. This finding enhanced the transformation rate in African oil palm that was previously difficult to achieve.

30.7 *Agrobacterium*-Mediated Plant Transformation

Over the years, many plants, including several important crops, have been genetically modified (transformed) using *Agrobacterium tumefaciens*-mediated transformation (Goedeke et al. 2007). *Agrobacterium*-mediated genetic engineering is currently the most widely used plant genetic engineering strategy; the first transformation among the monocots was accomplished in rice in 1994 (Hiei et al. 1994) and is now routinely used to transform other cereals, including maize (Ishida et al. 1996), barley (Tingay et al. 1997) and wheat (Cheng et al. 1997). Researcher familiarity with *Agrobacterium*-mediated transformation systems and their simplicity and independence from DNA delivery devices are an important rationale for the major shift toward using *Agrobacterium* for crop genetic engineering (Songstad et al. 1995).

Generally, *Agrobacterium*-mediated transformation is the preferred approach to deliver a foreign gene and compares with methods based upon direct gene transfer, because several advantages are provided; integration of a well-defined DNA sequence, typically low copy number and preferential integration into actively transcribed chromosomal regions (Birch 1997; Gheysen et al. 1998). Many approaches have been pursued in order to improve the efficiency of *Agrobacterium*-mediated transformation in recalcitrant monocot plant species, e.g. use of hypervirulent *Agrobacterium* strains, use of particular combinations of *Agrobacterium* and plasmids, optimization of co-culture media and conditions that increase the interaction of *Agrobacterium* with the plant cell (Cheng et al. 2004; Kumlehn et al. 2006; Roy et al. 2000; Somers et al. 2003).

Impressive progress has been achieved since 1983 in developing methods for genetically engineering a diverse range of plant species, in commercialization of transgenic traits in a number of crops that are planted on substantial areas throughout the world and in characterization of transgene locus structure. However, much remains to be learned about the mechanisms that result in the formation of transgene loci, why the main transformation systems differ in the proportions of complex transgene loci produced and how to routinely produce simple transgene loci (Finer 2010).

In the case of date palm, most published trials on *Agrobacterium*-mediated date palm transformation used GUS (β-glucuronidase) as a reporter gene, which is easy to assay. In addition, there has been no conclusive report on expression of economically-important genes in date palm to the present. In this regard, the first report on successful infection of date palm embryogenic callus with *Agrobacterium* and context, a system for *Agrobacterium*-mediated gene transfer to embryogenic callus of date palm was developed (Saker et al. 2009). This system involves callus production from shoot tip explants on callus induction medium (CIM), which contained basal MS salts, B5 vitamins, 30 g/l sucrose, 10 mg/l 2, 4-D, 3 mg/l 2ip, 170 mg/l KH_2PO_4 and 3 g/l activated charcoal, followed by mass propagation of the proliferated microcalli on MS medium supplemented with 0.4 mg/l NAA and 0.1 mg/l 2ip. Factors influencing transient expression of the GUS gene were evaluated following infection of date palm embryogenic callus with *Agrobacterium tumefaciens*. The obtained results indicated that high bacterial density (OD_{600} 1–1.5) and prolonged infection (2 h) gave the highest percentage of GUS-expressing calli. A further improvement of transient GUS expression was obtained through desiccation generated by exposing the embryogenic calli to a gentle air flow at 45°C, followed by rehydration in *Agrobacterium* suspension, significantly enhanced transient GUS expression. In contrast, sonication pre-treatment of the calli negatively affected the frequency of GUS expression. The results presented indicate that *Agrobacterium*-mediated gene transfer to date palm cells can be achieved, and thus represents a first step towards the establishment of stable genetic transformation of this agro-economically important monocot crops. Figure 30.2 shows the different steps of *Agrobacterium*-mediated transformation of date palm (Saker et al. 2009).

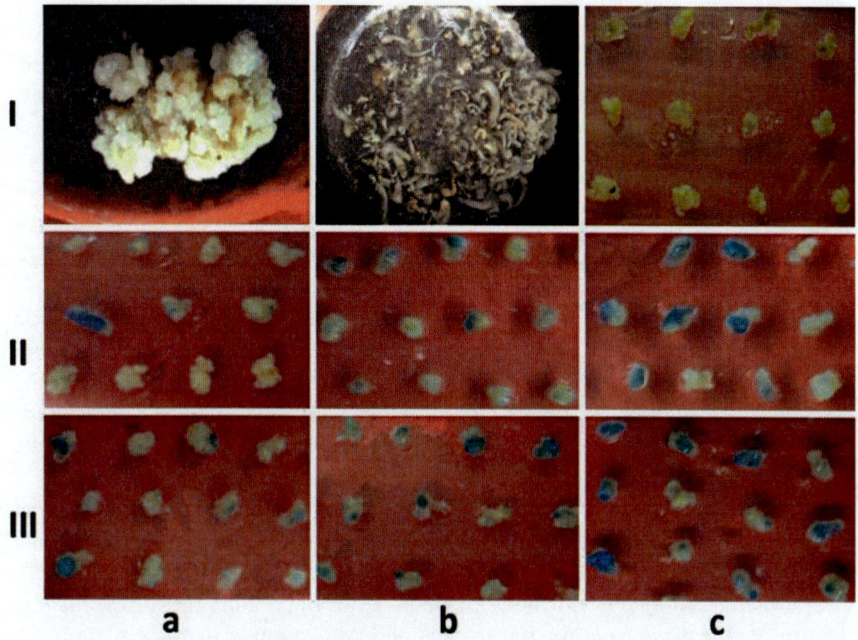

Fig. 30.2 *Agrobacterium*-mediated transformation of embryogenic callus of date palm (*source*: Saker et al. 2009) **I**: Shows embryogenic callus, and suspension of somatic embryos used in the experiment and negative control of GUS assay (from left to right). **II, III**): Histological detection of *gus* expression after 3 days of co-cultivation with Agrobacterium strain AGL1s harbors the binary plasmid pSB159. II, III represents incubation periods of 1 and 2 hrs respectively, while a, b and c represnt various Agrobacterium cell density (OD600) 0.5, 1 and 1.5 respectively

30.8 Direct Gene Transfer: Particle Bombardment

Direct DNA delivery methods were initially based on protoplasts as target cells and employed polyethylene glycol (PEG) and electroporation-mediated delivery of DNA (Paszkowski et al. 1984; Potrykus et al. 1987; Vain et al. 1995). Klein et al. (1988) reported the development of a microprojectile bombardment (gene gun) technology for delivery of DNA into intact plant cells, and this was followed by other reports. This development resulted in production of the first fertile transgenic plants in soybean (McCabe et al. 1988), followed by maize (Fromm et al. 1990; Gordon-Kamm et al. 1990), wheat (Vasil et al. 1992), oats (Somers et al. 1992) and barley (Wan and Lemaux 1994).

As mentioned above, the date palm tree is a target host of many serious pests such as red palm weevil *Rhynchophorus ferrugineus* (order Coleoptera) and bayoud disease caused by *Fusarium oxysporum* (Ajlan et al. 2000; Salama and Saker 2002). The surface application of pesticides to control the weevil is ineffective as the larvae persist in the trunk of the tree. Similarly the use of chemical fungicide is ineffective in bayoud management, as the fungus is soil borne. Moreover, the application of conventional breeding to produce new date palm breeds resistant to pests is a time-consuming and tedious endeavor. It usually takes up to 30 years to make three back-crosses, and to obtain the first offshoot (Saker and Moursy 2003). Accordingly there is great demand for the development of a transformation system for date palm, as a first step toward engineering new date palm breeds with auto-resistance against serious pests. This strategy proved to be effective and promising in other crops (Barton et al. 1987; Boller et al. 1983; Cho et al. 1995; Kramer and Muthukrishnan 1997; Lin et al. 1995; Morton et al. 2000; Schroeder et al. 1995; Valencia et al. 2000).

To avoid host range of *Agrobacterium*, Ghulam et al. (1998) addressed genetic transformation of oil palm using particle bombardment. They reported that hundreds of genetically transformed African oil palm trees carrying and expressing economically important genes were produced using a particle bombardment gene delivery system. They also studied biological parameters affecting microprojectile bombardment delivery of DNA into oil palm embryogenic calli.

New selection systems for isolation of transgenic tissue cultures and plants, and optimization of DNA delivery constraints combined with improvements in plant tissue culture techniques and microprojectile bombardment protocols, led to efficient transformation systems for a large number of important crops. Nevertheless the use of *Agrobacterium* for transformation is still limited in some plants, e.g. monocotyledons, due to the lack of available and efficient transformation and regeneration systems.

Pioneering trials on date palm transformation using particle bombardment was reported using the Iranian cv. Khorma (Habashi et al. 2008). Similarly, Saker et al. (2007) optimized some factors affecting transformation of the Egyptian cv. Sewi using particle bombardment. A construct harboring a cholesterol oxidase gene, which renders plants resistance to insect attack, was introduced into embryogenic date palm callus using PDS1000/He particle bombardment system. Three calli colonies out of 200 putative transformants microcalli colonies, which have been bombarded with DNA-coated particles, gave positive GUS expression. The successful integration of GUS gene in GUS positive clones was verified by PCR. The reported system involves the establishment of embryogenic callus cultures from shoot tip explants, followed by shooting of the embryogenic callus with DNA coated particles under optimized physical conditions. The developed system involved bombardment of embryogenic date palm callus as described by Klein et al. (1988), with minor modifications. Briefly, particles were prepared for bombardment as follows: 60 mg of particles were placed in a microcentrifuge tube, vortexed vigorously in 1 ml of 100% ethanol for 5 min, soaked at room temperature for 15 min and centrifuged for 15 min at 10,000 rpm. The supernatant was decanted and the particles washed three times in sterile distilled water, suspended in 2 ml of 50% (v/v) sterilized glycerol and divided into 125 µl aliquots with 3.75 mg of particles per aliquot. Nine µg of plasmid DNA was added to a microcentrifuge tube containing (3.75 mg of) particles in 125 µl, vortexed and 125 µl of 2.5 M CaC12 were added and vortexed. Fifty µl of 0.1 M spermidine (freshly prepared and filter sterilized 0.1 M spermidine solution) was added and vortexed for 4 min, kept on ice for 15 min, centrifuged for 20 s at 1,000 rpm and the supernatant removed. A half ml of 75% ethanol was added, vortexed for 2 min, centrifuged for 20 s at 1,000 rpm and supernatant removed, and 0.5 ml of 100% ethanol added, responded and 6 μl of the homogenous solution used per shot. The negative controls were bombarded with uncoated particles in an identical fashion. The shelf carrying the target tissues was fixed at three different flight positions (7, 10 and 13 cm), the chamber vacuum was fixed at 25 Hg and three different rupture-disc pressures (900, 1,100, 1,300 and 1,500 psi) were tested.

Selection of putative transformants was carried out on a selective medium containing 100 mg/l kanamycin and for rapid screening of putative transgenic calli, GUS expression was assayed using fluorescence detection method. Briefly, embryogenic micro calli colonies were squeezed in wells of ELISA plate in 100 μl GEP solution, then 100 μl MUG solution was added and incubated over night at 37°C, visualized under UV light (positive samples have a strong fluorescence). GEP solution (100 ml) consists of 50 mM Na_2HPO_4, 10 mM DTT or B- Mercaptoethanol, 1 mM Na 2 EDTA, 0.1% Na – Lauroylsarcosine and 0.1% Triton X 100. MUG solution (100 ml) consists of 20 mg MUG in 50 ml GEP Solution (Jefferson et al. 1987). The developed system can be seen as a first step toward production of transgenic date palm with auto resistance to serious pests.

In conclusion, the published protocols on direct gene transfer using practical bombardment were relayed on the main protocol described by Klein et al. (1988), with minor modifications. The overall objective of the two retrieved reports on date palm transformation was to establish a regeneration/transformation system for some local palm genotypes using particle bombardment. Identification of major parameters controlling DNA delivery by particle bombardment of date palm embryogenic calli and somatic embryos were investigated. Osmotic conditioning of explants before and after bombardment (osmotic and non-osmotic), type of explants (embryogenic callus and somatic embryo) and different bombardment parameters like acceleration pressure (900 and 1,100 psi), bombardment distance (6 and 9 cm) and gold particle size (1 and 0.6 μm) were revealed to be the most significant factors. Results indicated that transient GUS expression increased when 1.0 μm gold particle, 1,100-psi rupture disk, 6 cm bombardment distance, osmotic medium and mature somatic embryos were used (Habashi et al. 2008; Mousavi et al. 2009; Saker et al. 2007).

Preliminary investigations of physical factors influencing gene delivery using a biolistic gun indicated that flight distance of microprojectiles and applied pressure were the most effective parameters. Published data indicated that the highest percentage of explants showed positive GUS expression and was recorded following shooting of explants with microprojectile at a distance of 7 cm. Contrarily, when the distance increased, i.e. to 10 and 13 cm, a marked decrease in the percentage of explants showing positive GUS expression was recorded (Saker et al. 2007). Therefore, a distance of about 10 cm was suggested to be near the optimal for date palm embryogenic callus transformation using biolistic device PDS-1000-He (Bio-Rad) because short flight distance (7 cm) reduced the transient GUS expression and caused tissue dislocation which may be associated with high tissue damage as explained by (Parveez et al. 1997, 1998). In this context, other reports have shown similar results (King and Kasha 1994).

Published data also indicated that acceleration pressure around 1,000–1,100 psi gave the highest values of explant survival and percentage of explants showing positive transient GUS expression (Habashi et al. 2008; Sanford et al. 1993). Small (900 psi) and high (1,500 psi) acceleration pressure is associated with notable decrease in the percentage of explant survival on selective media and positive GUS expression (Saker et al. 2007). Increasing the applied helium pressure up to 1,500 psi dramatically reduces the GUS expression. This can be attributed to dislocation of target tissue and/or damage to cells as a result of very high velocity and force of microprojectile (Parveez et al. 1997, 1998; Ritala et al. 1993). In conclusion, optimal physical conditions affecting successful delivery of DNA into plant tissue were reported to be cell and tissue type dependent (Iida et al. 1990). Also interaction among factors affects the successful delivery process, for instance a microprojectile of small size will need lower force to penetrate the tissue while larger-size particles will need higher force and accordingly high helium pressure and short flight distance (Li et al. 1994). Not only physical factors but also cell, tissue and plant type affect the successful delivery of DNA as mentioned by Iida et al. (1990).

30.9 Date Palm Genomics

Recently, Joel Malek, director of the genomics laboratory and his team of laboratory assistants at Qatar University, increased the publicly-available knowledge of the date palm gene space nearly a thousand fold using an approach that takes advantage of the lower repetitive DNA in the date palm gene regions. The genome contained about 500 million base pairs, twice the size anticipated. To produce the draft map, the researchers used a next generation sequencing approach, which offers data quality between that of the expressed sequence tag (EST) method and the traditional whole-genome mapping method. The quality of their work is comparable to the versions of other plant draft sequences, such as the rice and papaya genomes.

The date palm genome is available at the website (http://qatar-weill.cornell.edu/research/datepalmGenome/index.html) as a resource for those interested in date palm genetics. The results obtained from this project may provide a starting point for researchers doing transformation, genetic and genomic studies of date palm and could revolutionize this field of study.

30.10 Conclusion

Considering the importance of date palm, it must be said that it lags behind other crops in the breadth of research scope of in-depth studies, whereas the development of plant transformation technology is impacting crop improvement options in unprecedented ways. Conversely, the transgenic technology promises greater efficiency and precision in plant breeding; to be effective, its application needs to be carefully planned. In the area of micropropagation, research on commercial production, scale up, to produce propagules with low-cost efficiency is an important issue. The current price of *in vitro*-derived date palm has not met consumer expectations (Al-Khayri 2005). Nevertheless, there were some successful endeavors for large-scale propagation of oil palm (Gorret et al. 2004). In date palm (Fki et al. 2003; 2005; Saker et al. 2006b), more research efforts should be directed towards *in vitro* multiplication of date palm via direct shoot organogenesis and further studies on industrial production of somatic embryos and artificial seed production are also assured. More effort should be directed towards *in vitro* selection of somaclonal variants and induced mutations as a promising tool for genetic improvement of date palm (Jain 2007). Genomic information makes it possible to identify female seeds at an early stage and would make sexual propagation of the trees much more viable. Additionally, the genome yields information about enzymes, which control traits such as fruit sweetness and rate of ripening; the more we understand about the genes and their functions, the closer we come to being able to manipulate the genome to control characteristics like resistance to disease and fruit quality.

Further studies on molecular markers are also needed to develop a robust, economically feasible and precise method(s) for the detection of sex and genetic stability in tissue culture-derived plants at early stages of development, especially since it was reported that 78% of tissue culture-derived date palm plants cultured in Saudi Arabia during 1992 and 1993 were unable to set seed (Al-Wasel 2001).

Date palm transformation is a powerful approach to plant improvement. However, it is not a solution to all problems. Several biosafety concerns associated with transgenic crops are real. By thoughtful choice of traits, strategies and government policies, however, engineered date palm can be developed and deployed to achieve national agricultural goals. However, lessons extracted from the literature search presented in this review indicate that date palm transformation is still in its infancy and the few published or unpublished reports have focused on optimization of some factors affecting either particle bombardment or *Agrobacterium*-mediated transformation. Greater efforts on date palm transformation are needed to benefit from the limitless applications of gene transfer technology in the improvement of date palm.

Acknowledgement The author express his thanks and appreciation to Dr. Tarek Kapeil (Botany Dept., Faculty of Science, Cairo University) and A. Researcher Mai A. Allam (Plant Biotechnology Dept., National Research Center, Egypt) for their so kind cooperation and providing me with information on origin of date palm and date palm tissue culture.

References

Aaouine M (1998) The application of biotechnology to date palm. In: Ives C, Bedford B (eds.) Agricultural biotechnology in international development. CABI, Wallingford, pp 133–146

Abdullah R, Chari C, Yap WSP, Yeun LH (2003) Transgenic oil palm with stably integrated *CpTI* gene confers resistance to bagworm larvae. In: Vasil IK (ed.) Plant biotechnology 2002 and beyond. Kluwer, Dordrecht, pp 163–165

Abdullah R, Alizah Z, Wee Y et al (2005) Immature embryo: a useful tool for oil palm genetic transformation studies. Electr J Biotechnol 8. ISSN 0717 3458. http://www.ejbiotechnology.info/content/vol8/issue1/full/1/

Ajlan AM, Shawir MS, Abo-El-Saad MM et al (2000) Laboratory evaluation of certain organophosphorus insecticides against the red palm weevil, *Rhynchophorus ferrugineus* (Olivier). Sci J King Faisal Univ Basic Appl Sci 1:15–26

Al-Khalifah N, Askari E (2003) Molecular phylogeny of date palm (*Phoenix dactylifera* L.) cultivars from Saudi Arabia by DNA fingerprinting. Theor Appl Genet 107:1266–1270

Al Khayri JM (2003) *In vitro* germination of somatic embryos in date palm: effect of auxin concentration and strength of MS salts. Curr Sci 84:5–10

Al-Khayri JM (2005) Date palm *Phoenix dactylifera* L. In: Jain SM, Gupta PK (eds.) Protocol for somatic embryogenesis in woody plants. Springer, Dordrecht, pp 309–319

Al-Khayri JM (2007) Date palm *Phoenix dactylifera* L. micropropagation. In: Jain SM, Häggman H (eds.) Protocols for micropropagation of woody trees and fruits. Springer, Dordrecht, pp 509–526

Al-Wasel AS (2001) Somaclonal variations in tissue culture-derived date palm (*Phoenix dactylifera*) trees. In: The second international conference on date palm, Al-Ain, 25–27 Mar 2001, p 591

Barton K, Whiteley H, Young S (1987) *Bacillus thuringiensis*–endotoxin in transgenic *Nicotiana tabacum* provides resistance to lepidopteran insects. Plant Physiol 85:1103–1109

Beetham PR, Kipp PR, Sawycky ZL (1999) A tool for functional plant genomics: chimeric RNA/DNA oligonucleotides cause *in vivo* gene-specific mutations. Proc Nat Acad Sci USA 96:8774–8778

Bekheet SA, Taha HS, Saker MM (2001) Factors affecting *in vitro* multiplication of date palm. Biol Plant 44:431–433

Bekheet SA, Taha HS, Saker MM, Moursy HA (2002) A synthetic seed system of date palm through somatic embryogenesis encapsulation. Ann Agric Sci Ain Shams Univ Cairo 47:325–337

Bhansali R, Kaul R, Dass H (1988) Mass cloning of date palm plantlets through repetitive somatic embryogenesis. J Plant Anat Morph 5:73–79

Birch RG (1997) Plant transformation: problems and strategies for practical application. Ann Rev Plant Physiol Plant Mol Biol 48:297–326

Boller T, Gehri A, Mauch F, Vogeli U (1983) Chitinase in bean leaves: induction by ethylene, purification, properties and possible function. Planta 157:22–31

Bouguedoura N (1991) Connaissance de la morphogenèse du palmier dattier (*Phoenix dactylifera* L.). Etude in situ et in vitro du développement morphogénétique des appareils végétatif et reproducteur. Thèse de Doctorat, Université des Sciences et de la Technologie Houari Boumediene, Alger

Bouguedoura N, Michaux-Ferriere N, Bompar JL (1990) Comportement *in vitro* de bourgeons axillaires de type indetermine du palmier dattier (*Phoenix dactylifera*). Can J Bot 68:2004–2009

Carpenter JB, Klotz LJ (1966) Diseases of the date palm. Date Growers Inst Rep 43:15–21

Chaibi N, Ben Abdallah A, Harzallah H, Lepoivre P (2002) Potentialités androgénétiques du palmier dattier *Phoenix dactylifera* L. et culture *in vitro* d'anthères. Biotechnol Agron Soc Environ 6:201–207

Cheng M, Fry JE, Peng S et al (1997) Genetic transformation of wheat mediated by *Agrobacterium tumefaciens*. Plant Physiol 115:971–980

Cheng M, Lowe BA, Spencer TM et al (2004) Factors influencing *Agrobacterium*-mediated transformation of monocotyledonous species. In Vitro Cell Dev Biol Plant 40:31–45

Cho H, Choi K, Yamashita M et al (1995) Introduction and expression of the *Streptomyces* cholestrol oxidase gene (ChoA), a potent insecticidal protein active against boll weevil larvae, into tobacco cells. Appl Microbiol Biotechnol 44:133–138

De Block M, Herrera-Estrella L, Van Montagu M et al (1984) Expression of foreign genes in regenerated plants and in their progeny. EMBO J 3:1681–1689

Djerbi M (1988) Les maladies du palmier dattier. Projet régional de lutte contre le Bayoud, FAO, Alger

Dransfield J, Uhl NW, Asmussen CB et al (2008) Genera palmarum. The evolution and classification of palms. Royal botanic gardens, Kew

El Bellaj M, El Jaafari S, El Hadrami I (2000) IAA-oxidase: regulator and potential marker of somatic embryogenesis in date palm (*Phoenix dactylifera* L.). Cah Agron 9:193–195

El Hadrami E, El Hadrami A (2009) Breeding date palm. In: Jain SM, Priyadarshan PM (eds.) Breeding plantation tree crops: tropical species. Springer, New York, pp 191–216

El-Hennawy H, Wally Y (1978) Date palm (*Phoenix dactylifera* L.), bud differentiation *in vitro*. Egypt J Hortic 5:81–82

Elshibli S, Korpelainen H (2008) Microsatellite markers reveal high genetic diversity in date palm (*Phoenix dactylifera* L.) germplasm from Sudan. Genetics 134:251–260

Finer JJ (2010) Plant nuclear transformation. In: Kempken F, Jung C (eds.) Genetic modification of plants: biotechnology in agriculture and forestry. Springer, Berlin

Fki L (2005) Application des suspensions cellulaires embryogènes au clonage et à l'amélioration {*in vitro*} du palmier dattier. Faculté des Sciences de Sfax-Tunisie

Fki L, Masmoudi R, Drira N, Rival A (2003) An optimized protocol for plant regeneration from embryogenic suspension cultures of date palm, *Phoenix dactylifera* L., cv. Deglet Nour. Plant Cell Rep 21:517–524

Fromm MJ, Morrish F, Armstrong C et al (1990) Inheritance and expression of chimeric genes in the progeny of transgenic maize plants. Bio/Tech 8:833–839

Gheysen G, Angenon G, Van Montagu M (1998) *Agrobacterium*-mediated plant transformation: a scientifically intriguing story with significant applications. In: Lindsey K (ed.) Transgenic plant research. Harwood Academic, Amsterdam, pp 1–33

Ghulam Kadir AP, Chowdhury MKU, Saleh NM (1998) Physical parameters affecting transient GUS gene expression in oil palm (*Elaeis guineensis Jacq.*) using the biolistic device. Ind Crops Prod 6:41–50

Goedeke S, Hensel G, Kapusi E et al (2007) Transgenic barley in fundamental research and biotechnology. Transgenic Plant J 1:104–117

Gordon-Kamm WJ, Spencer TM, Mangano ML et al (1990) Transformation of maize cells and regeneration of fertile transgenic plants. Plant Cell 2:603–618

Gorret N, bin Rosli SK, Oppenheim SF et al (2004) Bioreactor culture of oil palm (*Elaeis guineensis*) and effects of nitrogen source, inoculum size, and conditioned medium on biomass production. J Biotechnol 108:253–263

Habashi AA, Kaviani M, Mousavi A, Khoshkam S (2008) Transient expression of β-glucuronidase reporter gene in date palm (*Phoenix dactylifera* L.) embryogenic calli and somatic embryos via microprojectile bombardment. J Food Agric Environ 6:160–163

Hiei Y, Ohta S, Komari T, Kumashiro T (1994) Efficient transformation of rice (*Oryza sativa* L.) mediated by *Agrobacterium* and sequence analysis of the boundaries of the T-DNA. Plant J 6:271–282

Horsch RB, Fraley RT, Rogers SG et al (1984) Inheritance of functional foreign genes in plants. Science 223:496–498

Horsch RB, Fry JE, Hoffmann NL et al (1985) A simple and general method for transferring genes into plants. Science 227:1229–1231

Horsch RR, Fraley S, Rogers J et al (1987) *Agrobacterium*-mediated transformation of plants. In: Green CE, Somers DA, Hackett WP, Biesboer DD (eds.) Plant tissue and cell culture. Alan R. Liss, New York, pp 317–329

Iida A, Seki M, Kamada M et al (1990) Gene transfer into cultured plant cells by DNA-coated gold particles accelerated by a pneumatic particle gun. Theor Appl Genet 80:813–816

Ishida Y, Saito H, Ohta YS et al (1996) High efficiency transformation of maize)*Zea mays* L.) mediated by *Agrobacterium tumefaciens*. Nat Biotechnol 14:745–750

Jain SM (2002) A review of induction of mutations in fruits of tropical and subtropical regions. Acta Hortic 575:295–302

Jain SM (2006) Radiation-induced mutations for developing bayoud disease resistant date palm in North Africa. Proceedings of the international workshop on true-to-typeness of date palm tissue culture-derived plants, Morocco, 23–25 May 2005, pp 31–41

Jain SM (2007) Recent advances in date palm tissue culture and mutagenesis. Acta Hortic 736:205–211

Jefferson RA, Kavanagh TA, Bevan MW (1987) GUS Fusions: ß-glucuronidase as a sensitive and versatile gene fusion marker in higher plants. EMBO J 6:3901–3909

King SP, Kasha KJ (1994) Optimizing somatic embryogenesis and particle bombardment of barley (*Hordeum vulgare L.*) immature embryos. In Vitro Cell Dev Biol 30:117–123

Klein TM, Gradziel T, Fromm ME, Sanford JC (1988) Factors influencing gene delivery into *Zea mays* cells by high-velocity microprojectiles. Bio/Tech 6:559–563

Kramer KJ, Muthukrishnan S (1997) Insect chitinases: molecular biology and potential use as biopesticides. Insect Biotechnol Mol Biol 27:887–900

Kumlehn J, Serazetdinova L, Hensel G et al (2006) Genetic transformation of barley (*Hordeum vulgare* L.) via infection of androgenetic pollen cultures with *Agrobacterium tumefaciens*. Plant Biotechnol J 4:251–261

Laville E (1973) Les maladies du dattier. In: Munier P (ed.) Le palmier dattier. G.P. Maisonneuve and Larose, Paris, pp 95–108

Lee M-P, Yeun L-H, Abdallah R (2006) Expression of *Bacillus thuringiensis* insecticidal protein gene in transgenic oil palm. Electr J Biotechnol [online] 9. ISSN 0717–3458. http://www.ejbiotechnology.info/content/vol9/issue3/full/3/index.html

Li YH, Trembley FM, Seguin A (1994) Transient transformation of pollen and embryogenic tissues of white spruce (*Picea glauca* Moench) resulting from microprojectile bombardment. Plant Cell Rep 13:661–665

Lin W, Anuratha CS, Datta K et al (1995) Genetic engineering of rice for resistance to sheath blight. Biotechnol 13:686–691

Louvet J, Toutain G (1973) Recherches sur les fusarioses VIII. Nouvelles observations sur la fusariose du palmier dattier et précisions concernant la lutte. Ann Phytopathol 4:35–52

Malençon G (1934) Nouvelles observations concernant l'étiologie du Bayoud. C R Acad Sci Paris 19:1259–1262, (Abstract in Rev App Mycol 13: 505)

McCabe DE, Swain WF, Martinell BJ, Christou P (1988) Stable transformation of soybean (*Glycine max*) by particle acceleration. Bio/Tech 6:923–926

Morton R, Schroder H, Bateman K et al (2000) Bean α-amylase inhibitor 1 in transgenic peas (*Pisum stavium*) provides complete protection from pea weevil (*Bruchus pisorum*) under field conditions. PNAS 97:3820–3825

Mousavi M, Mousavi A, Habashi AK et al (2009) Optimization of physical and biological parameters for transient expression of uidA gene in embryogenic callus of date palm (*Phoenix dactylifera* L.) via particle bombardment. Afr J Biotechnol 8:3721–3730

Parveez GKA, Chowdhury MKU, Norihan MS (1997) Biological parameters affecting transient *GUS* gene expression in oil palm (*Elaeis guineensis* Jacq.) embryogenic calli via microprojectile bombardment. Ind Crops Prod 8:17–27

Parveez GKA, Chowdhury MKU, Norihan MS (1998) Physical parameters affecting transient GUS gene expression in oil palm (*Elaeis guineensis* Jacq.) using the biolistic device. Ind Crops Prod 6:41–50

Paszkowski J, Shillito RD, Saul M et al (1984) Direct gene transfer to plants. EMBO J 3:2717–2722

Potrykus I (1990) Gene transfer to cereals: an assessment. Bio/Tech 8:535–542

Potrykus I, Paszkowski J, Saul MW et al (1987) Direct gene transfer to plants: facts and future. In: Green CE, Somers DA, Hackett WT, Biesboer DD (eds.) Plant tissue and cell culture. Alan R. Liss, New York, pp 289–302

Poulain C, Rhiss A, Beauchesne G (1979) Multiplication vegetative en culture *in vitro* du palmier-dattier (*Phoenix dactylifera L.)*. C R Seanc Acad Agric Fr 11:1151–1154

Purseglove JW (1972) Tropical crops: monocotyledonous 2. Wiley, New York

Reynolds JF, Murashige T (1979) Asexual embryogenesis in callus cultures of palms. In Vitro 15:383–387

Ritala M, Monnonen L, Aspegren K et al (1993) Stable transformation of barley tissue culture by particle bombardment. Plant Cell Rep 12:435–440

Roy M, Jain RK, Rohila JS, Wu R (2000) Production of agronomically superior transgenic rice plants using *Agrobacterium*-transformation methods: present status and future perspectives. Curr Sci 79:954–960

Saker M, Moursy H (2003) Transgenic date palm: a new era in date palm biotechnology. In: Proceeding of the international conference on date palm, King Saud University, Qaseem, 16–19 Sept

Saker M, Ghareeb HA (2007) Factors influencing transient expression of *Agrobacterium*–mediated transformation of GUS gene in embryogenic callus of date palm. Fourth symposium on date palm in Saudi Arabia. King Faisal Univ, Al-Hassa, 5–8 May 2007

Saker M, Bekheet S, Taha H et al (2000) Detection of somaclonal variations in tissue culture-derived date palm plants using isozyme analysis and RAPD fingerprints. Biol Plant 43:347–351

Saker M, Adawy S, Mohamed A et al (2006a) Monitoring of cultivar identity in tissue culture-derived date palms using RAPD and AFLP analysis. Biol Plant 50:198–204

Saker M, Allam MA, Abd EL-Zaher H et al. (2006b)RAPD analysis of semi-dry Egyptian date palm during somatic embryogenesis. First Egyptian-Jordanian conference on biotechnology, 11–14 Dec 2006, pp 92–103

Saker M, Allam MA, Goma AH et al (2007) Optimization of some factors affecting genetic transformation of semi-dry Egyptian date palm cultivar (Sewi) using particle bombardment. J Genet Eng Biotechnol 5:1–6

Saker M, Ghareeb H, Kumlehn J (2009) Factors influencing transient expression of Agrobacterium-mediated transformation of GUS gene in embryogenic callus of date palm. Adv Hortic Sci 23:150–157

Salama HSM, Saker M (2002) DNA fingerprints of three different forms of the red palm weevil collected from Egyptian date palm orchards. Arch Phytopathol Plant Prot 35:99–306

Sanford J, Smith FD, Russel A (1993) Optimizing the biolistic process for different biological applications. Method Enzymol 217:483–509

Schroeder E, Gollasch S, Moore A et al (1995) Bean [alpha]-Amylase inhibitor confers resistance to the pea weevil *(Bruchus pisorum)* in transgenic peas (*Pisum sativum)*. Plant Physiol 107:1233–1239

Shaheen MA (1990) Propagation of date palm through tissue culture: a review and an interpretation. Ann Agric Sci Ain Shams Univ Cairo 35:895–909

Sharma HC, Sharma KK, Seetharama N, Ortiz R (2000) Prospects for using transgenic resistance to insects in crop improvement. Electr J Biotechnol 3. http://www.ejbiotechnology.info/content/vol3/issue2/full/3/

Sharma HC, Crouch JH, Sharma KK et al (2002) Applications of biotechnology for crop improvement: prospects and constraints. Plant Sci 63:381–395. http://dx.doi.org/10.1016/S0168-9452(02)00133-4

Somers DA, Rines HW, Gu W et al (1992) Fertile, transgenic oat plants. Bio/Tech 10:1589–1594

Somers DA, Samac DA, Olhoft PM (2003) Recent advances in legume transformation. Plant Physiol 131:892–899

Songstad DD, Somers DA, Griesbach RJ (1995) Advances in alternative DNA delivery techniques. Plant Cell Tissue Organ Cult 40:1–15

Tackholm V, Drar M 1973 (1950) Flora of Egypt, vol II, Otto Koeltz Antiquariat, Reprint

Tingay S, McElroy D, Kalla R et al (1997) *Agrobacterium tumefaciens*-mediated barley transformation. Plant J 11:1369–1376

Tisserat B (1979) Propagation of date palm (*Phoenix dactylifera* L.) *in vitro*. J Exp Bot 30:1275–1283

Tisserat B (1982) Factors involved in the production of plantlets from date palm callus cultures. Euphytica 31:201–214

Toutain G (1967) Le palmier dattier: culturer et production. Alawam 15:37–45
Vain P, De Buyser J, Bui Trang V et al (1995) Foreign gene delivery into monocotyledonous species. Biotechnol Adv 13:653–671
Valencia A, Bustillo AE, Ossa GE, Chrispeels MJ (2000) Alpha-amylases of the coffee berry borer (*Hypothenemus hampei*) and their inhibition by two plant amylase inhibitors. Insect Biotechnol Mol Biol 30:207–213
Vasil V, Castillo AM, Fromm ME, Vasil IK (1992) Herbicide resistant fertile transgenic wheat plants obtained by microprojectile bombardment of regenerable embryogenic callus. Bio/Tech 10:667–674
Wan Y, Lemaux PG (1994) Generation of large numbers of independently transformed fertile barley plants. Plant Physiol 104:37–48
Zaid A, Hughes H (1995) Water loss and polyethylene glycol-mediated acclimatization of *in vitro* grown seedlings of 5 cultivars of date palm (*Phoenix dactylifera* L.) plantlets. Plant Cell Rep 14:385–388
Zaid A, Tisserat B (1983) *In vitro* shoot tip differentiation in *Phoenix dactylifera* L. Date Palm J 2:163–182
Zaid A, Arias-Jiménez E (2002) Date palm cultivation. Rev 1. Plant production and protection paper 156, FAO, Rome
Zhu TD, Peterson DJ, Tagliani L et al (1999) Targeted manipulation of maize genes *in vivo* using chimeric RNA/DNA oligonucleotides. Proc Nat Acad Sci USA 96:8768–8773

Part V
Metabolites and Industrial Biotechnology

Chapter 31
Secondary Metabolites of Date Palm

A. El Hadrami, F. Daayf, and I. El Hadrami

Abstract Higher plants such as date palm accumulate a wide range of different chemicals in their tissue. These can conveniently be divided into two types of compounds: products of primary plant metabolism such as proteins, fats and carbohydrates, which guarantee the primary functions of growth and development; and products of secondary metabolism such as phenolics. Phenolic compounds are widely distributed in the plant kingdom. These organic compounds are not directly involved in primary metabolic processes of growth and development but are important constituents of resistance/tolerance to stress. These include lignins and other phenolics, which strengthen mechanically the cell walls while tannins, flavonoids, and some simple phenolics serve as defenses against herbivores, pathogens, ultraviolet radiation and other abiotic stress. In addition, some phenolics have allelopathic activity and may adversely influence the growth of neighboring plants. Other compounds such phytosterols, and alpha-tocopherols are also important in many growth and development processes. All these metabolites have been attributed to have nutritional and health-benefit value in the human diet. This chapter provides an overview on the work carried out on secondary metabolites in date palm, with special emphasis on the metabolic pathways involved, the regulation, potential functions, and health and nutritional benefits.

A. El Hadrami (✉)
OMEX Agriculture Inc. Canada, P.O. Box 301, 290 Agri Park Road, Oak Bluff, Manitoba, R0G 1N0, Canada
e-mail: abdele@omexcanada.com

F. Daayf
Department of Plant Science, University of Manitoba, 222 Agriculture Building, Winnipeg MB R3T 2 N2, Canada

I. El Hadrami
Department of Biology, Faculty of Sciences Semlalia, Laboratory of Biotechnologies, Cadi Ayyad University, B.P. 2390, 40 000, Marrakech, Morocco

S.M. Jain et al. (eds.), *Date Palm Biotechnology*,
DOI 10.1007/978-94-007-1318-5_31, © Springer Science+Business Media B.V. 2011

Keywords Alpha-tocopherol • Beta-sitosterol • Health benefits • Nutrition • Pathways • Phenolics • *Phoenix dactylifera* L. • Phytosterols • Secondary metabolites

31.1 Introduction

Date palm secondary metabolites comprise a number of phytochemicals that do not directly contribute to the plant growth and development, neither to photosynthesis, nutrient uptake, reproduction, nor to other functions ensured by primary metabolism (carbohydrates, proteins and fats). Secondary metabolites are very diverse and have served in the past as taxonomic characters for botanical classification of the species, and in studying cultivar diversity (Harborne and Williams 2000). These metabolites are often classified based on their chemical composition (carbon, hydrogen, oxygen and nitrogen: CHON), structure (i.e., benzene ring), solubility in water or other organic solvents, and most importantly on the basis of the pathway that leads to their biosynthesis i.e. phenylpropanoid, terpenes and saponins (Daayf et al. 2006; El Hadrami and Daayf 2002; Harborne and Williams 2000; Macheix et al. 1990).

Most secondary metabolites are highly bioactive and often are synthesized *de novo* in response to a given stress situation or freed from a non-active storage form that involves glycosylation, methylation or other conjugation forms (El Hadrami et al. 1997; Ramos et al. 1997). Upon synthesis and/or accumulation of these compounds they are thought to play a number of protective and antioxidant roles (J'Aiti et al. 2009). They also have been assigned numerous health-benefit functions of date palm fruit (Biglari et al. 2008).

Given the hostile environment under which date palm cultivation thrives, it is clear that secondary metabolites play an important role in conferring a protective effect against UV-damage, and tolerance to drought and salinity (El Hadrami and El Hadrami 2009). For instance, pigments such as the terpenoid carotenes, phenolics and particularly flavonoids are known for their protective effect against oxidative stress triggered by UV. These secondary metabolites have also been proven to be deleterious to plant pathogens and pests. These encompass direct toxicity, transient or permanent inhibition of the growth and development of microbes, and many repulsive properties against herbivores and pests.

Some of the accumulated secondary metabolites have the ability, when taken up or ingested by pathogens and pests, to cause serious alterations to the growth and development of these organisms, including blockage of ion channels, inhibition of enzymatic activity, cell division, motion coordination and many other vital processes. Tannins, for example, are known for their anti-nutritional properties towards insects and pests because of their ability to chelate micronutrients and essential enzymatic co-factors i.e. iron, leading to protein precipitation (Haslam 1989). On the other hand, terpenoid carotenes, some terpenes and phenolics, especially flavonoids, color the plant flowers and attract pollinators. Some of the flavonoids are even an essential component during the fertilization process (Buchanan et al. 2000).

Recent developments have shown that these compounds are able to transit through the vascular system of the plant (Buer et al. 2007), ensuring either a chemical signaling or the protection of sites remote from the infected/attacked location, preventing the spread of infection. They also play a role in scavenging harmful free radicals, hence reducing damage to the invaded tissues (Andersen and Markham 2006; Treutter 2006). Secondary metabolites contribute to phytohormonal homeostasis, especially those involved in triggering defense-related responses such as salicylic, jasmonic and abscissic acids (El Hadrami et al. 2009, 2011; J'Aiti et al. 2009).

Secondary metabolites also have an important ecological function, providing the date palm with a competitive advantage for water and nutrient resources over other cohabitant species (Treutter 2006). Upon biosynthesis and accumulation, secondary metabolites play a major role in the allelopathic effects that either reduce growth and development of neighboring species or restrict their accessibility to the resources. Flavonoids, for instance, are known for their ability to regulate root development and for chelating essential micronutrients (Andersen and Markham 2006; Daayf and Lattanzio 2008; El Hadrami and Daayf 2002)

In recent years, secondary metabolites of date palm have received special attention given their health-benefit claims and potential use in the booming industries of functional foods and nutraceuticals (Biglari et al. 2008). Phenolics, for example, are thought to exhibit an array of health benefits as antioxidants or anti-aging products, in preventing cancer, lowering cholesterol levels, and in stimulating the cardiovascular and immune systems (Andersen and Markham 2006; Macheix et al. 1990; Treutter 2006). A number of studies are under way to unveil more properties and in establishing procedures and protocols to economically and efficiently incorporate these date-derived products in the diet.

31.2 Date Palm Secondary Metabolism Pathways

Date palm shares many commonalities with other monocots and with dicot species in terms of the pathways that lead to the synthesis and accumulation of secondary metabolites. Simplistically, three main pathways can be distinguished in this plant. The terpenes, composed almost entirely of carbon and hydrogen (CH) (Fig. 31.1a), are secondary metabolites that have mevalonic acid as the main precursor. Terpenes present a high diversity in their chemical structure starting from the simple one (Fig. 31.1a). Phenolics are derived from the shikimic acid pathway (Herrmann 1995; Fig. 31.1b) with simple erythrose and pentose phosphate and aromatic amino acid i.e. phenylalanine or tyrosine as precursors. This category of compounds contains at least one benzene ring and is formed mainly of hydrogen and oxygen atoms (CHO).

31.2.1 Pathway Architecture and Biosynthesis

The majority if not all phenylpropanoids derive from the *t*-cinnamic acid (Dixon and Paiva 1995). The various classes of phenylpropanoids are interconnected; some of the connections are in common with several monocot and dicot species.

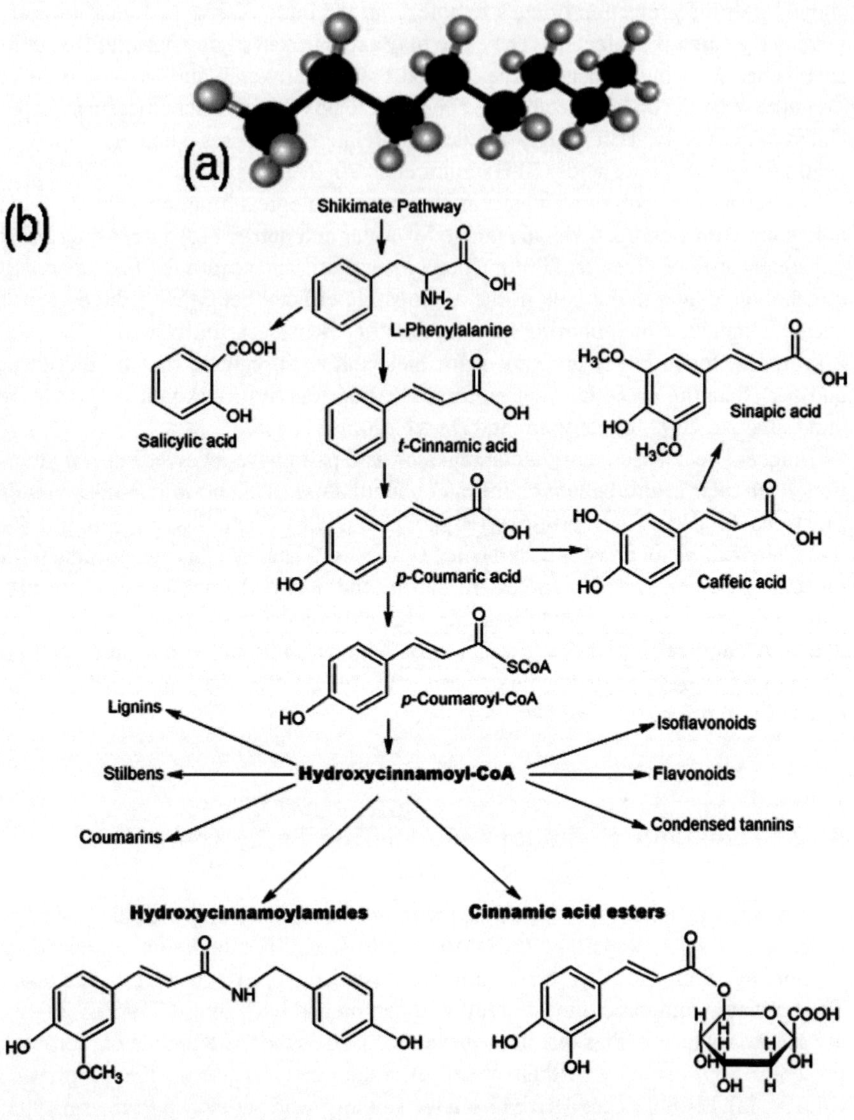

Fig. 31.1 Chemical structure of terpenes (a) and a simplified schematic for the metabolic pathway leading to phenolics (b)

The basic form of phenylpropanoids abundant in date palm is represented by C_6-C_3 molecules derived from cinnamate through hydroxylation, methylation or dehydration reactions. These include simple hydroxycinnamates such as *p*-coumarates, caffeiates, ferulates and sinapates.

More elaborated forms of phenylpropanoids are derived from the flavonol skeleton (C_6-C_3-C_6) involving enzymes such as the chalcone synthase, which catalyzes the condensation *p*-coumaroyl-CoA with three molecules of malonyl-CoA (Harborne and Williams 2000). The product of this condensation leads to the biosynthesis of other flavonoids including flavones, flavanones, flavanols, anthocyanins and proanthocyanidins (condensed tannins). Along the way, structural modifications such as hydroxylation, glycosylation, methylation, acylation, enylation, or sulfation may occur on all these phenylpropanoid nuclei.

31.2.2 Nature of the Compounds and Chemodiversity

In the date palm, three main families of phenylpropanoids can be detected in different tissue at various stages of growth and development, and in response to diverse stress. These include hydroxycinnamates, flavonols, flavan-3-ols, flavan-3,4-diols and proanthocyanidins.

31.2.2.1 Hydroxycinnamates

Hydroxycinnamates represent the first compounds found in the phenylpropanoid pathway (Ralph 2010). They result from hydroxylation, methylation or dehydration of the *t*-cinnamic acid. These compounds are very abundant in date palm tissue and contribute to a series of processes related to resistance/tolerance to biotic or abiotic stress. The major hydroxycinnamates found are derivatives of *p*-coumaric, caffeic, ferulic and sinapic acids.

The simple hydroxycinnamic acids are rarely free in the tissue but can be detected conjugated to other molecules, sugars and cell wall carbohydrates. For instance, date palm roots were shown to contain a high level of caffeoylshikimic acids (Fig. 31.2), as well as esters of sinapates and ferulates (El Hadrami 1995; El Hadrami et al. 1998; El Hassni et al. 2004; J'Aiti et al. 2009).

Hydroxycinnamates can also be found as conjugates with various amines in response to abiotic or biotic stress such as defense against bayoud. The phenolamide derivative of sinapic acid (I2) was found to play a major role in date-palm resistance to *Fusarium oxysporum* f. sp. *albedinis* (Foa) (Daayf et al. 2003; El Hadrami et al. 1997; Ramos et al. 1997).

31.2.2.2 Flavonoids

Flavonoids (Fig. 31.3) are a large group of secondary metabolites that have been subdivided into 12 subclasses, which include flavonols, chalcones, flavones, flavanones, flavanols, flavandiols, proanthocyanidins, aurones, anthocyanidins,

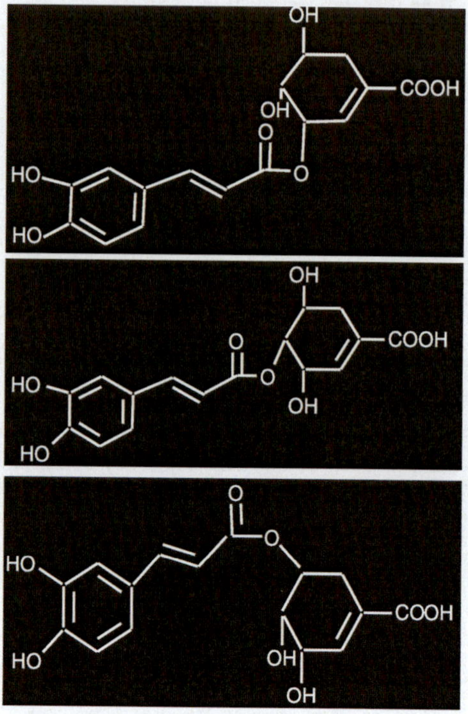

Fig. 31.2 Chemical structure of the isomers (3, 4 and 5) of caffeoylshikimic acid, the most abundant of the phenolics in date palm tissue

dihydroflavonols, dihydrochalcones, and iso- and neo-flavones (Andersen and Markham 2006; Treutter 2006). This subdivision is based on the oxidation state and the substitution residues found at the C_2-C_3 positions. All of these classes are not yet well studied in date palm and some of them are rather tissue-specific or differentially-abundant in one cultivar as compared to others.

Flavonoids biosynthesis in date palm, as in many other plant species, can be either constitutive or induced (Dixon and Paiva 1995; Macheix et al. 1990). Constitutive biosynthesis is under the control of many processes of growth and development such as leaf formation and the transition from the vegetative to the flowering stage (Andersen and Markham 2006; Heller and Forkman 1994; Treutter 2006). However, induced biosynthesis depends on a number of stimuli i.e. UV, heat, salinity, drought and injuries including pathogen attacks (Dixon and Paiva 1995). Most of the synthesized and accumulated flavonoids are stored in a conjugated form i.e., glycoside that is not or is less bioactive. The conjugated forms also allow long-distance distribution of these molecules from the site where they are synthesized to the sites where they need to be active (Buer and Djordjevic 2009; Buer et al. 2007). This property allows these molecules to contribute to local and systemic resistance/tolerance to stress (Arfaoui et al. 2007). The non-conjugated forms (aglycones) are biologically the most active and usually cleaved from their sugar or other conjugate moieties upon detection of stress conditions.

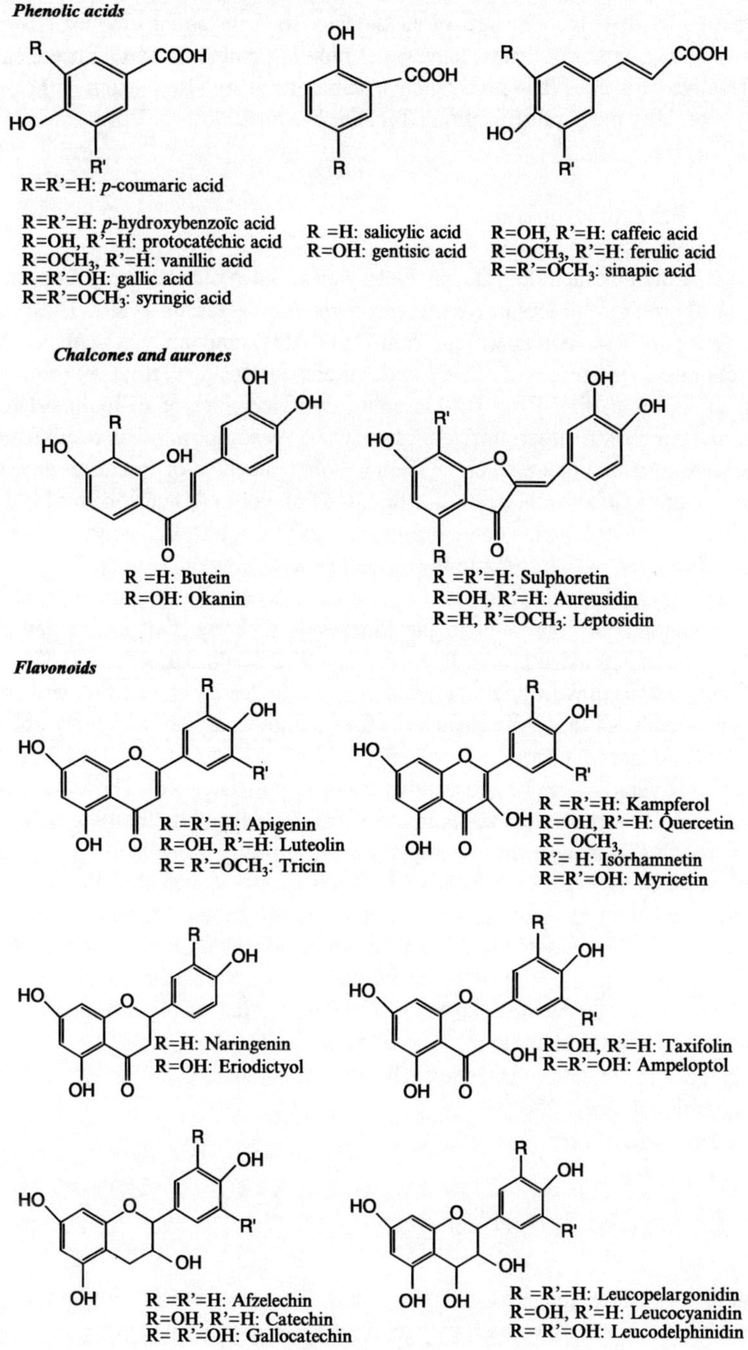

Fig. 31.3 Chemical structures of important phenolic compounds

Flavonoids in date palm are often detected in substantial amounts in major sampled tissue including fruits, leaves, germinating embryos, and embryogenic or proliferative calli. Free flavonoids such as kaempferol are also present in the pollen grain, especially during fertilization (Buchanan et al. 2000).

31.2.2.3 Proanthocyanidins

Condensed tannins including flavan-3-ols, flavan-3,4-diols and proanthocyanidins (Fig. 31.3) are more or less polymerized polyphenols (Haslam 1989). These molecules have a high molecular weight (500–3,000 kDa), and are water soluble. Their basic chemical structure is C_6-C_3-C_6 with an orthodiphenolic (Ring A) (phloroglucinol) to which another Ring B, that could be mono-, di-, or tri-hydroxylated, is attached through a tricarbon bridge. Polymerized proanthocyanidins result from the condensation of monomers such as flavan-3-ols (catechins or initiation units) and flavan-3,4-diols (leucoanthocyanidins or extension units) (Andersen and Markham 2006; Haslam 1989). Such a condensation occurs through C_4-C_8 bounds, and exceptionally through C_4-C_6, leading to the molecule branching.

Proanthocyanidins biosynthesis is a complex process. The major steps leading to these compounds are common to the flavonoids pathway. Out of the flavonoids C_6-C_3-C_6 nucleus, chalcone and flavonone synthases and isomerases (CHS, CHI, FS, FI), as well as dihydroflavonol reductases (DFR) lead to the synthesis of flavan-3-ols and flavan-3,4-diols, the basic units for proanthocyanidins and other polymerized and condensed tannins.

Proanthocyanidins can be constitutive or induced in date palm. These compounds exhibit a low direct toxicity and their action spectra depend on their concentrations (Haslam 1989). However, these compounds have *o*-dihydroxyphenyl groups capable of establishing hydroxylic bonds with ionic metals, hence chelating co-factors essential for pathogen enzymes for example (Andersen and Markham 2006). The *o*-diphenols of compounds also, in the presence of oxidases such as polyphenols oxidases and peroxidases, can be converted to highly-reactive *o*-quinones. The action of various tansferases can lead to the interaction of these *o*-quinones with nucleophilic radicals such as amino- (–NH_2) or thiols (–SH), preventing the spread of pathogens in the case of infection (Daayf and Lattanzio 2008; Daayf et al. 2006; El Hadrami and Daayf 2002).

31.2.3 Regulatory Mechanisms

Secondary metabolism, in general, and phenylpropanoid biosynthesis and accumulation, in particular, are highly regulated. The content of the constitutive compounds, as well as the balance between free- and conjugated forms, are kept under transcriptional, post-transcriptional and translational control. Conjugated forms are often cleaved upon reception of a signal freeing aglycones that are highly reactive.

For instance, during the interaction of date palm with *Fusarium oxysporum* f. sp. *albedinis*, the causal agent of bayoud disease, an up-regulation of the PAL as well as other downstream enzymes involved in phenylpropanoid pathways is often recorded (Daayf et al. 2003; El Hadrami et al. 1997). These lead to an increase in the biosynthesis and/or accumulation of phenolics such as hydroxycinnamates (El Hadrami et al. 1997; Ramos et al. 1997). These compounds are, at least in part, quickly used by oxidases such as polyphenol oxidase and peroxidases, which contribute to the strengthening of the cell wall and the restriction of pathogen spread (El Hassni et al. 2007; J'Aiti et al. 2009).

Induced compounds, on the other hand, are synthesized *de novo* and accumulate at a precise time in specific tissue in response to stimuli such as infection. For example, an accumulation of amides conjugates of sinapic acid were shown to play a major role in date palm defenses against bayoud (El Hadrami et al. 1997, 1998; El Hassni et al. 2007; J'Aiti et al. 2009; Ramos et al. 1997). Some stress conditions as well as micronutrients (i.e. boron levels) have been shown to play an important role in regulating the nature and content of the accumulated phenolics in dates (Al-Yousif et al. 1994).

For a number of years, investigations of date palm secondary metabolites relied on the use of time- and resource-consuming methods that led to the identification of several compounds, and families of compounds, such as phenolics, thought to play an important role in plant defense and/or tolerance to stress (El Hadrami 1995; El Hadrami et al. 1998; Ramos et al. 1997). To date the metabolic pathways have been identified mainly based on similarities with other plant species. The recent availability of the sequence of the entire genome (Al-Dous et al. 2009) will help identify the genes and enzymes involved and draw the particularities of the species. This will also open the door to understanding the interplays between pathways and to reveal the regulatory mechanisms.

31.2.4 Phenolic Content in Date Palm Tissue

The soluble phenolic contents in date palm are more or less higher dependant on tissue, cultivar, the growth and development stages and conditions of culture. The amounts can be rich and very high concentrations during the initiation of calli destined for somatic embryogenesis or organogenesis. The phenolic accumulation leads to tissue browning that constitutes one among other limiting factors of date palm micropropagation. In the roots and leaves, the contents are very different depending on the cultivar but, in general, roots accumulate less than leaves.

Date fruits were also found to be rich in phenolics (217–343 mg of ferulic acid equivalents/100 g). Date palm cultivars exhibit distinct levels and profiles of phenolic acids such as gallic, protocatechuic, *p*-hydroxybenzoic, vanillic, caffeic, syringic, *p*-coumaric, ferulic and *o*-coumaric acid. Other studies have reported an abundance of flavonoids (Biglari et al. 2008).

Dates represent a high energy source besides being rich in dietary fibers (Biglari et al. 2008; El Hadrami et al. 2010). Their carbohydrates content range is 44–88%, while proteins represent 2.3–5.6% and fats 0.2–0.5%. Dates are also a good source of essential minerals (916 mg per 100 g FW) and vitamins (C, B1, B2, A and riboflavin). The expectation of using dates as a functional food and in nutraceuticals will have to overcome the price of this fruit. Dates are high in price and among the most expensive fruits worldwide (about 1–20 Euros/kg depending on the cultivar).

31.2.5 Biological Activity and Roles

Phenolic compounds, particularly flavonoids, are among the most potent plant antioxidants (Andersen and Markham 2006; Daayf and Lattanzio 2008; El Hadrami and Daayf 2002; Macheix et al. 1990). They can form complexes with reactive metals such as iron, zinc and copper, reducing their absorption (Daayf and Lattanzio 2008). At first glance, this may seem to be a negative side effect (reducing nutrient absorption), but excess levels of such elements (metal cations) can promote the generation of free radicals and contribute to repairing the oxidative damage of cell membranes and cellular DNA. In addition, this contributes in preventing pathogens from accessing this micronutrient and achieving essential physiological and metabolic processes. Phenolic compounds also function as potent free radical scavengers, where they can neutralize free radicals before they can cause cellular damage (Andersen and Markham 2006; Daayf and Lattanzio 2008; El Hadrami and Daayf 2002).

Phenolics are also known as an important component during the establishment of date palm calli. A high production of these compounds combined with the activity of various oxidases often lead to browning (Fig. 31.4), which requires the use of activated charcoal to minimize its impact on the growth and development process. The excess of production of these compounds has also been detrimental to producing protoplasts or culturing anthers (El Hadrami and El Hadrami 2009).

In date palm, explants derived from young offshoots turn brown and produce abundant exudates that stain the culture medium a few days after culture (Fig. 31.5). Analysis of phenolic extracts showed that explants contained a high level of caffeoylshikimic acids ranging from 190 to 430 $\mu g \cdot g^{-1}$ FW, depending on the cultivar (El Hadrami 1995). It has been shown that these dominant hydroxycinnamic acids decreased largely during callogenesis induction in several tested cultivars (El Hadrami 1995).

To limit tissues browning that cause an appreciable loss of culture viability, activated charcoal and polyvinylpyrrolidone are commonly used in palm tissue culture to trap phenols and oxidized phenols. Activated charcoal used only at 150 $mg \cdot l^{-1}$ improved growth rate (callogenesis) and decreased tissue and medium browning evaluated after 40 and 80 days of culture. Polyvinylpyrrolidone incorporated into culture medium at 2 $g \cdot l^{-1}$ enhanced the percentage of explant reactivity. In addition, this treatment had no great impact on the growth rate but facilitated the trapping of phenols exuded from calli and enhanced medium browning. During the first months

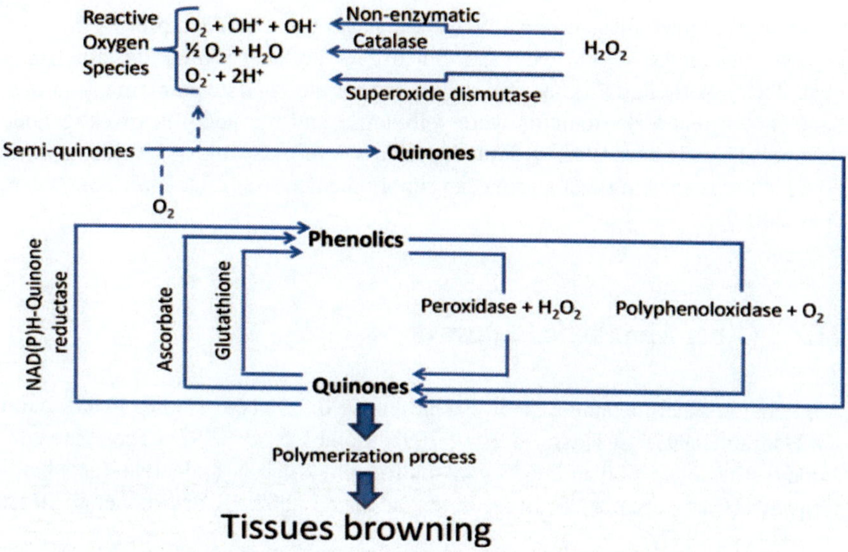

Fig. 31.4 Schematic representation of the mechanisms involved in tissue browning

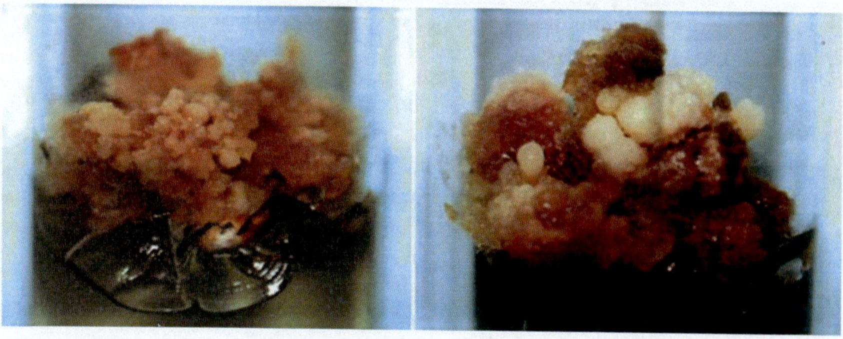

Fig. 31.5 Callogenesis in date palm showing browning that can be limited with activated charcoal

of culture, tissue browning was higher on control and polyvinylpyrrolidone media as compared with charcoal medium (El Hadrami 1995).

In addition, El Hadrami (1995) has shown that, depending on the culture medium, the phenolic accumulation was qualitatively different during callogenesis induction. On charcoal, calli accumulated appreciable amounts of monomeric and oligomeric flavanols (proanthocyanidins) as well as caffeoylshikimic acids (El Hadrami 1995). Calli from control and polyvinylpyrrolidone-treated media accumulated other phenolics represented by sinapic, ferulic and flavonoids derivatives (El Hadrami 1995). A lower amount of caffeoylshikimic acids was accumulated in these two treatments in comparison with charcoal (El Hadrami 1995).

Moreover, phenolics, especially certain flavonols (i.e. kaempferol), are key components of the stigma exudates that trigger pollen germination (Buchanan et al. 2000). Although this has still to be demonstrated in date palm, many studies have shown that CHS mutants were self-sterile and the addition of exogenous flavonols helps restore fertility. Mutant pollen was also capable of fertilizing wild-type plants, suggesting that stigma can supply the necessary flavonols (Buchanan et al. 2000).

31.3 Other Metabolic Pathways

A number of other metabolic pathways have more or less been studied in date palm (El Hadrami 1995; El Hadrami et al. 1998; Ramos et al. 1997). These include, among others, the polyamines/phenolamides, phytosterols and alpha-tocopherol pathways (Daayf et al. 2003; El Hadrami and Daayf 2002; El Hadrami et al. 1998; Ramos et al. 1997)

31.3.1 Chemistry and Content

31.3.1.1 Polyamines

Polyamines are a group of metabolites that play a number of important roles in regulating various processes of cell proliferation and differentiation at the cell level; and also growth and development at the plant level. There are three ubiquitous polyamines: putrescine, spermidine and spermine (Fig. 31.6). Evidence also has been showing that this category of metabolites is involved in programmed cell death triggered by injuries or pathogen attacks. Among these putrescine, spermidine and spermine, and their conjugated forms (mainly phenolamides), represent the most commonly detected compounds. In date palm, polyamines are often detected in calli and other *in vitro* cultured tissue (El Hadrami 1995). They seem to play a role in promoting cell proliferation and preventing browning and cell death. The latter comes from the fact that programmed cell death includes the activation of endogenous calcium- and magnesium-dependent endonucleases, leading to fragmentation of the chromosomal DNA. Tissue rich in polyamines does not exhibit such a phenomenon. However, tissue having cells with low polyamine content tend to have a destabilized DNA, exposed to nucleases. Polyamines also have been detected as a key component when date palm seedlings were exposed to salt and drought stresses.

The contents of polyamines in date palm are tissue-dependent and regulated by the stage and situation of growth and development. Variations among cultivars also have been detected (El Hadrami 1995).

(1) $NH_2(CH_2)_4NH_2$ (2) $NH_2(CH_2)_3NH(CH_2)_4NH_2$ (3) $NH_2(CH_2)_3NH(CH_2)_4NH(CH_2)_3NH_2$

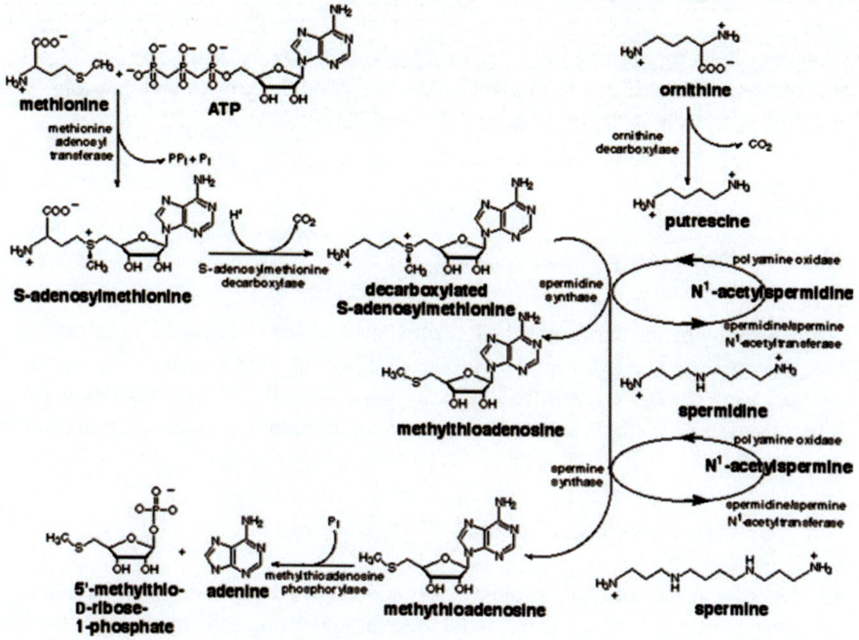

Fig. 31.6 Linear formulas of ubiquitous polyamines: (*1*) putrescine; (*2*) spermidine; (*3*) spermine and the pathway steps that lead to their biosynthesis

31.3.1.2 Phytosterols

Phytosterols are metabolites structurally related to steroids, which are widely distributed in animals and plants. Steroids belong to a large group of compounds known as terpenoids or isoprenoids. Terpenes are formed by the polymerization of isoprene units, and steroids are triterpenes or triterpenoids. The steroid family includes compounds such as the phytosterols and vitamin D.

Analyses of date palm cultivars for the presence of phytosterols indicate the abundance of these secondary metabolites in shoot tips and pollen grains (Bennett et al. 1996; Mahran et al. 1976; Mossa et al. 1986; Zaki et al. 1993). Calli and somatic embryos also were shown to contain a substantial amount of phytosterols. The lowest contents were detected in germinating embryos and leaves of cv. Zaghloul, as well as in the leaves of cv. Sewi. Overall, cv. Zaghloul was richer in phytosterols than Sewi regardless of the tissue tested. Among the tissue tested for phytosterols, pollen was the richest and roots the poorest.

Analysis using thin layer chromatography revealed a number of phytosterols including cholesterol, beta-sitosterol and stigmasterol, especially in the pollen grains (Bennett et al. 1996; Mahran et al. 1976; Mossa et al. 1986; Zaki et al. 1993).

31.3.1.3 Alpha-Tocopherols

In dates, alpha-tocopherol contents were reported to vary from 25% to 38.9%, according to the cultivar. The content depends also on the growth and development conditions, as well as the stage of fruit development.

31.3.2 Bioactivity

The bioactivity of secondary metabolites often depends on their nature and reactivity, biosynthesis triggers and the amount produced. Some of these metabolites have wide spectra of effects and are involved in many physiological and metabolic processes. Others have specific effects and are particularly produced in response to given stimuli.

31.3.2.1 Polyamines/Phenolamides

As in many other plant species, the exact functions of polyamines/phenolamides in date palm have not yet been identified. However, these metabolites are required for a number of cellular processes such as replication, transcription and translation. Their cationic nature and flexible charges make them exert specific interactions with other cell constituents, ions and molecules.

Polyamines/phenolamides intracellular contents are highly regulated, and a balance is maintained between their biosynthesis and degradation. For instance, in the pathway of spermidine and spermine, two enzymes, spermidine synthase and spermine synthase, are constitutively expressed while two other up-stream ones are highly regulated with a short turnover i.e. ornithine decarboxylase and *S*-adenosylmethionine decarboxylase.

The biosynthesis of polyamines/phenolamides is stimulated by a number of endo- and exogenous stimuli. Polyamine deficiency may lead to a reduction in cell proliferation. This importance of polyamines to the process of cell proliferation has been demonstrated by external supplementation to polyamine-deficient tissues in growth. Their catabolism, on the other hand, is regulated by the activity of enzymes such as the spermidine/spermine *N*-acetyl transferases which are tightly-regulated by the constitutively expressed polyamine oxidase. Other regulatory mechanisms include transport systems for remote utilization.

31.3.2.2 Phytosterols

Phytosterols are important in many physiological and metabolic processes, especially plant fertilization and germination of pollen grains (Buchanan et al. 2000).

31.3.2.3 Alpha-Tocopherols

Alpha-tocopherol is a low molecular weight lipophilic secondary metabolite known for its antioxidant abilities. It helps protect cell membranes from oxidative damage. For instance, exogenous application of antioxidant substances such as alpha-tocopherol, enhanced the ability of date palm seedlings to tolerate environmental stress and improved growth performance (Awad et al. 2005).

31.4 Secondary Metabolites in Dates

Phenolics represent one of the most important biochemical markers of date fruit development and ripening. The nature and content of phenolics increases from the transformation of the ovary walls until the later stages of ripening (tamar stage) (Fayadh and Al-Showiman 1990). This is due to the hormonal homeostasis that occurs in the fruit during growth and development. Phenolics may directly or indirectly affect fruit ripening as they are closely involved in the biosynthesis and regulation of the implicated hormones. *O*-Diphenols such as quercetin, rutin and caffeic acid derivatives exert inhibiting activities towards IAA-oxidase while monophenols such as vanillic, *p*-hydroxybenzoic and *p*-coumaric acids are known to be stimulators of the activity of this enzyme. On the other hand, chlorogenic acid, an *o*-diphenols, is proven to be an auxin protector. The ratio of mono-/*o*-di-phenols seems to play an important role in date-fruit growth and development, as it is the case in many other fruits (Macheix et al. 1990).

Phenolics are also among the main constituents that add taste and nutritional values to dates. For instance; color, astringency and the flavor of the dates are closely related to the nature and content of phenolics in the fruit tissue. Organoleptic properties of the fruit are highly dependent on enzymatic and/or non-enzymatic browning, which completes and defines the color of ripe dates. Quinones, as oxidation by-products of phenolics through the activity of polyphenol oxidases and peroxidases, are highly reactive molecules that form various complexes with carbohydrates and proteins, making some essential changes in the texture and flavor of the fruit. Browning in dates is a physiological process that accompanies date maturation. It is related to several abiotic and biotic stresses of the environment where date palms grow. It is also affected by the treatment and technological processes that follow fruit harvest, drying, packaging and transformation.

Date fruit astringency, especially at the kimri stage (Fayadh and Al-Showiman 1990), is very high and is ascribed to a high content of flavan-3-ols and caffeoylshikimic acids. Diffusion of other phenolics from the skin into the pulp during fruit maturation also is suspected to affect taste and flavor. This has yet to be determined in dates but is well known in other drupes such as grapes and olives.

31.4.1 Nutritional Value

Secondary metabolites such as phenolics, phytosterols and alpha-tocopherol are important in the functional properties of products extracted from date palm fruit. They contribute to the resistance to oxidation, taste and aroma, as well as color, hence preserving the health benefits of dates. Besbes et al. (2005) compared the phenolics, phytosterols and alpha-tocopherol contents and profiles in the oil extracted from two date palm cvs. Deglet Noor and Allig, and reported that phenolics were major components of resistance to oxidation. Their contents were more than twice as high in Deglet Noor (520.8 mg/kg) than in Allig (220.3 mg/kg). The authors attributed this property to hydroxytyrosol detected at 10.2% and 8.1% in Deglet Noor and Allig, respectively.

In the same study, Besbes et al. (2005) quantified date palm phytosterols to be 3,000 and 3,500 mg/kg for Allig and Deglet Noor, respectively. Beta-sitosterol was shown to be the predominant phytosterol, accounting for over 75% of the seed oil in both cultivars tested. As for alpha-tocopherol, its content varied from 25% to 38.9%.

31.4.2 Health Benefit Claims

Phenolics are well known for their powerful antioxidant properties and health benefits. Mansouri et al. (2005) investigated the content and nature of the phenolics of ripening dates produced by seven cultivars and assessed their antioxidant activity. Total phenolic content ranged from 2.5 to 8.4 mg gallic acid equivalents per 100 g of dates. The evaluation of the antioxidant activity revealed an antiradical efficiency of 0.08–0.22. A significant correlation ($R^2=0.975$) was established between phenolic contents and the antiradical efficiency among all cultivars tested. The main phenolics found in all cultivars were *p*-coumarates, ferulates and sinapates, as well as some derivatives of *t*-cinnamic acid and isomers of 5-*O*-caffeoylshikimate. Other flavonoids were also identified, including flavones, flavanones and flavonol glycosides.

In other studies, dates were found to be a high source of antioxidant constituents, total antioxidants (8,212–12,543 μmol of Trolox equivalents/g), and phenolics (217–343 mg of ferulic acid equivalents/100 g). Biglari et al. (2008) analyzed the antioxidant activities using three different methods, namely the Trolox equivalent antioxidant capacity (TEAC) method; 2,2′-azinobis (3-ethylbenzothiazoline-6-sulphonic acid) radical cation (ABTS·+) assays; and the ferric reducing/antioxidant power method (FRAP assay), as well as the total phenolic and flavonoid contents. The authors concluded that there is strong potential use of dates as an antioxidant functional food.

Other health benefits also are ascribed to phenolics, giving a added value to dates and increasing an interest in them as functional food and nutraceuticals. Al-Farsi et al. (2007) reported that the compositional and functional characteristics of date fruits, syrups, and their by-products, such as press cake and seed, make them a good source of natural antioxidants and could potentially be used directly as functional food or incorporated as an ingredient in functional food.

Phytosterols are lipid-like complexes of unsaturated alcohols. They are closely-related in terms of their structure to animal cholesterol but have an extra ethyl group on the side chain. Unlike cholesterol, phytosterols are beneficial to human health. They are thought to be involved in lowering the blood levels of low-density lipoprotein LDL, the so-called *bad* cholesterol. Phytosterols are also involved in blocking the absorption of dietary cholesterol into the bloodstream and in inhibiting the re-absorption of cholesterol from bile acids in the digestive process, thus reducing the amount of cholesterol returned to the bloodstream. Other health benefits may include positive effects for patients with autoimmune diseases. For instance, phytosterols were shown to reduce inflammation and suppress over-reactive immune systems in patients suffering from rheumatoid arthritis. In addition, the use of concentrated extracts of phytosterols has been shown to be effective against certain forms of cancer (i.e. prostate) and to enhance the insulin production of pancreatic cells.

Alpha-tocopherol (vitamin E) is often used orally to treat deficiencies, and in preventing cardiovascular diseases, diabetes and its complications, and benign prostatic hyperplasia. This vitamin is also administered against angina, thrombophlebitis, intermittent claudication, hypertension, and to prevent ischemia-reperfusion injury after coronary artery bypass surgery. Alpha-tocopherol is claimed to reduce the risks of various cancers, Alzheimer's, and Parkinson's diseases and other dementias. Vitamin E is also used against allergies, asthma and other respiratory problems, as well as digestive or circulatory diseases. Additionally, vitamin E is used topically against dermatitis, aging skin, granuloma annulare and in preventing skin ulceration often caused as a consequence of chemotherapeutic drugs use.

31.4.3 Role in Resistance/Tolerance to Stress

Several studies have shown the involvement of constitutive and induced hydroxycinnamates in the resistance of date palm to bayoud disease (Daayf et al. 2003; El Hadrami et al. 1997; El Hassni et al. 2004; J'Aiti et al. 2009; Ramos et al. 1997). A few of these compounds were revealed to exert a phytoalexin-like activity (El Hadrami et al. 1997; Ramos et al. 1997). These compounds were characterized as phenolamides (Ramos et al. 1997). Phenolamide derivatives, especially with a hydroxycinnoyl component, are known for their involvement with resistance/tolerance to biotic and abiotic stress (Edreva et al. 2007; El Hadrami and Daayf 2009). For instance, these compounds are believed to contribute to salt and drought tolerance. Calli treated with salt were shown to produce a high content of phenolamides, which helps prevent damage due to NaCl. These compounds are also believed to regulate the dedifferentiation process, the intracellular osmotic pressure between the cytoplasm and the vacuole, preserving cellular integrity. In addition, they exhibit a regulating role in cell proliferation and differentiation (El Hadrami and Daayf 2009).

Alpha-tocopherol is known for its antioxidant qualities and protective effect of membranes against oxidative damage. Exogenously treated date palm seedlings with alpha-tocopherol exhibited a certain level of tolerance to environmental stress while showing an improved growth and development (Awad et al. 2005).

31.5 Genomic Approaches

When the entire date palm genome sequence is publicly available, it will be possible to explore the gene expression on a larger scale including genomics. This will permit a global examination of spatio-temporal expression of genes and networks of effectors/genes in response to different signals. This technology was first developed for thale cress (*Arabidopsis thaliana* L.) and is being used for many other plants, especially those with an extensive genetic background i.e. alfalfa (*Medicago sativa)*, to determine the complexity of the organism's responses and signal transduction. Utilizing date palm cultivars with differential responses to bayoud, the genomics approach will provide novel clues to understanding the mechanism of resistance and hopefully provide new insights concerning the transfer of this character from resistant cultivars to susceptible, high-yield, commercial ones. Similar information will also be used to improve the nutritional value of dates.

31.6 Proteomic Approaches

The use of proteomics may allow characterizing the proteome or the potential set of proteins encoded for and by the genome (Wasinger et al. 1995). This includes protein localizations and their expression and interactions, depending on the physiological situations of the organism, cell or subcellular compartment. Several methods have been described recently concerning the protein purification and detection by mass spectrometry (Neubauer et al. 1997), protein localization (Ferrando et al. 2000, 2001; Farràs et al. 2001), and microarray techniques (Zhu et al. 2001). To the best of my knowledge, very little is known concerning proteomics in date palm. Although the potential role of phenolics, polyamines and other compounds such as ascorbic and jasmonic acids in date-palm resistance to bayoud fungus, the molecular mechanisms underlying their precise effects are not yet known. Determination of the transacting protein factors of these compounds may open a new and exciting area of research to study the function during the transcription process of genes. It is well known that many metabolic reactions are catalyzed by very complex sequentially-acting enzymes ordered in terms of structures and organization. These aspects should be studied in date palm.

A recent study in date palm using the proteomics approach established proteome maps (pH 5.0–8.0) of somatic embryos versus zygotic embryos, using 2-D PAGE combined with mass spectrometry analysis (Sghaier-Hammami et al. 2009a, b). The authors used embryos derived from cv. Deglet Noor and revealed qualitative and quantitative differences among the somatic and zygotic embryos. The study also revealed that the level of total proteins was higher in zygotic embryos as compared to their somatic counterparts. Differences were denoted between the two sets of embryos, especially in terms of enzymes controlling the glycolysis pathways. In addition, zygotic embryos showed an abundance of storage and stress-related proteins.

31.7 Metabolomics and Potential in Dates

Date palm research in metabolomics is not well documented, despite the fact that this discipline has been in development for over three decades. Most of the published data describe the diversity in terms of metabolites that can accumulate, in part, in response to several biotic and abiotic stresses. As compared to genomics, transcriptomics or proteomics, the goals of metabolomics are to determine the entire set of metabolites synthesized/accumulated by a plant, tissue or organ, when placed under particular conditions that allow the gathering of a clear view on its physiological status. This approach is of a great help in understanding gene function and regulation. It also allows one to determine and evaluate the impact of the treatment by chemical compounds or organic/biologic material, as well as to understand some of the mechanisms involved during the interactions of plants with pathogens. Unlike other well-genetically characterized and studied plant species, date palm is still lagging behind in terms of its genomics, proteomics and especially metabolomic aspects. Recent advances in genome sequencing (Al-Dous et al. 2009) will certainly help, in the years to come, to overcome previous challenges.

31.8 Transgenic Approaches

The powerful tools of modern genetics are now an essential part of our understanding of gene expression and function. It is currently conceivable, with some yet to overcome technical hurdles and public acceptance, to precisely map and manipulate genes that are implicated in specific growth and development processes or in responses to various stress conditions. This will improve and accelerate tremendously the breeding of traits of interest into high-yielding commercial cultivars. In each plant species, thousands of genes have tentatively been associated with resistance to stress conditions or ascribed to specific processes of growth and development. Depending on the species, these genes have been annotated and sub-cellulary localized under various environments. In date palm, this is still a slow process having had the entire genome sequence available only recently (Al-Dous et al. 2009). In addition, translating these molecular findings into an improved crop, ready for use by farmers, remains to be seen, especially in a crop like date palm that is still at its infancy in terms of molecular and the -omics era of genetics. Nonetheless, a number of modifications can be expected to improve the nutritional value of dates, disease and pest resistance, tolerance to several stresses and numerous other traits. For instance, an exciting prospect is the transformation of the male date palm to produce not only pollen but fruit as well.

31.9 Conclusion and Prospective

Over the past few decades, date palm secondary metabolism studies have led to the discovery of functional compounds and the identification of their metabolic biosynthesis pathways. The lack of genetic data has hampered the progress of these studies but the current release of the entirely sequenced genome will help identify some of the genes encoding various enzymes involved in the catalysis of the biosynthesis of these compounds, as well as the regulatory mechanisms. This, along with the great efforts been made in initiating, establishing and maintaining *in vitro* cultures for mass production, will soon lead to the production of useful secondary metabolites, and to metabolically engineer new derivatives. It will then be a matter of scaling-up *in vitro* multiplication operations to produce and purify these secondary metabolites in response to the demand of the flourishing industries of functional foods and nutraceuticals or other uses. In addition, the current developments of new -omic approaches, along with progress in terms of genetic transformation, hold the promise of a bright future for this economically and historically important crop.

Acknowledgments The work on date palm has been supported by several research grants to Professor Ismail El Hadrami (IFS/AUF/PRAD/CNRST/AI Morocco-Tunisia/AI Morocco-France/UCAM).

References

Al-Dous EK, George B, Salameh YM et al (2009) Qatar researchers sequence draft version of date palm genome. URL://qatar-weill.cornell.edu/research/datepalmGenome/download.html

Al-Farsi M, Alasalvar C, Al-Abid M et al (2007) Compositional and functional characteristics of dates, syrups, and their by-products. Food Chem 104:943–947

Al-Yousif FS, Al-Whaibi MH, El-Hirweris SO (1994) Influence of boron concentrations on some metabolites of date palm and sorghum seedlings. J Plant Nutr 17:1037–1052

Andersen ØM, Markham KR (eds.) (2006) Flavonoids: chemistry, biochemistry, and applications. Taylor & Francis, London

Arfaoui A, El Hadrami A, Mabrouk Y et al (2007) Treatment of chickpea with *Rhizobium* isolates enhances the expression of phenylpropanoid defense-related genes in response to infection by *Fusarium oxysporum* f. sp. *ciceris*. Plant Phys Biochem 45:470–479

Awad AM, Soaud AA, El-Konaissi MS et al (2005) Effect of elemental sulfur, some antioxidants and growth regulators on tolerance ability of *in-vitro* produced plantlets, and nutrient uptake, yield and fruit quality of mature date palm trees – part I. Tolerance ability of *in-vitro* produced plantlets. The sixth annual research conference at UAE University, 24–26 Apr 2005

Bennett RD, Ko ST, Heftmann E (1996) Isolation of estrone and cholesterol from the date palm. Phytochemistry 5:231

Besbes S, Bleckner C, Deroanne C et al (2005) Date seed oil: phenolic, tocopherol, and sterol profiles. J Food Lipids 11:251–265

Biglari F, AlKarkhi AFM, Easa AM (2008) Antioxidant activity and phenolic content of various date palm (*Phoenix dactylifera* L.) fruits from Iran. Food Chem 107:1636–1641

Buchanan B, Gruissem W, Jones R (eds.) (2000) Biochemistry and molecular biology of plants. Wiley, Hoboken

Buer CS, Djordjevic MA (2009) Architectural phenotypes in the transparent testa mutants of *Arabidopsis thaliana*. J Exp Bot 60:751–763

Buer CS, Muday GK, Djordjevic MA (2007) Long distance transport flavonoids. Plant Phys 145:478–490

Daayf F, Lattanzio V (eds.) (2008) Recent advances in polyphenol research, vol 1. Wiley InterScience, New York

Daayf F, El Bellaj M, El Hassni M et al (2003) Elicitation of soluble phenolics in date palm (*Phoenix dactylifera*) callus by *Fusarium oxysporum* f. sp. *albedinis* culture medium. Env Exp Bot 49:41–47

Daayf F, El Hadrami A, Adam LR, Balance GM (2006) Polyphenols communications. University of Manitoba, Winnipeg

Dixon RA, Paiva NL (1995) Stress-induced phenylpropanoid metabolism. Plant Cell 7:1085–1097

Edreva AM, Velikova VB, Tsonev TD (2007) Phenylamides in plants. Russ J Plant Phys 54:287–301

El Hadrami I (1995) L'embryogenèse somatique chez *Phoenix dactylifera* L. quelques facteurs limitants et marqueurs biochimiques. Thesis Doctorat d'Etat, Faculté des Sciences Semlalia, Université Cadi Ayyad, Marrakech

El Hadrami I, Daayf F (2002) Polyphenols communications 2002

El Hadrami A, Daayf F (2009) Priming canola resistance to blackleg with weakly aggressive isolates leads to an activation of hydroxycinnamates. Can J Plant Pathol 31(4):393–406

El Hadrami I, El Hadrami A (2009) Breeding date palm. In: Jain SM, Priyadarshan PM (eds.) Breeding plantation tree crops. Springer, New York, pp 191–216

El Hadrami I, Ramos T, El Bellaj M et al (1997) A sinapic derivative as an induced defense compound of date palm against *Fusarium oxysporum* f. sp. *albedinis*, the agent causing bayoud disease. J Phytopathol 145:329–333

El Hadrami I, El Bellaj M, El Idrissi A et al (1998) Biotechnologies végétales et amelioration du palmier dattier (*Phoenix dactylifera* L.), pivot de l'agriculture oasienne marocaine. Cah Agri 7:463–468

El Hadrami A, El Hadrami I, Daayf F (2009) Suppression of induced plant defense responses by fungal pathogens. In: Bouarab K, Brisson N, Daayf F (eds.) Molecular-plant microbe interactions CABI: Wallingford, UK, Chapter 10, pp 231–268

El Hadrami I, Daayf F, El Hadrami A (2010) On the socio-economical importance and traditional, health benefits, functional food and nutraceutical uses of dates. In: Manickavasagan A et al. (eds.) Dates – production, processing, marketing, food and medicine

El Hadrami A, Adam LR, Daayf F (2011) Biocontrol treatments confer protection against Verticillium dahliae infection of potato by inducing antimicrobial metabolites. Molecular Plant-Microbe Interactions 24(3):328–335

El Hassni M, J'Aiti F, Dihazi A et al (2004) Enhancement of defense responses against bayoud disease by treatment of date palm seedlings with a hypoaggressive *Fusarium oxysporum* isolate. J Phytopath 152:1–8

El Hassni M, El Hadrami A, Daayf F et al (2007) Biological control of bayoud disease in date palm: selection of microorganisms inhibiting the causal agent and inducing defense reactions. Envir Exper Bot 59:224–234

Fayadh JM, Al-Showiman SS (1990) A review on chemical composition of date palm (*Phoenix dactylifera* L.). J Chem Soc Pak 12:84–103

Ferrando A, Farràs R, Jásik J, Schell J, Koncz C (2000) Intron-tagged epitope: A tool for facile detection and purification of proteins expressed in Agrobacterium-transformed cells. Plant J 22:553–560

Ferrando A, Koncz-Kálmán Z, Farràs R, Tiburcio AF et al (2001) Detection of *in vivo* protein interactions between Snf1-related kinase subunits with intron-tagged epitope-labelling in plant cells. Nucleic Acids Res 29:3685–3693

Harborne JB, Williams CA (2000) Advances in flavonoid research since 1992. Phytochem 55:481–504

Haslam E (1989) Plant polyphenols – vegetable tannins revisited. Cambridge University Press, Cambridge

Heller W, Forkman G (1994) In: Harborne JB (ed) The flavonoids: advances in research since 1986. Chapman & Hall, New York

Herrmann KM (1995) The shikimate pathway: early steps in the biosynthesis of aromatic compounds. Plant Cell 7:907–919

J'Aiti F, Verdeil JL, El Hadrami I (2009) Effect of jasmonic acid on the induction of polyphenoloxidase and peroxidase activities in relation to date palm resistance against *Fusarium oxysporum* f. sp. *albedinis*. Phys Mol Plant Path 74:84–90

Macheix JJ, Fleuriet A, Billot J (1990) Fruit phenolics. CRC Press, Boca Raton

Mahran GH, Abdel-Wahab SM, Attia AM (1976) A photochemical study of date palm Pollen. Planta Medica 29:171–175

Mansouri A, Embared G, Kokkalou E, Kefalas P (2005) Phenolic profile and antioxidant activity of the Algerian ripe date palm fruit (*Phoenix dactylifera*). Food Chem 89:411–420

Mossa JS, Hifnawy MS, Mekkawi AG (1986) Phytochemical and biological investigations on date seeds (*Phoenix dactylifera* L.) produced in Saudi Arabia. Arab Gulf J Sci Res 4:495–507

Neubauer G, Gottschalk A, Fabrizio P, Seraphin B et al (1997) Identification of the proteins of the yeast U1 small nuclear ribonucleoprotein complex by mass spectrometry. Proc Natl Acad Sci USA 94:385–390

Ralph J (2010) Hydroxycinnamates in lignification. Phytochem Rev 9:65–83

Ramos T, El Bellaj M, El Idrissi-Tourane A et al (1997) Les phénolamides des rachis de palmes, composants de la réaction de défense du palmier dattier vis-à-vis de *Fusarium oxysporum* f. sp. *albedinis,* agent causal du bayoud. J Phytopath 145:487–493

Sghaier-Hammami B, Drira N, Jorrín-Novo JV (2009a) Comparative 2-DE proteomic analysis of date palm (*Phoenix dactylifera* L.) somatic and zygotic embryos. J Prot 73:161–177

Sghaier-Hammami B, Valledor L, Drira N, Jorrin-Novo JV (2009b) Proteomic analysis of the development and germination of date palm (*Phoenix dactylifera* L.) zygotic embryos. Proteomics 9:2543–2554

Wasinger VC, Cordwell SJ, Cerpa-Poljak A, Yan JX et al (1995) Progress with gene-product mapping of the mollicutes: Mycoplasma genitalium. Electrophoresis 16(7):1090–1094

Treutter D (2006) Significance of flavonoids in plant resistance: a review. Environ Chem Lett 4:147–157

Zaki AK, Schmidt J, Hammouda FM, Adam G (1993) Steroid constituents from pollen grains of *Phoenix dactylifera*. Planta Medica 59:A613–A614

Zhu H, Bilgin M, Bangham R, Hall D et al (2001) Global analysis of protein activities using proteome chips. Science 293: 2101–2105

Chapter 32
Industrial Biotechnology: Date Palm Fruit Applications

S.M. Aleid

Abstract Biotechnology can help find ways to utilize surplus dates as raw materials for the production of value-added products such as medical and industrial ethanol, bakery yeast, single-cell protein as a fodder yeast, citric acid, date flavored probiotic fermented dairy products, etc. Date juice is one of the richest foodstuffs in neutral compounds such as monosaccharides, disaccharide, mineral salts and vitamins. These substances are considered essential elements for the growth of microorganisms, especially yeasts. Another product that could be produced from date is bakery yeast. Claims are made that there are no technological constraints in using date extract for bakery yeast production. Dates are a good potential substrate for single-cell protein (SCP) production due to their sugar content. SCP can be used in animal feed without problems including for cattle, fish, poultry and rabbits. Dates present a good vehicle to transport probiotics both for their micro-architecture and for the presence of nutrients. Dates contain some micronutrients such as vitamins and minerals which might enhance the growth of microbial flora in probiotic dairy products. Addition of date syrup to probiotic yogurt enhanced the bifidobacterial count of the product. There is an increasing concern about the loss of economic value resulting from the large surplus of dates. This indicates the urgent need for biotechnology to expand and diversify the manufacture of date products.

Keywords Bakery yeast • Date flavoric • Probiotic fermented dairy products • Ethanol • Fodder yeast

S.M. Aleid (✉)
Date Palm Research Center of Excellence, King Faisal University,
Al-Hassa, Saudi Arabia
e-mail: seid@kfu.edu.sa

S.M. Jain et al. (eds.), *Date Palm Biotechnology*,
DOI 10.1007/978-94-007-1318-5_32, © Springer Science+Business Media B.V. 2011

32.1 Introduction

Dates can be utilized as a substrate for fermentation processes to manufacture different products such as alcohol, bakery yeast, organic acids, antibiotics and others. Date palm fruits are found to contain carbohydrates (44–88%), fats (0.2–0.4%), protein (2.3–5.6%), fiber (6.4–11.5%), minerals and vitamins (Al-Shahib and Marshall 2003). Carbohydrates in dates are mostly in the form of fructose and glucose, which are easily absorbed by the human body (Al-Farsi et al. 2005). Interestingly, dates contain higher concentrations of protein as compared to other major cultivated fruits such as apples, oranges, bananas and grapes (containing 0.3%, 0.7%, 1.0% and 1.0% protein, respectively) (Al-Showiman 1998). Twenty-three different amino acids are found in date protein, many of which are not contained in the most popular fruits. Several studies in the literature concluded that the aqueous extracts of dates have potent antioxidant and antimutagenic activity (Mansouri et al. 2005; Mohamed and Al-Okabi 2004). Dates are reported to have the second highest antioxidant activity among fruits commonly consumed in China (Guo et al. 2003). Al-Farsi et al. (2005) found that dates are a high source of antioxidants, anthocyanins, carotenoids and phenols. Antioxidants have received increased attention by nutritionists and medical researchers for their potential effects in the prevention of chronic and degenerative diseases such as cancer, cardiovascular diseases and to slow aging (Kaur and Kapoor 2001; Young and Woodside 2001). The most effective antioxidants in this respect appear to be the flavonoids and phenols. Because of their metal-chelating and radical-scavenging properties, phenols were considered effective inhibitors of lipid peroxidation (Mansouri et al. 2005). Furthermore, Al-Shahib and Marshall (2003) concluded that, in many ways, dates may be considered an almost ideal food, providing a wide range of essential nutrients and potential health benefits.

Based on the above compositional properties of date fruits, they are considered one of the most appropriate substances for production of value-added products through fermentation technology such as bakery yeast, single-cell protein as a fodder yeast, medical and industrial ethanol, and date flavored probiotic fermented dairy products, all of which are addressed in this chapter.

32.1.1 Date Fruit Production and Consumption

Dates are cultivated mainly in warmer regions of Asia and Africa. The fruit is also grown in some parts of Europe and the USA. It is now estimated that annually about seven million mt of dates are produced worldwide. The production period of the main supplying countries including Egypt, Iran, Saudi Arabia, UAE, Iraq and Tunisia etc. is from July to November. Dates are harvested and marketed at three stages of their development. The choice of harvesting at one or another stage depends on cultivar characteristics, climatic conditions and market demand

(Pakistan Horticulture Development and Export Board 2008). The world date fruit export market is about 0.42 million mt per annum (2005). Iran is on the top of the list with a 28% share, while Pakistan with 20% occupies second of the world's top ten exporting countries. Saudi Arabia and Tunisia both with almost 12% share are at par in their export performance securing third position in the year 2005. The world date imports are about 0.63 million mt per annum (2005). India is the largest importer with market share of about 38% while France and UK are the second and third largest importers with shares of 4% and 2.5%, respectively, in 2005 (FAOSTAT 2005).

The edible stages of ripening of date fruit can be divided into three main stages: khalal stage – physiological mature, hard and crisp, moisture contents 50–85%, yellowish in color; rutab stage – partially browned, reduced moisture contents 30–35%, softened; tamar stage – color from amber to dark brown, moisture contents reduced below 25–10%, texture soft pliable to firm. In conventional date processing, dry or soft dates are eaten as whole fruit, pitted and stuffed, or chopped and used in a great variety of ways: as ingredients in cereals, puddings, breads, cakes, cookies, ice cream and confectionaries. The pitting may be done by crushing and sieving the fruit or, more sophisticatedly, by piercing the seed out of the whole fruit. The calyces may also be mechanically removed. Surplus dates are processed into cubes, paste, spread, powder (date sugar), jam, jelly, juice, syrup, vinegar or alcohol. De-coloured and filtered date juice yields a clear invert sugar solution (Morton 1987).

32.1.2 Date Fruits as Raw Materials

Good quality raw material in volume is important to the development and modernization of the date industry. An important element is the formation of rural collection centers located in the main date-production areas. The purpose of the collection center is twofold. It provides a central location to create a buying hub and initial grading and processing of dates that have export potential and can be shipped to the central processing facilities for processing of lower-grade fruit and inferior-quality dates into value-added products such as syrup, paste, alcohol, vinegar etc. Secondly, it is a site from which agricultural extension agents can work to educate and assist farmers with improved practices in the care and harvesting of dates (Agland Investment Services Inc. 2008). Date fruits usually are fumigated and placed into cold storage. Dates are washed, graded for size and quality on belts, and then channeled to conveyor lines for further processing. The machines should have the advantage of being able to handle different date cultivars. A large area of the plant floor will be devoted to heavy usage of hand labor and fruit will again pass through an inspection belt prior to bulk packing (Fig. 32.1). End-products may include pressed date blocks, pitted dates, chopped/diced dates and date paste.

Fig. 32.1 An inspection belt for dates grading for size and quality prior to bulk packing (*Source*: Dates Processing Factory, Ministry of Agriculture, Al-Hassa, Saudi Arabia, 2002)

32.1.3 Improving Date Fruit Quality for Industrial Uses

In the season of tamar stage date harvest, some industries receive the fruits in amounts that far exceed immediate market capacity. Thus, most tamar dates are stored and then released into the market according to demand. Since quality parameters are affected by storage, it is very important to understand the effect of storage conditions on the different characteristics and acceptability of the date fruit to consumers (Ismail et al. 2008). There are several inherent constituents of dates, each of which in its own way takes part in the formation of the fruit. Due to genetic differences and growth conditions dates exhibit, perhaps more than other fruits, a wide variation in their final appearance and quality as one can perceive. Moreover, fruit quality, apart from these inherent properties, is also determined by exterior influences. Dates are classified in terms of the degree of insect infestation, defects, presence of foreign matter (sand, dust, debris) and pesticide residue (Saleem 2005). A quality profile of dates involves an evaluation of four aspects. First, color, shape, size, taste, texture, pit/flesh ratio and uniformity in color and size of the fruit. Second, moisture, sugar and fiber content. Third, defects of the fruits, which may include discoloration, broken skin, sunburn, blemishes, shrivel deformity etc. and fourth, presence of insect infestation, foreign matter, pesticide residues, molds and decay. A number of countries have formulated and applied date standards at the national level for both locally produced and imported dates. In an effort to arrive at global standards for dates the Codex Alimentarius Commission of the joint FAO/WHO Food Standards Program formulated a proposal for date standards intended to be the basis for worldwide application, subject to acceptance by the respective governments (Saleem 2005). Design of machines and the processes to harvest, handle and store agricultural materials and to convert these materials into food and feed requires an understanding of their physical properties (Keramat et al. 2008).

Dry dates are attacked by moths and beetles. Dates usually are subject to insect infestation during storage, resulting in high economic loss if disinfestation treatments are not applied. In order to store dates for a long period (several months to 1 year), the fruits must be thoroughly cleaned of any pests (eggs, pupas, larva or adults). This is done by fumigation, either in the field under various kinds of plastic sheets, or at the packaging warehouse in special sealed rooms. Infestation of dates with moths (almond moth, meal moth), beetles (sap beetle, saw-toothed grain beetle, flour beetle), rats, mice and ants leads to contamination and loss of volume. To comply with USA and European standards, for example, European markets require that the growers document the quality control processes used; especially a report concerning treatment against insects. Such a report must include a list of the materials permitted for use and approved by an official agent, in addition to the timetable of spraying with details of materials used, the date, concentration, number of days before harvesting and the level of residue of pesticides (Glasner et al. 2002).

32.1.4 Industrial Uses of Dates

Industrialization of dates has focused mainly on conventional processes, such as pitting, packaging, date pastes and animal feed. Biotechnological industrial processes using dates as raw materials are highly flexible and can accept most date cultivars. However, the most important factors to be considered in selecting date cultivars suitable for the production process are the sugar content, price per ton and storage life of the dates. As yet, there is no integrated date-processing industry, despite the early realization of the importance of dates as a source of many useful value-added products (Capital Advisory group 2004a). Even though date fruit is marketed all over the world as a high-value confectionery and fruit crop (Zaid 2006) and the production of dates has been increased many fold with modern biotechnological approaches, the processing industries have not been developed to keep pace. There is enormous industrial potential for fresh dates and date products with better quality attributes. Date processing industries are producing various date products like date paste, date syrup, date honey, date jam, date vinegar, etc. (Ahmed et al. 2005). The industrialization of dates is a highly demanding need (Capital Advisory group 2004a). The following are the advantages listed therefore:

- Availability of a consistent supply of raw materials (dates), taking into consideration adequate storage.
- Socioeconomic changes in date producing countries such as shifts in food habits and consumption patterns, which have led to a substantial surplus of fresh dates.
- Producing new products from dates will generate economic value and improve return.

- Date processing industries could use second- or third-grade dates that are not easily marketable.
- Generate more income for farmers by utilizing their production in manufacturing.

32.1.5 Economic Feasibility for Industrial Uses

Industrial projects utilizing dates as raw materials should focus on buying the less expensive cultivars that are not generally preferred for direct local consumption. To further control pricing and availability issues, annual supply contracts with date farmers are essential. Dates of an industrial processing grade (off-grade) could be purchased in bulk at low price. However, prices are relevant to the immediate post-harvest months and fluctuate at other seasons. Therefore, it is very important for industrial date processing projects to establish an efficient date collection and procurement mechanism. The trend of higher supply and declining demand will cause a downturn in date prices which might favor the industrial processor (Capital Advisory group 2004a).

32.2 Bakery Yeast Production from Dates

Dates are reputed to make a good potential substrate for bakery yeast production, serving mainly as a source of carbon and energy for the yeast. Molasses now is the dominant raw material for bakery yeast production worldwide. It is mainly used as a source of carbon and energy for the yeast in addition to providing some essential vitamins and minerals. Currently all bakery yeasts produced and used commercially in the world are strains of the species *Saccharomyces cerevisiae* (Barnett et al. 2000). The dry matter of the yeast cell is mainly composed of 40–54% raw protein (proteins, amino acids, nucleic acids and nucleotides), 39% carbohydrates (glycogen, trehalose, mannans and glycans), 7% lipids (neutral fats, sterines and phospholipids) and 6–10% (potassium, phosphorus, magnesium, sulfur, magnesium, sodium; smaller amounts of silicon, calcium, chlorine and iron and other trace elements) and ash (P_2O_5 and K_2O). In addition the yeast cell contains other components in smaller amounts such as vitamins, especially the B complex group, (about 480 mg/100 g yeast dry matter), of which D-biotin, D-pantothenic acid and m-Inositol are essential growth factors. The optimum growth temperature and pH for *S. cerevisiae* are 30°C and 4.5, respectively. It is facultatively anaerobic, i.e. it is able to grow aerobically and anaerobically. Under aerobic conditions, the yeast completely oxidizes sugars to CO_2 and produces 38 moles ATP (adenosine triphosphate) from 1 mole of glucose used for energy production. In this way a yield of about 0.5 g yeast/g sugar consumed is obtained. If the yeast grows anaerobically, it produces only 2 ATP moles from 1 mole glucose used for energy production. Hence the amount of biomass produced is much lower (maximum of 0.1 g yeast/g sugar), and the yeast produces high amounts of ethanol (about 0.5 g ethanol/g sugar consumed). A phenomenon unique to *S. cerevisiae* is the condition termed *aerobic respiration* which is the result of metabolic regulation known as the Crabtree Effect (Bailey and Ollis 1986). Due to this effect, if the sugar concentration in the growth medium exceeds 0.1 g/l, the yeast will start to ferment the sugars and produce ethanol, hence greatly reducing the biomass yield.

The role of the yeast in bread making is the rising of the dough to produce the characteristic loaf preferred by consumers. Dough rising occurs as a result of the gases produced by the yeast as it grows within the dough. During growth, the yeast metabolizes sugars in the dough with the help of a special enzyme system and produces alcohol and CO_2. The leavening power of the yeast depends on its activity and viability; hence the yeast used must be fully active with a high viable cell count. Furthermore, the leavening power of any yeast strain depends on its genetic makeup and on the process of production, and also on the storage conditions before use (Pyler 1988). The most important function of bakery yeast in bread making is to leaven the dough during the fermentation process by producing CO_2 via the alcoholic fermentation of sugars. Furthermore, the yeast produces desirable flavor and aroma compounds as products of secondary metabolism (Evans 1990), thus enhancing the characteristic flavor and aroma of baked products.

32.2.1 Dates as a Substrate for Bakery Yeast Production

Bakery yeast can be produced from substrates that contain suitable sources of carbon, energy, nitrogen, minerals and essential vitamins. Dates are said to be a good substrate for bakery yeast production. Their carbohydrate content is mainly sugars amounting to 65–87% of their dry matter. Date sugars are mainly glucose and fructose, which are easily assimilable to most microorganisms (Sawaya 1986). The protein content of dates is 1–3%. This is a low amount and hence a suitable source of nitrogen, in the form of inorganic salts, has to be added to the date substrate for bakery yeast production. Dates also contain vitamins important for yeast growth such as B_1 (0.75 mg/100 g), B_2 (0.2 mg/100 g) and nicotinic acid (0.33–2.2 ml/100 g). Also, the important minerals in 100 g dates are: potassium (650–750 mg), magnesium (50–58 mg), sulfur (43–51 mg), phosphorus (59–64 mg), iron (1.3–2 mg), calcium (58–68 mg) and chloride (268–290 mg) (Aleid et al. 2009). Commercial bakery yeasts produced from strains of *Saccharomyces cerevisiae* have the following average chemical composition: 47% C; 32% O_2; 6% H_2; 7.7% N_2; 2% K; 1.2% P; 1% S; 0.2% Mg; 0.1% Na and other trace elements. In addition, the yeast cells contain small amounts of vitamin B complex, of which D-Pantothenic acid, D-Biotin and m-Inositol are essential because the yeast cells cannot synthesize them (Bronn 1990). To produce 1 kg of yeast about 3 mg D-Biotin, 150 mg D-Pantothenic acid and 2 g m-Inositol are needed. These elements and compounds must be provided in the production medium in sufficient quantities and in metabolizable forms. If dates

Table 32.1 Date syrup and molasses compared as substrates for bakery yeast production

Nutrient	Date syrup	Beet molasses[a]	Sugarcane molasses[a]
Sugars (%)	80	50	50
N (%)	0.13	0.5	0.1
P (%)	0.11	0.03	0.09
K (%)	1.5	3.0	3.0
Mg (%)	0.08	0.01	0.3
Biotin (ppm)	2.73	0.05	2.0
Pantothenic acid (ppm)	240	80	25
m-Inositol (ppm)	0	6,500	4,000

[a] *Source*: Bronn (1990)

are used as a substrate for production, their sugars will act as a source of carbon and energy. According to the date fruit chemical composition given above, 1 mt of dates used as a carbon and energy source will yield about 325–435 kg of active dry yeast. The nitrogen content of dates is insufficient to produce such quantities of yeast and a suitable nitrogen source has to be added. The contents of dates in terms of other elements and essential vitamins must be determined and any deficiencies remedied (Aleid et al. 2009).

When comparing date and molasses as substrates for bakery yeast production, with regard to its nutrient contents, date syrup compares favorably with molasses which is the conventional substrate for bakery yeast production worldwide (Table 32.1). Date syrup contains much more sugar, biotin and pantothenic acid than molasses, about similar content of nitrogen, phosphorus and magnesium, about half the content of potassium (but still enough for bakery yeast production) and much less m-inositol. Compounds toxic to bakery yeast detected in date syrup include formic acid at 3.06%, acetic acid at 2.38% and propionic acid at 0.68% (total acids 6.12%), but no detectable amounts of the toxicants nitrite, sulfite and butyric acid. Formic acid becomes toxic to the yeast when its concentration in the medium exceeds 0.25%, whereas the toxicity level of the other two acids is in excess of 3.0% for the sum of the two (Aleid et al. 2009).

32.2.2 Production Process of Bakery Yeast

Bakery yeast is propagated under optimal conditions of temperature, pH, aeration and nutrient supply to give maximum yields of time, space and raw materials. The best fermentation process for bakery yeast production from strains of *Saccharomyces cerevisiae* is the fed-batch process, so that the Crabtree effect is avoided. Usually, part of the mineral medium and a small amount of the substrate is added to the fermentor, then the inoculum is added and the process started. The rest of the mineral medium and the substrate are fed to the fermentor at such a rate that the concentration of sugar in the fermentor does not exceed about 0.1 g/l. Continuous aeration and stirring is necessary to ensure the transfer of sufficient quantities of oxygen and nutrients to the growing yeast cells. Also, cooling is necessary to remove heat generated by the metabolic activity of the growing yeast culture and to maintain the fermentation temperature at about 30°C. When the fermentation process is completed, the final cell concentration in the fermentor is 4–5% by weight. Yeast cells are harvested by filtration or centrifugation and processed to the final product (Aleid et al. 2009).

32.2.3 *Bakery Yeast Production from Dates*

A few investigations into the production of bakery yeast from date extracts have been conducted (AlObaidi et al. 1985, 1987; Bassat 1971; Mohammed et al. 1986; Mudhaffer 1978). Comparisons were made between date extract and molasses. Positive findings were reported and claims were made that there are no technological constraints to using date extract for bakery yeast production. Nancib et al. (1997) used waste products from date in the production of bakery yeast from strains of *Saccharomyces cerevisiae*. They used a semi-synthetic fermentation medium containing sugars extracted from the date coat (freshly part), nitrogenous compounds extracted from seed hydrolysate, 6.0 g/l KH_2PO_4; 1.0 g/l date seed lipid; 0.6 g/l date seed ash and 1.0 g/l ammonium nitrate. Although they described this medium as satisfactory for bakery yeast production, yields obtained were very low with a maximum of 0.6 g/l biomass concentration in the fermentation medium compared to the optimum of about 40 g/l expected for an economical production. Khan et al. (1995) used Saudi Arabian dates in the production of bakery yeast. They propagated six different strains of *S. cerevisiae* in fermentation media containing Sefry Beesha cv. date extract (with 60% sugars) in place of molasses, in addition to 2 g/l ammonium sulfate and 50 μg/l biotin. Yields were also meager with a maximum of 10.7 g/l biomass concentration in the fermentation medium from 50 g/l sugar, representing a yield of only 42.8% of the theoretical. Date extract as a carbon and energy source for the propagation of bakery yeast on a pilot-plant scale, in comparison with molasses, was investigated by AlObaidi et al. (1986). Results showed that higher productivity of bakery yeast was observed when date extract was used. It was concluded from their study that date extract holds promise as a source of carbon and energy for the production of bakery yeast, although the average yields were only 47%. None of the authors discussed the Crabtree effect as a major technological problem encountered with bakery yeast propagation.

Aleid et al. (2009) used substrates from pure date syrup and pure molasses for the propagation of the bakery yeast strain *Saccharomyces cerevisiae*. All runs were fed-batch processes, at pH 4.5, 30°C, 8 g/l inoculum size and sugar concentration in all substrates of 200 g/l. The overall biomass yield from pure date syrup substrate was significantly lower than the yields from pure molasses substrates. Reduced yields could be attributed to the effect of yeast toxic organic acids contained in date syrup at high concentrations. The performance of the bakery yeast, propagated on date syrup as a fermentation substrate, in an Arabic bread test was excellent (Fig. 32.2).

Fig. 32.2 Bakery yeasts propagated on date syrup (**a**): fermentation vessel, (**b**): compressed yeast, (**c**): fermented bread dough, (**d**): Arabic bread. (Source: Aleid et al. 2009)

32.3 Single Cell-Protein Production from Dates

The technology of single-cell protein (SCP) production was established in the 1970s (Martin 1997). SCP is produced from bacteria, yeasts, molds and algae using different substrates as sources of carbon and energy such as food crops, by-products of agriculture and industry, wastes and also sunlight and atmospheric CO_2 (Israelidis 1987). Abduljabbar et al. (2008) reported about a yeast and a bacterium used for the production of SCP from ethanol, kerosene and gas oil. They found that optimum substrate concentrations during propagation were 0.5–4%. These are very low concentrations and will not yield more than about 2% biomass concentration in the bioreactor, hence the economic feasibility of the process is doubtful. No citations on SCP production from dates were found in the literature.

SCP is mainly considered a protein source and hence it is used to replace protein concentrates in animal feeds. SCP contains 50–70% protein, about 30% carbohydrates, 6% lipids and 8% minerals. In addition SCP is rich in vitamins, especially the B-complex group (Hamad 1986; Herbert 1976; Robinson 1986). Other advantages that support the use of SCP as animal feed include short production time (a few days compared to months for crops; years for animals), small land areas, no seasonality and use of cheap raw materials which are usually wastes that contaminate the environment.

32.3.1 Organisms Used for SCP Production

Desirable characteristics of the production organism include freedom from toxicity and pathogenicity, high content of protein with a well-balanced amino acid composition, thermo-tolerance, no growth factors needed, high yields and high growth rate (Hamad 1986).

32.3.1.1 Bacteria

Bacteria can be used for SCP production. The advantages are many e.g. they have high growth rates; are more stable in adverse growth conditions such as high temperatures; can utilize a wide range of substrates as sources of carbon, energy and nitrogen; and have high protein content. However, the main disadvantages of bacteria are: the cell wall makes digestion difficult, the high content of nucleic acids and the small size makes separation difficult (Holts 1994; Madigan et al. 1997).

32.3.1.2 Yeasts

Yeasts like *Candida utilis, Kluyveromyces marxianus* and *Saccharomyces cerevisiae* are used for SCP production. Yeasts used in SCP production have the advantages of better digestibility, lower nucleic acid content and easier handling during harvest (Hamad 1986; Ray 1996). On the other hand they have lower protein content than bacteria, lower growth rates and are less thermo-tolerant.

32.3.1.3 Algae

Algae are also used for SCP production, but less frequently. Algae are produced in bonds exposed to sun where they perform photosynthesis. Production costs are low, but much water and sunshine are needed.

32.3.2 Substrates for SCP Production

SCP can be produced from substrates that contain suitable sources of carbon, energy, nitrogen, minerals and vitamins Substrates range from food materials like grains and dates; by-products like molasses; alkanes, whey and wastes like sulfite liquor; residues of foods, plants and animals; and waste water. Some substrates contain easily-metabolized carbon and energy sources such as mono and disaccharides in molasses, dates and whey. Others contain complex carbohydrates such as starches in grains and cellulose in plant residues. Only a limited number of microorganisms are able to metabolize complex carbohydrates. In most cases such substrates need some treatments before use for SCP production. The use of wastes for SCP production serves two goals. First, it is possible to get rid of these environment-polluting wastes and, second, the process results in obtaining valuable products of commercial use (Allison 1975; Einsele 1975; Oura 1983). The yields on substrates are about 0.5 mt dry biomass per 1 mt carbohydrate (Allison 1975; Einsele 1975; Oura 1983). In most cases the substrates used for SCP production are low in nitrogen content and a suitable nitrogen source must be added. Usually inorganic nitrogen salts are added e.g. ammonia, ammonium salts, nitrates, etc.; sometimes urea is used. The process is therefore an upgrading of such inorganic nitrogenous compounds to the highly valuable organic nitrogen, the proteins.

32.3.2.1 Dates as a Substrate for SCP

Dates are a good potential substrate for SCP production. Their carbohydrate content is mainly sugars, amounting to 65–87% dry matter. The sugars are sucrose, glucose and fructose, which are easily assailable to most microorganisms (Aleid 2006; Sawaya 1986). This means that 1 mt of dates dry matter can produce up to 435 kg dry SCP. The protein content of dates is 1–3%; a low amount and hence inorganic nitrogen has to be added to the date substrate for SCP production (70 kg ammonium phosphate per mt of dates). The vitamin content of dates includes: thiamine (B_1), 0.75 mg/100 g; riboflavin (B_2), 0.2 mg/100 g and nicotinic acid (niacin, B_5), 0.33–2.2. The content of some important minerals (in 100 g dates) is: K, 650–750 mg; Mg, 50–58 mg; S, 43–51 mg; P, 59–64 mg; Fe, 1.3–2 mg; Ca, 58–68 mg and CI 268–290 mg).

32.3.2.2 Chemical Composition of Date Substrate

Date syrup for the production substrate should contain nutrients needed by yeasts such as sugars, protein, minerals, D-Biotin, D-Pantothenic acid and m-Inositol. As shown in Table 32.2, date syrup is deficient in P, K, Mg and proteins, it serves mainly as carbon and energy source. As a result, the deficient minerals need to be supplied to the yeast in the mineral medium in form of $(NH_4)_2SO_4$, KH_2PO_4, and $MgSO_4$ salts, and in case of the *Saccharomyces cerevisiae* strain, inositol needs to be added to the medium (Aleid et al. 2010).

32.3.3 Process of SCP Production

SCP is produced in fermentors using the fed-batch or continuous fermentation processes (Brauer 1985; Bronn 1990; Roels 1983). At first the substrate and the media are prepared by dilution, mixing, sterilization and purification if needed. Also the fermentor is cleaned and sterilized as needed. The process begins with the addition of a starter culture of the production organism to the fermentor containing some medium and substrate. The rest of the medium and substrate are then fed to the

Table 32.2 Date syrup contents from nutrients needed by yeasts

Chemicals	Date syrup[a]	Contents in *Saccharomyces cerevisiae*[b]	Contents in *Candida utilis*
P (g/kg)	1.13	14.10	70 (g/kg) (Total ash)[c]
K (g/kg)	14.88	20	
Mg (g/kg)	0.79	2	
Sugars (%)	80	30	–
Fructose	41		
Glucose	38		
Sucrose	1		
Crude protein (%)	2	50	46
Pantothenic acid (ppm)	240	150	Not needed[d]
Biotin (ppm)	2.73	0.30	Not needed
m-Inositol (ppm)	0	2,000	Not needed

[a]Aleid et al. (2010)
[b]Bronn (1990)
[c]Hamad (1986)
[d]Barnett et al. (2000)

fermentor. In the continuous fermentation process, feeding and harvest continue simultaneously and the process goes on as long as it is in a steady-state and no contamination of foreign microorganisms occurs. In the fed batch-process, feeding continues until a certain broth volume in the fermentor is reached, after which the process is stopped and the biomass harvested. Continuous aeration is needed because the process is mostly aerobic. Also cooling is necessary because large amounts of heat are produced during microbial growth. Stirring is needed to intensify cooling and air transport to the microorganisms. The final biomass concentration in the fermentor is about 4% on a dry weight basis. The biomass is harvested by filtration, centrifugation or sedimentation. The biomass is then dried to about 95% dry matter in dryers or under the sun. Drying increases the shelf-life by killing the cells of the production organism and preventing the growth of contaminants. According to use, the product is dried into powder, granules or flakes. Finally the product is packed in suitable containers and sent to the market.

For the assessment of the safety and nutritional value of SCP, factors such as nutrient composition, amino acid profile, vitamin and nucleic acid content as well as palatability, allergies and gastrointestinal effects should be taken into consideration (Litchfield 1968). Also, long-term feeding trials should be undertaken for toxicological effects and carcinogenesis.

32.3.4 SCP Production from Dates

Aleid et al. (2010) described using date syrup as the production substrate for single-cell protein and steps taken to formulate a suitable substrate from it. Fermentation process optimization experiments for both *Candida utilis* and *Saccharomyces*

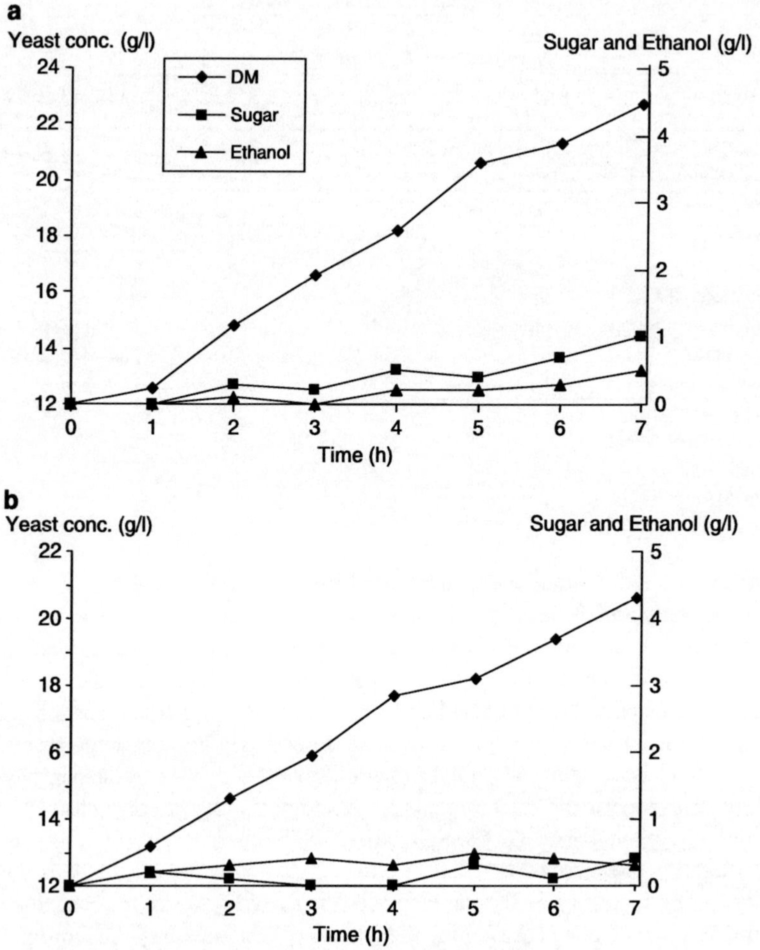

Fig. 32.3 Yeast dry matter (DM), sugar and ethanol of yeasts propagated on date syrup. (**a**) Propagation of *Candida utilis* at (12.5–42 g sugar/h) feeding rate and (12 g/l) inoculum size, (**b**) Propagation of *Saccharomyces cerevisiae* at (11–35.5 g sugar/h) feeding rate and (12 g/l) inoculum size. (Source: Aleid et al. 2010)

cerevisiae were conducted. The best process conditions for the propagation of *C. utilis* and *S. cerevisiae* were found to be at a high substrate feeding rate (12.5–42 g sugar/h) and medium substrate feeding rate (11–35.5 g sugar/h), respectively, with inoculum size of (12 g/l) for both microorganisms (Fig. 32.3).

In both propagation regimes the yeasts gave biomass yields on sugar of more than 90% of the theoretical, and the protein content of the biomass was about 50%. The amino acid composition of the proteins was high-quality containing most of the essential amino acids, especially lysine and the sulfur containing amino acids.

The biomass, especially that of *S. cerevisiae*, contained many essential vitamins of the B group. Also, the biomass of both yeasts contained high amounts of many important nutrient minerals such as Ca, Mg, Fe and Z, and the fatty acids of the yeasts were mostly of the preferred unsaturated types.

32.3.5 The Problem of Nucleic Acids

About 70–80% of the total cell nitrogen is represented by amino acids while the remainder occurs as nucleic acids. This concentration of nucleic acids is higher than other conventional proteins and is characteristic of all fast-growing organisms. This has two implications for the nutritional value of SCP. For use of SCP in animal feeds the major implication is simply that nucleic acid is not protein and essentially dilutes the protein, although there are at least some possibilities of physiological effects (Bull et al. 1997). As far as the potential use of SCP in human food is concerned, nucleic acids are undesirable because their digestion leads to unacceptably high levels of uric acid in the blood, sometimes resulting in gout disease (Edozien et al. 1970; White et al. 1964). Uric acid is a product of purine metabolism.

The bulk of the nucleic acid in microorganisms is RNA, which has a critical role in protein synthesis. Thus, it may be anticipated that the faster the rate of protein synthesis in a particular cell, the higher the nucleic acid content (Bull et al. 1997). The removal or reduction of nucleic acid content of various SCPs is achieved with one of the following treatments (Zee and Simard 1974): (a) chemical treatment with NaOH; (b) treatment of cells with 10% NaCl; (c) thermal shock. These methods aim to reduce the RNA content from about 7% to 1%, which is considered within acceptable levels. Thus SCP is treated with various methods in order to kill the cells, improve digestibility and reduce the nucleic acid content.

32.3.6 Metabolizable Energy of Yeasts Produced from Dates

The true metabolizable energy of *Saccharomyces cerevisiae* and *Candida utilis* yeasts are much higher than that of soybean meal, 2,485 kcal/kg (NRC 1984) and comparable to the energy in corn, 3,470 kcal/kg (NRC 1994) or higher as in case of the C. *utilis*. This is not surprising since date syrup used as a substrate to produce the SCP is recognized for its high energy value. The energy values are useful in formulating animal feeds. The true metabolizable energy may be more accurate than the apparent metabolizable energy, since it takes into account the metabolic and endogenous energy losses (sloughed intestinal cells, hormones, enzymes and endogenous urinary energy). However there is some criticism of this procedure in that poultry may refuse the ingredients and therefore synergism between ingredients cannot be accommodated (Leeson and Summer 2001).

32.3.7 *Uses of SCP*

SCP can be used as human food and animal feed. Its problem as human food lies in its high nucleic acid content. The purine bases of nucleic acids are degraded in the human body to uric acid. Increased levels of this acid in the blood and urine can cause gout or kidney stones. Through special treatments, the nucleic acid content of SCP can be reduced to a safe level of 3–4% (Scrimshaw 1975; Stringer 1982). SCP can be used in animal feed without problems including for cattle, fish, poultry and rabbits. SCP can also be used to replace milk powder in the diet of young stock like calves and lambs (Shacklady 1975).

32.3.7.1 Single Cell Protein for Poultry

Most of the initial work in the use of SCP as a protein source for poultry was done using SCP derived from petroleum materials and gas oil or normal paraffins as substrates (Plavnik et al. 1981). However, more recently increasing attention has focused on using bacteria and methane or oxygenated compounds (methanol and ethanol) in making SCP.

SCP in the Diet of Laying Hens

It has been found that feeding hens (10 months old) a mixture containing fish meal and paprin (single cell protein containing up to 4% hydrocarbon) had no significant affect on the number and weight of eggs. Ageev et al. (1984) used a paprin derived from yeast (*Candida* sp.), grown in N-Paraffins with broiler chickens and layers. They found that the addition of 3–4% paprin to the basal diet increased the livability rate, improved feed efficiency and slightly increased egg yield (by 3%). Paprin manufactured nowadays may contain 0.1% residual hydrocarbon, 0.4% DM, 56% protein, and is rich in B_{12} and all amino acids except methionine, they further added. Single-cell protein made in Taiwan was the focus of Taiwanese researchers. Lee and Chen (1983) fed poultry an appetite or gave restricted amounts of feed with or without SCP produced in Taipei. They found that egg production and final body weight were significantly greater with unrestricted feeding than with restricted. Also egg shell thickness and strength were significantly greater when SCP was added to both diets. When single cell protein was fed to layers, no depression in egg production was observed and the nucleic acid content was not affected in tissue and eggs (Al-Ani 1985). Layers performance was optimized when single-cell protein was fed at 2.5% of dry matter or 10% of dietary protein (FAO 2006). Single-cell protein was fed (0%, 5%, 10%, and 15%) levels to poultry; the best growth occurred at the 5% and 10% levels (Al-Shadeedl 1988). Najib (1996) studied the effect of incorporating yeast culture of *Saccharomyces cerevisiae* in the Saudi Baladi and White leghorn layers diet, and found that the best performance in terms of egg production, egg weight, feed conversion and feed intake was achieved when 0.3% yeast was fed to the Baladi poultry and 0.2% to the White leghorn. The possibility that SCP might harm the chicken's health was studied by Itakura et al. (1983) who reported that routine microscopic and macroscopic pathological examinations were made on 58 laying hens aged 21 weeks, which never had SCP in their diet, and on 59 birds given 11.0% pruteen (a protein produced by methanol-utilizing bacteria). No significant differences were found between groups in the occurrence of disorders.

SCP in the Diet of Broilers

In a series of studies by Plavnik et al. (1981), two types of SCP were tested; a pruteen (protein concentrate produced from methanol-utilizing bacteria) and a Lavera-type yeast utilizing the normal paraffins of heavy gas oil (LA). The inclusion of 90–150 gm of pruteen and LA per kg of broiler chicken diet depressed growth rate. However, this result was encountered by the addition of arginine. The reason for this depression could be due to lower feed intake, they further added. Similar weight gain depression was observed on the growth of broilers at the finishing period, fed the same above diets (Bornstein et al. 1981). Asif (1981) used a protein biomass from marine algae in feeding broiler chickens. He found that inclusion of 0.4% dried protein additive from marine algae improved average body weight by 8.5%, feed conversion by 6.3% and produced higher quality carcasses. Ergul and Vogt (1983), replaced fishmeal with bacterial bio-protein in broilers fed with high proportions of cotton seed and sunflower oil meals. They found no significant differences among groups; however, final body weight and feed efficiency were higher with 2% and 4% bacterial bioprotein. In trials with young quail, Krauss (1982) overcame the problem encountered by feeding SCP (probion) to the birds by supplementing the diet with methionine and arginine. He stated that methionine was the first limiting amino acid and arginine was the second. The feed value of Iranian SCP for broilers was determined by Pirmohammadi et al. (1999). They found that SCP concentration in the diet (0%, 3%, 6% or 9%) significantly affected performance and that a quadratic regression existed between SCP concentration and body weight gain or feed intake, while a linear regression was found between SCP concentration or feed consumption. As the dietary level of single-cell protein increases in broiler diet the gain, feed conversion and intake decreased (Jassim et al. 1986; Pirmohammadi et al. 1999). Schøyen et al. (2007) partially replaced soybean meal with basic bacterial protein (BBP) meal or autolyzed bacterial protein (AUT) meal. They concluded that 6% of either basic or autolyzed bacterial protein can replace soybean meal in diets for broiler chickens without impairing growth performance, and the basic bacterial protein seemed to be a slightly better substitute than the autolyzed protein. Nucleic acid content of the SCP has been a problem with humans. Therefore a study was conducted by Jassim et al. (1986) to determine the effect of using SCP on the level of nucleic acid of broilers. They found no significant differences due to diet on the nucleic acid content of tissue.

32.3.7.2 SCP in the Diet of Freshwater Fish

Aquaculture production has expanded worldwide at an annual rate of 15% and is predicted to continue to grow at this rate for at least another decade. Demands, therefore, on traditional fish feed ingredients, mainly fish meal and oil, which are finite global resources, are increasing (Hardy 2000). The proportion of global fish meal production utilized in fish feed has increased substantially over the past 10 years. In 1989, aquaculture was a minor consumer of fish meal, using approximately 10% of annual production (Barlow 1989). According to Hardy (2000) fish meal consumption by the aquaculture industry in 2000 was estimated to be 2.12 million mt. To be more specific, the fish meal used in tilapia feed amounted to 55,000 mt in 2000 and was predicted to increase by 35% to 74,000 mt in 2010. Suitable alternative feed ingredients will have to be utilized to provide the essential nutrients and energy needed to sustain the growth of aquaculture production.

Rumsey et al. (1991) found that the protein quality of *Saccharomyces* yeast (bakery or brewery) in diets of rainbow trout is improved by a treatment to disrupt the cell walls, thereby making the protein more available. When yeast cell walls are disrupted, 50% of the protein in rainbow trout diets can be supplied by bakery yeast with equivalent growth and feed conversion ratio to a control diet with protein supplied by casein and gelatine. Tests with fish have been conducted primarily using salmonids with variable results but during the last few years SCP have also been used to replace fishmeal in the diets of trout, Nile tilapia and Mozambique tilapia, producing comparable results (Barnes et al. 2006; McLean and Craig 2004; Olvera-Novoa et al. 2002; Schneider et al. 2004). Barnes et al. (2006) recently tried dietary yeast culture supplementation to initial rearing of rainbow trout. In general, yeast SCP could be substituted for about 25–40% of the fishmeal in standard diets (Spinelli 1980). Schneider et al. (2004) evaluated some plant-originated protein diets along with the single-cell protein diet as a fishmeal replacement on a 15% w/w basis in Nile tilapia. A mixture of plant feedstuffs with torula yeast (*Candida utilis*) has been evaluated by Olvera-Novoa et al. (2002) to substitute animal protein in diets for tilapia (*Oreochromis mossambicus*) fry, whereas McLean and Craig (2004) evaluated the growth of Nile tilapia fed with different levels of yeast-based diets. Shacklady (1974) incorporated up to 30% of alkane-grown yeasts (produced by British Petroleum) in the feed of trout with very good results.

It is necessary to develop new ways of using industrial residues, transforming them into protein for animal diets; besides the benefit of cost reduction it can minimize the pollutant impact of these residues in rivers and lakes (Medri et al. 2000). There are abundant date residues and wastes that could be utilized as substrates to grow yeast for producing SCP for animal feed because, according to Hardy (2000), among the noncommercial items, only single-cell protein (SCP), krill meal and possibly leaf-protein concentrate have a chance of becoming commercial commodities in feed formulations. It appears that there is great potential for the use of SCP in this field.

32.4 Ethanol Production from Dates

Biotechnology can help find ways to utilize surplus dates as a raw material for the production of value-added products such as medical and industrial ethanol, bakery yeast, fodder yeast, citric acid, vinegar, etc. Date juice is one of the richest foodstuffs in neutral compounds such as monosaccharides, disaccharide, mineral salts and vitamins. These substances are considered essential elements for the growth of microorganisms, especially yeasts (ALBassam 2001). If dates, which contain about 60–80% w/w sugar, are used as a substrate for medical and industrial ethanol production, 1 mt of dates should produce 300–400 kg of ethanol, i.e. about 380–500 l. In addition two by-products of potential economic value are produced, namely, carbon dioxide which can be used in the food industry and stillage which can be used as animal feed. Production of ethanol in the USA was about 20 million m^3 in 2007. Approximately 73% of world production of ethanol is used as fuel, 17% for the production of beverages and 10% in industry (Shapouri et al. 2006).

Pure ethyl alcohol is a colorless, mobile liquid of an agreeable odor. It boils at 78.3°C. Commercial alcohol contains about 90% pure ethyl alcohol and 10% water. This water cannot be entirely removed by fractional distillation, and to prepare anhydrous or absolute alcohol the commercial product must be allowed to stand over some dehydration agent, such as caustic lime, anhydrous copper sulfate, etc., and then distilled. A solution of 70–85% ethyl alcohol is commonly used as a disinfectant, killing organisms by denaturing their proteins and dissolving their lipids. It is effective against most bacteria and fungi, and many viruses, but is ineffective against bacterial spores (Capital Advisory Group 2004b).

Ethanol can be produced through microbial fermentation from any raw material containing carbohydrates. Such raw materials include dates, molasses, whey, grains (corn, sorghum, barley etc.), potatoes, cassava, Jerusalem artichoke etc. (Cardona and Sanchez 2007; Coté et al. 2004; Jeffries 2005; Xiang-Yang and Zhang 2005). Raw materials that contain carbohydrates in the form of simple sugars such as dates and molasses are more suitable for ethanol production because the substrate made from them does not need much preparatory treatment since simple sugars can be utilized directly by the fermenting yeast without the need of hydrolysis. Raw materials containing carbohydrates in the form of starch, however, such as grains, potatoes etc., need extensive preparatory treatments of the substrate including hydrolysis of starch by the addition of enzymes and acids before the resulting sugars can be converted to ethanol by the yeast. *Saccharomyces cerevisiae* is the most frequently used microorganism for ethanol production, but other microbes such as *Zymomonas mobilis*, *Candida tropicalis* and *Kluyveromyces marxianus* are described in the literature (Anderson et al. 1986; Cazetta et al. 2007; Cot et al. 2007; Jamai et al. 2007). The optimum temperature of fermentation for *S. cerevisiae, Zymomonas mobilis*, and *C. tropicalis* is about 30°C, and that of *Kluyveromyces marxianus* about 40°C. The overall yield on substrate must be above 90% of the theoretical yield, and the minimum ethanol concentration in the mash must be at least 10% v/v so that the process can be economically feasible. This requires the use of microbes with high specificity for ethanol production and with high tolerance to ethanol and osmotic pressure (Cot et al. 2007). The theoretical yield equals 0.51 g ethanol and 0.49 g carbon dioxide per gram sugar assimilated. An ethanol yield of 90% of the theoretical is considered satisfactory from the economical point of view (Cot et al. 2007; Coté et al. 2004). If dates, which contain 80% w/w sugar, are used as a substrate, 1 mt of dates should produce a maximum of 400 kg of ethanol, i.e. about 500 l.

Several methods were used for the sugar-fermentation process such as the immobilized cell method for *Saccharomyces cerevisiae*. Results showed that the productivity of ethanol by the immobilized cells method is higher and more economical than the free-cell method (Hahn-Hagerda 1985). No significant scientific publications on the production of ethanol from dates were found cited in the literature. However, the ability of individual strains of *S. cerevisiae* and *Candida utilis* to utilize date juice has been studied by ALBassam (2001). Results showed that *S. cerevisiae* has a high ability to metabolize date juice for ethanol production. The data on optimization of physiological conditions of fermentation, pH, and temperature and sucrose concentration showed similar effects on immobilized and free-cell of *S. cerevisiae* and *C. utilis*, in batch and immobilized fermentation. When fermentation was carried out at pH 4.5 and 30°C with *S. cerevisiae* using a substrate of 22 g/L of monosaccharides and disaccharides found in date juice, a maximum yield of 12.8% ethanol was obtained with *S. cerevisiae* in the free-cell method, and 13.4%,w/v, ethanol when those cells where immobilized by sodium alginate respectively (ALBassam 2001). In addition, immobilized fermentation has large economical advantages. Thus yeast cells keep their activity and fermentation viability for more than 3 months (Mohite and Sivaraman 1984).

32.4.1 *Major Steps of Ethanol Production from Dates*

32.4.1.1 Substrates

A substrate suitable for ethanol production can be extracted from dates. Date fruits are pitted and the flesh heated with an equal amount of water at 80°C for 30 min. The mixture is filtered to remove impurities to obtain a clear extract. Finally the extract is concentrated by heating. The concentration process yields a concentrated product, a date substrate, which contains the following nutrients. About 80% sugars, mainly in form of fructose (41%); glucose (38%); a small amount of sucrose (1%); about 2% crude protein; 1.13 g/kg phosphorus; 14.88 g/kg potassium; 0.79 g/kg magnesium; 240 ppm pantothenic acid and 2.73 ppm biotin. Therefore, date substrate used as a carbon and energy source will provide about 400 kg of ethanol (about 500 l). The substrate usually is diluted with water to a sugar concentration of 20%, pH adjusted to 4.5, heated to the boiling point and cooled for clarification, filtration to remove precipitates and then sterilized at 121°C for 30 min.

32.4.1.2 Preparation of the Mineral Medium

Salts are dissolved in water in the bioreactor, the pH adjusted to 4.5 and then sterilized at 121°C for 30 min.

32.4.1.3 Preparation of Inoculum from Yeast Strains

Yeast is propagated in the bioreactor containing suitable medium, harvested by filtration as a cake and kept under refrigeration to be used as inoculum.

32.4.1.4 Fermentation Under Controlled Conditions

Fermentation can be accomplished by two methods: the batch method, where the mineral medium and the whole substrate are added to the bioreactor at the beginning of fermentation and the process continues to its end; and the fed-batch method where only the mineral medium is placed in the bioreactor while the substrate is added at graduated quantities over a certain period of time until the mash is fully fermented before the distillation process is started. Samples can be taken from the bioreactor every hour and the concentrations of ethanol, sugars and yeast dry matter determined to ensure the end of the fermentation process.

32.4.1.5 Distillation

Fermented mash contains about 10% v/v alcohol, as well as all the non-fermentable solids from the raw material (dates) and the yeast cells. The mash will then be pumped to a continuous-flow, multi-column distillation system where the alcohol is removed from the solids and water. The alcohol will leave the top of the final column at about 96% strength, and the residue mash, stillage, transferred from the base of the column to the co-product processing area (Global Business Network 2007).

32.4.1.6 Dehydration

Alcohol from the top of the column will then pass through a dehydration system where the remaining water is removed. Most ethanol plants use a molecular sieve to capture the last bit of water in the ethanol. The alcohol product at this stage is called anhydrous (pure, without water) ethanol and is approximately 200 proof.

32.4.1.7 Denaturation

Ethanol to be used for fuel is then denatured with a small amount (2–5%) of some product, like gasoline, to make it unfit for human consumption.

32.4.2 Co-products

There are two main co-products created in the production of ethanol: carbon dioxide and distiller's stillage. Carbon dioxide is given off in great quantities during fermentation and many ethanol plants collect the carbon dioxide, clean it of any residual alcohol, compress it and sell it for use to carbonate beverages or in flash-freezing meat. Distiller's stillage, wet and dried, is high in protein and other nutrients and is a highly-valued livestock feed ingredient. Ethanol production is a no-waste process that adds value to the raw material by converting it into more valuable products (Global Business Network 2007).

32.5 Utilization of Dates in Probaiotic Dairy Products

Fermented dairy products containing probiotic cultures are healthful. The ability of probiotic bacteria to support the immune system could be important to the elderly or other people with compromised immune function, especially children (Parodi 1999). Yogurt is most often flavored with fruit preserves or other ingredients (Potter and Hotchkiss 1995). Flavored yogurts are made by adding fruit concentrates or flavors to cultured milk before or after incubation (Keating and White 1990). Dates are of higher nutritional value and do not need the addition of sugar or colors and flavors, compared to other fruits.

32.5.1 Use of Dates in the Manufacture of Some Dairy Products

Date palm products were used in the manufacture of some dairy products. Yousif et al. (1996) used date syrup in the manufacture of date juice milk drink. Hamad et al. (1983) used date syrup as a sweetener in ice cream. Hashim (2001) studied the characteristics and acceptability of yogurt containing date palm products. Results indicated that addition of 10–20% of date paste with or without 5% date syrup did not affect yogurt acidity, protein or fat contents, but increased total solids significantly. Also, addition of 15% date paste with 5% date syrup provided yogurt with desired sensory quality.

The effect of fortification with date fibers on quality was investigated (Hashim et al. 2009). Fortification of fresh yogurt with 1.5%, 3.0% and 4.5% date fiber did not cause significant changes in yogurt acidity, although the pH was increased. Yogurt fortified with date fiber had firmer texture (higher hardness values) and darker color. Sensory ratings and acceptability of yogurt decreased significantly when increasing date fibers to 4.5%. Fortifying yogurt with 3% date fibers produced acceptable yogurt with beneficial health effects. Gad et al. (2010) used date fruit as a part of water (v/v) used in reconstituting skimmed milk powder in processing yogurt with 14% total solids. Results showed that yogurt enriched with 10% dates had a significant sweetness, recorded the highest antioxidant values, higher in HCl-soluble minerals and folate concentration compared to plain yogurt. It could be concluded that numerous health benefits beyond its nutritional value have been associated with consuming yogurt enriched with 10% date fruit.

32.5.2 Background of Probiotics

Tissier (1900) found that bifidobacteria are dominant in the gut flora of breast-fed babies and he observed clinical benefits from treating diarrhea in infants with bifidobacteria. The claimed effect was that bifidobacteria displaced the proteolytic bacteria causing the disease. Cheplin and Rettger (1920) demonstrated that Metchnikoff's *Bulgarian Bacillus*, scientifically, *Lactobacillus delbrueckii* subsp. *bulgaricus*, could not live in the human intestine, and the fermented food phenomena petered out. Certain strains of *L. acidophilus* were found to be very active when implanted in the human digestive tract.

The term *probiotics* was first introduced in 1953 by Kollath (Hamilton-Miller et al. 2003). In contrast to antibiotics, probiotics were defined as microbial- derived factors that stimulate the growth of other microorganisms. Fuller (1989) suggested a definition of probiotics which has been widely used: *a live microbial feed supplement which beneficially affects the host animal by improving its intestinal microbial balance*. Fuller's definition emphasizes the requirement of viability for probiotics and introduces the aspect of a beneficial effect on the host. FAO/WHO (2001) defines probiotics as *live microorganisms which when administered in adequate amounts confer a health benefit on the host.*

32.5.3 Impact of Probiotic Foods in Human Health

Probiotics are lactic acid bacteria, *Bifidobacterium*, used in the most effective species combinations. The most traditional form of their consumption is through cultured dairy products (Maity and Misra 2009). Probiotics are gaining importance because of their healthful role and innumerable benefits, e.g. treating lactose intolerance, hypercholesterol problem, and cardiac diseases and managing cardiac problems like atherosclerosis and arteriosclerosis. Consumption of cultured milk produced using lactic lactobacilli and probiotics has been shown to have a protecting effect against carcinogens (Zhang and Ohta 1991). Consumption of probiotic cultured dairy products has also been shown to reduce the risk of diarrhea, one of the most common causes of sickness in young children. Also these products reduce the risk of developing a type of diarrhea which commonly occurs as a result of antibiotic treatments (Guandalini et al. 2000).

Consumption of probiotic products is found to have favorable effects on immune function. These products have been shown to enhance the number of IGA-producing plasma cells as well as increasing IGA levels in mice and humans (Perdigon et al. 1991). Probiotic products have been also shown to reduce the incidence of childhood eczema by half (Rautavaara et. al. 2002). In addition, probiotic consumption by young children may beneficially affect immune system development and prevent allergies (Majamaa and Isolauri 1997). The antioxidant activity of several species of probiotic bacteria found in fermented milk can significantly affect human health. Sour milk products have shown important improvement of the overall antioxidant activity of blood, prolonged resistance of lipoproteins fractions to oxidation and reduced level of peroxide lipoproteins and oxidized LDL cholesterol (Fitzgerald and Murray 2006).

32.5.4 Probiotic Dairy Products

Classic yogurt is produced by the thermophilic protosymbiotic culture of *Streptococcus thermophilus* and *Lactobacillus delbrueckii* subsp. *bulgaricus*. Yogurt milk is produced with a thermophilic culture of *S. thermophilus* and a *Lactobacillus* species such as *L. acidophilus*. The thermophilic starter culture fermentation is usually carried out at 40–45°C. The time needed for fermentation may be as short as 2.5 h for classic yogurt starter culture; this fast fermentation is mainly the result of the protosymbiosis. Rapid acidification and a short time are needed to prepare classic yogurt starter culture, and heat treatment is not required. Yogurt milk, on the other hand, requires about 6–8 h for fermentation, mainly because of the use of *L. acidophilus* as the lactobacillus component of the starter. In any case, a pH<4.8 is necessary to guarantee formation of a stable gel from coagulated milk protein. This is especially important for natural-set yogurt. As a result of the method used in manufacturing, stirred yogurt and drink yogurt are well-suited to the addition of probiotics after fermentation. Probiotics can be added easily during stirring of the product just before filling of the final containers. For natural-set yogurt, probiotic bacteria must be present during fermentation because fermentation takes place in the final containers and subsequent stirring would destroy the product's texture (Robinson 1995).

Sweet acidophilus milk and sweet milk are probiotic dairy products based on unfermented milk. Both are produced by adding concentrated probiotic bacteria to intensively heat-treated milk. Heat treatment is necessary to achieve sufficient microbiological stability during storage of the final product. *Lactobacillus acidophilus* plus bifidobacteria are added to sweet acidophilus milk. In contrast, acidophilus milk (fermented) is produced by fermentation with *L. acidophilus*. Again, intensive heat treatment before fermentation, yielding almost sterile milk, is necessary for successful fermentation because *L. acidophilus* acidifies slowly and thus can be readily out-competed by contaminating bacteria (Teuber et al. 1994).

For the manufacture of yogurt, probiotic lactobacilli can even be used as starter cultures because they meet the legal requirements. However, such manufacture is a compromise between full expression of the potential health properties of the probiotic strain and the technologic suitability of the strain. The probiotic strain must meet, not only the criteria for good survival, but also the criteria for fermentation and harmonious interaction with the *Streptococcus thermophilus* starter strain used. This could mean that the strain with the best combination of functional and technologic properties is the one used, not the strain with the best health properties. Whereas the coagulation of milk proteins is a consequence of acid production in yogurt, coagulation in cheese is achieved through the proteolytic action of rennet.

Less rennet is added for fresh cheese (cheeses that do not undergo ripening) than for ripened cheese. As an example, cottage cheese manufacture will be described. Milk is usually inoculated with a mesophilic starter culture and incubated at 20–30°C for a relatively short period before rennet is added. Incubation proceeds until the curd forms. The curd is cut to allow expelling of whey from the coagulated casein. Expelling is reinforced by raising the temperature of the whey-coagulum mixture to 50–55°C for 1–2 h. During this time the coagulum particles shrink (because of further loss of whey) and become firmer. After the whey is drained off, the coagulum is washed in clear water at 7–10°C and then at 2°C to remove residual lactose. Finally, cream and salt (and spices for some products) are added in desired concentrations and the mixed products are poured into retail containers (Robinson 1995).

32.5.5 *Probiotic Yogurt Flavored with Dates*

Aljasass et al. (2010) produced date flavored probiotic stirred yogurt from both fresh cow milk and reconstituted whole milk powder as follows. Fresh whole cow milk (about 11% total solid and 3% fat) was heated at 80°C for 10 min, cooled to 40°C, then 4% skim milk powder added to raise the total solids level up to 15%. In order to stabilize the resultant product and improve the firmness and protect the product against a whey-off flavor, a 1% food grade gelatin was added to the yogurt milk. Heating was continued to 60°C and then 1% gelatin was added. (The gelatin was dissolved in hot water before addition) After addition of gelatin solution, the heating process for yogurt milk was continued to 80°C for 10 min and then the milk rapidly cooled to 40°C. The milk was then inoculated with 2% of yogurt mixed culture of 1:1 *Streptococcus thermophilus* and *Lactobaccilus bulgaricus* and then divided into two equal portions. The first portion was inoculated with 5% of *Bifidobacterium infantis*, while the second portion was inoculated with 5% of *B. angulatum*. Both portions were incubated at 37°C until complete coagulation occurred (about 4 h) and then the curd was cooled to 4°C overnight. After this, each portion was subdivided into five parts and then these parts were mixed with date spread at levels of 0.0, 5.0%, 10.0%, 15.0% and 20.0%, respectively. The resultant *date flavored probiotic stirred yogurt*, from all five treatments, was stored at 4°C for 10 days.

32.5.5.1 Effect of Date Syrup on Microbial Flora in Probiotic Yogurt

The viable bifidobacterial count for the date syrup flavored probiotic yogurt was higher than the minimum concentration of probiotic required for beneficial affect to the end of the storage period. No significant differences in total bacterial counts of the probiotic yogurt were observed (Fig. 32.4), For example, at day 0 the total count of *Bifidobacterium infantis* in date flavored probiotic stirred yogurt made from fresh cow milk was over 9.7 logs CFU/g in all samples treated with 5%, 10%, 15% and 20% date dips.

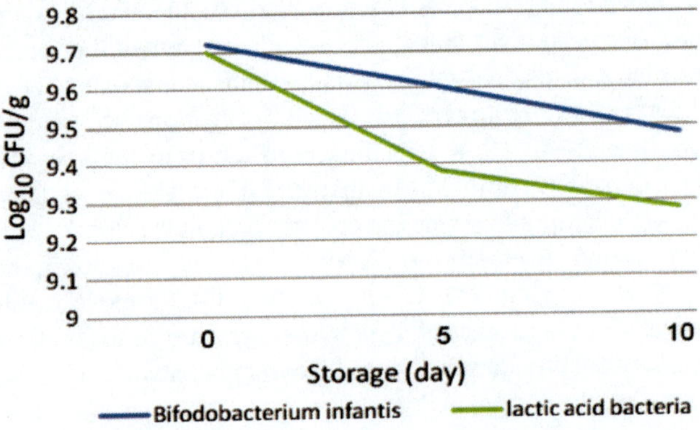

Fig. 32.4 Changes in *Bifodobacterium infantis* and lactic acid bacteria count (log CFU/g) of flavored probiotic yogurt treated with 15% date syrup made from fresh cow milk during cold storage (1±5°C). (Source: (Aljasass et al. 2010)

Moreover it was noted that the addition of date syrup to the probiotic yogurt enhanced the bifidobacterial count of the product of all treatments and improved their survival during the cool storage period up to 10 days (Aljasass et al. 2010). This could be explained on the basis that date syrup may contain some micronutrients, such as vitamins and minerals, which might enhance the growth of bifidobacteria (Gad et al. 2010). Meanwhile the viable bifidobacterial count for the product of all treatments was higher than the minimum concentration of probiotic required for beneficial effect up to the end of the storage period. Raipulis et al. (2005) reported that consumption should be more than 100 g/day of bio-yogurt containing more than 10^6 CFU/gm; other scientists suggested a minimum viable number of 10^6 CFU/g, but recommended 10^8 CFU/g to be more effective (Shah 2000). The results of this study indicated that the resultant date flavored probiotic yogurt from all treatments contained viable numbers of bifido bacteria more than the minimum number reported by above studies.

An increase in total lactic bacteria in the flavored probiotic stirred yogurt containing date syrup at the 0 time and during storage period is noted. This could be due to the higher simple sugars content which stimulates the lactic acid bacteria. The coliform bacteria, mold and yeast counts were not detected in any sample either during production or refrigerated storage at 5±1°C for 10 days. This could be attributed to the effect of the probiotic cultures which protected the products from the spoilage organisms (Anukam et al. 2006) and also to the high hygienic precautions taken during the production, handling and storage of the yogurt. Similar results were reported by Al-Otaibi (2009) who did not detect coliform bacteria, mold or yeast in probiotec fermented milks at up to 3 weeks of cold storage.

32.5.5.2 Effect of Date Syrup on Quality of Probiotic Yogurt

The addition of concentrated date syrup to the probiotic yogurt resulted in a remarkable increase in the total calorie content of the products. This increase was in proportion to the level of concentrated date dips added. The total calories of date flavored probiotic yogurt made from fresh cow milk were increased by 29.3, 51.9, 75.4, and 92.7 when the concentrated date syrup was added at 5%, 10%, 15% and 20%, respectively (Aljasass et al. 2010). Hashim (2001) reported that addition of 10–20% date paste increased total solids of the resultant yogurt, but did not significantly affect fat and protein contents of the products. Gad et al. (2010) used dates as a part of water (v/v) used in reconstituting skim milk powder in processing yogurt with 14% total solids. They reported that this treatment enhanced the nutritive value of the resultant yogurt. Addition of date syrup to the probiotic yogurt resulted in a remarkable increase in total solids, total carbohydrates and total calorie contents and this relationship was date syrup-level dependent. However this addition did not affect total protein, fat and acidity contents or pH values of the flavored probiotic yogurt. This could be explained on the basis that addition of the concentrated date syrup to the probiotic yogurt was carried out after the complete fermentation process, and so this addition did not affect the fermentation by the bacterial cultures (Aljasass et al. 2010). Hashim (2001), reported that the addition of 10–20% date paste to plain yogurt did not affect acidity and pH of the resultant yogurt. The type of milk used, culture strains and storage period did not affect these constituents.

Color characteristics showed a decrease in the lightness values (L) and an increase in yellowness and redness values (a, b). Sensory evaluation indicated that addition of date syrup improved the sensory quality of the probiotic yogurt, whereas products contained date syrup up to 20% were acceptable, but that the level of 15% gained the highest acceptability. All products remained of acceptable quality up to the end of 10 days of cold storage. No significant differences in the sensory quality of the product occurred using the two bifids. Strains were observed, but the products made from reconstituted whole dried milk gained higher scores than those made from fresh cow milk, up to the end of the storage period (Aljasass et al. 2010).

32.5.6 Fortification of Dates with Probiotics

Dates present a good vehicle to transport probiotics both for their micro-architecture and for the presence of nutrients. Probiotic date products could attract more consumers including children showing no preference toward milk consumption and individuals who are in need of low-cholesterol diets. Al-Otaibi and Saleh (2010) incorporated probiotic microorganisms (*Bifidobacterium lactis* Bb-12, *B. longum* Bb-46 and *Lactobacillus acidophilus* La-5) with dates to produce a multi-functional product, enhancing the inherent functionality of dates plus the added functionality of probiotics.

32.5.6.1 Preparation of Probiotic Dates

As an ingredient, probiotic fermented milk could be prepared as follows. Whole cow milk is pasteurized at 65 °C for 30 min and cooled to 43°C, then inoculated

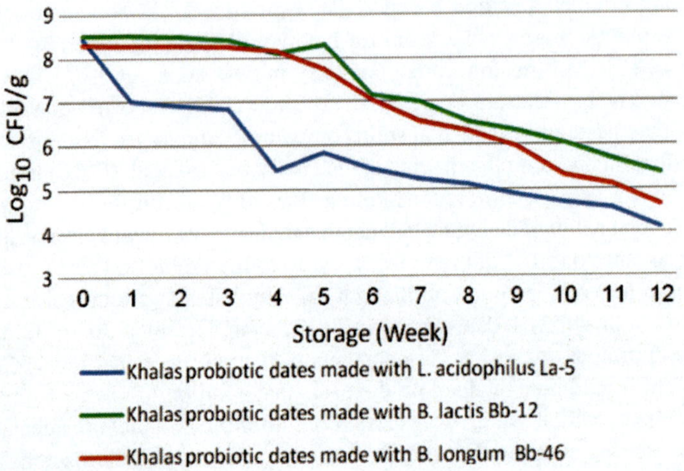

Fig. 32.5 Changes in viability of probiotic bacteria in probiotic dates during refrigerated storage period. (Source: Al-Otaibi and Saleh 2010)

with 0.07% (w/v) of lyophilized *Bifidobacterium lactis* Bb-12, *B. longum* Bb-46 or *Lactobacillus acidophilus* La-5. After incubation, all products are stored at 5 ± 1°C. Dates (Khalas cv.) are washed in water, seed removed, soaked in boiling water for 1 min and dried by forced air. The dates are milled to obtain date paste. For the probiotic date preparation, a formula could be used with the following ingredients: (33% date paste, 17% probiotic fermented milk, 25% full cream milk powder, 12.5% biscuit powder and 12.5% nut powder). The ingredients are mixed thoroughly, formed into small balls and coated with chocolate. The product can be stored at 5°C for 12 weeks. For quality assurance, the probiotic bacteria counts and pH values can be determined weekly during storage (Al-Otaibi and Saleh 2010).

32.5.6.2 Viability of Microbial Flora in Probiotic Dates

Viable counts of *Lactobacillus acidophilus*, *Bifodobacterium lactis* and *B. longum* in Khalas cv. probiotic dates during regenerated storage for 12 weeks are presented in (Fig. 32.5). All probiotic bacteria were capable of surviving in dates. The initial viable counts of all probiotic bacteria in the date product were above 8 $\log_{10}$ cfu/g. The viability in general decreased gradually during the storage period. The viable count of *L. acidophilus* remained above 6 $\log_{10}$ cfu/g until the third week of storage in Khalas probiotic date. Moreover, the viable count of *B. lactis* and *B. longum* remained above 6 $\log_{10}$ cfu/g until the eighth and tenth weeks of storage. For maximum health benefits, the minimum number of probiotic organisms in a food product should be 10^6 cfu/g. Therefore, the viability of the probiotic bacteria is the most important factor during refrigerated storage (Al-Otaibi and Saleh 2010).

The viability of probiotic microorganisms are dependent on the level of oxygen in products, oxygen permeation of the package, fermentation time and storage temperature (Shah 2000). The viability of probiotic bacteria is also affected by inhibitory substances such as lactic acid produced during production and cold storage. The loss of viability of probiotic bacteria during storage of probiotic dates may be due to cold storage and the presence of antibacterial substances in the dates such as phenolic compounds, which have antibacterial activity against gram positive and gram negative bacteria (Pereira et al. 2007; Regnault-Roger et al. 1987).

32.5.6.3 Changes in pH Values of Probiotic Dates

The initial pH value is 6.8 for cv. Khalas and probiotic dates. This pH value is suitable for growth and activity of bifiobacteria. Shah (2006) reported that the optimum pH for the growth of bifiobacteria is 6.0–7.0 with virtually no growth below pH 4.5–5.0, or above pH 8.0–8.5. Optimum growth occurs at 37–41°C, maximum growth is 43–45°C, while minimum growth temperature is 25–28°C. Probiotic dates exhibited a slight decrease in pH during storage, to 6.1 at the end of 12 weeks storage, possibly due to the low acidifying activity of the probiotic bacteria as a result of the low moisture content of probiotic dates and cold storage. The probiotic dates containing either *Lactobacillus acidophilus* or bifidobacteria showed almost similar patterns of pH values over 12 weeks of storage (Al-Otaibi and Saleh 2010).

32.5.6.4 Quality of Probiotic Dates

The addition of probiotic fermented milk to the mixture improved the taste, appearance and color of probiotic date. There were significant differences in the smell and odor between probiotic dates and ordinary dates. Luckow and Delahunty (2004) and Luckow et al. (2006) reported that the incorporation of probiotic bacteria in fruits caused off-flavors. However, this finding is not true with probiotic dates. From quality evaluation, it could be concluded that *Lactobacillus acidophilus*, *Bifidobacterium lactis* and *B. longum* could be used as probiotic cultures for the production of healthy and acceptable probiotic dates (Al-Otaibi and Saleh 2010).

32.6 Conclusion and Prospective

Integrated facilities seriously need to be considered within industrial complexes to diversify downstream products from dates, including date liquid sugar, date juice, medical alcohol, vinegar, bakery and fodder yeasts, animal feed etc. to minimize cost and to maximize return from industrial projects. Moreover, interested investors should not take risks in using cheaper and unproven technology providers. Furthermore, the production process should allow for the use of a wide variety of dates and therefore permit both flexibility of production and procurement. Utilizing dates and their by-products in industrial biotechnology will have positive economic, social and environmental impacts. Lignocellulosic feedstocks are considered an alternative (second generation raw material) as they are available in large quantities. However the high costs for conversion of the polysaccharides (cellulose and hemicellulose) into fermentable sugars inhibit their use. Dates, in fact, have readily assimilable forms of sugars such as glucose and fructose which could be directly utilized by fermentation microorganisms. Moreover, due to the use of biotechnology techniques, new and improved microorganisms can be created that convert date biomass components (such as sugar and fibers) into end-products or intermediates that can be utilized in the food or feed industries.

References

Abduljabbar AH, Raouf SR, Alhelu J (2008) Petroleum single cell protein production. Eng Technol J 27:468–469

Ageev V, Kupina L, Dolbeneva E et al (1984) An important protein source for poultry. Ptitsevodstovo 3:19–22

Agland Investment Services, Inc (2008) Pre-feasibility report – date processing. Agland Investment Services, Larkspur. Prepared for: USAID/Iraq/Inma Agribusiness Program

Ahmed J, Ramaswamy HS, Khan AR (2005) Effect of water activity on glass transition of date pastes. J Food Eng 66:253–258

Al-Ani TA (1985) Effect of using single-cell protein on the productive efficiency and nucleic acid content in laying hens eggs and tissues (in Iraq). Baghdad University, College of Agriculture, Baghdad

ALBassam R (2001) Industrial ethanol production using juice of dates in a fixed cell process. In: Second international conference on date palms, Al-Ain, 25–26 Mar pp 850–860

Aleid SM (2006) Chromatographic separation of fructose from date syrup. Int J Food Sci Nutr 57:83–96

Aleid SM, Hamad SH, Al-jasass FM (2009) Use of dates in the production of baker's yeast. Final report, Project # APR-24-27, King Abdulaziz City for Science and Technology, Riyadh

Aleid SM, Al-jasass FM, Hamad SH et al (2010) Production of single cell protein from surplus dates and its use as a protein source in the diets of poultry and fish. Second annual report. Project # APR-27-96, King Abdulaziz City for Science and Technology, Riyadh

Al-Farsi M, Alaslvar C, Morris A et al (2005) Comparison of antioxidant activity, anthocyanins, carotenoids and phenolics of three native fresh and sun-dried date (*Phoenix dactylifera* L.) varieties grown in Oman. J Agr Food Chem 53:7592–7599

Aljasass FM, Aleid SM, Elneshwy AA (2010) Utilization of dates in the manufacture of new probiotic dairy food. First annual report, Date Palm Research Center, King Faisal University, Al-Ahsa, Project No. PR3

AlKhateeb AA, Dinar HM (2002) Date palm in Saudi Arabia: Cultivation, production and processing. King Faisal University Press. Alhassa, Saudi Arabia. p 145 (in Arabia)

Allison K (1975) Protein from oil. The PB-alkane process, Mikrobielle Proteingewinnung, Verlag Chemie, Weinheim

AlObaidi ZS, Al-Hilli MA, Al Hakkak TS (1985) Optimization of propagation medium for baker's yeast using date extract and molasses.1: determination of the optimum concentration of the macro elements. J Agric Water Resour Res Iraq 5:97–108

AlObaidi ZS, Mohamed NA, Hasson NA et al (1986) Semi-industrial production of baker's yeast using date extract and molasses. J Agric Water Resour Res 5:162–174

AlObaidi ZS, Aziz GM, Al-Hakkak TS et al (1987) Optimization of propagation medium for baker's yeast using date extract and molasses. Determination of the optimum concentration of micro-elements and vitamins. Date Palm J 5:64–78

Al-Otaibi MM (2009) Evaluation of some probiotic fermented milk products from al-Ahsa markets, Saudi Arabia. Am J Food Technol 4:1–8

Al-Otaibi MM, Saleh FA (2010) Fortification of some dates with probiotic fermented milk, Project No. 110003, King Faisal University, College of Agricultural and Food Sciences, Dept Food and Nutrition Science, Al-Hassa

Al-Shadeedl MJB (1988) The effect of different levels of single cell protein in broiler ration and bird densities on body weight and feed conversion ratio, Baghdad University, College of Agriculture, Baghdad

Al-Shahib W, Marshall RJ (2003) The fruit of the date palm: its possible use as the best food for the future. Int J Food Sci Nutr 54:247–259

Al-Showiman SS (1998) Date, food and health. Dar Al-Khareji Press, Qassim, Saudi Arabia

Anderson PJ, McNeil K, Watson K (1986) High-efficiency carbohydrate fermentation to ethanol at temperatures above 40°C by *Kluyveromyces marxianus* var. *marxianus* isolated from sugar mills. *Appl Environ Microbiol*, pp 1314–1320

Anukam K, Osazuwa E, Ahonkhai I (2006) Augmentation of antimicrobial metronidazole therapy of bacterial vaginosis with oral probiotic *Lactobacillus rhamnosus* GR-1 and *Lactobacillus reuteri* RC-14: randomized, double-blind, placebo controlled trial. Microbes Infect 8:1450–1454

Asif M (1981) Use of protein biomass from marine algae in feeding of broiler chickens. Sbornik Nauchnykh Rabot Leningradskogo Veterinarnogo Instituta 67:88–92

Bailey JE, Ollis DF (1986) Biochemical engineering fundamentals, 2nd edn. McGraw-Hill, New York

Barlow S (1989) Fishmeal-world outlook to the year 2000. Fish Farm 43:40–41

Barnes ME, Durben DJ, Reeves SG et al (2006) Dietary yeast culture supplementation improves initial rearing of McConaughy strain rainbow trout. Aquac Nutr 12:388–394

Barnett JA, Payne RW, Yarrow D (2000) Yeasts: characteristics and identification, 3rd edn. Cambridge University Press, Cambridge

Bassat FF (1971) Manufacturing of dates products, Baghdad

Bornstein S, Plavnik I, Lipstein B (1981) Evaluation of methanol-grown bacteria and hydrocarbon-grown yeast as sources of protein for poultry: performance of broilers during the finishing period. Br Poult Sci 22:141–152

Brauer H (1985) Biotechnology, vol 2. Verlag Chemie, Weinheim

Bronn W (1990) Technology of baker's yeast production (in German). Technical University of Berlin, Berlin

Bull AT, Ellwood DC, Ratledge C (1997) Microbial technology: current state and future prospects. Twenty-Ninth Symposium of the society for general microbiology, University of Cambridge, London

Capital Advisory Group (2004a) Pre-feasibility study: date liquid sugar manufacturing facility, Riyadh

Capital Advisory Group (2004b) Pre-feasibility study: medical alcohol and vinegar from dates, Riyadh

Cardona CA, Sanchez OJ (2007) Fuel ethanol production: process design trends and integration opportunities. Bioresour Technol 98:2415–2457

Cazetta ML, Celligoi MAPC, Buzato JB, Scarmino IS (2007) Fermentation of molasses by *Zymomonas mobilis*: effect of temperature and sugar concentration on ethanol production. Biores Technol 98:2824–2828

Cheplin HA, Rettger LF (1920) Studies on the transformation of the intestinal flora, with special reference to the implantation of *Bacillus acidophilus*: II. Feeding experiments on man. Proc Nat Acad Sci USA 6:704–705

Cot M, Loret M-O, François J, Benbadis L (2007) Physiological behaviour of *Saccharomyces cerevisiae* in aerated fed-batch fermentation for high level production of bioethanol. FEMS Yeast Res 7:22–32

Coté A, Brown WA, Cameron D et al (2004) Hydrolysis of lactose in whey permeate for subsequent fermentation to ethanol. J Dairy Sci 87:1608–1620

Edozien JC, Udo UV, Young VR et al (1970) Effects of high levels of yeast feeding on uric acid metabolism of young men. Nature 228:180

Einsele A (1975) Biomass from higher n-alkanes, single cell protein II. MIT Press, Cambridge

Ergul M, Vogt H (1983) Replacement of fishmeal by bacterial bioprotein in broiler feeds with high proportions of cottonseed and sunflower oilmeals. Landbauforschung Volkenrode 33:79–84

Evans IH (1990) Yeast strains for baking. In: Spencer JFT, Spencer DM (eds.) Yeast technology. Springer, Berlin, pp 13–45

FAO/WHO (2001) Health and nutritional properties in food including powder milk with live lactic acid bacteria. Report of a Joint FAO/ WHO Expert consultation on evaluation of health and nutritional properties of probiotics in food including powder milk with lactic acid bacteria. http://www.who.int/foodsafety/publication/fs_management/probiotics/en/index.html

FAO (2006) Animal Feed Resources Information System. http://fao.org

FAOSTAT (2005). http://faostat.fao.org

Fitzgerald RK, Murray BA (2006) Bioactive peptides and lactic fermentations. Int J Dairy Technol 59:118–125

Fuller R (1989) Probiotics in man and animals. J Appl Bacteriol 66:365–378

Gad AS, Kholif AM, Sayed AF (2010) Evaluation of the nutritional value of functional yogurt resulting from combination of date palm syrup and skim milk. Am J Food Technol 5:250–259

Glasner B, Botes A, Zaid A et al (2002) Date harvesting, packinghouse management and marketing aspects. In: Zaid A (ed.) Date palm cultivation. FAO, Rome, pp 237–267

Guandalini S, Pensabene L, Zikrei MA (2000) *Lactobacillus* GG administered in oral rehydration solution to children with acute diarrhea: a multicenter European trial. J Pediatr Gastroenterol Nutr 24:399–404

Guo C, Yang J, Wei J et al (2003) Antioxidant activities of peel, pulp and seed fractions of common fruits as determined by FRAP assay. Nutr Res 23:719–726

Hahn-Hagerda B (1985) Comparison between immobilized *Kluyveromyces fragilis* and *Saccharomyces cerevisiae* co-immobilized with b-galactosidase, with respect to continuous ethanol production from concentrated whey permeate. Biotech Bioeng 27:914–916

Hamad AM, Mustafa MI, Al-Kahtani S (1983) Possibility of utilizing date syrup as sweetening and flavoring agent in ice cream making. In: Proceedings of first symposium on date palm in Saudi Arabia, King Faisal University, Al-Hassa, pp 544–550

Hamad SH (1986) Screening of yeasts associated with food from Sudan and their possible use for single cell protein and ethanol production. PhD thesis, Technical University of Berlin, Germany

Hamilton-Miller JMT, Gibson GR, Bruck W (2003) Some insights into the derivation and early uses of the word 'probiotic'. Br J Nutr 90:845

Hardy RW (2000) New developments in aquatic feed ingredients, and potential of enzyme supplements. In: Cruz-Suárez LE, Ricque-Marie D, Tapia-Salazar M et al. (eds.) Avances en nutrición acuícola V. Memorias del V simposium international de nutrición acuícola, Mérida, Mexico, 19–22 Nov 2000, pp 216–226

Hashim IB (2001) Characteristics and acceptance of yoghurt containing date palm products. In: Second international conference on date palms, Al-Ain, United Arab Emirates, 25–26 Mar, pp 842–849

Hashim IB, Khalil AH, Afifi HS (2009) Quality characteristics and consumer acceptance of yogurt fortified with date fiber. J Dairy Sci 92:5403–5407

Herbert D (1976) Stoichiometric aspects of microbial growth. In: Methods in microbiology, vol 5B. Academic, London

Holts JG (1994) Bergey's manual of determinative bacteriology, 9th edn. Williams and Wilkins, Baltimore

Ismail B, Haffar I, Baalbakic R et al (2008) Physico-chemical characteristics and sensory quality of two date varieties under commercial and industrial storage conditions. LWT Food Sci Technol 41:896–904

Israelidis JC (1987) Nutrition-single cell protein, twenty years later. Presented at First B.I.O. International conference Athens, Food Technology Institute, Southeastern College, Athens, May 1987. Website. http://www. biopolitics.gr/HTML/PUBS/VOL1/isreali.htm

Itakura C, Emi Y, Takeehi M et al (1983) Pathology of hens fed a diet containing methanol utilizing bacteria. Jpn Poult Sci 20:302–306

Jassim AH, Hussain NA, Muhssin AA et al (1986) The effect of using single cell protein in productive performance, level of nucleic acids and histopathological changes broilers. J Agric Water Resour Res Anim Prod 5:275–283

Jamai L, Ettayebi K, El Yamani J, Ettayebi M (2007) Production of ethanol from starch by free immobilized Candida tropicalis in the presence of α-amylase. Biores Technol 98:2765–2770

Jeffries TW (2005) Ethanol fermentation on the move. Nat Biotechnol 23:40–41

Kaur C, Kapoor HC (2001) Antioxidants in fruits and vegetables-the millenniums health. Int J Food Sci Technol 36:703–725

Keating K, White CH (1990) Effect of alternative sweeteners in plain and fruit flavored yoghurt. J Dairy Sci 73:54–62

Keramat Jahromi M, Jafari A, Rafiee S et al (2008) Changes in physical properties of date fruit (cv. Shahani) during three edible stages of ripening. Am-Eur J Agric Environ Sci 3:132–136

Khan JA, Abulnaja KO, Kumosani TA et al (1995) Utilization of Saudi date sugars in production of baker's yeast. Bioresour Technol 53:63–66

Krauss S (1982) Characterization of imbalances of nutrients and additives in single cell protein (probion) in poultry. Thesis, Ludwig-Maximillians-Universitat

Lee PK, Chen SC (1983) Effect of adding Lih Feng Jiuen, the microorganism feed supplement, to laying diets on egg production, feed efficiency, and egg quality of leghorn hens. Taiwan J Vet Med Anim Husb 41:25–35

Leeson S, Summers JD (2001) Scott's nutrition of the chicken, 4th edn. University Books, Guelph

Litchfield JH (1968) The production of fungi. In: Mateles RI, Tannenbaum SR (eds.) Single cell protein. MIT Press, Cambridge

Luckow T, Delahunty C (2004) Which juice is "healthier"? A consumer study of probiotic non-dairy juice drinks. Food Qual Prefer 15:751–759

Luckow T, Sheehan V, Fitzgerald G et al (2006) Exposure, health information and flavour-masking strategies for improving the sensory quality of probiotic juice. Appetite 47:315–323

Madigan MT, Martinko JM, Parker J (1997) Brock biology of microorganisms. Prentice Hall International, Herts

Maity TK, Misra AK (2009) Probiotics and human health: synoptic review. Afr J Food Agric Nutr Dev 9:1778–1796

Majamaa H, Isolauri E (1997) Probiotics: a novel approach in the management of food allergy. J Allergy Clin Immunol 99:179–185

Mansouri A, Embared G, Kokkalou E et al (2005) Phenolic profile and antioxidant activity of the Algerian ripe date palm fruit (*Phoenix dactylifera*). Food Chem 89:411–420

Martin AM (1997) Bioconversion of waste materials to industrial products, 2nd edn. Blackie Academic and Professional, p 576

McLean E, Craig SR (2004) Growth performance of Nile tilapia fed an organically certified yeast-based alternative protein source. In: Proceedings of the 5th international conference on recirculating aquaculture, Roanoke, 22–25 July, pp 580–586

Medri V, Pereira GV, Leonhardt JH (2000) Growth of Nile tilapia *Oreochromis niloticus* fed with different levels of alcohol yeast. Rev Bras Biol 60:113–121

Mohamed DA, Al-Okabi S (2004) In vivo evaluation of antioxidant and anti-inflammatory activity of different extracts of date fruits in adjuvant arthritis. Pol J Food Nutr Sci 13:397–402

Mohammed NA, Al-Obaidi ZS, Hasson NA, Jassem MA (1986) Semi-industrial production of baker's yeast using date extract and molasses. J Agric Water Resour Res Iraq 5:147–162 (in Arabic)

Mohite V, Sivaraman H (1984) Continuous of sweet sorghum juice to ethanol using immobilized yeast cells. Biotechnol Bioeng 26:1126–1127

Morton JF (1987) Date. In: Fruits of warm climates. Julia F Morton, Miami, pp 5–11
Mudhaffer S (1978) Studies on the conditions of growth of baker's yeast, grown in solution of molasses or debis (date juice). Nahrung 22:285–289
Najib H (1996) Effect of incorporating yeast culture *Saccharomyces cervisiae* in to the Saudi Baladi and White Leghorn layer's diet. J Appl Anim Res 10:181–186
Nancib N, Nancib A, Boudrant J (1997) Use of waste date products in the formation of baker's yeast biomass by *Saccharomyces cerevisiae*. Bioresour Technol 60:67–71
Olvera-Novoa MA, Martinez-Palacio CA, Olivera-Castillo L (2002) Utilization of torula yeast (*Candida utilis*) as a protein source in diets for tilapia (*Oreochromis mossambicus* Peters) fry. Aquac Nutr 8:257–264
Oura E (1983) Biomass from carbohydrates. In: Biotechnology, vol 3. Verlag Chemie, Weinheim
Pakistan Horticulture Development and Export Board (2008) Dates marketing strategy
Parodi PW (1999) The role of intestinal bacteria in the causation and prevention of cancer: modulation by diet and probiotics. Aust J Dairy Technol 54:103–121
Perdigon G, Alvarez S, Holgado AP (1991) Immunoadjuvant activity of oral *Lactobacillus casei*: influence of dose on the secretory immune response and protective capacity in intestinal infections. J Dairy Res 58:485–496
Pereira AP, Isabel CFR, Ferreira IC et al (2007) Phenolic compounds and antimicrobial activity of olive (*Olea europaea* L. Cv. Cobrançosa) leaves. Molecule 12:1153–1162
Pirmohammadi R, Shivazad M, Yoisef-Hakimi AA (1999) Feeding value of Iranian produced single cell protein for broilers. Iran J Agric Sci 30:277–286
Plavnik I, Bornstein S, Hurwitz S (1981) Evaluation of methanol-grown bacteria and hydrocarbon-grown yeast as sources of protein for poultry: studies with young chicks. Br Poult Sci 22:123–140
Potter NN, Hotchkiss JH (1995) Milk and milk products in food science, 5th edn. Chapman and Hall, New York
Pyler EJ (1988) Baking science and technology, vol II, 3rd edn. Sosland Publishing, Merriam, pp 1071–1076
Raipulis J, Toma MM Semjonovs P (2005) The effect of probiotics on the genotoxicity of furazolidone. Int J Food Microbiol 102:343–347
Rautavaara S, Kalliomaki M, Isolauri E (2002) Probiotics during pregnancy and breast-feeding might confer immunomodulatory protection against astopic disease in the infant. J Allergy Clin Immunol 109:119–121
Ray B (1996) Fundamental food microbiology. CRC Press, Boca Raton
Regnault-Roger C, Hadidane R, Biard JF et al (1987) High performance liquid and thin-layer chromatography determination of phenolic acids in palm (*Phoenix dactylifera*). Food Chem 25:61–71
Robinson RK (1986) Modern dairy technology, vol 1, Advances in milk processing. Elsevier, London
Robinson RK (1995) A colour guide to cheese and fermented milks. Chapman and Hall, London/New York
Roels JA (1983) Energetics and kinetics in biotechnology. Elsevier, Amsterdam
Rumsey GL, Hughes SG, Smith RR et al (1991) Digestibility and energy values of intact, disrupted, and extracts from brewers dried yeast fed to rainbow trout (*Oncorhynchus mykiss*). Anim Feed Sci Technol 33:185–193
Saleem SA (2005) Aspects of ripening of Dhakki dates (*Phoenix dactylifera* L.) and post harvest stability employing hurdle technology. PhD Dissertation, Department of Food Science and Technology, Gomal University, Pakistan
Sawaya WN (1986) Dates of Saudi Arabia, Regional Agriculture and Water Research Center, Ministry of Agriculture and Water, Riyadh
Schøyen HF, Hetland H, Rouvinen-Watt K et al (2007) Growth performance and ileal and total tract amino acid digestibility in broiler chickens fed diets containing bacterial protein produced on natural gas. Poult Sci 86:87–93

Schneider O, Amirkolaie AK, Vera-Cartas J et al (2004) Digestibility, faeces recovery, and related carbon, nitrogen and phosphorus balances of five feed ingredients evaluated as fishmeal alternatives in Nile tilapia, *Oreochromis niloticus* L. Aquac Res 35:1370–1379

Scrimshaw NS (1975) Single cell protein for human consumption, single cell protein II. MIT Press, Cambridge

Shacklady CA (1974) Response of livestock and poultry to SCP. In: Davis P (ed.) Single cell Protein. Proceedings of the International symposium held in Rome, Italy, November. Academic Press, London

Shacklady CA (1975) Value of SCP for animals, single cell protein II. MIT Press, Cambridge

Shah NP (2000) Effect of milk derived bioactives: an overview. Br J Nutr 84:3–10

Shah NP (2006) Probiotics and fermented milks. In: Chandan RC, White CH, Kilara A, Hui YH (eds.) Manufacturing yogurt and fermented milks. Blackwell Publishing, Oxford, pp 352–354

Shapouri H, Salassi M, Nilson JF (2006) The economic feasibility of ethanol production from sugar in the United States, US Department of Agriculture Report, Washington, DC

Spinelli J (1980) Unconventional feed ingredients for fish feed. In: Fish feed technology. UNDP/FAO, Rome

Stringer DA (1982) Industrial development and evaluation of new protein sources, microorganisms. Proc Nutr Soc 41:289–300

Teuber GA, Krusch U, Lembke J (1994) Biotechnological procedures for production of food and feeding-stuffs. In: Präve P, Faust U, Sittig W, Sukatsch DA (eds.) Handbuch der biotechnologie. R Oldenbourg Verlag, Munich, pp 479–540

Tissier H (1900) Recherchers sur la flora intestinal normal et pathologique du nourisson. Thesis, University of Paris, Paris

White A, Handler P, Smith EL (1964) Principles of biochemistry, 3rd edn. McGraw Hill, New York

Xiang-Yang GE, Zhang W-G (2005) A shortcut to the production of high ethanol concentration from Jerusalem artichoke tubers. Food Technol Biotechnol 43:241–246

Young IS, Woodside JV (2001) Antioxidant in health and disease. J Clin Pathol 54:176–186

Yousif AK, Alghamdi AS, Ahmed A et al (1996) Processing and evaluation of date juice milk drink. Egypt J Dairy Sci 24:277–288

Zaid A (2006) World date industry: situation, challenges and opportunities. International conference on date palm production and processing technology, Muscat

Zee JA, Simard RE (1974) Simple process for the reduction in the nucleic acid content in yeast. Appl Microbiol 29:59–62

Zhang XB, Ohta Y (1991) In vitro binding of mutagenic pyrolyzates to lactic acid bacterial cells in human gastric juice. J Dairy Sci 74:752–757

Chapter 33
Date Palm as a Source of Bioethanol Producing Microorganisms

N. Gupta and H. Kushwaha

Abstract Expanding world population, increasing energy demand, depleting reserves of fossil fuels and increasing effects of pollution from these fuels demand more ecofriendly alternatives which can substitute for fossil fuel (petrol, diesel, coal etc.). Ethanol derived from biomass has the potential to be a substitute of fossil fuel which is renewable, non-toxic, biodegradable and more ecofriendly. The three major classes of feedstocks for ethanol production are sugars, starches and lignocelluloses. Date palm (*Phoenix dactylifera*) sap is highly nutritive and has high sugar content which varies from 60% to 70%; it is also a very good source of fermentation microorganisms. Palm-wine fermentation is always alcoholic-lactic-acetic acid fermentation, involving mainly yeasts and lactic acid bacteria. Currently research is being directed to develop metabolically and genetically engineered *Saccharomyces* strains and other ethanol-fermenting microflora that have the potential to utilize a wide range of substrates including pentose and hexose sugars, an ability for direct and efficient ethanol production from cellulosic materials and to tolerate ethanol stress. Thus *Saccharomyces* strains from date palm sap could be genetically modified to overcome the constraints in the path of higher yield ethanol production.

Keywords Biofuels • Bioethanol • Biomass • Fermentation • *Phoenix dactylifera*

N. Gupta (✉)
Department of Microbiology and Biotechnology, Faculty of Science, M S University, Baroda, Gujarat, India

Department of Molecular Biology and Genetic Engineering, College of Basic Sciences and Humanities, G.B. Pant University of Agriculture and Technology, Pantnagar, Uttarakhand 263 145, India
e-mail: nidhigupta_2411@yahoo.co.in

H. Kushwaha
Department of Molecular Biology and Genetic Engineering, College of Basic Sciences and Humanities, G.B. Pant University of Agriculture and Technology, Pantnagar, Uttarakhand 263 145, India

S.M. Jain et al. (eds.), *Date Palm Biotechnology*,
DOI 10.1007/978-94-007-1318-5_33, © Springer Science+Business Media B.V. 2011

33.1 Introduction

Rapid depletion of energy resources, based on non-renewable fuels, is of deep concern worldwide, especially in the developing countries. This is mainly due to increased transportation, modernization and industrialization which ultimately lead to environmental pollution resulting in health hazards and ecological imbalances. Hence, there is a dire need to search for an alternative fuel, which is more ecofriendly and enhances quality of life. Global energy usage is projected to nearly double in the next two decades, and biological fuel production might serve as a sustainable, carbon-neutral energy source compatible with current engine technology. Conversion of biomass to biofuels has been the subject of intense research efforts since the 1970s. This work has recently gained significant political and scientific momentum owing to concerns about climate change, global energy security and petroleum supply. Today, biomass covers about 10% of the world's primary energy demand. Plant biomass is an abundant and renewable source of energy-rich carbohydrates which can be efficiently converted into biofuels by microbes. Today bioethanol is one of the major products available commercially.

The world's largest producers of bio-ethanol are Brazil (sugarcane ethanol) and the United States (corn ethanol). Brazil is largest exporter of ethanol; delivering 70% of worldwide supply and the USA is that country's largest client as it imported 1.74 billion liters in 2006, which represents 58% of Brazil's ethanol exports (Wust 2007). In an effort to offset increases in consumption and to limit the fossil fuel-related negative impacts on the environment, the US Department of Energy has established the goal of supplanting 30% gasoline consumption with cellulosic ethanol by 2030 (Herrera 2006). Similarly, a European Union Directive of 2003 aims to replace 5.75% of all gasoline and diesel transport fuels with biofuels by 2010 (Schubert 2006). A few countries are actively involved in generating biofuel from date palm at a very large scale by setting up various companies aimed at ethanol production. *Oman Green Energy Company Makes Ethanol From Date Palm, Plans Large Refinery, 100 Ethanol Pumps By 2010*, reads a news account (Gulf News 2007). In Algeria, Algerian Biotech Company aims at production of biomethanol from dates.

Owing to physical and political limitations on arable land, it is believed that future biofuels will, of necessity, originate from abundantly available lignocellulosic biomass. There are many advantages to using bioresource derived ethanol as a liquid transportation fuel. Bioethanol blended with gasoline extends crude oil utilization, reduces reliance on oil imports and helps to mitigate increasing oil prices. The higher oxygen content of ethanol results in relatively cleaner combustion and has long been used as an additive in gasoline to reduce urban smog and other environmental pollution problems. Therefore, ignition improvers, glow-plug, surface ignition and pilot injection are applied to promote self-ignition by using diesel-bioethanol blended fuel (Kim et al. 2005).

The most popular blend for light-duty vehicles is known as E85, and contains 85% bioethanol and 15% gasoline. In Brazil, bioethanol for fuel is derived from sugarcane and is used either pure or blended with gasoline in a mixture called gasohol (24% bioethanol, 76% gasoline) (de Oliveria et al. 2005). In several states of the United States, a small amount of bioethanol (10% by volume) is added to gasoline, known as gasohol or E10. Blends having higher concentrations of bioethanol in gasoline are also used, e.g. in flexible-fuel vehicles that can operate on blends of up to 85% bioethanol-E85 (Malca and Freire 2006). Some countries have exercised biofuel program involving both forms: bioethanol-gasoline blend program, e.g. the United States (E10 and for Flexible Fuel Vehicle (FFV) E85), Canada (E10 and for FFV E85), Sweden (E5 and for FFV E85), India (E5), Australia (E10), Thailand (E10), China (E10), Columbia (E10), Peru (E10), Paraguay (E7) and Brazil (E20, E25 and FFV any blend) (Kadiman 2005).

33.2 Major Feedstocks and Bioethanol Production Processes

The three major classes of feedstocks for ethanol production are sugars (e.g., molasses, cane juice), starches (corn, wheat, cassava) and lignocelluloses (rice straw, wheat straw, bagasse, wood, energy crops) (Fig. 33.1). Starch and sugar-based ethanol is often referred to as a first-generation biofuel. Even though the production of ethanol from starch represents the most convenient and technically advanced option for bioenergy in the USA, it would result in severe competition between energy and food supplies. Lignocellulosic feedstock can be acquired from either dedicated biomass crops or forestry and agricultural residuals (Boerjan 2005; Sims et al. 2006). The key obstacle for transitioning from starch-based to lignocellulosic biofuels is the

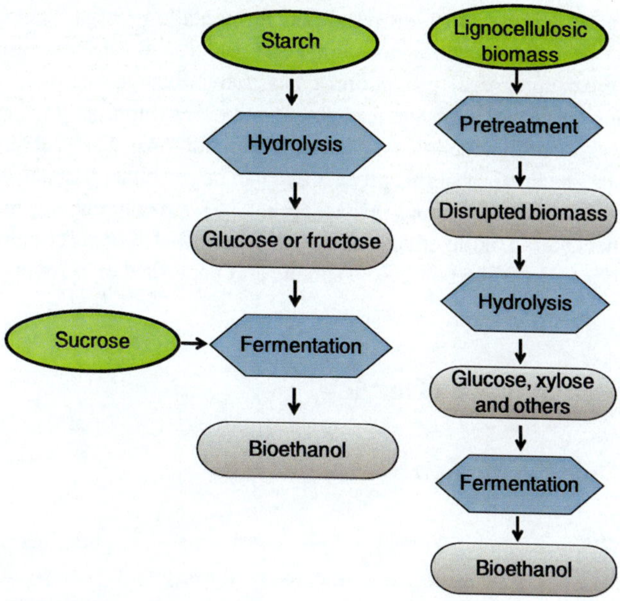

Fig. 33.1 Platform for fermentation process by major feedstocks of ethanol *A* Sucrose, *B* Starch and *C* Lignocellulosic material

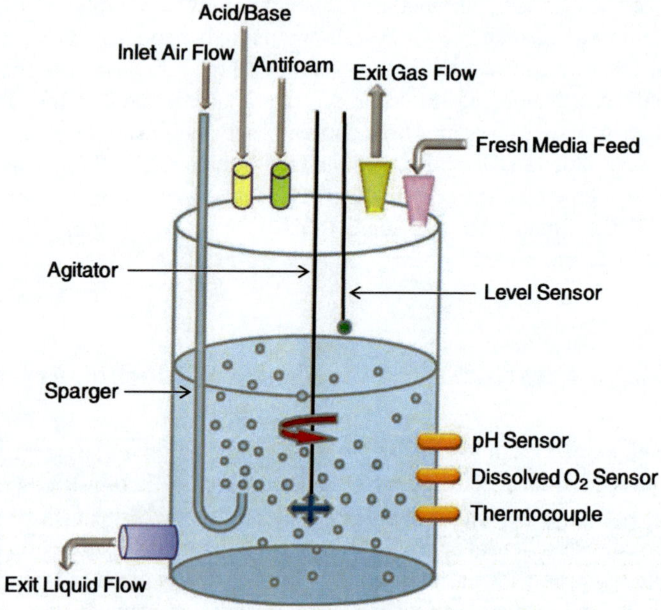

Fig. 33.2 Structure of fermentation tank

complicated structure of the cell wall, which is resistant to breakdown and represents a recalcitrance problem. Current processes for lignocellulosic biomass include pretreatment, saccharification (hydrolysis) and fermentation (Ragauskas et al. 2006). Improvement or replacement of these processes is crucial for increasing efficiency and for decreasing biofuel production costs. Obviating pretreatment, along with simultaneous saccharification and fermentation, are two important factors that would decrease the cost of lignocellulosic ethanol production (Ragauskas et al. 2006).

Ethanol production is a simple process that can be run either as a batch reactor in a confined space or as a continuous process (Fig. 33.2). An entire plant can in fact fit in one's own backyard making ethanol a very attractive fuel source for communities or even countries that wish to be self-sustainable and not reliant on foreign resources.

33.3 Date Palm: An Overview

33.3.1 Date Production

Date palm (*Phoenix dactylifera* L.), a diploid with 2n = 36, is a member of the monocot family Arecaceae classified as a dioecious tall evergreen. Date palm is the most cultivated palm in the arid and semi-arid regions of the world. Dates constitute part of a popular subsistence among the populace of the Middle Eastern peninsula.

There are estimated 90–100 million trees worldwide and mainly concentrated between latitudes 10–30° North, in arid regions of the Middle East and North Africa, where it is thought to have been cultivated for several thousands of years. According to the FAOSTAT database (http://faostat.org), world date fruit production has risen from 2,659,406 mt in 1980 to 7,109,974 mt in 2008. The major producers of date palm are Egypt, Iran, Saudi Arabia, United Arab Emirates, Pakistan, Algeria, Iraq, Sudan, Oman and Libya.

33.3.2 *Date Chemical Composition*

The development of date fruits is divided into three final stages: khalal, rutab and tamar. Khalal stage dates are immature with hard texture, yellow, red or pink in color, total soluble solids (TSS) of 30–45° brix, astringent and in some cultivars edible; rutab stage dates soften at the tip of the fruit, with TSS of 55–60 brix, are free of astringency and edible; tamar stage dates are fully ripe with TSS of 60–84° brix and edible (Pareek 1985). Dates are generally harvested at the tamar stage, that is after the development of TSS of 60–70° brix. Date fruit at the tamar stage contains moisture ranging of 10–22%; total sugars 62–75%; protein 2.2–2.7%; fiber 5–8%; fat 0.4– 0.7%, ash 3.5–4.2%; total acidity 0.06–0.20% and ascorbic acid of 30.0–50.0 mg%, on a dry weight basis (Baraem et al. 2006; El-Sharnouby et al. 2007; FAO 1962).

The fruit of the date palm is composed of a fleshy pericarp and seed. Pits (seeds) of date palm are a waste product of many date fruit processing plants producing pitted dates, date powders, date syrup, date juice, chocolate-coated dates and date confectionery. Date palm female trees bear fruits at 3–5 years and are fully mature at 12 years. The fruit is a nutritious source of sugar, minerals, and vitamins. Appropriately called the *palm of life*, for over 5,000 years the date palm has provided food, ornament and material for shelter, fiber and fuel in a harsh environment where relatively few other plants are able to grow. Mature date palms are highly desirable landscape subjects because they are plentiful, relatively inexpensive, uniform in size and habit and highly ornamental.

33.3.3 *Date Utilization*

Dates are known to be rich in carbohydrates (80%) but quite low in protein (2–3%) (Al-Hooti et al. 1997). Dates are an excellent source of simple sugars, minerals and vitamins (El-Shaarawy et al. 1989) and its fiber content reaches about 8% (FAO 1962; Lambiote 1982). The flesh of a fully ripe date (tamar), consist of two-third sugars and one-quarter water, the rest being mainly cellulose, pectin, ash and vitamins (FAO 1962). The date is considered as a nutritious fruit as research has indicated the clear contribution of dates to human health when consumed with other food constituents (Lambiote 1982). There are at least 15 minerals in dates. The percentage of each mineral in dried dates varies from 0.1 to 916 mg/100 g date depending on the mineral. In many varieties, potassium can be found at a concentration as high as 0.9% in the flesh while it is as high as 0.5% in some seeds. Other minerals and salts that are found in various proportions include boron, calcium, cobalt, copper, fluorine, iron, magnesium, manganese, potassium, phosphorous, sodium and zinc. Additionally, the seeds contain aluminum, cadmium, chloride, lead and sulphur in various proportions. Dates contain elemental fluorine that is useful in protecting teeth against decay. Selenium, another element believed to help prevent cancer and important in immune function, is also found in dates. The protein in dates contains 23 types of amino acids, some of which are not present in the most popular fruits such as oranges, apples and bananas. Dates contain at least six vitamins including a small amount of vitamin C, and vitamins B_1 thiamine, B_2 riboflavin, nicotinic acid (niacin) and vitamin A. The fruits have an important therapeutic role in glycemic and lipid control of diabetic patients. Dates have also been identified as having antioxidant and antimutagenic properties, and were found to reduce heart disease and cancer.

33.4 Indigenous Ethanol Producing Microflora of Date Palm Sap

Since the principal constituent of the date fruit is sugar and its total sugar content at harvest ranges from 70% to 80%. High sugar content also is present in date palm sap and could be used as a good source of fermentative microorganisms. In almost all tropical locations in Asia where palm trees grow, the sap obtained from the decapitated inflorescence of various palm species is fermented to produce an alcoholic beverage called palm wine or toddy. There is an art in binding the flower spathes, pounding them to cause the sap to flow properly by cutting the spathe tip and collecting the sap into the earthen pitchers which contain yeasts and bacteria in the left-over toddy from the previous lots. The fermentation starts as soon as the sap flows into the pitcher.

Palm wine is either consumed fresh as it is brought down from the tree or fermented for up to 24 h. The freshly harvested sap is generally a dirty brown sweet liquid having 10–18% w/w sugar, a pH of 7.0–7.4 and traces of ethanol, which after fermentation results in the formation of a product containing as much as 9% (by volume, v/v) ethanol and pH of 4.0–5.5 (Joshi et al. 1999; Steinkraus 1996).

Palm wine fermentation is always alcoholic-lactic-acetic acid fermentation, involving mainly yeasts and lactic acid bacteria. In the fermenting sap, *Saccharomyces cerevisiae* is invariably present but lactic acid bacteria such as *Lactobacillus plantarum, L. mesenteroides* or other species of bacteria like *Zymomonas mobilis* and *Acetobacter* spp. vary. The other yeast types include *Schizosaccharomyces pombe, Saccharomyces chevalieri, S. exiguus, Candida* spp.; *Saccharomycodes ludwigii, S. pombe, Saccharomyces cerevisiae, Kodamaea ohmeri* and *Hanseniaspora occidentalis* which are characterized as maximum ethanol producers in toddy

(Joshi et al. 1999). The yeasts, especially *Saccharomyces* spp., are largely responsible for the characteristic aroma of palm wine (Uzochukwu et al. 1999). During fermentation, there is continuous effervescence as a result of the production of carbon dioxide. A yeasty odor develops and after a couple of hours yeasts form sediment at the bottom of the container.

Palm wine is a good source of B vitamins. Recently Gupta et al. (2009) reported the occurrence of high ethanol producing microorganisms (*Saccharomyces* sp.) with faster growth rate in date palm sap. Various indigenous strains of *Saccharomyces* sp. were isolated from date palm sap and were evaluated for alcohol dehydrogenase (ADH) enzyme activity, ethanol production and alcohol tolerance limits. Alcoholic contents in juice samples fermented with different yeast strains varied considerably (8.9–12.5%, v/v) as determined by GLC. Yeast cultures showed varied *in vitro* ethanol tolerance (3–12%). Isolate SCP-1 was found superior showing 12.5% ethanol production, high ADH enzyme activity (4.38 units/ml) and higher alcohol tolerance maintaining cell viability at 12% ethanol in YPD medium up to 48 h (Gupta et al. 2009).

33.5 Constraints in Bioethanol Production

33.5.1 *Lack of Proper Substrate Utilization for Biofuel Production*

Bioethanol production from plant biomass has received considerable attention recently in order to mitigate global warming and demands for petroleum. Currently, bioethanol is produced mainly from sugar-containing or starchy biomass such as sugarcane and corn as the raw material. As sugar-containing and starchy biomass is used for food and animal feed, there arises competition for its use as both food and fuel. Due to this competition, lignocellulosic bioethanol production has been eagerly researched worldwide. Lignocellulosic biomass, such as woods and agricultural residues, is an attractive feedstock for bioethanol production because of its large amount of potentially-fermentable sugars. The main structural components of lignocellulosic biomass are cellulose, hemicellulose and lignin. Of these, only cellulose and hemicellulose can be used as raw materials to produce ethanol by fermentation of carbohydrates obtained by chemical or enzymatic hydrolysis (saccharification). The main component of lignocellulosic hydrolysates is glucose, a hexose sugar derived from cellulose and hemicellulose. Although the proportion of monosaccharides in hemicellulose hydrolysates varies depending on the raw material and the hydrolysis procedure (Hendriks and Zeeman 2009; Lee 1997), they all contain both pentose sugars, such as D-xylose and L-arabinose and hexose sugars. D-xylose is the second most abundant carbohydrate and its content is particularly high in grass and hardwood. Thus, a substantial number of the hydrolysates obtained from lignocellulosic biomass contain xylose, requiring an economic conversion of biomass into ethanol through xylose utilization.

The microorganisms for hexose sugars including glucose, mannose, and galactose is *Saccharomyces cerevisiae*, yeast with high ethanol productivity, high tolerance to ethanol and tolerance to inhibitory compounds present in the hydrolysate of lignocellulosic biomass (Olsson and Hahn-Hagerdal 1993; Olsson and Nielsen 2000). Although *S. cerevisiae* can grow well even at relatively low pH, which prevents contamination by other bacteria, native strains are unable to utilize xylose for growth or fermentation. Instead, it metabolizes D-xylulose, an isomerization product of D-xylose (Chiang et al. 1981; Wang and Schneider 1980; Wang et al. 1980). Some yeast strains have been reported to ferment xylose into ethanol (Jeffries 1981; Schneider et al. 1981; Slininger et al. 1982), but the rate and yield of ethanol production from xylose in these xylose-utilizing yeast strains are considerably low as compared to their glucose fermentation. Therefore, genetic engineering and/or adaptation may be promising methods to develop sufficient pentose sugar fermentation in *S. cerevisiae*.

For the choice of the fermenting microorganism, complete substrate utilization, inhibitor tolerance and ethanol productivity are important aspects. The yeast *Saccharomyces cerevisiae* satisfies the last two conditions (Almeida et al. 2007; Bettiga et al. 2009; Piskur et al. 2006), while metabolic engineering is required to obtain strains able to ferment lignocellulosic biomass (Olofsson et al. 2008), which mainly consists of pentose and hexose, where the ratio depends on the source of lignocellulose. Since a limited number of microorganisms can ferment pentose, in the cases that pentose is the main sugar such as in corn stover and bagasse, usually a pentose-fermenting microorganism such as *Pichia stipitis* is used for cultivation. Lignocellulose is a complex polymer consisting of fibrous bundles of crystalline cellulose encased in a polymeric matrix of hemicellulose and lignin. Although compositions vary, this material can be generally regarded as being composed of 50% cellulose, 25% hemicellulose and 25% lignin (and extractables). For bioconversion, the carbohydrate portion must be solubilized while the lignin and residues can be used to provide energy for ethanol purification. The glycosidic linkages in hemicellulose are readily hydrolyzed by dilute acids at elevated temperatures to yield syrup containing xylose and arabinose for agricultural residues and hardwoods or mannose, xylose and glucose for softwoods. However, no natural organisms have been found which can efficiently and rapidly convert xylose and arabinose into ethanol. For the complete fermentation of the pentose fraction in lignocellulose hydrolyzate, a strain combining L-arabinose and D-xylose utilization is desired. Fermentation of either L-arabinose or D-xylose by recombinant *S. cerevisiae* has been demonstrated.

33.5.2 *Ethanol Fermentation Utilizing Related Substrates and Ethanol Extraction: Lack of Suitable Technology*

Lignocellulose is the most plentiful renewable biomass produced by photosynthesis and its annual production was estimated in 1×10^{10} millions tons worldwide (Sanchez and Cardona 2008). The potential for using lignocellulosic materials in bioethanol production is well recognized. However, the task of hydrolyzing lignocellulose to fermentable monosugars is still technically problematic because the linear polymer has a strong crystallinity and is usually surrounded by lignin, which reduces accessibility to hydrolytic enzymes. Many pretreatment techniques have been used to increase the hydrolysis of lignocellulosic biomass; for instance, dilute acid, ammonia recycle percolation, lime, steam explosion (Hendriks and Zeeman 2009), alkaline and acidic wet oxidation (Varga et al. 2004). Current processes for lignocellulosic biomass include pretreatment, saccharification (hydrolysis) and fermentation. Some options proposed to reduce the cost of the conversion of lignocellulose to ethanol, include: eliminating pretreatment, increasing cellulose hydrolysis yield, enhancing the enzyme activity to reduce its consumption, and improving the fermentation process both in yield and specificity. This conversion includes two processes: (i) hydrolysis of cellulose in the lignocellulosic materials to fermentable reducing sugars and (ii) fermentation of the sugars to ethanol. The hydrolysis is usually catalyzed by cellulase enzymes and the fermentation is carried out by yeast or bacteria. The factors that have been identified to affect the hydrolysis of cellulose include porosity, i.e., accessible surface area of the waste materials, cellulose fiber crystallinity and lignin and hemicellulose content (McMillan 1994). The presence of lignin and hemicellulose makes the access of cellulose enzymes to cellulose difficult, thus reducing the efficiency of the hydrolysis. Removal of lignin and hemicellulose, reduction of cellulose crystallinity and increase of porosity in pretreatment processes can significantly improve the hydrolysis (McMillan 1994).

33.5.3 Increased Ethanol Concentration During Fermentation

One of the most common stresses that yeast cells encounter during fermentation is the increased ethanol concentration. Yeast cells have developed appropriate mechanisms to deal with several types of damages caused by increased ethanol concentration. First, ethanol increases the fluidity of the plasma membrane and destroys the normal membrane structures. In response, yeast cells may change membrane compositions to antagonize membrane fluidization and stabilize plasma membrane. Specifically, it has been shown that the levels of unsaturated fatty acids (UFAs) (You et al. 2003), and ergosterol (Daum et al. 1998; Swan and Watson 1998), increase in response to the high concentration of ethanol. Furthermore, the addition of some types of amino acids (Hu et al. 2005; Takagi et al. 2005), and inositol (Kelley et al. 1988), can enhance ethanol tolerance when provided as a supplement, most likely through enhanced membrane stability. Second, the expression of factors that stabilize and/or repair denatured proteins in yeast cells, such as trehalose and induced heat shock proteins (HSPs), have been revealed to correlate with the capabilities to tolerate alcohol (Swan and Watson 1998; Vianna et al. 2008). Third, some candidate proteins involved in the expression of stress-related genes like the zinc finger protein (MacPherson et al. 2006) and the recently reported alcohol sensitive ring/PHD finger 1 protein (Asr1p) (Betz et al. 2004), also play a role in ethanol tolerance in

Saccharomyces cerevisiae. Lastly, the global transcription machinery engineering (gTME) technology can reprogram gene transcription and then improve glucose/ethanol tolerance of yeast cells (Alper et al. 2006).

33.6 Strain Improvement Through Mutagenesis and Recombinant DNA Technology

A number of genetic and molecular biological mechanisms based advanced technologies could be used for strain improvement for ethanol production. Improvement in the yeast strain should address the following topics.

33.6.1 Utilization of a Wide Range of Substrates

Sustainable and economically viable manufacturing of bioethanol from lignocellulose raw material is dependent on the availability of a robust ethanol producing microorganism, able to ferment all sugars present in the feedstock, including the pentose sugars L-arabinose and D-xylose. *Saccharomyces cerevisiae* is a robust ethanol producer, but needs to be engineered to achieve pentose sugar fermentation. *S accharomyces cerevisiae*, which is one of the most prominent ethanol-producing microorganisms utilizing hexose, has been found unable to utilize xylose due to lack of the key enzymes in the xylose-metabolising pathway (Meinander et al. 1999). Thus, the efficient utilization of xylose in hemicellulose in addition to glucose in cellulose by a recombinant xylose-fermenting *S. cerevisiae* strain would offer an opportunity to reduce the production cost of bioethanol significantly (Chandrakant and Bisaria 2000). To date, numerous studies regarding the metabolic engineering of *S. cerevisiae* for xylose utilization have been reported, and many reviews have already addressed the current advancement in metabolic engineering of xylose-fermenting strains and factors which affect xylose metabolism in yeasts (Almeida and Hahn-Hagerdal 2009; Aristidou and Penttila 2000; Chu and Lee 2007; Dien et al. 2003; Gong et al. 1999; Hahn-Hagerdal et al. 2001, 2007a,b; Jeffries 2006; Jeffries and Jin 2004; Jeffries and Shi 1999; van Maris et al. 2006, 2007).

33.6.2 Direct and Efficient Ethanol Production from Cellulosic Materials

Lignocellulosic biomass shows significant potential as a future substrate for bioethanol production. An optimal microorganism for lignocellulose-based bioethanol production must be able to hydrolyze sugar polymers, utilize all fermentable compounds, and convert them into ethanol at a high rate, yield and productivity.

No single microorganism is known to possess all of these characteristics. For example the traditional ethanol producer, *Saccharomyces cerevisiae*, cannot hydrolyze cellulose and hemicelluloses. Moreover, it cannot utilize many compounds which are generated during hydrolysis of lignocellulosic biomass, such as xylose, arabinose, glucuronic acids and cello-oligosaccharides. Another problem that arises during biomass pretreatment in a lignocellulosic process is the formation of several substances which inhibit yeast growth, such as furans, organic acids, phenols and inorganic salts.

Since the yeast *Saccharomyces cerevisiae* cannot utilize cellulosic materials, these materials must undergo saccharification to glucose before ethanol production can take place. Various cellulose and β-glucosidase genes have been expressed in *S. cerevisiae* with the aim of direct ethanol production from cellulose (Okada et al. 1998). Fermentation of cellulose to ethanol by recombinant yeast cells has, however, not been successful, although some recombinant yeast strains are able to assimilate soluble cellooligosaccharides (2–6 glucose units) as carbon sources (Cho et al. 1999; Murai et al. 1998; van Rensburg et al. 1998). Recently, yeast strains displaying various proteins on the cell surface have been developed by using genetic engineering techniques (Boder and Wittrup 1997; Kondo et al. 2002; Murai et al. 1998; Nakamura et al. 2001; Schreuder et al. 1996; Ueda and Tanaka 2000). Yeast strains which codisplay cellulolytic enzymes on the cell surface through cell surface engineering could mimic cellulosome, which is a multienzyme complex consisting of cellulase and hemicellulase organized on the bacterial cell surface (Bayer et al. 1994; Shoham et al. 1999). The most promising strategy for converting cellulose to ethanol in yeast is certainly the concerted heterologous expression of all types of cellulolytic enzymes to maximize their synergies (Baker et al. 1998; Fujita et al. 2004; Zhang and Lynd 2004).

Lignocellulosic plant biomass requires chemical pretreatment, exposing the polysaccharides (cellulose and hemicelluloses) to enzymatic hydrolysis and fermentation. Current pretreatment processes, which usually rely on high temperatures, acid hydrolysis and/or high pressure (Chandra et al. 2007; Galbe and Zacchi 2007; Lin and Tanaka 2006) form several degradation products with various inhibitory effects on yeast fermentation (Klinke et al. 2003, 2004). These substances fall into the following classes: carboxylic acids, furans, phenolic compounds and inorganic salts (Zaldivar et al. 2001). To overcome the various inhibitory substances, metabolic, genetic and evolutionary and gene disruption strategies were used.

33.6.3 Yeast Strains Tolerant of Ethanol Stress

Eukaryotic cells will encounter stresses during phases of their growth and reproductive cycles. As a result, signatures of stress response mechanisms are widespread in the genomes, cytoplasms and plasma membranes of cells. For example, in yeast, during the process of biomass propagation, yeast cells are dynamically exposed to a mixed and interrelated group of known stresses such as osmotic, oxidative, thermic, ethanol tolerance and/or starvation. These stress conditions can dramatically affect the population dynamics and industrial fermentation, including ethanol production (Estruch 2000; Hohmann 2002). One of the most common stresses that yeast cells encounter during fermentation is the increased ethanol concentration.

Many researchers have attempted to analyze the mechanism of ethanol stress tolerance and to improve the ethanol stress tolerance of yeast. It was reported that yeast cells adapt to high concentration of ethanol by changing the components in the cytoplasmic membrane to maintain the membrane fluidity (Alexandre et al. 1994; Chi and Arneborg 1999; Chi et al. 1999; You et al. 2003). Additionally, it was reported that increasing intracellular accumulation of trehalose by deletion or expression of antisense for acid trehalase gene *ATH1* improved ethanol stress tolerance (Jung and Park 2005; Kim et al. 1996). Also it was reported (Takagi et al. 2005) that enhancement of intracellular proline accumulation improved ethanol tolerance of yeast used for Japanese rice wine (sake) brewing. On the other hand, genome-wide analysis are also useful to identify the genes responsible for ethanol stress tolerance of yeast except for the genes related to the characteristics described above. To analyze the transcriptional response upon ethanol stress and identify the genes responsible for ethanol tolerance, DNA microarray analysis have been performed (Alexandre et al. 2001; Ogawa et al. 2000). Moreover, genome-wide screening of ethanol-sensitive mutants has been carried out. The genes responsible for ethanol stress tolerance were identified by transposon mutagenesis (Takahashi et al. 2001). The knockout mutant library of *Saccharomyces cerevisiae*, which is now commercially available, has also been used for genome-wide screening of the genes responsible for ethanol stress tolerance (Fujita et al. 2006; Kubota et al. 2004; van Voorst et al. 2006). In these approaches, mutants showing ethanol sensitivity are screened, and the deleted genes are determined as the genes required for growth under high ethanol concentration conditions.

33.7 Conclusions and Prospective

Although date palm sap is the rich source of microorganisms, it could also serve as fermentation substrate. Recently lignocellulosic biomass represents the most prospective feedstock for ethanol production. The availability and low cost of a wide range of lignocellulosic materials offer many possibilities for the development of bioindustries that could support the growth of the international biofuel market and contribute to the reduction of greenhouse gas emissions worldwide. Fermentative microflora from date palm thus could be utilized for fermentation utilizing ethanol fermenting major feedstocks. These microorganisms could be modified utilizing genetic and metabolic engineering techniques for higher ethanol yield. The current research tendencies toward improving fuel ethanol production are thus linked to the nature of employed raw materials, processing steps and related process engineering issues.

References

Alexandre H, Rousseaux I, Charpentier C (1994) Relationship between ethanol tolerance, lipid composition and plasma membrane fluidity in *Saccharomyces cerevisiae* and *Kloeckera apiculata*. FEMS Microbiol Lett 124:17–22

Alexandre H, Ansanay-Galeote V, Dequin S et al (2001) Global gene expression during short-term ethanol stress in *Saccharomyces cerevisiae*. FEBS Lett 498:98–103

Al-Hooti S, Sidhu JS, Qabazard H (1997) Physicochemical characteristics of five date fruit cultivars grown in the United Arab Emirates. Plant Foods Hum Nutr 50:101–113

Almeida JR, Hahn-Hagerdal B (2009) Developing *Saccharomyces cerevisiae* strains for second generation bioethanol: improving xylose fermentation and inhibitor tolerance. Int Sugar J 111:172–180

Almeida JR, Modig T, Petersson A et al (2007) Increased tolerance and conversion of inhibitors in lignocellulosic hydrolysates by *Saccharomyces cerevisiae*. J Chem Technol Biotechnol 82:340–349

Alper H, Moxley J, Nevoigt E et al (2006) Engineering yeast transcription machinery for improved ethanol tolerance and production. Science 314:1565–1568

Aristidou A, Penttila M (2000) Metabolic engineering applications to renewable resource utilization. Curr Opin Biotechnol 11:187–198

Baker JO, Ehrman CI, Adney WS et al (1998) Hydrolysis of cellulose using ternary mixtures of purified cellulases. Appl Biochem Biotechnol 70–72:395–403

Baraem I, Imad H, Riad B et al (2006) Physico-chemical characteristics and total quality of five date varieties grown in the United Arab Emirates. Int J Food Sci Technol 41:919–926

Bayer EA, Morag E, Lamed R (1994) The cellulosome–a treasure-trove for biotechnology. Trends Biotechnol 12:378–386

Bettiga M, Gorwa-Grauslund MF, Hahn-Hagerdal B (2009) Metabolic engineering in yeasts. In: Smolke C (ed) The metabolic pathway engineering handbook. Taylor & Francis CRC Press, London, pp 22.21–22.46

Betz C, Schlenstedt G, Bailer SM (2004) Asr1p, a novel yeast ring/PHD finger protein, signals alcohol stress to the nucleus. J Biol Chem 279:28174–28181

Boder ET, Wittrup KD (1997) Yeast surface display for screening combinatorial polypeptide libraries. Nat Biotechnol 15:553–557

Boerjan W (2005) Biotechnology and the domestication of forest trees. Curr Opin Biotechnol 16:159–166

Chandra RP, Bura R, Mabee WE et al (2007) Substrate pretreatment: the key to effective enzymatic hydrolysis of lignocellulosics? Adv Biochem Eng Biotechnol 108:67–93

Chandrakant P, Bisaria VS (2000) Simultaneous bioconversion of glucose and xylose to ethanol by *Saccharomyces cerevisiae* in the presence of xylose isomerase. Appl Microbiol Biotechnol 53:301–309

Chi Z, Arneborg N (1999) Relationship between lipid composition, frequency of ethanol-induced respiratory deficient mutants, and ethanol tolerance in *Saccharomyces cerevisiae*. J Appl Microbiol 86:1047–1052

Chi Z, Kohlwein SD, Paltauf F (1999) Role of phosphatidylinositol (PI) in ethanol production and ethanol tolerance by a high ethanol producing yeast. J Ind Microbiol Biotechnol 22:58–63

Chiang LC, Gong CS, Chen LF et al (1981) D-xylulose fermentation to ethanol by *Saccharomyces cerevisiae*. Appl Environ Microbiol 42:284–289

Cho KM, Yoo YJ, Kang HS (1999) Integration of endo/exoglucanase and β-glucosidase genes into the yeast chromosomes for direct conversion of cellulose to ethanol. Enzyme Microb Technol 25:23–30

Chu BC, Lee H (2007) Genetic improvement of *Saccharomyces cerevisiae* for xylose fermentation. Biotechnol Adv 25:425–441

Daum G, Lees ND, Bard M et al (1998) Biochemistry, cell biology, and molecular biology of lipids of *Saccharomyces cerevisiae*. Yeast 14:1471–1510

de Oliveria MED, Vaughan BE, Rykiel EJ Jr (2005) Ethanol as fuel: energy, carbon dioxide balances, and ecological footprint. BioScience 55:593–602
Dien BS, Cotta MA, Jeffries TW (2003) Bacteria engineered for fuel ethanol production: current status. Appl Microbiol Biotechnol 63:258–266
El-Shaarawy M, Mesallam MI, El-Nakhal AS et al (1989) Studies on extraction of dates. In: Proceedings second symposium on date palm, King Faisal University, Al-Hassa, Saudi Arabia, March 3–6, pp 259–271
El-Sharnouby GA, Al-wesali MS, Al-Shathri AA (2007) Effect of some drying methods on quality of palm date fruits powder. Fourth symposium on date palm in Saudi Arabia, King Faisal University, 5–8 May
Estruch F (2000) Stress-controlled transcription factors, stress-induced genes, and stress tolerance in budding yeast. FEMS Microbiol Rev 24:469–486
FAO (1962) Dates handling, processing and packing. FAO, Rome
Fujita Y, Ito J, Ueda M et al (2004) Synergistic saccharification, and direct fermentation to ethanol, of amorphous cellulose by use of an engineered yeast strain codisplaying three types of cellulolytic enzyme. Appl Environ Microbiol 70:1207–1212
Fujita K, Matsuyama A, Kobayashi Y et al (2006) The genome wide screening of yeast deletion mutants to identify the genes required for tolerance to ethanol and other alcohols. FEMS Yeast Res 6:744–750
Galbe M, Zacchi G (2007) Pretreatment of lignocellulosic materials for efficient bioethanol production. Adv Biochem Eng Biotechnol 108:41–65
Gong CS, Cao NJ, Du J et al (1999) Ethanol production from renewable resources. Adv Biochem Eng Biotechnol 65:207–241
Gulf News (2007) Tapping green alternative, June 23
Gupta N, Dubey A, Tewari L (2009) High efficiency alcohol tolerant *Saccharomyces* isolates of *Phoenix dactylifera* for bioconversion of sugarcane juice into bioethanol. JSIR 68:401–405
Hahn-Hagerdal B, Wahlbom CF, Gardonyi M et al (2001) Metabolic engineering of *Saccharomyces cerevisiae* for xylose utilization. Adv Biochem Eng Biotechnol 73:53–84
Hahn-Hagerdal B, Karhumaa K, Fonseca C et al (2007a) Towards industrial pentose fermenting yeast strains. Appl Microbiol Biotechnol 74:937–953
Hahn-Hagerdal B, Karhumaa K, Jeppsson M et al (2007b) Metabolic engineering for pentose utilization in *Saccharomyces cerevisiae*. Adv Biochem Eng Biotechnol 108:147–177
Hendriks ATWM, Zeeman G (2009) Pretreatments to enhance the digestibility of lignocellulosic biomass. Bioresour Technol 100:10–18
Herrera S (2006) Bonkers about biofuels. Nat Biotechnol 24:755–760
Hohmann S (2002) Osmotic stress signaling and osmoadaptation in yeasts. Microbiol Mol Biol Rev 66:300–372
Hu CK, Bai FW, An LJ (2005) Protein amino acid composition of plasma membranes affects membrane fluidity and thereby ethanol tolerance in a self-flocculating fusant of *Schizosaccharomyces pombe* and *Saccharomyces cerevisiae*. Sheng Wu Gong Cheng Xue Bao 21:809–813
Jeffries TW (1981) Conversion of xylose to ethanol under aerobic conditions by Candida tropicalis. Biotech Lett 3:213–218
Jeffries TW (2006) Engineering yeasts for xylose metabolism. Curr Opin Biotechnol 17:320–326
Jeffries TW, Jin YS (2004) Metabolic engineering for improved fermentation of pentoses by yeasts. Appl Microbiol Biotechnol 63:495–509
Jeffries TW, Shi NQ (1999) Genetic engineering for improved xylose fermentation by yeasts. Adv Biochem Eng Biotechnol 65:117–161
Joshi VK, Sandhu DK, Thakur NS (1999) Fruit based alcoholic beverages. In: Joshi VK, Pandey A (eds.) Biotechnology: food fermentation. Educational Publishers, Ernakulam, pp 647–744
Jung YJ, Park HD (2005) Antisense-mediated inhibition of acid trehalase (*ATH1*) gene expression promotes ethanol fermentation and tolerance in *Saccharomyces cerevisiae*. Biotechnol Lett 27:1855–1859
Kadiman OK (2005) Crops: beyond foods. In: Proceedings first international conference of crop security. Malang, Indonesia, September 20–23

Kelley MJ, Bailis AM, Henry SA et al (1988) Regulation of phospholipid biosynthesis in *Saccharomyces cerevisiae* by inositol; inositol is an inhibitor of phosphatidylserine synthase activity. J Biol Chem 263:18078–18085

Kim J, Alizadeh P, Harding T et al (1996) Disruption of the yeast *ATH1* gene confers better survival after dehydration, freezing, and ethanol shock: potential commercial applications. Appl Environ Microbiol 62:1563–1569

Kim H, Choi B, Park S et al (2005) Engine performance and emission characteristics of CRDI diesel engine equipped with the WCC and the DOC. Using ethanol blended diesel fuel. In: Proceedings 15th international symposia on alcohol fuels (ISAF XV), San Diego. September 26–28,p 30

Klinke HB, Olsson L, Thomsen AB et al (2003) Potential inhibitors from wet oxidation of wheat straw and their effect on ethanol production of *Saccharomyces cerevisiae*: wet oxidation and fermentation by yeast. Biotechnol Bioeng 81:738–747

Klinke HB, Thomsen AB, Ahring BK (2004) Inhibition of ethanol producing yeast and bacteria by degradation products produced during pre-treatment of biomass. Appl Microbiol Biotechnol 66:10–26

Kondo A, Shigechi H, Abe M et al (2002) High-level ethanol production from starch by a flocculent *Saccharomyces cerevisiae* strain displaying cell-surface glucoamylase. Appl Microbiol Biotechnol 58:291–296

Kubota S, Takeo I, Kume K et al (2004) Effect of ethanol on cell growth of budding yeast: genes that are important for cell growth in the presence of ethanol. Biosci Biotechnol Biochem 68:968–972

Lambiote B (1982) Some aspects of the role of dates in human nutrition. In: Proceedings first international symposium on date palm, King Faisal University, Saudi Arabia, March 23–25, pp 573–579

Lee J (1997) Biological conversion of lignocellulosic biomass to ethanol. J Biotechnol 56:1–24

Lin S, Tanaka S (2006) Ethanol fermentation from biomass resources: current state and prospects. Appl Microbiol Biotechnol 69:627–642

MacPherson S, Larochelle M, Turcotte B (2006) A fungal family of transcriptional regulators: the zinc cluster proteins. Microbiol Mol Biol Rev 70:583–604

Malca J, Freire F (2006) Renewability and life-cycle energy efficiency of bioethanol and bioethyl tertiary butyl ether (bioETBE): assessing the implications of allocation. Energy 31:3362–3380

McMillan JD (1994) Pretreatment of lignocellulosic biomass. In: Himmel ME, Baker JO, Overend RP (eds.) Enzymatic conversion of biomass for fuels production. American Chemical Society, Washington, DC, pp 292–324

Meinander NQ, Boels I, Hahn-Hagerdal B (1999) Fermentation of xylose/glucose mixtures by metabolically engineered *Saccharomyces cerevisiae* strains expressing XYL1 and XYL2 from Pichia stipitis with and without overexpression of TALl. Bioresour Technol 68:79–87

Murai T, Ueda M, Kawaguchi T et al (1998) Assimilation of cellooligosaccharides by cell surface-engineered yeast expressing β-glucosidase and carboxymethylcellulase from *Aspergillus aculeatus*. Appl Environ Microbiol 64:4857–4861

Nakamura Y, Shibasaki S, Ueda M et al (2001) Development of novel whole-cell immunoadsorbents by yeast surface display of the IgG-binding domain. Appl Microbiol Biotechnol 57:500–505

Ogawa Y, Nitta A, Uchiyama H et al (2000) Tolerance mechanism of the ethanol-tolerant mutant of sake yeast. J Biosci Bioeng 90:312–320

Okada H, Sekiya T, Yokoyama K et al (1998) Efficient secretion of *Trichoderma reesei* cellobiohydrolase II in *Schizosaccharomyces pombe* and characterization of its products. Appl Microbiol Biotechnol 49:301–308

Olofsson K, Bertilsson M, Liden G (2008) A short review on SSF–an interesting process option for ethanol production from lignocellulosic feedstocks. Biotechnol Biofuels 1:7

Olsson L, Hahn-Hagerdal B (1993) Fermentative performance of bacteria and yeast in lignocellulose hydrolysates. Process Biochem 28:249–257

Olsson L, Nielsen J (2000) The role of metabolic engineering in the improvement of *Saccharomyces cerevisiae:* utilization of industrial media. Enzyme Microb Technol 26:785–792

Pareek OP (1985) Date palm. In: Bose TK (ed.) Fruits of India tropical and subtropical. Naya Prokash, Calcutta, pp 662–675

Piskur J, Rozpedowska E, Polakova S et al (2006) How did *Saccharomyces* evolve to become a good brewer? Trends Genet 22:183–186

Ragauskas AJ, Williams CK, Davison BH et al (2006) The path forward for biofuels and biomaterials. Science 311:484–489

Sanchez OJ, Cardona CA (2008) Trends in biotechnological production of fuel ethanol from different feedstocks. Bioresour Technol 99:5270–5295

Schneider H, Wang PY, Chan YK et al (1981) Conversion of D-xylose into ethanol by the yeast *Pachysolen tannophilus*. Biotechnol Lett 3:89–92

Schreuder MP, Mooren ATA, Toschka HY et al (1996) Immobilizing proteins on the surface of yeast cells. Trends Biotechnol 14:115–120

Schubert C (2006) Can biofuels finally take center stage? Nat Biotechnol 24:777–784

Shoham Y, Lamed R, Bayer EA (1999) The cellulosome concept as an efficient microbial strategy for the degradation of insoluble polysaccharides. Trends Microbiol 7:275–281

Sims REH, Astley H, Bernhard S et al (2006) Energy crops: current status and future prospects. Glob Change Biol 12:2054–2076

Slininger PJ, Bothast RJ, Van Cauwenberge JE et al (1982) Conversion of D-xylose to ethanol by the yeast *Pachysolen tannophilus*. Biotechnol Bioeng 24:371–384

Steinkraus KH (1996) Handbook of indigenous fermented foods, 2nd edn. Marcel Dekker, New York

Swan TM, Watson K (1998) Stress tolerance in a yeast sterol auxotroph: role of ergosterol, heat shock proteins and trehalose. FEMS Microbiol Lett 169:191–197

Takagi H, Takaoka M, Kawaguchi A et al (2005) Effect of L-proline on sake brewing and ethanol stress in *Saccharomyces cerevisiae*. Appl Environ Microbiol 71:8656–8662

Takahashi T, Shimoi H, Ito K (2001) Identification of genes requires for growth under ethanol stress using transposon mutagenesis in *Saccharomyces cerevisiae*. Mol Genet Genom 265:1112–1119

Ueda M, Tanaka A (2000) Genetic immobilization of proteins on the yeast cell surface. Biotechnol Adv 18:121–140

Uzochukwu S, Balogh E, Tucknot OG et al (1999) Role of palm wine yeasts and bacteria in palm wine aroma. J Food Sci Technol 36:301–304

Van Maris AJ, Abbott DA, Bellissimi E et al (2006) Alcoholic fermentation of carbon sources in biomass hydrolysates by *Saccharomyces cerevisiae*: current status. Antonie Leeuwenhoek 90:391–418

Van Maris AJ, Winkler AA, Kuyper M et al (2007) Development of efficient xylose fermentation in *Saccharomyces cerevisiae*: xylose isomerase as a key component. Adv Biochem Eng Biotechnol 108:179–204

Van Rensburg P, Van Zyl WH, Pretorius IS (1998) Engineering yeast for efficient cellulose degradation. Yeast 14:67–76

van Voorst F, Houghton-Larsen J, Jonson L et al (2006) Genome-wide identification of genes required for growth of *Saccharomyces cerevisiae* under ethanol stress. Yeast 23:351–359

Varga E, Klinke HB, Reczey K et al (2004) High solid simultaneous saccharification and fermentation of wet oxidized corn stover to ethanol. Biotechnol Bioeng 88:567–574

Vianna CR, Silva CL, Neves MJ et al (2008) *Saccharomyces cerevisiae* strains from traditional fermentations of Brazilian cachaca: trehalose metabolism, heat and ethanol resistance. Antonie Leeuwenhoek 93:205–217

Wang PY, Schneider H (1980) Growth of yeasts on D-xylulose. Can J Microbiol 26:1165–1168

Wang PY, Shopsis C, Schneider H (1980) Fermentation of a pentose by yeasts. Biochem Biophys Res Commun 94:248–254

Wust C (2007) Thanks giving in the car tank. Fuel from biomass can replace the oil–if better technology is used. Spiegel 8:104–111

You KM, Rosenfield CL, Knipple DC (2003) Ethanol tolerance in the yeast *Saccharomyces cerevisiae* is dependent on cellular oleic acid content. Appl Environ Microbiol 69:1499–1503
Zaldivar J, Nielsen J, Olsson L (2001) Fuel ethanol production from lignocellulose: a challenge for metabolic engineering and process integration. Appl Microbiol Biotechnol 56:17–34
Zhang YH, Lynd LR (2004) Toward an aggregated understanding of enzymatic hydrolysis of cellulose: noncomplexed cellulase systems. Biotechnol Bioeng 88:797–824

Index

S.M. Jain et al. (eds.), *Date Palm Biotechnology*,
DOI 10.1007/978-94-007-1318-5, © Springer Science+Business Media B.V. 2011

C

D

N

O

P

T